T0100531

Advances in the Physics of Particles and Nuclei
Volume 31

For further volumes:
http://www.springer.com/series/6938

Advances in the Physics of Particles and Nuclei

The series *Advances in the Physics of Particles and Nuclei* (APPN) is devoted to the archiving, in printed high-quality book format, of the comprehensive, long shelf-life reviews published in *The European Physical Journal A* and *C*. APPN will be of benefit, in particular, to those librarians and research groups who have chosen to have only electronic access to these journals. Occasionally, original material in review format and refereed by the series' editorial board will also be included.

Advances in the Physics of Particles and Nuclei

Volume 31

Edited by

Douglas H. Beck
Dieter Haidt
John W. Negele

Volume 31

Contributions to this Volume:

Douglas H. Beck
Department of Physics
University of Illinois at Urbana-Champaign
1110 West Green Street
Urbana, IL 61801-3080
USA
e-mail: dhbeck@illinois.edu

Dieter Haidt
DESY
Notkestraße 85
22603 Hamburg
Germany
e-mail: dieter.haidt@desy.de

John W. Negele
William A. Coolidge Professor of Physics
Massachusetts Institute of Technology
Center for Theoretical Physics
77 Massachusetts Ave. 6-315
Cambridge MA 02139
USA
e-mail: negele@mitlns.mit.edu

Originally published in Eur. Phys. J. C (2011) 71: 1534 and Eur. Phys. J. C (2011) 71: 1627

ISSN 1868-2146 e-ISSN 1861-440X
ISBN 978-3-642-23305-0 e-ISBN 978-3-642-23306-7
DOI 10.1007/978-3-642-23306-7
Springer Heidelberg Dordrecht London New York

Printed on acid-free paper

Springer is part of Springer Science+Business Media (www.springer.com)

Table of Contents

Heavy quarkonium: progress, puzzles, and opportunities

N. Brambilla[1,*,†], **S. Eidelman**[2,3,*,†], **B.K. Heltsley**[4,a,*,†], **R. Vogt**[5,6,*,†], **G.T. Bodwin**[7,†], **E. Eichten**[8,†], **A.D. Frawley**[9,†], **A.B. Meyer**[10,†], **R.E. Mitchell**[11,†], **V. Papadimitriou**[8,†], **P. Petreczky**[12,†], **A.A. Petrov**[13,†], **P. Robbe**[14,†], **A. Vairo**[1,†], **A. Andronic**[15], **R. Arnaldi**[16], **P. Artoisenet**[17], **G. Bali**[18], **A. Bertolin**[19], **D. Bettoni**[20], **J. Brodzicka**[21], **G.E. Bruno**[22], **A. Caldwell**[23], **J. Catmore**[24], **C.-H. Chang**[25,26], **K.-T. Chao**[27], **E. Chudakov**[28], **P. Cortese**[16], **P. Crochet**[29], **A. Drutskoy**[30], **U. Ellwanger**[31], **P. Faccioli**[32], **A. Gabareen Mokhtar**[33], **X. Garcia i Tormo**[34], **C. Hanhart**[35], **F.A. Harris**[36], **D.M. Kaplan**[37], **S.R. Klein**[38], **H. Kowalski**[10], **J.-P. Lansberg**[39,40], **E. Levichev**[2], **V. Lombardo**[41], **C. Lourenço**[42], **F. Maltoni**[43], **A. Mocsy**[44], **R. Mussa**[16], **F.S. Navarra**[45], **M. Negrini**[20], **M. Nielsen**[45], **S.L. Olsen**[46], **P. Pakhlov**[47], **G. Pakhlova**[47], **K. Peters**[15], **A. D. Polosa**[48], **W. Qian**[49,14], **J.-W. Qiu**[12,50], **G. Rong**[51], **M.A. Sanchis-Lozano**[52], **E. Scomparin**[16], **P. Senger**[15], **F. Simon**[23,53], **S. Stracka**[41,54], **Y. Sumino**[55], **M. Voloshin**[56], **C. Weiss**[28], **H.K. Wöhri**[32], **C.-Z. Yuan**[51]

[1]Physik-Department, Technische Universität München, James-Franck-Str. 1, 85748 Garching, Germany
[2]Budker Institute of Nuclear Physics, Novosibirsk 630090, Russia
[3]Novosibirsk State University, Novosibirsk 630090, Russia
[4]Cornell University, Ithaca, NY 14853, USA
[5]Physics Division, Lawrence Livermore National Laboratory, Livermore, CA 94551, USA
[6]Physics Department, University of California at Davis, Davis, CA 95616, USA
[7]High Energy Physics Division, Argonne National Laboratory, 9700 South Cass Avenue, Argonne, IL 60439, USA
[8]Fermi National Accelerator Laboratory, P.O. Box 500, Batavia, IL 60510, USA
[9]Physics Department, Florida State University, Tallahassee, FL, 32306-4350, USA
[10]Deutsches Elektronen-Synchrotron DESY, Hamburg, Germany
[11]Indiana University, Bloomington, IN 47405, USA
[12]Physics Department, Brookhaven National Laboratory, Upton, NY 11973-5000, USA
[13]Department of Physics and Astronomy, Wayne State University, Detroit, MI 48201, USA
[14]Laboratoire de l'Accélérateur Linéaire, IN2P3/CNRS and Université Paris-Sud 11, Centre Scientifique d'Orsay, BP 34, 91898 Orsay Cedex, France
[15]GSI Helmholtzzentrum für Schwerionenforschung, 64291 Darmstadt, Germany
[16]INFN Sezione di Torino, Via P. Giuria 1, 10125 Torino, Italy
[17]Department of Physics, The Ohio State University, Columbus, OH 43210, USA
[18]Institut für Theoretische Physik, Universität Regensburg, 93040 Regensburg, Germany
[19]INFN Sezione di Padova, Via Marzolo 8, 35131 Padova, Italy
[20]Università di Ferrara and INFN Sezione di Ferrara, Via del Paradiso 12, 44100 Ferrara, Italy
[21]Institute of Nuclear Physics, Polish Academy of Sciences, Kraków, Poland
[22]Università di Bari and INFN Sezione di Bari, Via Amendola 173, 70126 Bari, Italy
[23]Max Planck Institute for Physics, München, Germany
[24]Department of Physics, Lancaster University, Lancaster, LA1 4YB, UK
[25]CCAST (World Laboratory), P.O. Box 8730, Beijing 100190, China
[26]Institute of Theoretical Physics, Chinese Academy of Sciences, Beijing 100190, China
[27]Department of Physics, Peking University, Beijing 100871, China
[28]Thomas Jefferson National Accelerator Facility, 12000 Jefferson Ave., Newport News, VA 23606, USA
[29]Clermont Université, Université Blaise Pascal, CNRS-IN2P3, LPC, BP 10448, 63000 Clermont-Ferrand, France
[30]University of Cincinnati, Cincinnati, OH 45221, USA
[31]Laboratoire de Physique Théorique, Unité mixte de Recherche, CNRS, UMR 8627, Université de Paris-Sud, 91405 Orsay, France
[32]LIP, Av. Elias Garcia 14, 1000-149 Lisbon, Portugal
[33]SLAC National Accelerator Laboratory, Stanford, CA 94309, USA
[34]Department of Physics, University of Alberta, Edmonton, Alberta, Canada T6G 2G7
[35]Institut für Kernphysik, Jülich Center for Hadron Physics, and Institute for Advanced Simulation, Forschungszentrum Jülich, 52425 Jülich, Germany
[36]Department of Physics and Astronomy, University of Hawaii, Honolulu, HI 96822, USA
[37]Illinois Institute of Technology, Chicago, IL 60616, USA
[38]Lawrence Berkeley National Laboratory, Berkeley, CA 94720, USA
[39]IPNO, Université Paris-Sud 11, CNRS/IN2P3, Orsay, France
[40]Centre de Physique Théorique, École Polytechnique, CNRS, 91128 Palaiseau, France
[41]INFN Sezione di Milano, Via Celoria 16, 20133 Milano, Italy
[42]CERN, 1211 Geneva 23, Switzerland
[43]Center for Cosmology, Particle Physics and Phenomenology, Université Catholique de Louvain, 1348 Louvain-la-Neuve, Belgium

[44] Department of Math and Science, Pratt Institute, 200 Willoughby Ave, ARC LL G-35, Brooklyn, NY 11205, USA
[45] Instituto de Física, Universidade de São Paulo, C.P. 66318, 05315-970 São Paulo, SP, Brazil
[46] Department of Physics & Astronomy, Seoul National University, Seoul, Korea
[47] Institute for Theoretical and Experimental Physics, Moscow 117218, Russia
[48] INFN Sezione di Roma, Piazzale Aldo Moro 2, 00185 Roma, Italy
[49] Department of Engineering Physics, Tsinghua University, Beijing 100084, China
[50] C.N. Yang Institute for Theoretical Physics, Stony Brook University, Stony Brook, NY 11794-3840, USA
[51] Institute of High Energy Physics, Chinese Academy of Sciences, Beijing 100049, China
[52] Instituto de Física Corpuscular (IFIC) and Departamento de Física Teórica, Centro Mixto Universitat de Valencia-CSIC, Doctor Moliner 50, 46100 Burjassot, Valencia, Spain
[53] Excellence Cluster 'Universe', Technische Universität München, Garching, Germany
[54] Dipartimento di Fisica, Università di Milano, 20133 Milano, Italy
[55] Department of Physics, Tohoku University, Sendai, 980-8578, Japan
[56] William I. Fine Theoretical Physics Institute, School of Physics and Astronomy, University of Minnesota, 116 Church Street SE, Minneapolis, MN 55455, USA

Abstract A golden age for heavy-quarkonium physics dawned a decade ago, initiated by the confluence of exciting advances in quantum chromodynamics (QCD) and an explosion of related experimental activity. The early years of this period were chronicled in the Quarkonium Working Group (QWG) CERN Yellow Report (YR) in 2004, which presented a comprehensive review of the status of the field at that time and provided specific recommendations for further progress. However, the broad spectrum of subsequent breakthroughs, surprises, and continuing puzzles could only be partially anticipated. Since the release of the YR, the BESII program concluded only to give birth to BESIII; the B-factories and CLEO-c flourished; quarkonium production and polarization measurements at HERA and the Tevatron matured; and heavy-ion collisions at RHIC have opened a window on the deconfinement regime. All these experiments leave legacies of quality, precision, and unsolved mysteries for quarkonium physics, and therefore beg for continuing investigations at BESIII, the LHC, RHIC, FAIR, the Super Flavor and/or Tau–Charm factories, JLab, the ILC, and beyond. The list of newly found conventional states expanded to include $h_c(1P)$, $\chi_{c2}(2P)$, B_c^+, and $\eta_b(1S)$. In addition, the unexpected and still-fascinating $X(3872)$ has been joined by more than a dozen other charmonium- and bottomonium-like "XYZ" states that appear to lie outside the quark model. Many of these still need experimental confirmation. The plethora of new states unleashed a flood of theoretical investigations into new forms of matter such as quark–gluon hybrids, mesonic molecules, and tetraquarks. Measurements of the spectroscopy, decays, production, and in-medium behavior of $c\bar{c}$, $b\bar{b}$, and $b\bar{c}$ bound states have been shown to validate some theoretical approaches to QCD and highlight lack of quantitative success for others. Lattice QCD has grown from a tool with computational possibilities to an industrial-strength effort now dependent more on insight and innovation than pure computational power. New effective field theories for the description of quarkonium in different regimes have been developed and brought to a high degree of sophistication, thus enabling precise and solid theoretical predictions. Many expected decays and transitions have either been measured with precision or for the first time, but the confusing patterns of decays, both above and below open-flavor thresholds, endure and have deepened. The intriguing details of quarkonium suppression in heavy-ion collisions that have emerged from RHIC have elevated the importance of separating hot- and cold-nuclear-matter effects in quark–gluon plasma studies. This review systematically addresses all these matters and concludes by prioritizing directions for ongoing and future efforts.

Contents

[a] e-mail: bkh2@cornell.edu
[*] Editors
[†] Section coordinators

1 Introduction

Heavy quarkonium is a multiscale system which can probe all regimes of quantum chromodynamics (QCD). At high energies, a perturbative expansion in the strong-coupling constant $\alpha_s(Q^2)$ is possible. At low energies, nonperturbative effects dominate. In between, the approximations and techniques which work at the extremes may not succeed, necessitating more complex approaches. Hence heavy quarkonium presents an ideal laboratory for testing the interplay between perturbative and nonperturbative QCD within a controlled environment. To do so in a systematic manner requires the intersection of many avenues of inquiry: experiments in both particle physics and nuclear physics are required; perturbative and lattice QCD calculations must be performed in conjunction with one another; characteristics of confinement and deconfinement in matter must be confronted; phenomenology should be pursued both within the Standard Model and beyond it. Above all, experiments must continue to provide measurements which constrain and challenge all aspects of QCD, and theory must then guide experiment toward the next important observables.

Effective field theories[1] (EFTs) describing quarkonium processes have continued to develop and now provide a unifying description as well as solid and versatile tools yielding well-defined predictions. EFTs rely on higher-order perturbative calculations and lattice simulations. Progress on both fronts has improved the reach and precision of EFT-based predictions, enabling, e.g., the increasingly precise determinations of several fundamental parameters of the Standard Model (i.e., α_s, m_c, and m_b).

Several experiments operating during this era, primarily BABAR at SLAC and Belle at KEK), CLEO-III and CLEO-c at CESR, CDF and DØ at Fermilab, and BESII and BESIII at IHEP have, in effect, operated as quarkonium factories, vastly increasing the available data on quarkonia spectra and decays. Over the same period, investigations of quarkonium production in fixed target experiments at Fermilab and CERN, HERA-B at DESY, and PHENIX and STAR at RHIC have vastly increased the knowledge base for cold- and hot-medium studies. The resulting variety of collision types, energy regimes, detector technologies, and analysis techniques has yielded quarkonium-related physics programs that are both competitive and complementary. Taken together, the experimental programs provide the confirmations and refutations of newly observed phenomena that are crucial for sustained progress in the field as well as the breadth and depth necessary for a vibrant quarkonium research environment.

The Quarkonium Working Group (QWG) was formed in 2002 as a dedicated and distinct effort to advance quarkonium studies by drawing sometimes disparate communities together in this common cause. QWG activities bring experts in theory and experiment together to discuss the current status and progress in all the relevant subfields. Subsequent participant interactions are intended to synthesize a consensus of progress and priorities going forward. Periodic QWG meetings have been effective in achieving this function. The exhaustive *CERN Yellow Report* [1], the first document produced by QWG detailing the state of quarkonium physics and suggestions for future efforts, was released in

[1] EFTs such as HQEFT, NRQCD, pNRQCD, SCET, ..., are described elsewhere in this article.

2004 to embody such a synthesis. Since that report appeared, much has been accomplished in theory and experiment, warranting an updated review.

This review provides a comprehensive exploration of heavy-quarkonium physics applicable to the landscape of 2010, with particular emphases on recent developments and future opportunities. The presentation is organized into five broad and frequently overlapping categories:

- *Spectroscopy* (Sect. 2), which focuses on the existence, quantum numbers, masses, and widths of heavy-quarkonium (or quarkonium-like) bound states;
- *Decay* (Sect. 3), an examination of the patterns and properties of quarkonia transitions and decays, with special attention given to the decay dynamics and exclusive final-state branching fractions;
- *Production* (Sect. 4), the study of heavy-quarkonium creation in e^+e^-, $p\bar{p}$, ℓp, γp, and pp collisions;
- *In medium* (Sect. 5), the investigation of deconfinement and formation of quark–gluon plasma in heavy-ion collisions via measurement of quarkonium suppression;
- *Experimental outlook* (Sect. 6), the status and physics reach of new and planned experimental facilities.

Below we briefly introduce and motivate each of these sections.

Heavy quarkonium *spectroscopy* examines the tableau of heavy-quark bound states, thereby providing the starting point for all further investigations. Which states exist? Why? What are their masses, widths, and quantum numbers? Which states should exist but have not yet been observed? Does QCD fully explain the observed terrain? If not, why? New experimental and theoretical efforts over the last decade have provided some answers to these questions, while also raising new ones. Some long-anticipated states have, at last, been measured (e.g., $h_c(1P)$, $\eta_c(2S)$, and $\eta_b(1S)$), while many unanticipated states (e.g., $X(3872)$ and $Y(4260)$) also appeared. Does the underestimation of the $\eta_b(1S)$ hyperfine splitting by some QCD calculations indicate faults in application of theory, inaccuracy of measurements, or the presence of new physics? Have we observed mesonic molecules? Tetraquarks? Quark–gluon hybrids? How would we know if we had? How many of the new states are experimental artifacts? Do $X(3872)$ decay patterns comport with those of any conventional quarkonium? Is $X(3872)$ above or below $D^{*0}\bar{D}^0$ threshold? Is the e^+e^- hadronic cross section enhancement near 10.86 GeV simply the $\Upsilon(5S)$ resonance or does $\Upsilon(5S)$ overlap with a new Y_b state, as suggested by recent dipion transition data? These questions, among many others, animate continuing theoretical and experimental spectroscopic investigations.

For states away from threshold, theory provides a description, at the level of the binding-energy scale, in the form of an EFT called pNRQCD. Precise and accurate calculation of the $\eta_b(1S)$ hyperfine splitting remains a challenge for both perturbative and lattice calculations. With one exception, no EFT description has yet been constructed nor have the appropriate degrees of freedom been clearly identified for most new states close to threshold. The exception is $X(3872)$, which displays universal characteristics due to its proximity to $D^{*0}\bar{D}^0$ threshold, thus prompting a plethora of calculations based on a single elegant formalism. Spectroscopy has advanced from both direct and EFT-formulated lattice calculations. In general, however, the threshold regions remain troublesome for the lattice as well as EFTs, excited-state lattice calculations have been only recently pioneered, and the full treatment of bottomonium on the lattice remains a challenge.

A substantial challenge in the realm of quarkonium *decay* is for theory to keep pace with the large number of new measurements. These include increasingly precise measurements of prominent decay modes (e.g., dilepton branching fractions and widths of J/ψ and Υ, branching fractions for and dynamical descriptions of dipion transitions from $\psi(2S)$ and $\Upsilon(nS)$), and first measurements or important refinements of previously low-statistics results (e.g., $J/\psi \to 3\gamma$; $J/\psi \to \gamma\eta_c(1S)$; $\Upsilon(4S) \to \pi^+\pi^-\Upsilon(1S, 2S, 3S)$), and the burgeoning lists of exclusive hadronic decay modes (e.g., $\eta_c(1S)$ and χ_{bJ}). Some previously puzzling situations (e.g., theory–experiment disagreements for higher-order multipoles in $\psi(2S) \to \gamma\chi_{cJ}$, $\chi_{cJ} \to \gamma J/\psi$) have been resolved by improved measurements while others (e.g., the $\rho\pi$ puzzle, suppressed $\psi(2S)$ and $\Upsilon(1S)$ decays to $\gamma\eta$) remain. Has the two-peak dipion mass structure in $\Upsilon(3S) \to \pi^+\pi^-\Upsilon(1S)$ been explained? What exactly is the source of the distorted photon lineshape in $J/\psi \to \gamma\eta_c(1S)$? Does the $\psi(3770)$ have non-$D\bar{D}$ decay modes summing to more than $\sim 1\%$? Our review of decays details new measurements and addresses these and related questions.

For a quarkonium with a small radius, an EFT description of radiative magnetic dipole transitions has been recently obtained, replacing the now-outdated model description; its extension to electric dipole transitions and to states with larger radius is needed. Steady improvement in NRQCD inclusive decay-width calculations has taken place in higher-order expansions in the velocity and strong-coupling constant as well as in the lattice evaluation of matrix elements. Predictions match measurements adequately at the level of ratios of decay widths. Further improvements would require the lattice calculation or data extraction of the NRQCD matrix elements and perturbative resummation of large contributions to the NRQCD matching coefficients. The new data on hadronic transitions and hadronic decays pose interesting challenges to the theory.

The pioneering measurements of quarkonium *production* at the Tevatron were carried out in the early 1990s. Soon after, NRQCD factorization became the standard tool for theoretical calculations. Since then, the Tevatron, B-factories, and HERA have all performed important measurements, some of which have given rise to inconsistencies, puzzles, and new challenges for both theory and experiment. Among these are apparent inconsistencies in quarkonium polarization at the Tevatron between Run I and Run II for the J/ψ, between CDF and DØ for the Υ, and between experiment and NRQCD factorization predictions for both. At least as surprising was the observation at the B-factories that close to 60% of e^+e^- collisions that contain a J/ψ also include a charm meson pair. Photoproduction measurements at HERA revealed discrepancies with LO NRQCD factorization predictions. In response to these and other challenges, the theory of quarkonium production has progressed rapidly.

NRQCD factorization is the basis for much of the current theoretical work on quarkonium production. Factorization in exclusive quarkonium production has recently been proven to all orders in perturbation theory for both double-charmonium production in e^+e^- annihilation and B-meson decays to charmonium and a light meson. NRQCD factorization for inclusive quarkonium production has been shown to be valid at NNLO. However, an all-orders proof of factorization for inclusive quarkonium production remains elusive. This is a key theoretical issue, as a failure of factorization at any order in perturbation theory would imply that there are large, non-factorizing contributions, owing to the presence of soft-gluon effects.

Corrections to hadroproduction have been calculated at NLO, and, in the case of the color-singlet channel, partially at NNLO, even though just a few years ago these calculations were thought to be barely possible. The new calculations show that, because of kinematic enhancements, higher-order corrections can be orders of magnitude larger than the Born-level contributions. In the case of double-charmonium production in e^+e^- collisions, relativistic and perturbative corrections increased the predicted cross sections by more than a factor of four, bringing them into agreement with experiment. New NRQCD factorization calculations of quarkonium photoproduction to NLO at HERA have also moved predictions into agreement with experiment. The importance of higher-order corrections has raised the issue of the convergence of the perturbation series. New methods to address this issue are on the horizon.

New observables have been proposed that may help us to understand the mechanisms of quarkonium production. For example, alternative methods for obtaining information about the polarization of produced quarkonia have been suggested. The associated production of quarkonia may also be an important tool in understanding new states. The production characteristics of the $X(3872)$ may shed light on its exotic nature. The improved theoretical landscape will soon be confronted with the first phase of running at the LHC, where charmonium and bottomonium production will be measured with high statistics in a greatly extended kinematic range.

The study of quarkonium *in medium* has also undergone crucial development. The large datasets from heavy-ion collisions at RHIC suggest that the quark–gluon plasma is actually more like a liquid than a plasma. The suppression of quarkonium production in a hot medium was proposed as a clean probe of deconfined matter. However, the use of quarkonium yields as a diagnostic tool of the hot medium has turned out to be quite challenging. Indeed, quarkonium production was already found to be suppressed by cold-nuclear-matter effects in proton–nucleus collisions. Such effects require dedicated experimental and theoretical attention themselves. In high-energy environments such as at heavy-ion colliders, where more than one $Q\overline{Q}$ pair may be produced in a collision, coalescence of Q and $\overline{Q}$ can lead to secondary quarkonium production, requiring understanding of the transport properties of the medium to separate primary and secondary quarkonium production. The interpretation of in-medium hot-matter effects requires understanding the $Q\overline{Q}$ interaction in terms of finite temperature (T) QCD. The successful Hard Thermal Loop effective theory integrates over the hardest momenta proportional to T for light-quark and gluon observables. To extend the Hard Thermal Loop theory to heavy quarkonium at finite temperature, the additional scales introduced by the bound state must be taken into account. Recently there has been significant progress in constructing a perturbative EFT description of quarkonium at finite T, resulting in a clearly defined potential. This potential displays characteristics that are considerably different from the phenomenological, lattice-inspired description used up to now with well-defined phenomenological implications, as we further discuss. The higher energy of the heavy-ion collisions at the LHC will expand the study of quarkonium in media to bottomonia production and suppression. These studies will be crucial for arriving at a uniform description of heavy quarkonia in cold and hot nuclear matter.

Lastly, we turn our attention to a discussion of the *experimental outlook* in the near term as well as longer-term prospects. For the LHC, the future is now, with the first quarkonium data presented this year. While the preliminary data are encouraging, the full potential for LHC quarkonium studies is still to come. There is a future in low-energy quarkonium hadroproduction studies as well, including two experiments at GSI in Darmstadt, Germany. $\overline{\text{P}}$ANDA will make precision spectroscopy studies in $\bar{p}p$ and $\bar{p}A$ interactions, while the CBM detector will make fixed-target studies of pA and AA interactions to further the understanding of quarkonium production and suppression in high baryon-density matter. Quarkonium physics goals at the currently running BESIII, as well as at proposed super flavor and tau–charm factories are also discussed. Measurements of quarkonium photoproduction offer important insight into the gluon generalized parton distribution (GPD) in nuclei, the role of color correlations, and the color-dipole nature of quarkonia undergoing elastic scattering at high energies. These investigations can be performed at JLab, CERN, and the EIC in the medium term at lower energies, whereas higher energy studies will have to await the ENC, EIC, or LHeC. Important top-quark measurements with high precision can be performed at a future e^+e^- linear collider (ILC or CLIC) in the region just above $t\bar{t}$ threshold. Overall, an extremely active and ambitious future lies ahead for the study of heavy quarkonia with new facilities.

2 Spectroscopy[2]

Spectroscopy is, in part, bump-hunting in mass spectra. Of late, progress has occurred mostly at e^+e^- colliding beam facilities (BES at BEPC, CLEO at CESR, BABAR at PEP-II, Belle at KEKB, KEDR at VEPP-4M), but other venues have gotten into the game as well, including E835 at Fermilab ($\bar{p}p$ gas-jet target) and CDF and DØ at the Tevatron $p\bar{p}$ collider. Tevatron searches target inclusive production of a fully reconstructed state, and can best succeed when the presence of leptons (e.g., $J/\psi \to \mu^+\mu^-$) or displaced vertices (e.g., B-decay) can suppress backgrounds and when there is no need to observe photons. The main strength of e^+e^- colliders is the capability to obtain large datasets at or near charmonium and/or bottomonium vector state masses with well-known initial-state quantum numbers and kinematics. Modern e^+e^- detectors feature precision charged particle trackers, electromagnetic calorimeters, Cherenkov-radiation imagers or time-of-flight taggers, and muon filters, which together allow measurement of the individual decay remnants: γ, $e^\pm$, $\mu^\pm$, $\pi^\pm$, $K^\pm$, $p(\bar{p})$. These capabilities in e^+e^- collisions are exploited using the following techniques.

Full-event reconstruction Datasets taken on-resonance at vector quarkonium masses allow full reconstruction of cascade transitions involving another state (e.g., $\psi(2S) \to \pi^0 h_c$, $h_c \to \gamma\eta_c(1S)$ or $Y(4260) \to \pi^+\pi^- J/\psi$).

Inclusive spectra One or more final-state particles are selected in each event. The measured four-momenta are then used to search directly for mass peaks, or indirectly via missing mass; i.e., the mass recoiling against the particle(s) selected. Two examples are an inclusive photon or π^0 momentum spectrum to identify transitions (e.g., $\Upsilon(3S) \to \gamma\eta_b$ or $\psi(2S) \to \pi^0 h_c(1P)$), which typically have small signals on very large backgrounds. In the continuum reaction $e^+e^- \to XJ/\psi$ with $J/\psi \to \ell^+\ell^-$ (double-charmonium production), the unmeasured particle(s) X can be identified via peaks in the missing-mass spectrum.

Energy scan Scans in e^+e^- center-of-mass energy ($\sqrt{s}$) can map out vector resonances via either inclusive hadronic-event counting (R) and/or exclusive final states (e.g., $D\bar{D}^*$). This does not use machine time efficiently, however, because accelerators work best when operated at a single energy for a long time so tuning can optimize the instantaneous luminosity. Competing priorities usually limit the duration of such scans.

$\gamma\gamma$-fusion The process $e^+e^- \to e^+e^-\gamma^{(*)}\gamma^{(*)} \to e^+e^-X$ allows searches for a large range of masses for X, but X is restricted to having spin-0 or 2 and positive C-parity (e.g., $\eta_c(2S)$ or $\chi_{c2}(2P)$). The outgoing e^+e^- tend to escape the detector at small angles, leaving X with very small momentum transverse to the beamline.

ISR Initial-state radiation (ISR) allows access to all vector states with masses below the $\sqrt{s}$ value of the e^+e^- collision. The effective luminosity per unit of radiative photon energy (or the mass recoiling against it) is well known, allowing for well-normalized exposures at all masses. The ISR photon tends to escape the detector at small angles, leaving the recoiling state with small momentum transverse to the beam but large momentum along it. While the rate for ISR production at any fixed $\sqrt{s}$ is small per unit luminosity, factory-sized datasets at BABAR and Belle make this a viable tool (e.g., $e^+e^- \to \gamma Y(4260)$, $e^+e^- \to \gamma D\bar{D}^*$), and the simultaneous exposure of all allowed masses recoiling against the ISR photon contrasts with the discrete points available from a direct e^+e^- scan.

B-decays Large B-factory datasets at the $\Upsilon(4S)$ make it possible to utilize two-body kinematics to search for exclusive decays of B-mesons (e.g., $B \to KZ_i^+$).

It is worth emphasizing here that the key tenet of experimental science is that discoveries must be reproducible and *verified by independent parties* as a prerequisite for general acceptance. This is no small point in a period such as the present when the world has been bombarded with more than one new state found per year. It is worth pondering spectra which initially were thought to indicate new

[2] Contributing authors: N. Brambilla†, B.K. Heltsley†, A.A. Petrov†, G. Bali, S. Eidelman, U. Ellwanger, A. Gabareen Mokhtar, X. Garcia i Tormo, R. Mussa, F.S. Navarra, M. Nielsen, S.L. Olsen, P. Pakhlov, G. Pakhlova, A.D. Polosa, M.A. Sanchis-Lozano, Y. Sumino, and M. Voloshin.

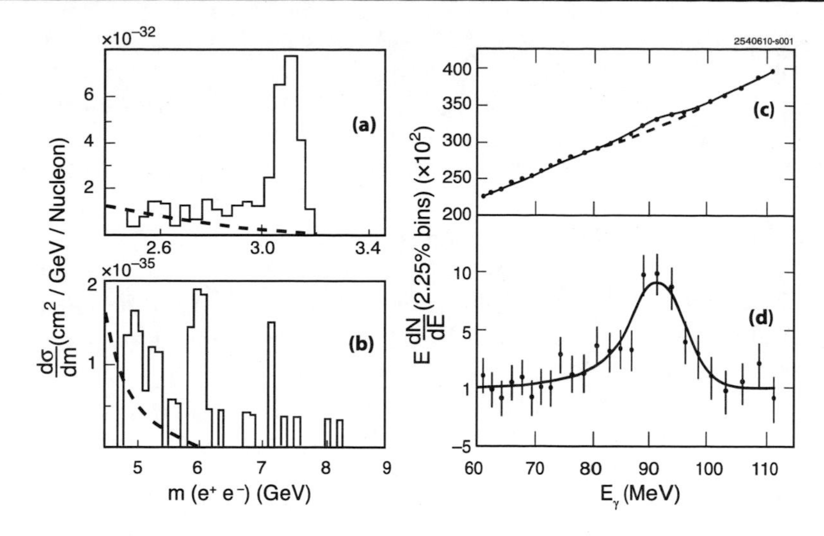

Fig. 1 (**a**), (**b**) Dielectron mass distributions from inclusively selected proton-on-beryllium collisions [2], in which *dashed curves* indicate estimated backgrounds. (**c**) Inclusive photon energy distribution in $\psi(2S)$ decays [5] (*points with error bars*), showing a fit (*solid line*) to a signal and a smooth background (*dashed line*). Part (**d**) is the same as (**c**) but with the smooth backgrounds subtracted. Both peaks were later refuted. Adapted from [2, 5] with kind permission, copyrights (1976, 1982), respectively, The American Physical Society

states in the heavy-quarkonium mass region but were later proven otherwise by independent measurements. Such occurrences happen to the best of experiments and most often can be attributed to fluctuations and/or extremely subtle systematic effects, not overt analysis blunders. Figure 1 highlights two such examples. Figures 1(a) and (b) show dielectron mass distributions observed [2] inclusively in 400 GeV proton collisions on beryllium. A J/ψ peak is clearly seen in (a) while there is an apparent peak near 6 GeV in (b). The authors estimated a 2% probability that the 6 GeV peak was due to background fluctuations, explicitly cautioned that confirmation was needed, and suggested that the name "Υ" be given to this or any subsequently confirmed high-mass dilepton peak. The 6 GeV phenomenon was not confirmed several months later in a dimuon version [3] of the same experiment. The same authors discovered the true $\Upsilon(1S)$ shortly thereafter [4]. Figures 1(c) and (d) show an inclusively selected photon-energy distribution in $\psi(2S)$ decays [5]. The size of the peak near 91 MeV represents a branching fraction of 0.8% and has statistical significance of $>6\sigma$. The peak position corresponds to a mass recoiling against the photon of 3594 ± 5 MeV and a width $\Gamma < 8$ MeV. The result stood as an $\eta_c(2S)$ candidate for twenty years. It was finally refuted [6] as having $\mathcal{B} < 0.2\%$ at 90% CL (confidence level). Incidents such as these (and many others) have led many experiments to adopt more stringent criteria and procedures for establishing signals. These include requiring a threshold of "5σ" statistical significance for claiming "observation", allowing systematic variations to reduce the reported significance, tuning of selection criteria on small subsamples of data not used in the signal search, and the intentional obscuring of signal regions until cuts are frozen (so-called "blind" analysis). *However, every potential signal deserves independent confirmation or refutation.*

This section will first focus on recent measurements: What is the current status of each state? How were the measurements performed? Which need verification? Which are in conflict? Then the theoretical issues will be addressed.

2.1 Conventional vectors above open-flavor threshold

Here we describe recent measurements relevant to the determinations of mass, width, and open-charm content of the four known vector charmonia above open-charm threshold. These states were first observed thirty years ago in e^+e^- annihilation as enhancements in the total hadronic cross section [7–11]. No update of their parameters was made until 2005, when a combined fit to the Crystal Ball [12] and BES [13] R-measurements was performed by Seth [14]. Even more recently, BES [15] reported new parameter values for the ψ resonances. A plethora of open charm cross section measurements has become available and is discussed in what follows. Finally, recent studies of resonant structures just above the open-bottom threshold are described.

2.1.1 Vectors decaying to open charm

The total cross section for hadron production in e^+e^- annihilation is usually parametrized in terms of the ratio R, defined as

$$R = \frac{\sigma(e^+e^- \to \text{hadrons})}{\sigma(e^+e^- \to \mu^+\mu^-)}, \tag{1}$$

where the denominator is the lowest-order QED cross section,

$$\sigma\left(e^+e^- \to \mu^+\mu^-\right) = \frac{4\pi\alpha^2}{3s}. \tag{2}$$

Away from flavor thresholds, measured R values are consistent with the three-color quark model predictions plus terms governed by QCD and the running of $\alpha_s(Q^2)$. Resonant states in the vicinity of flavor thresholds can be studied with fits of measured R distributions. As a part of a study of open charm cross sections in the region from 3.97–4.26 GeV, CLEO [16] published radiatively corrected R-values as shown in Fig. 2. These are in good agreement with earlier measurements [12, 13], which are also shown, demonstrating that in this energy range R values are reasonably well-vetted experimentally.

The extraction of resonance parameters from such R measurements, however, has evolved in complexity, causing systematic movement in some of the parameters over time. The latest BES [15] fit to their R-scan data is more sophisticated than previous efforts and includes the effects of interference and relative phases, as shown in Fig. 3 and Table 1. To take into account interference, BES relied on model predictions for branching fractions of ψ states into all possible two-body charm meson final states. Thus the measured parameters from this fit still include some model uncertainties which are difficult to estimate. Other systematic uncertainties are estimated using alternative choices and combinations of Breit–Wigner forms, energy dependence of the full width, and continuum charm background. It was found that the results are sensitive to the form of the energy-dependent total width but are not sensitive to the form of background.

In a separate analysis, BES [19] fit their R data from 3.65–3.90 GeV, finding a 7σ preference for two interfering lineshapes peaked near 3763 and 3781 MeV relative to a single such shape for the $\psi(3770)$, although other sources for the observed distortion of a pure D-wave Breit–Wigner are possible (see also Sect. 3.4.4). A very recent preliminary analysis of KEDR [20] e^+e^- scan data near the $\psi(3770)$ applies an extended vector dominance model and includes interference with the tail of the $\psi(2S)$ resonance, concluding that the latter interference causes a significant shift upward in the fitted peak of the $\psi(3770)$ as compared to most previous fits, including those of BES. The KEDR measurements are not consistent with the two-peak distortion seen by BES.

For determination of the resonance parameters in the open charm region, inclusive hadronic cross section measurements appear not to supply enough information to deter-

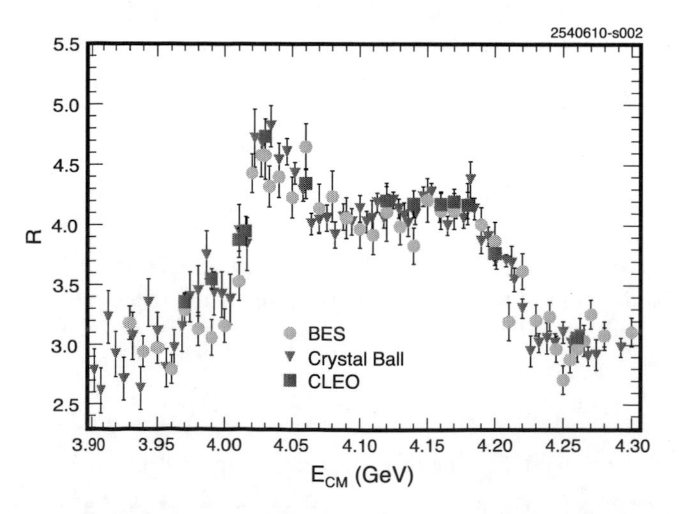

Fig. 2 Measurements of R, including radiative corrections, in the open charm region. From Crystal Ball [12], BES [13], and CLEO [16]. Adapted from [16] with kind permission, copyright (2009) The American Physical Society

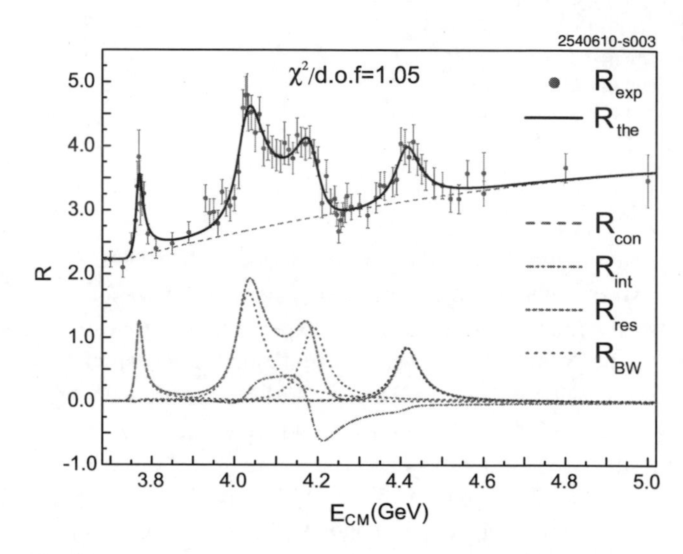

Fig. 3 From BES [15], measured R values from data (*dots with error bars*) and *curves* showing the best fit (*solid*) and the contributions of its components: continuum background (*long dashed*), individual resonance (*dotted*), interference (*dash-dot-dot*), and the summation of the nonbackground curves (*short dashed*). Adapted from [15] with kind permission, copyright (2008) Elsevier

Table 1 The resonance parameters of the high-mass charmonia from the BES global fit [15] together with the values from PDG04 [17], Seth [14], and PDG08 [18]

Resonance	m (MeV)	Γ_{tot} (MeV)	δ (°)	Ref.
$\psi(3770)$	3769.9 ± 2.5	23.6 ± 2.7		PDG04 [17]
	3771.1 ± 2.4	23.0 ± 2.7		Seth [14]
	3772.0 ± 1.9	30.4 ± 8.5	0	BES [15]
	3772.92 ± 0.35	27.3 ± 1.0		PDG08 [18]
$\psi(4040)$	4040 ± 1	52 ± 10		PDG04 [17]
	4039 ± 1.0	80 ± 10		Seth [14]
	4039.6 ± 4.3	84.5 ± 12.3	130 ± 46	BES [15]
$\psi(4160)$	4159 ± 20	78 ± 20		PDG04 [17]
	4153 ± 3	103 ± 8		Seth [14]
	4191.7 ± 6.5	71.8 ± 12.3	293 ± 57	BES [15]
$\psi(4415)$	4415 ± 6	43 ± 15		PDG04 [17]
	4421 ± 4	62 ± 20		Seth [14]
	4415.1 ± 7.9	71.5 ± 19.0	234 ± 88	BES [15]

Table 2 From *BABAR* [28], ratios of branching fractions for the $\psi(4040)$, $\psi(4160)$ and $\psi(4415)$ resonances. The first error is statistical, the second systematic. Theoretical expectations are from models denoted $^3P_0$ [31], C^3 [32], and $\rho K\rho$ [33]

State	Ratio	Measured	$^3P_0$	C^3	$\rho K\rho$
$\psi(4040)$	$D\bar{D}/D\bar{D}^*$	$0.24 \pm 0.05 \pm 0.12$	0.003		0.14
	$D^*\bar{D}^*/D\bar{D}^*$	$0.18 \pm 0.14 \pm 0.03$	1.0		0.29
$\psi(4160)$	$D\bar{D}/D^*\bar{D}^*$	$0.02 \pm 0.03 \pm 0.02$	0.46	0.08	
	$D\bar{D}^*/D^*\bar{D}^*$	$0.34 \pm 0.14 \pm 0.05$	0.011	0.16	
$\psi(4415)$	$D\bar{D}/D^*\bar{D}^*$	$0.14 \pm 0.12 \pm 0.03$	0.025		
	$D\bar{D}^*/D^*\bar{D}^*$	$0.17 \pm 0.25 \pm 0.03$	0.14		

mine the relative strength of different decay channels. More data and more reliable physical models appear to be needed in order to make further progress. The PDG-supplied parameters in Table 1 bypass these issues and provide parameters under the simplest of assumptions, which may or may not turn out to be correct.

Detailed studies of the open-charm content in the charmonium region were not undertaken until large datasets were obtained by CLEO at discrete energy points and by the B-factory experiments using radiative returns to obtain a continuous exposure of the mass region. The picture that has emerged is complex due to the many thresholds in the region, nine of which are two-body final states using allowed pairs of D^0, D^+, D^{*0}, D^{*+}, D_s^+, and D_s^{*+}. Moreover, distinguishing genuine two-body from "multibody" decays (e.g., $D\bar{D}^*$ from $D^0D^-\pi^+$) poses a challenge. Experimentally, the data are consistent where measurements overlap; significant discrepancies with some predictions mean that theoretical work remains.

Exclusive e^+e^- cross sections for hadronic final states containing charm mesons were studied by several groups. In the $\sqrt{s} = 3.7$–5 GeV energy region, Belle [21–26], and *BABAR* [27–29] used initial-state radiation (ISR) to reach the charmonium region, and CLEO used its large data sample taken at the $\psi(3770)$ peak [30] and its scan over $\sqrt{s} = 3.97$–4.26 GeV [16]. Some of these results can be seen in Table 2 and Figs. 4 and 5. Measurements of the neutral to charged $D\bar{D}$ cross section ratio at the $\psi(3770)$ peak show consistency but the PDG08 [18] world average, 1.260 ± 0.021, is dominated by the CLEO [30] value. The $D\bar{D}$ cross sections across the entire charm energy range from Belle [21] and *BABAR* [27] appear in Fig. 4 and are consistent with one another. Both observe a structure in the ISR $D\bar{D}$ cross section (Figs. 4(a) and (b)), known as $G(3900)$, which must be taken into account to describe both the $D\bar{D}$ cross section and R in the region between $\psi(3770)$ and $\psi(4040)$. The $G(3900)$ is not considered to be a specific $c\bar{c}$ bound state, as it is qualitatively consistent with a prediction from a coupled-channel model [34]. The D^+D^{*-}

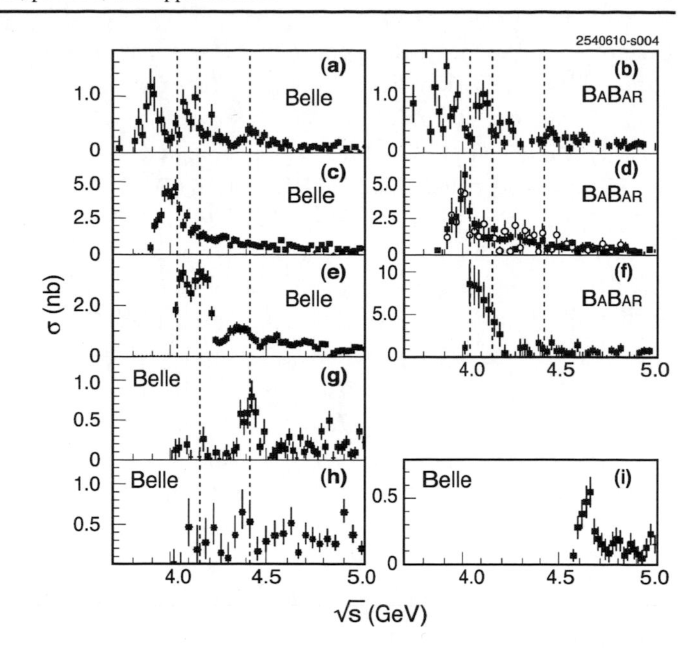

Fig. 4 Measured e^+e^- exclusive open-charm meson- or baryon-pair cross sections for $\sqrt{s} = 3.7$–5.0 GeV from Belle and *BABAR*, showing **(a)** $D\bar{D}$ [21]; **(b)** $D\bar{D}$ [27]; **(c)** D^+D^{*-} [22]; **(d)** $D\bar{D}^*$ for $D = D^0$ (*solid squares*) and $D = D^+$ (*open circles*) [28]; **(e)** $D^{*+}D^{*-}$ [22]; **(f)** $D^*\bar{D}^*$ [28]; **(g)** $D^0D^-\pi^+$ [23]; **(h)** $D^0D^{*-}\pi^+$ [24]; **(i)** $\Lambda_c^+\Lambda_c^-$ [25]. *Vertical dashed lines* indicate ψ masses in the region. Adapted from [21–25, 27, 28] with kind permission, copyrights (2008, 2007, 2008, 2009, 2008, 2007, 2008), respectively, The American Physical Society

cross sections from Belle [22] and *BABAR* [28] exhibit a single broad peak near threshold whereas $D^{*+}D^{*-}$ results [22, 28] feature several local maxima and minima across this energy range. The $e^+e^- \to \Lambda_c^+\Lambda_c^-$ cross section measured by Belle [25], shown in Fig. 4(i), exhibits a substantial enhancement just above threshold near 4.6 GeV (addressed below).

BABAR [28, 29] performed unbinned maximum-likelihood fits to the $D\bar{D}$, $D\bar{D}^*$, $D^*\bar{D}^*$, and $D_s^{(*)+}D_s^{(*)-}$ spectra. The expected ψ signals were parametrized by P-wave relativistic Breit–Wigner (RBW) functions with their parameters fixed to the PDG08 values [18]. An interference between the resonances and the nonresonant contributions was required in the fit. The computed ratios of the branching fractions for the ψ resonances to nonstrange open-charm meson pairs and the quark model predictions are presented in Table 2. The *BABAR* results deviate from some of the theoretical expectations. The *BABAR* [29] cross sections for $D_s^{(*)+}D_s^{(*)-}$ production show evidence for $\psi(4040)$ in $D_s^+D_s^-$ and $\psi(4160)$ in $D_s^{*+}D_s^-$, and are consistent with the CLEO [16] results where they overlap.

The $e^+e^- \to D^0D^-\pi^+$ cross section measured by Belle [23] is shown in Fig. 4(g) and exhibits an unambiguous $\psi(4415)$ signal. A study of the resonant structure shows clear signals for the $\bar{D}_2^*(2460)^0$ and $D_2^*(2460)^+$ mesons and constructive interference between the neutral $D^0\bar{D}_2^*(2460)^0$ and the charged $D^-D_2^*(2460)^+$ decay amplitudes. Belle performed a likelihood fit to the $D\bar{D}_2^*(2460)$ mass distribution with a $\psi(4415)$ signal parametrized by an S-wave RBW function. The significance of the signal is $\sim 10\sigma$ and the peak mass and total width are in good agreement with the PDG06 [35] values and the BES fit results [15]. The branching fraction for $\psi(4415) \to D\bar{D}_2^*(2460) \to D\bar{D}\pi^+$ was found to be between 10% and 20%, depending on the $\psi(4415)$ parametrization. The fraction of $D\bar{D}_2^*(2460) \to D\bar{D}\pi^+$ final states composed of nonresonant $D^0D^-\pi^+$ was found to be <22%. Similarly, the $D^0D^{*-}\pi^+$ content of $\psi(4415)$, shown in Fig. 4(h), has been determined by Belle [24]; a marginal signal is found (3.1σ), and its branching fraction was limited to <10.6%. Belle [26] has also reported a spectrum of $e^+e^- \to D_s^{(*)+}D_s^{(*)-}$ cross sections from $\sqrt{s} = 3.8$–5 GeV using ISR from a data sample of 967 fb^{-1} in 40 MeV bins; the values are consistent with but higher-statistics and more finely binned than those of BABAR [29].

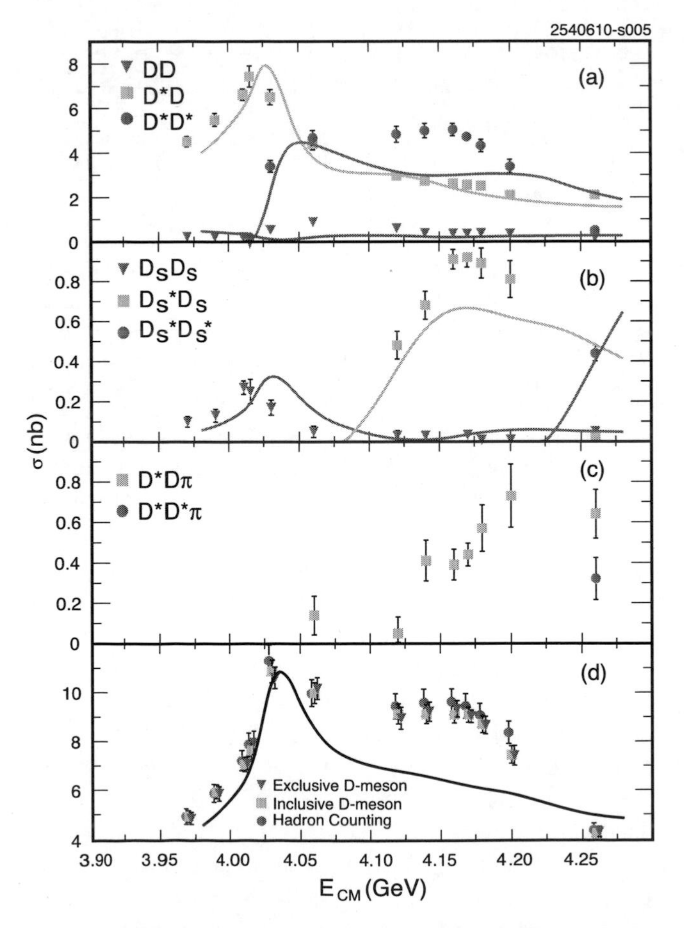

Fig. 5 From CLEO [16], cross sections, without radiative corrections, for e^+e^- annihilation into: (**a**)–(**c**) the exclusive open-charm meson-pairs shown; and (**d**) for two methods of open-charm counting added onto the extrapolated *uds* cross section compared to an all-flavor hadron-counting cross section. *Solid lines* show updated potential model predictions [34]. Adapted from [16] with kind permission, copyright (2009) The American Physical Society

The CLEO exclusive cross sections [16] in the top three frames of Fig. 5 are not directly comparable to those from BABAR and Belle as they are not radiatively corrected, but generally seem to reflect consistency. The updated potential model predictions of Eichten [16, 34] shown in Fig. 5 fail to describe many features of the data. The CLEO total cross section determinations, shown in Fig. 5(d), reveal that, within the measurement accuracy of 5–10%, two- and three-body modes with open charm saturate the yield of all multihadronic events above the extrapolated *uds* contribution.

2.1.2 Vectors decaying to open bottom

The current generation of B-factories have scanned the energy range above open bottom threshold. BABAR [36] performed a comprehensive low-luminosity (25 pb^{-1} per point), high-granularity ($\approx$5 MeV steps) scan between 10.54 and 11.2 GeV, followed by an eight-point scan, 0.6 fb^{-1} total, in the proximity of the $\Upsilon(6S)$ peak. Belle [37] acquired $\approx$30 pb^{-1} for just nine points over 10.80–11.02 GeV, as well as 8.1 fb^{-1} spread over seven additional points more focused on the $\Upsilon(5S)$ peak. The BABAR scan is shown in Fig. 6. Both scans suggest instead that the simple Breit–Wigner parametrization, previously used to model the peaks observed in the CLEO [38] and CUSB [39] scans, is not adequate for the description of the complex dynamics in the proximity of the $B^{(*)}\bar{B}^{(*)}$ and $B_s^{(*)}\bar{B}_s^{(*)}$ thresholds. Data points on $R_b = \sigma(b\bar{b})/\sigma(\mu\mu)$ are better modeled assuming a flat $b\bar{b}$ continuum contribution which interferes constructively with the $5S$ and $6S$ Breit–Wigner resonances, and a second flat contribution which adds incoherently. Such fits strongly alter the PDG results on the $5S$ and $6S$ peaks,

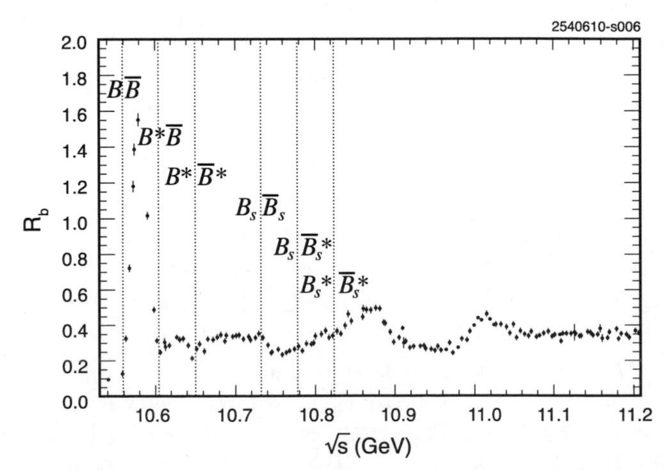

Fig. 6 From BABAR [36], measured values of the hadronic cross section attributable to b-flavored states, normalized to the point muon-pair cross section, from a scan of the center-of-mass energy region just below the $\Upsilon(4S)$ to above the $\Upsilon(6S)$. *Dashed vertical lines* indicate the various $B\bar{B}$ mass thresholds. Adapted from [36] with kind permission, copyright (2009) The American Physical Society

Table 3 New *BABAR* and Belle results on masses and widths of the $\Upsilon(5S)$ and $\Upsilon(6S)$ resonances, compared to PDG averages. The angle ϕ parametrizes the phase of interfering continuum

Υ	m (MeV)	Γ (MeV)	ϕ (rad)	Ref.
5S	10876 ± 2	43 ± 4	2.11 ± 0.12	*BABAR* [36]
	10879 ± 3	46^{+9}_{-7}	$2.33^{+0.26}_{-0.24}$	Belle [37]
	10865 ± 8	110 ± 13	–	PDG08 [18]
6S	10996 ± 2	37 ± 3	0.12 ± 0.07	*BABAR* [36]
	11019 ± 8	79 ± 16	–	PDG08 [18]

as shown in Table 3. Strong qualitative agreement is observed between the experimental behavior of the R_b ratio and the theory predictions based on the coupled-channel approach [40].

Additional insight can be provided by the exclusive decomposition of the two-body (i.e., $B\bar{B}$, $B\bar{B}^*$, $B^*\bar{B}^*$) and many-body decay modes. Results from e^+e^- collisions have been given by Belle [41] using a data sample of 23.6 fb^{-1} acquired at the $\Upsilon(5S)$. Charged B-mesons were reconstructed in two decay channels, $K^\pm J/\psi$ and $D^0\pi^\pm$ (with $J/\psi\to l^+l^-$ and $D^0\to K\pi, K\pi\pi\pi$). Neutral B mesons were reconstructed in $K^{*0}J/\psi$ and $D^\pm\pi^\mp$, with $D^\pm\to K^\pm\pi^\pm\pi^\mp$. The B^* mesons were reconstructed via their radiative transition. Belle observes a large fraction (about 16.4% of the total $b\bar{b}$ pairs) from 3- and 4-body decay modes, i.e., $B^{(*)}\bar{B}^{(*)}\pi$, $B^{(*)}\bar{B}^{(*)}\pi\pi$. A significant fraction of these events can actually be expected from ISR production of $\Upsilon(4S)$. Theory predictions on multibody decays at $\Upsilon(5S)$ range from 0.03% [42] to 0.3% [43].

2.2 Newly found conventional quarkonia

Table 4 lists properties of new conventional heavy-quarkonium states. The h_c is the ${}^1P_1$ state of charmonium, singlet partner of the long-known χ_{cJ} triplet ${}^3P_J$. The $\eta_c(2S)$ is the first excited state of the pseudoscalar ground state $\eta_c(1S)$, lying just below the mass of its vector counterpart, $\psi(2S)$. The first B-meson seen that contains charm is the B_c. The ground state of bottomonium is the $\eta_b(1S)$. And the $\Upsilon(1D)$ is the lowest-lying D-wave triplet of the $b\bar{b}$ system. All fit into their respective spectroscopies roughly where expected. Their exact masses, production mechanisms, and decay modes provide guidance to their descriptions within QCD.

2.2.1 *Observation of $h_c(1P)$*

Two experiments reported $h_c(1P)$ sightings in 2005, with CLEO [44, 45] reporting an observation at $>6\sigma$ in the isospin-forbidden decay chain $e^+e^-\to\psi(2S)\to\pi^0h_c$, $h_c\to\gamma\eta_c(1S)$, and E835 [48] found 3σ evidence in $p\bar{p}\to h_c$, $h_c\to\gamma\eta_c(1S)$, $\eta_c(1S)\to\gamma\gamma$. CLEO [46] later updated its measurements with a larger dataset, refining its mass measurement to a precision of just over 0.2 MeV, finding a central value slightly more accurate than that of E835, which has an uncertainty of just under 0.3 MeV. CLEO utilized two detection methods. The first was a semi-inclusive selection that required detection of both the transition π^0 and radiative photon but only inferred the presence of the $\eta_c(1S)$ through kinematics. The second employed full reconstruction in fifteen different $\eta_c(1S)$ decay modes, five of them previously unseen. The two methods had some statistical and almost full systematic correlation for the mass measurement because both rely on the π^0 momentum determination. As the parent $\psi(2S)$ has precisely known mass and is produced nearly at rest by the incoming e^+e^- pair, the mass of the $h_c(1P)$ is most accurately determined by fitting the distribution of the mass recoiling against the π^0, as shown for the exclusive analysis in Fig. 7. CLEO's two methods had comparable precision and gave consistent masses within their uncorrelated uncertainties. Statistical uncertainties from the numbers of signal (background) events in the exclusive (inclusive) analysis are larger than the systematic errors attributable to calorimeter energy resolution. The E835 measurement relies on knowledge of the initial center-of-mass energy of the $p\bar{p}$ for each event during a scan of the $h_c(1P)$ mass region as well as upon reconstruction of all three photons with kinematics consistent with the production and decay hypothesis. Unlike the CLEO result, backgrounds are negligible. Mass measurement accuracy was limited equally by statistics (13 signal events with a standard deviation in center-of-mass energy of 0.07 MeV) and systematics of $\bar{p}$ beam energy stability. Using a sample of 106M $\psi(2S)$, in 2010 BESIII [47] reported a mass result using the $\pi^0\gamma$ inclusive method, matching CLEO's precision. The spin-averaged centroid of the triplet states, $\langle m(1^3P_J)\rangle\equiv[m(\chi_{c0})+3m(\chi_{c1})+5m(\chi_{c2})]/9$, is expected to be near the $h_c(1P)$ mass, making the hyperfine mass splitting, $\Delta m_{\text{hf}}[h_c(1P)]\equiv\langle m(1^3P_J)\rangle-m[h_c(1P)]$, an important measure of the spin–spin interaction. The h_c-related quantities are summarized in Table 5; mass measurements are consistent. It could be a coincidence [46] that $\Delta m_{\text{hf}}[h_c(1P)]_{\text{exp}}\approx0$, the same as the lowest-order perturbative QCD expectation, because the same theoretical assumptions lead to the prediction

$$\frac{m(\chi_{c1})-m(\chi_{c0})}{m(\chi_{c2})-m(\chi_{c1})}=\frac{5}{2},\tag{3}$$

whereas measured masses [18] yield a value of 2.1, 20% smaller than predicted.

2.2.2 *Observation of $\eta_c(2S)$*

The search for a reproducible $\eta_c(2S)$ signal has a long and checkered history. There were hints in early $e^+e^-\to c\bar{c}$

Table 4 New *conventional* states in the $c\bar{c}$, $b\bar{c}$, and $b\bar{b}$ regions, ordered by mass. Masses m and widths Γ represent the weighted averages from the listed sources. Quoted uncertainties reflect quadrature summation from individual experiments. In the Process column, the decay mode of the new state claimed is indicated in parentheses. Ellipses (...) indicate inclusively selected event topologies; i.e., additional particles not required by the Experiments to be present. For each Experiment a citation is given, as well as the statistical significance in number of standard deviations (#σ), or "(np)" for "not provided." The Year column gives the date of first measurement cited, which is the first with significance of $>5\sigma$. The Status column indicates that the state has been observed by at most one (NC!-needs confirmation) or at least two independent experiments with significance of $>5\sigma$ (OK). The state labelled $\chi_{c2}(2P)$ has previously been called $Z(3930)$

State	m (MeV)	Γ (MeV)	J^{PC}	Process (mode)	Experiment (#σ)	Year	Status
$h_c(1P)$	3525.45 ± 0.15	0.73 ± 0.53	1^{+-}	$\psi(2S) \to \pi^0(\gamma\eta_c(1S))$	CLEO [44–46] (13.2)	2004	OK
		(<1.44)		$\psi(2S) \to \pi^0(\gamma \ldots)$	CLEO [44–46] (10.0), BES [47] (18.6)		
				$p\bar{p} \to (\gamma\eta_c) \to (\gamma\gamma\gamma)$	E835 [48] (3.1)		
				$\psi(2S) \to \pi^0(\ldots)$	BESIII [47] (9.5)		
$\eta_c(2S)$	3637 ± 4	14 ± 7	0^{-+}	$B \to K(K_S^0 K^- \pi^+)$	Belle [49] (6.0)	2002	OK
				$e^+e^- \to e^+e^-(K_S^0 K^- \pi^+)$	*BABAR* [50] (4.9), CLEO [51] (6.5), Belle [52] (6)		
				$e^+e^- \to J/\psi\ (\ldots)$	*BABAR* [53] (np), Belle [54] (8.1)		
$\chi_{c2}(2P)$	3927.2 ± 2.6	24.1 ± 6.1	2^{++}	$e^+e^- \to e^+e^-(D\bar{D})$	Belle [55] (5.3), *BABAR* [56] (5.8)	2005	OK
B_c^+	6277.1 ± 4.1	–	0^-	$\bar{p}p \to (\pi^+ J/\psi) \ldots$	CDF [57] (8.0), DØ [58] (5.2)	2007	OK
$\eta_b(1S)$	9390.7 ± 2.9	?	0^{-+}	$\Upsilon(3S) \to \gamma + (\ldots)$	*BABAR* [59] (10), CLEO [60] (4.0)	2008	NC!
				$\Upsilon(2S) \to \gamma + (\ldots)$	*BABAR* [61] (3.0)		
$\Upsilon(1^3D_2)$	10163.8 ± 1.4	?	2^{--}	$\Upsilon(3S) \to \gamma\gamma(\gamma\gamma\Upsilon(1S))$	CLEO [62] (10.2)	2004	OK
				$\Upsilon(3S) \to \gamma\gamma(\pi^+\pi^-\Upsilon(1S))$	*BABAR* [63] (5.8)		

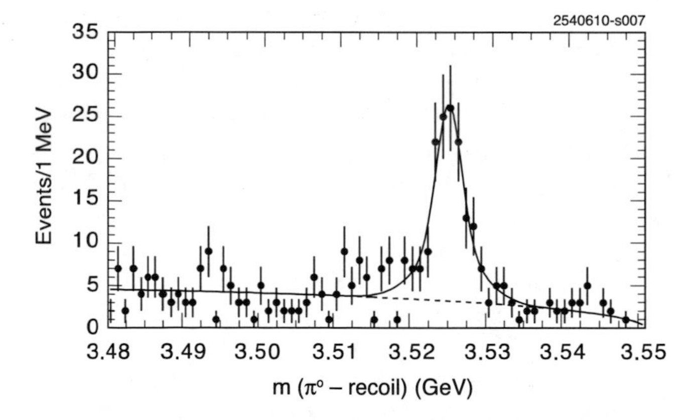

Fig. 7 From CLEO [46], the mass recoiling against the π^0 in the $\psi(2S) \to \pi^0 h_c$, $h_c \to \gamma\eta_c(1S)$ *exclusive* sample in which the π^0, γ, and $\eta_c(1S)$ are all explicitly reconstructed in the detector. Adapted from [46] with kind permission, copyright (2008) The American Physical Society

Table 5 Width and mass measurements of the $h_c(1P)$. $\langle m(1^3P_J)\rangle$ and $\Delta m_{\rm hf}$ are defined in the text

Quantity	Value (MeV)	Ref. (χ^2/d.o.f.)
Width	$0.73 \pm 0.45 \pm 0.28$	BES [47]
	<1.44@90% CL	BES [47]
Mass	$3525.8 \pm 0.2 \pm 0.2$	E835 [48]
	$3525.28 \pm 0.19 \pm 0.12$	CLEO [46]
	$3525.40 \pm 0.13 \pm 0.18$	BES [47]
	3525.45 ± 0.15	Avg[3] (2.2/2)
$\langle m(1^3P_J)\rangle$	3525.30 ± 0.07	PDG08 [18]
$\Delta m_{\rm hf}$	-0.15 ± 0.17	

data for a purported $\eta_c(2S)$ with mass near 3455 MeV in $\psi(2S) \to \gamma\gamma J/\psi$ events [64] and in inclusive radiative $\psi(2S)$ decays [65, 66]. A possible signal [67] near 3591 MeV was reported in 1978 in $\psi(2S) \to \gamma\gamma J/\psi$. Crystal Ball ruled out that result in 1982 [5] and also reported an $\eta_c(2S)$ signal in *inclusive* radiative $\psi(2S)$

[3] A note concerning tables in this section: where the label "Avg" is attached to a number, it signifies an inverse-square-error-weighted average of values appearing directly above, for which all statistical and systematic errors were combined in quadrature without accounting for any possible correlations between them. The uncertainty on this average is inflated by the multiplicative factor S if $S^2 \equiv \chi^2/\text{d.o.f.} > 1$.

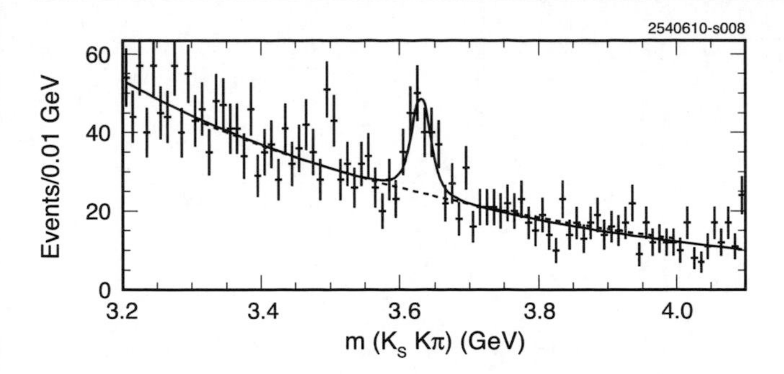

Fig. 8 From *BABAR* [50], the $K\bar{K}\pi$ invariant-mass distribution from selected $e^+e^- \to e^+e^-K_S^0K^+\pi^-$ events, data (*points with error bars*) overlaid with a fit (*solid line*) having two components, a smooth background (*dashed line*) and an $\eta_c(2S)$ signal. Adapted from [50] with kind permission, copyright (2004) The American Physical Society

decays with a mass of 3592 ± 5 MeV. The latter result persisted, in limbo, unconfirmed and unrefuted, for twenty years, until Belle [49] found a signal in $B \to K\eta_c(2S)$ in the exclusive $\eta_c(2S) \to K_S^0K^-\pi^+$ decay mode (a favorite all-charged final state for $\eta_c(1S)$), at $3654 \pm 6 \pm 8$ MeV. Since then measurements of $\eta_c(2S)$ in that mass region have been reported by *BABAR* [50] (see Fig. 8), CLEO [51], and Belle [52] in $\gamma\gamma$-fusion to $K\bar{K}\pi$ final states and by *BABAR* [53] and Belle [54] in double charmonium production.

With this plethora of independent measurements in three different production mechanisms and two methods of mass reconstruction (fully reconstructed exclusive decay to $K\bar{K}\pi$ and missing mass), it might have been reasonable to expect clarity and cohesion to have emerged. However, complete experimental unity eludes us because, while the mass values are all in the same vicinity, when averaged they have a PDG S-factor of 1.7, the factor by which the weighted-average uncertainty is inflated. The two most precise measurements, both from $\gamma\gamma \to \eta_c(2S)$, disagree by 2.5σ; the two least precise, both by Belle, disagree by 2.2σ; the two double-charmonium results disagree by 1.6σ; and the two *BABAR* results disagree by 1.4σ. There are no easily identifiable outliers to discard. The lesson here may be that statistics in all the methods utilized are hard to come by, or that background shapes are more complicated than assumed, or that these measurements have been plagued by extraordinary bad luck. In any case, further exploration is clearly merited. CLEO [68] attempted to find exclusive $\eta_c(2S)$ decays in radiative $\psi(2S)$ decays, guided by the success of such methods for $\eta_c(1S)$ [69], but found no clear signals in its sample of 25M $\psi(2S)$.

Just prior to submission of this article, Belle [70] announced a preliminary observation of $\eta_c(2S)$, produced in two-photon fusion, in three new decay modes ($3(\pi^+\pi^-)$, $K^+K^-2(\pi^+\pi^-)$, and $K_S^0K^-\pi^+\pi^-\pi^+$). These modes will offer more concrete avenues of approach to $\eta_c(2S)$ in order to better measure its properties.

Table 6 Properties of the $\chi_{c2}(2P)$ (originally $Z(3930)$)

Quantity	Value	Ref. (χ^2/d.o.f.)
Mass (MeV)	$3929 \pm 5 \pm 2$	Belle [55]
	$3926.7 \pm 2.7 \pm 1.1$	*BABAR* [56]
	3927.2 ± 2.6	Avg[3] (0.14/1)
Width (MeV)	$29 \pm 10 \pm 2$	Belle [55]
	$21.3 \pm 6.8 \pm 3.6$	*BABAR* [56]
	24.1 ± 6.1	Avg[3] (0.37/1)
$\Gamma_{\gamma\gamma} \times \mathcal{B}(D\bar{D})$	$0.18 \pm 0.05 \pm 0.03$	Belle [55]
(keV)	$0.24 \pm 0.05 \pm 0.04$	*BABAR* [56]
	0.21 ± 0.04	Avg[3] (0.46/1)

2.2.3 Observation of $\chi_{c2}(2P)$

In 2005 Belle [55] observed an enhancement in the $D\bar{D}$ mass spectrum from $e^+e^- \to e^+e^-D\bar{D}$ events with a statistical significance of 5.3σ. Properties are shown in Table 6. It was initially dubbed the $Z(3930)$, but since has been widely[4] (if not universally) accepted as the $\chi_{c2}(2P)$. The analysis selects fully reconstructed D meson pairs with at most one π^0 and at most six pions/kaons per event, using the decays $D^0 \to K^-\pi^+$, $K^-\pi^+\pi^0$, and $K^-\pi^+\pi^+\pi^-$, and $D^+ \to K^-\pi^+\pi^+$. The outgoing e^+e^- were presumed to exit the detector at small angles. This $\gamma\gamma$-fusion signature was enforced by requiring small transverse momentum with respect to the beam direction in the e^+e^- center-of-mass frame and restricting the $D\bar{D}$ longitudinal momentum to kinematically exclude $e^+e^- \to \gamma D\bar{D}$. Figure 9 shows the resulting $D\bar{D}$ mass and angular distributions; the latter are consistent with the spin-2, helicity-2 hypothesis but disagree with spin-0. *BABAR* [56] confirmed the Belle observation in $\gamma\gamma$-fusion with significance of 5.8σ and found properties consistent with those from Belle.

2.2.4 Observation of B_c^+

Unique among mesons is the B_c^+ because it is the lowest-lying (and only observed) meson composed of a heavy quark and a heavy antiquark of different flavors. As such, its mass, lifetime, decay, and production mechanisms garner attention so as to constrain and cross-check QCD calculations similar to those used for other heavy quarkonia. Since its mass is well above 6 GeV, production of pairs $B_c^+B_c^-$ at the e^+e^- B-factories, which take most of their data near the $\Upsilon(4S)$, has not been possible. Although a hint from OPAL [72] at

[4] Lattice calculations [71] suggest that the $\chi_{c2}(2P)$ (i.e., the $2\,^3P_2$ $c\bar{c}$ state) and the $1\,^3F_2$ state could be quite close in mass, so that perhaps the $Z(3930)$ is *not* the $2\,^3P_2$ but rather the $1\,^3F_2$.

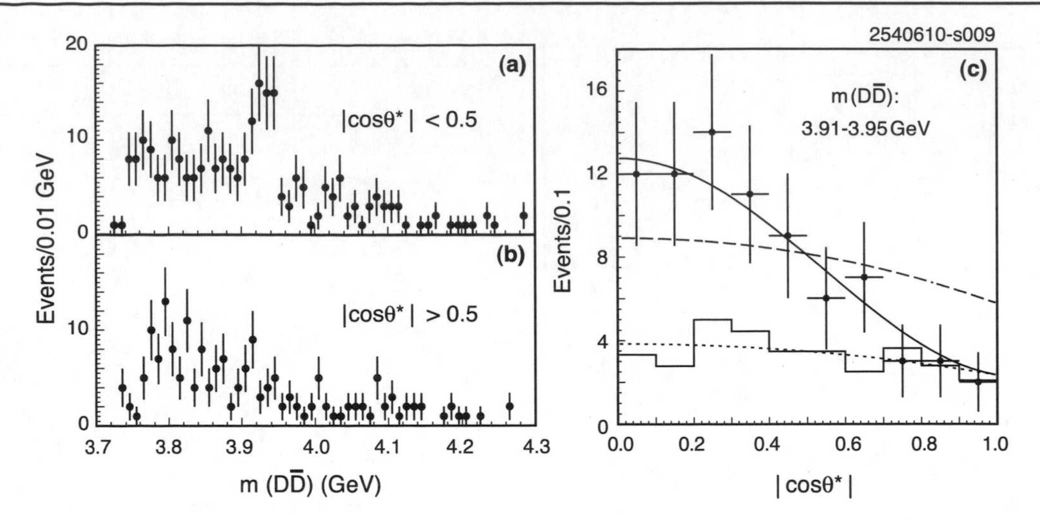

Fig. 9 From the Belle [55] observation of $\chi_{c2}(2P) \to D\bar{D}$. (**a**) and (**b**) $m(D\bar{D})$ distributions from selected $e^+e^- \to e^+e^- D\bar{D}$ events, for the $|\cos\theta^*|$ regions indicated, where θ^* is the polar angle of a D momentum vector in the $\gamma\gamma$ center-of-mass frame. Part (**c**) shows the corresponding $|\cos\theta^*|$ distributions for the $m(D\bar{D})$ region indicated, from data (*points with error bars*) and background (*solid line histogram*). Also shown are expected distributions for the spin-2 (helicity-2) (*solid curve*) and spin-0 (*dashed curve*) hypotheses, both of which include background (*dotted curve*). Adapted from [55] with kind permission, copyright (2006) The American Physical Society

LEP and then suggestive evidence from CDF [73] for the existence of B_c^+ were published in 1998, it was not until a decade later that two confirming observations in excess of 5σ significance were made in Run II at the Tevatron. Both mass measurements used the decay chain $B_c^+ \to J/\psi\pi^+$, $J/\psi \to \mu^+\mu^-$, and obtain for mass and statistical significance the values

$$m(B_c^+) = 6275.6 \pm 2.9 \pm 2.5\ \text{MeV} \quad (8\sigma)\ \text{CDF [57]}$$
$$= 6300 \pm 14 \pm 5\ \text{MeV} \quad (5.2\sigma)\ \text{DØ [58]}. \qquad (4)$$

The CDF B_c^+ mass plot is shown in Fig. 10. Their weighted average (Table 4) is about 2σ lower than the lattice QCD prediction [74] of $6304 \pm 12^{+18}_{-0}$ MeV. The only observed decay modes for B_c^+ are $J/\psi\pi^+$ and $J/\psi\ell^+\nu_\ell$. The semileptonic mode has been used by both CDF [75, 76] and DØ [77] to measure the B_c^+ lifetime. Their results are consistent with each other and have a weighted average [76] of 0.46 ± 0.04 ps. (See also Sect. 4.7 for discussion of B_c^+.)

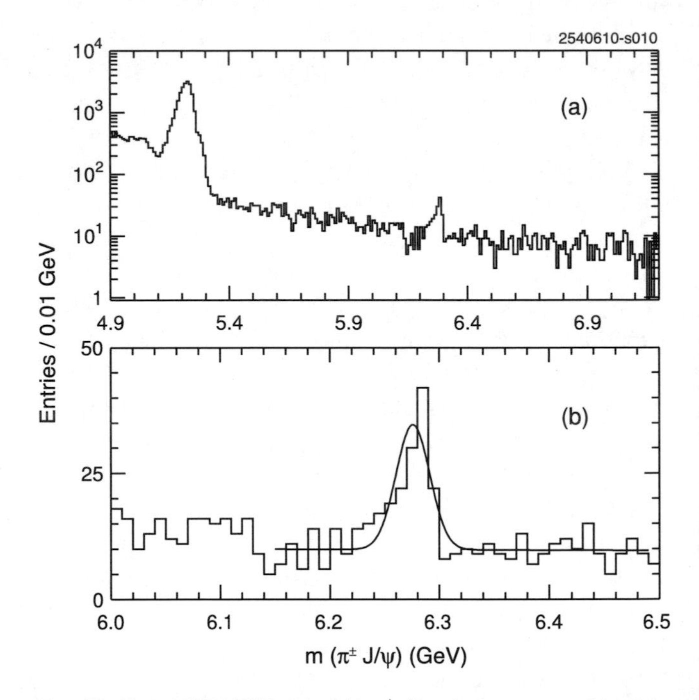

Fig. 10 From CDF [57], (**a**) $J/\psi\pi^+$ invariant-mass combinations from selected $\bar{p}p \to \pi^+ J/\psi X$ events. The bump near 5.2 GeV is due to $B^+ \to K^+ J/\psi$ decays with a pion-mass assignment for the kaon. (**b**) As in (**a**), but zoomed in on the 6.0–6.5 GeV mass region; the *solid curve* indicates the projection of the B_c^+ maximum-likelihood fit to the data. Adapted from [57] with kind permission, copyright (2008) The American Physical Society

2.2.5 *Observation of* $\eta_b(1S)$

Nonobservation of the bottomonium ground state was an annoying thorn in the side of heavy-quarkonium spectroscopy until 2008, when BABAR [59, 61] succeeded in observing the $\eta_b(1S)$ where previous efforts had failed. The hyperfine mass-splitting of singlet-triplet states, $\Delta m_{\rm hf} \equiv m(1^3S_1) - m(1^1S_0)$, probes the spin-dependence of bound-state energy levels, and, once measured, imposes constraints on theoretical descriptions. The η_b remained elusive for a variety of reasons. Branching fractions for transitions from the $\Upsilon(nS)$ states are small and no low-multiplicity, high-rate "golden" decay modes analogous to $\eta_c(1S) \to K\bar{K}\pi$ appear to exist for η_b. This left inclusive $\Upsilon(nS) \to \gamma\eta_b$ as the first line of attack.

BABAR's success was mainly due to large data samples obtained just prior to shutdown of the experiment. For the express objective of η_b-discovery (among others), *BABAR* accumulated 122M $\Upsilon(3S)$ and 100M $\Upsilon(2S)$ decays, compared to CLEO (9M $\Upsilon(2S)$ and 6M $\Upsilon(3S)$) and Belle (11M $\Upsilon(3S)$). Even with such large data samples and a high-performance cesium iodide crystal calorimeter, *BABAR*'s task was far from trivial: the expected photon line was buried under a sea of π^0-decay photons even after all photon candidates that combine with any other photon to form a π^0 were vetoed. The η_b photon line was also obscured by two other physics processes, each inducing structure in E_γ, the photon energy in the $\Upsilon(nS)$ rest frame. The η_b photon line lies in the high-energy tail of the three Doppler-smeared and merged $\chi_{bJ}(nP) \to \gamma\Upsilon(1S)$ peaks and adjacent to that of the radiative return process, $e^+e^- \to \gamma\Upsilon(1S)$. *BABAR* introduced a method to suppress nonresonant "continuum" photons and thereby enhance experimental signal-squared-to-background ratio (S^2/B), noting that such backgrounds tend to follow initial parton (jet) directions, whereas the η_b decay products will have direction uncorrelated with that of the transition photon. The angle θ_T was defined to be the angle between each transition photon candidate and the thrust axis [78] of the rest of the event. (The thrust axis is the direction that maximizes the sum of absolute values of momenta projected upon it, and, on a statistical basis, follows the axis of two-jet events.) The thrust angle associated with each candidate radiative photon was calculated and required to satisfy $|\cos\theta_T| < 0.7$, the criterion found by *BABAR* to maximize S^2/B. The analysis extracted a signal by fitting the E_γ distribution to four components: an empirically determined smooth background, merged χ_{bJ} peaks, a monochromatic ISR photon line, and an η_b signal. The resulting E_γ spectrum from the *BABAR* [59] $\Upsilon(3S)$ analysis is shown in Fig. 11, with an η_b signal of significance of $>10\sigma$. A few months after this discovery, *BABAR* [61] announced confirmation of their signal with a nearly identical analysis of their $\Upsilon(2S)$ data, albeit with smaller significance (3.0σ). To avoid bias, these analyses established procedures while "blind" to the η_b signal region in E_γ.

Initially, there was some worry that the *BABAR* results were in mild conflict with earlier nonobservation upper limits from a CLEO [79] analysis, which had as its primary focus a detailed study of the dipole transitions $\Upsilon(nS) \to \gamma\chi_{bJ}(mP)$. However, CLEO [60] later corrected errors and omissions in that analysis and announced new results consistent with but less precise than *BABAR*'s, including 4σ evidence for $\Upsilon(3S) \to \gamma\eta_b$ and a larger upper limit on $\mathcal{B}(\Upsilon(2S) \to \gamma\eta_b)$. In addition to including the initially omitted ISR peak in the fit to E_γ and assuming a more reasonable width, $\Gamma(\eta_b) = 10$ MeV, for the signal, CLEO exploited an E_γ resolution slightly better than *BABAR*, parametrized the observed photon lineshape more accurately than before, and added a new twist to the *BABAR*-inspired thrust-angle restriction. Instead of simply rejecting a high-background region of thrust angle, CLEO accumulated three

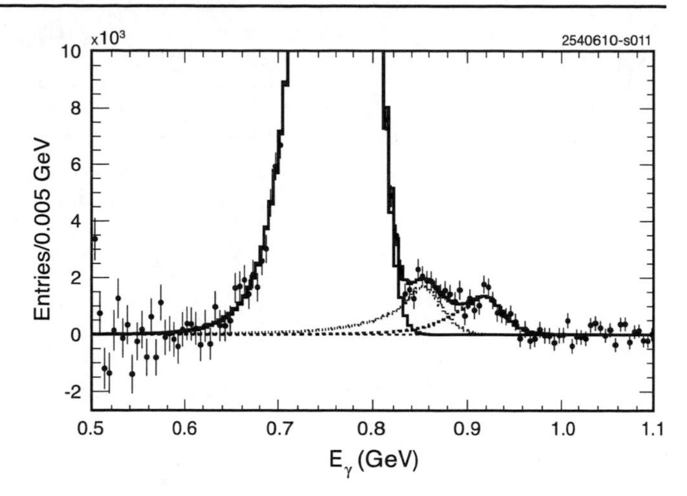

Fig. 11 From *BABAR* [59], the inclusive photon energy spectrum in the e^+e^- center-of-mass frame $\Upsilon(3S)$ data after subtraction of the smooth background. The *solid curve* shows the best fit, and the *peaks* correspond to, from left to right, $\chi_{b1,2} \to \gamma\Upsilon(1S)$, ISR production of $\Upsilon(1S)$, and $\Upsilon(3S) \to \gamma\eta_b$. Adapted from [59] with kind permission, copyright (2008) The American Physical Society

E_γ distributions, one for each of three $|\cos\theta_T|$ bins, the middle one being $0.3 < |\cos\theta_T| < 0.7$. Hence no statistical power was wasted by throwing any data away, and an improved S^2/B in the combined $|\cos\theta_T| < 0.3$ and 0.3–0.7 bins relative to <0.7 was exploited. A *BABAR*-like fit to the measured E_γ distributions in all three $|\cos\theta_T|$ bins simultaneously extracted the η_b signal. CLEO left the photon selection criteria from the original analysis unchanged and quoted a final mass, rate, and significance which were each the mean from an ensemble of fits with reasonable confidence levels, not on any arbitrarily chosen individual fit. The fit ensemble contained many variations, each specifying a different background parametrization, E_γ range, and/or logarithmic or linear E_γ scale.

Tables 4 and 7 summarize the experimental η_b results, which together yield

$$\Delta m_{\text{hf}}\big[\eta_b(1S)\big]_{\text{exp}} = 69.6 \pm 2.9 \text{ MeV}. \tag{5}$$

Belle is poised to search for η_b using its 11M $\Upsilon(3S)$ events and recently augmented 160M $\Upsilon(2S)$ dataset.

Theoretical predictions for $\eta_b(1S)$ hyperfine splitting are discussed Sects. 2.5.3, 2.6.2, 2.7, 2.8.1, and 2.10.1.

2.2.6 *Search for $h_b(1P)$*

A preliminary analysis from *BABAR* [80] describes two searches for $h_b(1P)$ in a sample of 122M $\Upsilon(3S)$ decays. The first search employs a method similar to the CLEO $h_c(1P)$ inclusive search (see Sect. 2.2.1) by selecting $\Upsilon(3S)$ decays with both a soft π^0 and a radiative photon, looking for the decay chain $\Upsilon(3S) \to \pi^0 h_b(1P)$, $h_b(1P) \to \gamma\eta_b(1S)$. With the radiative photon restricted

Table 7 Measured $\eta_b(1S)$ properties. The value quoted for the weighted average of Δm_{hf} includes all three measurements

Quantity	$\Upsilon(2S) \to \gamma\eta_b$	$\Upsilon(3S) \to \gamma\eta_b$	Ref. (χ^2/d.o.f.)
E_γ	$610.5^{+4.5}_{-4.3} \pm 1.8$	$921.2^{+2.1}_{-2.8} \pm 2.4$	*BABAR* [59, 61]
(MeV)	–	$918.6 \pm 6.0 \pm 1.8$	CLEO [60]
$m(\eta_b)$	$9392.9^{+4.6}_{-4.8} \pm 1.8$	$9388.9^{+3.1}_{-2.3} \pm 2.7$	*BABAR* [59, 61]
(MeV)	–	$9391.8 \pm 6.6 \pm 2.0$	CLEO [60]
Δm_{hf}	$67.4^{+4.8}_{-4.5} \pm 1.9$	$71.4^{+2.3}_{-3.1} \pm 2.7$	*BABAR* [59, 61]
(MeV)	–	$68.5 \pm 6.6 \pm 2.0$	CLEO [60]
		69.6 ± 2.9	Avg[3] (0.6/2)
$\mathcal{B} \times 10^4$	$4.2^{+1.1}_{-1.0} \pm 0.9$	$4.8 \pm 0.5 \pm 1.2$	*BABAR* [59, 61]
	<8.4	$7.1 \pm 1.8 \pm 1.1$	CLEO [60]
	<5.1	<4.3	CLEO [79]

to the range allowed for the transition to $\eta_b(1S)$, the mass recoiling against the soft π^0 is plotted and scanned for a peak above a smooth background. *BABAR* sees a 2.7σ effect at $m(\pi^0 - \text{recoil}) = 9903 \pm 4 \pm 1$ MeV. In a second search, an upper limit of

$$\mathcal{B}\big(\Upsilon(3S) \to \pi^+\pi^- h_b(1P)\big) < 2.5 \times 10^{-4}$$
$$\text{for } 9.88 < m\big(h_b(1P)\big) < 9.92 \text{ GeV at 90\% CL} \quad (6)$$

is set.

2.2.7 Observation of $\Upsilon(1^3D_2)$

CLEO [62] made the first of two observations of $\Upsilon(1D)$, using the four-photon cascade shown in Fig. 12:

$$\begin{aligned}
&\Upsilon(3S) \to \gamma\chi_{bJ}(2P),\\
&\chi_{bJ}(2P) \to \gamma\Upsilon(1D),\\
&\Upsilon(1D) \to \gamma\chi_{bJ}(1P),\\
&\chi_{bJ}(1P) \to \gamma\Upsilon(1S),\\
&\Upsilon(1S) \to \ell^+\ell^-,
\end{aligned} \quad (7)$$

where $\ell^\pm \equiv e^\pm$ or $\mu^\pm$. The largest background source of four soft photons and an $\Upsilon(1S)$ is $\Upsilon(3S) \to \pi^0\pi^0\Upsilon(1S)$, which was suppressed by vetoing events with two photon-pairings that are both consistent with π^0 masses. The next-most pernicious background is the quite similar four-photon cascade through $\Upsilon(2S)$ instead of $\Upsilon(1D)$ (also shown in Fig. 12); the softer two photons overlap the signal photons within the experimental resolution. This latter background was suppressed by kinematically constraining each event to a $\Upsilon(1D)$ hypothesis with unknown $\Upsilon(1D)$ mass and including all J possibilities for the intermediate $\chi_{bJ}(nP)$ states,

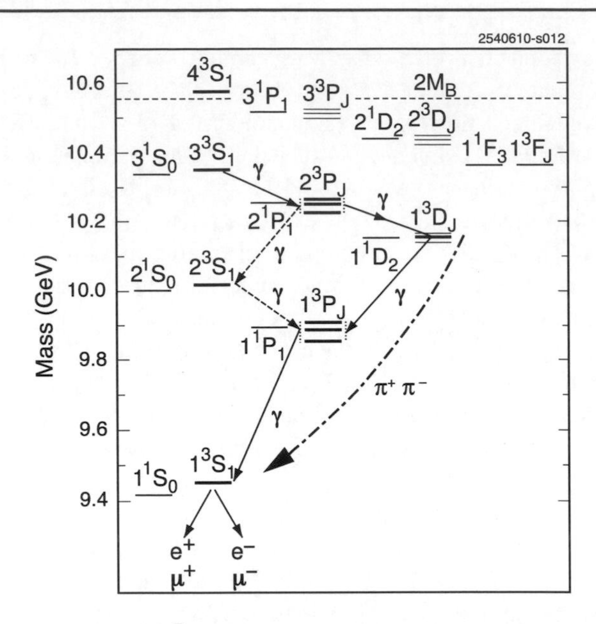

Fig. 12 Expected $b\bar{b}$ bound-state mass levels. The four-photon transition sequence from the $\Upsilon(3S)$ to the $\Upsilon(1S)$ via the $\Upsilon(1D)$ states is shown (*solid lines*). An alternative route for the four-photon cascade via the $\Upsilon(2S)$ state is also displayed (*dashed lines*). The hadronic dipion transition from $\Upsilon(1D)$ to $\Upsilon(1S)$ is indicated by the *dot-dash curve*. Adapted from [62] with kind permission, copyright (2004) The American Physical Society

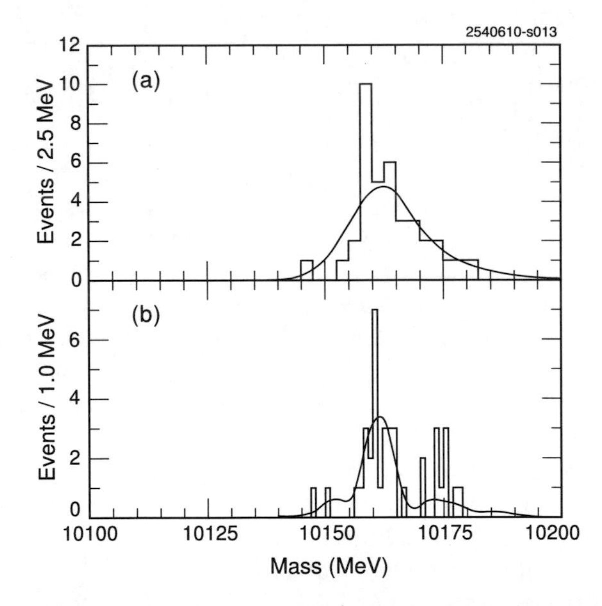

Fig. 13 From CLEO [62], distributions of (**a**) mass recoiling against the softest two photons, and (**b**) mass that produces the smallest $\chi^2(1D)$ (see text) per event, from $\Upsilon(3S) \to \gamma\gamma\gamma\gamma\ell^+\ell^-$ events, selected to be consistent with a four-photon cascade through $\chi_{bJ}(2P)$, $\Upsilon(1D)$, and $\chi_{bJ}(1P)$ to $\Upsilon(1S)$. The *solid line histogram* represents data, and the *curves* represent the CLEO fits. Adapted from [62] with kind permission, copyright (2004) The American Physical Society

and then requiring a good fit quality, $\chi^2(1D)$. Masses from the surviving candidates are shown in Fig. 13, with the mass recoiling against the softest two photons in (a) and the value obtained from the minimum $\chi^2(1D)$ combination in (b).

Both give consistent masses for an $\Upsilon(1^3D_2)$, and the latter has an inconclusive 1.9σ hint of a second peak 13 MeV above the primary one, which could be an indication of the corresponding $\Upsilon(1^3D_3)$ state. The observed 34.5 ± 6.4 signal events in the central peak correspond to a statistical significance of 10σ, most of which are attributed to cascades involving the $\chi_{b1}(nP)$ for both $n = 1$ and 2 and to production of an $\Upsilon(1^3D_2)$. The product branching fraction for the entire cascade was found to be $(2.5 \pm 0.7) \times 10^{-5}$. Upper limits on other possible decays relative to the four-photon cascade were also set to be <0.25 for $\Upsilon(1D) \to \eta\Upsilon(1S)$ and <1.2 for $\pi^+\pi^-\Upsilon(1D)$, both at 90% CL.

Belle has an $\Upsilon(3S)$ dataset slightly larger than CLEO and therefore could mount a comparable $\Upsilon(1D)$ search. BABAR has twenty times more $\Upsilon(3S)$ than CLEO, and therefore has the capability to search for other decay chains and to explore hyperfine mass structure of the allowed $\Upsilon(1D)$ spin states. While neither Belle nor BABAR has yet explored the four-photon cascade, BABAR has observed [63] $\Upsilon(1^3D_2)$ produced from a two-photon cascade from $\Upsilon(3S)$ decay as does CLEO, but then undergoing a charged dipion transition to the $\Upsilon(1S)$, which then decays to $\ell^+\ell^-$. The $\pi^+\pi^-\ell^+\ell^-$ invariant-mass distribution from such events, restricted to those with dilepton masses near that of the $\Upsilon(1S)$, is shown in Fig. 14. In addition to confirming the $\Upsilon(1^3D_2)$ signal at 5.8σ, the $\Upsilon(1^3D_1)$ and $\Upsilon(1^3D_3)$ states are seen at 1.8σ and 1.6σ, respectively. The $\Upsilon(1^3D_2)$ mass is somewhat larger than but consistent with the CLEO value, as shown in Table 8. The BABAR analysis also concludes, first, that the dipion invariant-mass distribution is in substantially better agreement with that predicted for a $\Upsilon(1^3D_J)$ than an S or P state, and second, that angular distributions of the $\Upsilon(1^3D_2)$ signal events are consistent with the quantum number assignments of $J = 2$ and $P = -1$.

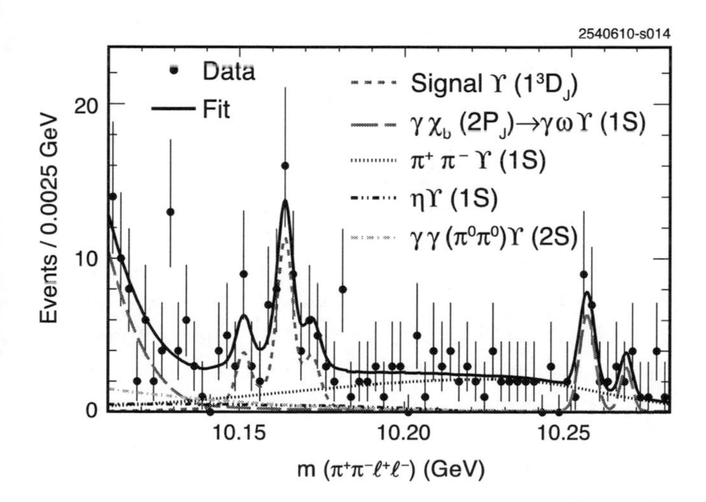

Fig. 14 From BABAR [63], the $\pi^+\pi^-\ell^+\ell^-$ invariant mass restricted to those with dilepton masses near that of the $\Upsilon(1S)$. *Curves* represent $\Upsilon(1^3D_J)$ signals only (*short dash*), the total fit (*solid*) which includes four backgrounds it (others). Adapted from [63] with kind permission, copyright (2010) The American Physical Society

Table 8 Measured mass values of the $\Upsilon(1^3D_2)$

Decay	Value (MeV)	Ref. (χ^2/d.o.f.)
$\Upsilon(3S) \to \gamma\gamma\gamma\gamma\Upsilon(1S)$	$10161.1 \pm 0.6 \pm 1.6$	CLEO [62]
$\Upsilon(3S) \to \gamma\gamma\pi^+\pi^-\Upsilon(1S)$	$10164.5 \pm 0.8 \pm 0.5$	BABAR [63]
Both of above	10163.8 ± 1.4	Avg[3] (3.1/1)

2.3 New unanticipated states

2.3.1 $X(3872)$, *the enduring exotic*

The $X(3872)$ occupies a unique niche in the menagerie of unanticipated states listed in Table 9 as both the first and the most intriguing. At this point it is widely studied, yet its interpretation demands much more experimental attention. Its apparent quantum numbers, mass, and decay patterns make it an unlikely conventional charmonium candidate, and no consensus explanation has been found.

In 2003, while studying $B^+ \to K^+\pi^+\pi^-J/\psi$, Belle [85] discovered an unexpected enhancement in the $\pi^+\pi^-J/\psi$ invariant-mass spectrum near 3872 MeV. This sighting of the X in B-decays was later confirmed by BABAR [87]. The $X \to \pi^+\pi^-J/\psi$ decay was also observed inclusively in prompt production from $\bar{p}p$ collisions at the Tevatron by both CDF [88–90] and DØ [91]. CDF [89] studied the angular distributions and correlations of the $\pi^+\pi^-J/\psi$ final state, finding that the dipion was favored to originate as a ρ^0, and that only J^{PC} assignments of 1^{++} and 2^{-+} explained their measurements adequately. Belle [92] reported evidence for the $\gamma J/\psi$ decay, which BABAR [97, 98] confirmed at 4σ significance. The radiative decay verifies the positive C-parity assignment of CDF. It also bolsters the 1^{++} assignment because a 2^{-+} state would have to undergo a high-order multipole transition which would be more strongly suppressed than the observed rates allow.

From the beginning, the proximity of the X mass to $D^{*0}\bar{D}^0$ threshold was conspicuous, and eventually decays to $D^{*0}\bar{D}^0$ were observed by BABAR [96] and Belle [95]. Interest in the relationship of X to $D^{*0}\bar{D}^0$ fueled improvements in measurements of its mass, as shown in Table 10, and of the D^0 mass, as shown in Table 11. The X mass measurements based upon the $\pi^+\pi^-J/\psi$ decay are consistent with one another. The world-average X mass, restricted to measurements using $\pi^+\pi^-J/\psi$ decays, is dominated by the CDF [90] inclusive result, illustrated in Fig. 15. The CDF systematic uncertainty on the mass was obtained from studies of $\psi(2S) \to \pi^+\pi^-J/\psi$ decays, which have a similar topology and a well-known $\psi(2S)$ mass to match. The measured mass discrepancy was extrapolated from the $\psi(2S)$ mass to the X mass to obtain error estimates. The world-average D^0 mass precision is dominated by a CLEO [119] measurement that uses the decay

Table 9 As in Table 4, but for new *unconventional* states in the $c\bar{c}$ and $b\bar{b}$ regions, ordered by mass. For $X(3872)$, the values given are based only upon decays to $\pi^+\pi^-J/\psi$. $X(3945)$ and $Y(3940)$ have been subsumed under $X(3915)$ due to compatible properties. The state known as $Z(3930)$ appears as the $\chi_{c2}(2P)$ in Table 4. See also the reviews in [81–84]

State	m (MeV)	Γ (MeV)	J^{PC}	Process (mode)	Experiment (#σ)	Year	Status
$X(3872)$	3871.52 ± 0.20	1.3 ± 0.6	$1^{++}/2^{-+}$	$B \to K(\pi^+\pi^-J/\psi)$	Belle [85, 86] (12.8), *BABAR* [87] (8.6)	2003	OK
		(<2.2)		$p\bar{p} \to (\pi^+\pi^-J/\psi) + \cdots$	CDF [88–90] (np), DØ [91] (5.2)		
				$B \to K(\omega J/\psi)$	Belle [92] (4.3), *BABAR* [93] (4.0)		
				$B \to K(D^{*0}\bar{D}^0)$	Belle [94, 95] (6.4), *BABAR* [96] (4.9)		
				$B \to K(\gamma J/\psi)$	Belle [92] (4.0), *BABAR* [97, 98] (3.6)		
				$B \to K(\gamma\psi(2S))$	*BABAR* [98] (3.5), Belle [99] (0.4)		
$X(3915)$	3915.6 ± 3.1	28 ± 10	$0/2^{?+}$	$B \to K(\omega J/\psi)$	Belle [100] (8.1), *BABAR* [101] (19)	2004	OK
				$e^+e^- \to e^+e^-(\omega J/\psi)$	Belle [102] (7.7)		
$X(3940)$	3942^{+9}_{-8}	37^{+27}_{-17}	$?^{?+}$	$e^+e^- \to J/\psi(D\bar{D}^*)$	Belle [103] (6.0)	2007	NC!
				$e^+e^- \to J/\psi\ (\ldots)$	Belle [54] (5.0)		
$G(3900)$	3943 ± 21	52 ± 11	1^{--}	$e^+e^- \to \gamma(D\bar{D})$	*BABAR* [27] (np), Belle [21] (np)	2007	OK
$Y(4008)$	4008^{+121}_{-49}	226 ± 97	1^{--}	$e^+e^- \to \gamma(\pi^+\pi^-J/\psi)$	Belle [104] (7.4)	2007	NC!
$Z_1(4050)^+$	4051^{+24}_{-43}	82^{+51}_{-55}	?	$B \to K(\pi^+\chi_{c1}(1P))$	Belle [105] (5.0)	2008	NC!
$Y(4140)$	4143.4 ± 3.0	15^{+11}_{-7}	$?^{?+}$	$B \to K(\phi J/\psi)$	CDF [106, 107] (5.0)	2009	NC!
$X(4160)$	4156^{+29}_{-25}	139^{+113}_{-65}	$?^{?+}$	$e^+e^- \to J/\psi(D\bar{D}^*)$	Belle [103] (5.5)	2007	NC!
$Z_2(4250)^+$	4248^{+185}_{-45}	177^{+321}_{-72}	?	$B \to K(\pi^+\chi_{c1}(1P))$	Belle [105] (5.0)	2008	NC!
$Y(4260)$	4263 ± 5	108 ± 14	1^{--}	$e^+e^- \to \gamma(\pi^+\pi^-J/\psi)$	*BABAR* [108, 109] (8.0)	2005	OK
					CLEO [110] (5.4)		
					Belle [104] (15)		
				$e^+e^- \to (\pi^+\pi^-J/\psi)$	CLEO [111] (11)		
				$e^+e^- \to (\pi^0\pi^0J/\psi)$	CLEO [111] (5.1)		
$Y(4274)$	$4274.4^{+8.4}_{-6.7}$	32^{+22}_{-15}	$?^{?+}$	$B \to K(\phi J/\psi)$	CDF [107] (3.1)	2010	NC!
$X(4350)$	$4350.6^{+4.6}_{-5.1}$	$13.3^{+18.4}_{-10.0}$	$0,2^{++}$	$e^+e^- \to e^+e^-(\phi J/\psi)$	Belle [112] (3.2)	2009	NC!
$Y(4360)$	4353 ± 11	96 ± 42	1^{--}	$e^+e^- \to \gamma(\pi^+\pi^-\psi(2S))$	*BABAR* [113] (np), Belle [114] (8.0)	2007	OK
$Z(4430)^+$	4443^{+24}_{-18}	107^{+113}_{-71}	?	$B \to K(\pi^+\psi(2S))$	Belle [115, 116] (6.4)	2007	NC!
$X(4630)$	4634^{+9}_{-11}	92^{+41}_{-32}	1^{--}	$e^+e^- \to \gamma(\Lambda_c^+\Lambda_c^-)$	Belle [25] (8.2)	2007	NC!
$Y(4660)$	4664 ± 12	48 ± 15	1^{--}	$e^+e^- \to \gamma(\pi^+\pi^-\psi(2S))$	Belle [114] (5.8)	2007	NC!
$Y_b(10888)$	10888.4 ± 3.0	$30.7^{+8.9}_{-7.7}$	1^{--}	$e^+e^- \to (\pi^+\pi^-\Upsilon(nS))$	Belle [37, 117] (3.2)	2010	NC!

chain $D^0 \to \phi K_S^0$, $\phi \to K^+K^-$, $K_S^0 \to \pi^+\pi^-$, and is limited by statistics. Despite all these advances, the $D^{*0}\bar{D}^0$ mass threshold test remains ambiguous, with $m[X(3872)] - [m(D^{*0}) + m(D^0)] = -0.42 \pm 0.39$ MeV. This limits the hypothetical $D^{*0}\bar{D}^0$ binding energy to be <0.92 MeV at 90% CL and does not foreclose the possibility that the $X(3872)$ is *above* $D^{*0}\bar{D}^0$ threshold. Further clarity here would require much more precise mass measurements for both the X and the D^0.

Both Belle and *BABAR* have reported $X(3872)$ signals in the $D^{*0}\bar{D}^0$ final state with branching fractions about ten times higher than for $\pi^+\pi^-J/\psi$. Both used $D^{*0} \to D^0\pi^0$ and $D^0\gamma$ decays, both selected and kinematically constrained a D^{*0} candidate in each event, and both performed unbinned maximum-likelihood fits to the $D^{*0}\bar{D}^0$ mass. (Belle's fit is two-dimensional, the second dimension being a B-meson-consistency kinematic variable; *BABAR* cuts on B-meson consistency.) Both results appear in Table 10. (An earlier Belle publication [94] used a dataset smaller by one-third than in [95], made no D^{*0}-mass constraint, and measured a mass value of $3875.2 \pm 0.7^{+0.3}_{-1.6} \pm 0.8$ MeV.) Belle [95] fit to a conventional Breit–Wigner signal shape convolved with a Gaussian resolution function. *BABAR* [96]

Table 10 $X(3872)$ mass and width measurements by decay mode and experiment. The χ^2/d.o.f. values given in parentheses refer to weighted averages of the masses only. The lines marked $(B^\pm)$ and (B^0) represent mass values quoted by *BABAR* in charged and neutral B-decays, respectively

Mode	Mass (MeV)	Width (MeV)	Ref. (χ^2/d.o.f.)
$\pi^+\pi^- J/\psi$	$3871.46 \pm 0.37 \pm 0.07$	1.4 ± 0.7	Belle [86]
$(B^\pm)$	$3871.4 \pm 0.6 \pm 0.1$	$1.1 \pm 1.5 \pm 0.2$	*BABAR* [87]
(B^0)	$3868.7 \pm 1.5 \pm 0.4$	–	*BABAR* [87]
	$3871.8 \pm 3.1 \pm 3.0$	–	DØ [88]
	$3871.61 \pm 0.16 \pm 0.19$	–	CDF [90]
	3871.52 ± 0.20	1.3 ± 0.6	Avg³ (2.1/4)
$D^{*0}\bar{D}^0$	$3875.1^{+0.7}_{-0.5} \pm 0.5$	$3.0^{+1.9}_{-1.4} \pm 0.9$	*BABAR* [96]
	$3872.9^{+0.6+0.4}_{-0.4-0.5}$	$3.9^{+2.8+0.2}_{-1.4-1.1}$	Belle [95]
	3874.0 ± 1.2	$3.5^{+1.6}_{-1.0}$	Avg³ (4.7/1)

Table 11 Mass measurements relevant to the $X(3872)$. We define $\delta m_0 \equiv m(D^{*0}) - m(D^0)$ and $\Delta m_{thr} \equiv m[X(3872)] - [m(D^{*0}) + m(D^0)]$

Quantity	Mass (MeV)	Ref. (χ^2/d.o.f.)
$m(D^0)$	$1864.6 \pm 0.3 \pm 1.0$	ACCMOR [118]
	$1864.847 \pm 0.150 \pm 0.095$	CLEO [119]
	$1865.3 \pm 0.33 \pm 0.23$	KEDR [120]
	1864.91 ± 0.16	Avg³ (1.2/2)
δm_0	142.12 ± 0.07	PDG08 [18]
$2m(D^0) + \delta m_0$	3871.94 ± 0.33	–
$m[X(3872)]$	3871.52 ± 0.20	Table 10 ($\pi^+\pi^- J/\psi$)
Δm_{thr}	-0.42 ± 0.39	–
	$\in [-0.92, 0.08]$	@90% CL

fit the data to an ensemble of MC samples, each generated with different plausible X masses and widths and assuming a purely S-wave decay of a spin-1 resonance. The *BABAR* X mass from $D^{*0}\bar{D}^0$ decays is more than 3 MeV larger than the world average from $\pi^+\pi^- J/\psi$, which engendered speculation that the $D^{*0}\bar{D}^0$ enhancement might be a different state than that observed in $\pi^+\pi^- J/\psi$, but the smaller value observed by Belle in $D^{*0}\bar{D}^0$ seems to make that possibility unlikely. The two X mass measurements using $D^{*0}\bar{D}^0$ decays are inconsistent by 2.2σ, and are 1.8σ and 4.7σ higher than the $\pi^+\pi^- J/\psi$-based mass. However, important subtleties pointed out by Braaten and co-authors [121, 122] appear to explain at least qualitatively why masses extracted in this manner are larger than in $\pi^+\pi^- J/\psi$.

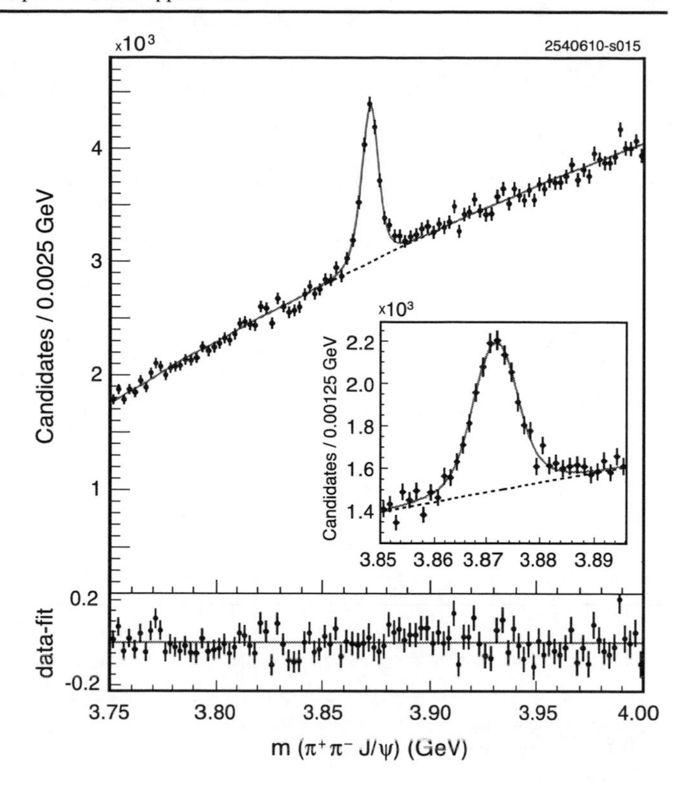

Fig. 15 From CDF [90], the $\pi^+\pi^- J/\psi$ invariant-mass distribution for $X(3872)$ candidates, showing the data (*points*), the projection of the unbinned maximum-likelihood fit (*solid line*) and its smooth background component (*dashed line*), and the inset which enlarges the peak region with finer binning. The lower panel shows residuals of the data with respect to the fit. Adapted from [90] with kind permission, copyright (2009) The American Physical Society

Measuring the X mass with the $D^{*0}\bar{D}^0$ decay is considerably more challenging than with $\pi^+\pi^- J/\psi$ for several reasons [121, 122]. If conceived as a bound or virtual $D^{*0}\bar{D}^0$ state [123], the X lineshape in this decay mode is determined by the binding energy, the D^{*0} natural width, and the natural width of the X itself, which is at least as large as the D^{*0} width [121]. Because the binding energy of the X is less than 1 MeV, whether or not its mass *peak* is below $D^{*0}\bar{D}^0$ threshold, substantial fractions of the *lineshape* will lie both above and below that threshold. The portion of the X lineshape below $D^{*0}\bar{D}^0$ threshold, by definition, cannot decay to $D^{*0}\bar{D}^0$. However, $D^0\bar{D}^0\pi^0$ and $D^0\bar{D}^0\gamma$ final states are possible from decays of a bound, effectively off-shell, D^{*0}, as there is adequate phase space available above $D^0\bar{D}^0\pi^0$ threshold. Due to imperfect experimental resolution, these final states are indistinguishable from $D^{*0}\bar{D}^0$ even though the D^{*0} decay products have masses below that of D^{*0}. Furthermore, the analysis procedure which mass-constrains a D^{*0} candidate in each event distorts the purported X mass distribution for below-threshold decays. Conversely, that portion of the X lineshape above $D^{*0}\bar{D}^0$ threshold can, of course, decay to $D^{*0}\bar{D}^0$, but the $D^{*0}\bar{D}^0$ mass distribution should, by definition, be exactly zero below threshold. Therefore the kinematic constraint on the reconstructed

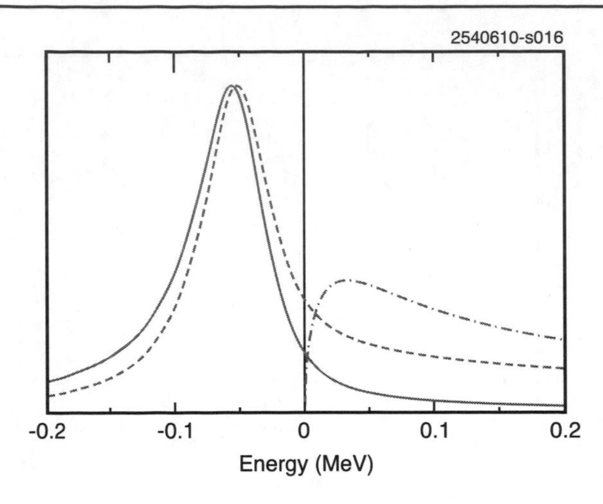

Fig. 16 From Braaten and Stapleton [122], the $X(3872)$ lineshapes extracted from a fit to the Belle [95] $D^{*0}\bar{D}^0$ events, unfolding the effects of experimental resolution, for $\pi^+\pi^-J/\psi$ (*solid curve*), $D^0\bar{D}^0\pi^0$ (*dashed*), and, when always constraining one $D^0\pi^0$ pairing per event to a D^{*0} mass, $D^{*0}\bar{D}^0$ (*dot-dashed*). The *horizontal axis* is the invariant mass of the decay products relative to $D^{*0}\bar{D}^0$ threshold, and the *solid* and *dashed curves* are normalized so as to have the same peak height. Adapted from [122] with kind permission, copyright (2010) The American Physical Society

$D^0\pi^0$ to the D^{*0}-mass, as carried out by Belle and BABAR, results in a broad $D^{*0}\bar{D}^0$ mass peak above threshold that should not be misconstrued as the true X lineshape: neither the mass nor width results from $D^{*0}\bar{D}^0$ reflect the true mass or width of the X. Rather, the lineshapes for $\pi^+\pi^-J/\psi$, $D^{*0}\bar{D}^0$, and $D^0\bar{D}^0\pi^0$ (and $D^0\bar{D}^0\gamma$) final states are related but slightly different from one another, as shown in Fig. 16. More data and more sophisticated analyses are required to fully exploit what $D^0\bar{D}^0\pi^0$ and $D^0\bar{D}^0\gamma$ decays can reveal about the nature of the X.

Branching-fraction-related measurements for $X(3872)$ appear in Table 12. Note that $X \to D^{*0}\bar{D}^0$ decays are an order of magnitude more prevalent than $X \to \pi^+\pi^-J/\psi$, and that experimental information concerning the radiative decay $X \to \gamma\psi(2S)$ has recently become murky; Belle's preliminary upper limit [99] is inconsistent with the BABAR [98] measurement. Belle [92] found 4.3σ evidence for $X \to \pi^+\pi^-\pi^0 J/\psi$ in B-decays, with the 3π invariant mass clustered near the kinematic endpoint, which is almost one full Γ_ω below the ω mass peak. This suggests the decay $X \to \omega J/\psi$, $\omega \to \pi^+\pi^-\pi^0$ on the ω low-side tail. Despite this apparent phase-space suppression, the rate for $X \to \omega J/\psi$ was found to be comparable to that of $\pi^+\pi^-J/\psi$. In 2010 BABAR [93] reported corroborating evidence (4.0σ) of an $X(3872)$ peak in $B^{0,+} \to (J/\psi\pi^+\pi^-\pi^0)K^{0,+}$ decays using their full data sample, also finding a decay rate comparable to that of $\pi^+\pi^-J/\psi$, as shown in Table 12. Their analysis was able to identify the three-pion decay as coming from an ω-meson decay by weighting the entries based on the pion opening angles in the ω rest-frame: phase-space

Table 12 For $X(3872)$, measured branching fractions and products thereof, in units of 10^{-6}: $\mathcal{B}_{B^+} \equiv \mathcal{B}(B^+ \to K^+X)$,
$\mathcal{B}_{\gamma 1} \equiv \mathcal{B}(B^+ \to K^+X) \times \mathcal{B}(X \to \gamma J/\psi)$,
$\mathcal{B}_{\gamma 2} \equiv \mathcal{B}(B^+ \to K^+X) \times \mathcal{B}(X \to \gamma\psi(2S))$,
$\mathcal{B}_0 \equiv \mathcal{B}(B^0 \to KX) \times \mathcal{B}(X \to f)$,
$\mathcal{B}_+ \equiv \mathcal{B}(B^+ \to KX) \times \mathcal{B}(X \to f)$, and
$\mathcal{B}_{0+} \equiv (\mathcal{B}_0 + \mathcal{B}_+)/2$ for final state f.
Branching fraction ratios are defined as:
$R_{0+} \equiv \mathcal{B}(B^0 \to K^0X)/\mathcal{B}(B^+ \to K^+X)$,
$r_{DD\pi} \equiv \mathcal{B}_{0+}(D^0\bar{D}^0\pi^0)/\mathcal{B}_{0+}(\pi^+\pi^-J/\psi)$,
$r_\omega \equiv \mathcal{B}(X \to \omega J/\psi)/\mathcal{B}(X \to \pi^+\pi^-J/\psi)$,
$r_{\gamma 1} \equiv \mathcal{B}(X \to \gamma J/\psi)/\mathcal{B}(X \to \pi^+\pi^-J/\psi)$, and
$r_{\gamma 2} \equiv \mathcal{B}(X \to \gamma\psi(2S))/\mathcal{B}(X \to \pi^+\pi^-J/\psi)$

What	Mode	Value	Ref. (χ^2/d.o.f.)
$\mathcal{B}_{B^+}$		<320	BABAR [124]
$\mathcal{B}_{\gamma 1}$	$\gamma J/\psi$	$1.8 \pm 0.6 \pm 0.1$	Belle [92]
		$2.8 \pm 0.8 \pm 0.1$	BABAR [98]
		2.2 ± 0.5	Avg[3] (1.0/1)
$\mathcal{B}_{\gamma 2}$	$\gamma\psi(2S)$	<3.4	Belle [99]
		0.8 ± 2.0[a]	Belle [99]
		$9.5 \pm 2.7 \pm 0.6$	BABAR [98]
		3.8 ± 4.1	Avg[3] (6.4/1)
$\mathcal{B}_{\gamma 2}/\mathcal{B}_{\gamma 1}$	$\gamma\psi(2S)/\gamma J/\psi$	4.3 ± 1.6	Values above
$\mathcal{B}_0$	$D^{*0}\bar{D}^0$	$167 \pm 36 \pm 47$	BABAR [96]
$\mathcal{B}_+$		$222 \pm 105 \pm 42$	BABAR [96]
$\mathcal{B}_0$		$97 \pm 46 \pm 13$	Belle [95]
$\mathcal{B}_+$		$77 \pm 16 \pm 10$	Belle [95]
$\mathcal{B}_{0+}$		90 ± 19	Avg[3] (3.6/3)
$\mathcal{B}_{0+}/\mathcal{B}_{B^+}$	$D^{*0}\bar{D}^0$	>28%	Above
$\mathcal{B}_0$	$\pi^+\pi^-J/\psi$	$3.50 \pm 1.90 \pm 0.40$	BABAR [87]
$\mathcal{B}_+$		$8.40 \pm 1.50 \pm 0.70$	BABAR [87]
$\mathcal{B}_0$		$6.65 \pm 1.63 \pm 0.55$	Belle [86]
$\mathcal{B}_+$		$8.10 \pm 0.92 \pm 0.66$	Belle [86]
$\mathcal{B}_{0+}$		7.18 ± 0.97	Avg[3] (4.9/3)
$\mathcal{B}_{0+}/\mathcal{B}_{B^+}$	$\pi^+\pi^-J/\psi$	>2.2%	Above
R_{0+}	$\pi^+\pi^-J/\psi$	$0.82 \pm 0.22 \pm 0.05$	Belle [86]
	$\pi^+\pi^-J/\psi$	$0.41 \pm 0.24 \pm 0.05$	BABAR [87]
	$D^{*0}\bar{D}^0$	$1.26 \pm 0.65 \pm 0.06$	Belle [95]
	$D^{*0}\bar{D}^0$	$1.33 \pm 0.69 \pm 0.43$	BABAR [96]
	Both	0.70 ± 0.16	Avg[3] (3.0/3)
r_{D^*D}	$D^{*0}\bar{D}^0$	12.5 ± 3.1	Ratio of avgs
$\mathcal{B}_0$	$\omega J/\psi$	$3.5 \pm 1.9 \pm 0.7$	Belle [92]
$\mathcal{B}_+$		$8.5 \pm 1.5 \pm 1.7$	Belle [92]
$\mathcal{B}_0$		$6 \pm 3 \pm 1$	Belle [93]
$\mathcal{B}_+$		$6 \pm 2 \pm 1$	Belle [93]
$\mathcal{B}_{0+}$		5.8 ± 1.2	Avg[3] (2.7/3)

Table 12 (*Continued*)

What	Mode	Value	Ref. (χ^2/d.o.f.)
r_ω	$\omega J/\psi$	$1.0 \pm 0.4 \pm 0.3$	Belle [92]
		0.8 ± 0.3	*BABAR* [93]
		0.85 ± 0.26	Avg[3] (0.1/1)
$r_{\gamma 1}$	$\gamma J/\psi$	0.31 ± 0.08	Values above
$r_{\gamma 2}$	$\gamma \psi(2S)$	0.53 ± 0.57	Values above
$\mathcal{B}_0$	$\pi^+\pi^0 J/\psi$	-5.7 ± 4.9	*BABAR* [125]
$\mathcal{B}_+$		2.0 ± 3.8	*BABAR* [125]
$\mathcal{B}_{0+}$		-0.9 ± 3.7	Avg[3] (1.5/1)

[a]Belle only quotes an upper limit for this preliminary result. From the information presented in [99], we have extracted an approximate central value and error for this table

weighting results in no net signal. In a comparison of the observed $m(3\pi)$ mass distribution to that of MC simulations, *BABAR* also found that the inclusion of one unit of orbital angular momentum in the $J/\psi\omega$ system, with its consequent negative parity, substantially improves the description of the data. Hence the $X(3872)$ quantum number assignment [89] of $J^{PC} = 2^{-+}$ is preferred somewhat over the 1^{++} hypothesis in the *BABAR* analysis, leading *BABAR* to conclude that the $X(3872)$ can be interpreted as an $\eta_{c2}(1D)$ charmonium state [81, 126]. However, the 1^{++} assignment cannot be ruled out as unlikely by this analysis, just less likely than 2^{-+}. In addition, it has been shown [127] that a 2^{-+} assignment is not consistent with other properties of the $X(3872)$.

BABAR [125] searched for a charged partner state in the decay $X^+ \to \rho^+ J/\psi$, finding the results in Table 12. The average from charged and neutral B-decays should be compared with the isospin-symmetry prediction, which is double the rate for $\rho^0 J/\psi$. These rates disagree by more than 4σ, making it most likely that the X is an isosinglet. The *BABAR* [124] upper limit on $\mathcal{B}_{B^+} \equiv \mathcal{B}(B^+ \to K^+ X)$ permits an inferred lower limit on $\mathcal{B}(X \to \pi^+\pi^- J/\psi)$, which, when combined with the relative rates of $D^{*0}\bar{D}^0$ and $\pi^+\pi^- J/\psi$, yields $2.2\% < \mathcal{B}(X \to \pi^+\pi^- J/\psi) < 10.5\%$ and $28\% < \mathcal{B}(X \to D^{*0}\bar{D}^0) < 94.2\%$. Belle [86] has studied the question of whether or not the X, like conventional charmonia, tends to be produced more strongly in $B^0 \to K^{*0} X$ relative to nonresonant (NR) $B^0 \to (K^+\pi^-)_{\rm NR} X$. Using $X \to \pi^+\pi^- J/\psi$ decays, they limit the $K^{*0}/(K^+\pi^-)_{\rm NR}$ ratio to be <0.5 at 90% CL, contrasted with ratios closer to 3 for other charmonium states.

The possibility that the X enhancement in $\pi^+\pi^- J/\psi$ is composed of two different narrow states, X_L and X_H, was addressed by CDF [90]. By analyzing the observed lineshape, X_L and X_H were found to have masses closer than 3.2 (4.3) MeV for a relative production fraction of unity (20%). Any mass difference between the X appearing in charged and neutral B-decays has also been limited to <2.2 MeV, as detailed in Table 13.

Table 13 For $X(3872)$, Δm_{0+} (in MeV), the difference between the $X(3872)$ mass obtained from neutral and charged B decays; and Δm_{LH} (in MeV), the difference in mass of two X states produced with equal strength in $p\bar{p}$ collisions

Quantity	Mode	Value	Ref. (χ^2/d.o.f.)
Δm_{0+}	$\pi^+\pi^- J/\psi$	$2.7 \pm 1.6 \pm 0.4$	*BABAR* [87]
		$0.18 \pm 0.89 \pm 0.26$	Belle [86]
		0.79 ± 1.08	Avg[3] (1.8/1)
		<2.2 at 90% CL	Above
Δm_{LH}	$\pi^+\pi^- J/\psi$	< 3.2 at 90% CL	CDF [90]

Taking the totality of experimental information on the $X(3872)$ at face value, the X is a narrow resonant structure with the most probable quantum numbers $J^{PC} = 1^{++}$ and $I = 0$, and has mass within 1 MeV of $D^{*0}\bar{D}^0$ threshold. It may have comparable decay rates to $\gamma\psi(2S)$ and (often-slightly-below-threshold) $D^{*0}\bar{D}^0$, but has an order-of-magnitude smaller rate to both $\omega J/\psi$ and $\rho^0 J/\psi$. Decays to $\gamma J/\psi$ occur at roughly a quarter of the $\gamma\psi(2S)$ rate. If there are two components of the observed enhancements, they must be closer in mass than a few MeV. It is produced and observed in Tevatron $p\bar{p}$ collisions with a rate similar to conventional charmonia, and at the B-factories in $B \to KX$ decays. Unlike conventional charmonia, $B \to K^* X$ is suppressed with respect to $B \to K\pi X$.

The summarized properties of $X(3872)$ do not comfortably fit those of any plausible charmonium state. Prominent decays to $D^{*0}\bar{D}^0$ and proximity to $D^{*0}\bar{D}^0$ mass threshold naturally lead to models [123] which posit the X to be either a weakly bound molecule of a D^{*0} and a $\bar{D}^0$ slightly below threshold or a virtual one slightly unbound. Accommodating the large radiative decay rates and substantial $\pi^+\pi^- J/\psi$ rate in such models creates challenges because the D^{*0} and $\bar{D}^0$ would be spatially separated by large distances, suppressing the probability of overlap for annihilation. This has led to the hypothesis of mixing with a charmonium state having the same quantum numbers. Models of a tightly bound diquark–diantiquark system $cu\bar{c}\bar{u}$ feature two neutral ($cd\bar{c}\bar{d}$, $cd\bar{c}\bar{s}$) and one charged ($cu\bar{c}\bar{d}$) partner state, which are limited by the corresponding mass-difference and null search measurements. Better understanding of the $X(3872)$ demands more experimental constraints and theoretical insight.

2.3.2 Unconventional vector states

$Y(4260)$, $Y(4360)$, $Y(4660)$, and $Y(4008)$ The first observation of an unexpected vector charmonium-like state was made by *BABAR* [108] in ISR production of $Y(4260) \to$

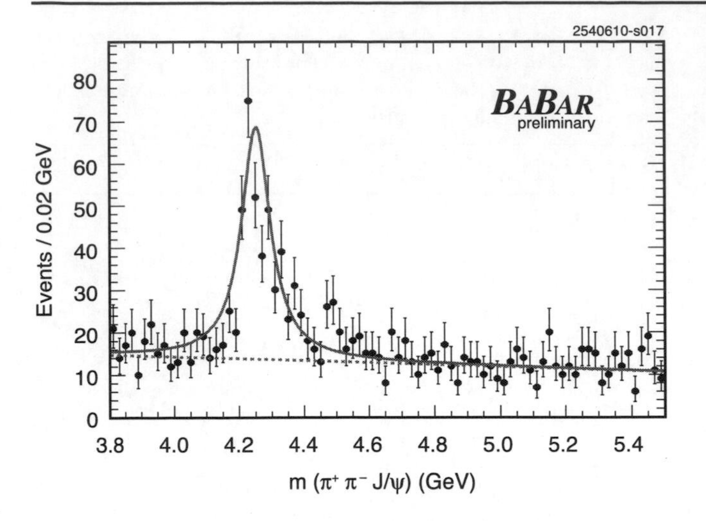

Fig. 17 From *BABAR* [109], the invariant mass of $\pi^+\pi^- J/\psi$ candidates produced in initial-state radiation, $e^+e^- \to \gamma_{\rm ISR}\,\pi^+\pi^- J/\psi$. *Points with error bars* represent data, and the *curves* show the fit (*solid*) to a signal plus a linear background (*dashed*)

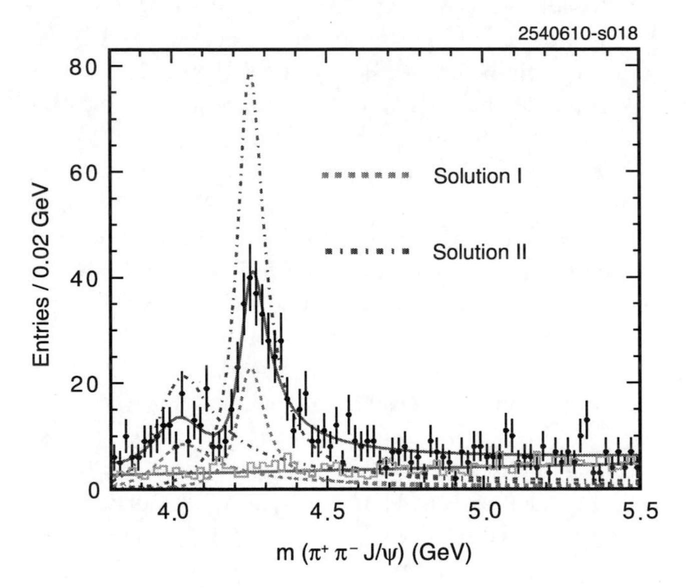

Fig. 18 From Belle [104], the invariant mass of $\pi^+\pi^- J/\psi$ candidates produced in initial-state radiation, $e^+e^- \to \gamma_{\rm ISR}\,\pi^+\pi^- J/\psi$, with J/ψ-sidebands already subtracted, unlike Fig. 17. *Points with error bars* represent data, the *solid curve* shows the best fits to the data to two resonances including interference with a floating phase, and the *dashed* and *dashed-dot curves* show the two pairs of individual resonance contributions for the two equally probable best-fit phases. Adapted from [104] with kind permission, copyright (2007) The American Physical Society

$\pi^+\pi^- J/\psi$, which was later updated [109] with twice the data, as shown in Fig. 17. CLEO [110] and Belle [104] confirmed the *BABAR* result, but Belle also found a smaller, broader structure at 4008 MeV, as seen in Fig. 18. Aside from the lower-mass state, for which the updated *BABAR* [109] analysis placed an upper limit, the three sets of measurements were quite consistent in mass and width, as shown in Table 14, but only roughly so in strength. *BABAR* [113] found

Table 14 Measured properties of the decay $Y(4260) \to \pi^+\pi^- J/\psi$, including its mass m, width Γ, and branching fraction $\mathcal{B}$. The Belle [104] single-resonance fit result is quoted to allow for comparison to the other two

Quantity	Value	Ref. (χ^2/d.o.f.)
m	$4259 \pm 8^{+2}_{-6}$	*BABAR* [109]
(MeV)	4263 ± 6	Belle [104]
	$4284^{+17}_{-16} \pm 4$	CLEO [110]
	4263 ± 5	Avg[3] (1.8/2)
Γ	$88 \pm 23^{+6}_{-4}$	*BABAR* [109]
(MeV)	126 ± 18	Belle [104]
	$73^{+39}_{-25} \pm 5$	CLEO [110]
	108 ± 15	Avg[3] (2.4/2)
$\mathcal{B} \times \Gamma_{ee}$	$5.5 \pm 1.0^{+0.8}_{-0.7}$	*BABAR* [109]
(eV)	9.7 ± 1.1	Belle [104]
	$8.9^{+3.9}_{-3.1} \pm 1.8$	CLEO [110]
	8.0 ± 1.4	Avg[3] (6.1/2)

Table 15 Measured properties of the two enhancements found in the $\pi^+\pi^-\psi(2S)$ mass distribution: $Y(4360)$ and $Y(4660)$. Liu et al. [128] performed a binned maximum-likelihood fit to the combined Belle and *BABAR* cross section distributions (Fig. 19)

Quantity	Value	Ref. (χ^2/d.o.f.)
m	4324 ± 24	*BABAR* [113]
(MeV)	$4361 \pm 9 \pm 9$	Belle [114]
	4353 ± 15	Avg[3] (1.8/1)
	$4355^{+\ 9}_{-10} \pm 9$	Liu [128]
Γ	172 ± 33	*BABAR* [113]
(MeV)	$74 \pm 15 \pm 10$	Belle [114]
	96 ± 42	Avg[3] (6.8/1)
	$103^{+17}_{-15} \pm 11$	Liu [128]
m	$4664 \pm 11 \pm 5$	Belle [114]
(MeV)	$4661^{+9}_{-8} \pm 6$	Liu [128]
Γ	$48 \pm 15 \pm 3$	Belle [114]
(MeV)	$42^{+17}_{-12} \pm 6$	Liu [128]

one more apparent enhancement, $Y(4360)$, in $\pi^+\pi^-\psi(2S)$, which Belle [114] measured with somewhat larger mass and smaller width, as seen in Table 15. Belle also found a second structure near 4660 MeV, as seen in Fig. 19. (A combined fit [128] to Belle and *BABAR* $\pi^+\pi^-\psi(2S)$ data found them to be consistent with one other.) Because dipion transitions between vector quarkonia are commonplace for charmonium and bottomonium, it was natural, then, that the first inclination was to ascribe the Y's to excited vector charmonia.

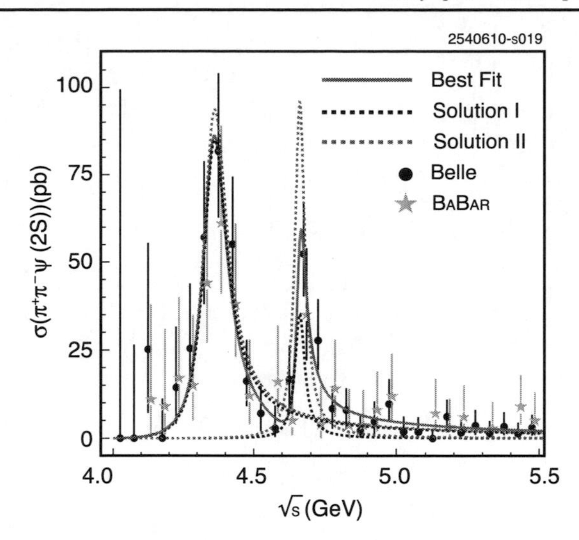

Fig. 19 From [128], the $\pi^+\pi^-\psi(2S)$ cross section as a function of $\sqrt{s}$, showing the result of a binned maximum-likelihood fit of combined Belle and *BABAR* data, The *solid circles* and *stars* show the Belle and *BABAR* data, respectively. The *solid curve* shows the best fits to the data to two resonances including interference with a floating phase, and the *dashed curves* show the two pairs of individual resonance contributions for the two equally probable best-fit phases. Adapted from [128] with kind permission, copyright (2008) The American Physical Society

A number of additional features of these states contradicted this hypothesis, however. Only one, $Y(4660)$, is remotely near a predicted 1^{--} $c\bar{c}$ state (3^3D_1). The $Y(4260)$ and $Y(4360)$ did not show up in inclusive hadronic cross section (R) measurements as seen in Fig. 3, as would be expected of such states. (There is no fine-grained R-scan data near $Y(4660)$.)

A comparison of the measured $\pi^+\pi^-J/\psi$ and total hadronic cross sections in the $\sqrt{s} \simeq 4260$ MeV region yields a lower bound for $\Gamma(Y \to \pi^+\pi^-J/\psi) > 508$ keV at 90% CL, an order of magnitude higher than expected for conventional vector charmonium states [129]. Charmonium would also feature dominant open charm decays, exceeding those of dipion transitions by a factor expected to be $\gtrsim 100$, since this is the case for $\psi(3770)$ and $\psi(4160)$. As summarized in Table 16, no such evidence has been found, significantly narrowing any window for either charmonia or, in some cases, quark–gluon hybrid interpretations. CLEO [111] studied direct production of $Y(4260)$ in e^+e^- collisions; verified the production cross section; and identified the only non-$\pi^+\pi^-J/\psi$ decay mode seen so far, $\pi^0\pi^0J/\psi$, occurring at roughly half of the $\pi^+\pi^-J/\psi$ rate.

Any explanation for these vector states will have to describe their masses, widths, and manifest reluctance to materialize in open charm or unflavored light meson final states. The dipion invariant-mass spectra exhibit curious structures, as seen for $Y(4260)$ in Fig. 20 [109], $Y(4360)$ in Fig. 21(a) [114], and $Y(4660)$ in Fig. 21(b) [114]. The first shows a distinctly non-phase-space double-hump structure

Table 16 Upper limits at 90% CL on the ratios $\sigma(e^+e^- \to Y \to T)/\sigma(e^+e^- \to Y \to \pi^+\pi^-J/\psi)$ at $E_{\text{c.m.}} = 4.26$ GeV (CLEO [16]) and $\mathcal{B}(Y \to T)/\mathcal{B}(Y \to \pi^+\pi^-J/\psi)$ (for $Y(4260)$), and $\mathcal{B}(Y \to T)/\mathcal{B}(Y \to \pi^+\pi^-\psi(2S))$ (for $Y(4360)$ and $Y(4660)$), from *BABAR* [27–29] and Belle [22], where T is an open charm final state

T	$Y(4260)$	$Y(4360)$	$Y(4660)$
$D\bar{D}$	4.0 [16], 7.6 [27]		
$D\bar{D}^*$	45 [16], 34 [28]		
$D^*\bar{D}^*$	11 [16], 40 [28]		
$D\bar{D}^*\pi$	15 [16], 9 [22]	8 [22]	10 [22]
$D^*\bar{D}^*\pi$	8.2 [16]		
$D_s^+D_s^-$	1.3 [16], 0.7 [29]		
$D_s^+D_s^{*-}$	0.8 [16], 44 [29]		
$D_s^{*+}D_s^{*-}$	9.5 [16]		

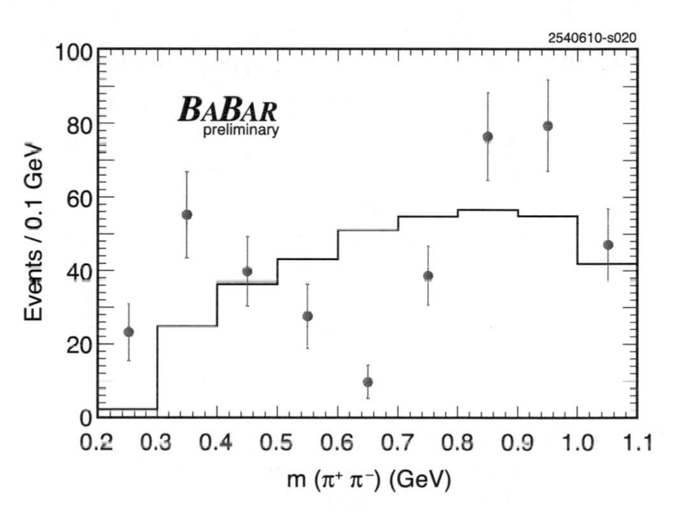

Fig. 20 From *BABAR* [109], the dipion invariant-mass distribution of ISR-produced $Y(4260) \to \pi^+\pi^-J/\psi$ decays, where *points* represent data and the *line histogram* is phase-space MC simulation

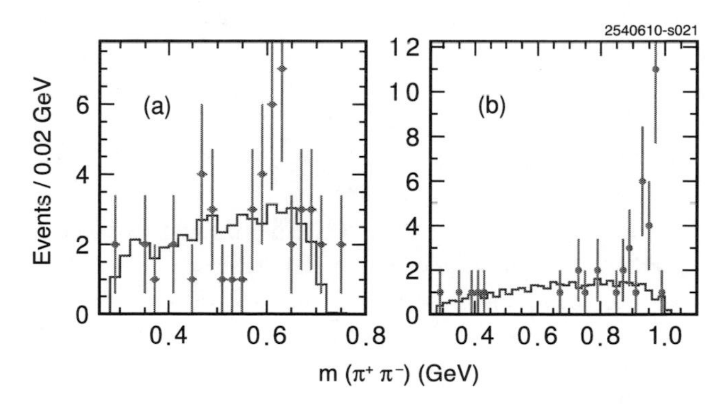

Fig. 21 From Belle [114], the dipion invariant-mass distribution of ISR-produced (**a**) $Y(4360) \to \pi^+\pi^-\psi(2S)$ and (**b**) $Y(4660) \to \pi^+\pi^-\psi(2S)$, where *points* represent data and the *line histograms* show phase-space MC simulations. Adapted from [114] with kind permission, copyright (2007) The American Physical Society

which is qualitatively confirmed by Belle [104], the second exhibits a plurality of events at higher masses, and the third indicates a quite dominant $f_0(980)$ component.

$X(4630)$ The $e^+e^- \to \Lambda_c^+\Lambda_c^-$ cross section was measured by Belle [25] using ISR and partial reconstruction (Fig. 4(i)). A clear peak is evident near the threshold, and corresponds to

$$\mathcal{B}\big(X(4630) \to e^+e^-\big) \times \mathcal{B}\big(X(4630) \to \Lambda_c^+\Lambda_c^-\big) = (0.68 \pm 0.33) \times 10^{-6}. \quad (8)$$

The nature of this enhancement remains unclear. Although both mass and width of the $X(4630)$ (see Table 9) are consistent within errors with those of the $Y(4660)$, this could be coincidence and does not exclude other interpretations.

$Y_b(10888)$ A recent Belle scan above the $\Upsilon(4S)$ was motivated by an earlier observation [117] of anomalously large $\pi^+\pi^-\Upsilon(nS)$ $(n = 1, 2, 3)$ cross sections near the $\Upsilon(5S)$ peak energy. These new data allowed independent determinations of the $\Upsilon(5S)$ lineshape and that of the $\pi^+\pi^-\Upsilon(nS)$ enhancement [37]. A simultaneous fit to all three measured $\pi^+\pi^-\Upsilon(nS)$ cross sections to a single Breit–Wigner function represents the data well; this lineshape has somewhat[5] higher mass and narrower width (see Table 9) than does the $\Upsilon(5S)$ resonance measured with loosely selected hadronic events in the same experiment. This suggests that the enhancement $\pi^+\pi^-\Upsilon(nS)$ could be a 1^{--} Y_b state distinct from $\Upsilon(5S)$ and perhaps of a similar origin as $Y(4260)$. The relevant cross sections and lineshapes are shown in Fig. 22. See also the discussion in Sect. 3.3.11.

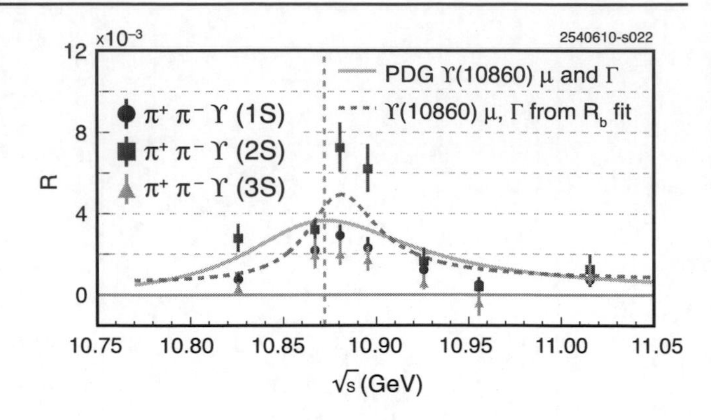

Fig. 22 From Belle [37], results of an energy scan near the $\Upsilon(5S)$. The *points with error bars* indicate the three sets of cross sections for the $\pi^+\pi^-\Upsilon(nS)$ $(n = 1, 2, 3)$, normalized to the point cross section. The *solid* and *dashed curves* show the $\Upsilon(5S)$ lineshape from PDG08 [18] and Belle [37], respectively

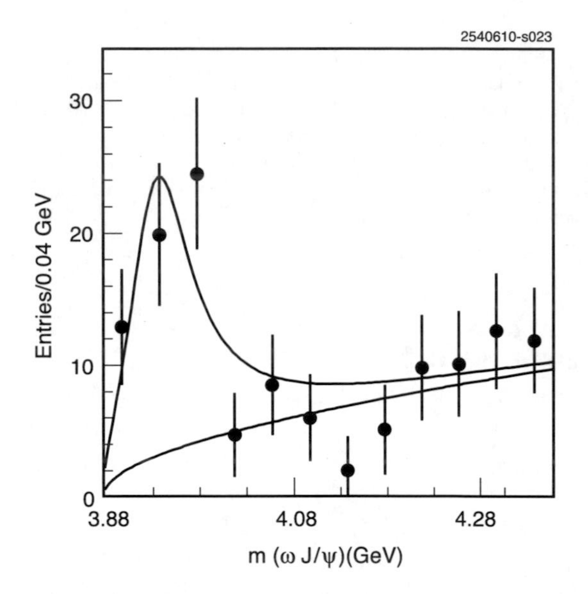

Fig. 23 From Belle [100], the $\omega J/\psi$ mass distribution in $B^+ \to K^+\omega J/\psi$ decays. The *upper curve* is the total fit function, the *lower one* is the contribution of the phase-space-like threshold function. Adapted from [100] with kind permission, copyright (2005) The American Physical Society

2.3.3 Other positive C-parity states

Of the multitude of new charmonium-like states, a puzzling cluster of them from different production mechanisms and/or decay chains gather near a mass of 3940 MeV ($Z(3930)$, $Y(3940)$, $X(3940)$, $X(3915)$) and have positive C-parity. Four others ($Y(4140)$, $X(4160)$, $Y(4274)$, and $X(4350)$) also have $C = +$ and have related signatures. Definitive determination of whether some of these are distinct from others and whether any can be attributed to expected charmonia requires independent confirmations, more precise mass and branching fraction measurements, and unambiguous quantum number assignments.

$Z(3930)$, $Y(3940)$, and $X(3915)$ Some of this fog is clearing as more measurements appear. As described previously, the state decaying to $D\bar{D}$, previously known as $Z(3930)$, has been identified as the $\chi_{c2}(2P)$. The $Y(3940) \to \omega J/\psi$ enhancement initially found by Belle [100] in $B^+ \to K^+Y(3940)$ decays is shown in Fig. 23. It was confirmed by BABAR [101] with more statistics, albeit with somewhat smaller mass, as shown in Fig. 24. But Belle [102] also found a statistically compelling (7.7σ) resonant structure $X(3915)$ in $\gamma\gamma$ fusion decaying to $\omega J/\psi$, as seen in Fig. 25. As the higher-mass Belle $B \to K\,Y(3940)$ ($\to \omega J/\psi$) sighting shares the same production and decay signature as that of BABAR's $Y(3940)$, which has mass and width consistent with the $X(3915)$, the simplest interpretation is that the

[5]When Belle [117] adds the cross sections from $\pi^+\pi^-\Upsilon(nS)$ events at each energy scan point to the loosely selected hadronic events for a fit to a single resonance, the quality of the fit degrades by 2.0σ (where systematic uncertainties are included) relative to the hadronic events alone. That is, the $\pi^+\pi^-\Upsilon(nS)$ and hadronic events are consistent with a single enhancement within two standard deviations.

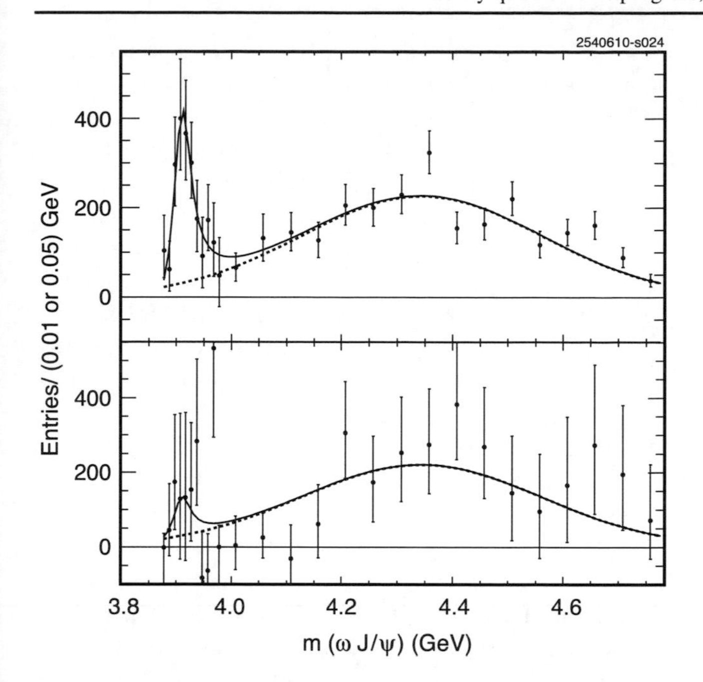

Fig. 24 From *BABAR* [101], the $\omega J/\psi$ mass distribution in $B^+ \to K^+\omega J/\psi$ (*upper*) and $B^0 \to K^0\omega J/\psi$ (*lower*) decays. The *solid* (*dashed*) *curve* represents the total fit (background) function. Adapted from [101] with kind permission, copyright (2008) The American Physical Society

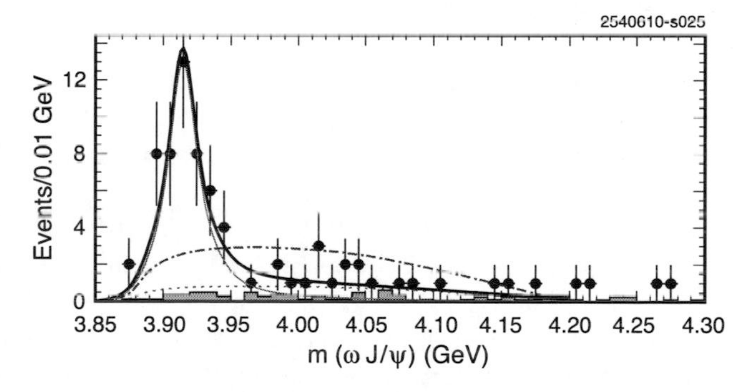

Fig. 25 From Belle [102], the $\omega J/\psi$ invariant-mass distribution for the $\gamma\gamma \to \omega J/\psi$ data (*points*) and background from scaled non-$\omega J/\psi$ sidebands (*shaded*). The *bold solid*, *thinner solid*, and *dashed curves* are the total, resonance, and background contributions, respectively. The *dot-dashed curve* is the best fit with no resonance included. Adapted from [102] with kind permission, copyright (2009) The American Physical Society

$Y(3940)$ and $X(3915)$ are the same state, and that the latter name should prevail as the mass is closer to 3915 MeV. It is this reasoning which motivates grouping them together in Tables 9 and 17. The $X(3915)$ clearly has $C = +$, but J^P remains to be determined.

What more can be gleaned from the existing measurements? The production rate and total width determinations summarized in Table 17 are useful in testing whether the $X(3915)$ behaves like other charmonia. Existing measurements for product branching fractions in $B^+ \to K^+\psi$ decays, where $\psi = \eta_c(1S)$, J/ψ, χ_{cJ}, $h_c(1P)$, $\eta_c(2S)$, and

Table 17 Measured properties of the $X(3915) \to \omega J/\psi$ (subsuming what has previously been called $Y(3940)$). Here $\mathcal{B}_B \times \mathcal{B}_X \equiv \mathcal{B}(B^+ \to K^+X) \times \mathcal{B}(X \to \omega J/\psi)$ and $\Gamma_{\gamma\gamma} \times \mathcal{B}_X \equiv \Gamma(X \to \gamma\gamma) \times \mathcal{B}(X \to \omega J/\psi)$

Quantity	Value	Decay	Ref. (χ^2/d.o.f.)
m	$3942 \pm 11 \pm 13$	$B^+ \to K^+X$	Belle [100]
(MeV)	$3914.6^{+3.8}_{-3.4} \pm 2.0$	$B^+ \to K^+X$	*BABAR* [101]
	$3915 \pm 3 \pm 2$	$\gamma\gamma \to X$	Belle [102]
	3915.6 ± 3.1	Both	Avg[3] (2.7/2)
Γ	$87 \pm 22 \pm 26$	$B^+ \to K^+X$	Belle [100]
(MeV)	$34^{+12}_{-8} \pm 5$	$B^+ \to K^+X$	*BABAR* [101]
	$17 \pm 10 \pm 3$	$\gamma\gamma \to X$	Belle [102]
	28 ± 10	Both	Avg[3] (4.6/2)
$\mathcal{B}_B \times \mathcal{B}_X$	$7.1 \pm 1.3 \pm 3.1$	$B^+ \to K^+X$	Belle [100]
(10^{-5})	$4.9^{+1.0}_{-0.9} \pm 0.5$	$B^+ \to K^+X$	*BABAR* [101]
	5.1 ± 1.0	$B^+ \to K^+X$	Avg[3] (0.4/1)
$\Gamma_{\gamma\gamma} \times \mathcal{B}_X$	$61 \pm 17 \pm 8$	$\gamma\gamma \to X\ (0^+)$	Belle [102]
(keV)	$18 \pm 5 \pm 2$	$\gamma\gamma \to X\ (2^+)$	Belle [102]

$\psi(2S)$, indicate that such states appear with branching fractions of 10^{-3} or below. If the $X(3915)$ behaved at all similarly, this would, in turn, imply that $\Gamma(X(3915) \to \omega J/\psi) \gtrsim 1$ MeV, whereas $\psi(2S)$ and $\psi(3770)$ have hadronic transition widths at least ten times smaller. Similarly, the $\gamma\gamma$-fusion results in Table 17 imply a much larger $\Gamma(X(3915) \to \gamma\gamma)$ than is typical of the few-keV two-photon widths of $\chi_{c0,2}$, unless $\omega J/\psi$ completely dominates its width (which would also be surprising for a $c\bar{c}$ state). In agreement with this pattern, $X(3915)$ does not appear to have prominent decays to $\gamma J/\psi$ [97], $D\bar{D}$ [130], or $D\bar{D}^*$ [95]. Hence any conventional $c\bar{c}$ explanation for $X(3915)$ would likely have trouble accommodating these quite *un*charmonium-like features. More data and more analysis, especially of angular distributions, are necessary to firmly establish these conclusions.

$X(3940)$ The situation for masses near 3940 MeV gets even messier when Belle's analyses of resonances in $e^+e^- \to J/\psi(\ldots)$ [54] and $e^+e^- \to J/\psi D\bar{D}^*$ [103] are considered. The former, as shown in Fig. 26, which examined mass recoiling against a J/ψ in inclusive production, was confirmed by the latter partial reconstruction analysis with the mass and width shown in Table 9. The latter analysis reconstructed the J/ψ and a D or D^*, and then used the missing-mass spectrum to isolate events consistent with the desired topology. The event was then kinematically constrained to have the requisite missing mass (D^* or D), improving resolution on the missing momentum. That four-momentum was then combined with that of the reconstructed $D^{(*)}$ to form the $D\bar{D}^*$ invariant mass, as illustrated

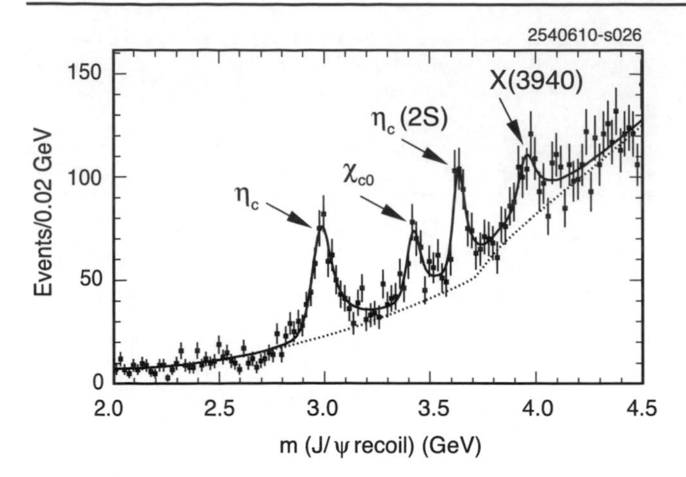

Fig. 26 From Belle [54], the distribution of mass recoiling against the J/ψ in $e^+e^- \to J/\psi(\dots)$ events (*points with error bars*). Results of the fit are shown by the *solid curve*, the *dashed curve* corresponds to the expected background distribution. Adapted from [54] with kind permission, copyright (2007) The American Physical Society

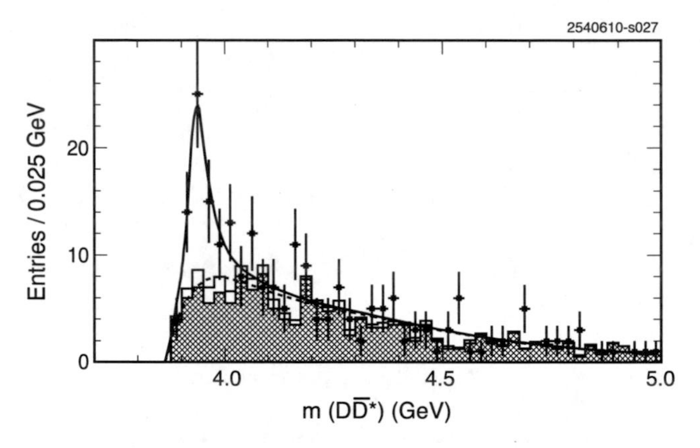

Fig. 27 From Belle [103], the distribution of $D\bar{D}^*$ mass recoiling against the J/ψ (*points with error bars*). Results of the full (background only) fit are shown by the *solid* (*dashed*) *curve*, and the *hatched histogram* is from the scaled $D^{(*)}$-sidebands. Adapted from [103] with kind permission, copyright (2008) The American Physical Society

in Fig. 27, yielding a signal for $X(3940)$ with a significance of 6.0σ. From mass measurements alone, this state appears distinct from the $X(3915)$ by 3.1σ. Bolstering this notion, explicit searches for a state with the appropriate mass in B-decays (to $D\bar{D}^*$ [95]) or double-charmonium production (decaying to $\omega J/\psi$ [103]) were negative, yielding incompatible $\omega J/\psi/D\bar{D}^*$ relative branching fractions for the two production mechanisms.

$X(4160)$ Belle [103] extended the $e^+e^- \to J/\psi D\bar{D}^*$ analysis to also search for resonances decaying to $D^*\bar{D}^*$, and found a broad enhancement $X(4160)$ just above $D^*\bar{D}^*$ threshold with mass and width shown in Table 9. Double vector-charmonium production, which can occur through two intermediate virtual photons instead of one, has not yet been observed, so it seems unlikely that this state could be $\psi(4160)$. And if $e^+e^- \to J/\psi\, X(4160)$ is produced via annihilation, its C-parity would necessarily be positive.

$Y(4140)$, $Y(4274)$, and $X(4350)$ New measurements from CDF [106, 107] indicate at least one more $C = +$ state seen in B decays and decaying to two vectors, one being a J/ψ, near threshold. In inclusively selected $B^+ \to K^+\phi J/\psi$ decays, two enhancements in the $\phi J/\psi$ mass spectrum, with masses and widths as shown in Table 9, are $Y(4140)$ and $Y(4274)$. The analysis requires both that the final state be kinematically consistent with B-decay, but also uses the CDF particle identification system to require three charged kaons in the decay. The signal significances are 5.0σ for $Y(4140)$ and 3.1σ for $Y(4274)$. Both remain unconfirmed. However, Belle [112] searched for production of $Y(4140)$ in two-photon fusion, $e^+e^- \to e^+e^-Y(4140)$, $Y(4140) \to \phi J/\psi$, and found no evidence for it, obtaining a limit of $\Gamma_{\gamma\gamma} \times \mathcal{B}(\phi J/\psi) < 40$ eV for $J^P = 0^+$ and < 5.9 eV for 2^+ at 90% CL. In that same analysis, Belle reported a 3.2σ enhancement, $X(4350) \to \phi J/\psi$, with mass and width as in Table 9 and a production rate measured to be $\Gamma_{\gamma\gamma} \times \mathcal{B}(\phi J/\psi) = 6.7^{+3.2}_{-2.4} \pm 1.1$ eV for $J^P = 0^+$ and $1.5^{+0.7}_{-0.6} \pm 0.3$ eV for 2^+.

2.3.4 *Charged exotic mesons: the Z's*

The charmonium-like *charged* Z states, seen by Belle in $Z^- \to \psi(2S)\pi^-$ and $\chi_{c1}\pi^-$ in $B \to Z^-K$ decays, are of special interest. If these states are mesons, they would necessarily have a minimal quark substructure of $c\bar{c}u\bar{d}$ and therefore be manifestly exotic. In a manner similar to the first unearthing of $X(3872)$, Z^- states were found in exclusively reconstructed B-decays in which a conventional charmonium state is a decay product of the Z^-. Here, a single charged pion accompanies either a $\psi(2S)$ or χ_{c1} (compared to a $\pi^+\pi^-$ pair accompanying a J/ψ in the case of $X(3872)$). The $\psi(2S)$ is found via either $\psi(2S) \to \pi^+\pi^-J/\psi$ followed by $J/\psi \to \ell^+\ell^-$ or from direct dileptonic decay, $\psi(2S) \to \ell^+\ell^-$. The χ_{c1} is tagged by its decay $\chi_{c1} \to \gamma J/\psi$, $J/\psi \to \ell^+\ell^-$. Statistics are gained in the $Z(4430)^-$ analysis by using both charged and neutral B decays, combining each Z^- candidate with either a neutral ($K^0_S \to \pi^+\pi^-$) or charged kaon candidate, if present. In all cases, background not from B-decays is small after the usual B-selection criteria on expected B-candidate energy and mass have been applied, and is well-estimated using appropriately scaled sidebands in those variables. Each experiment finds consistency in various subsets of its own data (e.g., from charged and neutral B mesons and in different $\psi(2S)$ and J/ψ decay modes) and thereby justifies summing them for final results. Belle and BABAR have comparable mass resolution and statistical power for studying these decays.

Belle found [115] the first Z^- state by observing a sharp peak near $M(\psi(2S)\pi^-) = 4430$ MeV with statistical significance of $> 6\sigma$. The largest backgrounds are $B \to \psi(2S)K_i^*$, $K_i^* \to K\pi^-$, where $K_i^* \equiv K^*(892)$ or $K_2^*(1430)$. Hence $K\pi^-$ mass regions around $K^*(892)$ and $K_2^*(1430)$ were excised in this Belle analysis. However, as *interference* between different partial waves in the $K\pi^-$ system can produce fake "reflection" peaks in the $M(\psi(2S)\pi^-)$ distribution, further attention is warranted. In the kinematically allowed $K\pi^-$ mass range for this three-body B-decay, only S-, P- and D-partial waves in $K\pi^-$ are significant. Belle found that no combination of interfering $L = 0, 1, 2$ partial waves can produce an artificial mass peak near 4430 MeV without also producing additional, *larger* structures nearby in $M(\psi(2S)\pi^-)$. Such enhancements are absent in the Belle data, ruling out such reflections as the origin of the apparent signal.

In recognition of the role $K\pi^-$ dynamics can play in the background shape and in response to the BABAR non-confirming analysis described below, Belle released a second analysis [116] of $Z(4430)^-$ in their data. Here Belle modeled the $B \to \psi(2S)\pi^- K$ process as the sum of several two-body decays, with a $B \to Z(4430)^- K$ signal component and, for background, $B \to \psi(2S)K_i^*$, where K_i^* denotes all of the known $K^* \to K\pi^-$ resonances that are kinematically accessible. Results of this second analysis are depicted in Fig. 28, which is an $m^2(\psi(2S)\pi^-)$ Dalitz plot projection with the prominent K^* bands removed. The data (points with error bars) are compared to the results of the fit with (solid histogram) or without (dashed histogram) the $Z(4430)^-$ resonance; the former can be seen to be strongly favored over the latter, a 6.4σ effect. This Dalitz plot fit yields the mass and width shown in Table 9 as well as the product branching fraction, $\mathcal{B}(B^0 \to Z^- K) \times \mathcal{B}(Z^- \to \psi(2S)\pi^-) = (3.2^{+1.8}_{-0.9}{}^{+9.6}_{-1.6}) \times 10^{-5}$. These values for the mass, width, and rate are consistent with the corresponding measurements reported in the initial Belle publication, but have larger uncertainties, which is indicative of the larger set of systematic variations in both signal and background properties that were considered. In the default fit, the $Z(4430)^-$ resonance was assumed to have zero spin. Variations of the fit included a $J = 1$ assignment for the $Z(4430)^-$, models with hypothetical $K^* \to K\pi^-$ resonances with floating masses and widths, and radically different parametrizations of the $K\pi^-$ S-wave amplitude.

The corresponding BABAR search [131] added a decay mode $Z(4430)^- \to J/\psi\pi^-$ that was not considered in either Belle study. Its inclusion increases statistics for two purposes: first, potentially for more signal, since it is entirely reasonable to expect this mode to occur with at least as large a branching fraction as the discovery mode; and second, to study the $K\pi^-$ resonance structure in the background, since the J/ψ modes contain about six times more

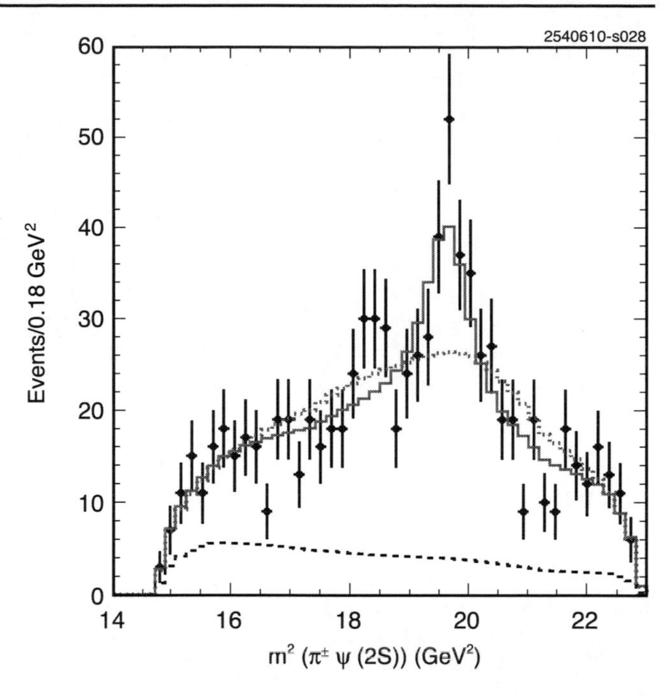

Fig. 28 From Belle [116], for $B \to K\pi^-\psi(2S)$ candidates, the data *points* show the $m^2(\psi(2S)\pi^-)$ projection of the Dalitz plot with the K_i^* bands removed. The *solid (dotted) histogram* shows the corresponding projections of the fits with (without) a $Z(4430)^- \to \psi(2S)\pi^-$ resonance term. The *dashed histogram* represents non-B-decay background estimated from energy-difference sidebands. Adapted from [116] with kind permission, copyright (2009) The American Physical Society

events than those with the $\psi(2S)$. The BABAR analysis exploits this more copious B-decay mode with exhaustive studies of $K\pi^-$ partial wave dynamics, including fine-grained determination of angular distributions and selection efficiencies over all regions of the Dalitz plot. The data were fit with floating S-, P-, and D-wave intensities. For both $J/\psi\pi^- K$ and $\psi(2S)\pi^- K$ samples, good fits are obtained, as shown in the projections in Fig. 29. Hence both $J/\psi\pi^-$ and $\psi(2S)\pi^-$ mass distributions from BABAR are well-represented by simulations with no extra resonant structure near $M(\psi(2S)\pi^-) = 4430$ MeV. At the same time, if the BABAR $M(\psi(2S)\pi^-)$ distribution with a $K^*(892)$ and $K_2^*(1430)$ veto (as done by Belle) is fit for the presence of a $Z(4430)^-$ at the same mass and width found by Belle, a signal with 2σ statistical significance is found, indicating a statistical consistency in the corresponding Belle and BABAR mass distributions. This latter finding was verified with a direct bin-by-bin comparison between the Belle and BABAR $M(\psi(2S)\pi^-)$ distributions after the $K^*(892)$ and $K_2^*(1430)$ veto: the two samples were found to be statistically equivalent. That is, while no statistically significant signal for $Z(4430)^-$ in the BABAR data has been found, neither does the BABAR data refute the positive Belle observation of $Z(4430)^-$.

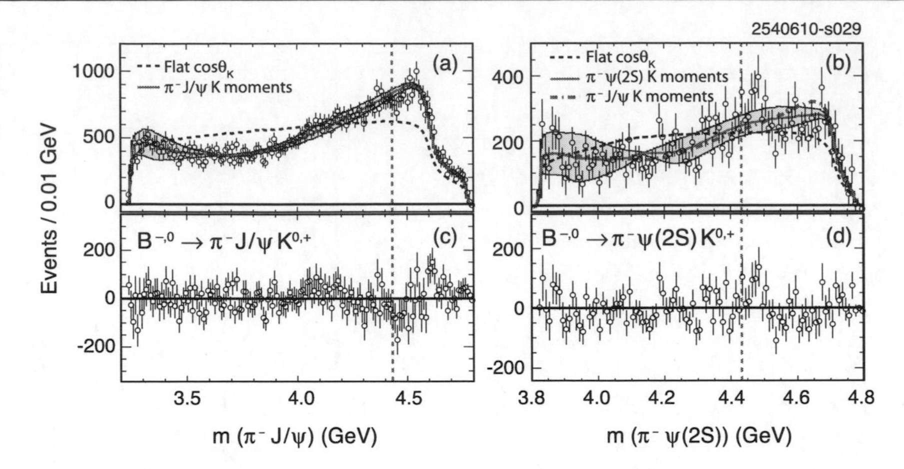

Fig. 29 From *BABAR* [131], the $\psi\pi^-$ mass distributions for (**a**) $B \to J/\psi\pi^- K$ and (**b**) $B \to \psi'\pi^- K$. The *points* show the data (integrated over all $K\pi^-$ regions) after efficiency correction and background subtraction. The *dashed curves* show the $K\pi^-$ reflection expected for a uniform decay angle ($\cos\theta_K$) distribution, while the *solid curves* show the result when accounting for the measured angular variation. The *shaded bands* represent the effect of statistical uncertainty on the normalized moments. In (**b**), the *dot-dashed curve* indicates the result expected if the $K\pi^-$ properties measured for $J/\psi\pi^- K$ are used. The *dashed vertical lines* indicate the value of $m(\psi\pi^-) = 4433$ MeV. In (**c**) and (**d**) appear the residuals (*data-solid curve*) for (**a**) and (**b**), respectively, after the *solid curves* are subtracted from the data. Adapted from [131] with kind permission, copyright (2009) The American Physical Society

Belle has also found [105] signals, dubbed $Z_1(4050)^-$ and $Z_2(4250)^-$, in the $\chi_{c1}\pi^-$ channel, again using $B \to Z^- K$ decays. Here the kinematically allowed mass range for the $K\pi^-$ system extends beyond the $K_3^*(1780)$ F-wave resonance. Thus S-, P-, D- and F-wave terms for the $K\pi^-$ system are all included in the model. A Dalitz fit with a single resonance in the $Z^- \to \chi_{c1}\pi^-$ channel is favored over a fit with only K_i^* resonances and no Z^- by $>10\sigma$. Moreover, a fit with two resonances in the $\chi_{c1}\pi^-$ channel is favored over the fit with only one Z^- resonance by 5.7σ. Fitted mass values appear in Table 9. The product branching fractions have central values similar to that for the $Z(4430)^-$ but with large errors. Figure 30 shows the $m(\chi_{c1}\pi^-)$ projection of the Dalitz plot with the K^* bands excluded and the results of the fit with no $Z^- \to \chi_{c1}\pi^-$ resonances and with two $Z^- \to \chi_{c1}\pi^-$ resonances.

Thus, although two experiments have explored these new states, we are left in the less than satisfying situation of three claims of definitive observations and one nonobservation which does not exclude the positive measurement. If any or all of the three charged Z^- states reported by Belle are in fact meson resonances, they would be "smoking guns" for exotics. It is therefore crucial that these states be confirmed or refuted with independent measurements. In particular, *BABAR* should search for $Z_1(4050)^-$ and $Z_2(4250)^-$ in the $\chi_{c1}\pi^-$ channel and Belle should search for $Z(4430)^-$ in the $J/\psi\pi^-$ channel. That the purported $Z(4430)^-$ might decay copiously to $\psi(2S)\pi^-$ but barely or not at all to $J/\psi\pi^-$ is a theoretical puzzle worth addressing. The DØ and CDF experiments at the Tevatron could also search for inclusive production of $Z(4430)^-$.

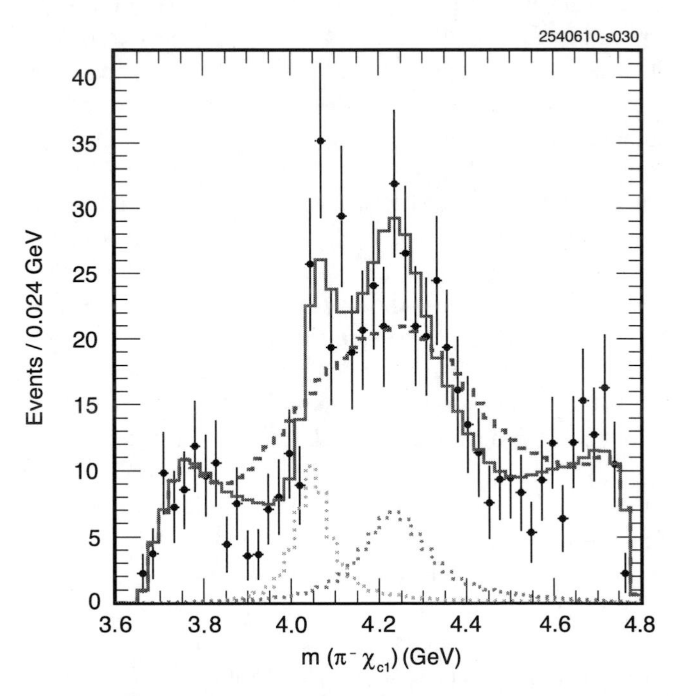

Fig. 30 From Belle [105], for $B \to K\pi^-\chi_{c1}$ candidates, the data *points* show the $m(\pi^-\chi_{c1})$ projection of the Dalitz plot with the K^* bands removed. The *solid* (*dashed*) *histogram* shows the corresponding projection of the fit with (without) the two $Z_i^- \to \chi_{c1}\pi^-$ resonance terms. The *dotted histograms* show the fitted contributions of the two resonances. Adapted from [105] with kind permission, copyright (2008) The American Physical Society

2.4 Characteristics of quarkonium systems

Heavy quarkonia are systems composed of two heavy quarks, each having mass m much larger than the QCD confinement scale $\Lambda_{\rm QCD}$. Because the system is nonrelativistic, quarkonium is characterized by the heavy-quark bound-state velocity, $v \ll 1$ ($v^2 \sim 0.3$ for $c\bar{c}$, $v^2 \sim 0.1$ for $b\bar{b}$, $v^2 \sim 0.01$ for $t\bar{t}$) and by a hierarchy of energy scales: the mass m (hard scale, H), the relative momentum $p \sim mv$ (soft scale, S), and the binding energy $E \sim mv^2$ (ultrasoft scale, US). For energy scales close to Λ_{QCD}, perturbation theory breaks down and one has to rely on nonperturbative methods. Regardless, the nonrelativistic hierarchy of scales,

$$m \gg p \sim 1/r \sim mv \gg E \sim mv^2, \tag{9}$$

where r is the typical distance between the heavy quark and the heavy antiquark, also persists below the scale Λ_{QCD}. Since $m \gg \Lambda_{\text{QCD}}$, $\alpha_s(m) \ll 1$, and phenomena occurring at the scale m may be always treated perturbatively. The coupling may also be small if $mv \gg \Lambda_{\text{QCD}}$ and $mv^2 \gg \Lambda_{\text{QCD}}$, in which case $\alpha_s(mv) \ll 1$ and $\alpha_s(mv^2) \ll 1$, respectively. This is likely to happen only for the lowest charmonium and bottomonium states (see Fig. 31). Direct information on the radius of the quarkonia systems is not available, and thus the attribution of some of the lowest bottomonia and charmonia states to the perturbative or the nonperturbative soft regime is at the moment still ambiguous. For $t\bar{t}$ threshold states even the ultrasoft scale may be considered perturbative.

This hierarchy of nonrelativistic scales separates quarkonia [1] from heavy-light mesons, the latter of which are characterized by just two scales: m and Λ_{QCD} [132, 133]. This makes the theoretical description of quarkonium physics more complicated. All quarkonium scales get entangled in a typical amplitude involving a quarkonium observable, as illustrated in Fig. 32. In particular, quarkonium annihilation and production take place at the scale m, quarkonium binding takes place at the scale mv (which is the typical momentum exchanged inside the bound state), while very low-energy gluons and light quarks (also called ultrasoft degrees of freedom) are sufficiently long-lived that a bound state has time to form and therefore are sensitive to the scale mv^2. Ultrasoft gluons are responsible for phenomena like the Lamb shift in QCD. The existence of several scales complicates the calculations. In perturbative calculations of loop diagrams the different scales get entangled, challenging our abilities to perform higher-order calculations. In lattice QCD, the existence of several scales for quarkonium sets requirements on the lattice spacing ($a < 1/m$) and overall size ($La > 1/(mv^2)$) that are challenging to our present computational power.

However, it is precisely the rich structure of separated energy scales that makes heavy quarkonium particularly well suited to the study of the confined region of QCD, its interplay with perturbative QCD, and of the behavior of the perturbation series in QCD: heavy quarkonium is an ideal probe of confinement and deconfinement. Quarkonia systems with different radii have varying sensitivities to the Coulombic and confining interactions, as depicted in Fig. 33. Hence

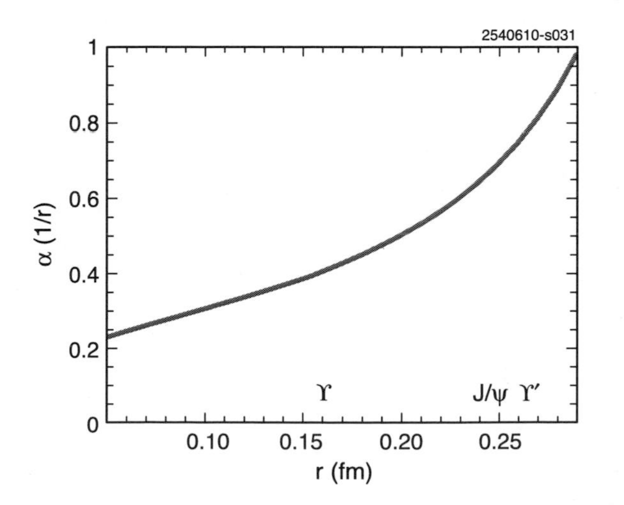

Fig. 31 The strong-coupling constant, α_s, at one loop, as a function of quarkonium radius r, with labels indication approximate values of mv for $\Upsilon(1S)$, J/ψ, and $\Upsilon(2S)$

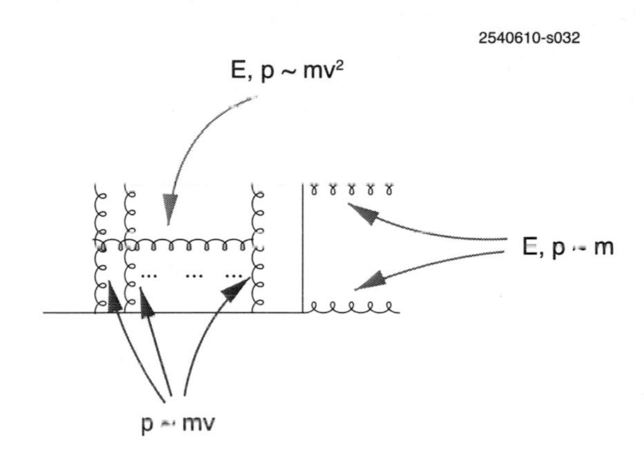

Fig. 32 Typical scales appearing in a quarkonium annihilation diagram

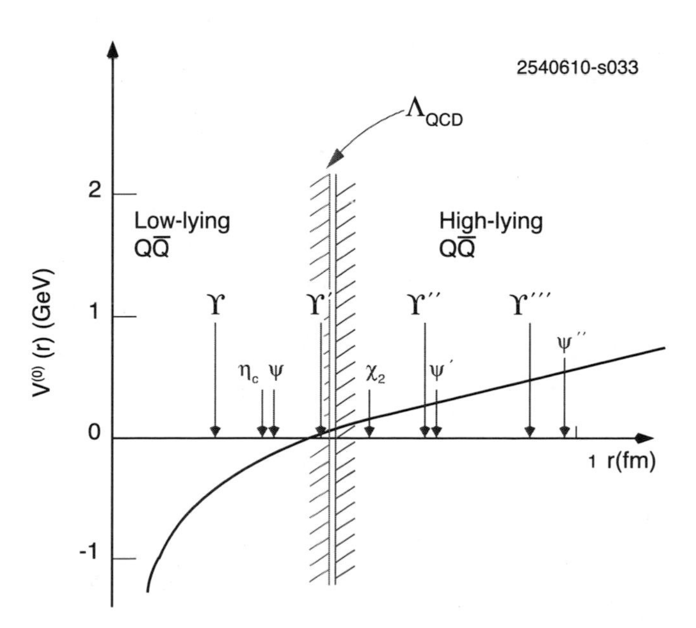

Fig. 33 Static $Q\overline{Q}$ potential as a function of quarkonium radius r

different quarkonia will dissociate in a medium at different temperatures, providing, e.g., a thermometer for the plasma, as discussed in Sect. 5.3.

2.5 Nonrelativistic effective field theories

The modern approach to heavy quarkonium is provided by Nonrelativistic Effective Field Theories (NR EFTs) [134]. The idea is to take advantage of the existence of a hierarchy of scales to substitute simpler but equivalent NR EFTs for QCD. A hierarchy of EFTs may be constructed by systematically integrating out modes associated with high-energy scales not relevant for the quarkonium system. Such integration is performed as part of a matching procedure that enforces the equivalence between QCD and the EFT at a given order of the expansion in v. The EFT realizes factorization between the high-energy contributions carried by the matching coefficients and the low-energy contributions carried by the dynamical degrees of freedom at the Lagrangian level. The Poincaré symmetry remains intact at the level of the NR EFT in a nonlinear realization that imposes exact relations among the EFT matching coefficients [135, 136].

2.5.1 Physics at the scale m: NRQCD

Quarkonium annihilation and production occur at the scale m. The suitable EFT is Nonrelativistic QCD [137, 138], which follows from QCD by integrating out the scale m. As a consequence, the effective Lagrangian is organized as an expansion in $1/m$ and $\alpha_s(m)$:

$$\mathcal{L}_{\text{NRQCD}} = \sum_n \frac{c_n(\alpha_s(m), \mu)}{m^n} \times O_n(\mu, mv, mv^2, \ldots), \quad (10)$$

where O_n are the operators of NRQCD that are dynamical at the low-energy scales mv and mv^2, μ is the NRQCD factorization scale, and c_n are the Wilson coefficients of the EFT that encode the contributions from the scale m and are nonanalytic in m. Only the upper (lower) components of the Dirac fields matter for quarks (antiquarks) at energies lower than m. The low-energy operators O_n are constructed out of two or four heavy-quark/antiquark fields plus gluons. The operators bilinear in the fermion (or antifermion) fields are the same ones that can be obtained from a Foldy–Wouthuysen transformation of the QCD Lagrangian. Four-fermion operators have to be added. Matrix elements of O_n depend on the scales μ, mv, mv^2 and Λ_{QCD}. Thus operators are counted in powers of v. The imaginary part of the coefficients of the four-fermion operators contains the information on heavy-quarkonium annihilation. The NRQCD heavy-quarkonium Fock state is given by a series of terms, where the leading term is a $Q\overline{Q}$ in a color-singlet state, and the first correction, suppressed in v, comes from a $Q\overline{Q}$ in an octet state plus a gluon. Higher-order terms are subleading in increasing powers of v. NRQCD is suitable for spectroscopy studies on the lattice. The latest results on charmonium, bottomonium, and B_c spectroscopy are reported in Sect. 2.6.2. For the latest results on NRQCD inclusive decay amplitudes, see Sect. 3.2.1.

2.5.2 Physics at the scales mv, mv^2: pNRQCD

Quarkonium formation occurs at the scale mv. The suitable EFT is pNRQCD [139, 140], which follows from NRQCD by integrating out the scale $mv \sim r^{-1}$. The soft scale mv may or may not be larger than the confinement scale Λ_{QCD} depending on the radius of the quarkonium system. When $mv^2 \sim \Lambda_{\text{QCD}}$, we speak about weakly coupled pNRQCD because the soft scale is perturbative and the matching of NRQCD to pNRQCD may be performed in perturbation theory. When $mv \sim \Lambda_{\text{QCD}}$, we speak about strongly coupled pNRQCD because the soft scale is nonperturbative and the matching of NRQCD to pNRQCD is not possible in perturbation theory. Below we will review recent results and applications of the two EFTs.

2.5.3 $mv \gg \Lambda_{\text{QCD}}$: weakly coupled pNRQCD

The effective Lagrangian is organized as an expansion in $1/m$ and $\alpha_s(m)$, inherited from NRQCD, and an expansion in r [140]:

$$\mathcal{L}_{\text{pNRQCD}} = \int d^3r \sum_n \sum_k \frac{c_n(\alpha_s(m), \mu)}{m^n} \times V_{n,k}(r, \mu', \mu) r^k \times O_k(\mu', mv^2, \ldots), \quad (11)$$

where O_k are the operators of pNRQCD that are dominant at the low-energy scale mv^2, μ' is the pNRQCD factorization scale and $V_{n,k}$ are the Wilson coefficients of the EFT that encode the contributions from the scale r and are nonanalytic in r. The degrees of freedom that make up the operators O_k are $Q\overline{Q}$ states, color-singlet S, color-octet $O_a T^a$, and (ultrasoft) gluons. The operators are defined in a multipole expansion. In the equations of motion of pNRQCD, we may identify $V_{n,0} = V_n$ with the $1/m^n$ potentials that enter the Schrödinger equation and $V_{n,k\neq 0}$ with the couplings of the ultrasoft degrees of freedom that provide corrections to the Schrödinger equation. Since the degrees of freedom that enter the Schrödinger description are, in this case, both $Q\overline{Q}$ color singlet and $Q\overline{Q}$ color octets, both singlet and octet potentials exist.

The bulk of the interaction is contained in potential-like terms, but non-potential interactions, associated with the propagation of low-energy degrees of freedom are, in general, present as well and start to contribute at NLO in the multipole expansion. They are typically related to nonperturbative effects. Matrix elements of $O_{n,k}$ depend on the scales μ', mv^2 and Λ_{QCD}.

If the quarkonium system is small, the soft scale is perturbative and the potentials can be *entirely* calculated in perturbation theory [134]. They are renormalizable, develop a scale dependence, and satisfy renormalization group equations that eventually allow resummation of potentially large logarithms.

Progress in the perturbative calculation of the potentials

- $Q\overline{Q}$ singlet and octet potentials

There has been much progress in the computation of the perturbative corrections to the potential between a static quark and a static antiquark in a color-singlet configuration. The subleading logarithmic term coming from ultrasoft effects has been computed [141]. The resummation of those subleading ultrasoft logarithms has also been performed [142], along with a comparison of the static energy to lattice data [142]. These calculations confirm that perturbation theory can reproduce the lattice very accurately at short distances (up to 0.25 fm) once the cancelation of the leading renormalon singularity is implemented.

The three-loop correction to the static potential is now completely known: the fermionic contributions to the three-loop coefficient [143] first became available, and, more recently, the remaining purely gluonic term has been obtained [144, 145]. The value found for the n_f independent three-loop coefficient is much lower than the previous, widely used Padé estimate and within the range obtained by comparing to lattice data [142]. The future implementation of the three-loop result may improve the precision of some mass and strong-coupling determinations. In particular, the recently obtained theoretical expression [146] for the complete QCD static energy at next-to-next-to-next-to-leading-logarithmic accuracy (NNNLL) has been used to determine $r_0 \Lambda_{\overline{\mathrm{MS}}}$ by comparison with available lattice data, where r_0 is the lattice scale and $\Lambda_{\overline{\mathrm{MS}}}$ is the QCD scale, obtaining

$$r_0 \Lambda_{\overline{\mathrm{MS}}} = 0.622^{+0.019}_{-0.015} \tag{12}$$

for the zero-flavor case. This extraction was previously performed at the NNLO level (including an estimate at NNNLO) in [147]. The same procedure can be used to obtain a precise evaluation of the unquenched $r_0 \Lambda_{\overline{\mathrm{MS}}}$ value after short-distance unquenched lattice data for the $Q\overline{Q}$ exist.

At three-loop order a violation of the Casimir scaling for the static QQbar potential is found, see [148].

The static octet potential is known up to two loops [149]. Relativistic corrections to the static singlet potential have been calculated over the years and are summarized in [134].

- QQQ and QQq static potentials

The three-quark static potential has been calculated in perturbation theory at next-to-leading order in the singlet, octet, and decuplet channels [150, 151]. Mixing between the octet representations has been found already at tree level. At next-to-next-to-leading order, the subset of diagrams producing three-body forces has been identified in Coulomb gauge and their contribution to the potential calculated. Combining it with the contribution of the two-body forces, which may be extracted from the quark–antiquark static potential, the complete next-to-next-to-leading order of the three-quark static potential in the color-singlet channel has been obtained in [150]. These results may be important for accurate calculations of the lowest QQQ states as well as for comparison and study of the QQQ lattice static energies in the domain of small interquark distance.

The same pNRQCD description is also possible for QQ states, which are relevant for doubly charmed baryons [151, 152]. The QQ antitriplet static potential, relevant for the QQq states, has been calculated at next-to-next-to-leading order [150]. First lattice calculations of the QQq potential have also become available [153–155].

Progress on the lowest spectra calculation For systems with small radii, precision calculations are possible. In such cases, quarkonium may become a benchmark for our understanding of QCD, in particular for the transition region between perturbative and nonperturbative QCD, and for the precise determination of Standard Model parameters (e.g., heavy-quark masses, α_s, as described in Sect. 2.8). When the soft scale is perturbative, the energy levels are given by the expectation value of the perturbative potentials, calculated to the appropriate order of the expansion in α_s plus nonperturbative terms that are not part of the potential, which start to contribute to the energy levels at order $m\alpha_s^5$. They enter energy levels and decay calculations in the form of local or nonlocal electric and magnetic condensates [134, 156–159]. A precise and systematic knowledge of such nonperturbative, purely glue-dependent objects is still lacking. It is therefore important that these condensates be fixed, either by dedicated lattice determinations or extracted from the data. Within pNRQCD it is possible to relate the leading electric and magnetic nonlocal correlators to the gluelump masses and to some existing lattice (quenched) determinations [134].

However, since the nonperturbative contributions are suppressed in the power counting, it is possible to obtain precise determinations of the masses of the lowest quarkonium resonances within purely perturbative calculations, in the cases in which the perturbative series is convergent (i.e., after the appropriate subtractions of renormalons have been performed), and large logarithms are resummed [160–163].

Once the potentials are known, energy levels can be computed. The lowest heavy-quarkonium states are suitable to obtain precise determinations of the b and c mass. Such determinations are competitive with those coming from different systems and different approaches, as has been discussed at length [1]. An update of recent result on mass extractions is given in Sect. 2.8.

Once the quark masses have been obtained, it is possible to obtain the energy levels of the lowest resonances. However, which quarkonium state belongs to which regime is an open issue and no clear-cut method exists to decide this *a priori*, since we lack a direct method to determine the quarkonium radius [164, 165]. Typically the lowest states $\Upsilon(1S)$, η_b, B_c, and possibly J/ψ and η_c are assumed to be in the weakly coupled regime. The S-wave energy levels are known in perturbation theory at next-to-next-to-next-to-leading order ($m\alpha_s^5$) [156, 157, 166–169].

A prediction of the B_c mass has been obtained [170]. The NNLO calculation with finite charm mass effects [171, 172] predicts a mass of 6307(17) MeV, in agreement with the CDF measurement [173] and the lattice determination [74]. A NLO calculation reproduces, in part, the $\chi_{bJ}(1P)$ fine splitting [174]. The same procedure seems to work at NNLO even for higher-mass bottomonium states (i.e., measured masses match the predictions within the respective theory errors, which are larger for higher-mass states) [171]. Including logs resummation at NLL, it is possible to obtain a prediction [175]

$$\Delta m_{\rm hf}\big[\eta_b(1S)\big] = 41 \pm 11(\text{th})^{+9}_{-8}(\delta\alpha_s)\ \text{MeV}, \tag{13}$$

in which the second error comes from the uncertainty in α_s. This value is consistent with another perturbative prediction [176] but both undershoot the average experimental value from (5) by about 2σ. This discrepancy with experiment remains a challenge for theory. (There are further discussions of the $\eta_b(1S)$ hyperfine splitting in Sects. 2.2.5, 2.6.2, 2.7, 2.8.1, and 2.10.1, the last of which offers the possibility of new physics becoming manifest in $\eta_b(nS)$ mass splittings.) Similar calculations yield a predicted B_c hyperfine separation [177]

$$m(B_c^*) - m(B_c) = 50 \pm 17^{+15}_{-12}\ \text{MeV}. \tag{14}$$

2.5.4 *$mv \sim \Lambda_{\rm QCD}$: strongly coupled pNRQCD*

When $mv \sim \Lambda_{\rm QCD}$ the soft scale is nonperturbative and the matching from NRQCD to pNRQCD cannot be performed in perturbation theory. Then the potential matching coefficients are obtained in the form of expectation values of gauge-invariant Wilson-loop operators. In this case, heavy-light meson pairs and heavy hybrids develop a mass gap of order $\Lambda_{\rm QCD}$ with respect to the energy of the $Q\overline{Q}$ pair, the second circumstance being apparent from lattice simulations. Thus, away from threshold, the quarkonium singlet field S is the only low-energy dynamical degree of freedom in the pNRQCD Lagrangian (neglecting ultrasoft corrections coming from pions and other Goldstone bosons). This pNRQCD Lagrangian may be expressed as [134, 178, 179]:

$$\mathcal{L}_{\rm pNRQCD} = S^\dagger\left(i\partial_0 - \frac{\mathbf{p}^2}{2m} - V_S(r)\right)S. \tag{15}$$

The singlet potential $V_S(r)$ can be expanded in powers of the inverse of the quark mass; static, $1/m$ and $1/m^2$ terms were calculated long ago [178, 179]. They involve NRQCD matching coefficients (containing the contribution from the hard scale) and low-energy nonperturbative parts given in terms of static Wilson loops and field-strength insertions in the static Wilson loop (containing the contribution from the soft scale). In this regime of pNRQCD, we recover the quark potential singlet model. However, here the potentials are calculated in QCD by nonperturbative matching. Their evaluation requires calculations on the lattice or in QCD vacuum models.

Recent progress includes new, precise lattice calculations of these potentials obtained using the Lüscher multi-level algorithm (see Sect. 2.6.3 for more details).

Then, away from threshold, all the masses can be obtained by solving the Schrödinger equation with such potentials. Some applications of these results to the spectrum calculation are ongoing [180].

A trivial example of application of this method is the mass of the $h_c(1P)$. The lattice data show a vanishing long-range component of the spin–spin potential. Thus the potential appears to be entirely dominated by its short-range, delta-like, part, suggesting that the $^1P_1$ state should be close to the center-of-gravity of the ^3P_J system. Indeed, the measurements described in Sect. 2.2.1 and summarized in Table 5 are consistent with this expected value.

If we explicitly consider light quarks, each quarkonium state may develop a width due to decay through pion emission. The neavy-light states develop a mass gap of order $\Lambda_{\rm QCD}$ with respect to quarkonium which can be absorbed into the definition of the potentials or of the (local or nonlocal) condensates [181].

2.6 Lattice QCD spectrum calculations

In quarkonia, the ultrasoft scale mv^2 is often of a similar size as the scale $\Lambda_{\rm QCD}$, where nonperturbative effects become important. For all charmonium and many bottomonium states the soft scale $mv \sim 1/r$ is not much larger than $\Lambda_{\rm QCD}$ either. The nonperturbative contributions can be evaluated via computer simulations of lattice-regularized QCD (Lattice QCD), where the lattice spacing a provides a hard ultraviolet cut-off on the available momenta in a Euclidean space-time volume.

Light sea quarks are particularly expensive to simulate numerically since the computational effort increases as a large inverse power of the corresponding pseudoscalar mass m_{PS}. The spatial lattice extent, La, should be much larger than m_{PS}^{-1} to control finite-size effects, necessitating large volumes. Only very recently have the first simulations near the physical pion mass $m_{PS} \approx m_\pi$ become possible [182–185]. In the first reference a variant of the staggered fermion action was applied. These fermions are usually only defined for n_f mass-degenerate fermions, where n_f is a multiple of four. This restriction was circumvented by replacing the determinant of the $n_f = 8 + 4$ staggered action by a fourth root that may then correspond to $n_f = 2 + 1$. However, it remains controversial whether this *ad hoc* prescription recovers a unitary, local quantum field theory and thus the correct continuum limit. Moreover, there are claims that the additional so-called taste symmetry cannot be completely restored in the continuum limit [186, 187], some aspects of which have been refuted [188], and others of which have yet to be clearly established [189].

The lattice regularization of QCD is not unique, and many different discretized actions can be constructed that should yield the same results after removing the cut-off (continuum limit extrapolation: $a \to 0$). While the Wilson action is subject to $O(a)$ discretization effects, other actions are $O(a)$ improved, e.g., chiral actions fulfilling the Ginsparg–Wilson relation, staggered actions, twisted-mass QCD or the nonperturbatively improved Sheikholeslami–Wilson (clover) action. Ideally, lattice simulations are repeated at several values of the lattice spacing, $a \ll \Lambda_{QCD}^{-1}$, and the results then extrapolated to the continuum limit. Lattice artifacts will be large if physics at scales $q \not\ll a^{-1}$ becomes important, spoiling the continuum limit extrapolation. In this respect, the charm quark mass, m_c, and, even more so, the bottom quark mass, m_b, pose challenges. By exploiting or ignoring the multiscale nature of quarkonium systems, different routes can be taken.

One way to proceed is to integrate out m as well as mv and to evaluate the resulting potential nonperturbatively in lattice simulations of pNRQCD [139, 140]. Subsequently, the energy levels can be obtained by solving a Schrödinger equation. This directly relates lattice simulations to QCD vacuum models and to potential models. However, at present this approach is only semi-quantitative since no calculations of the matching of lattice pNRQCD, where rotational symmetry is broken, to QCD is available.

Another approach, which is more common, is to integrate out only m and to simulate NRQCD [137, 138] on the lattice [190]. In this case, the lattice spacing, a^{-1}, plays the role of the soft matching scale, μ, used above. Unless this is restricted to $m \gtrsim a^{-1} \gtrsim mv$, the NRQCD matching coefficients will become uncontrollably large, and the continuum limit cannot be taken. Therefore a must be sufficiently fine for discretization effects to be smaller than the neglected higher-order terms in the heavy-quark expansion. This restricts the applicability of NRQCD methods to systems containing b quarks.

Relativistic heavy-quark actions can also be used because the inequality $a^{-1} \gg m$ usually still holds so that spin-averaged level-splittings can be obtained. To a lesser extent, reliable results on the fine structure can be achieved as well. The Fermilab effective field theory interpretation of the heavy-quark action [191, 192] smoothly connects the region $m \sim a^{-1}$ with the continuum limit $a^{-1} \gg m$, allowing charm and bottom systems to be treated in the same set up.

The following sections survey the present state-of-the-art. We start with results obtained using a relativistic heavy-quark action, continue with NRQCD, and conclude with lattice pNRQCD results.

2.6.1 Relativistic heavy-quark actions

One way of limiting the computational cost of small lattice spacings, i.e., of a large number of lattice points, is the use of anisotropic actions, with a temporal lattice spacing, a_τ, smaller than the spatial lattice spacing, $a_\sigma = \xi a_\tau$, where $\xi > 1$. The spatial lattice extent, $L_\sigma a_\sigma$, still needs to be sufficiently large to accommodate the quarkonium state, which has a size of order $r \simeq (mv)^{-1}$. In the presence of light sea quarks, one would additionally wish to realize $L_\sigma a \gg m_{PS}^{-1}$. With sufficiently large a_σ, it is possible to limit the number of lattice points, $\propto L_\sigma^3$. It is then easy to realize $a_\tau < m^{-1}$, where m is the particle mass. Naively, such simulations are cheaper by a factor ξ^3 relative to the isotropic case. This method was successfully explored in the quenched approximation [193, 194], and is reviewed in [1].

At tree level, it can be arranged that the lattice spacing errors are $O[(ma_\tau)^n]$, where $n \in \{1, 2\}$ depends on the heavy-quark action, but care is needed to ensure it [195]. One-loop corrections may lead to $O[\alpha_s(ma_\sigma)^n]$ terms: to the extent that $\alpha_s\xi^n$ is small, the leading-order lattice effects can be regarded as $O[(ma_\tau)^n]$. The anisotropy parameter ξ must be determined consistently for the quark and gluon contributions to the QCD action. Within the quenched approximation, where the feedback of quark fields onto the gluons is neglected, this problem factorizes. The tuning is much harder to achieve and numerically more costly in QCD with sea quarks. Nevertheless such a program was pursued very successfully by the Hadron Spectrum Collaboration [196]. It should be noted, however, that, in this case, the $O(a)$ improvement is *not* nonperturbatively accurate.

Such anisotropic configurations have been employed to calculate electromagnetic transition rates from excited charmonium states, high-spin states, and exotics [71, 197, 198] with small volumes. The lattice spacing and sea quark mass dependencies will be investigated in the near future. The same holds for an exploratory study of the Regensburg group, using the isotropic clover action for $n_f = 2$ sea quarks (generated by the QCDSF Collaboration [199]) and charm quarks, using a lattice cut-off $a^{-1} \approx 1.7$ GeV [200].

While the spin-averaged splittings are in qualitative agreement with experiment, the fine structure is underestimated due to unphysically heavy sea quarks and the missing continuum limit extrapolation. Another study with improved Wilson fermions was performed by PACS-CS with $n_f = 2+1$ sea quark flavors at the physical π mass [183], focusing only on very few J^{PC} channels. In this case, the J/ψ-$\eta_c(1S)$ splitting is underestimated by about 10% relative to experiment, which can probably be attributed to the finite lattice spacing. Note that valence quark annihilation channels were omitted in all these studies.

To this end, mixing effects with noncharmed mesons and flavor-singlet contributions to the spectrum were evaluated [201] and found to be smaller than 10 MeV in the pseudoscalar channel. The latter effect was also investigated by the MILC Collaboration [202], with similar results. It should be noted that valence quark annihilation diagrams were neglected in all other unquenched studies. Finally, mixing effects between radially excited $c\bar{c}$ states and four-quark molecules (or tetraquarks) were investigated at a sea pion mass, $m_{\mathrm{PS}} < 300$ MeV, at $a^{-1} \approx 2.4$ GeV [203]. Indications of attraction and mixing were found in the 1^{++} channel, supporting the interpretation of the $X(3872)$ as a $D^{*0}\bar{D}^0$ bound state with a $\chi_{c1}(2P)$ admixture. Other charmed tetraquark studies [204, 205] were performed in the quenched approximation and did not take into account valence-annihilation diagrams. A recent review on the tetraquark topic is presented in [206].

At present, all lattice studies of heavy-quarkonium spectroscopy in which a continuum limit is taken are based on $n_f \approx 2+1$ configurations generated by the MILC Collaboration [207] using the so-called AsqTad improved staggered sea-quark action [208]. Together with the Fermilab Lattice Collaboration, charmonium and bottomonium spectra were investigated using the clover action for the heavy quarks at four different lattice spacings ranging from $(1.2\ \mathrm{GeV})^{-1}$ down to $(2.2\ \mathrm{GeV})^{-1}$, and various light-quark masses [209]. Only S- and P-waves were studied. After extrapolating to the physical limit, all spin-averaged splittings, with the exception of the charmonium $2\overline{S} - 1\overline{S}$ splitting, which is muddled with threshold effects, are in agreement with experiment. Moreover, the $1S$ and $1P$ fine structures are compatible with the experimental values. The continuum limit extrapolation was essential to achieve this. The same combination of AsqTad sea-quark action and clover charm-quark action was also used in a recent calculation of singly and doubly charmed baryons [210].

2.6.2 *NRQCD lattice calculations*

As discussed above, lattice NRQCD is a suitable method for bottomonia studies, for which $am_b > 1$ and $\Lambda_{\mathrm{QCD}}/(vm_b) \ll 1$. However, the precision of the results for fine-structure splittings is limited here by the fact that the QCD matching coefficients are typically taken at tree level only in such lattice simulations.

The HPQCD and UKQCD Collaborations presented calculations [211] of the bottomonium spectrum using the $n_f = 2+1$ MILC gauge configurations [207] described above, expanding to order-v^4 in the heavy-quark velocity v. Another lattice calculation for bottomonium [212] that closely followed the methods of [211] was later released, but using the $n_f = 2+1$ configurations provided by the RBC and UKQCD Collaborations [213]) with the chiral domain-wall sea-quark action for a single lattice spacing $a^{-1} \approx 1.7$ GeV, also expanding to order v^4. Both calculations [211, 212] found agreement with experiment for the spin-averaged splittings. The spin-dependent splitting was seen to be systematically underestimated in [212], possibly due to the omission of relativistic, radiative, and discretization corrections (see Sect. 2.7 for more on $\Delta m_{\mathrm{hf}}[\eta_b(1S)]$). The same method was then used to calculate the spectrum of other states containing one or two b-quarks, including baryons [212, 214]. Again, the underestimation of the experimental fine structure might be due to either the coarse lattice spacing or the imprecise matching between lattice NRQCD and QCD.

Large scale simulations of the spectrum of Υ states and B_c, which include sea-quark mass and continuum limit extrapolations, were performed by the HPQCD Collaboration [215, 216]. They employed an NRQCD action for the b quark and combined this with the relativistic HISQ (highly improved staggered quark) action for the charm quark. Their findings are very similar to those of [209]; it will be very interesting to perform a detailed comparison between results from a relativistic b-quark action and NRQCD on the same ensemble of gauge configurations.

The mass of the triply heavy baryon Ω_{bbb} has been calculated [217] in lattice QCD with $2+1$ flavors of light sea quarks. The b-quark is implemented with improved lattice NRQCD. Gauge-field ensembles from both the RBC/UKQCD and MILC Collaborations with lattice spacings in the range from 0.08–0.12 fm are used. The final result for the mass, which includes an electrostatic correction, is

$$m(\Omega_{bbb}) = 14.371 \pm 0.004\ (\mathrm{stat.}) \pm 0.011\ (\mathrm{syst.}) \pm 0.001\ (\mathrm{exp.})\ \mathrm{GeV}. \qquad (16)$$

The hyperfine splitting between the physical $J = \frac{3}{2}$ state and a fictitious $J = \frac{1}{2}$ state is also presented [217].

2.6.3 *pNRQCD lattice calculations*

Another approach is offered by pNRQCD [178, 179]. Unfortunately, lattice pNRQCD also suffers from the fact that the matching coefficients between lattice NRQCD and QCD are only known at tree level. pNRQCD bridges the gap between QCD and potential models. Implemented on the lattice, it amounts to calculating static potentials and spin- and velocity-dependent corrections. Recent and not so recent progresses in this area include the calculation of string breaking in the static sector [218], which offers one entry point into the study of threshold states, the calculation of potentials in baryonic static-static-light systems [155, 219], and the calculation of interaction energies between two static-light mesons [220]. New very precise results on all leading relativistic corrections to the static $Q\overline{Q}$ potential have also been calculated [221–223].

2.7 Predictions for the $\eta_b(1S)$ mass

The calculation described above in Sect. 2.5.3 with result in (13) gives a numerical result for $\Delta m_{\text{hf}}[\eta_b(1S)]$ (41 ± 14 MeV) that is typical of perturbative calculations (e.g., 44 ± 11 MeV is given by [176]). These values are somewhat smaller than those obtained from lattice NRQCD as described in Sect. 2.6.2 above (e.g., 61 ± 14 MeV in [211] and 52 ± 1.5(stat.) MeV in [212]). However, it has been argued [224] that additional short-range corrections of $\delta^{\text{hard}}\Delta m_{\text{hf}}[\eta_b(1S)] \approx -20$ MeV would lower these unquenched lattice results to the level of the perturbative predictions.

Very recently a newer lattice prediction [225] performed at order v^6 in the NRQCD velocity expansion has been obtained at tree level for the NRQCD matching coefficients and with domain-wall actions for sea quarks, including the spin splittings, and based on the RBC/UKQCD gauge-field ensembles. This approach [225] addresses the concerns in [224] (namely, that radiative contributions in the calculation are missing because the NRQCD matching coefficients are calculated at tree level) by calculating appropriate *ratios* of spin splittings and thereafter normalizing to a measured value. With this method, and using the experimental result for the $1P$ tensor splitting as input, a $1S$ bottomonium hyperfine splitting of

$$\begin{aligned}\Delta m_{\text{hf}}\big[\eta_b(1S)\big] &= 60.3 \pm 5.5 \text{ (stat.)} \\ &\quad \pm 5.0 \text{ (syst.)} \\ &\quad \pm 2.1 \text{ (exp.) MeV} \\ &= 60.3 \pm 7.7 \text{ MeV}\end{aligned} \tag{17}$$

is determined. This value is slightly smaller (1.1σ) than but consistent with the experimental measurements ((5) and Table 7), and somewhat larger (1.2σ) than but consistent with earlier pQCD calculations [175, 176] (see (13)). However, this still leaves the weakly coupled pNRQCD (perturbative) and experimental values 2.0σ apart, and it is not clear why; the lattice NRQCD lies in between. More study and better precision in the predictions are required to gain confidence in both perturbative and nonperturbative calculations. (See further discussions of the $\eta_b(1S)$ hyperfine splitting in Sects. 2.2.5, 2.5.3, 2.8.1, and 2.10.1.)

2.8 Standard Model parameter extractions

Given the progress made in the effective field theories formulation, in the calculation of high order perturbative contributions and in the lattice simulations, quarkonium appears to be a very suitable system for precise determination of Standard Model parameters like the heavy-quark masses and α_s. Below we report about recent determinations. For a review and an introduction to the procedure and methods used in this section, see [1].

2.8.1 α_s determinations

Below we review several extractions of the strong-coupling constant related to observables in heavy-quarkonium physics. All values for α_s are quoted in the $\overline{\text{MS}}$ scheme with $n_f = 5$, unless otherwise indicated.

α_s from quarkonia masses on the lattice A precise determination of α_s from lattice simulations has been presented by the HPQCD Collaboration [182]. The mass difference $m[\Upsilon(2S)] - m[\Upsilon(1S)]$, the masses $m[\Upsilon(1S)]$, $m[\eta_c(1S)]$, and light meson masses were used to tune the bare parameters. Several short-distance quantities, mainly related to Wilson loops, were computed to obtain the original result [182]

$$\alpha_s(m_{Z^0}) = 0.1183 \pm 0.0008. \tag{18}$$

An independent implementation of a similar lattice-based approach, using the same experimental inputs as [182], yields [226]

$$\alpha_s(m_{Z^0}) = 0.1192 \pm 0.0011. \tag{19}$$

The HPQCD Collaboration updated their original result of (18), superseding it with [227]

$$\alpha_s(m_{Z^0}) = 0.1184 \pm 0.0006. \tag{20}$$

This value is not independent of the result in (19). The two results in (18) and (19) are expected to be nearly identical, as they use the same inputs and different calculations of the same theoretical effects.

Another new determination [227, 228] of α_s uses moments of heavy-quark correlators calculated on the lattice and continuum perturbation theory. The same references also provide a determination of the c- and b-quark masses, described in Sect. 2.8.2. The result is [227]

$$\alpha_s(m_{Z^0}) = 0.1183 \pm 0.0007. \tag{21}$$

These extractions of α_s, based on lattice calculations and quarkonia masses, are the most precise individual determinations among the many methods and measurements available [229], and will thus tend to dominate any average over different α_s determinations such as that in [229].

α_s *from quarkonium radiative decays* The CLEO [230] measurement of the inclusive radiative decay $\Upsilon(1S) \to \gamma X$ (see Sect. 3.2.2), together with a theoretical description of the photon spectrum [231], has made it possible to obtain a precise determination of α_s from $\Upsilon(1S)$ decays [232]. A convenient observable is the parton-level ratio

$$R_\gamma \equiv \frac{\Gamma(V \to \gamma g g)}{\Gamma(V \to g g g)} \tag{22}$$

for decays of heavy vector meson V. The wave function of $V = \Upsilon(1S)$ at the origin and the relativistic corrections cancel at order v^2 in this ratio. Furthermore, one also needs to include color-octet contributions in the decay rates, requiring an estimation of the color-octet NRQCD matrix elements. The two color-octet matrix elements that appear in the numerator of R_γ at order v^2, $O_8(^1S_0)$ and $O_8(^3P_0)$, also appear in the denominator, which includes $O_8(^3S_1)$, thus decreasing the theoretical uncertainty associated with the estimation of those matrix elements in the α_s extraction. The theoretical expression for R_γ at order v^2 is

$$R_\gamma = \frac{36}{5}\frac{e_b^2\alpha}{\alpha_s}\frac{1 + C_{\gamma i}\mathcal{R}_i}{1 + C_i\mathcal{R}_i}. \tag{23}$$

Here e_b is the b-quark charge, α is the fine-structure constant, and $C_{\gamma i}\mathcal{R}_i$ and $C_i\mathcal{R}_i$ represent the order v^2 corrections to the numerator and denominator, respectively. The v^2 corrections [232] account for radiative, relativistic, and octet effects. Experimental values of R_γ for $V = \Upsilon(1S, 2S, 3S)$ appear in Table 24. The value of R_γ for $\Upsilon(1S)$ obtained by CLEO [230] using the Garcia–Soto (GS) QCD calculation [231] for the photon spectrum[6] is

$$R_\gamma\big[\Upsilon(1S)\big]_{\rm exp} = 0.0245 \pm 0.0001 \pm 0.0013, \tag{24}$$

where the first error is statistical and the second is systematic. The value of α_s obtained from (23) using (24) as input is [232]

$$\begin{aligned} &\alpha_s(m_{\Upsilon(1S)}, n_f = 4) = 0.184^{+0.015}_{-0.014}, \\ &\alpha_s(m_{Z^0}) = 0.119^{+0.006}_{-0.005}. \end{aligned} \tag{25}$$

The recent lattice [233] and continuum [234] estimates of the octet matrix elements are used to obtain this result. The experimental systematic uncertainty from (24) dominates the error in α_s shown in (25).

There are also CLEO measurements of $R_\gamma[J/\psi]$ [235] and $R_\gamma[\psi(2S)]$ [236] (see Sect. 3.2.2 and Table 25). One could, in principle, extract α_s in the same way as for $\Upsilon(1S)$ above. However, the relativistic and octet corrections are more severe than for the $\Upsilon(1S)$, so terms of higher order than v^2 may not be small enough to ignore. The effects due to the proximity of $\psi(2S)$ to open-charm threshold are difficult to estimate [237].

α_s *from bottomonium hyperfine splitting* The observation of $\eta_b(1S)$ by BABAR [59, 61] and CLEO [60] allows α_s to be extracted from the singlet-triplet hyperfine mass splitting,

$$\Delta m_{\rm hf}\big[\eta_b(1S)\big] \equiv m\big[\Upsilon(1S)\big] - m\big[\eta_b(1S)\big]. \tag{26}$$

The theoretical expression used for the hyperfine splitting includes a perturbative component and a nonperturbative one. The perturbative component is given by the expression of [175], which includes order α_s corrections to the leading order (LO) term,

$$\Delta m_{\rm hf}\big[\eta_b(1S)\big]_{\rm LO} = \frac{C_F^4\alpha_s^4 m_b}{3}, \tag{27}$$

and resummation of the logarithmically enhanced corrections up to the subleading logarithms (which are of the form $\alpha_s^n \ln^{n-1}\alpha_s$). The nonperturbative part is parametrized in terms of the dimension-four gluon condensate, which is fixed according to [238].

Using the average experimental value in Table 7 and (5) the resulting value of α_s is [239]

$$\alpha_s(m_{Z^0}) = 0.125 \pm 0.001 \pm 0.001 \pm 0.001, \tag{28}$$

where the first error is experimental, the second is associated with the gluon condensate and the third accounts for the b-quark mass uncertainty. This α_s value is slightly more than 3σ higher than the updated HPQCD [227] lattice result in (20).

2.8.2 *Determinations of m_b and m_c*

Below we review recent extractions of the heavy-quark masses related to observables in heavy-quarkonium physics.

Sum rules

- Moments

The determination of the heavy-quark masses from a sum-rule analysis requires theoretical predictions for the $n^{\rm th}$ moments of the cross section for heavy-quark production in e^+e^- collisions. The theoretical expression for the moments is related to derivatives of the vacuum polarization function

[6] A theoretical description of the photon spectrum is needed to extrapolate to the experimentally inaccessible, low-energy part of the spectrum.

at $q^2 = 0$. Four-loop $[\mathcal{O}(\alpha_s^3)]$ results for the vacuum polarization function have appeared for the first moment [240–242], second moment [243], third moment [244], and approximate results for higher moments [245, 246]. All those four-loop results are used in the most recent low-momentum sum-rule determinations of the heavy-quark masses reported below.

- Low-n sum rules

The most recent determination [247, 248] of the c- and b-quark masses using low-momentum sum rules incorporates four-loop results for the derivatives of the vacuum polarization function along with the most recent experimental data. The results are [248]

$$m_c^{\overline{\rm MS}}(3\ {\rm GeV}) = 0.986 \pm 0.013\ {\rm GeV}, \tag{29}$$

$$m_c^{\overline{\rm MS}}(m_c^{\overline{\rm MS}}) = 1.279 \pm 0.013\ {\rm GeV}, \tag{30}$$

and

$$m_b^{\overline{\rm MS}}(10\ {\rm GeV}) = 3.610 \pm 0.016\ {\rm GeV}, \tag{31}$$

$$m_b^{\overline{\rm MS}}(m_b^{\overline{\rm MS}}) = 4.163 \pm 0.016\ {\rm GeV}. \tag{32}$$

For a critical discussion of the error attached to these determinations and a new (preliminary) mass determination using low-momentum sum rules see [249].

- Large-n sum rules

A determination of the b-quark mass using nonrelativistic (large-n) sum rules, including resummation of logarithms, has been performed [250]. It incorporates next-to-next-to leading order results along with the complete next-to-leading logarithm resummation (and partial next-to-next-to-leading logarithm resummation). Including logarithm resummation improves the reliability of the theoretical computation. The value of the $\overline{\rm MS}$ mass is [250]

$$m_b^{\overline{\rm MS}}(m_b^{\overline{\rm MS}}) = 4.19 \pm 0.06\ {\rm GeV}. \tag{33}$$

The c-quark mass has also been determined from a nonrelativistic sum-rules analysis [251], with the result

$$m_c^{\overline{\rm MS}}(m_c^{\overline{\rm MS}}) = 1.25 \pm 0.04\ {\rm GeV}. \tag{34}$$

- Alternative approaches

A determination of the c- and b-quark masses which uses moments at $q^2 \neq 0$ and includes the dimension-six gluon condensate (also determined from the sum rules) has been reported [252]:

$$m_c^{\overline{\rm MS}}(m_c^{\overline{\rm MS}}) = 1.260 \pm 0.018\ {\rm GeV}, \tag{35}$$

$$m_b^{\overline{\rm MS}}(m_b^{\overline{\rm MS}}) = 4.220 \pm 0.017\ {\rm GeV}, \tag{36}$$

which employ an estimate of the four-loop contribution to the $q^2 \neq 0$ moments.

Quark masses from the lattice A determination of the b-quark mass in full (unquenched) lattice QCD, using one-loop matching to continuum QCD, finds [211]

$$m_b^{\overline{\rm MS}}(m_b^{\overline{\rm MS}}) = 4.4 \pm 0.3\ {\rm GeV}. \tag{37}$$

The c-quark mass was calculated by comparing lattice determinations of moments of heavy-quark correlators to four-loop continuum perturbation theory [227, 228]. A b-quark mass calculation is also included in [227]. Due to the use of continuum, rather than lattice, perturbation theory, a higher-order perturbative calculation can be used, achieving very precise results [227]:

$$m_c^{\overline{\rm MS}}(3\ {\rm GeV}) = 0.986 \pm 0.006\ {\rm GeV}, \tag{38}$$

$$m_c^{\overline{\rm MS}}(m_c^{\overline{\rm MS}}) = 1.273 \pm 0.006\ {\rm GeV}, \tag{39}$$

and

$$m_b^{\overline{\rm MS}}(10\ {\rm GeV}) = 3.617 \pm 0.025\ {\rm GeV}, \tag{40}$$

$$m_b^{\overline{\rm MS}}(m_b^{\overline{\rm MS}}) = 4.164 \pm 0.023\ {\rm GeV}. \tag{41}$$

2.8.3 m_t determination

Determination of the top-quark mass m_t at a future e^+e^- linear collider from a $t\bar{t}$ lineshape measurement (see Sect. 6.13) requires good theoretical knowledge of the total $t\bar{t}$ production cross section in the threshold region. The threshold regime is characterized by $\alpha_s \sim v$. The N^{k}LO result includes corrections of order $\alpha_s^n v^m$ with $n + m = k$. The N^{2}LO result has been known for some time now. Several contributions to the N^{3}LO result have been calculated. Those include corrections to the Green functions and wave function at the origin [166, 167, 253–255]; matching coefficients of the effective theory currents [256, 257]; electroweak effects in NRQCD [258, 259]; and corrections to the static potential (see Sect. 2.5.3).

Renormalization-group-improved expressions, which sum terms of the type $\alpha_s \ln v$, are also necessary to reduce the normalization uncertainties of the cross section and improve the reliability of the calculation. Those resummations, which were originally only done in the framework of velocity NRQCD, have now been calculated within pNRQCD [260]. The terms at next-to-next-to-leading logarithmic accuracy are not yet completely known [261]. Consistent inclusion of all effects related to the instability of the top quark is needed. Some recent progress in this direction has been made [262–264]. It is expected that a linear collider will provide an m_t determination with uncertainties at the level of 100 MeV (see Sect. 6.13). For comparison, the Tevatron Electroweak Working Group has reported a best-current value of $m_t = 173.1 \pm 1.3$ GeV [265].

At the LHC, top quarks will be produced copiously. It has been pointed out [266] that in the threshold region of $t\bar{t}$ production at LHC, a significant amount of the (remnant of) color-singlet $t\bar{t}$ resonance states will be produced, unlike at the Tevatron where the color-octet $t\bar{t}$ states dominate. In fact, there appears the $1S$ peak in the $t\bar{t}$ invariant-mass distribution below $t\bar{t}$ threshold, even after including the effects of initial-state radiation and parton-distribution function, and the position of this $1S$ peak is almost the same as that in e^+e^- collisions [267, 268]. Namely, theoretically there is a possibility of extracting the top-quark mass with high accuracy from this peak position, although experimentally it is quite challenging to reconstruct the $t\bar{t}$ invariant mass with high accuracy.

Recently, a theoretical framework to compute the fully differential cross sections for top-quark production and its subsequent decays at hadron colliders has been developed, incorporating the bound-state effects which are important in the $t\bar{t}$ threshold region [269]. A Monte Carlo event generator for LHC has been developed and various kinematical distributions of the decay products of top quarks have been computed. In particular, it was found that a bound-state effect deforms the (bW^+)–$(\bar{b}W^-)$ double-invariant-mass distribution in a correlated manner, which can be important in the top event reconstruction.

2.9 Exotic states and states near or above threshold

For states *away* from threshold, it has been shown that appropriate EFTs to describe the quarkonium spectrum can be constructed. In particular, in pNRQCD, the relevant degrees of freedoms are clearly identified: the leading order description coincides with the Schrödinger equation, the potentials are the pNRQCD matching coefficients, and the energy levels are calculable in a well-defined procedure. *Close* to threshold, the situation changes drastically [270, 271]. As described earlier in this section, the region close to and just above threshold is presently the most interesting, with a wealth of newly discovered states. Most new states do not fit potential-model expectations. This is to be expected, as we have seen that a potential model description of quarkonium (strongly coupled pNRQCD) emerges only for binding energies smaller than $\Lambda_{\rm QCD}$. Since the open heavy-flavor threshold is at the scale $\Lambda_{\rm QCD}$ in HQET, a potential model description of states above that threshold cannot provide a reasonable approach. On the other hand, from a QCD point of view, a plethora of new states are expected. NRQCD is still a good EFT for states close to and just above threshold, at least when their binding energies remain much smaller than the heavy-flavor mass. The heavy quarks move slowly in these states, and the static limit should remain a good starting point.

Below we examine how things change close to threshold and the new degrees of freedom that emerge. First, Sect. 2.9.1 considers the case in which there are only quarkonium and gluonic excitations. Away from threshold, the gluonic excitations have been integrated out to obtain strongly coupled pNRQCD and the QCD nonperturbative potentials are the pNRQCD matching coefficients. Close to threshold the gluonic excitations no longer develop a gap with respect to quarkonium and they have to be considered as dynamical degrees of freedom. Next, Sect. 2.9.2 considers the situation with dynamical ultrasoft light quarks and we discuss all the new degrees of freedom that may be generated. No QCD based theory description is yet possible in this situation, apart from systems like the $X(3872)$ that display universal characteristics and may be treated with EFTs methods. Models for the description of states close to threshold just pick up some of the possible degrees of freedom and attribute to them some phenomenological interaction. These models will be described and their predictions contrasted. Third, Sect. 2.9.6 will summarize the predictions of sum rules, a method that allows calculation of the masses of the states once an assumption on the operator content is made. Lastly, all the new unconventional states will be summarized along with possible interpretations in Sect. 2.9.7.

2.9.1 Gluonic excitations

First, consider the case without light quarks. Here the degrees of freedom are heavy quarkonium, hybrids and glueballs. In the static limit, at and above the $\Lambda_{\rm QCD}$ threshold, a tower of hybrid static energies (i.e., of gluonic excitations) must be considered on top of the $Q\overline{Q}$ static singlet energy [272, 273]. The spectrum has been thoroughly studied on the lattice [274]. At short distances, it is well described by the Coulomb potential in the color-singlet or in the color-octet configurations. At short distances, the spectrum of the hybrid static energies is described in the leading multipole expansion of pNRQCD by the octet potential plus a mass scale, which is called *gluelump mass* [140, 275]. At large distances the energies rise linearly in r. The first hybrid excitation plays the role of the open heavy-flavor threshold, which does not exist in this case. If the Born–Oppenheimer approximation is viable, many states built on each of the hybrid potentials are expected. Some of these states may develop a width if decays to lower states with glueball emission (such as hybrid $\to$ glueball $+$ quarkonium) are allowed. The states built on the static potential (ground state) are the usual heavy-quarkonium states.

Consider, for example, the $Y(4260)$, for which many interpretations have been proposed, including a charmonium hybrid [276–278]. If the $Y(4260)$ is interpreted as a charmonium hybrid, one may rely on the heavy-quark expansion and on lattice calculations to study its properties. Decays into $D^{(*)}\bar{D}^{(*)}$ should be suppressed since they are forbidden at leading order in the heavy-quark expansion [277]

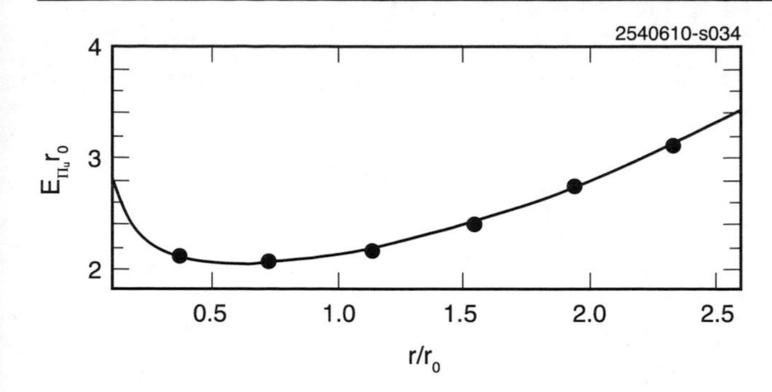

Fig. 34 The hybrid static potential Π_u at short and intermediate distances, $r_0 \approx 0.5$ fm. The *solid circles* are the lattice data [274] and the *smooth curve* traces out (43)

(see also [279]). This is in agreement with the upper limit on $Y \to D\bar{D}$ reported by BABAR (see Table 16). The quantum numbers of the $Y(4260)$ are consistent with those of a pseudoscalar 0^{-+} fluctuation $|\phi\rangle$, belonging to the family of mv^2 fluctuations around the gluonic excitation between a static quark and a static antiquark, with quantum numbers 1^{+-}, also known as Π_u,

$$|Y\rangle = |\Pi_u\rangle \otimes |\phi\rangle. \quad (42)$$

It is suggestive that, according to lattice calculations [274], Π_u is the lowest gluonic excitation between a static quark and a static antiquark above the quark–antiquark color singlet. $|\phi\rangle$ is a solution to the Schrödinger equation with a potential that is the static energy of Π_u. Fitting the static energy of Π_u at short and intermediate distances, one finds

$$E_{\Pi_u} r_0 = \text{constant} + 0.11\frac{r_0}{r} + 0.24\left(\frac{r}{r_0}\right)^2, \quad (43)$$

as illustrated in Fig. 34. Solving the corresponding Schrödinger equation,

$$m_Y = (2 \times 1.48 + 0.87 + 0.53)\ \text{GeV} = 4.36\ \text{GeV}, \quad (44)$$

where 1.48 GeV is the charm mass in the RS scheme [280] and 0.87 GeV is the Π_u gluelump mass in the same scheme [275].

2.9.2 Spectrum with light quarks

Once light fermions have been incorporated into the spectrum, new gauge-invariant states appear beside the heavy quarkonia, hybrids, and glueballs. On the one hand, we have the states with no heavy-quark content. Due to chiral symmetry, there is a mass gap, of $O(\Lambda_\chi)$, between the Goldstone bosons, which are massless in the chiral limit, and the rest of the spectrum. The Goldstone bosons are considered as ultrasoft degrees of freedom and $\Lambda_\chi \sim \Lambda_{\text{QCD}}$, so that away from threshold the rest of the spectrum should be integrated out. Besides these, there are also bound states made of one heavy quark and light quarks, i.e., the $\overline{Q}q$–$Q\overline{q}$ system. The energy of this system is, according to the HQET counting rules, $m_{\overline{Q}q} + m_{Q\overline{q}} = 2m + 2\overline{\Lambda}$. Therefore, since the heavy-light binding energy $\overline{\Lambda} \sim \Lambda_{\text{QCD}}$, away from threshold these states have to be integrated out. Close to threshold the situation is different. In this case, there is no mass gap between the heavy quarkonium and the creation of a $\overline{Q}q$–$Q\overline{q}$ pair. Thus, for study of heavy quarkonium near threshold, these degrees of freedom must be included in the spectrum, even if the mixing between the heavy quarkonium and the $\overline{Q}q - Q\overline{q}$ is expected to be small, being suppressed in the large Nc counting. Summarizing, light fermions contribute within this picture in three ways:

- *Hard light fermions,* which are encoded into the matching coefficients of the NRQCD Lagrangian and obtained from the computation of perturbative Feynman diagrams at the scale m.
- *Soft light fermions,* a term that denotes, in a generic way, all the fermions that are incorporated in the potentials. It is expected that their main effects can be simulated by a variation of the value of the parameters in the potentials. They can be evaluated nonperturbatively via unquenched lattice calculation of the potentials.
- *Ultrasoft light fermions,* which are the ones that will become pions and, since they are also ultrasoft degrees of freedom, they should be incorporated in the effective Lagrangian together with the heavy quarkonium.

So the general picture is as follows: The inclusion of light quarks does not remove any states predicted in the no-light-quarks scenario, but the availability of decays via pion emission does increase the width of each such state in the spectrum. Moreover, in addition to the regular quarkonium states, new states built using the light-quark quantum numbers may form. States made of two heavy and light quarks include those built on pairs of heavy-light mesons ($D\bar{D}$, $B\bar{B}$, ...), like hadronic molecular states [33, 281]; molecular states composed of the usual quarkonium states (built on the static potential); and light hadrons (hadro-quarkonium [282]); pairs of heavy-light baryons [283]; tetraquark states [284]; and likely many others. It would be particularly interesting to have the spectrum of tetraquark potentials, or at least their ground states, from lattice QCD, since a tetraquark interpretation of some of the newly discovered states has been advocated [285, 286] (see Sect. 2.9.4). If, again, the Born-Oppenheimer approximation is a reasonable approach, many states built on each of the tetraquark potentials may be expected, many developing (large) widths due to decays by emission of a pion (or other light hadron).

How these different kinds of states "talk to each other" is an important issue [287]. Results on crosstalk of the static potential with a pair of heavy-light mesons on the lattice have recently been reported (see Sect. 2.6). This explains why, from the QCD point of view, so many states of a new nature appear in this region of the spectrum. However, a systematic QCD description of these states has not yet been developed. For the time being, models are developed in order to obtain more detailed information on these systems. Exceptional cases, e.g., those for which the state is extremely close to a threshold (e.g., $X(3872)$), allow for an effective field theory treatment [288] (see also Sect. 2.9.3).

Results from some of the above-mentioned models follow: There are differences among models involving four-quark fields, two heavy and two light. Given four-quark fields of the type, e.g., $c\bar{c}q\bar{q}$ (where q represents a generic light quark), three quark-pair configurations are possible. All of them have been exploited in the literature. However, the resulting models are not equivalent because different dynamics are attributed to different configurations. Due to the absence of further theoretical input from QCD, many tetraquark studies rely on phenomenological models of the tetraquark interaction, unless some special hierarchy of dynamical scales may be further exploited on the top of the nonrelativistic and perturbative expansions discussed so far. In [289, 290], it is assumed that

$$X \sim (c\bar{c})^8_{S=1}(q\bar{q})^8_{S=1}, \tag{45}$$

i.e., that the dominant Fock-space component of the $X(3872)$ contains a $c\bar{c}$ pair and a $q\bar{q}$ pair in a color-octet configuration with spin 1. Calculations have been based on a phenomenological interaction Hamiltonian. In [285], it is assumed that

$$X \sim (cq)^{\bar{3}}_{S=1}(\bar{c}\bar{q})^3_{S=0} + (cq)^{\bar{3}}_{S=0}(\bar{c}\bar{q})^3_{S=1}. \tag{46}$$

Here the clustering of quark pairs in tightly bound color-triplet diquarks is not induced by a scale separation as it would happen in baryons with two heavy quarks [151], but is a dynamical assumption of the model. In [291–294], it is assumed that

$$\begin{aligned} X &\sim (c\bar{q})^1_{S=0}(q\bar{c})^1_{S=1} + (c\bar{q})^1_{S=1}(q\bar{c})^1_{S=0} \\ &\sim D^0\bar{D}^{*0} + D^{*0}\bar{D}^0, \end{aligned} \tag{47}$$

i.e., that the dominant Fock-space component of the $X(3872)$ is a $D^0\bar{D}^{*0}$ and $D^{*0}\bar{D}^0$ molecule. Small short-range components of the type

$$(c\bar{c})^1_{S=1}(q\bar{q})^1_{S=1} \simeq J/\psi\,\rho\,(\text{or }\omega) \tag{48}$$

are included as well. Predictions depend on the phenomenological Hamiltonian, which typically contains short-range ($\sim 1/\Lambda_{\rm QCD}$), potential-type interactions among the quarks and long-range ($\sim 1/m_\pi$) one-pion exchange. In [288, 295–297], it is assumed not only that the $X(3872)$ is a $D^0\bar{D}^{*0}$ and $\bar{D}^0 D^{*0}$ molecule but also that it is loosely bound, i.e., that the following hierarchy of scales is realized:

$$\Lambda_{\rm QCD} \gg m_\pi \gg \frac{m_\pi^2}{m_{D^0}} \approx 10\ \text{MeV} \gg E_{\rm b}. \tag{49}$$

Indeed, the binding energy, $E_{\rm b}$, which may be estimated from $m_X - (m_{D^{*0}} + m_{D^0})$, is, as Table 13 shows, very close to zero, i.e., much smaller than the natural scale m_π^2/m_{D_0}. Systems with a short-range interaction and a long scattering length have universal properties that may be exploited: in particular, production and decay amplitudes factorize into short-range and long-range component, where the latter depends only on a single parameter, the scattering length (see Sect. 2.9.3).

2.9.3 Molecular states

Loosely bound hadronic molecules The $X(3872)$ resonance (see Sect. 2.3.1) cannot be easily explained as a standard charmonium excitation [298]. The close proximity of its mass to $D^{*0}\bar{D}^0$ threshold suggested that it could be a good example of a hadronic molecule with $J^{PC} = 1^{++}$ quantum numbers. A $D^{*0}\bar{D}^0$ molecule would be characterized by an extremely small binding energy, as small as $E_{\rm b} \approx 0.1$ MeV. Indeed, the $B \to KX$ Belle production mechanism allows the formation of a nearly-at-rest $D^{*0}\bar{D}^0$ system which could be very weakly bound. Several other bound states with similar properties have also been discovered.

The threshold proximity of many of the new states implies that, regardless of the binding mechanism, there should be a significant component of a hadronic molecule in the wave function of the state. The small binding energy of this molecular component indicates that the $D^{*0}\bar{D}^0$ scattering length, a, is unnaturally large. This leads to some simplifications in the description of the properties of the molecule, as its binding energy, $E_{\rm b}$, and wave function are in fact largely determined by a, a phenomenon known as low-energy *universality*. In the limit of a very shallow bound state, the scattering length is

$$a = \frac{1}{\sqrt{2m_{\rm r}E_{\rm b}}}, \tag{50}$$

where

$$m_{\rm r} = \frac{m_{D^0}m_{D^{*0}}}{m_{D^0} + m_{D^{*0}}} \tag{51}$$

is the reduced mass. Clearly, the scattering length $a \simeq 10$ fm is much larger than the natural length scale R for a molecular state bound by pion exchange, $R \simeq 1/m_\pi = 1.5$ fm. There is a universal prediction for a wave function of the S-wave molecular state,

$$\psi(r)_{\text{mol}} = \frac{1}{\sqrt{2\pi a}} \frac{e^{-\frac{r}{a}}}{r}, \tag{52}$$

where r is the separation of the constituents in the molecule's rest frame, which is correct up to terms of order R/a. A proper investigation of near-threshold resonances is needed to address the appearance of this new length scale, the possibility of distinguishing between an "elementary" particle ($q\bar{q}$, hybrid, or compact tetraquark) and a composite state (hadronic molecule), and how to estimate the admixture of the composite. It was suggested by Weinberg [299–301] that the admixture fraction can be determined model-independently for such near-threshold bound states. In this scheme, the admixture of a nonmolecular component is parametrized in terms of a single parameter, $0 \leq \xi^2 \leq 1$, which measures the probability of finding the molecular component in the physical wave function of the state of interest. For $\xi < 1$, (52) becomes

$$\psi(r)_{\text{mol}} = \frac{\xi^2}{\sqrt{2\pi a}} \frac{e^{-\frac{r}{a}}}{r}. \tag{53}$$

Accordingly, the expression for the scattering length, (50), is then

$$a = \frac{2\xi^2}{1+\xi^2} \frac{1}{\sqrt{2m_r E_b}}. \tag{54}$$

The expression acquires corrections of order R. A simultaneous measurement of both binding energy and scattering length can extract the value of the parameter ξ, and the nature of the state becomes an observable. When this formalism [299–301] was applied to the deuteron, it was shown that, indeed, the deuteron is a proton–neutron molecule. The method just described can only be applied if the particles forming the molecule are in a relative S-wave and if the state studied is sufficiently close to threshold, i.e., if $k \simeq \sqrt{2m_r E_b}$ is the smallest momentum scale in the problem. The approach was generalized [302–304] to include inelastic channels, as well as an above-threshold resonance. It is stressed in [302, 303] that the relevant quantity to be studied is the effective coupling constant squared, or, equivalently, the residue at the bound-state pole that parametrizes the coupling strength of this state to the relevant continuum channel, which can be shown to be proportional to ξ^2 with a known factor of proportionality. Thus this coupling constant, which is an observable, measures the amount of molecular admixture in the sense defined in [299–301]. A related approach uses pole-counting [305], which studies the structure of the near-threshold singularities of the scattering amplitude. It appears that the state is mostly elementary if there are two nearby poles in the scattering amplitude, whereas composite particle corresponds to a single, near-threshold pole. While these methods provide a diagnostic tool for identifying near-threshold molecular states, they do not provide information on the binding mechanism. Some of these states might be interpreted as hadrocharmonia, discussed in detail in Sect. 2.9.5.

The analysis sketched above has been applied to various states. Evidence supporting the identification of $Y(4660)$ as a $\psi(2S)f_0(980)$ [306] bound system has been found. In addition, it has been proposed that $X(3872)$ [33, 281, 307, 308] is a $D^{*0}\bar{D}^0$ bound system, and that $D_s(2317)$ and $D_s(2460)$ [309–312] are bound states of KD and KD^*, respectively. One may use unitarization schemes to investigate corresponding states *not* located near thresholds; e.g., the $Y(4260)$ was suggested to be a $J/\psi f_0(980)$ bound system [313], which would make it a close relative of $Y(4660)$.

Molecules and effective field theories Effective field theory (EFT) techniques can be used to study the dynamical properties of a threshold molecular state independent of any particular model. This is possible due to the multitude of scales present in QCD. The small binding energy suggests that this state can play the role of the deuteron in meson–antimeson interactions. Thus methods similar to those developed for the deuteron may be employed, with the added benefit of heavy-quark symmetry. A suitable effective Lagrangian describing the $X(3872)$ contains only heavy-meson degrees of freedom with interactions approximated by local four-boson terms constrained only by the symmetries of the theory. While the predictive power of this approach is somewhat limited, several model-independent statements can be made. For instance, existence of $D^{*0}\bar{D}^0$ molecule does not necessarily imply existence of $D^{*0}\bar{D}^{*0}$ or $D^0\bar{D}$ molecular states [314]. First steps towards the development of a systematic EFT for the $X(3872)$ have been taken [314–317].

Effective field theories can be used to study formation [318] and decays [319] of $X(3872)$ and other molecular states. In particular, it can be employed to study line-shapes [121, 320]. Those studies reveal that the spectral shape of the resonances located near thresholds is the relevant observable. This is a direct consequence of the importance of the effective coupling constant for the nature of the state. In the case of the $Y(4660)$, the spectrum shows a visible deviation from a symmetric distribution [306], which, in the molecular picture, can be traced to the increasing phase space available for the $\psi(2S)f_0(980)$ system. (Alternatively, the asymmetry might originate from interference of the resonance signal with that of the lower-lying $Y(4360)$ [321]—see also the discussion in Sect. 2.9.4.) Only the mass of the Y and the overall normalization were left as free parameters in a fit to the experimental mass spectrum; the width can be calculated from the effective coupling constant under the assumption that the $Y(4660)$ is indeed a $\psi(2S)f_0(980)$ bound system. This fit [306] gives a mass of $m_Y = (4665^{+3}_{-5})$ MeV. From this fit, the effective coupling constant was found to be in the range 11–14 GeV. This, in turn, allowed a prediction of the width of $\Gamma_Y = (60 \pm 30)$ MeV. The current quality of data allows for additional decay channels, e.g., $\Lambda_c^+\Lambda_c^-$. To double check that the analysis is sensible, a second fit to the experimental mass spectrum is performed in which the effective coupling is allowed to float in addition to m_Y and the overall normalization. This second fit calls for a coupling constant of 13 GeV. This result is interpreted in [306] as strong evidence in favor of a molecular interpretation for the $Y(4660)$. Under this interpretation, employing heavy-spin symmetry allows one to predict a close relative, Y_η, to the $Y(4660)$, namely a bound state of $\eta_c(2S)$ and $f_0(980)$ [322]. The mass difference between this state and $Y(4660)$ is predicted to match that between $\psi(2S)$ and $\eta_c(2S)$ up to corrections of order $(\Lambda_{\text{QCD}}/m_c)^2$, which gives $m_{Y_\eta} = 4616^{+5}_{-6}$ MeV. The width and spectral shape in the $\eta_c(1S)\pi\pi$ channel are predicted to be equal to those of the $Y(4660)$, respectively. Further systematic studies are necessary to put these conjectures on firmer ground. This model-independent scheme has been extended to states with one unstable constituent [323]. Thus the method can now be applied to many more states in the spectrum. Measurements with higher statistics are needed to test these predictions. For example, an improved spectral shape measurement for $Y(4660)$ could determine if the large predicted coupling to $\psi(2S)f_0(980)$ is present. Another test is a search for the decay $B^+ \to K^+Y_\eta$, which has been estimated [322] to have a branching fraction of $\sim 10^{-3}$. After accounting for the possibility of final-state interactions in this picture, it has also been proposed [324] that the $X(4630)$, which decays to $\Lambda_c^+\Lambda_c^-$, could be the same state as the $\psi(2S)f_0(980)$ molecule $Y(4660)$, which could be tested with measurements of $B^+ \to K^+\Lambda_c^+\Lambda_c^-$ decays.

$X(3872)$ as a $D^{*0}\bar{D}^0$ molecule Analyses of published $X(3872)$ data in both the $D^{*0}\bar{D}^0$ and $\pi^+\pi^-J/\psi$ channels have shed light on the nature of the X. As for the $Y(4660)$, the lineshape contains the important information. One approach [325], using then-existing data, concluded that the $X(3872)$ is indeed generated by nonperturbative $D^{*0}\bar{D}^0$ interactions, which are not sufficiently strong to form a bound state, but only to produce a virtual state very close to $D^{*0}\bar{D}^0$ threshold. A different approach [320] stressed that, if the $X(3872)$ is a bound state, there will be a resonant $D^0\bar{D}^0\pi^0$ peak *below* the $D^{*0}\bar{D}^0$ threshold attributable to the nonzero width of the D^{*0}. This latter approach, also using the initial lineshape measurements, identified the X as a molecule, although a virtual state was not excluded. Later, using the same formalism as [325] and additional data that had become available, a fit to the measured lineshape found [326] that a significant admixture of a compact component inside the X wave function is required. As discussed in Sect. 2.3.1 and illustrated in Fig. 16, it was pointed out [122] that the experimental lineshapes for these $D^0\bar{D}^0\pi^0$ events had been generated by constraining the measured particle momenta so that the $D^0\pi^0$ (or $\bar{D}^0\pi^0$) candidates would have the D^{*0} mass. If the $X(3872)$ were a loosely bound molecule, such an analysis procedure would cause the resonant invariant-mass peak of $D^0\bar{D}^0\pi^0$ that is actually *below* $D^{*0}\bar{D}^0$ threshold to erroneously appear *above* the threshold and broadened. Taking this effect into account, the analysis [122] found the data to be consistent with the identification of the $X(3872)$ as a bound state with mass below $D^{*0}\bar{D}^0$ threshold. If future measurements have more statistics in all relevant decay modes of $X(3872)$ and/or improve upon measured mass resolutions, while simultaneously avoiding the kinematic constraint to the D^{*0} mass, more definitive statements on the molecular nature of the $X(3872)$ could be made.

Another avenue for studying the molecular nature of $X(3872)$ could become available at the LHC. In particular, an EFT description [327] of $X(3872)$ scattering off D^0 or D^{*0} mesons has been developed, and proposed for testing with either B_c or $B\bar{B}$ decays with final-state interactions.

Questioning the $X(3872)$ molecular interpretation The production cross section of $X(3827)$ at the Tevatron is a potential discriminant for its molecular interpretation. Neither CDF nor DØ have reported such a measurement because it is not a trivial one. However, based on published and unpublished-but-public CDF documents, the product of the cross section of $X(3872)$ and its branching fraction into $\pi^+\pi^-J/\psi$ can be estimated. The inclusive, prompt[7] production rate of $X \to \pi^+\pi^-J/\psi$ relative to $\psi(2S) \to \pi^+\pi^-J/\psi$, both with $J/\psi \to \mu^+\mu^-$, for the same transverse momentum (p_T) and rapidity (y) restrictions, *assuming equal selection efficiencies*, is estimated to be [328]

$$\frac{\sigma(p\bar{p} \to X + \text{any})_{\text{prompt}} \times \mathcal{B}(X \to \pi^+\pi^-J/\psi)}{\sigma(p\bar{p} \to \psi(2S) + \text{any})_{\text{prompt}}} = (4.7 \pm 0.8)\%. \tag{55}$$

This value is used in conjunction with a CDF [329] measurement of the absolute cross section for inclusive $\psi(2S)$ production as a function of p_T for central rapidity to obtain [328]

$$\sigma(p\bar{p} \to X + \text{any})_{\text{prompt}} \times \mathcal{B}\big(X \to \pi^+\pi^-J/\psi\big) = (3.1 \pm 0.7)\text{ nb} \quad \text{for } p_T > 5\text{ GeV}/c \text{ and } |y| < 0.6, \tag{56}$$

[7]The adjective *prompt* refers to X particles that are produced by QCD interactions and not by the weak decays of b-hadrons.

assuming X and $\psi(2S)$ have the same rapidity distribution. Since the unknown branching fraction satisfies $\mathcal{B}(X \to \pi^+\pi^- J/\psi) < 1$, (56) also provides a lower limit on the prompt X production cross section for the transverse momentum and rapidity restrictions given.

The large magnitude of the prompt production cross sections measured in $p\bar{p}$ collisions at the Tevatron came as a surprise to many. The original Monte Carlo (MC) studies, based in part on the generators HERWIG [330] and PYTHIA [331], suggested that formation of loosely bound $D^{*0}\bar{D}^0$ molecules in this environment fall far short of the observed rates. Further theoretical work coupled with MC studies [328] reinforced this viewpoint, which was then challenged by an independent examination [332] of the issues involved. Both approaches allow formation of a molecule if its constituents, after their initial production by the underlying generator, have relative momentum up to a value $k_{\max}$. The two approaches differ markedly in the values of $k_{\max}$ that are permitted. In [328], $k_{\max}$ is chosen to be comparable to the binding momentum $k_b = \sqrt{2m_r E_b}$, whereas [332] argues that a value larger by an order of magnitude, and correspondingly even larger prompt production cross section, is more appropriate due to constituent rescattering effects.

Further justification for the choices in [332] was given in [333], in which deuteron production is taken as a case study to judge the efficacy of the arguments. In addition, it allowed tuning of the underlying generator for production of the molecule constituents in the required pairs. It is argued that the prescription in [328] for the value of $k_{\max}$ used in MC generation, as applied to deuterons, is flawed for both fundamental and empirical reasons. Fundamentally, a loosely bound S-wave molecule does not satisfy the minimum uncertainty principle, $\Delta r \Delta k \sim 1$. Instead, it maximizes the uncertainty, satisfying $\Delta r \Delta k \gg 1$ with $\Delta r \sim k_b^{-1}$ and $\Delta k \gg k_b$. Empirically, the MC technique in [328] underpredicts the CLEO [334] measurements of antideuterons in $\Upsilon(1S)$ decays (see Sect. 3.4.5).

A rebuttal to [333] has been made [335]. It argues that the deuteron and $X(3872)$ are not comparable. First, the discrepancy obtained between data and MC using a small $k_{\max}$ (based mostly on the uncertainty principle, which limits the reasonable variations of k, and therefore $k_{\max}$, to be of order k_b, not an order of magnitude larger) is modest for the deuteron (factors of 2–3) compared to the $X(3872)$ (factor of $\approx$300). Moreover, MC studies with PYTHIA have found considerably better agreement, within a factor of 2–3, between the estimated and observed cross sections of the deuteron. Discrepancies of this size in e^+e^- collisions at LEP were considered reasonably close to the measurements and do not justify rejecting the MC altogether [336]. Finally, contends [335], the deuteron is a system qualitatively different from a $D^{*0}\bar{D}^0$ molecule because the D^0 is spinless and cannot participate in spin interactions. Conversely, spin interactions play an important role in the determination of the deuteron binding: the spin-singlet deuteron and its isospin partner, the dineutron, are not bound.

Much of the motivation for the $X(3872)$ molecular interpretation, which assumes S-wave binding, no longer applies if its quantum numbers are found to be 2^{-+}, as preferred by a recent BABAR [93] analysis. A $D^{*0}\bar{D}^0$ 2^{-+} state would require a relative P-wave. It is unlikely [335] that π-exchange could bind such a state, given that, even in an S-wave configuration, it is not clear that the attraction is sufficiently strong. Even if such a state exists, there remains the further problem that unless spin-dependent forces prevent the binding, one should expect partner states with 0^{-+} and (J^{PC}-exotic) 1^{-+}, for which there is no experimental evidence. A P-wave 2^{-+} molecule would also imply the existence of an extremely narrow, more deeply bound S-wave 1^{++} molecule. Alternatively, forming a 2^{-+} S-wave bound state would require a different molecule type such as $D_2 D$ or $D_1 D^*$, which would require not only an immense binding energy of some 500 MeV, but also loss of the appealing connection between the mass of the $X(3872)$ and $D^{*0}\bar{D}^0$ threshold.

On the basis of a study by Cho and Wise [337], it is difficult to reconcile the observed $X(3872)$ prompt production cross section with the expectations for $1\,^1D_2$ standard charmonium. The integrated prompt cross section found using the Cho and Wise gluon-fragmentation function in a $1\,^1D_2$ state is

$$\sigma\big(p\bar{p} \to 1\,^1D_2 + \text{all}\big) = 0.6\ \text{nb}, \qquad (57)$$

some 50 and 120 times smaller than the estimated experimental cross section [335]. As for the mass of the $1\,^1D_2$ charmonium, there are a number of studies available in the literature [335]. A hadron–string calculation, with results that agree very well with previous determinations of charmonium and bottomonium levels, is also proposed in [335]. Most of these calculations indicate that the $X(3872)$ has the most difficulty matching quarkonium levels. In this model, while all charmonium and bottomonium levels agree with data within $\sim$10 MeV (excluding the $1\,^3D_1$, which departs from the experimental mass by 40 MeV), the predicted $1\,^1D_2$, the would-be $X(3872)$, falls short of the measured value by 80 MeV. The mass mismatch and production cross section jeopardize a 2^{-+} charmonium interpretation of the $X(3872)$. Clarification of the $X(3872)$ decay modes and relative branching fractions would help disentangle the possible explanations. The prominent radiative decay mode for a D-wave charmonium $X(3872)$ is expected to be $X(3872) \to h_c\gamma \to J/\psi\pi^0\gamma$ while the $\eta_c(1S)\pi\pi$ channel should have the highest rate among hadronic modes.

The reader is also directed to Sects. 2.9.4 and 4.2.2 for more information on the $X(3872)$.

2.9.4 Tetraquark states

The nonstandard decay patterns of $X(3872)$ suggested other theoretical interpretations as well. Could the $X(3872)$ be a pointlike hadron resulting from the binding of a diquark and an antidiquark? This idea was discussed in Maiani et al. [285], following one interpretation of pentaquark baryons (antidiquark–antidiquark–quark states) proposed by Jaffe and Wilczek [338] and the recent discussion of light scalar mesons in terms of tetraquarks by 't Hooft et al. [339, 340] (see also the review [341] and references therein). Some considerations by 't Hooft on an open-string description of baryons [342] were also sources of inspiration.

Diquarks A spin-zero diquark operator in the attractive anti-triplet color (greek subscripts α, β, γ) channel, anti-symmetric in flavor (latin subscripts i, j, k) can be written as:

$$[qq]_{i\alpha} = \epsilon_{ijk}\epsilon_{\alpha\beta\gamma}\bar{q}_c^{j\beta}\gamma_5 q^{k\gamma} \tag{58}$$

where the subscript c denotes charge conjugation. The Fermi statistics of light quarks is respected in (58). Spin-one diquarks (the so-called 'bad' ones) can be conceived but they are believed to have a smaller binding energy with respect to spin-zero candidates (see, e.g., [343]). A 'bad' diquark (spin-one) operator can be written as:

$$[qq]_\alpha^{ijr} = \epsilon_{\alpha\beta\gamma}\left(\bar{q}_c^{i\beta}\gamma^r q^{j\gamma} + \bar{q}_c^{j\beta}\gamma^r q^{i\gamma}\right) \tag{59}$$

which is a **6** in flavor space (and has three spin components) as required by Fermi statistics ($r = 1, 2, 3$). Both represent positive parity states, 0^+ and 1^+, respectively. Similarly one can construct 0^- and 1^- operators, $\bar{q}_c q$ and $\bar{q}_c\gamma^r\gamma_5 q$. The latter are identically zero in the 'single-mode configuration': quarks that are unexcited with respect to one another. In fact, the most solid tetraquark candidates are scalar mesons made up of 'good' diquarks. The spin-zero light diquarks are very effective at reducing the number of expected four-quark states. A $qq\bar{q}\bar{q}$ multiplet should contain $81 = \mathbf{3}\otimes\mathbf{3}\otimes\bar{\mathbf{3}}\otimes\bar{\mathbf{3}}$ particles for three quark flavors. But if the diquark degrees of freedom are the relevant ones, the number of states is reduced to $9 = \mathbf{3}\otimes\bar{\mathbf{3}}$ for a diquark that behaves as an antiquark and an antidiquark as a quark (see (58)). In the case of light scalar mesons this represents a way of encrypting the exoticity (we would have 9 light scalar mesons even if their structure were $q\bar{q}$ [339, 340]). Spin-one diquarks of (59) would enlarge the flavor structure as they are $\mathbf{6}_f$ operators.

Tetraquarks The tetraquark model provides fertile ground for investigations of heavy-flavored states. One of the features of the diquark–antidiquark model proposed in [285], which could also be considered a drawback, is the proliferation of predicted states. Another is the paucity of insight from *selection rules* that could explain why many of these states are not observed (for a recent account, see, e.g., [344]). It is quite possible that those states are waiting to be discovered.

A tetraquark in the diquark–antidiquark incarnation is a state like $[qq]_{i\alpha}[\bar{q}\bar{q}]^{j\alpha}$, if spin-zero diquarks are concerned, as is the case for the tetraquark interpretation of light scalar mesons. To use a notation making flavor explicit one can write $[q_1q_2][\bar{q}_3\bar{q}_4]$. There is no real distinction between 'good' and 'bad' diquarks once one of the quarks in the bound state is heavy: spin–spin interactions between quarks are $1/m$ suppressed and $m\to\infty$ with respect to the light-quark mass scale. In other words, one can expect that tetraquarks like $[cq][\bar{c}\bar{q}']$ have the same chances to be formed by spin-one or spin-zero diquarks. Moreover, the considerations of Fermi statistics made above are no longer valid here. This enlarges the spectrum of predicted states.

There is also the question discussed in Sect. 2.9.3 of whether an $X(3872)$ tetraquark state can better match observed cross sections at the Tevatron. If only spin-zero diquarks were allowed, a $J^{PC} = 1^{++}$ $X(3872)$ could not be described as a tetraquark. For a 2^{-+} $X(3872)$, however, the tetraquark interpretation is still viable. Known problems with the proliferation of states are shifted to lower-mass scales [335].

The $[cq][\bar{c}\bar{q}']$ states should also appear in *charged* combinations. Although there is unconfirmed evidence of a $Z(4430)^-$ state (see Sect. 2.3.4) decaying into charmonium plus charged pion, there is yet no evidence of charged, almost degenerate partners, of the $X(3872)$. Another example of a possible tetraquark meson is the $Y(4260)$, a 1^{--} resonance decaying into $\pi\pi J/\psi$, as described in Sect. 2.3.2. The dipion-mass distribution is consistent with there being a substantial $f_0(980)$ component in the $Y(4260)$ decay, suggesting an exotic tetraquark structure $[cs][\bar{c}\bar{s}]$. The tetraquark model suggests a $Y(4260)\to D_s^+D_s^-$ decay mode [345], a mode for which the experimental upper limit at 90% CL on branching fraction, <1.3 relative to $\pi^+\pi^-J/\psi$ (see Table 16), is not a particularly stringent one.

As stated above, the main drawback of the tetraquark model is the proliferation of predicted particles. For example, using a naive constituent diquark model, in the hidden-strange and hidden-charm sector one can predict a quite complex pattern of states [344]. But since no such states near thresholds, such as $\phi J/\psi$ or $f_0(980)J/\psi$, are predicted, this model cannot account for the unconfirmed evidence (3.8σ significance) for a new resonance, $Y(4140)$, decaying to $\phi\,J/\psi$ near threshold, as reported by CDF [106] (see Table 9 and Sect. 2.3.3). The naive constituent diquark model [344] does predict, however, a 0^{-+} state decaying to $\phi J/\psi$ at about 4277 MeV. The measured $\phi J/\psi$ mass spectrum reported by CDF [106] does show an intriguing enhancement near this mass, but the statistical significance reported by CDF for this structure, $<3\sigma$, leaves the possibility that it is an artifact or a fluctuation. If this peak becomes more significant with more data, it could bolster the constituent diquark approach.

Constituent quark models can only give rough estimates of the expected mass values. In contrast, the most striking prediction of the model are particles decaying to charmonia plus charged pions or ρ-mesons. One or more of the unconfirmed $Z(4430)^+$, $Z_1(4050)^-$, and $Z_2(4250)^-$ (see Table 9 and Sect. 2.3.4) could be examples of such particles. If any of these were confirmed by CDF (or by LHC experiments), there would be a much stronger argument in favor of the tetraquark model than any mass spectrum determination of neutral candidates.

Baryonia Assuming an *open-string* hadron picture of diquark–antidiquark tetraquarks, compelling evidence for their existence should be found in experimental searches for narrow structures coupled preferentially to a baryon and an anti baryon. It has been proposed [321] that there is compelling experimental evidence for a single vector baryonium candidate, Y_B, that explains *two* unconfirmed states reported by Belle [25, 114]. These states, the $X(4630)$, observed in the decay to $\Lambda_c^+\Lambda_c^-$, and $Y(4660)$, which decays to $\pi^+\pi^-\psi(2S)$ (see Table 9 and Sect. 2.3.2), have reported masses that differ by only two standard deviations and com patible widths. They can also be fit well by a single resonance [321]

$$\begin{aligned} m(Y_B) &= 4661 \pm 9\ \text{MeV}, \\ \Gamma(Y_B) &= 63 \pm 23\ \text{MeV}. \end{aligned} \tag{60}$$

The mass of the Y_B is significantly above the decay threshold and a straightforward four-quark interpretation explains its decay modes. The ratio of branching ratios found,

$$\frac{\mathcal{B}(Y_B \to \Lambda_c^+\Lambda_c^-)}{\mathcal{B}(Y_B \to \psi(2S)\pi^+\pi^-)} = 25 \pm 7, \tag{61}$$

highlights a strong affinity of Y_B to the baryon–antibaryon decay mode. The phase space involved in the decays in (61) are rather similar because the $\pi^+\pi^-$ pair results from an $f_0(980)$ decay (see Fig. 21 in Sect. 2.3.2). As the $f_0(980)$ can be identified as a diquark–antidiquark particle (see 't Hooft et al. [339, 340]), the Y_B can also be interpreted [321] as such an exotic state. A $[cd][\bar{c}\bar{d}]$ assignment for Y_B naturally explains [321] the ratio found in (61); in the baryon–antibaryon mode, a string of two heavy quarks with angular momentum excitation $\ell \neq 0$ would break:

$$[qq]\sim\sim\sim[\bar{q}\bar{q}] \to [qq]\sim\sim\sim q + \bar{q}\sim\sim\sim[\bar{q}\bar{q}]. \tag{62}$$

While baryonic decays of tetraquarks should be the most favorable according to the string-color picture, but these are typically phase-space forbidden.

The $Y(4360)$ (see Sect. 2.3.2) could be the radial ground state $(1P)$ of the $Y_B(4660)$ $(2P)$. These two states both decay into $\pi^+\pi^-\psi(2S)$ rather than $\pi^+\pi^-J/\psi$, a puzzling characteristic that awaits explanation [321]. An alternative molecular interpretation of the $Y(4660)$ can be found in [306].

The $Y(2175)$, observed by BABAR [346], BES [347], and Belle [348], is another interesting baryonium-like candidate, albeit in the light-quark sector. Assuming that $Y(2175)$ is a four-quark meson [349], it should preferentially decay into $\Lambda\bar{\Lambda}$. Because the $\Lambda\bar{\Lambda}$ threshold is ≈ 2231 MeV, this decay proceeds through the high-mass tail of the $Y(2175)$. The BABAR data are consistent with this hypothesis [349].

What about doubly charged particles? A diquark–antidiquark open-charm $[cu][\bar{d}\bar{s}]$ composition, denoted here by A^{++}, could exist [350] and decay, e.g., into D^+K^+. It is very unlikely that a loosely bound molecule of this kind could be produced. There has not been a search for a doubly charged particle like A^{++} close to the D^+K^+ mass. Theory is still not able to reliably predict the A^{++} rate within the tetraquark model.

Counting quarks in heavy-ion collisions Heavy-ion collisions also provide a means for definitively determining the quark nature of, e.g., the $X(3872)$. The nuclear modification ratios R_{AA} and R_{CP} (see Sect. 5.5 and (184) and (185), respectively) of the $X(3872)$ and its anisotropy coefficient, v_2 (see Sect. 5.5), which can be measured by ALICE, could be useful tools in this task. In the recombination picture, the $X(3872)$ is expected to be produced with rates similar to charm mesons and baryons. Thus the soft part of the spectrum, where recombination is more effective, can be highly populated with X's. On the other hand, the fragmentation functions of a tetraquark-X are different from those of a $D^{*0}\bar{D}^0$ molecule since the D fragmentation functions are the standard ones. This effect could be studied in a manner similar to those described for light scalar mesons in [351].

To summarize, we refer to Table 18, which lists the most significant tetraquark candidates. The Y_B and $Z(4430)^+$ are the most likely. Experimental study of the $Y(4260) \to D_s^+D_s^-$ decay, responsible for most of the width in the tetraquark model, would be an important discriminant to assess its nature. More work on the tetraquark picture has been done recently by Ali et al. [352, 353] and Ebert et al. [354].

Table 18 For significant tetraquark candidates, their spin-parity (J^{PC}), decay modes, and quark content

4q candidate	J^{PC}	Decay Modes	$([qq][\bar{q}\bar{q}])_{nJ}$
$Y_B(4660)$	1^{--}	$\Lambda_c^+\Lambda_c^-$, $\pi\pi\psi(2S)$	$([cd][\bar{c}\bar{d}])_{2P}$
$Z(4430)^+$	1^{+-}	$\pi^+\psi(2S)$	$([cu][\bar{c}\bar{d}])_{2S}$
$Y(4260)$	1^{--}	$\pi\pi\,J/\psi$, $D_s^+D_s^-$ (?)	$([cs][\bar{c}\bar{s}])_{1P}$
$X(3872)$	1^{++}	$\rho^0 J/\psi$, $\omega J/\psi$, $D\bar{D}\pi$	$([cu][\bar{c}\bar{u}])_{1S}$

2.9.5 Hadrocharmonium

The decay pattern for six of the new states ($Y(4260)$, $Y(4360)$, $Y(4660)$, $Z_1(4050)^-$, $Z_2(4250)^-$, and $Z(4430)^+$) could be interpreted [355] as an indication of an intact charmonium state within a more complex hadronic structure. These states have only been observed decaying to a single preferred charmonium state accompanied by one or more light mesons (see Table 9). Decays into a different, apparently nonpreferred, charmonium resonance with the same quantum numbers (e.g., J/ψ instead of $\psi(2S)$ for $Y(4660)$) or open-charm hadrons have not been observed. In some cases there are meaningful experimental upper bounds on such decays, as shown in Table 16. One explanation [355] is that each such state consists of its preferred charmonium embedded in a shell of light-quark and gluon matter, i.e., a compact charmonium is bound inside a spatially large region of excited light matter. The observed decays can then be viewed as the de-excitation of the light-hadronic matter into light mesons and liberation of the compact charmonium. This structure is referred to as hadrocharmonium [355], or more generally, hadroquarkonium.

The picture of a hadroquarkonium mesonic resonance is quite similar to the much-discussed nuclear-bound quarkonium. The primary difference is that, instead of a nucleus, an excited mesonic resonance provides the large spatial configuration of light-quark matter. For the quarkonium to remain intact inside hadroquarkonium, the binding has to be relatively weak. In complete analogy with the treatment of charmonium binding in nuclei (in terms of the QCD multipole expansion) [356–359], the interaction between a compact, colorless quarkonium and the soft, light matter can be described [282] by an effective Hamiltonian proportional to the quarkonium chromopolarizability α,

$$H_{\text{eff}} = -\frac{1}{2}\,\alpha\,\vec{E}^a\cdot\vec{E}^a, \tag{63}$$

where $\vec{E}^a$ is the operator of the chromoelectric field. The chromopolarizability can be estimated from the transition $\psi(2S)\to\pi\pi J/\psi$, giving [360] $\alpha^{(12)}\approx 2\ \text{GeV}^{-3}$. The average of the gluonic operator over light-hadron matter (h) with mass m_h can be found using the conformal anomaly relation in QCD:

$$\langle h|\frac{1}{2}\vec{E}^a\cdot\vec{E}^a|h\rangle \geq \frac{8\pi^2}{9}m_h, \tag{64}$$

which provides an estimate of the strength of the van der Waals-type quarkonium–light hadrons interaction. The likelihood of binding charmonium in light-hadronic matter depends on the relation between the mass m_h and the spatial extent of h [282]. In a particular model [361, 362] of mesonic resonances based on an AdS/QCD correspondence, it can be proven [363] that a heavy quarkonium does form a bound state inside a sufficiently excited light-quark resonance. The decay of such a bound state into open heavy-flavor hadrons is suppressed in the heavy-quark limit as $\exp(-\sqrt{\Lambda_{\text{QCD}}/m})$, where m is the heavy-quark mass. This is consistent with the nonobservation of Y and Z decays into charm meson pairs. However, it is not clear whether the charm quark is heavy enough for the heavy-quark limit to be applicable.

If the Y and Z resonances are hadrocharmonia, the following is expected [282].

- Bound states of J/ψ and $\psi(2S)$ with light nuclei and with baryonic resonances should exist, e.g., baryocharmonium decaying into $pJ/\psi(+\text{pions})$.
- Resonances containing χ_{cJ} that decay into χ_{cJ} + pion(s) should also exist. The as-yet-unconfirmed $Z_1(4050)^-$ and $Z_2(4250)^-$ states reported by Belle [105] are candidates (see Table 9 and Sect. 2.3.4).
- Decays of hadrocharmonia candidates to nonpreferred charmonium states, e.g., $Y(4260)\to\pi^+\pi^-\psi(2S)$, or $Y(4360)\to\pi^+\pi^- J/\psi$, should be suppressed relative to preferred charmonia.
- Resonances containing excited bottomonia such as $\Upsilon(3S)$, $\chi_b(2P)$, and/or $\Upsilon(1D)$ should exist in the mass range 11–11.5 GeV.

2.9.6 QCD sum rules

QCD sum rules (QCDSR) [238, 364–366] provide a method to perform QCD calculations of hadron masses, form factors and decay widths. The method is based on identities between two- or three-point correlation functions, which connect hadronic observables with QCD fundamental parameters, such as quark masses, the strong-coupling constant, and the quantities which characterize the QCD vacuum, i.e., the condensates. In these identities (*sum rules*), the phenomenological side (which contains information about hadrons) is related to the QCD or OPE (operator product expansion) side, where the information about the quark content is introduced. Since the correlation functions are written in terms of well-defined quark currents, the method is effective in establishing the nature of the exotic states (molecule, tetraquark, hybrid, etc.).

In principle, QCDSR allows first-principle calculations. In practice, however, in order to extract the result, it is necessary to make expansions, truncations, and other approximations that may reduce the power of the formalism and introduce large errors. In addition, the convergence of the method often critically depends upon the decay channel.

Formalism QCD sum rule calculations of hadron masses are based on the correlator of two hadronic currents:

$$\Pi(q) \equiv i \int d^4x e^{iq\cdot x} \langle 0| T\big[j(x) j^\dagger(0)\big] |0\rangle, \tag{65}$$

where $j(x)$ is a current with the appropriate quantum numbers. The phenomenological and OPE correlation functions must then be identified. The Borel transformation [238, 364–366], which converts the Euclidean four-momentum squared, Q^2, into the variable M^2 (where M is the *Borel mass*), is applied to improve the overlap between the two sides of this identity. Working in Euclidean space is necessary to avoid singularities in the propagators in (65). More precisely, it is necessary to be in the *deep Euclidean* region, i.e., $Q^2 = -q^2 \gg \Lambda_{\text{QCD}}$. The Borel transform is well-defined in this region and a good OPE convergence is obtained, dominated by the perturbative term.

After equating the two sides of the sum rule, assuming quark-hadron duality [238, 364–366] and making a Borel transform, the sum rule can be written as

$$\lambda e^{-m^2/M^2} = \int_{s_{\text{min}}}^{s_0} ds \rho^{\text{OPE}}(s), \tag{66}$$

where m is the mass of the particle, M is the Borel mass, ρ is the spectral density obtained from the OPE side (taking the imaginary part of the correlation function), s_0 is the parameter which separates the pole (particle) from the continuum (tower of excitations with the same quantum numbers), and s_{min} is determined by kinematical considerations. The parameter λ represents the coupling of the current to the hadron. Solving (66) for the mass, a function which is approximately independent of M should be obtained. In practice, the result depends on the Borel mass and a value of M must be chosen within a domain called the *Borel window*. In order to determine the Borel window, the OPE convergence and the pole contribution are examined: the minimum value of M is fixed by considering the convergence of the OPE, while the maximum value of M is determined by requiring that the pole contribution be larger than the continuum contribution. As pointed out in [367], it becomes more difficult for tetraquarks to simultaneously satisfy pole dominance and OPE convergence criteria. Reasonably wide Borel windows in which these two conditions are satisfied can exist only for heavy systems. Increasing the number of quark lines in a given system, the OPE convergence becomes gradually more problematic. For example, when a change is made from a meson (two quark lines) to a baryon (three quark lines), the perturbative term goes from a single loop to a double loop, suppressed by a factor of π^2 with respect to the single loop. At the same time, higher-order quark condensates become possible, while the nonperturbative corrections grow larger. As a consequence, larger Borel masses must be used to obtain convergence. However, at higher Borel masses the correlation function is dominated by the continuum contribution. With more quark lines, it becomes difficult to find a Borel window where both OPE convergence and pole dominance are satisfied. When a heavy quark is present, its mass provides a hard scale, which helps to make the OPE convergent at lower Borel masses. In summary: the calculations seem to indicate that it is more difficult to keep a larger number of quarks together with small spatial separation.

QCDSR and $X(3872)$ The $X(3872)$ (see Sect. 2.3.1 and Tables 9, 10, 11, 12, and 13) has quantum numbers $J^{PC} = 1^{++}$ or 2^{-+}. It decays with equal strength into $\pi^+\pi^- J/\psi$ and $\pi^+\pi^-\pi^0 J/\psi$, indicating strong isospin-violation, which is incompatible with a $c\bar{c}$ state. Its mass and the isospin-violation could be understood in several four-quark approaches. However, this state has a decay width of less than 2.2 MeV, which is sometimes difficult to accommodate. In order to discuss four-quark configurations in more detail, a distinction between a *tetraquark* and a *molecule* will be made: the former is simply a combination of four quarks with the correct quantum numbers, whereas the latter is a combination of two meson-like color-neutral objects. This separation can easily be made at the start of a calculation when the current is chosen. However, performing a Fierz transformation on the currents will mix tetraquarks with molecules. Having this ambiguity in mind, this notation will be used to clearly refer to the employed currents. The treatment given below applies only for the $J^{PC} = 1^{++}$ assignment for $X(3872)$.

A current can be constructed for the X based on diquarks in the color-triplet configuration with symmetric spin distribution: $[cq]_{S=1}[\bar{c}\bar{q}]_{S=0} + [cq]_{S=0}[\bar{c}\bar{q}]_{S=1}$ (see (46)). Therefore the corresponding lowest-dimension interpolating operator for describing X_q as a tetraquark state is given by

$$\begin{aligned} j_\mu^{(q,\text{di})} = \frac{i\epsilon_{abc}\epsilon_{dec}}{\sqrt{2}} \big[& \big(q_a^T C\gamma_5 c_b\big)\big(\bar{q}_d \gamma_\mu C \bar{c}_e^T\big) \\ & + \big(q_a^T C\gamma_\mu c_b\big)\big(\bar{q}_d \gamma_5 C \bar{c}_e^T\big)\big], \end{aligned} \tag{67}$$

where q denotes a u or d quark and c is the charm quark. We can also construct a current describing X_q as a molecular $D^{*0}\bar{D}^0$ state:

$$j_\mu^{(q,\text{mol})}(x) = \frac{1}{\sqrt{2}} \Big[\big(\bar{q}_a(x)\gamma_5 c_a(x) \bar{c}_b(x)\gamma_\mu q_b(x)\big)$$

$$- \left(\bar{q}_a(x)\gamma_\mu c_a(x)\bar{c}_b(x)\gamma_5 q_b(x)\right)\Big]. \quad (68)$$

The currents in (67) [368] and (68) [369] have both been used. In each case it was possible to find a Borel window where the pole contribution is bigger than the continuum contribution and with a reasonable OPE convergence. On the OPE side, the calculations were done to leading order in α_s including condensates up to dimension eight. The mass obtained in [368] considering the allowed Borel window and the uncertainties in the parameters was $m_X = (3.92 \pm 0.13)$ GeV, compatible with the measured value. For the current in (68), the OPE convergence and pole contribution yield a similar Borel window, resulting in a predicted mass [369] of $m_X = (3.87 \pm 0.07)$ GeV, also consistent with the measured value and more precise than that obtained with the tetraquark current.

In principle, we might expect a large partial decay width for the decay $X \to \rho^0 J/\psi$. The initial state already contains the four necessary quarks and no rules prohibit the decay. Therefore this decay is allowed, similar to the case of the light scalars σ and κ studied in [370], with widths of order 400 MeV. The decay width is essentially determined by the $X J/\psi V$ ($V = \rho, \omega$) coupling constant. The decay width was computed using QCDSR [371] with $g_{X J/\psi V}$ evaluated assuming that the $X(3872)$ is described by the tetraquark current, (67). The QCDSR calculation for the vertex $X(3872) J/\psi V$ is based on the evaluation of the three-point correlation function, which is a straightforward extension of (65) to the case of three currents representing the three particles in the decay. In [372] the current representing the X was given by

$$j_\alpha^X = \cos\theta j_\alpha^{(u,\mathrm{di})} + \sin\theta j_\alpha^{(d,\mathrm{di})}, \quad (69)$$

with $j_\alpha^{(q,\mathrm{di})}$ given in (67). This mixing between diquarks with different light flavors was first introduced [285] to explain the decay properties of the $X(3872)$. Using $\theta \approx 20°$ it is possible to reproduce the measured ratio r_ω given in Table 12. With the same angle, the $X J/\psi \omega$ coupling constant was calculated with QCDSR [371] and found to be $g_{X\psi\omega} = 13.8 \pm 2.0$. This value is much bigger than the estimate of [285] and leads to a large partial decay width of $\Gamma(X \to J/\psi(n\pi)) = (50 \pm 15)$ MeV. A similar value was also obtained [372] using a molecular current similar to (68). Therefore it is not possible to explain the small width of the $X(3872)$ from a QCDSR calculation if the X is a pure four-quark state. In [372], the $X(3872)$ was treated as a mixture of a $c\bar{c}$ current with a molecular current, similar to the mixing considered in [373] to study the light scalar mesons:

$$J_\mu^q(x) = \sin\alpha j_\mu^{(q,\mathrm{mol})}(x) + \cos\alpha j_\mu^{(q,2)}(x), \quad (70)$$

with $j_\mu^{(q,\mathrm{mol})}(x)$ given in (68) and

$$j_\mu^{(q,2)}(x) = \frac{1}{6\sqrt{2}} \langle\bar{q}q\rangle \left[\bar{c}_a(x)\gamma_\mu\gamma_5 c_a(x)\right]. \quad (71)$$

The introduction of the quark condensate, $\langle\bar{q}q\rangle$, ensures that $j_\mu^{(q,2)}(x)$ and $j_\mu^{(q,\mathrm{mol})}(x)$ have the same dimension.

It is not difficult to reproduce the experimental mass of the $X(3872)$ [368–370, 372]. This is also true for the current in (70) for a wide range of mixing angles, α, but, as observed in [372], it is not possible to match the measured value of r_ω. In order to reproduce the ratio r_ω given in Table 12, it is also necessary to consider a mixture of D^+D^{*-} and D^-D^{*+} components [285]. In this case the current is given by

$$j_\mu^X(x) = \cos\theta J_\mu^u(x) + \sin\theta J_\mu^d(x), \quad (72)$$

where $J_\mu^u(x)$ and $J_\mu^d(x)$ are given by (70). With this particular combination one obtains

$$\frac{\Gamma(X \to J/\psi\pi^+\pi^-\pi^0)}{\Gamma(X \to J/\psi\pi^+\pi^-)} \simeq 0.15 \left(\frac{\cos\theta + \sin\theta}{\cos\theta - \sin\theta}\right)^2, \quad (73)$$

exactly the same relation [285, 371] that imposes $\theta \sim 20°$ in (69) and compatible with the measured value of r_ω given in Table 12.

It was shown [372] that, with (72) and a mixing angle $\alpha = (9 \pm 4)°$ in (70), it is possible to describe the measured $X(3872)$ mass with a decay width of $\Gamma(X \to J/\psi(n\pi)) = (9.3 \pm 6.9)$ MeV, which is compatible with the experimental upper limit. The same mixing angle was used to evaluate the ratio

$$r_{\gamma 1} \equiv \frac{\Gamma(X \to \gamma J/\psi)}{\Gamma(X \to \pi^+\pi^- J/\psi)}, \quad (74)$$

obtaining $r_{\gamma 1} = 0.19 \pm 0.13$ [374], in excellent agreement with the measured value given in Table 12. Hence QCDSR calculations strongly suggest that the $X(3872)$ can be well described by a $c\bar{c}$ current with a small, but fundamental, admixture of molecular ($D\bar{D}^*$) or tetraquark $[cq][\bar{c}\bar{q}]$ currents. In connection with the discussion in Sect. 2.9.3, a possible $D_s^+ D_s^{*-}$ molecular state, such an X_s state, was also considered. The X_s mass obtained, 3900 MeV, was practically degenerate with the $X(3872)$. Therefore QCDSR indicate a larger binding energy for X_s than for the $X(3872)$, leading to a smaller mass than predicted in [344].

It is straightforward to extend the analysis done for the $X(3872)$ to the case of the bottom quark. Using the same interpolating field of (67) with the charm quark replaced by the bottom quark, the analysis done for $X(3872)$ was repeated [368] for an analogous X_b. Here there is also a good Borel window. The prediction for the mass of the state that couples to a tetraquark $[bq][\bar{b}\bar{q}]$ with $J^{PC} = 1^{++}$ current is

$m_{X_b} = (10.27 \pm 0.23)$ GeV. The central value is close to the mass of $\Upsilon(3S)$ and appreciably below the $B^*\bar{B}$ threshold at about 10.6 GeV. For comparison, the molecular model predicts a mass for X_b which is about 50–60 MeV below this threshold [33], while a relativistic quark model without explicit $(b\bar{b})$ clustering predicts a value about 133 MeV below threshold [286].

Summarizing, in QCDSR it is possible to satisfactorily explain all the $X(3872)$ properties with a mixture of a $\approx$97% $c\bar{c}$ component and a $\approx$3% meson molecule component. This molecular component must be a mixture of 88% $D^0 D^{*0}$ and 12% $D^+ D^{*-}$ [372]. These conclusions hold only for the quantum number assignment $J^{PC} = 1^{++}$.

QCDSR and the Y states The states $Y(4260)$, $Y(4360)$ and $Y(4660)$ do not easily fit in the predictions of the standard quark model. The $Y(4260)$ has a $\pi^+\pi^- J/\psi$ decay width of $\simeq$100 MeV, and no isospin-violating decay such as $Y \to J/\psi\pi^+\pi^-\pi^0$ has been observed. With these features, the Y is likely to be a meson molecule or a hybrid state. QCDSR calculations of the mass strongly suggest the $D^*\bar{D}_0 - \bar{D}^* D_0$ and $D\bar{D}_1 - \bar{D}D_1$ as favorite molecular combinations, where the symbols D, D^*, D_0 and D_1 represent the lowest-lying pseudoscalar, vector, scalar, and axial-vector charm messons, respectively. From now on we shall omit the combinations arising from symmetrization and use the short forms for these states, e.g., $D_0\bar{D}$.

The vector Y states can be described by molecular or tetraquark currents, with or without an $s\bar{s}$ pair. QCD sum rule calculations have been performed [375, 376] using these currents. A possible interpolating operator representing a $J^{PC} = 1^{--}$ tetraquark state with the symmetric spin distribution

$$[cs]_{S=0}[\bar{c}\bar{s}]_{S=1} + [cs]_{S=1}[\bar{c}\bar{s}]_{S=0} \tag{75}$$

is given by

$$j_\mu = \frac{\epsilon_{abc}\epsilon_{dec}}{\sqrt{2}}\big[\big(s_a^T C\gamma_5 c_b\big)\big(\bar{s}_d\gamma_\mu\gamma_5 C\bar{c}_e^T\big) + \big(s_a^T C\gamma_5\gamma_\mu c_b\big)\big(\bar{s}_d\gamma_5 C\bar{c}_e^T\big)\big]. \tag{76}$$

This current has good OPE convergence and pole dominance in a given Borel window. The result for the mass of the state described by the current in (76) is $m_Y = (4.65 \pm 0.10)$ GeV [375], in excellent agreement with the mass of the $Y(4660)$ meson, lending credence to the conclusion [375] that the $Y(4660)$ meson can be described with a diquark–antidiquark tetraquark current with a spin configuration given by scalar and vector diquarks. The quark content of the current in (76)) is also consistent with the experimental dipion invariant-mass spectra, which give some indication that the $Y(4660)$ has a well-defined dipion intermediate state consistent with $f_0(980)$.

Replacing the strange quarks in (76) by a generic light quark q, the mass obtained for a 1^{--} state described with the symmetric spin distribution

$$[cq]_{S=0}[\bar{c}\bar{q}]_{S=1} + [cq]_{S=1}[\bar{c}\bar{q}]_{S=0} \tag{77}$$

is $m_Y = (4.49 \pm 0.11)$ GeV [375], which is slightly larger than but consistent with the measured $Y(4360)$ mass.

The Y mesons can also be described by molecular-type currents. A $D_{s0}(2317)\bar{D}_s^*(2110)$ molecule with $J^{PC} = 1^{--}$ could also decay into $\psi(2S)\pi^+\pi^-$ with a dipion-mass spectrum consistent with $f_0(980)$. A current with $J^{PC} = 1^{--}$ and a symmetric combination of scalar and vector mesons is

$$j_\mu = \frac{1}{\sqrt{2}}\big[(\bar{s}_a\gamma_\mu c_a)(\bar{c}_b s_b) + (\bar{c}_a\gamma_\mu s_a)(\bar{s}_b c_b)\big]. \tag{78}$$

The mass obtained in [375] for this current is $m_{D_{s0}\bar{D}_s^*} = (4.42 \pm 0.10)$ GeV, which is in better agreement with $Y(4360)$ than $Y(4660)$.

To consider a molecular $D_0\bar{D}^*$ current with $J^{PC} = 1^{--}$, the strange quarks in (78) must be replaced with a generic light quark q. The mass obtained with such current is [375] $m_{D_0\bar{D}^*} = (4.27 \pm 0.10)$ GeV, in excellent agreement with the $Y(4260)$ mass. In order to associate this molecular state with $Y(4260)$, a better understanding of the dipion invariant-mass spectra in $Y(4260) \to \pi^+\pi^- J/\psi$ is needed. From the measured spectra, it seems that the $Y(4260)$ is consistent with a nonstrange molecular state $D_0\bar{D}^*$. Using a D_0 mass of $m_{D_0} = 2352 \pm 50$ MeV, the $D_0\bar{D}^*$ threshold is $\sim$4360 MeV, 100 MeV above the 4.27 ± 0.10 GeV quoted above, indicating the possibility of a bound state.

A $J^{PC} = 1^{--}$ molecular current can also be constructed with pseudoscalar and axial-vector mesons. A molecular $D\bar{D}_1$ current was used in [376]. The mass obtained with this current is $m_{D\bar{D}_1} = (4.19 \pm 0.22)$ GeV. Thus, taking the mass uncertainty into account, the molecular $D\bar{D}_1$ assignment for the $Y(4260)$ is also viable, in agreement with a meson-exchange model [377]. The $D\bar{D}_1$ threshold is $\approx$4285 MeV, close to the $Y(4260)$ mass, indicating the possibility of a loosely bound molecular state.

Summarizing, the Y states can be understood as charmonium hybrids, tetraquark states, and a $D_0\bar{D}^*$ or $D\bar{D}_1$ molecular state for $Y(4260)$. Also possible are a tetraquark state with two axial $[cs]$ P-wave diquarks, or two scalar $[cs]$ P-wave diquarks for $Y(4360)$.

$Z(4430)^+$ A current describing the $Z(4430)^+$ as a $D^* D_1$ molecule with $J^P = 0^-$ is [378]

$$j = \frac{1}{\sqrt{2}}\big[(\bar{d}_a\gamma_\mu c_a)\big(\bar{c}_b\gamma^\mu\gamma_5 u_b\big) + (\bar{d}_a\gamma_\mu\gamma_5 c_a)\big(\bar{c}_b\gamma^\mu u_b\big)\big]. \tag{79}$$

This current corresponds to a symmetric $D^{*+}\bar{D}_1^0 + \bar{D}^{*0}D_1^+$ state with positive G-parity, consistent with the observed decay $Z(4430)^+ \to \pi^+\psi(2S)$. The mass obtained in a QCDSR calculation using such a current is [378] $m_{D^*D_1} = (4.40 \pm 0.10)$ GeV, in an excellent agreement with the measured mass.

To check if the $Z(4430)^+$ could also be described as a diquark–antidiquark state with $J^P = 0^-$, the current [379]

$$j_{0^-} = \frac{i\epsilon_{abc}\epsilon_{dec}}{\sqrt{2}}\left[\left(u_a^T C\gamma_5 c_b\right)\left(\bar{d}_d C\bar{c}_e^T\right) - \left(u_a^T C c_b\right)\left(\bar{d}_d\gamma_5 C\bar{c}_e^T\right)\right] \quad (80)$$

was used to obtain the mass $m_{Z_{(0^-)}} = (4.52 \pm 0.09)$ GeV [379], somewhat larger than, but consistent with, the experimental value. The result using a molecular-type current is in slightly better agreement with the experimental value. However, since there is no one-to-one correspondence between the structure of the current and the state, this result cannot be used to conclude that the $Z(4430)^+$ is favored as a molecular state over a diquark–antidiquark state. To get a measure of the coupling between the state and the current, the parameter λ, defined in (66), is evaluated as $\lambda_{D^*D_1} \simeq 1.5\lambda_{Z_{(0^-)}}$. This suggests that a physical particle with $J^P = 0^-$ and quark content $c\bar{c}u\bar{d}$ has a stronger coupling to the molecular D^*D_1-type current than with that of (80).

A diquark–antidiquark interpolating operator with $J^P = 1^-$ and positive G-parity was also considered [379]:

$$j_\mu^{1^-} = \frac{\epsilon_{abc}\epsilon_{dec}}{\sqrt{2}}\left[\left(u_a^T C\gamma_5 c_b\right)\left(\bar{d}_d\gamma_\mu\gamma_5 C\bar{c}_e^T\right) + \left(u_a^T C\gamma_5\gamma_\mu c_b\right)\left(\bar{d}_d\gamma_5 C\bar{c}_e^T\right)\right]. \quad (81)$$

In this case the Borel stability obtained is worse than for the Z^+ with $J^P = 0^-$ [379]. The mass obtained is $m_{Z_{(1^-)}} = (4.84 \pm 0.14)$ GeV, much larger than both the measured value and that obtained using the $J^P = 0^-$ current. Thus, it is possible to describe the $Z(4430)^+$ as a diquark–antidiquark or molecular state with $J^P = 0^-$, and $J^P = 1^-$ configuration is disfavored.

It is straightforward to extend the D^*D_1-molecule analysis to the bottom quark. Using the same interpolating field of (79) but replacing the charm quark with bottom, an investigation of the hypothetical Z_b is performed [378]. The OPE convergence is even better than for $Z(4430)^+$. The predicted mass is $m_{Z_{B^*B_1}} = (10.74 \pm 0.12)$ GeV, in agreement with that of [380]. For the analogous strange meson Z_s^+, considered as a pseudoscalar $D_s^*D_1$ molecule, the current is obtained by replacing the d quark in (79) with an s quark. The predicted mass is [378] $m_{D_s^*D_1} = (4.70 \pm 0.06)$ GeV, larger than the $D_s^*D_1$ threshold of ~4.5 GeV, indicating that this state is probably very broad and therefore might be difficult to observe.

Table 19 Summary of QCDSR results [381]. The labels 1, 3, and $\bar{3}$ refer to singlet, triplet and anti-triplet color configurations, respectively. The symbols S, P, V, and A refer to scalar, pseudoscalar, vector and axial-vector $q\bar{q}$, qq or $\bar{q}\bar{q}$ combinations, respectively

State	Configuration	Mass (GeV)
$X(3872)$	$[cq]_{\bar{3}}[\bar{c}\bar{q}]_3$ (S + A)	3.92 ± 0.13
	$[c\bar{q}]_1[\bar{c}q]_1$ (P + V)	3.87 ± 0.07
	$[c\bar{c}]_1(A)+$	
	$[c\bar{q}]_1[\bar{c}q]_1$ (P + V)	3.77 ± 0.18
$Y(4140)$	$[c\bar{q}]_1[\bar{c}q]_1$ (V + V)	4.13 ± 0.11
	$[c\bar{s}]_1[\bar{c}s]_1$ (V + V)	4.14 ± 0.09
$Y(4260)$	$[c\bar{q}]_1[\bar{c}q]_1$ (S + V)	4.27 ± 0.10
	$[c\bar{q}]_1[\bar{c}q]_1$ (P + A)	4.19 ± 0.22
$Y(4360)$	$[cq]_{\bar{3}}[\bar{c}\bar{q}]_3$ (S + V)	4.49 ± 0.11
	$[c\bar{s}]_1[\bar{c}s]_1$ (S + V)	4.42 ± 0.10
$Y(4660)$	$[cs]_{\bar{3}}[\bar{c}\bar{s}]_3$ (S + V)	4.65 ± 0.10
$Z(4430)$	$[c\bar{d}]_1[\bar{c}u]_1$ (V + A)	4.40 ± 0.10
	$[cu]_{\bar{3}}[\bar{c}\bar{d}]_3$ (S + P)	4.52 ± 0.09

Summarizing, the $Z(4430)^+$ has been successfully described by both a D^*D_1 molecular current and a diquark–antidiquark current with $J^P = 0^-$.

Table 19 summarizes the QCDSR results for the masses and corresponding quark configurations of the new states [381]. Masses obtained with QCDSR in the molecular approach can be found in [382] (they agree with those in Table 19 from [381] apart from small discrepancies, which should be addressed). The principal input parameters used to obtain the values in Table 19 are

$$\begin{aligned} &m_c(m_c) = (1.23 \pm 0.05)\ \text{GeV}, \\ &\langle\bar{q}q\rangle = -(0.23 \pm 0.03)^3\ \text{GeV}^3, \\ &\langle\bar{q}g\sigma.Gq\rangle = m_0^2\langle\bar{q}q\rangle, \\ &m_0^2 = 0.8\ \text{GeV}^2, \\ &\langle g^2G^2\rangle = 0.88\ \text{GeV}^4. \end{aligned} \quad (82)$$

Uncertainties in Table 19 come from several sources: quark masses and α_s are varied by their errors around their central values in (82); the condensates are taken from previous QCDSR analyses and their uncertainties propagated; particle masses are extracted for a range of Borel masses throughout the Borel window; and in the case of mixing, the masses and widths are computed for several values of the mixing angle centered on its optimal value.

Some of the states discussed above can be understood as both tetraquark and molecular structures. This freedom will be reduced once a comprehensive study of the decay width is performed. At this point, it appears that the only decay width calculated with QCDSR is in [372]. The next challenge for the QCDSR community is to understand the existing data on decays. Explaining these decays will impose severe constraints in the present picture of the new states [372].

2.9.7 Theoretical explanations for new states

Table 20 lists the new states with their properties and proposed theoretical explanations. The theoretical hypothesis list is far from exhaustive.

2.10 Beyond the Standard Model

2.10.1 Mixing of a light CP-odd Higgs and $\eta_b(nS)$ resonances

This section explores the possibility that the measured $\eta_b(1S)$ mass (Table 7 and (5)) is smaller than the predictions (see Sects. 2.5.3, 2.6.2, 2.7, and 2.8.1) due to mixing with a CP-odd Higgs scalar A, and predictions for the spectrum of the $\eta_b(nS)$-A system and the branching fractions into $\tau^+\tau^-$ as functions of m_A are made. Such mixing can cause masses of the η_b-like eigenstates of the full mass matrix to differ considerably from their values in pure QCD [389–391]. Thus the mass of the state interpreted as the $\eta_b(1S)$ can be smaller than expected if m_A is slightly above 9.4 GeV. The masses of the states interpreted as $\eta_b(2S)$ and $\eta_b(3S)$ can also be affected. Furthermore, all $\eta_b(nS)$ states can acquire non-negligible branching ratios into $\tau^+\tau^-$ due to their mixing with A.

A relatively light, CP-odd Higgs scalar can appear, e.g., in nonminimal supersymmetric extensions of the Standard Model (SM) as the NMSSM (Next-to-Minimal Supersymmetric Standard Model); see [392] and references therein. Its mass must satisfy constraints from LEP, where it could have been produced in $e^+e^- \to Z^* \to ZH$ and $H \to AA$ (where H is a CP-even Higgs scalar). For $m_A >$ 10.5 GeV, where A would decay dominantly into $b\bar{b}$, and $m_H <$ 110 GeV, corresponding LEP constraints are quite strong [393]. If $2m_\tau < m_A <$ 10.5 GeV, A would decay dominantly into $\tau^+\tau^-$ and values for m_H down to ~86 GeV are allowed [393] even if H couples to the Z boson with the strength of a SM Higgs boson. A possible explanation of an excess of $b\bar{b}$ events found at LEP [394, 395] provides additional motivation for a CP-odd Higgs scalar with a mass below 10.5 GeV. However, a recent (but preliminary) analysis from ALEPH has found no further evidence of such an excess.

The masses of all 4 physical states (denoted by η_i, $i = 1\ldots4$) as functions of m_A are shown together with the uncertainty bands in Fig. 35 [396]. By construction, $m_{\eta_1} \equiv m_{\rm obs}[\eta_b(1S)]$ is constrained to the measured value. For clarity the assumed values for $m_{\eta_b^0(nS)}$ are indicated as horizontal dashed lines. For m_A not far above 9.4 GeV, the effects of the mixing on the states $\eta_b^0(2S)$ and $\eta_b^0(3S)$ are negligible, but for larger m_A the spectrum can differ considerably from the standard one.

Now turning to the tauonic branching ratios of the η_i states induced by their A-components, assuming $\Gamma_{\eta_b^0(1S)} \sim$ 5–20 MeV, the predicted branching ratio of $\eta_b(1S) \to \tau^+\tau^-$ is compatible with the BABAR upper limit of 8% at 90% CL [397]. For the heavier η_i states, the corresponding branching fractions vary with m_A as shown in Fig. 36, where $\Gamma_{\eta_b^0(1S)} \sim 10$ MeV and $\Gamma_{\eta_b^0(2S)} \sim \Gamma_{\eta_b^0(3S)} \sim 5$ MeV. With larger (smaller) total widths, these branching fractions would be smaller (larger).

The predicted masses and branching fractions in Figs. 35 and 36, respectively, together with an accurate test of lepton-universality-breaking in Υ decays [396, 398], can play an important role both in the experimental search for excited $\eta_b(nS)$ states and subsequent interpretation of the observed spectrum. Such comparisons would test the hypothesis of η_b mixing with a light CP-odd Higgs boson.

2.10.2 Supersymmetric quarkonia

As the top quark is too heavy to form QCD bound states (it decays via the electroweak transition $t \to bW$), the heaviest quarkonia that can be formed in the Standard Model are restricted to the bottom-quark sector. This, however, does not preclude the existence of the heavier quarkonium-like structures in theories beyond the Standard Model.

In particular, there exists some interest in detection of bound states of a top *squark*, $\tilde{t}$, i.e., *stoponium*. From a practical point of view, observation of such a state would allow precise determination of squark masses, since stoponium decays via annihilation into SM particles do not have missing energy signatures with weakly interacting lightest supersymmetric particles (LSPs). From a theoretical point of view such a scenario could be interesting because it could generate correct relic abundance for neutralino Dark Matter and baryon number asymmetry of the universe [399].

As it turns out, some supersymmetric models allow for a relatively light top squark that can form bound states. One condition for such an occurrence involves forbidding two-body decay channels

$$\tilde{t}_1 \to b\tilde{C}_1, \qquad \tilde{t}_1 \to t\tilde{N}_1, \tag{83}$$

where $\tilde{t}_1$ is the lighter mass eigenstate of a top squark, and $\tilde{C}_1$ and $\tilde{N}_1$ are the lightest chargino and neutralino states, respectively. These decay channels can be kinematically forbidden if $m_{\tilde{t}_1} - m_{\tilde{C}_1} < m_b$ and $m_{\tilde{t}_1} - m_{\tilde{N}_1} < m_t$, which can be arranged in SUSY models (although not in

Table 20 As in Table 9, new *unconventional* states in the $c\bar{c}$, $b\bar{c}$, and $b\bar{b}$ regions, ordered by mass, with possible interpretations (which do not apply solely to the decay modes listed alongside). References are representative, not necessarily exhaustive. The QCDSR notation is explained in the caption to Table 19

State	m (MeV)	Γ (MeV)	J^{PC}	Modes	Interpretation	Reference(s)
$X(3872)$	3871.52 ± 0.20	1.3 ± 0.6	$1^{++}/2^{-+}$	$\pi^+\pi^- J/\psi$	$D^{*0}\bar{D}^0$ molecule (bound)	[121, 122] [383–385]
				$D^{*0}\bar{D}^0$	$D^{*0}\bar{D}^0$ unbound	
				$\gamma J/\psi, \gamma\psi(2S)$	if 1^{++}, $\chi_{c2}(2P)$	[71]
				$\omega J/\psi$	if 2^{-+}, $\eta_{c2}(1D)$	[81, 93, 126]
					charmonium + mesonic-molecule mixture	[381]
					QCDSR: $[cq]_{\bar{3}}[\bar{c}\bar{q}]_3$ (S + A)	[381]
					QCDSR: $[c\bar{q}]_1[\bar{c}q]_1$ (P + V)	[381]
					QCDSR: $[c\bar{c}]_1$(A) + $[c\bar{q}]_1[\bar{c}q]_1$ (P + V)	[381]
$X(3915)$	3915.6 ± 3.1	28 ± 10	$0, 2^{?+}$	$\omega J/\psi$	$D^{*+}D^{*-} + D^{*0}\bar{D}^{*0}$	[386]
$Z(3930)$	3927.2 ± 2.6	24.1 ± 6.1	2^{++}	$D\bar{D}$	$\chi_{c2}(2P)$ (i.e., 2^3P_2 $c\bar{c}$)	
					1^3F_2 $c\bar{c}$	[71]
$X(3940)$	3942^{+9}_{-8}	37^{+27}_{-17}	$?^{?+}$	$D\bar{D}^*$		
$G(3900)$	3943 ± 21	52 ± 11	1^{--}	$D\bar{D}$	Coupled-channel effect	[34]
$Y(4008)$	$4008^{+121}_{-\ 49}$	226 ± 97	1^{--}	$\pi^+\pi^- J/\psi$		
$Z_1(4050)^+$	4051^{+24}_{-43}	82^{+51}_{-55}	?	$\pi^+\chi_{c1}(1P)$	hadrocharmonium	[282, 355]
$Y(4140)$	4143.0 ± 3.1	$11.7^{+9.1}_{-6.2}$	$?^{?+}$	$\phi J/\psi$	QCDSR: $[c\bar{q}]_1[\bar{c}q]_1$ (V + V)	[381]
					QCDSR: $[c\bar{s}]_1[\bar{c}s]_1$ (V + V)	[381]
					$D_s^{*+}D_s^{*-}$	[386]
$X(4160)$	4156^{+29}_{-25}	139^{+113}_{-65}	$?^{?+}$	$D\bar{D}^*$		
$Z_2(4250)^+$	$4248^{+185}_{-\ 45}$	$177^{+321}_{-\ 72}$	?	$\pi^+\chi_{c1}(1P)$	hadrocharmonium	[282, 355]
$Y(4260)$	4263 ± 5	108 ± 14	1^{--}	$\pi^+\pi^- J/\psi$	charmonium hybrid	[276–278]
				$\pi^0\pi^0 J/\psi$	$J/\psi f_0(980)$ bound state	[313]
					$D_0\bar{D}^*$ molecular state	[375]
					$[cs][\bar{c}\bar{s}]$ tetraquark state	[345, 381]
					hadrocharmonium	[282, 355]
					QCDSR: $[c\bar{q}]_1[\bar{c}q]_1$ (S + V)	[381]
					QCDSR: $[c\bar{q}]_1[\bar{c}q]_1$ (P + A)	[381]
$Y(4274)$	$4274.4^{+8.4}_{-6.7}$	32^{+22}_{-15}	$?^{?+}$	$B \to K(\phi J/\psi)$	(see $Y(4140)$)	
$X(4350)$	$4350.6^{+4.6}_{-5.1}$	$13.3^{+18.4}_{-10.0}$	$0, 2^{++}$	$\phi J/\psi$		
$Y(4360)$	4353 ± 11	96 ± 42	1^{--}	$\pi^+\pi^-\psi(2S)$	hadrocharmonium	[282, 355]
					crypto-exotic hybrid	[71]
					$Y_B(4360) = [cd][\bar{c}\bar{d}](1P)$, baryonium	[321]
					QCDSR: $[cq]_{\bar{3}}[\bar{c}\bar{q}]_3$ (S + V)	[381]
					QCDSR: $[c\bar{s}]_1[\bar{c}s]_1$ (S + V)	[381]
$Z(4430)^+$	4443^{+24}_{-18}	$107^{+113}_{-\ 71}$	?	$\pi^+\psi(2S)$	$D^{*+}\bar{D}_1^0$ molecular state	[378, 387]
					$[cu][\bar{c}\bar{d}]$ tetraquark state	[379]
					hadrocharmonium	[282, 355]
					QCDSR: $[c\bar{d}]_1[\bar{c}u]_1$ (V + A)	[381]
					QCDSR: $[cu]_{\bar{3}}[\bar{c}\bar{d}]_3$ (S + P)	[381]
$X(4630)$	$4634^{+\ 9}_{-11}$	92^{+41}_{-32}	1^{--}	$\Lambda_c^+\Lambda_c^-$	$Y_B(4660) = [cd][\bar{c}\bar{d}](2P)$, baryonium	[321]
					$\psi(2S)f_0(980)$ molecule	[324]
$Y(4660)$	4664 ± 12	48 ± 15	1^{--}	$\pi^+\pi^-\psi(2S)$	$\psi(2S)f_0(980)$ molecule	[306]
					$[cs][\bar{c}\bar{s}]$ tetraquark state	[375]
					hadrocharmonium	[282, 355]
					$Y_B(4660) = [cd][\bar{c}\bar{d}](2P)$, baryonium	[321]
					QCDSR: $[cs]_{\bar{3}}[\bar{c}\bar{s}]_3$ (S + V)	[381]
$Y_b(10888)$	10888.4 ± 3.0	$30.7^{+8.9}_{-7.7}$	1^{--}	$\pi^+\pi^-\Upsilon(nS)$	$\Upsilon(5S)$	[388]
					b-flavored $Y(4260)$	[37, 117]

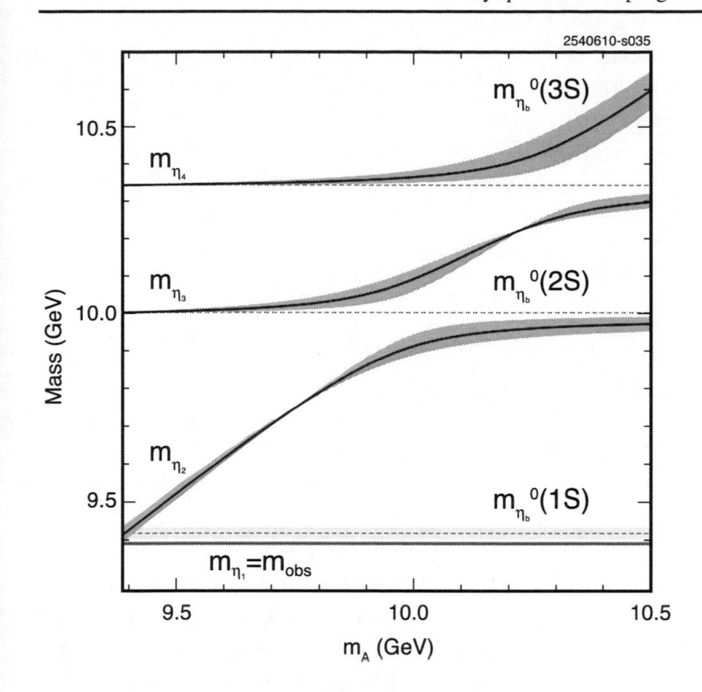

Fig. 35 The masses of all eigenstates as function of m_A

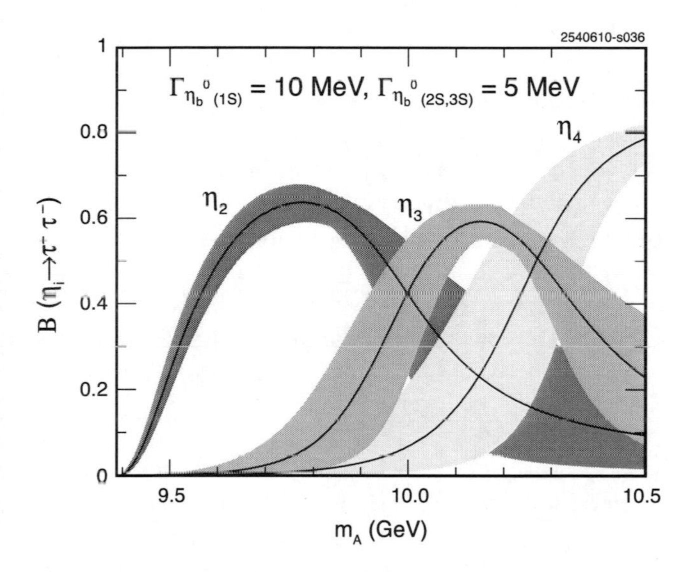

Fig. 36 The branching ratios into $\tau^+\tau^-$ for the eigenstates η_2, η_3 and η_4 as functions of m_A

mSUGRA). Other decay channels are either flavor-violating or involve more than two particles in the final state and are therefore suppressed enough to allow formation of stoponia. Since top squarks are scalars, the lowest-energy bound state $\eta_{\tilde{t}}$ has spin zero and mass predicted in the range of 200–800 GeV [400–402].

The $\eta_{\tilde{t}}$ spin-zero state can be most effectively produced at the LHC in the $gg \to \eta_{\tilde{t}}$ channel. Possible decay channels of $\eta_{\tilde{t}}$ include $\gamma\gamma$, gg, ZZ, WW, as well as hadronic and leptonic final states. It is the $\gamma\gamma$ decay channel that has recently received most attention.

The production cross section for a stoponium $\eta_{\tilde{t}}$ detected in the $\gamma\gamma$ channel can be written as [402]

$$\begin{aligned}\sigma(pp \to \eta_{\tilde{t}} \to \gamma\gamma) &= \frac{\pi^2}{8m^3_{\eta_{\tilde{t}}}}\mathcal{B}(\eta_{\tilde{t}} \to gg)\Gamma(\eta_{\tilde{t}} \to \gamma\gamma) \\ &\quad\times \int_\tau^1 dx\frac{t}{x}g\left(x, Q^2\right)g\left(\tau/x, Q^2\right), \end{aligned} \tag{84}$$

where $g(x, Q^2)$ is the gluon parton distribution function, and $\mathcal{B}(\eta_{\tilde{t}} \to gg) = \Gamma(\eta_{\tilde{t}} \to gg)/\Gamma_{tot}$ is the branching fraction for $\eta_{\tilde{t}} \to gg$ decay. Note that (84) contains $\eta_{\tilde{t}}$ decay widths into $\gamma\gamma$ and gg channels that depend on the value of the $\eta_{\tilde{t}}$ wave function at the origin,

$$\begin{aligned}\Gamma(\eta_{\tilde{t}} \to \gamma\gamma) &= \frac{4}{3}\frac{8}{9}\frac{4\pi\alpha^2}{m^2_{\eta_{\tilde{t}}}}\left|\psi_{\eta_{\tilde{t}}}(0)\right|^2, \\ \Gamma(\eta_{\tilde{t}} \to gg) &= \frac{9}{8}\frac{\alpha_s}{\alpha}\Gamma(\eta_{\tilde{t}} \to \gamma\gamma). \end{aligned} \tag{85}$$

These decay widths have recently been evaluated taking into account one-loop QCD corrections [403, 404]. These next-to-leading order QCD corrections to $\Gamma(\eta_{\tilde{t}} \to \gamma\gamma)$ significantly decrease [403] the rate while NLO corrections to the production cross section tend to increase it [404], leaving the product in (84) largely unchanged from the leading-order prediction.

It appears that, if possible, observation of stoponium would require significant statistics at the LHC. For example, for stoponium masses on the order of or less than 300 GeV, an integrated luminosity of 10 fb^{-1} [402] is required. Larger mass values would require even greater statistics. Thus discovery of stoponium decays could only improve determinations of the squark mass, which would probably be already available using other methods. The possibility of learning about bound-state QCD dynamics in a regime where calculations are under better theoretical control makes this study worthwhile.

2.10.3 Invisible decays of ψ and Υ

Measurements of invisible, meaning undetected, decay rates of ψ and Υ resonances can discover or place strong constraints on dark matter scenarios where candidate dark matter constituents are lighter than the b-quark [405].

According to the SM, invisible decays of the $\Upsilon(1S)$ can proceed via $b\bar{b}$ annihilation into a $\nu\bar{\nu}$ pair with a branching fraction $\mathcal{B}(\Upsilon(1S) \to \nu\bar{\nu}) \simeq 10^{-5}$ [406], well below current experimental sensitivity. However, models containing low-mass dark matter components might enhance such invisible decay modes up to observable rates [407–410]. Interestingly, from the astroparticle and cosmological side, DAMA and CoGeNT experiments have reported the observation of events compatible with a Light Dark-Matter (LDM) candidate with a mass inside the interval [5, 10] GeV [411]. Such a mass range for LDM constituents is especially attractive in certain cosmological scenarios, e.g., the Asymmetric Dark Matter model where the cosmological dark matter density arises from the baryon asymmetry of the universe and is expected to have an LDM mass of order [1–10] GeV (see [412, 413] and references therein).

Searches for the $\Upsilon(1S) \to$ invisible mode have been carried out by Argus [414], CLEO [415], Belle [416], and BABAR [417] using the cascade decay, $\Upsilon(2,3S) \to \pi^+\pi^-\Upsilon(1S)$, in which the dipion tags the invisibly decaying $\Upsilon(1S)$. The most recent experimental data from BABAR [417] yield an upper limit at the 90% CL of

$$\mathcal{B}\big(\Upsilon(1S) \to \text{invisible}\big) < 3 \times 10^{-4} \quad \text{at 90\% CL,} \tag{86}$$

which is only one order of magnitude above the SM expectation. This bound only applies to LDM candidates with masses less than half the $\Upsilon(1S)$ mass. The corresponding limit from BES [418] is

$$\mathcal{B}(J/\psi \to \text{invisible}) < 7.2 \times 10^{-4} \quad \text{at 90\% CL.} \tag{87}$$

However, the invisible decay mode induced by scalar or pseudoscalar mediators (like CP-odd or CP-even Higgs bosons) actually vanishes, independent of the character of the dark matter candidate (either scalar, Dirac or Majorana fermion), if the decaying resonance has $J^{PC} = 1^{--}$ [419]. In order to get a non-vanishing decay rate, a new vector U-boson associated with gauging an extra $U(1)$ symmetry (i.e., a new kind of interaction) would be required [419]. Hence a different type of search has been performed by looking at the decay $\Upsilon(1S) \to \gamma +$invisible, which can proceed via a light scalar or pseudoscalar Higgs mediator decaying into a LDM pair. As discussed in Sect. 2.10.1, a light CP-odd Higgs boson A_1 mixing with η_b resonances could naturally become the mediator of the decay into undetected dark matter particles: $\Upsilon \to \gamma\ A_1(\to$ invisible). The result for $\Upsilon(3S)$ decays from BABAR [420] is

$$\mathcal{B}\big(\Upsilon(3S) \to \gamma + \text{invisible}\big) < (0.7\text{–}30) \times 10^{-6}$$
$$\quad \text{at 90\% CL, for } s_{\text{inv}}^{1/2} < 7.8 \text{ GeV,} \tag{88}$$

where s_{inv} denotes the invariant-mass squared of the hypothetical LDM pair. The corresponding limit from CLEO [421] for J/ψ decays is

$$\mathcal{B}(J/\psi \to \gamma + \text{invisible}) < (2.5\text{–}6.5) \times 10^{-6}$$
$$\quad \text{at 90\% CL, for } s_{\text{inv}}^{1/2} < 960 \text{ MeV.} \tag{89}$$

Experimental systematic effects make improving limits from searches of this type at higher mass difficult. The energy of the final-state photon gets progressively smaller for larger invisible mass; at low energy, the energy resolution and number of fake photon candidates are all typically less favorable than at higher photon energy. Nevertheless, such improvements, if achieved, would provide important constraints on theoretical possibilities by reaching higher LDM masses.

3 Decay[8]

3.1 Radiative transitions

An electromagnetic transition between quarkonium states, which occurs via emission of a photon, offers the distinctive experimental signature of a monochromatic photon, a useful production mechanism for discovery and study of the lower-lying state, and a unique window on the dynamics of such systems. Below we first review the status and open questions regarding the relevant theoretical frameworks and tools, and then describe important measurements of charmonium and bottomonium electromagnetic transitions.

3.1.1 Theoretical status

The nonrelativistic nature of heavy quarkonium may be exploited to calculate electromagnetic transitions. Nonrelativistic effective field theories provide a way to systematically implement the expansion in the relative heavy-quark velocity, v. Particularly useful are nonrelativistic QCD (NRQCD) coupled to electromagnetism [137, 138], which follows from QCD (and QED) by integrating out the heavy-quark mass scale m, and potential NRQCD coupled to electromagnetism [134, 139, 140, 422], which follows from NRQCD (and NRQED) by integrating out the momentum-transfer scale mv.

Electromagnetic transitions may be classified in terms of electric and magnetic transitions between eigenstates of the leading-order pNRQCD Hamiltonian. The states are classified in terms of the radial quantum number, n, the orbital angular momentum, l, the total spin, s, and the total angular momentum, J. In the nonrelativistic limit, the spin dependence of the quarkonium wave function decouples from the spatial dependence. The spatial part of the wave function, $\psi(x)$, can be expressed in terms of a radial wave function, $u_{nl}(r)$, and the spherical harmonics, Y_{lm}, as $\psi(x) = Y_{lm}(\theta,\phi)u_{nl}(r)/r$. The spatial dependence of the electromagnetic transition amplitudes reduces to expectation values of various functions of quark position and momentum between the initial- and final-state wave functions [1].

[8]Contributing authors: E. Eichten†, R.E. Mitchell†, A. Vairo†, A. Drutskoy, S. Eidelman, C. Hanhart, B. Heltsley, G. Rong, and C.-Z. Yuan.

Magnetic transitions flip the quark spin. Transitions that do not change the orbital angular momentum are called magnetic dipole, or M1, transitions. In the nonrelativistic limit, the spin-flip transition decay rate between an initial state $i = n\ {}^{2s+1}l_J$ and a final state $f = n'\ {}^{2s'+1}l_{J'}$ is:

$$\Gamma(i \xrightarrow{\text{M1}} \gamma + f) = \frac{16}{3}\alpha e_Q^2 \frac{E_\gamma^3}{m_i^2}(2J'+1)\mathrm{S}_{if}^{\mathrm{M}}|\mathcal{M}_{if}|^2, \quad (90)$$

where e_Q is the electrical charge of the heavy quark Q ($e_b = -1/3$, $e_c = 2/3$), α the fine-structure constant, $E_\gamma = (m_i^2 - m_f^2)/(2m_i)$ is the photon energy, and m_i, m_f are the masses of the initial- and final-state quarkonia, respectively. The statistical factor $\mathrm{S}_{if}^{\mathrm{M}} = \mathrm{S}_{fi}^{\mathrm{M}}$ reads

$$\mathrm{S}_{if}^{\mathrm{M}} = 6(2s+1)(2s'+1) \times \begin{Bmatrix} J & 1 & J' \\ s' & l & s \end{Bmatrix}^2 \begin{Bmatrix} 1 & \frac{1}{2} & \frac{1}{2} \\ \frac{1}{2} & s' & s \end{Bmatrix}^2. \quad (91)$$

For $l = 0$ transitions, $S_{if}^{\mathrm{M}} = 1$. For equal quark masses m, the overlap integral $\mathcal{M}_{if}$ is given by

$$\mathcal{M}_{if} = (1+\kappa_Q) \times \int_0^\infty dr u_{nl}(r) u'_{n'l}(r) j_0\left(\frac{E_\gamma r}{2}\right), \quad (92)$$

where j_n are spherical Bessel functions and κ_Q is the anomalous magnetic moment of a heavy quarkonium $Q\bar{Q}$. In pNRQCD, the quantity $1+\kappa_Q$ is the Wilson coefficient of the operator $S^\dagger \sigma \cdot e_Q \mathbf{B}^{\mathrm{em}}/(2m)S$, where $\mathbf{B}^{\mathrm{em}}$ is the magnetic field and S is a $Q\bar{Q}$ color-singlet field.

Electric transitions do not change the quark spin. Transitions that change the orbital angular momentum by one unit are called electric dipole, or E1, transitions. In the nonrelativistic limit, the spin-averaged electric transition rate between an initial state $i = n\ {}^{2s+1}l_J$ and a final state $f = n'\ {}^{2s'+1}l'_{J'}$ ($l = l' \pm 1$) is

$$\Gamma(i \xrightarrow{\text{E1}} \gamma + f) = \frac{4}{3}\alpha e_Q^2 E_\gamma^3 (2J'+1)\mathrm{S}_{if}^{\mathrm{E}}|\mathcal{E}_{if}|^2, \quad (93)$$

where the statistical factor $\mathrm{S}_{if}^{\mathrm{E}} = \mathrm{S}_{fi}^{\mathrm{E}}$ is

$$\mathrm{S}_{if}^{\mathrm{E}} = \max(l, l') \begin{Bmatrix} J & 1 & J' \\ l' & s & l \end{Bmatrix}^2. \quad (94)$$

The overlap integral $\mathcal{E}_{if}$ for equal quark masses m is given by

$$\mathcal{E}_{if} = \frac{3}{E_\gamma}\int_0^\infty dr u_{nl}(r) u_{n'l'}(r) \times \left[\frac{E_\gamma r}{2} j_0\left(\frac{E_\gamma r}{2}\right) - j_1\left(\frac{E_\gamma r}{2}\right)\right]. \quad (95)$$

Since the leading-order operator responsible for the electric transition does not undergo renormalization, the electric transition rate does not depend on a Wilson coefficient, analogous to the case of the quarkonium magnetic moment appearing in the magnetic transitions.

If the photon energy is smaller than the typical inverse radius of the quarkonium, we may expand the overlap integrals in $E_\gamma r$, generating electric and magnetic multipole moments. At leading order in the multipole expansion, the magnetic overlap integral reduces to $\mathcal{M}_{if} = \delta_{nn'}$. Transitions for which $n = n'$ are called *allowed* M1 transitions, transitions for which $n \neq n'$ are called *hindered* transitions. Hindered transitions happen only because of higher-order corrections and are suppressed by at least v^2 with respect to the allowed ones. At leading order in the multipole expansion the electric overlap integral reduces to

$$\mathcal{E}_{if} = \int_0^\infty dr u_{nl}(r) r u_{n'l'}(r). \quad (96)$$

Note that E1 transitions are more copiously observed than allowed M1 transitions, because the rates of the electric transitions are enhanced by $1/v^2$ with respect to the magnetic ones. Clearly, the multipole expansion is always allowed for transitions between states with the same principal quantum numbers ($E_\gamma \sim mv^4$ or $mv^3 \ll mv$) or with contiguous principal quantum numbers ($E_\gamma \sim mv^2 \ll mv$). For transitions that involve widely separated states, the hierarchy $E_\gamma \ll mv$ may not be realized. For example, in $\Upsilon(3S) \to \gamma\eta_b(1S)$, we have $E_\gamma \approx 921$ MeV, which is smaller than the typical momentum transfer in the $\eta_b(1S)$, about 1.5 GeV [175], but may be comparable to or larger than the typical momentum transfer in the $\Upsilon(3S)$. On the other hand, in $\psi(2S) \to \gamma\chi_{c1}$, we have $E_\gamma \approx 171$ MeV, which is smaller than the typical momentum transfer in both the $\psi(2S)$ and the χ_{c1}.

Beyond the nonrelativistic limit, (90) and (93) get corrections. These are radiative corrections counted in powers of $\alpha_s(m)$ and relativistic corrections counted in powers of v. These last ones include proper relativistic corrections of the type $(mv)^2/m^2$, recoil corrections and, for weakly coupled quarkonia, also corrections of the type $\Lambda_{\mathrm{QCD}}/(mv)$. Finally, we also have corrections of the type $E_\gamma/(mv)$ that involve the photon energy. In the charmonium system, $v^2 \approx 0.3$, and corrections may be as large as 30%. Indeed, a negative correction of about 30% is required to bring the nonrelativistic prediction of $\mathcal{B}(J/\psi \to \gamma\eta_c(1S))$, which is about 3%, close to the experimental value, which is about 2%. We will see that this is actually the case. In the bottomonium system, $v^2 \approx 0.1$ and corrections may be as large as 10%.

For a long time, corrections to the electromagnetic transitions have been studied almost entirely within phenomenological models [31, 423–435] (a sum rule analysis appears in [436]). We refer to reviews in [1, 81]; a textbook presentation can be found in [437]. In contrast to models, the effective field theory approach allows a systematic and rigorous treatment of the higher-order corrections. The use of EFTs for electromagnetic transitions was initiated in [422], in which a study of magnetic transitions was performed. The results of that analysis may be summarized in the following way.

- The quarkonium anomalous magnetic moment κ_Q does not get contributions from the scale mv: it is entirely determined by the quark anomalous magnetic moment. Since the quark magnetic moment appears at the scale m, it is accessible by perturbation theory: $\kappa_Q = 2\alpha_s(m)/(3\pi) + \mathcal{O}(\alpha_s^2)$. As a consequence, κ_Q is a small positive quantity, about 0.05 in the bottomonium case and about 0.08 in the charmonium one. This is confirmed by lattice calculations [438] and by the analysis of higher-order multipole amplitudes (see Sect. 3.1.6).
- QCD does not allow for a scalar-type contribution to the magnetic transition rate. A scalar interaction is often postulated in phenomenological models.

The above conclusions were shown to be valid at any order of perturbation theory as well as *non*perturbatively. They apply to magnetic transitions from any quarkonium state. For ground state magnetic transitions, we expect that perturbation theory may be used at the scale mv. Under this assumption, the following results were found at relative order v^2.

- The magnetic transition rate between the vector and pseudoscalar quarkonium ground state, including the leading relativistic correction (parametrized by α_s at the typical momentum-transfer scale $m_i\alpha_s/2$) and the leading anomalous magnetic moment (parametrized by α_s at the mass scale $m_i/2$), reads

$$\Gamma(i \to \gamma + f) = \frac{16}{3}\alpha e_Q^2 \frac{E_\gamma^3}{m_i^2} \times \left[1 + \frac{4}{3}\frac{\alpha_s(m_i/2)}{\pi} - \frac{32}{27}\alpha_s^2(m_i\alpha_s/2)\right], \quad (97)$$

in which $i = 1^3 0_1$ and $f = 1^1 0_1$. This expression is not affected by nonperturbative contributions. Applied to the charmonium and bottomonium case it gives $\mathcal{B}(J/\psi \to \gamma\eta_c(1S)) = (1.6 \pm 1.1)\%$ (see Sect. 3.1.2 for the experimental situation) and $\mathcal{B}(\Upsilon(1S) \to \gamma\eta_b(1S)) = (2.85 \pm 0.30) \times 10^{-4}$ (see Sect. 3.1.8 for some experimental perspectives).

- A similar perturbative analysis, performed for hindered magnetic transitions, mischaracterizes the experimental data by an order of magnitude, pointing either to a breakdown of the perturbative approach for quarkonium states with principal quantum number $n > 1$, or to large higher-order relativistic corrections.

The above approach is well suited to studying the lineshapes of the $\eta_c(1S)$ and $\eta_b(1S)$ in the photon spectra of $J/\psi \to \gamma\eta_c(1S)$ and $\Upsilon(1S) \to \gamma\eta_b(1S)$, respectively. In the region of $E_\gamma \ll m\alpha_s$, at leading order, the lineshape is given by [439]

$$\frac{d\Gamma}{dE_\gamma}(i \to \gamma + f) = \frac{16}{3}\frac{\alpha e_Q^2}{\pi}\frac{E_\gamma^3}{m_i^2} \times \frac{\Gamma_f/2}{(m_i - m_f - E_\gamma)^2 + \Gamma_f^2/4}, \quad (98)$$

which has the characteristic asymmetric behavior around the peak seen in the data (compare with the discussion in Sect. 3.1.2).

No systematic analysis is yet available for relativistic corrections to electromagnetic transitions involving higher quarkonium states, i.e., states for which Λ_{QCD} is larger than the typical binding energy of the quarkonium. These states are not described in terms of a Coulombic potential. Transitions of this kind include magnetic transitions between states with $n > 1$ and all electric transitions, $n = 2$ bottomonium states being on the boundary. Theoretical determinations rely on phenomenological models, which we know do not agree with QCD in the perturbative regime and miss some of the terms at relative order v^2 [422]. A systematic analysis is, in principle, possible in the same EFT framework developed for magnetic transitions. Relativistic corrections would turn out to be factorized in some high-energy coefficients, which may be calculated in perturbation theory, and in Wilson-loop amplitudes similar to those that encode the relativistic corrections of the heavy-quarkonium potential [179]. At large spatial distances, Wilson-loop amplitudes cannot be calculated in perturbation theory but are well suited for lattice measurements. Realizing the program of systematically factorizing relativistic corrections in Wilson-loop amplitudes and evaluating them on the lattice, would, for the first time, produce model-independent determinations of quarkonium electromagnetic transitions between states with $n > 1$. These are the vast majority of transitions observed in nature. Finally, we note that, for near-threshold states such as $\psi(2S)$, intermediate meson loops may provide important contributions [440], which should be systematically accounted for.

Higher-order multipole transitions have been observed in experiments (see Sect. 3.1.6), Again, a systematic treatment is possible in the EFT framework outlined above, but has not yet been realized.

3.1.2 Study of $\psi(1S,2S) \to \gamma\eta_c(1S)$

Using a combination of inclusive and exclusive techniques, CLEO [69] has recently measured

$$\mathcal{B}\big(J/\psi \to \gamma\eta_c(1S)\big) = (1.98 \pm 0.09 \pm 0.30)\%,$$
$$\mathcal{B}\big(\psi(2S) \to \gamma\eta_c(1S)\big) = (0.432 \pm 0.016 \pm 0.060)\%. \tag{99}$$

The lineshape of the $\eta_c(1S)$ in these M1 transitions was found to play a crucial role. Because the width of the $\eta_c(1S)$ is relatively large, the energy dependence of the phase space term and the matrix element distort the lineshape (see (98)). Indeed, the photon spectrum measured by CLEO shows a characteristic asymmetric behavior (see Fig. 37). The theoretical uncertainty in this lineshape represents the largest systematic error in the branching ratios. Both M1 transitions are found to be larger than previous measurements due to a combination of a larger $\eta_c(1S)$ width and the first accounting for the pronounced asymmetry in the lineshape. This process has also been recently measured by KEDR [441].

The distortion of the $\eta_c(1S)$ lineshape also has implications for the mass of the $\eta_c(1S)$. As of 2006, there was a 3.3σ discrepancy between $\eta_c(1S)$ mass measurements made from $\psi(1S,2S) \to \gamma\eta_c(1S)$ (with a weighted average of 2977.3 ± 1.3 MeV) and from $\gamma\gamma$ or $p\overline{p}$ production (averaging 2982.6 ± 1.0 MeV). The CLEO [69] analysis suggests that the solution to this problem may lie in the lineshape of the $\eta_c(1S)$ in the M1 radiative transitions. When no distortion in the lineshape is used in the CLEO fit to $J/\psi \to \gamma\eta_c(1S)$, the resulting $\eta_c(1S)$ mass is consistent with other measurements from M1 transitions (2976.6 ± 0.6 MeV, statistical error only); however, when a distorted lineshape is taken into account, the mass is consistent with those from $\gamma\gamma$ or $p\overline{p}$ production (2982.2 ± 0.6 MeV, statistical error only). Recent measurements of the $\eta_c(1S)$ mass in $\gamma\gamma$ production are consistent with this general picture. An $\eta_c(1S)$ mass of $2986.1 \pm 1.0 \pm 2.5$ MeV is reported in a Belle analysis of $\gamma\gamma \to h^+h^-h^+h^-$, where $h = \pi, K$ [442], consistent with the higher $\eta_c(1S)$ mass. Also, BABAR measures a mass of $2982.2 \pm 0.4 \pm 1.6$ MeV in $\gamma\gamma \to K_S K^{\pm}\pi^{\mp}$ [443].

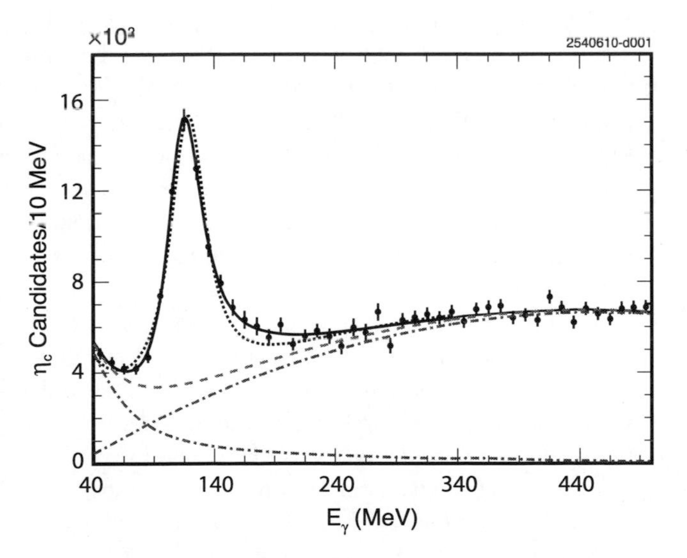

Fig. 37 From CLEO [69], the photon energy spectrum from $J/\psi \to \gamma\eta_c(1S)$. The $\eta_c(1S)$ is reconstructed in 12 different exclusive decay modes. The *dotted curve* represents a fit with a relativistic Breit–Wigner function; the *solid curve* uses a relativistic Breit–Wigner function distorted by the energy dependence of the phase-space term and the matrix element. The *dash-dotted curves* show two components of the background, which when summed become the *dashed curve*. Adapted from [69] with kind permission, copyright (2009) The American Physical Society

3.1.3 Observation of $h_c(1P) \to \gamma\eta_c(1S)$

The decay chain $\psi(2S) \to \pi^0 h_c(1P)$, $h_c(1P) \to \gamma\eta_c(1S)$ was first observed by CLEO [45, 46] using 24.5 million $\psi(2S)$ events, and later confirmed with higher statistics by BESIII [47] using 106 million $\psi(2S)$. While the mass difference of the $h_c(1P)$ and $\chi_{cJ}(1P)$ states is a measure of the hyperfine splitting in the $1P$ $c\bar{c}$ system, the product branching fraction can be used to glean information about the size of the E1 transition $h_c(1P) \to \gamma\eta_c(1S)$. The product branching fraction $\mathcal{B}(\psi(2S) \to \pi^0 h_c(1P)) \times \mathcal{B}(h_c(1P) \to \gamma\eta_c(1S))$ was measured to be

$$(4.19 \pm 0.32 \pm 0.45) \times 10^{-4} \quad \text{CLEO [46]},$$
$$(4.58 \pm 0.40 \pm 0.50) \times 10^{-4} \quad \text{BESIII [47]}, \tag{100}$$

where both CLEO and BESIII used an inclusive technique (requiring reconstruction of just the π^0 and the transition photon and imposing appropriate kinematic constraints), but CLEO also utilized a fully exclusive technique (in addition to the π^0 and γ, reconstructing the $\eta_c(1S)$ in multiple exclusive decay channels). BESIII has also measured $\mathcal{B}(\psi(2S) \to \pi^0 h_c(1P))$ (see Sect. 3.3.3), allowing extraction of

$$\mathcal{B}\big(h_c(1P) \to \gamma\eta_c(1S)\big) = (54.3 \pm 6.7 \pm 5.2)\%. \tag{101}$$

As part of the same study, CLEO has measured the angular distribution of the transition photon from $h_c(1P) \to \gamma\eta_c(1S)$ (Fig. 38). Fitting to a curve of the form $N(1 + \alpha\cos^2\theta)$, and combining the results from the inclusive and exclusive analyses, it was found that $\alpha = 1.20 \pm 0.53$, consistent with $\alpha = 1$, the expectation for E1 transitions.

3.1.4 Nonobservation of $\psi(2S) \to \gamma\eta_c(2S)$

After years of false alarms, the $\eta_c(2S)$ was finally observed in B-decays and two-photon fusion (see Sect. 2.2.2). In an attempt to both discover new decay modes and to observe it in a radiative transition, CLEO [68] modeled an $\eta_c(2S)$ analysis after its effort on the $\eta_c(1S)$, wherein a

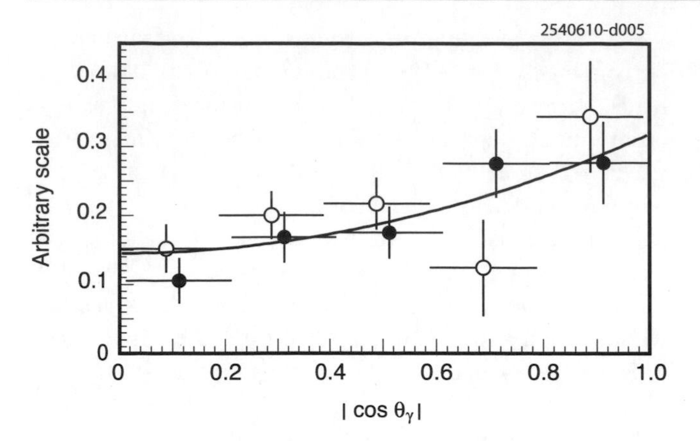

Fig. 38 From CLEO [46], the distribution of the photon polar angle in the e^+e^- center-of-mass frame from the transition sequence $\psi(2S)\to\pi^0 h_c(1P)$, $h_c(1P)\to\gamma\eta_c(1S)$. *Open circles* represent data from an inclusive $\eta_c(1S)$ decays and *solid circles* data from exclusive $\eta_c(1S)$ decays (see text). The *solid curve* represents a fit of both inclusive and exclusive data to $N(1+\alpha\cos^2\theta)$, from which α was found to be 1.20 ± 0.53. Adapted from [46] with kind permission, copyright (2008) The American Physical Society

systematic study of many exclusive hadronic decay modes aided in measuring the lineshapes and branching fractions in $\psi(1S,2S)\to\gamma\eta_c(1S)$ transitions [69]. Eleven modes, which were chosen based in part upon success in finding similar $\eta_c(1S)$ decays, were sought in the exclusive decay chain $\psi(2S)\to\gamma\eta_c(2S)$, $\eta_c(2S)\to$ hadrons in CLEO's 26 million $\psi(2S)$ sample. One of the modes sought was the dipion transition $\eta_c(2S)\to\pi^+\pi^-\eta_c(1S)$, which used proven hadronic decay modes of the $\eta_c(1S)$. No $\eta_c(2S)$ signals were found, and eleven product branching-fraction upper limits were set. None but one of these products can be used to directly set a limit on the transition because none but one have a measured $\eta_c(2S)$ branching fraction. Using the BABAR [444] branching fraction for $\eta_c(2S)\to KK\pi$ allows CLEO to set an upper limit

$$\mathcal{B}\big(\psi(2S)\to\gamma\eta_c(2S)\big)<7.6\times10^{-4}. \tag{102}$$

This value is greater than a phenomenological limit obtained [68] by assuming the matrix element is the same as for $J/\psi\to\gamma\eta_c(1S)$ and correcting the measured J/ψ branching fraction by the ratio of total widths and phase-space factors, $(3.9\pm1.1)\times10^{-4}$.

3.1.5 $\psi(2S)\to\gamma\gamma J/\psi$ through χ_{cJ}

One component of the CLEO [445, 446] $\psi(2S)\to XJ/\psi$ branching-fraction analysis described in Sect. 3.3.2 addresses the $\gamma\gamma J/\psi$ final states that proceed through the doubly radiative decays $\psi(2S)\to\gamma\chi_{cJ}$, $\chi_{cJ}\to\gamma J/\psi$. The resulting product branching fractions, measured using CLEO's dataset of 27M $\psi(2S)$ decays, are shown in Table 21. These inputs are used to determine $\mathcal{B}(\chi_{cJ}\to\gamma J/\psi)$ branching fractions, which are calculated by the Particle Data Group [447] from world averages of the quantities in Table 21 and those for $\psi(2S)\to\gamma\chi_{cJ}$.

Table 21 Results from CLEO [445, 446] for the $\psi(2S)\to\gamma\gamma J/\psi$ branching fractions through either χ_{cJ} intermediate states or a nonresonant (nr) channel

Final state	$\mathcal{B}$ (%)
$\gamma(\gamma J/\psi)_{\chi_{c0}}$	$0.125\pm0.007\pm0.013$
$\gamma(\gamma J/\psi)_{\chi_{c1}}$	$3.56\pm0.03\pm0.12$
$\gamma(\gamma J/\psi)_{\chi_{c2}}$	$1.95\pm0.02\pm0.07$
$\gamma(\gamma J/\psi)_{\text{nr}}$	≤0.1

A substantial difference between the original [445] and final [446] CLEO analyses, aside from an eightfold increase in statistics, is the treatment of the transition through χ_{c0}, which has the smallest rate of the three. The primary systematic challenge for this mode is dealing with its small $\gamma J/\psi$ branching fraction relative to that of χ_{c1}. In the energy spectrum of the lower energy photon in such decays, there are peaks near 128, 172, and 258 MeV, corresponding to the intermediate χ_{c2}, χ_{c1}, and χ_{c0} states, respectively (see Fig. 4 of [446]). However, due to nonzero photon energy-measurement resolution and nonzero natural widths of the χ_{cJ}, the three peaks overlap one another. In particular, the high-side tail of the lower-energy photon's spectrum for the transition through χ_{c1} has a significant contribution in the χ_{c0} energy region (relative to the small χ_{c0} signal), a fact which introduces subtleties to the analysis. Because of the proximity of the photon lines to one another and the large disparity of rates, the measured product branching fraction for the χ_{c0} transition is sensitive to $\Gamma(\chi_{c1})$ and to the detailed lineshape in these decays, i.e., some of the apparent χ_{c0} signal is actually feedacross from χ_{c1}. The second CLEO analysis [446] implemented the E_γ^3-weighting in its MC simulation of the transition, as required from phase-space considerations for E1 decays given in (93). The inclusion of the E_γ^3 factor significantly increases the contribution of χ_{c1} near the χ_{c0} peak (relative to not using this factor). However, even if the MC simulation of these decays allows for a χ_{c1} natural width of up to 1 MeV (about 2.5σ higher than the nominal value [18]) and uses the expected E1-transition energy dependence of the lineshape, the region in the lower-energy photon spectrum in the valley between the χ_{c1} and χ_{c0} peaks showed an excess of events in the CLEO data over the number expected. One hypothesis for filling that deficit suggested by CLEO is the *nonresonant* (nr) decay $\psi(2S)\to\gamma\gamma J/\psi$. CLEO estimated that a branching fraction of $\approx$0.1% or smaller could be accommodated by the data. However, CLEO did *not* claim observation of this nonresonant decay mode due to a combination of limited statistics, uncertainties about χ_{cJ} widths,

Table 22 From CLEO [450], normalized magnetic dipole (M2) amplitudes from an analysis of $\psi(2S) \to \gamma\chi_{cJ}$, $\chi_{cJ} \to \gamma J/\psi$ decays. For the $J = 2$ values, the electric octupole (E3) moments were fixed to zero

Decay	Quantity	Value (10^{-2})
$\psi(2S) \to \gamma\chi_{c1}$	$b_2^{J=1}$	$2.76 \pm 0.73 \pm 0.23$
$\psi(2S) \to \gamma\chi_{c2}$	$b_2^{J=2}$	$1.0 \pm 1.3 \pm 0.3$
$\chi_{c1} \to \gamma J/\psi$	$a_2^{J=1}$	$-6.26 \pm 0.63 \pm 0.24$
$\chi_{c2} \to \gamma J/\psi$	$a_2^{J=2}$	$-9.3 \pm 1.6 \pm 0.3$

and possible dynamical distortion of the lineshape, all matters deserving further attention. In fact, such a distortion was later observed in radiative transitions from $\psi(2S)$ to $\eta_c(1S)$ (see Sect. 3.1.2), raising the importance of verifying the assumed photon lineshapes with data for any future analysis.

A preliminary analysis of nonresonant $\psi(2S) \to \gamma\gamma J/\psi$ decays from BESIII has been released [448, 449], which uses a dataset of 106 M $\psi(2S)$ decays, nearly four times that of CLEO. A statistically significant signal is claimed, and a branching fraction of $\approx 0.1\%$ is measured, compatible with the CLEO upper bound. BESIII also raised the issue of how interference between the resonant $\gamma\gamma J/\psi_{\chi_{cJ}}$ final states might affect the lower-energy photon spectrum, and what the effect might be from interference between the nonresonant and resonant channels.

3.1.6 Higher-order multipole amplitudes

The radiative decays $\psi(2S) \to \gamma\chi_{c1,2}$ and $\chi_{c1,2} \to \gamma J/\psi$ are dominated by electric dipole (E1) amplitudes. However, they are expected to have a small additional contribution from the higher-order magnetic quadrupole (M2) amplitudes. Previous measurements of the relative sizes of the M2 amplitudes have disagreed with theoretical expectations. CLEO [450] has recently revisited this issue with a high-statistics analysis of the decay chains $\psi(2S) \to \gamma\chi_{cJ}$; $\chi_{cJ} \to \gamma J/\psi$; $J/\psi \to l^+l^-(l = e, \mu)$ for $J = 1, 2$. Starting with 24×10^6 $\psi(2S)$ decays, CLEO observes approximately 40 000 events for $J = 1$ and approximately 20 000 events for $J = 2$, significantly larger event samples than previous measurements. Using an unbinned maximum-likelihood fit to angular distributions, CLEO finds the normalized M2 admixtures in Table 22. For the quoted $J = 2$ measurements, the electric octupole (E3) moments were fixed to zero. As shown in Fig. 39, these new measurements agree well with theoretical expectations when the anomalous magnetic moment of the charm quark is assumed to be zero and the mass of the charm quark is assumed to be 1.5 GeV.

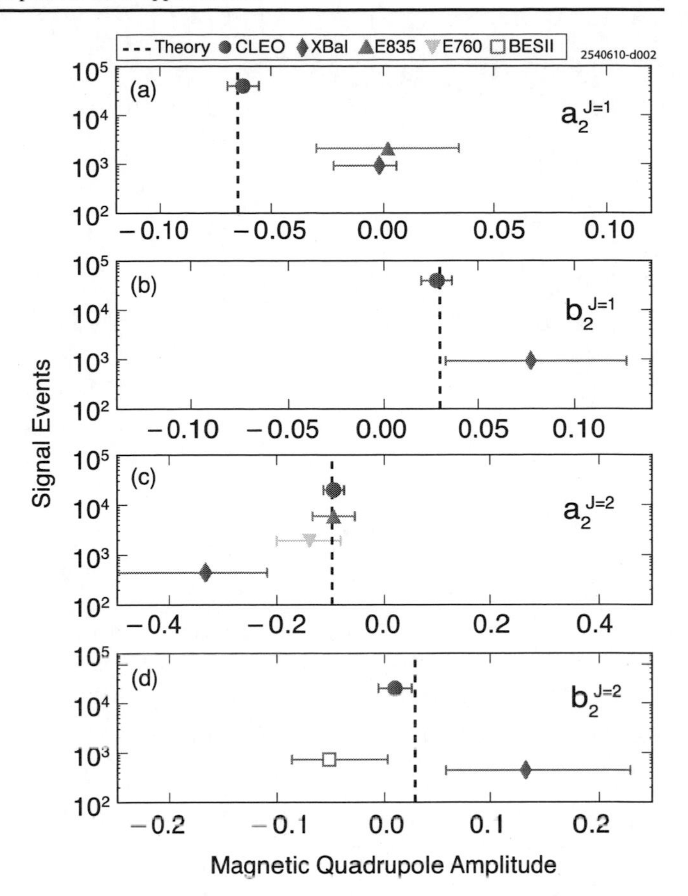

Fig. 39 A compilation of measurements of normalized magnetic dipole amplitudes from $\chi_{c1} \to \gamma J/\psi$ ($a_2^{J=1}$), $\psi(2S) \to \gamma\chi_{c1}$ ($b_2^{J=1}$), $\chi_{c2} \to \gamma J/\psi$ ($a_2^{J=2}$), and $\psi(2S) \to \gamma\chi_{c2}$ ($b_2^{J=2}$). The *solid circles* represent data from CLEO [450], which show consistency with predictions (*dashed vertical lines*), unlike some earlier measurements. The nonrelativistic theoretical expectations are calculated with an anomalous magnetic moment of the charm quark of zero and an assumed 1.5 GeV charm quark mass. Adapted from [450] with kind permission, copyright (2009) The American Physical Society

3.1.7 Observation of $\psi(3770) \to \gamma\chi_{cJ}(1P)$

The existence of the $\psi(3770)$ has long been established, and it has generally been assumed to be the 1^3D_1 charmonium state with a small admixture of 2^3S_1. However, because it predominantly decays to $D\bar{D}$, its behavior as a state of charmonium has gone relatively unexplored in comparison to its lighter partners. The charmonium nature of the $\psi(3770)$ is especially interesting given the unexpected discoveries of the X, Y, and Z states, opening up the possibility that the $\psi(3770)$ could include more exotic admixtures. The electromagnetic transitions, $\psi(3770) \to \gamma\chi_{cJ}$, because they are straightforward to calculate assuming the $\psi(3770)$ is the $^3D_1$ state of charmonium, provide a natural testing ground for the nature of the $\psi(3770)$ [31, 451, 452].

CLEO has observed these transitions in two independent analyses. In the first [453], the χ_{cJ} were reconstructed exclusively in the decay chain $\psi(3770) \to \gamma\chi_{cJ}$, $\chi_{cJ} \to$

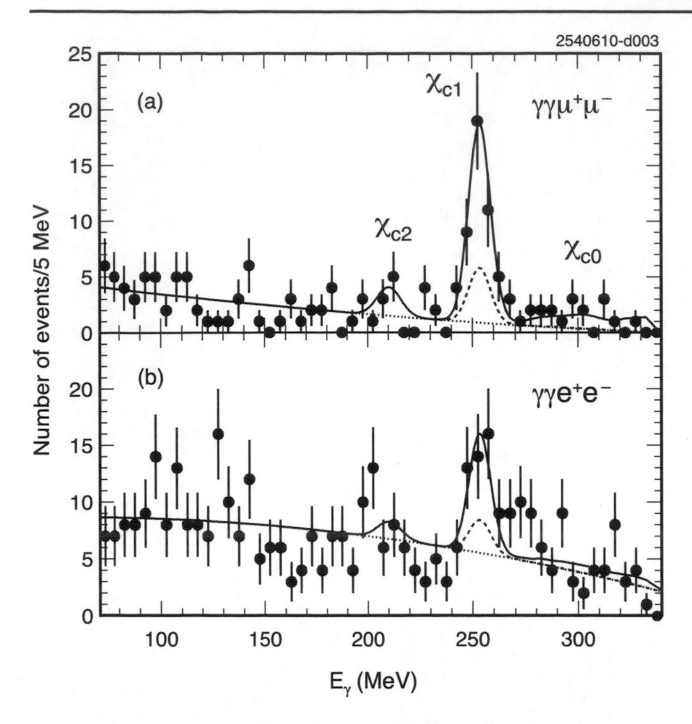

Fig. 40 From CLEO [453], the energy of the transition (lower energy) photon from $\psi(3770) \to \gamma\chi_{cJ}$ found when reconstructing $\chi_{cJ} \to \gamma J/\psi$ and requiring the J/ψ decay to (**a**) $\mu^+\mu^-$ or (**b**) e^+e^-. *Solid circles* represent data, the *dotted curve* shows the smooth fitted background, the *dashed curve* shows the sum of the smooth background fit and an estimated contribution from the tail of the $\psi(2S)$ (events individually indistinguishable from signal), and the *solid curve* is a result of a fit of the data to all background and signal components. Background saturates the data at the χ_{c2} and χ_{c0}, but a significant χ_{c1} signal is obtained. Adapted from [453] with kind permission, copyright (2006) The American Physical Society

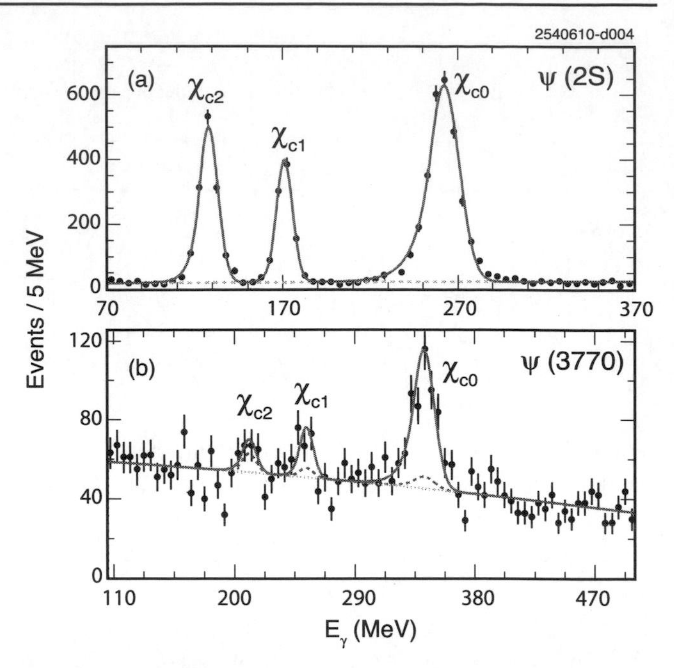

Fig. 41 From CLEO [454], the energy of the transition photon from (**a**) $\psi(2S)$, or (**b**) $\psi(3770)$ decaying to $\gamma\chi_{cJ}$ when the χ_{cJ} are reconstructed in exclusive hadronic modes. *Short-dashed curves* represent fits to the smooth background. The *long-dashed curve* in (**b**) represents estimated background from the tail of the $\psi(2S)$, for which events are individually indistinguishable from signal. A significant $\psi(3770) \to \gamma\chi_{c0}$ signal is obtained. Adapted from [454] with kind permission, copyright (2006) The American Physical Society

$\gamma J/\psi$, $J/\psi \to \ell^+\ell^-$, with results depicted in Fig. 40. In the second [454], the χ_{cJ} were reconstructed in several exclusive hadronic modes and then normalized to $\psi(2S) \to \gamma\chi_{cJ}$ using the same exclusive modes, with results shown in Fig. 41. Due to differing relative rates of the χ_{cJ} decay modes employed, the first method has more sensitivity to the transitions to $\chi_{c1,2}$ whereas the second is more suited to χ_{c0}. Combining the results of the two analyses, the partial widths of $\psi(3770) \to \gamma\chi_{cJ}$ were found to be 172 ± 30 keV for $J = 0$, 70 ± 17 keV for $J = 1$, and <21 keV at 90% CL for $J = 2$. These measurements are consistent with relativistic calculations assuming the $\psi(3770)$ is the $^3D_1$ state of charmonium.

3.1.8 Observation of $\Upsilon(2S, 3S) \to \gamma\eta_b(1S)$

The recent discovery of the $\eta_b(1S)$ state by BABAR [59] has, through a measurement of the $\eta_b(1S)$ mass, given us our first measurement of the $1S$ hyperfine splitting in the bottomonium system. This is obviously an important accomplishment for spectroscopy as it provides a window into the spin–spin interactions within the $b\bar{b}$ system. However, in addition to its contributions to spectroscopy, the observation of the decays $\Upsilon(2S, 3S) \to \gamma\eta_b(1S)$ has resulted in our first measurements of M1 radiative transition rates in the bottomonium system. A large range of theoretical predictions have been made for these rates [455], especially for the hindered transitions, to which the experimental measurement brings much-needed constraints.

Using 109 million $\Upsilon(3S)$ decays and 92 million $\Upsilon(2S)$ decays, BABAR [59, 61] has measured

$$\begin{aligned} \mathcal{B}\big(\Upsilon(3S) \to \gamma\eta_b(1S)\big) &= (4.8 \pm 0.5 \pm 0.6) \times 10^{-4}, \\ \mathcal{B}\big(\Upsilon(2S) \to \gamma\eta_b(1S)\big) &= \big(4.2^{+1.1}_{-1.0} \pm 0.9\big) \times 10^{-4}. \end{aligned} \tag{103}$$

Both measurements assume an $\eta_b(1S)$ width of 10 MeV. The large systematic errors in the branching fractions are due to the difficulty in isolating the small $\eta_b(1S)$ signal from other nearby photon lines ($\chi_{bJ}(2P, 1P) \to \gamma\Upsilon(1S)$ and $\Upsilon(3, 2S) \to \gamma\Upsilon(1S)$) and from the large background in the energy spectrum of inclusive photons.

In addition to the M1 transition rates, the energy dependence of the matrix elements is also of interest. In the case of charmonium (see below), this energy dependence can introduce a nontrivial distortion of the $\eta_c(1S)$ lineshape which can artificially pull the mass measurement several MeV from its true value. It is expected that the same distortion mechanism will hold in the bottomonium system. This effect must then be understood if M1 transitions are to be used for precision $\eta_b(1S)$ mass measurements.

Studying the $\eta_b(1S)$ lineshape in M1 transitions will require a large reduction in background levels. One possibility would be to study exclusive $\eta_b(1S)$ decays. Using exclusive $\eta_b(1S)$ decays could also allow a measurement of $\mathcal{B}(\Upsilon(1S) \to \gamma\eta_b(1S))$, the allowed M1 transition, since background levels in the $\Upsilon(1S)$ inclusive photon energy spectrum are likely prohibitively large.

3.2 Radiative and dileptonic decays

Here we review theoretical status and experimental results for radiative and dileptonic decays of heavy quarkonia. The simplest parton-level decay of any heavy-quarkonium vector state occurs through annihilation into a virtual photon and thence into dilepton or quark–antiquark pairs. The latter can be difficult to isolate from ggg decay at the charmonium and bottomonium mass scales, but fortunately is known to have a rate proportional to $R \equiv \sigma(e^+e^- \to \text{hadrons})/\sigma(e^+e^- \to \mu^+\mu^-)$, a quantity that is well measured with off-resonance data and which has a well understood energy dependence. Conversely, dilepton pairs have distinctive experimental signatures for which most modern detectors are optimized, offering the prospect of high precision. This high precision is quite useful in studies of both production and decays of vector charmonium and bottomonium states. Dileptonic widths also offer relative and absolute measures of wave function overlap at the origin. For all these reasons, decays to $\ell^+\ell^-$ are heavily studied and used to characterize the most basic features of each vector state.

The simplest three-body decays of vector quarkonia are to $\gamma\gamma\gamma$, γgg, and ggg, and their relative rates should reflect directly upon the value of α_s at the relevant mass scale: naively, $R_\gamma \equiv \mathcal{B}(\gamma gg)/\mathcal{B}(ggg) \simeq \alpha/\alpha_s(m)$. Although the $\gamma\gamma\gamma$ final state is experimentally straightforward to isolate, its rate is exceedingly small. Conversely, while γgg and ggg decays are abundant, distinguishing them from each other and from transitions involving final-state hadrons is quite challenging. Indeed, ggg decays cannot be effectively differentiated on an event-to-event basis from quarkonium annihilation into light-quark–antiquark pairs nor from γgg final states with soft photons. Experimental study of $\gamma\gamma\gamma$, γgg, and ggg quarkonium decays has progressed substantially, but has not yet entered the realm of precision.

The simplest decay or production mechanism of any scalar or tensor quarkonium state is to and from a pair of photons. Two-photon decay offers clean experimental signatures and a measure of the frequently small two-photon branching fraction, whereas production via two-photon fusion in e^+e^- collisions offers the prospect of determination of the diphotonic width. As with vector states and three photons, the two-photon coupling provides quite basic information about scalar and tensor quarkonia.

For completeness, we mention that radiative decays offer convenient production mechanisms for scalar and tensor states such as light-quark or hybrid mesons, glueball candidates, and non-Standard-Model Higgs or axion searches. Treatment of such decays is beyond the scope of this review, but some examples can be found in Sect. 3.2.4.

3.2.1 Theoretical status

Quarkonium annihilation happens at the heavy-quark mass scale m. Processes that happen at the scale m are best described by NRQCD. The NRQCD factorization formula for the quarkonium annihilation width into light hadrons or photons or lepton pairs reads [138, 456]

$$\Gamma_{H\text{-annih.}} = \sum_n \frac{2\,\mathrm{Im}\,c_n}{m^{(d_{O_n}-4)}} \langle H|O_n^{\text{4-fermion}}|H\rangle, \qquad (104)$$

where $O_n^{\text{4-fermion}}$ are four-fermion operators, c_n are their Wilson coefficients, d_{O_n} their dimensions, and $|H\rangle$ is the state that describes the quarkonium in NRQCD. The Wilson coefficients c_n are series in powers of the strong-coupling constant α_s, evaluated at the heavy-quark mass, and the matrix elements $\langle H|O_n^{\text{4-fermion}}|H\rangle$ are counted in powers of the heavy-quark velocity v. The matrix elements live at the scale mv: they are nonperturbative if $mv \gtrsim \Lambda_{\text{QCD}}$, while they may be evaluated in perturbation theory if $mv^2 \gtrsim \Lambda_{\text{QCD}}$.

Substantial progress has been made in the evaluation of the factorization formula at order v^7 [457, 458], in the lattice evaluation of the NRQCD matrix elements $\langle H|O_n^{\text{4-fermion}}|H\rangle$ [233], and in the data of many hadronic and electromagnetic decays (see [1] and subsequent sections). As discussed in [1], the data are clearly sensitive to NLO corrections in the Wilson coefficients c_n (and presumably also to relativistic corrections). For an updated list of ratios of P-wave charmonium decay widths, see Table 23.

Table 23 Comparison of measured χ_{cJ} decay-width ratios (using PDG10 [447]) with LO and NLO determinations [1], assuming $m_c = 1.5$ GeV and $\alpha_s(2m_c) = 0.245$, but without corrections of relative order v^2. LH ≡ light hadrons

Ratio	PDG	LO	NLO
$\frac{\Gamma(\chi_{c0}\to\gamma\gamma)}{\Gamma(\chi_{c2}\to\gamma\gamma)}$	4.5	3.75	5.43
$\frac{\Gamma(\chi_{c2}\to \text{LH}) - \Gamma(\chi_{c1}\to \text{LH})}{\Gamma(\chi_{c0}\to\gamma\gamma)}$	450	347	383
$\frac{\Gamma(\chi_{c0}\to \text{LH}) - \Gamma(\chi_{c1}\to \text{LH})}{\Gamma(\chi_{c0}\to\gamma\gamma)}$	4200	1300	2781
$\frac{\Gamma(\chi_{c0}\to \text{LH}) - \Gamma(\chi_{c2}\to \text{LH})}{\Gamma(\chi_{c2}\to \text{LH}) - \Gamma(\chi_{c1}\to \text{LH})}$	8.4	2.75	6.63
$\frac{\Gamma(\chi_{c0}\to \text{LH}) - \Gamma(\chi_{c1}\to \text{LH})}{\Gamma(\chi_{c2}\to \text{LH}) - \Gamma(\chi_{c1}\to \text{LH})}$	9.4	3.75	7.63

In [232], the high precision of data and matrix elements has been exploited to provide a new determination of α_s from

$$\frac{\Gamma(\Upsilon(1S)\to\gamma\,\mathrm{LH})}{\Gamma(\Upsilon(1S)\to\mathrm{LH})}: \tag{105}$$

$$\alpha_s(m_{\Upsilon(1S)}) = 0.184^{+0.015}_{-0.014}, \tag{106}$$

implying

$$\alpha_s(m_{Z^0}) = 0.119^{+0.006}_{-0.005}. \tag{107}$$

The NRQCD factorization formulas for electromagnetic and inclusive hadronic decay widths lose their predictive power as soon as we go to higher orders in v, due to the rapid increase in the number of nonperturbative matrix elements [1, 457, 458]. Quarkonia, with typical binding energies much smaller than Λ_{QCD}, conservatively include all quarkonia above the ground state. Matrix elements of these states are inherently nonperturbative and may be evaluated on the lattice [233]; few, however, are known. A way to reduce the number of these unknown matrix elements is to go to the lower-energy EFT, pNRQCD, and to exploit the hierarchy $mv \gg mv^2$. In pNRQCD, NRQCD matrix elements factorize into two parts: one, the quarkonium wave-function or its derivative at the origin, and the second, gluon-field correlators that are universal, i.e., independent of the quarkonium state. The pNRQCD factorization has been exploited for P-wave and S-wave decays in [181].

Quarkonium ground states have typical binding energy larger than or of the same order as Λ_{QCD}. Matrix elements of these states may be evaluated in perturbation theory with the nonperturbative contributions being small corrections encoded in local or nonlocal condensates. Many higher-order corrections to spectra, masses, and wave functions have been calculated in this manner [157], all of them relevant to the quarkonium ground state annihilation into light hadrons and its electromagnetic decays. For some recent reviews about applications, see [270, 459]. In particular, $\Upsilon(1S)$, $\eta_b(1S)$, J/ψ, and $\eta_c(1S)$ electromagnetic decay widths at NNLL have been evaluated [260, 460]. The ratios of electromagnetic decay widths were calculated for the ground state of charmonium and bottomonium at NNLL order [460], finding, e.g.,

$$\frac{\Gamma(\eta_b(1S)\to\gamma\gamma)}{\Gamma(\Upsilon(1S)\to e^+e^-)} = 0.502 \pm 0.068 \pm 0.014. \tag{108}$$

A partial NNLL-order analysis of the absolute widths of $\Upsilon(1S)\to e^+e^-$ and $\eta_b(1S)\to\gamma\gamma$ can be found in [260].

As the analysis of $\Gamma(\Upsilon(1S)\to e^+e^-)$ of [260] illustrates, for this fundamental quantity there may be problems of convergence of the perturbative series. Problems of convergence are common and severe for all the annihilation observables of ground state quarkonia and may be traced back to large logarithmic contributions, to be resummed by solving suitable renormalization group equations, and to large $\beta_0\alpha_s$ contributions of either resummable or nonresummable nature (these last ones are known as renormalons). Some large $\beta_0\alpha_s$ contributions were successfully treated [461] to provide a more reliable estimate for

$$\frac{\Gamma(\eta_c(1S)\to\mathrm{LH})}{\Gamma(\eta_c(1S)\to\gamma\gamma)} = (3.26 \pm 0.6) \times 10^3, \tag{109}$$

or $(3.01 \pm 0.5) \times 10^3$ in a different resummation scheme. A similar analysis could be performed for the $\eta_b(1S)$, which combined with a determination of $\Gamma(\eta_b(1S)\to\gamma\gamma)$ would then provide a theoretical determination of the $\eta_b(1S)$ width. At the moment, without any resummation or renormalon subtraction performed,

$$\frac{\Gamma(\eta_b(1S)\to\mathrm{LH})}{\Gamma(\eta_b(1S)\to\gamma\gamma)} \simeq (1.8\text{–}2.3) \times 10^4. \tag{110}$$

Recently a new resummation scheme has been suggested for electromagnetic decay ratios of heavy quarkonium and applied to determine the $\eta_b(1S)$ decay width into two photons [462]:

$$\Gamma\big(\eta_b(1S)\to\gamma\gamma\big) = 0.54 \pm 0.15\ \mathrm{keV}. \tag{111}$$

Substituting (111) into (110) gives $\Gamma(\eta_b(1S)\to\mathrm{LH}) = 7$–16 MeV.

3.2.2 *Measurement of $\psi, \Upsilon \to \gamma gg$*

In measurements of the γgg rate from J/ψ [235], $\psi(2S)$ [236], and $\Upsilon(1S, 2S, 3S)$ [230], CLEO finds that the most effective experimental strategy to search for γgg events is to focus solely upon those with energetic photons (which are less prone to many backgrounds), then to make the inevitable large subtractions of ggg, $q\bar{q}$, and transition backgrounds on a statistical basis, and finally to extrapolate the radiative photon energy spectrum to zero with the guidance of both theory and the measured high-energy spectrum. The most troublesome background remaining is from events with energetic $\pi^0 \to \gamma\gamma$ decays which result in a high-energy photon in the final state. One of several methods used to estimate this background uses the measured *charged* pion spectra and the assumption of isospin invariance to simulate the resulting photon spectrum with Monte Carlo techniques; another measures the exponential shape of the photon-from-π^0 distribution at low photon energy, where γgg decays are few, and extrapolates to the full energy range. Backgrounds to γgg from transitions require the input of the relevant branching fractions and their uncertainties. The rate for ggg decays is then estimated as that fraction of decays that remains after all dileptonic, transition, and $q\bar{q}$ branching fractions are subtracted, again requiring

Table 24 Measured values of R_γ, as defined in (22), assuming the Garcia–Soto (GS) [231] or Field [463] models of the direct photon spectrum from Υ decays, and their averages, from CLEO [230]. The uncertainty on each average includes a component to account for model dependence (which dominates the other uncertainties); statistical errors are negligible

State	R_γ (%)		
	GS	Field	Average
$\Upsilon(1S)$	2.46 ± 0.13	2.90 ± 0.13	2.70 ± 0.27
$\Upsilon(2S)$	3.06 ± 0.22	3.57 ± 0.22	3.18 ± 0.47
$\Upsilon(3S)$	2.58 ± 0.32	3.04 ± 0.32	2.72 ± 0.49

Table 25 Measured values of R_γ from CLEO, as defined in (22), and the respective absolute branching fractions from PDG10 [447]

State	R_γ (%)	$\mathcal{B}(\gamma gg)$ (%)	$\mathcal{B}(ggg)$ (%)	Source
J/ψ	13.7 ± 1.7	8.8 ± 0.5	64.1 ± 1.0	[235, 447]
$\psi(2S)$	9.7 ± 3.1	1.02 ± 0.29	10.6 ± 1.6	[236, 447]
$\Upsilon(1S)$	2.70 ± 0.27	2.21 ± 0.22	81.7 ± 0.7	[230, 447]
$\Upsilon(2S)$	3.18 ± 0.47	1.87 ± 0.28	58.8 ± 1.2	[230, 447]
$\Upsilon(3S)$	2.72 ± 0.49	0.97 ± 0.18	35.7 ± 2.6	[230, 447]

input of many external measurements and their respective uncertainties. Not surprisingly, the relative errors on the results of 10–30% are dominated by the systematic uncertainties incurred from background-subtraction methods, photon-spectrum model dependence, and external branching fractions.

The CLEO measurements of the observable R_γ defined in (22) are shown in Tables 24 and 25. It should be noted that the uncertainties in R_γ for $V = \Upsilon(1S, 2S, 3S)$ are partially correlated with one another because of shared model dependence and analysis systematics. In addition, the *shape* of the measured direct photon spectrum from J/ψ is quite similar to that of the Υ's, in contrast to the $\psi(2S)$, for which the spectrum appears to be softer. Absolute values for the branching fractions have been calculated with input of other world-average branching fractions in PDG10 [447], and are reproduced in Table 25. See Sect. 2.8.1 for discussion of extraction of α_s from R_γ measurements.

3.2.3 Observation of $J/\psi \to \gamma\gamma\gamma$

Orthopositronium, the $^3S_1$ e^+e^- bound state, decays to 3γ almost exclusively and has long been used for precision tests of QED [464]. The rate of its analog for QCD, three-photon decay of vector charmonium, in particular that of the J/ψ, acts as a probe of the strong interaction [355]. Due to similarities at the parton level, relative rate measurements of the branching fractions for $J/\psi \to 3\gamma$, $J/\psi \to \gamma gg$, $J/\psi \to 3g$, and $J/\psi \to l^+l^-$ provide crucial grounding

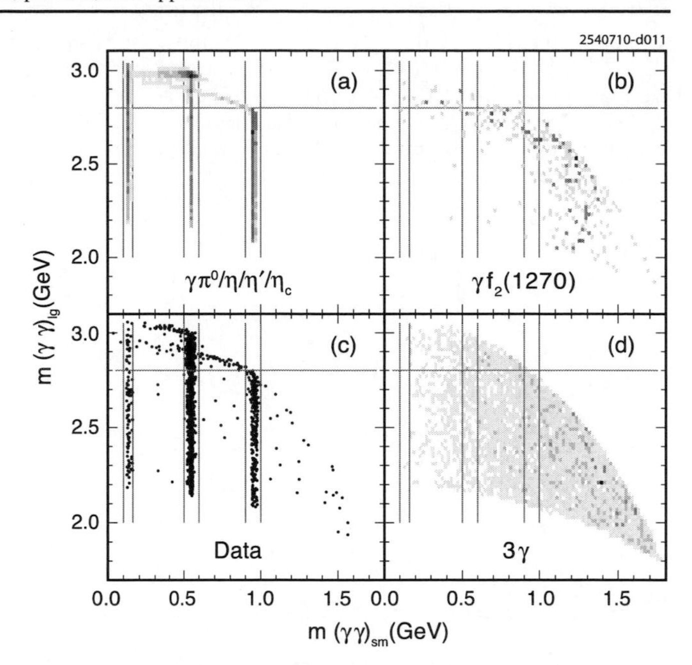

Fig. 42 From CLEO [469], distributions of the largest vs. the smallest diphoton mass combination per event in $\psi(2S) \to \pi^+\pi^- J/\psi$, $J/\psi \to 3\gamma$ candidates. Parts (**a**), (**c**), and (**d**) are from MC simulations of $J/\psi \to \gamma\eta/\eta'/\eta_c(1S) \to 3\gamma$, $\gamma f_2(1270) \to \gamma\pi^0\pi^0 \to 5\gamma$, and the signal process, $J/\psi \to 3\gamma$, respectively. In these three MC plots *darker shading* of bins signifies relatively larger event density than *lighter shades*. Part (**b**) shows the CLEO data, where each *solid circle* indicates a single event. *Solid lines* demarcate regions excluded from the $J/\psi \to 3\gamma$ event selection. 38 data events populate the signal region, of which $24.2^{+7.2}_{-6.0}$ were estimated to be from $J/\psi \to 3\gamma$, with the remainder due to background, mostly various sources of $J/\psi \to \gamma\pi^0\pi^0$. Adapted from [469] with kind permission, copyright (2008) The American Physical Society

for QCD predictions [355, 465, 466]. Previous searches for the quarkonium decay to 3γ have yielded only upper limits: $\mathcal{B}(\omega \to 3\gamma) < 1.9 \times 10^{-4}$ at 95% CL [467] and $\mathcal{B}(J/\psi \to 3\gamma) < 5.5 \times 10^{-5}$ at 90% CL [468].

CLEO [469] performed a search for the all-photon decays of the J/ψ using tagged J/ψ mesons from 9.6×10^6 $\psi(2S) \to J/\psi\pi^+\pi^-$ decays. After excluding backgrounds from $J/\psi \to \gamma\eta/\eta'/\eta_c(1S) \to 3\gamma$ with restrictions on photon-pair masses, and minimizing $J/\psi \to \gamma\pi^0\pi^0$ backgrounds in which two of the photons are very soft by imposing stringent energy-momentum conservation via a kinematic fit, CLEO reported the first observation of the decay $J/\psi \to 3\gamma$ by finding 38 events, 12.8 of which were estimated to be background, mostly from various sources of $J/\psi \to \gamma\pi^0\pi^0 \to 5\gamma$ decays. The branching fraction is measured to be

$$\mathcal{B}(J/\psi \to 3\gamma) = (1.2 \pm 0.3 \pm 0.2) \times 10^{-5} \tag{112}$$

with statistical significance for the signal of 6.3σ. Diphoton mass plots are shown in Fig. 42 for data, signal MC, and two possible sources of background. The measured three-photon branching fraction lies between the zeroth-order pre-

Table 26 Branching fractions (in units of 10^{-4}) for charmonium decays to $\gamma(\pi^0/\eta/\eta')$ from CLEO [471] and PDG08 [18], the latter of which is dominated by BESII [470]. The rightmost column shows the difference between the two in units of standard error (σ). Upper limits are quoted at 90% CL. Entries in the last two rows include the effects of estimated continuum background and ignore (include) maximal destructive interference between $\psi(3770)$ and continuum sources

Mode	CLEO	PDG08	#σ
$J/\psi \to \gamma\pi^0$	$0.363 \pm 0.036 \pm 0.013$	$0.33^{+0.06}_{-0.04}$	0.4
$\to \gamma\eta$	$11.01 \pm 0.29 \pm 0.22$	9.8 ± 1.0	1.2
$\to \gamma\eta'$	$52.4 \pm 1.2 \pm 1.1$	47.1 ± 2.7	1.7
$\psi(2S) \to \gamma\pi^0$	<0.07	<54	–
$\to \gamma\eta$	<0.02	<0.9	–
$\to \gamma\eta'$	$1.19 \pm 0.08 \pm 0.03$	1.36 ± 0.24	−0.7
$\psi(3770) \to \gamma\pi^0$	<3	–	–
$\to \gamma\eta$	<0.2 (1.5)	–	–
$\to \gamma\eta'$	<0.2 (1.8)	–	–

dictions [465] for $\mathcal{B}_{3\gamma}/\mathcal{B}_{\gamma gg} \approx (\alpha/\alpha_s)^2/3$ and $\mathcal{B}_{3\gamma}/\mathcal{B}_{3g} \approx (\alpha/\alpha_s)^3$ and is consistent with both, but is a factor of $\approx$2.5 below that for $\mathcal{B}_{3\gamma}/\mathcal{B}_{ll} \approx \alpha/14$, all assuming $\alpha_s(m_{J/\psi}) \approx 0.3$. Upper limits on the branching fractions for $J/\psi \to \gamma\gamma$, 4γ, and 5γ were also determined, setting more stringent restrictions than previous experiments.

3.2.4 Nonobservation of $\psi(2S), \Upsilon(1S) \to \gamma\eta$

Both BESII [470] and CLEO [471] report studies on the exclusive final states $\gamma(\pi^0/\eta/\eta')$ from charmonium, BESII from J/ψ decays alone, and CLEO from decays of J/ψ, $\psi(2S)$, and $\psi(3770)$. Resulting branching fractions appear in Table 26, where it can be seen that the results from the two experiments are consistent with each other, and that precision has steadily improved. From the perspective of charmonium physics, the most striking feature of these numbers is the nonobservation of $\psi(2S) \to \gamma\eta$.

CLEO and BESII have consistent values of the ratio

$$r_n \equiv \frac{\mathcal{B}(\psi(nS) \to \gamma\eta)}{\mathcal{B}(\psi(nS) \to \gamma\eta')} \tag{113}$$

for J/ψ of $r_1 \approx 0.2$ within a few percent; the naive expectation would be that, for $\psi(2S)$, $r_2 \approx r_1$. Yet the CLEO result implies that $r_2 < 1.8\%$, or $r_2/r_1 < 8\%$, both at 90% CL. Specifically, the rate of $\gamma\eta$ relative to $\gamma\eta'$ from $\psi(2S)$ is *at least an order of magnitude smaller* than that from J/ψ. If instead we characterize the effect in terms of "the 12% rule" (Sect. 3.4.1), we note that, relative to its dileptonic width, $\psi(2S) \to \gamma\eta'$ is suppressed by a factor of five with respect to $J/\psi \to \gamma\eta'$, but $\psi(2S) \to \gamma\eta$ is suppressed by at least *two* orders of magnitude with respect to $J/\psi \to \gamma\eta$.

Do we see such transitions at expected rates in Υ decays? The CLEO [472] search for $\Upsilon(1S) \to \gamma\eta^{(\prime)}$ failed to find evidence for either pseudoscalar meson in radiative decays, setting the limits

$$\begin{aligned} &\mathcal{B}(\Upsilon(1S) \to \gamma\eta) < 1.0 \times 10^{-6}, \\ &\mathcal{B}(\Upsilon(1S) \to \gamma\eta') < 1.9 \times 10^{-6}. \end{aligned} \tag{114}$$

These values rule out the predictions of Chao [473], which are based on mixing of η, η', and $\eta_b(1S)$, but not sensitive enough to probe those of Ma [474], which uses a QCD factorization approach, nor those of Li [475], which posits a substantial two-gluon component in η and η' within a perturbative QCD framework.

What dynamical effect is present in $\psi(2S)$ decays that is absent in J/ψ decays that can explain this large of a suppression? Is it related to $\rho\pi$ suppression in $\psi(2S)$ decays relative to J/ψ (Sect. 3.4.1), another vector-pseudoscalar final state? Is there a connection between the suppression of $\gamma\eta$ and $\gamma\eta'$ in $\Upsilon(1S)$ decay and that of $\gamma\eta$ in $\psi(2S)$ decay? These questions remain unanswered.

Events selected from this charmonium analysis have been used to address physics questions other than those directly associated with $c\bar{c}$ bound states:

- The flavor content of η and η' mesons, which are commonly thought to be mixtures of the pure SU(3)-flavor octet and singlet states, with a possible admixture of gluonium [476–478]; if $J/\psi \to \gamma\eta^{(\prime)}$ occurs through $c\bar{c} \to \gamma gg$, which is expected to fragment in a flavor-blind manner, the mixing angle can be extracted from r_1 as defined in (113). The measured value [470] from charmonium is consistent with that obtained from other sources [479–481].
- The high-statistics sample of $J/\psi \to \gamma\eta'$ decays was also used by CLEO to perform the first simultaneous measurement of the largest five η' branching fractions [471], attaining improved precisions, and to improve the measurement precision of the η' mass [482].
- Although CLEO [471] only set limits for $\psi(3770)$ decays to these final states, clean signals for both $\gamma\eta$ and $\gamma\eta'$ final states were observed. However, these rates were seen to be consistent with that expected from continuum production as extrapolated from the bottomonium energy region using the only other measurement of $e^+e^- \to \gamma\eta^{(\prime)}$, by BABAR [483].
- Rosner [484] explores the implications of these measurements on production mechanisms for γP (P = pseudoscalar) final states, in particular the contribution of the vector dominance model (through $\rho\pi$) to $J/\psi \to \gamma\pi^0$.

3.2.5 Two-photon widths of charmonia

Considerable progress has been made with measurements of the two-photon width of the $\eta_c(1S)$ and $\chi_c(1P)$ states.

Two different approaches have been used. The first one uses the formation of a charmonium state in two-photon collisions followed by the observation of its decay products. In this case the directly measured quantity is the product of the charmonium two-photon width and the branching ratio of its decay to a specific final state. A two-photon width can then be computed from the product if the corresponding branching fraction has been measured. Such measurements were performed at Belle for the following decays: $\chi_{c0(2)} \to \pi^+\pi^-$, K^+K^- [485], $\eta_c(1S) \to p\bar{p}$ [486], $\chi_{c0(2)} \to K_S^0 K_S^0$ [487], $\chi_{c0(2)} \to \pi^0\pi^0$ [488, 489] and at CLEO for the $\chi_{c2} \to J/\psi\gamma$ decay [490]. Belle also applied this method to study the two-photon formation of the $\eta_c(1S)$, $\chi_{c0}(1P)$ and $\chi_{c2}(1P)$ via various final states with four charged particles ($2\pi^+2\pi^-, \pi^+\pi^-K^+K^-, 2K^+2K^-$) and quasi-two-body final states ($\rho\rho, \phi\phi, \eta\eta, \ldots$) [442, 491]. In the latter study [442], Belle also sought the two-photon production of the $\eta_c(2S)$ and did not find a significant signal in any four-body final state.

The second method is based on the charmonium decay into two photons. CLEO performed [492] such measurements using the reactions $\psi(2S) \to \gamma_1\chi_{cJ}$, $\chi_{cJ} \to \gamma_2\gamma_3$, where γ_1 is the least-energetic final-state photon in the $\psi(2S)$ center of mass frame. Clear signals were observed for the χ_{c0} and χ_{c2}, as shown in Fig. 43. (Two-photon decay of spin-one states is forbidden by the Landau–Yang theorem [493, 494].) Using the measured signal yield and the previously determined number of $\psi(2S)$ produced (24.6 million), the product of the branching fractions is determined as:

$$\begin{aligned} &\mathcal{B}(\psi(2S) \to \gamma\chi_{c0}) \times \mathcal{B}(\chi_{c0} \to \gamma\gamma) \\ &\quad = (2.17 \pm 0.32 \pm 0.10) \times 10^{-5}, \\ &\mathcal{B}(\psi(2S) \to \gamma\chi_{c2}) \times \mathcal{B}(\chi_{c2} \to \gamma\gamma) \\ &\quad = (2.68 \pm 0.28 \pm 0.15) \times 10^{-5}. \end{aligned} \tag{115}$$

World-average values for the $\psi(2S) \to \gamma\chi_{cJ}$ branching fractions and total widths of the $\chi_{c0(2)}$ states can then used to calculate the two-photon widths for the corresponding charmonia. In Table 27 we list these two-photon widths, which result from a constrained fit to all relevant experimental information from PDG08 [18]. Using those values, the experimental ratio becomes $\Gamma_{\gamma\gamma}(\chi_{c2})/\Gamma_{\gamma\gamma}(\chi_{c0}) = 0.22 \pm 0.03$. The LO and a NLO determination are shown in the first row of Table 23; some relevant theoretical issues are discussed in Sect. 3.2.1.

As part of the $J/\psi \to 3\gamma$ analysis described in Sect. 3.2.3, CLEO [469] failed to find evidence for the decay $\eta_c(1S) \to \gamma\gamma$. The product branching fraction for the decay chain $J/\psi \to \gamma\eta_c(1S)$, $\eta_c(1S) \to \gamma\gamma$ was measured to be $(1.2^{+2.7}_{-1.1} \pm 0.3) \times 10^{-6}$ ($< 6 \times 10^{-6}$ at 90% CL), which, using the CLEO [69] measurement of $\mathcal{B}(J/\psi \to \gamma\eta_c(1S)) =$ $(1.98 \pm 0.31)\%$ (discussed in Sect. 3.1.2), can be rewritten as

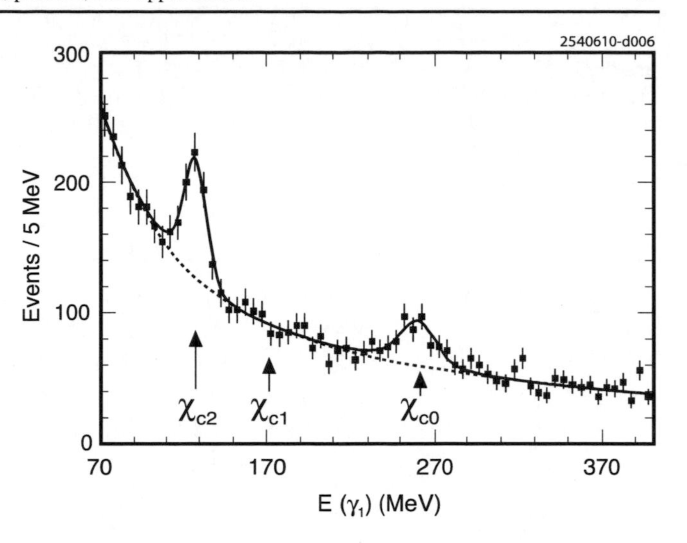

Fig. 43 From CLEO [492], the fitted spectrum for $E(\gamma_1)$ from the reaction $\psi(2S) \to \gamma_1\chi_{cJ}$, $\chi_{cJ} \to \gamma_2\gamma_3$, where γ_1 is the least energetic photon in the $\psi(2S)$ center-of-mass frame. The expected positions of $E(\gamma_1)$ from χ_{c0}, χ_{c1}, χ_{c2} are marked with *arrows*. The *dashed curve* represents a fit of the nonpeaking background to a polynomial. Adapted from [492] with kind permission, copyright (2008) The American Physical Society

Table 27 World-average values of the two-photon width for various charmonium states from PDG08 [18]

State	$\chi_{c0}(1P)$	$\chi_{c2}(1P)$	$\eta_c(1S)$
$\Gamma_{\gamma\gamma}$ (keV)	2.36 ± 0.25	0.515 ± 0.043	$7.2 \pm 0.7 \pm 2.0$

$$\begin{aligned} \mathcal{B}\big(\eta_c(1S) \to \gamma\gamma\big) &= \big(6.1^{+13.7}_{-5.6}\big) \times 10^{-5} \quad \text{(CLEO)} \\ &< 2.4 \times 10^{-4} \quad \text{at 90\% CL.} \end{aligned} \tag{116}$$

This central value is smaller than but consistent with the only other direct measurement of $\eta_c(1S) \to \gamma\gamma$ measured in $B^\pm \to K^\pm\eta_c(1S)$ decays by Belle [496], which obtains

$$\mathcal{B}\big(\eta_c(1S) \to \gamma\gamma\big) = \big(2.4^{+1.2}_{-0.9}\big) \times 10^{-4} \quad \text{(Belle)}, \tag{117}$$

for which the Belle product branching fraction was unfolded by PDG08 [18].

3.2.6 Dileptonic widths in the ψ family

BABAR [497], CLEO [498–500], and KEDR [501] have all performed measurements of dileptonic decays from the narrow members of the ψ resonance family with much-improved precisions. BABAR [497] pioneered the use of vector charmonia produced through initial-state radiation from e^+e^- collisions collected for other purposes at energies higher than the state being studied, in particular for the

Table 28 Measurements of $\Gamma(J/\psi \to e^+e^-) \times \mathcal{B}(J/\psi \to \ell^+\ell^-)$ and their relative accuracies δ; the weighted average and its (χ^2/d.o.f.) are shown assuming lepton universality. The two values from KEDR have some systematic uncertainties in common and are shown separately below the average

Experiment	$\ell^+\ell^-$	$\Gamma(J/\psi \to e^+e^-)\times$ $\mathcal{B}(J/\psi \to \ell^+\ell^-)$ (keV)	δ (%)
BABAR [497]	$\mu^+\mu^-$	$0.3301 \pm 0.0077 \pm 0.0073$	3.2
CLEO [498]	$\mu^+\mu^-$	$0.3384 \pm 0.0058 \pm 0.0071$	2.7
KEDR [501]	$\ell^+\ell^-$	$0.3321 \pm 0.0041 \pm 0.0050$	1.9
Avg (0.43/2)	$\ell^+\ell^-$	0.3334 ± 0.0047	1.4
KEDR [501]	$\mu^+\mu^-$	$0.3318 \pm 0.0052 \pm 0.0063$	2.5
KEDR [501]	e^+e^-	$0.3323 \pm 0.0064 \pm 0.0048$	2.4

J/ψ from $\sqrt{s} \approx 10.58$ GeV. CLEO followed *BABAR*'s lead, studying J/ψ [498] and $\psi(2S)$ [500] mesons from $\sqrt{s} = 3770$ MeV. KEDR [501] followed the more straightforward route, which requires a dedicated scan of the resonance. In both kinds of analyses the directly measured quantity is $\Gamma(J/\psi \to e^+e^-) \times \mathcal{B}(J/\psi \to \ell^+\ell^-)$. The dileptonic branching fraction can then be divided out from the result once to obtain Γ_{ee} and twice for Γ_{tot}. CLEO and *BABAR* used only $J/\psi \to \mu^+\mu^-$ decays because the e^+e^- final state has larger backgrounds from radiative Bhabha events, whereas KEDR used dimuons and dielectrons, obtaining the best precision to date. In all three cases $e^+e^- \to \ell^+\ell^-$ and its interference with the resonant signal must be addressed. The results are listed in Table 28, where it can be seen that the four measurements are consistent with one another and have combined precision of 1.4%.

CLEO [499] also provided an independent measurement of J/ψ dileptonic branching fractions using $\psi(2S) \to J/\psi\pi^+\pi^-$, $J/\psi \to \ell^+\ell^-$ decay chains, and normalizing to all J/ψ decays produced via the $\pi^+\pi^-$ transition by fitting the dipion recoil mass distribution. CLEO obtained

$$\begin{aligned}
&\mathcal{B}\big(J/\psi \to \mu^+\mu^-\big) = (5.960 \pm 0.065 \pm 0.050)\%, \\
&\mathcal{B}\big(J/\psi \to e^+e^-\big) = (5.945 \pm 0.067 \pm 0.042)\%, \\
&\frac{\mathcal{B}(J/\psi \to e^+e^-)}{\mathcal{B}(J/\psi \to \mu^+\mu^-)} = (99.7 \pm 1.2 \pm 0.6)\%, \\
&\mathcal{B}\big(J/\psi \to \ell^+\ell^-\big) = (5.953 \pm 0.070)\%,
\end{aligned} \tag{118}$$

consistent with and having uncertainties at least factor of two smaller than previous determinations. With this measurement, the assumption of lepton universality, and the weighted average from Table 28, we obtain $\Gamma_{ee}(J/\psi) = 5.60 \pm 0.10$ keV (1.8%) and $\Gamma_{\text{tot}}(J/\psi) = 94.1 \pm 2.6$ keV (2.7%), compared to relative uncertainties on these quantities before any of these measurements (PDG04 [17]) of 3.1% and 3.5%, respectively.

For $\psi(2S)$, CLEO [500] used the decays $\psi(2S) \to X_i J/\psi$, $J/\psi \to \ell^+\ell^-$, where $X_i = \pi^+\pi^-$, $\pi^0\pi^0$, and η,

Table 29 Measurements of $\Gamma_{ee}(\psi(2S))$ and their relative accuracies δ. The CLEO value quoted has been updated with branching fractions from [446] and the weighted-average with (χ^2/d.o.f.) of the top three measurements is given

Experiment	$\Gamma(\psi(2S) \to e^+e^-)$ (keV)	δ (%)
CLEO [446, 500]	2.407 ± 0.083	3.4
BES [502]	$2.330 \pm 0.036 \pm 0.110$	5.0
BES [503]	$2.388 \pm 0.037 \pm 0.096$	4.3
Avg (0.29/2)	2.383 ± 0.056	2.3
PDG04 [17]	2.12 ± 0.12	5.7

directly measuring the products $\mathcal{B}(\psi(2S) \to X_i J/\psi) \times \Gamma_{ee}[\psi(2S)]$. CLEO initially used $\mathcal{B}(\psi(2S) \to X_i J/\psi)$ values from [445] to extract $\Gamma_{ee}[\psi(2S)] = 2.54 \pm 0.11$ keV (4.3%), but if instead we use CLEO's updated values [446], described below in Sect. 3.3.2, we obtain $\Gamma_{ee}[\psi(2S)] = 2.407 \pm 0.083$ keV (3.4%). We then obtain the ratio $\Gamma_{ee}[\psi(2S)]/\Gamma_{ee}(J/\psi) = 0.43 \pm 0.02$, a quantity which might be more precisely predicted in lattice QCD than either Γ_{ee} alone [211], in which we have used our updated $\Gamma_{ee}[\psi(2S)]$ and our world-average $\Gamma_{ee}[J/\psi]$ from above.

BES [502, 503] studied the energy range $\sqrt{s} = 3.660$–3.872 GeV to determine the resonance parameters of $\psi(2S)$ and $\psi(3770)$. From the fit of the cross sections for $D^0\bar{D}^0$, D^+D^- and non-$D\bar{D}$ production the branching fractions and partial widths for $\psi(2S) \to e^+e^-$ and $\psi(3770) \to e^+e^-$ decays were determined. The results of these measurements of $\Gamma_{ee}[\psi(2S)]$ are shown in Table 29, where the improvement since 2004 can be observed.

Decays of the $\psi(2S)$ into τ-lepton pairs are less probable and therefore less studied. BES [504] used a sample of 14 million produced $\psi(2S)$ to measure the corresponding branching fraction. The result, $\mathcal{B}(\psi(2S) \to \tau^+\tau^-) = (3.08 \pm 0.21 \pm 0.38) \times 10^{-3}$, has better relative precision (14%) than previous measurements [505, 506]. Using part of their statistics KEDR measured $\Gamma(\psi(2S) \to e^+e^-) \times \mathcal{B}(\psi(2S) \to \tau^+\tau^-)$ to be 9.0 ± 2.6 eV [507], which, using the average from Table 29, implies a value $\mathcal{B}(\psi(2S) \to \tau^+\tau^-) = (3.8 \pm 1.1) \times 10^{-3}$, consistent with but considerably less precise than the result from BES.

CLEO [508] measured the hadronic cross section at a single energy point near the peak of the $\psi(3770)$, $\sqrt{s} = 3773$ MeV, taking interference between the final states of resonance decays and nonresonant e^+e^- annihilation into account. From the observed cross section, which is significantly smaller than some of the previous measurements [8, 11], $\Gamma_{ee}[\psi(3770)]$ is also obtained. In a scan over the $\psi(3770)$ energy region, 68 energy points in the range 3.650–3.872 GeV, BES [509] measured $R \equiv \sigma(e^+e^- \to \text{hadrons})/\sigma(e^+e^- \to \mu^+\mu^-)$, determining the parameters of the $\psi(3770)$ resonance, including the leptonic width. The results of the described measurements of $\Gamma_{ee}[\psi(3770)]$ are

Table 30 Measurements of $\Gamma_{ee}(\psi(3770))$ and their relative accuracies δ. The BES and CLEO results listed do not include the potential effect of interference with higher-mass ψ-states

Experiment	$\Gamma(\psi(3770) \to e^+e^-)$ (keV)	δ (%)
PDG04 [17]	0.26 ± 0.04	15.4
CLEO [508]	$0.203 \pm 0.003^{+0.041}_{-0.027}$	$^{+20.0}_{-13.3}$
BES [502]	$0.251 \pm 0.026 \pm 0.011$	11.2
BES [509]	$0.277 \pm 0.011 \pm 0.013$	6.1
PDG10 [447]	0.259 ± 0.016	6.2

Table 31 Γ_{ee} of the higher ψ states from PDG04 [17] and the BES [15] global fit

Resonance	Γ_{ee} (keV) from PDG04	Γ_{ee} (keV) from BES
$\psi(3770)$	0.26 ± 0.04	0.22 ± 0.05
$\psi(4040)$	0.75 ± 0.15	0.83 ± 0.20
$\psi(4160)$	0.77 ± 0.23	0.48 ± 0.22
$\psi(4415)$	0.47 ± 0.10	0.35 ± 0.12

shown in Table 30, where it can be seen that world-average uncertainty improved by more than a factor of two between 2004 and 2010.

Finally, BES [13] performed a global fit of R in the energy range 3.7–5.0 GeV, covering the four resonances, $\psi(3770)$, $\psi(4040)$, $\psi(4160)$, and $\psi(4415)$ [15]. Interference between the four ψ states was accounted for (which was not the case for the $\Gamma_{ee}[\psi(3770)]$ measurements in Table 30) and an energy-dependent width based on all accessible two-body decay channels was used. The results are shown in Table 31. It can be seen that the new results have larger uncertainties than previous ones, which ignored interference with higher-mass ψ states.

3.2.7 Dileptonic widths in the Υ family

CLEO [510–512] has made a systematic study of dileptonic decays of the narrow states in the Υ family. To determine dimuonic branching fractions, the quantities $\tilde{\mathcal{B}}_{\mu\mu} \equiv \Gamma_{\mu\mu}/\Gamma_{\text{had}}$ are measured for each $\Upsilon(nS)$, where $\Gamma_{\mu\mu}$ (Γ_{had}) is the rate for Υ decay to $\mu^+\mu^-$ (hadrons). Assuming lepton universality,

$$\mathcal{B}_{\mu\mu} = \frac{\Gamma_{\mu\mu}}{\Gamma_{\text{tot}}} = \frac{\tilde{\mathcal{B}}_{\mu\mu}}{1 + 3\tilde{\mathcal{B}}_{\mu\mu}}. \tag{119}$$

The results of this analysis, based on much larger data samples than available to previous experiments, are summarized in Table 32. While the result for the $\Upsilon(1S)$ is in good agreement with the world average, the CLEO results for the $\Upsilon(2S)$ and $\Upsilon(3S)$ are about 3σ larger than previous world averages. However, the CLEO values are confirmed by their proximity to $\tau^+\tau^-$ branching-fraction measurements, which are discussed below.

Table 32 Dimuonic branching fractions of the narrow Υ states from PDG04 [17] and CLEO [510]

Resonance	$\mathcal{B}_{\mu\mu}$ (%) from PDG04	$\mathcal{B}_{\mu\mu}$ (%) from CLEO
$\Upsilon(1S)$	2.48 ± 0.06	$2.49 \pm 0.02 \pm 0.07$
$\Upsilon(2S)$	1.31 ± 0.21	$2.03 \pm 0.03 \pm 0.08$
$\Upsilon(3S)$	1.81 ± 0.17	$2.39 \pm 0.07 \pm 0.10$

Table 33 From CLEO [511], measured values of $\Gamma_{ee}\Gamma_{\text{had}}/\Gamma_{\text{tot}}$ for the narrow Υ states

Resonance	$\Gamma_{ee}\Gamma_{\text{had}}/\Gamma_{\text{tot}}$ (keV)
$\Upsilon(1S)$	$1.252 \pm 0.004 \pm 0.019$
$\Upsilon(2S)$	$0.581 \pm 0.004 \pm 0.009$
$\Upsilon(3S)$	$0.413 \pm 0.004 \pm 0.006$

Table 34 From CLEO [512], measured τ-pair branching fractions of the narrow Υ states and ratios to corresponding dimuonic rates

Resonance	$\mathcal{B}_{\tau\tau}/\mathcal{B}_{\mu\mu}$	$\mathcal{B}_{\tau\tau}$ (%)
$\Upsilon(1S)$	$1.02 \pm 0.02 \pm 0.05$	$2.54 \pm 0.04 \pm 0.12$
$\Upsilon(2S)$	$1.04 \pm 0.04 \pm 0.05$	$2.11 \pm 0.07 \pm 0.13$
$\Upsilon(3S)$	$1.05 \pm 0.08 \pm 0.05$	$2.52 \pm 0.19 \pm 0.15$

In order to measure a quantity very close to Γ_{ee}, CLEO [511] performed dedicated scans to measure the integral of the Υ production cross section over incident e^+e^- energies to determine

$$\tilde{\Gamma}_{ee} \equiv \frac{\Gamma_{ee}\Gamma_{\text{had}}}{\Gamma_{\text{tot}}}, \qquad \tilde{\Gamma}_{ee} = \frac{m_\Upsilon^2}{6\pi^2}\int \sigma\bigl(e^+e^- \to \Upsilon \to \text{hadrons}\bigr)\, dE. \tag{120}$$

The resulting values, listed in Table 33, are consistent with, but more precise than, the PDG world averages.

CLEO [512] also addressed the third dileptonic width, $\Gamma_{\tau\tau}$ (making the first observation of the decay $\Upsilon(3S) \to \tau^+\tau^-$) and measured precise values of $\mathcal{B}_{\tau\tau}/\mathcal{B}_{\mu\mu}$ for $\Upsilon(nS)$, $n = 1, 2, 3$, as shown in Table 34. Using the CLEO values of $\mathcal{B}_{\mu\mu}$ [510] (discussed above) allowed reporting of the absolute $\mathcal{B}_{\tau\tau}$ values as well. The results obtained are consistent with the expectations from the Standard Model, and $\mathcal{B}_{\tau\tau}$ values for $\Upsilon(1S, 2S)$ have much-improved precision over previous measurements.

Table 35 Values of Γ_{tot} for the $\Upsilon(1S, 2S, 3S)$ from PDG04 [17], CLEO [510], and PDG08 [18]

$\Upsilon(nS)$	Γ_{tot} (keV)		
	PDG04	CLEO	PDG08
$\Upsilon(1S)$	53.0 ± 1.5	52.8 ± 1.8	54.02 ± 1.25
$\Upsilon(2S)$	43 ± 6	29.0 ± 1.6	31.98 ± 2.63
$\Upsilon(3S)$	26.3 ± 3.4	20.3 ± 2.1	20.32 ± 1.85

The total width of the resonances can be expressed as

$$\Gamma_{\text{tot}} = \frac{\tilde{\Gamma}_{ee}}{\mathcal{B}_{\mu\mu}\,(1 - 3\mathcal{B}_{\mu\mu})}, \tag{121}$$

where $\tilde{\Gamma}_{ee}$ is measured as in (120), and which, when combined with the more precise $\mathcal{B}_{\mu\mu}$ also described above, yields improved measurements of the total widths Γ_{tot}. The larger $\mathcal{B}_{\mu\mu}$ for $\Upsilon(2S, 3S)$ as determined by CLEO (validated by consistent values of $\mathcal{B}_{\tau\tau}$) leads to smaller and more precise $\Gamma_{\text{tot}}(2, 3S)$, as seen in Table 35.

Improved measurements of the $\Upsilon(4S)$ parameters were reported by BABAR [513]. Three scans of the energy range $\sqrt{s} = 10.518$–10.604 GeV were performed with 11, 7, and 5 energy points, respectively, with integrated luminosity of typically 0.01 fb^{-1} per point. This information was complemented by a large data sample of 76 fb^{-1} collected at the peak of the $\Upsilon(4S)$, from which the cross section at the peak was determined. The nominal $\sqrt{s}$ values of the scans were corrected using an energy calibration based on the dedicated run at the $\Upsilon(3S)$. A fit of the energy dependence allows a determination of the $\Upsilon(4S)$ parameters, among them mass, total, and electronic width, obtaining

$$\begin{aligned} m\big(\Upsilon(4S)\big) &= 10579.3 \pm 0.4 \pm 1.2 \text{ MeV}, \\ \Gamma_{\text{tot}}\big(\Upsilon(4S)\big) &= 20.7 \pm 1.6 \pm 2.5 \text{ MeV}, \\ \Gamma_{ee}\big(\Upsilon(4S)\big) &= 321 \pm 17 \pm 29 \text{ eV}, \end{aligned} \tag{122}$$

all of which dominate the PDG08 [18] world averages.

3.3 Hadronic transitions

3.3.1 Theoretical status

The general form for a hadronic transition is

$$\Phi_i \to \Phi_f + h \tag{123}$$

where Φ_i, Φ_f and h stand for the initial-state, final-state quarkonia, and the emitted light hadron(s). In the $c\bar{c}$ and $b\bar{b}$ systems, the mass $m_{\Phi_i} - m_{\Phi_f}$ varies from a few hundred MeV to slightly over a GeV, so the kinematically allowed final light hadron(s) h are dominated by single-particle ($\pi^0, \eta, \omega, \ldots$) or two-particle ($2\pi$ or $2K$) states. The low momenta of the light hadrons in these transitions allow the application of chiral Lagrangian methods. To date, over twenty hadronic transitions have been observed experimentally.

Hadronic transitions are important decay modes for low-lying heavy-quarkonium states. In fact, the first observed hadronic transition, $\psi(2S) \to \pi\pi J/\psi$, has a branching fraction recently measured by CLEO [446] to be $(52.7 \pm 1.3)\%$ (see Sect. 3.3.2). Calculating such transitions requires nonperturbative QCD. The standard approach is a QCD Multipole Expansion (QCDME) for gluon emission, which is modeled after the multipole expansion used for electromagnetic transitions (see Sect. 3.1.1).

Many contributed to the early development of the QCDME approach [158, 514, 515], but Yan [516] was the first to present a gauge-invariant formulation within QCD. For a heavy $Q\bar{Q}$ bound state, a *dressed* (*constituent*) quark is defined as

$$\tilde{\psi}(\mathbf{x}, t) \equiv U^{-1}(\mathbf{x}, t)\psi(x), \tag{124}$$

where $\psi(x)$ is the usual quark field and U is defined as a path-ordered exponential along a straight line from $\mathbf{X} \equiv (\mathbf{x}_1 + \mathbf{x}_2)/2$ (the center-of-mass coordinate of Q and $\bar{Q}$) to $\mathbf{x}$,

$$U(\mathbf{x}, t) = P \exp\left[ig_s \int_{\mathbf{X}}^{\mathbf{x}} \mathbf{A}(\mathbf{x}', t) \cdot d\mathbf{x}'\right]. \tag{125}$$

Gluon-field color indices have been suppressed. The *dressed* gluon field is defined by

$$\begin{aligned} \tilde{A}_\mu(\mathbf{x}, t) &\equiv U^{-1}(\mathbf{x}, t)A_\mu(x)U(\mathbf{x}, t) \\ &\quad - \frac{i}{g_s}U^{-1}(\mathbf{x}, t)\partial_\mu U(\mathbf{x}, t). \end{aligned} \tag{126}$$

Now we can make the QCD multipole expansion, in powers of $(\mathbf{x} - \mathbf{X}) \cdot \nabla$ operating on the gluon field in exact analogy with QED:

$$\begin{aligned} \tilde{A}_0(\mathbf{x}, t) &= A_0(\mathbf{X}, t) - (\mathbf{x} - \mathbf{X}) \cdot \mathbf{E}(\mathbf{X}, t) + \cdots, \\ \tilde{\mathbf{A}}(\mathbf{X}, t) &= -\frac{1}{2}(\mathbf{x} - \mathbf{X}) \times \mathbf{B}(\mathbf{X}, t) + \cdots, \end{aligned} \tag{127}$$

where $\mathbf{E}$ and $\mathbf{B}$ are color-electric and color-magnetic fields, respectively. The resulting Hamiltonian for a heavy $Q\bar{Q}$ system is then [516]

$$H_{\text{QCD}}^{\text{eff}} = H_{\text{QCD}}^{(0)} + H_{\text{QCD}}^{(1)} + H_{\text{QCD}}^{(2)}, \tag{128}$$

with $H_{\text{QCD}}^{(0)}$ taken as the zeroth-order Hamiltonian, even though it does not represent free fields but instead the sum of the kinetic and potential energies of the heavy quarks. We also define

$$H_{\text{QCD}}^{(1)} \equiv Q_a A_0^a(\mathbf{X}, t), \tag{129}$$

in which Q_a is the color charge of the $Q\bar{Q}$ system (zero for color-singlets), and

$$H^{(2)}_{\text{QCD}} \equiv -\mathbf{d}_a \cdot \mathbf{E}^a(\mathbf{X},\mathbf{t}) - \mathbf{m_a} \cdot \mathbf{B^a}(\mathbf{X},\mathbf{t}) + \cdots, \tag{130}$$

which is treated perturbatively. The quantities

$$\mathbf{d}^i_a = g_E \int d^3x \tilde{\psi}^\dagger (x-X)^i t_a \tilde{\psi} \tag{131}$$

and

$$\mathbf{m}^i_a = g_M/2 \int d^3x \tilde{\psi}^\dagger \epsilon^{ijk} (x-X)_j \gamma_k t_a \tilde{\psi} \tag{132}$$

are the color-electric dipole moment (E1) and the color-magnetic dipole moment (M1) of the $Q\bar{Q}$ system, respectively. Higher-order terms (not shown) give rise to higher-order electric (E2, E3, ...) and magnetic moments (M2, ...). Because $H^{(2)}_{\text{QCD}}$ in (130) couples color-singlet to octet $Q\bar{Q}$ states, the transitions between eigenstates $|i\rangle$ and $|f\rangle$ of $H^{(0)}_{\text{QCD}}$ are at least second-order in $H^{(2)}_{\text{QCD}}$. The leading-order term is given by

$$\begin{aligned}&\langle f\,h|H_2 \frac{1}{E_i - H^{(0)}_{\text{QCD}} + i\partial_0 - H^{(1)}_{\text{QCD}}}|i\rangle\\ &= \sum_{KL} \langle f\,h|H_2|KL\rangle \frac{1}{E_i - E_{KL}} \langle KL|H_2|i\rangle,\end{aligned} \tag{133}$$

where the sum KL is over a complete set of color-octet $Q\bar{Q}$ states $|K\,L\rangle$ with associated energy E_{KL}. Finally, a connection is made to the physical hadronic transitions in (123) by assuming factorization of the heavy-quark interactions and the production of light hadrons. For example, the leading order E1–E1 transition amplitude is:

$$\begin{aligned}&\mathcal{M}(\Phi_i \to \Phi_f + h)\\ &= \frac{1}{24} \sum_{KL} \frac{\langle f|d^{ia}_m|K\,L\rangle\langle K\,L|d^j_{ma}|i\rangle}{E_i - E_{KL}} \langle h|\mathbf{E}^{ai}\mathbf{E}^j_a|0\rangle.\end{aligned} \tag{134}$$

The allowed light-hadronic final state h is determined by the quantum numbers of the gluonic operator. The leading order term E1–E1 in (133) has $CP = ++$ and $L = 0, 2$ and hence couples to 2π and $2K$ in $I = 0$ states. Higher-order terms (in powers of v) couple as follows: E1–M1 in $O(v)$ with $(CP = --)$ couples to ω; E1–M1, E1–E2 in $O(v)$ and M1–M1, E1–M2 in $O(v^2)$ with $(CP = +-)$ couples to both π^0 (isospin-breaking) and η (SU(3)-breaking); and M1–M1, E1–E3, E2–E2 $(CP = ++)$ are higher-order corrections to the E1–E1 terms.

Applying this formulation to observed hadronic transitions requires additional phenomenological assumptions. Following Kuang and Yan [516, 517], the heavy $Q\bar{Q}$ bound states spectrum of $H^{(0)}_{\text{QCD}}$ is calculated by solving the state equation with a given potential model. The intermediate octet $Q\bar{Q}$ states are modeled by the Buchmueller–Tye quark-confining string (QCS) model [518]. Then chiral symmetry relations can be employed to parametrize the light-hadronic matrix element. The remaining unknown coefficients in the light-hadron matrix elements are set by experiment or calculated using a duality argument between the physical light-hadron final state and associated two-gluon final state. A detailed discussion of all these assumptions can be found in the QWG review [1].

For the most common transitions, $h = \pi_1 + \pi_2$, the effective chiral Lagrangian form is [519]

$$\begin{aligned}&\frac{g_E^2}{6} \langle \pi_1 \pi_2 | \mathbf{E}^a_i \mathbf{E}_{aj} | 0 \rangle\\ &= \frac{1}{\sqrt{(2\omega_1)(2\omega_2)}} \Big[C_1 \delta_{ij} q_1^\mu q_{2\mu}\\ &\quad + C_2 \Big(q_{1k} q_{2l} + q_{1l} q_{2k} - \frac{2}{3} \delta_{ij} q_1^\mu q_{2\mu} \Big) \Big].\end{aligned} \tag{135}$$

If the polarization of the heavy $Q\bar{Q}$ initial and final states is measured, more information can be extracted from these transitions and a more general form of (135) is appropriate [520].

Important *single* light-hadron transitions include those involving the η, π^0, or ω mesons. The general form of the light-hadronic factor for the η transition, which is dominantly E1–M2, is [521]

$$\frac{g_e g_M}{6} \langle \eta | \mathbf{E}^a_i \partial_i \mathbf{B}^a_j | 0 \rangle = i (2\pi)^{\frac{3}{2}} C_3 q_j. \tag{136}$$

The π^0 and η transitions are related by the structure of chiral symmetry-breaking [522]. Many more details for these and other transitions within the context of the Kuang–Yan model can be found in the review of Kuang [521].

A summary of all experimentally observed hadronic transitions and their corresponding theoretical expectations within the Kuang–Yan (KY) model is presented in Table 36. The experimental partial widths are determined from the measured branching fractions and total width of the initial state. If the total width is not well measured, the theoretically expected width is used, as indicated. The theory expectations are adjusted using the current experimental inputs to rescale the model parameters $|C_1|$ and $|C_2|$ in (135) and $|C_3|$ in (136).

The multipole expansion works well for transitions of heavy $Q\bar{Q}$ states below threshold [81]. Within the specific KY model a fairly good description of the rates for the two-pion transitions is observed. The partial width $\Gamma(\Upsilon(3S) \to \Upsilon(1S)\pi^+\pi^-)$ was predicted to be suppressed due to cancelations between the various QCS intermediate states [517], allowing nonleading terms, $O(v^2)$, to contribute significantly. The non-S-wave behavior of the $m_{\pi^+\pi^-}$ dependence in $\Upsilon(3S)$ decays, also observed in the $\Upsilon(4S) \to$

$\Upsilon(2S)\pi^+\pi^-$ transitions, may well reflect this influence of higher-order terms. Other possibilities are discussed in Sect. 3.3.11. For single light-hadron transitions some puzzles remain. For example, the ratio

$$\frac{\Gamma(\Upsilon(2S) \to \eta\Upsilon(1S))}{\Gamma(\psi(2S) \to \eta J/\psi(1S))} \tag{137}$$

is much smaller than expected from theory (see Sect. 3.3.6).

The situation is more complicated for above-threshold, strong open-flavor decays. The issues are manifest for $\Upsilon(5S)$ two-pion transitions to $\Upsilon(nS)$ ($n = 1, 2, 3$). First, states above threshold do not have sizes that are small compared to the QCD scale (e.g., $\sqrt{\langle r^2\rangle_{\Upsilon(5S)}} = 1.2$ fm), making the whole QCDME approach less reliable. Second, even within the KY model, the QCS intermediate states are no longer far away from the initial-state mass. Thus the energy denominator, $E_i - E_{KL}$ in (134), can be small, leading to large enhancements in the transition rates that are sensitive to the exact position of the intermediate states [528]. This is the reason for the large theory widths seen in Table 36. Third, a number of new states (see Sect. 2.3) that do not fit into the conventional $Q\bar{Q}$ spectra have been observed, implying additional degrees of freedom appearing in the QCD spectrum beyond naive-quark-model counting. Hence the physical quarkonium states have open-flavor meson-pair contributions and possible hybrid ($Q\bar{Q}g$) or tetraquark contributions. The effect of such terms on hadronic transitions is not yet understood [531]. A possibly related puzzle is the strikingly large ratio

$$R_\eta[\Upsilon(4S)] \equiv \frac{\Gamma(\Upsilon(4S) \to \Upsilon(1S)\eta)}{\Gamma(\Upsilon(4S) \to \Upsilon(1S)\pi^+\pi^-)} \approx 2.5\,. \tag{138}$$

This ratio is over a hundred times larger than one would expect within the KY model, which is particularly surprising, since the similarly defined ratio, $R_\eta[\Upsilon(2S)] \approx 10^{-3}$, is actually less than half of the KY model expectations (see Table 36) and the experimental upper bound on $R_\eta[\Upsilon(3S)]$ is already slightly below KY-model expectations. Much theoretical work remains in order to understand the hadronic transitions of the heavy $Q\bar{Q}$ systems above threshold.

Many of the new XYZ states (see Sect. 2.3) are candidates for so-called hadronic molecules. If this were the case, and they indeed owe their existence to nonperturbative interactions among heavy mesons, the QCDME needs to be extended by heavy-meson loops. These loops provide non-multipole, long-ranged contributions, so as to allow for the inclusion of their influence. However, if hadron loops play a significant, sometimes even nonperturbative, role above $\bar{D}D$ threshold, one should expect them to be at least of some importance below the lowest inelastic threshold. Correspondingly, one should expect to find some systematic deviations between quark-model predictions and data. By including

Table 36 Partial widths for observed hadronic transitions. Experimental results are from PDG08 [18] unless otherwise noted. Partial widths determined from known branching fractions and total widths. Quoted values assume total widths of $\Gamma_{\rm tot}(\chi_{b2}(2P)) = 138 \pm 19$ keV [523], $\Gamma_{\rm tot}(\chi_{b1}(2P)) = 96 \pm 16$ keV [523], $\Gamma_{\rm tot}(\Upsilon(1^3D_2)) = 28.5$ keV [524, 525] and $\Gamma_{\rm tot}(\Upsilon(5S)) = 43 \pm 4$ MeV [36]. Only the charged dipion transitions are shown here, but the corresponding measured $\pi^0\pi^0$ rates, where they exist, are consistent with a parent state of $I = 0$. Theoretical results are given using the Kuang and Yan (KY) model [517, 521, 526]. Current experimental inputs were used to rescale the parameters in the theory partial rates. ($|C_1| = 10.2 \pm 0.2 \times 10^{-3}$, $C_2/C_1 = 1.75 \pm 0.14$, $C_3/C_1 = 0.78 \pm 0.02$ for the Cornell case)

Transition	$\Gamma_{\rm partial}$ (keV) (Experiment)	$\Gamma_{\rm partial}$ (keV) (KY Model)
$\psi(2S)$		
$\to J/\psi + \pi^+\pi^-$	102.3 ± 3.4	input ($\lvert C_1\rvert$)
$\to J/\psi + \eta$	10.0 ± 0.4	input (C_3/C_1)
$\to J/\psi + \pi^0$	0.411 ± 0.030 [446]	0.64 [522]
$\to h_c(1P) + \pi^0$	0.26 ± 0.05 [47]	0.12–0.40 [527]
$\psi(3770)$		
$\to J/\psi + \pi^+\pi^-$	52.7 ± 7.9	input (C_2/C_1)
$\to J/\psi + \eta$	24 ± 11	
$\psi(3S)$		
$\to J/\psi + \pi^+\pi^-$	<320 (90% CL)	
$\Upsilon(2S)$		
$\to \Upsilon(1S) + \pi^+\pi^-$	5.79 ± 0.49	8.7 [528]
$\to \Upsilon(1S) + \eta$	$(6.7 \pm 2.4) \times 10^{-3}$	0.025 [521]
$\Upsilon(1^3D_2)$		
$\to \Upsilon(1S) + \pi^+\pi^-$	0.188 ± 0.046 [63]	0.07 [529]
$\chi_{b1}(2P)$		
$\to \chi_{b1}(1P) + \pi^+\pi^-$	0.83 ± 0.33 [523]	0.54 [530]
$\to \Upsilon(1S) + \omega$	1.56 ± 0.46	
$\chi_{b2}(2P)$		
$\to \chi_{b2}(1P) + \pi^+\pi^-$	0.83 ± 0.31 [523]	0.54 [530]
$\to \Upsilon(1S) + \omega$	1.52 ± 0.49	
$\Upsilon(3S)$		
$\to \Upsilon(1S) + \pi^+\pi^-$	0.894 ± 0.084	1.85 [528]
$\to \Upsilon(1S) + \eta$	$<3.7 \times 10^{-3}$	0.012 [521]
$\to \Upsilon(2S) + \pi^+\pi^-$	0.498 ± 0.065	0.86 [528]
$\Upsilon(4S)$		
$\to \Upsilon(1S) + \pi^+\pi^-$	1.64 ± 0.25	4.1 [528]
$\to \Upsilon(1S) + \eta$	4.02 ± 0.54	
$\to \Upsilon(2S) + \pi^+\pi^-$	1.76 ± 0.34	1.4 [528]
$\Upsilon(5S)$		
$\to \Upsilon(1S) + \pi^+\pi^-$	228 ± 33	
$\to \Upsilon(1S) + K^+K^-$	26.2 ± 8.1	
$\to \Upsilon(2S) + \pi^+\pi^-$	335 ± 64	
$\to \Upsilon(3S) + \pi^+\pi^-$	206 ± 80	

intermediate heavy-meson effects within the framework of QCDME [532, 533], improved agreement with the experimental data on dipion transitions in the ψ and Υ systems was obtained.

Alternatively, a nonrelativistic effective field theory (NREFT) was introduced [534] that allows one to study the effect of heavy-meson loops on charmonium transitions with controlled uncertainty. In this work, it was argued that the presence of meson loops resolves the long-standing discrepancy between, on the one hand, the values of the light-quark mass-differences extracted from the masses of the Goldstone bosons, and on the other, the ratio of selected charmonium transitions, namely $\psi(2S) \to J/\psi\pi^0/\psi(2S) \to J/\psi\eta$. NREFT uses the velocity of the heavy mesons in the intermediate state, $v \sim \sqrt{|m - 2m_D|/m_D}$, as expansion parameter. Thus, for transitions of states below $D\bar{D}$ threshold, the analytic continuation of the standard expression is to be used. For low-lying charmonium transitions, v is found to be of order 0.5. A typical transition via a D-meson loop may then be counted as

$$v^3/(v^2)^2 \times \text{vertex factors}. \tag{139}$$

For the transition between two S-wave charmonia, which decay into $D^{(*)}D^{(*)}$ via a P-wave vertex, the vertex factors scale as v^2. Thus the loop contributions appear to scale as order v, and, for values of the velocity small relative to those that can be captured by QCDME, are typically suppressed. However, in certain cases enhancements may occur. For example, for $\psi(2S) \to J/\psi\pi^0$ and $\psi(2S) \to J/\psi\eta$, flavor symmetry is broken, and therefore the transition matrix element needs to scale as δ, the energy scale that quantifies the degree of flavor-symmetry violation in the loop and which originates from the mass differences of charged and neutral D-mesons. However, if an energy scale is pulled out of the integral, this needs to be balanced by removing the energy scale v^2 from the power counting. Thus the estimate for the loops contribution scales as δ/v compared to the piece of order δ that emerges from QCDME. Hence, for certain transitions, meson loops are expected to significantly influence the rates.

The contributions of heavy-meson loops to charmonium decays follow a very special pattern. They are expected to be more important for charmonia pairs close to the lowest two-meson threshold. In addition, loops appear to be suppressed for transitions between S and P wave charmonia [535]. These conjectures can be tested experimentally by systematic, high-precision measurements of as many transitions as possible. Once the role of meson loops is established for transitions between conventional heavy quarkonia below or very close to open-flavor threshold, an extension from perturbative to a nonperturbative treatment might lead to a combined analysis of excited charmonia and at least a few of the XYZ states.

3.3.2 Branching fractions for $\psi(2S) \to XJ/\psi$

Precision measurements of the hadronic transitions $\psi(2S) \to XJ/\psi$ are important for a number of reasons beyond the obvious one of providing an accurate normalization and accounting for $\psi(2S)$ decays. First, they allow experimental comparisons with increasingly sophisticated theoretical calculations (see Sect. 3.3.1). Second, they can be used in comparisons with the analogous Υ transitions in the bottomonium system. Finally, the transitions provide a convenient way to access the J/ψ, where the transition pions in $\psi(2S) \to \pi^+\pi^-J/\psi$, for example, can be used in tagging a clean and well-normalized J/ψ sample. Using its full sample of 27 million $\psi(2S)$ decays, CLEO [446] measured these rates with substantially improved precision over previous measurements. The analysis measures $\mathcal{B}(\psi(2S) \to \pi^+\pi^-J/\psi)$ *inclusively*, selecting events based upon the mass recoiling against the transition dipion and placing no restriction on the J/ψ decay. Its uncertainty is dominated by a 2% systematic error on the produced number of $\psi(2S)$. Other transitions to J/ψ, the *exclusive* modes through $\pi^0\pi^0$, η, π^0, and $\psi(2S) \to \gamma\chi_{cJ} \to \gamma\gamma J/\psi$, as well as the *inclusive* rate for $\psi(2S) \to J/\psi$ + any, are all measured *relative* to the $\pi^+\pi^-$ transition and use $J/\psi \to \ell^+\ell^-$, thereby reducing systematic error from the number of $\psi(2S)$, which cancels in the ratios. These relative rates are measured with precision from 2–6%. The absolute branching fractions determined from this analysis are shown in Table 37 which are higher values than most previous measurements. The ratio of $\pi^0\pi^0$ to $\pi^+\pi^-$ transitions is consistent with one-half, the expectation from isospin invariance, which was not the case in some earlier analyses. See Sect. 3.1.5 for a discussion of the $\gamma\gamma J/\psi$ portion of this analysis, which addresses the product branching fractions $\mathcal{B}(\psi(2S) \to \gamma\chi_{cJ}) \times \mathcal{B}(\chi_{cJ} \to \gamma J/\psi)$ and whether any nonresonant such final states are present.

Table 37 Results from the branching-fraction analyses for $\psi(2S)$ and $\psi(3770)$ decays to XJ/ψ from CLEO [446, 500]

Quantity	$\psi(2S)$ (%)	$\psi(3770)$ (%)
$\mathcal{B}(\pi^+\pi^-J/\psi)$	$35.04 \pm 0.07 \pm 0.77$	$0.189 \pm 0.020 \pm 0.020$
$\mathcal{B}(\pi^0\pi^0J/\psi)$	$17.69 \pm 0.08 \pm 0.53$	$0.080 \pm 0.025 \pm 0.016$
$\mathcal{B}(\eta J/\psi)$	$3.43 \pm 0.04 \pm 0.09$	$0.087 \pm 0.033 \pm 0.022$
$\mathcal{B}(\pi^0 J/\psi)$	$0.133 \pm 0.008 \pm 0.003$	<0.028 at 90% CL
$\mathcal{B}(J/\psi + \text{any})$	$62.54 \pm 0.16 \pm 1.55$	–
$\frac{\mathcal{B}(\pi^0\pi^0J/\psi)}{\mathcal{B}(\pi^+\pi^-J/\psi)}$	50.47 ± 1.04	42 ± 17

3.3.3 Observation of $\psi(2S) \to \pi^0 h_c(1P)$

The hadronic transition $\psi(2S) \to \pi^0 h_c(1P)$ was first observed by CLEO for $h_c(1P) \to \gamma\eta_c(1S)$ [45, 46] and later seen for $h_c(1P) \to 2(\pi^+\pi^-)\pi^0$ [536] (see Sect. 3.4.2). However, those analyses could only measure product branching fractions. BESIII [47] has used its much larger 106 million $\psi(2S)$ sample to inclusively measure $\mathcal{B}(\psi(2S) \to \pi^0 h_c(1P))$ by observing a significant enhancement at the $h_c(1P)$ in the mass recoiling against a reconstructed π^0, with none of the $h_c(1P)$ decay products reconstructed. The result is $\mathcal{B}(\psi(2S) \to \pi^0 h_c(1P)) = (8.4 \pm 1.3 \pm 1.0) \times 10^{-4}$, in agreement with the range predicted in [527] (see Table 36).

3.3.4 Nonobservation of $\eta_c(2S) \to \pi^+\pi^-\eta_c(1S)$

The historically uncooperative nature of the $\eta_c(2S)$ has continued into the modern era, despite its discovery (Sect. 2.2.2). CLEO [68] failed not only to see $\eta_c(2S)$ in radiative transitions (Sect. 3.1.4), but also failed to observe the expected dipion transition to $\eta_c(1S)$, setting the upper limit

$$\mathcal{B}\big(\psi(2S) \to \gamma\eta_c(2S)\big) \times \mathcal{B}\big(\eta_c(2S) \to \pi^+\pi^-\eta_c(1S)\big) < 1.7 \times 10^{-4} \quad \text{at 90\% CL.} \tag{140}$$

3.3.5 Observation of $\psi(3770) \to XJ/\psi$

Hadronic transitions of the $\psi(3770)$ to J/ψ are sensitive to the relative sizes of the 2^3S_1 and 1^3D_1 admixture contained in the $\psi(3770)$, and thus, at least in principle, contain information about the nature of the $\psi(3770)$. BESII [537] first reported evidence for the transition $\psi(3770) \to \pi^+\pi^- J/\psi$ at $\sim 3\sigma$ significance using approximately 2×10^5 $\psi(3770)$ decays. CLEO [500], with roughly 2×10^6 $\psi(3770)$ decays, later observed the $\pi^+\pi^-$ transition at 11.6σ, and also found evidence for the $\pi^0\pi^0 J/\psi$ (3.4σ) and $\eta J/\psi$ (3.5σ) transitions. The CLEO results appear in Table 37. With more data, these hadronic transitions could be used to shed light on the $c\bar{c}$ purity of the $\psi(2S)$ and $\psi(3770)$ states [538].

3.3.6 Observation of $\Upsilon(2S) \to \eta\Upsilon(1S)$

The observation of $\Upsilon(2S) \to \eta\Upsilon(1S)$ using 9 million $\Upsilon(2S)$ decays collected with the CLEO III detector represents the first observation of a quarkonium transition involving the spin flip of a bottom quark. The transition rate is sensitive to the chromomagnetic moment of the b quark. Using three decay modes, $\eta \to \gamma\gamma$, $\pi^+\pi^-\pi^0$, and $3\pi^0$, CLEO [539] observes $13.9^{+4.5}_{-3.8}$ events of the form $\Upsilon(2S) \to \eta\Upsilon(1S)$ (Fig. 44). Using the difference in log-likelihood for fits with and without signal, it was determined that this corresponds to a $5.3\,\sigma$ observation. Correcting for acceptance and incorporating systematic errors, the observed number of events translates to a branching fraction of

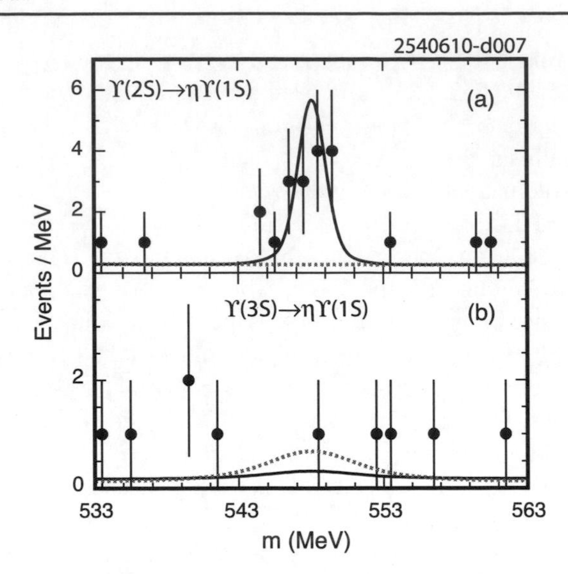

Fig. 44 From CLEO [539], the invariant mass of the η candidate for (**a**) $\Upsilon(2S) \to \eta\Upsilon(1S)$ and (**b**) $\Upsilon(3S) \to \eta\Upsilon(1S)$ for η-mesons exclusively reconstructed in the three decay modes $\eta \to \gamma\gamma$, $\pi^+\pi^-\pi^0$, and $3\pi^0$. In each case, the *solid curve* corresponds to the best fit to a flat background and signal. The *dotted curve* represents (**a**) the best-fit background, and (**b**) the 90% CL upper limit for a signal. Adapted from [539] with kind permission, copyright (2008) The American Physical Society

$$\begin{aligned} \mathcal{B}\big(\Upsilon(2S) \to \eta\Upsilon(1S)\big) &= \big(2.1^{+0.7}_{-0.6} \pm 0.3\big) \times 10^{-4}, \\ \mathcal{B}\big(\Upsilon(3S) \to \eta\Upsilon(1S)\big) &< 1.8 \times 10^{-4} \quad \text{at 90\% CL.} \end{aligned} \tag{141}$$

The $2S$ rate is a factor of four smaller than one would expect scaling from the analogous charmonium transition rate, $\psi(2S) \to \eta J/\psi$. Similarly, the $3S$ limit is already two times smaller than what one would expect from the same scaling. The interpretation of this pattern is still unclear.

3.3.7 Observation of $\chi_{b1,2}(2P) \to \omega\Upsilon(1S)$

In 2003, CLEO [540] reported the observation of the decay chain $\Upsilon(3S) \to \gamma\chi bJ(2P)$, $\chi bJ(2P) \to \omega\Upsilon(1S)$, for $J = 1, 2$, with $\omega \to \pi^+\pi^-\pi^0$ and $\Upsilon(1S) \to \ell^+\ell^-$, for which there is marginal phase space available, and measured

$$\begin{aligned} \mathcal{B}\big(\chi_{b1} \to \omega\Upsilon(1S)\big) &= \big(1.63^{+0.35+0.16}_{-0.31-0.15}\big)\%, \\ \mathcal{B}\big(\chi_{b2} \to \omega\Upsilon(1S)\big) &= \big(1.10^{+0.32+0.11}_{-0.28-0.10}\big)\%. \end{aligned} \tag{142}$$

The relative rates are comparable, in agreement with Voloshin's prediction [541] on the basis of S-wave phase-space factors and the $E1 * E1 * E1$ gluon configurations expected by the multipole expansion.

3.3.8 Dipion transitions from $\Upsilon(2S, 3S)$

The double-hump dipion invariant-mass distribution in $\Upsilon(3S) \to \Upsilon(1S)\pi\pi$ transitions has been thought to be at least puzzling, and frequently thought to be anomalous, indicative of either new physics or an intermediate scalar dipion resonance. However, a CLEO [542] analysis has offered an alternative to these characterizations; simply, that dipion–quarkonium dynamics expected to occur within QCD is responsible. Brown and Cahn [519] derived the general matrix element for dipion transitions from heavy vector quarkonia; it is constrained by PCAC and simplified by treating it as a multipole expansion [514, 516, 543]. This general matrix element has three terms: one proportional to $(m_{\pi\pi}^2 - 2m_\pi^2)$, one proportional to the product $E_1 E_2$ of the two pion energies in the parent rest frame, and a third which characterizes the transition requiring a b-quark spin-flip. Although the third chromomagnetic term is thought to be highly suppressed, and therefore ignorable, even the second term had generally been neglected prior to the CLEO analysis. Sensitivity to the second term, which is also proportional to $\cos\theta_X$, θ_X being the dipion-helicity angle, is greatest at low $m_{\pi\pi}$, where experiments using only charged pions can only reconstruct very few events: soft charged particles curl up in the detector magnetic field before reaching any tracking chambers. The notable aspect of the CLEO analysis is not only that it fits for all three terms with complex form factors $\mathcal{A}$, $\mathcal{B}$, and $\mathcal{C}$, respectively, but also that it performs a simultaneous fit to dipion transitions through both charged *and neutral* pion pairs. The latter subset of the data enhances sensitivity to the $\cos\theta_X$ dependence because even when neutral pions are slow, frequently both π^0-decay photons can be reconstructed in the calorimeter.

Results are shown in Fig. 45 and Table 38, from which the following conclusions are drawn:

- The CLEO data in all three Υ dipion transitions can be represented well by a two-term form of the matrix element, provided the form factors are allowed to be different for all three decays.
- Sensitivity of the CLEO data to the chromomagnetic term, i.e., to a nonzero value of $\mathcal{C}/\mathcal{A}$, is small because the functional dependencies of the $\mathcal{B}$ and $\mathcal{C}$ terms are quite similar; a much larger dataset of $\Upsilon(3S) \to \Upsilon(1S)\pi\pi$ is required to probe this component. None of the three Υ dipion transitions require a nonzero $\mathcal{C}/\mathcal{A}$; i.e., there is no evidence yet that spin-flips play a significant role in these decays. In the case of $\Upsilon(3S) \to \Upsilon(1S)\pi\pi$, where a slight sensitivity to this term exists, an upper limit of $|\mathcal{C}/\mathcal{A}| < 1.09$ at 90% CL is given.
- The dynamics in $\Upsilon(2S) \to \Upsilon(1S)\pi\pi$ are reproduced by the fits, but only if non zero values of $\mathcal{B}/\mathcal{A}$ are allowed. The data in this transition are described well without any complex component in $\mathcal{B}/\mathcal{A}$.
- The two-peak structure in $\Upsilon(3S) \to \Upsilon(1S)\pi\pi$ is reproduced without any new physics or any intermediate dipion resonances, $\mathcal{B}/\mathcal{A}$ is found to have a significant *complex* component, and the dipion-helicity form factor is nearly three times larger than that of the dipion-mass-dependent term.

Dubynskiy and Voloshin [520] comment further, on both the formalism and prospects for learning more from such decays. They cast doubt upon the possibility that $\mathcal{B}$ can have a complex component when $\mathcal{C} \equiv 0$, on general principles; conclude that $\mathcal{B}$ and $\mathcal{C}$ terms are degenerate unless $\mathcal{C}$ is set to zero; and suggest resolving the $\mathcal{B}$–$\mathcal{C}$ terms' degeneracy by including initial- and final-state polarization information in fits to experimental data.

CLEO [544] used the matrix elements determined above to obtain correct efficiencies in the first new measurement of these transition branching ratios in a decade. BABAR [545] followed suit soon after, using radiative returns from e^+e^- collisions at 10.58 GeV and the nearby continuum to the narrow Υ states as a source. Both groups used $\Upsilon \to \ell^+\ell^-$ decays; BABAR used only charged, fully reconstructed dipions, whereas CLEO used both charged and neutral, except for $\Upsilon(3S) \to \Upsilon(2S)\pi\pi$, where the low efficiency for charged-track reconstruction made the charged mode too difficult to normalize precisely. CLEO gained some statistical power by performing both exclusive ($\Upsilon \to \ell^+\ell^-$) and inclusive ($\Upsilon \to$ any) versions for each transition; in the latter case the signal was obtained by a fit to the invariant-mass recoiling against the dipion for a smooth background and signal term peaking at the appropriate Υ mass. BABAR extracted signals by fitting distributions of $\Delta m \equiv m(\pi^+\pi^-\ell^+\ell^-) - m(\ell^+\ell^-)$ for a smooth background and signal at the mass difference appropriate for the desired parent Υ, after cutting on $m(\ell^+\ell^-)$ around the mass appropriate for the desired daughter Υ. The combination of CLEO and BABAR branching fractions, as shown in Table 39, which are consistent with each other and previous measurements, made significant improvements to the branching-fraction precisions and dominate the PDG10 [447] world averages shown.

3.3.9 Observation of $\Upsilon(1^3D_J) \to \pi^+\pi^-\Upsilon(1S)$

The $\Upsilon(1^3D_J) \to \pi^+\pi^-\Upsilon(1S)$ measurements from BABAR [63] (see Sect. 2.2.7) represent the only available data on hadronic transitions of the $\Upsilon(1^3D_J)$ states. Partial rates are expected to be independent of J [526]. Using *predicted* [524] branching fractions for $\Upsilon(3S) \to \gamma\chi_{bJ'}(2P)$ and $\chi_{bJ'}(2P) \to \gamma\Upsilon(1^3D_J)$, BABAR quotes

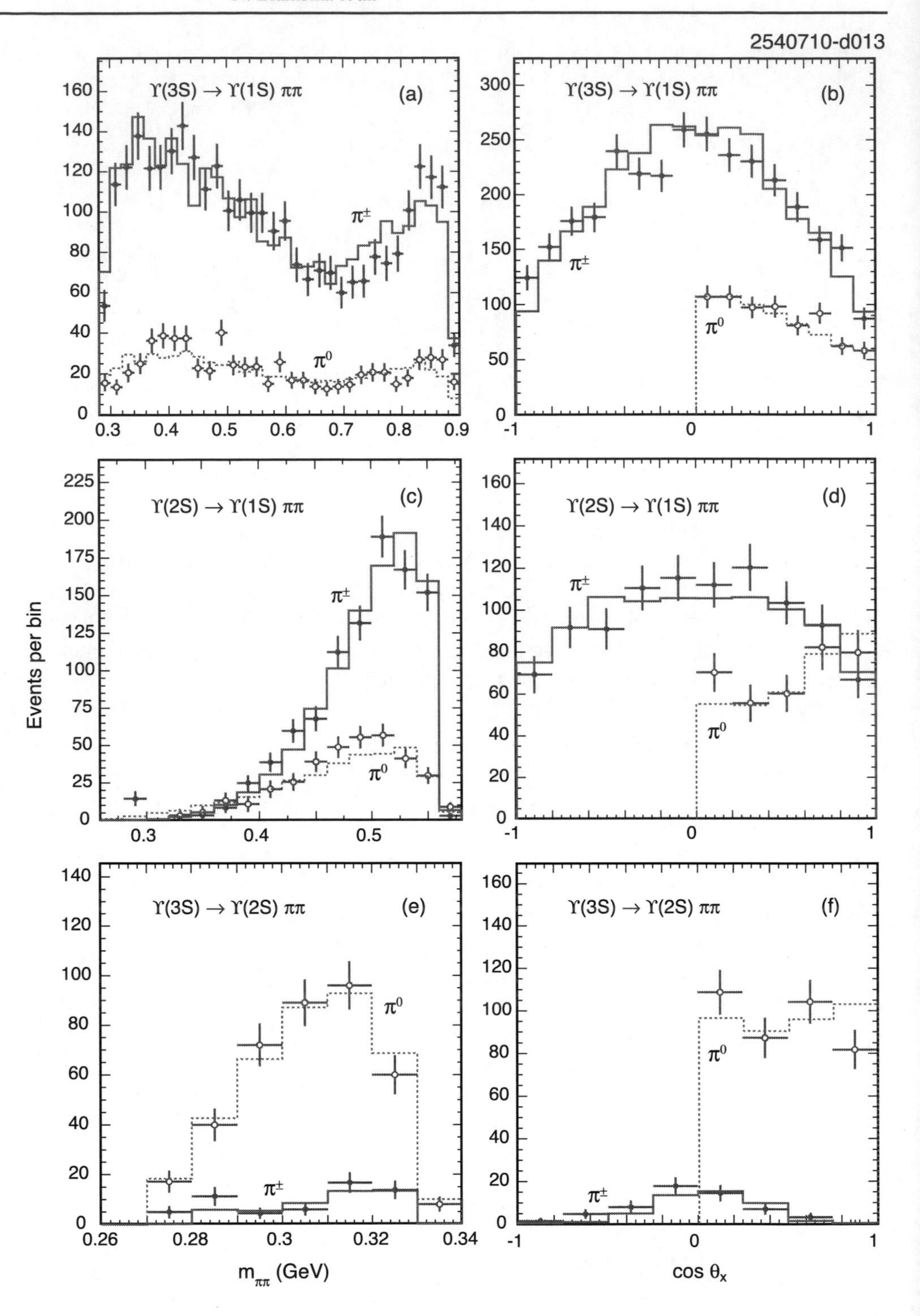

Fig. 45 From CLEO [542], (**a**), (**c**), (**e**) the dipion-mass distributions and (**b**), (**d**), (**f**) dipion-helicity angle $\cos\theta_X$ distributions for the transitions (**a**), (**b**) $\Upsilon(3S) \to \Upsilon(1S)\pi\pi$, (**c**), (**d**) $\Upsilon(2S) \to \Upsilon(1S)\pi\pi$, and (**e**), (**f**) $\Upsilon(3S) \to \Upsilon(2S)\pi\pi$, in which *closed* (*open*) *circles* represent the $\pi^+\pi^-$ ($\pi^0\pi^0$) data, and *solid* (*dotted*) *line histogram* the MC simulation for $\pi^+\pi^-$ ($\pi^0\pi^0$) transitions generated with the best-fit parameters in Table 38. Adapted from [542] with kind permission, copyright (2007) The American Physical Society

$$
\begin{aligned}
&\mathcal{B}\big(\Upsilon(1^3D_1) \to \pi^+\pi^-\Upsilon(1S)\big)\\
&\qquad = \big(0.42^{+0.27}_{-0.23} \pm 0.10\big)\% \; (< 0.82\%),\\
&\mathcal{B}\big(\Upsilon(1^3D_2) \to \pi^+\pi^-\Upsilon(1S)\big)\\
&\qquad = \big(0.66^{+0.15}_{-0.14} \pm 0.06\big)\%, \quad \text{and}\\
&\mathcal{B}\big(\Upsilon(1^3D_3) \to \pi^+\pi^-\Upsilon(1S)\big)\\
&\qquad = \big(0.29^{+0.22}_{-0.18} \pm 0.06\big)\% \; (< 0.62\%),
\end{aligned}
\tag{143}
$$

where upper limits are given at 90% CL and include systematic uncertainties. This hadronic $\Upsilon(1^3D_2)$ transition provides an important benchmark for comparing various theoretical predictions for partial rates [524, 526, 529]. Furthermore, comparing with the observed $\psi(3770) \to \pi^+\pi^- J/\psi$ transition may give insight into threshold effects in the $\psi(3770)$ state.

Table 38 Results from fits to the CLEO [542] $\Upsilon(mS) \to \pi\pi\Upsilon(nS)$, $(m, n) = (3, 2), (3, 1), (2, 1)$ transitions data for the complex form factors $(\mathcal{A}, \mathcal{B}, \mathcal{C})$ of the three terms, and the associated phase angle, $\delta_{\mathcal{BA}}$. The phase angles are quoted in degrees, and have a two-fold ambiguity of reflection in the real axis. The first three fits constrain the chromomagnetic coefficient to be zero ($\mathcal{C} \equiv 0$), but $\mathcal{C}$ floats in the fourth fit. The operators $\mathcal{R}$ and $\mathcal{I}$ denote real and imaginary parts, respectively

Transition	Quantity	Value
$\Upsilon(3S) \to \Upsilon(1S)\pi\pi$	$\Re(\mathcal{B}/\mathcal{A})$	-2.52 ± 0.04
	$\Im(\mathcal{B}/\mathcal{A})$	$\pm 1.19 \pm 0.06$
$(\mathcal{C} \equiv 0)$	$\lvert\mathcal{B}/\mathcal{A}\rvert$	2.79 ± 0.05
	$\delta_{\mathcal{BA}}$ (°)	$155(205) \pm 2$
$\Upsilon(2S) \to \Upsilon(1S)\pi\pi$	$\Re(\mathcal{B}/\mathcal{A})$	-0.75 ± 0.15
	$\Im(\mathcal{B}/\mathcal{A})$	0.00 ± 0.11
$(\mathcal{C} \equiv 0)$	$\lvert\mathcal{B}/\mathcal{A}\rvert$	0.75 ± 0.15
	$\delta_{\mathcal{BA}}$ (°)	180 ± 9
$\Upsilon(3S) \to \Upsilon(2S)\pi\pi$	$\Re(\mathcal{B}/\mathcal{A})$	-0.40 ± 0.32
$(\mathcal{C} \equiv 0)$	$\Im(\mathcal{B}/\mathcal{A})$	0.00 ± 1.1
$\Upsilon(3S) \to \Upsilon(1S)\pi\pi$	$\lvert\mathcal{B}/\mathcal{A}\rvert$	2.89 ± 0.25
	$\lvert\mathcal{C}/\mathcal{A}\rvert$	0.45 ± 0.40
($\mathcal{C}$ floats)		< 1.09 (90% CL)

3.3.10 *Dipion and η transitions from $\Upsilon(4S)$*

Even above open-bottom threshold, the $b\bar{b}$ vector bound state undergoes hadronic transitions, as first measured by *BABAR* [546], which was later updated [545]. The latter analysis is done in a manner similar to that described in Sect. 3.3.8, using mass difference windows around $\Delta m = m[\Upsilon(4S)] - m[\Upsilon(jS)]$, with $j = 1, 2$. Belle [547] reported a similar analysis, but only on $\Upsilon(4S) \to \pi^+\pi^-\Upsilon(1S)$. The two resulting branching fractions for the latter transition are consistent with one other; PDG10 reports (averaged for $\Upsilon(4S) \to \pi^+\pi^-\Upsilon(1S)$) branching fractions

$$\begin{aligned}
&\mathcal{B}\big(\Upsilon(4S) \to \pi^+\pi^-\Upsilon(1S)\big) \\
&\quad = (0.810 \pm 0.06) \times 10^{-4}, \\
&\mathcal{B}\big(\Upsilon(4S) \to \pi^+\pi^-\Upsilon(2S)\big) \\
&\quad = (0.86 \pm 0.11 \pm 0.07) \times 10^{-4}, \\
&\frac{\mathcal{B}(\Upsilon(4S) \to \pi^+\pi^-\Upsilon(2S))}{\mathcal{B}(\Upsilon(4S) \to \pi^+\pi^-\Upsilon(1S))} = 1.16 \pm 0.16 \pm 0.14.
\end{aligned} \tag{144}$$

Of particular note is that, while the dipion mass spectrum for $\Upsilon(4S) \to \pi^+\pi^-\Upsilon(1S)$ has a typical spectrum with a single peak, that of $\Upsilon(4S) \to \pi^+\pi^-\Upsilon(2S)$ appears to have a double-peak structure like $\Upsilon(3S) \to \pi\pi\Upsilon(1S)$, as seen in Fig. 46.

In the same analysis, *BABAR* [546] sought η transitions in the Υ system as well, via $\eta \to \pi^+\pi^-\pi^0$. While no evidence is found for such transitions from $\Upsilon(2S)$ or $\Upsilon(3S)$ (consistent with the CLEO results discussed in Sect. 3.3.6), a quite significant signal was observed for $\eta\Upsilon(1S)$ final states in the $\Upsilon(4S)$ data sample, which, if attributed fully to resonant production, corresponds to

$$\begin{aligned}
&\mathcal{B}\big(\Upsilon(4S) \to \eta\Upsilon(1S)\big) = (1.96 \pm 0.06 \pm 0.09) \times 10^{-4}, \\
&\frac{\mathcal{B}(\Upsilon(4S) \to \eta\Upsilon(1S))}{\mathcal{B}(\Upsilon(4S) \to \pi^+\pi^-\Upsilon(1S))} = 2.41 \pm 0.40 \pm 012.
\end{aligned} \tag{145}$$

The *BABAR* continuum sample taken just below $\Upsilon(4S)$ is only a small fraction of that taken on-$\Upsilon(4S)$, so even though no $\eta\Upsilon(1S)$ are observed in that sample, the possibility that the observed events are attributable to $e^+e^- \to \eta\Upsilon(1S)$ can only be excluded at the 2.7σ level.

Table 39 Branching fractions for bottomonium dipion transitions $\Upsilon(nS) \to \pi\pi\Upsilon(mS)$ for $n = 2, 3$ and $m < n$, as compiled by the Particle Data Group as indicated and as measured by CLEO and *BABAR*; the PDG10 [447] (*PDG08* [18]) numbers include (*do not include*) the CLEO [544] and *BABAR* [545] results

Transition	$\mathcal{B}$ (%)	Source
$\Upsilon(2S) \to \pi^+\pi^-\Upsilon(1S)$	18.8 ± 0.6	PDG08 [18]
	$18.02 \pm 0.02 \pm 0.61$	CLEO [544]
	$17.22 \pm 0.17 \pm 0.75$	*BABAR* [545]
	18.1 ± 0.4	PDG10 [447]
$\Upsilon(2S) \to \pi^0\pi^0\Upsilon(1S)$	9.0 ± 0.8	PDG08 [18]
	$8.43 \pm 0.16 \pm 0.42$	CLEO [544]
	8.6 ± 0.4	PDG10 [447]
$\Upsilon(3S) \to \pi^+\pi^-\Upsilon(1S)$	4.48 ± 0.21	PDG08 [18]
	$4.46 \pm 0.01 \pm 0.13$	CLEO [544]
	$4.17 \pm 0.06 \pm 0.19$	*BABAR* [545]
	4.40 ± 0.10	PDG10 [447]
$\Upsilon(3S) \to \pi^0\pi^0\Upsilon(1S)$	2.06 ± 0.28	PDG08 [18]
	$2.24 \pm 0.09 \pm 0.11$	CLEO [544]
	2.20 ± 0.13	PDG10 [447]
$\Upsilon(3S) \to \pi^+\pi^-\Upsilon(2S)$	2.8 ± 0.6	PDG08 [18]
	$2.40 \pm 0.10 \pm 0.26$	*BABAR* [545]
	2.45 ± 0.23	PDG10 [447]
$\Upsilon(3S) \to \pi^0\pi^0\Upsilon(2S)$	2.00 ± 0.32	PDG08 [18]
	$1.82 \pm 0.09 \pm 0.12$	CLEO [544]
	1.85 ± 0.14	PDG10 [447]

3.3.11 *Dipion transitions near $\Upsilon(5S)$*

As described in Sect. 2.3.2, Belle [117] reported first observation of apparent dipion transitions from the $\Upsilon(5S)$ to the lower-mass narrow Υ states. The caveat "apparent" is applied because a later analysis, also by Belle [37], and also

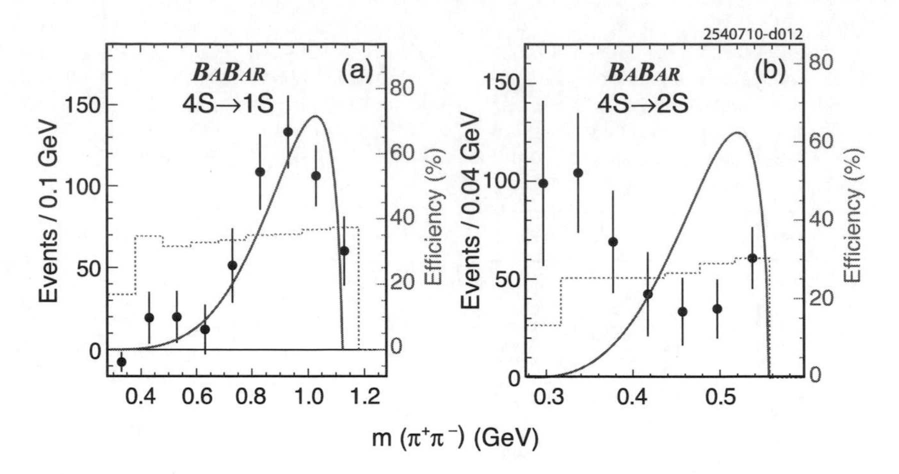

Fig. 46 From *BABAR* [546], efficiency-corrected dipion-mass distributions for (**a**) $\Upsilon(4S) \to \pi^+\pi^-\Upsilon(1S)$, and (**b**) $\Upsilon(4S) \to \pi^+\pi^-\Upsilon(2S)$, in which *solid circles* denote data, *dotted histograms* denote reconstruction and event selection efficiencies, which follow the scales on the *right-side axes*, and the *solid curves* represent theoretical predictions [546]. Experimental resolution on $m_{\pi^+\pi^-}$ is less than 5 MeV. Adapted from [546] with kind permission, copyright (2006) The American Physical Society

described in Sect. 2.3.2, suggests that these transitions may originate not from the enhancement in the hadronic cross section from e^+e^- collisions near $\sqrt{s} \approx 10.87$ GeV known as "$\Upsilon(5S)$", but from a *separate new state* Y_b close by in mass, much as $Y(4260)$ was observed via dipion transitions to lower-mass charmonium.

If interpreted as $\Upsilon(5S)$ transitions, the measured dipion decay widths are more than two orders of magnitude larger than those in the similar $\Upsilon(4S) \to \Upsilon(1S)\pi^+\pi^-$ [545, 547] (see Sect. 3.3.10), and $\Upsilon(2S,3S) \to \Upsilon(1S)\pi^+\pi^-$ (see Sect. 3.3.8) transitions. The reason for such large differences is not clear. Two theoretical ideas have been proposed. The first approach [548] assumes that the bottomonium transitions with two-pion emission come through intermediate virtual $B\bar{B}\pi(\pi)$ formation, followed by the recombination of the B pair into the Υ. The difference between the $\Upsilon(5S)$ and $\Upsilon(4S)$ decays results from a large imaginary part in the $\Upsilon(5S)$ decay amplitude appearing due to the positive difference between the $\Upsilon(5S)$ mass and the sum of the masses of the $B\bar{B}\pi\pi$ system. A similar approach was also used [549, 550], in which specific decay parameters were predicted with good accuracy. An alternative theoretical idea was proposed [388], in which the large $\Upsilon(5S) \to \Upsilon(1S)\pi^+\pi^-$ decay width was explained as a possible indication of the production of a $b\bar{b}g$ hybrid state Y_b with a mass close to the mass of the $\Upsilon(5S)$ resonance.

3.3.12 Observation of $\chi_{bJ}(2P) \to \pi\pi\chi_{bJ}(1P)$

The decay $\chi_{bJ}(2P) \to \pi\pi\chi_{bJ}(1P)$ ($J = 1,2$) was observed by CLEO [523] using 6 million decays of the $\Upsilon(3S)$. This is the only observed hadronic transition from the $\chi_{bJ}(nP)$ states aside from the surprisingly large $\chi_{bJ}(2P) \to \omega\Upsilon(1S)$; as such, it provides an important benchmark. The following decay chain was used in the search: $\Upsilon(3S) \to \gamma\chi_{bJ'}(2P)$; $\chi_{bJ'}(2P) \to \pi\pi\chi_{bJ}(1P)$; $\chi_{bJ}(1P) \to \gamma\Upsilon(1S)$; $\Upsilon(1S) \to \ell^+\ell^-$. Both charged and neutral pion pairs were sought, with results combined assuming isospin invariance. The data were not adequate to distinguish between allowed values of J' and J; hence it was assumed that $J' = J$. The $J = 0$ transition was inaccessible due to the smallness of the branching fractions in the decay chain. Assuming the $J = 2$ and $J = 1$ transitions have the same partial width, it was found that

$$\Gamma\big(\chi_b(2P) \to \pi\pi\chi_b(1P)\big) = 0.83 \pm 0.22 \pm 0.21 \text{ keV}. \quad (146)$$

This rate is consistent with theoretical expectations [517].

3.4 Hadronic decays

In general, the nonrelativistic quark model does a remarkable job also for decays of the lowest (below $\bar{D}D$ threshold) charmonia—see, e.g., [31]. However, there are some striking discrepancies where additional experimental as well as theoretical work is necessary, as will be outlined below—see, e.g., Sect. 3.4.1. Improved data, e.g., sensitive to lineshapes due to improved resolution and statistics, as well as data on additional transitions, should shed important light on the structure of the light charmonia.

As described at the end of Sect. 3.3.1, for some transitions heavy-meson loops might play a significant role. The natural question that arises asks what their influence on decays could be. Here, unfortunately, the NREFT described above is not applicable anymore, for the momenta of the final-state particles as well as the intermediate heavy-meson velocities get too large for a controlled expansion. However, those studies can be performed within phenomenological models [440, 551–553].

3.4.1 The 12% rule and $\rho\pi$ puzzle

From perturbative QCD (pQCD), it is expected that both J/ψ and $\psi(2S)$ decay into any exclusive light-hadron final state with a width proportional to the square of the wave function at the origin [554, 555]. This yields the pQCD "12% rule",

$$Q_h \equiv \frac{\mathcal{B}(\psi(2S) \to h)}{\mathcal{B}(J/\psi \to h)} = \frac{\mathcal{B}(\psi(2S) \to e^+e^-)}{\mathcal{B}(J/\psi \to e^+e^-)} \approx 12\%. \tag{147}$$

A large violation of this rule was first observed in decays to $\rho\pi$ and $K^{*+}K^- + c.c.$ by Mark II [556], and became known as *the $\rho\pi$ puzzle*. Since then, many two-body decay modes of the $\psi(2S)$ (and some multibody ones) have been measured by BES [557–563] and CLEO [564]; some decays obey the rule while others violate it to varying degrees.

The $\rho\pi$ mode is essential for this study—the recent measurements, together with the old information, show us a new picture of the charmonium decay dynamics [565]. With a weighted average of $\mathcal{B}(J/\psi \to \pi^+\pi^-\pi^0) = (2.00 \pm 0.09)\%$ from the existing measurements, and an estimation of

$$\frac{\mathcal{B}(J/\psi \to \rho\pi)}{\mathcal{B}(J/\psi \to \pi^+\pi^-\pi^0)} = 1.17 \times (1 \pm 10\%) \tag{148}$$

using the information given in [566], one gets $\mathcal{B}(J/\psi \to \rho\pi) = (2.34 \pm 0.26)\%$. This is substantially larger than the world average listed by PDG08 [18], which is $(1.69 \pm 0.15)\%$, from a simple average of many measurements. The branching fraction of $\psi(2S) \to \pi^+\pi^-\pi^0$ is measured to be $(18.1 \pm 1.8 \pm 1.9) \times 10^{-5}$ and $(18.8^{+1.6}_{-1.5} \pm 1.9) \times 10^{-5}$ at BESII [567] and CLEO [564], respectively. To extract the $\rho\pi$ component, however, the experiments make different choices, which in turn lead to different answers. BESII uses a partial wave analysis (PWA), while CLEO counts the number of events by applying a ρ mass cut. The branching fraction from BESII is $(5.1 \pm 0.7 \pm 1.1) \times 10^{-5}$, while that from CLEO is $(2.4^{+0.8}_{-0.7} \pm 0.2) \times 10^{-5}$. If we take a weighted average and inflate the resulting uncertainty with a PDG-like scale factor accounting for the disagreement, we obtain $\mathcal{B}(\psi(2S) \to \rho\pi) = (3.1 \pm 1.2) \times 10^{-5}$. With the results from above, one gets

$$Q_{\rho\pi} = \frac{\mathcal{B}(\psi(2S) \to \rho\pi)}{\mathcal{B}(J/\psi \to \rho\pi)} = (0.13 \pm 0.05)\%. \tag{149}$$

The suppression compared to the 12% rule is about two orders of magnitude.

There are enough measurements of $\psi(2S)$ and J/ψ decays for an extensive study of the "12% rule" [18, 557–564, 568–570], among which the Vector-Pseudoscalar (VP) modes, like the $\rho\pi$, have been measured with the highest priority. The ratios of the branching fractions are generally suppressed relative to the 12% rule for the non-isospin-violating VP and Vector-Tensor (VT) modes (i.e., excluding modes like $\omega\pi^0$ and $\rho^0\eta$), while Pseudoscalar-Pseudoscalar (PP) modes are enhanced. The multihadron modes and the baryon–antibaryon modes cannot be simply characterized, as some are enhanced, some are suppressed, and some match the expectation of the rule. Theoretical models, developed for interpreting specific modes, have not yet provided a solution for all of the measured channels. For a recent review, see [1, 571].

3.4.2 Observation of $h_c(1P) \to 2(\pi^+\pi^-)\pi^0$

CLEO reported the first evidence for an exclusive hadronic decay mode of the $h_c(1P)$ [536], previously only seen through its radiative transition to the $\eta_c(1S)$. Using 25.7×10^6 $\psi(2S)$ decays, CLEO performed a search for hadronic decays of the $h_c(1P)$ in the channels $\psi(2S) \to \pi^0 h_c(1P)$; $h_c(1P) \to n(\pi^+\pi^-)\pi^0$, where $n = 1, 2, 3$. Upper limits were set for the 3π and 7π decay modes, but evidence for a signal was found in the 5π channel with a significance of 4.4 σ. The 5π mass distribution from data is shown in Fig. 47(a); the corresponding spectrum from background Monte Carlo is shown in Fig. 47(b). The measured mass is consistent with previous measurements. The measured product branching fraction is

$$\begin{aligned}&\mathcal{B}\big(\psi(2S) \to \pi^0 h_c(1P)\big) \times \mathcal{B}\big(h_c(1P) \to 2(\pi^+\pi^-)\pi^0\big)\\&\quad = \big(1.88^{+0.48+0.47}_{-0.45-0.30}\big) \times 10^{-5}.\end{aligned} \tag{150}$$

This value is approximately 5% of that given in (100), indicating the total hadronic decay width of the $h_c(1P)$ is likely of the same order as its radiative transition width to the $\eta_c(1S)$.

3.4.3 $\chi_{cJ}(1P)$ hadronic decays

Precision in study of hadronic $\chi_{cJ}(1P)$ decays continues to improve with larger datasets, and is beginning to approach that achieved for J/ψ and $\psi(2S)$. Decays of the P-wave states provide information complementary to that from the S-wave states, which probe short-range processes. At CLEO and BESII the P-wave states are accessed through the predominant E1 radiative transitions $\psi(2S) \to \gamma\chi_{cJ}(1P)$. Recent measurements have also come from Belle through the process $\gamma\gamma \to \chi_{c0,2}(1P)$.

A large number of different hadronic decay modes of the $\chi_{cJ}(1P)$ states have recently been measured, many for the

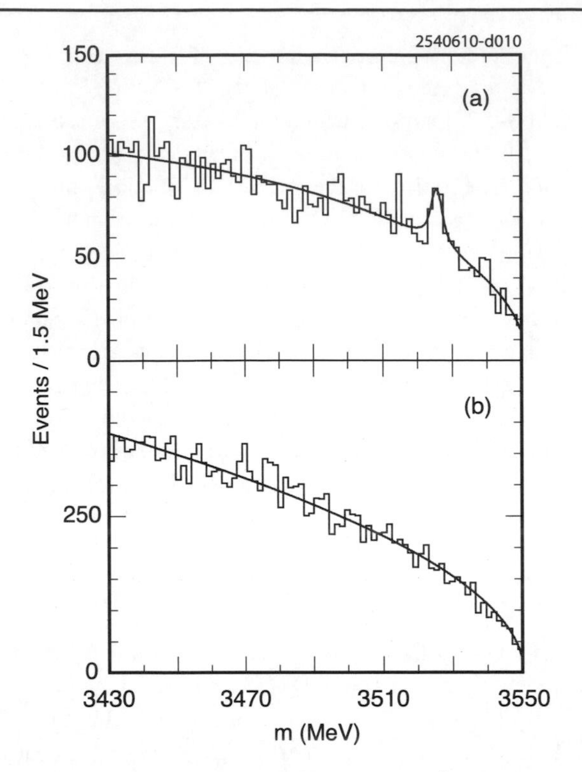

Fig. 47 From CLEO [536], the $2(\pi^+\pi^-)\pi^0$ mass distribution showing evidence for the $\psi(2S) \to \pi^0 h_c(1P)$ followed by $h_c(1P) \to 2(\pi^+\pi^-)\pi^0$. *Solid line histograms* show (**a**) data and (**b**) Monte Carlo simulation of backgrounds. *Solid curves* show the results of fits to the respective histograms for a smooth background and $h_c(1P)$ signal. Adapted from [536] with kind permission, copyright (2009) The American Physical Society

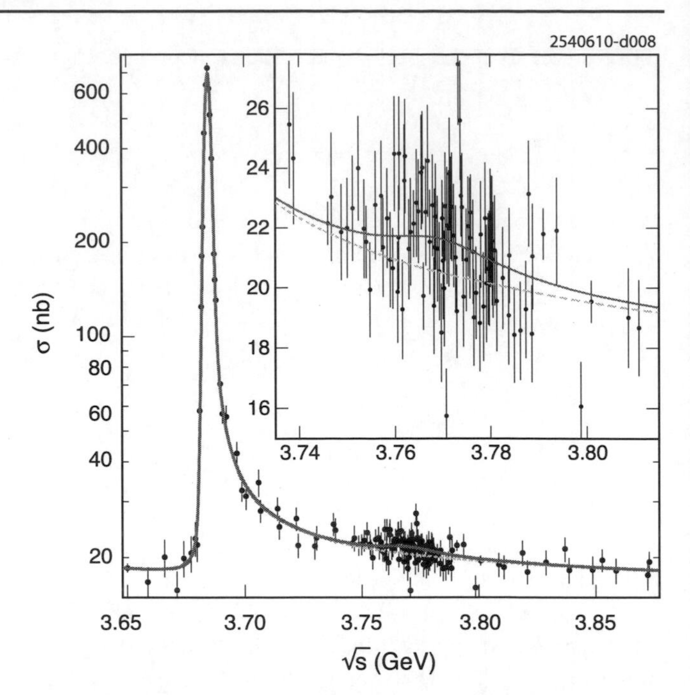

Fig. 48 From BES [503], the hadronic cross section versus $\sqrt{s}$, extracted from counting inclusively selected hadronic events with a charged kaon of energy 1.15–2.00 GeV, a range which excludes $D\bar{D}$ events. *Solid circles* represent data, the *dashed curve* represents the contributions from J/ψ, $\psi(2S)$, and continuum hadron production, and the *solid curve* the best fit to the data of expected background plus a floating $\psi(3770) \to$ non-$D\bar{D}$ component (see text). Adapted from [503] with kind permission, copyright (2008) Elsevier

first time. These include two-meson decays at BESII [572], CLEO [573], and Belle [485, 487–489, 491]; two-baryon decays at CLEO [574]; three-meson decays at CLEO [575] and BESII [576]; four-meson decays at BESII [577–580], CLEO [581], and BELLE [442]; as well as others like $4\pi p\bar{p}$ and $p\bar{n}\pi^0$ at BESII [563, 582].

More work is required on the theoretical front to understand the wide variety of hadronic $\chi_{cJ}(1P)$ decays now experimentally observed. There are indications that the pattern of decays may require the introduction of a color octet mechanism that includes contributions from the subprocess $c\bar{c}g \to q\bar{q}$ [456]. Hadronic decays of the $\chi_{cJ}(1P)$ may also provide insight into the relative contributions of singly and doubly OZI violating processes [583].

3.4.4 Non-$D\bar{D}$ $\psi(3770)$ hadronic decays

Experimental measurements of the branching fractions for $\psi(3770) \to$ non-$D\bar{D}$ provide important information about the nature of the $\psi(3770)$ resonance. While the $\psi(3770)$ is thought to be dominantly the 1^3D_1 state of charmonium, its dilepton width indicates it should also contain a substantial n^3S_1 component. The size of this component is related to the non-$D\bar{D}$-width of the $\psi(3770)$ [425, 452]. Several measurements of the cross sections for $\psi(3770) \to$ non-$D\bar{D}$ final states have recently been made, but there is a disagreement between those from BES [502, 503, 584, 585] and that of CLEO [508].

The most recent and precise result from BES [503] is $\mathcal{B}(\psi(3770) \to \text{non-}D\bar{D}) = (15.1 \pm 5.6 \pm 1.8)\%$. This was obtained by counting inclusively selected hadronic events with a charged kaon of energy 1.15–2.00 GeV, a range which excludes $D\bar{D}$ events, in the range of center-of-mass energies 3.650–3.872 GeV, and correcting the observed number for efficiency and integrated luminosity at each energy point. These cross sections were then fit for a smooth background and the expected lineshape of $\psi(3770)$ with floating normalization. Figure 48 shows these measured cross sections versus the center-of-mass energy together with the best fit, which indicates evidence for non-$D\bar{D}$ decays. Combining this result with another one, also by BES [585], which measured cross sections at just two points (3.773 GeV and 3.650 GeV) gives a 4.8σ signal significance for $\psi(3770) \to$ non-$D\bar{D}$ decays. Other measurements from BES—which include variations on the above-described techniques—are listed in Table 40.

CLEO [508] measured the cross section for $\psi(3770) \to$ non-$D\bar{D}$ by comparing the cross sections for $\psi(3770) \to$

Table 40 Measurements of branching fractions for $\psi(3770) \to D\bar{D}$ and non-$D\bar{D}$ at BES, CLEO, and as averaged by PDG10. Rates for $D^0\bar{D}^0$ and D^+D^- are constrained to sum to that of $D\bar{D}$, and the rates for $D\bar{D}$ and non-$D\bar{D}$ are constrained to sum to unity. Hence within each set of measurements, uncertainties are highly correlated

Final state	$\mathcal{B}$ (%)	Source
$D^0\bar{D}^0$	$49.9 \pm 1.3 \pm 3.8$	BES [502]
D^+D^-	$35.7 \pm 1.1 \pm 3.4$	BES [502]
$D\bar{D}$	$85.5 \pm 1.7 \pm 5.8$	BES [502]
non-$D\bar{D}$	$14.5 \pm 1.7 \pm 5.8$	BES [502]
$D^0\bar{D}^0$	$46.7 \pm 4.7 \pm 2.3$	BES [584]
D^+D^-	$36.9 \pm 3.7 \pm 2.8$	BES [584]
$D\bar{D}$	$83.6 \pm 7.3 \pm 4.2$	BES [584]
non-$D\bar{D}$	$16.4 \pm 7.3 \pm 4.2$	BES [584]
$D\bar{D}$	$86.6 \pm 5.0 \pm 3.6$	BES [503]
non-$D\bar{D}$	$13.4 \pm 5.0 \pm 3.6$	BES [503]
$D\bar{D}$	$103.3 \pm 1.4^{+4.8}_{-6.6}$	CLEO [508]
non-$D\bar{D}$	$-3.3 \pm 1.4^{+6.6}_{-4.8}$	CLEO [508]
$D\bar{D}$	93^{+8}_{-9}	PDG10 [447]
non-$D\bar{D}$	7^{+9}_{-8}	PDG10 [447]

Table 41 Measurements of non-$D\bar{D}$ cross sections for $\psi(3770)$ decays and the experimentally observed cross section for $\psi(3770)$ production at 3.773 GeV

Experiment	$\sigma^{\text{obs}}_{\text{non-}D\bar{D}}$ (nb)	$\sigma^{\text{obs}}_{\psi(3770)}$ (nb)
CLEO [508]	$-0.01 \pm 0.08^{+0.41}_{-0.30}$	$6.38 \pm 0.08^{+0.41}_{-0.30}$
BESII [502]	$1.14 \pm 0.08 \pm 0.59$	$7.18 \pm 0.20 \pm 0.63$
BESII [584]	$1.04 \pm 0.23 \pm 0.13$	$6.94 \pm 0.48 \pm 0.28$
BESII [503]	$0.95 \pm 0.35 \pm 0.29$	$7.07 \pm 0.36 \pm 0.45$
BESII [585]	$1.08 \pm 0.40 \pm 0.15$	–
MARKII [11]	–	9.1 ± 1.4

hadrons to that measured for $\psi(3770) \to D\bar{D}$ [30]. The former quantity is obtained by subtracting the hadronic cross section measured on the "continuum" at $\sqrt{s} = 3671$ MeV, as extrapolated to $\sqrt{s} = 3773$ MeV and corrected for interference with $\psi(2S)$ decays, from the hadronic cross section measured near the peak of the $\psi(3770)$, $\sqrt{s} = 3773$ MeV. The net non-$D\bar{D}$ cross section obtained by CLEO is smaller than those of BES and is consistent with zero. These non-$D\bar{D}$ cross sections are summarized in Table 41, and resulting branching fractions in Table 40. As the CLEO and BES results are in conflict, the PDG10 average in Table 40 averages between the two and inflates the combined uncertainty so as to be consistent with both.

To search for light-hadron decays of $\psi(3770)$, both BES [586–592] and CLEO [593–595] extensively studied

Table 42 Measurements from CLEO [595] of cross sections for $e^+e^- \to \pi^+\pi^-\pi^0$ and $e^+e^- \to$ VP channels on the continuum just below the $\psi(2S)$ at $\sqrt{s} = 3.671$ GeV and on the $\psi(3770)$ resonance at $\sqrt{s} = 3.773$ GeV

Final state	σ (pb) $\sqrt{s} = 3.671$ GeV	σ (pb) $\sqrt{s} = 3.773$ GeV
$\pi^+\pi^-\pi^0$	$13.1^{+1.9}_{-1.7} \pm 2.1$	$7.4 \pm 0.4 \pm 2.1$
$\rho\pi$	$8.0^{+1.7}_{-1.4} \pm 0.9$	$4.4 \pm 0.3 \pm 0.5$
$\rho^0\pi^0$	$3.1^{+1.0}_{-0.8} \pm 0.4$	$1.3 \pm 0.2 \pm 0.2$
$\rho^+\pi^-$	$4.8^{+1.5}_{-1.2} \pm 0.5$	$3.2 \pm 0.3 \pm 0.2$
$\omega\eta$	$2.3^{+1.8}_{-1.0} \pm 0.5$	$0.4 \pm 0.2 \pm 0.1$
$\phi\eta$	$2.1^{+1.9}_{-1.2} \pm 0.2$	$4.5 \pm 0.5 \pm 0.5$
$K^{*0}\bar{K}^0$	$23.5^{+4.6}_{-3.9} \pm 3.1$	$23.5 \pm 1.1 \pm 3.1$
$K^{*+}K^-$	$1.0^{+1.1}_{-0.7} \pm 0.5$	<0.6

various exclusive light-hadron decay modes for $\psi(3770) \to$ LH (LH $\equiv$ light hadron), but for only one channel was found to have a significant signal: CLEO [595] measured the branching fraction $\mathcal{B}(\psi(3770) \to \phi\eta) = (3.1 \pm 0.6 \pm 0.3) \times 10^{-4}$. This branching fraction is obtained by subtracting the extrapolated continuum cross section for $e^+e^- \to \phi\eta$ measured at $\sqrt{s} = 3671$ MeV from that measured at 3.773 GeV. CLEO's measurement explicitly ignored the possible interference among amplitudes for this final state from $\psi(3770)$, continuum, and $\psi(2S)$.

Although CLEO did not claim observations for other light hadron decay modes, some evidence for such decays can be found in the CLEO cross sections. Table 42 lists some cross sections for $e^+e^- \to$ LH measured at 3.773 and 3.671 GeV by CLEO [595]. Curiously, the final states $\pi^+\pi^-\pi^0$, $\rho\pi$, and $\omega\eta$ have *smaller* cross sections at $\sqrt{s} =$ 3773 MeV than at 3671 MeV, suggesting that an interference effect has come into play to reduce the observed cross section. If this is the cause, this would imply $\psi(3770)$ branching fractions for these modes of order 10^{-4}–10^{-3}. It is also noteworthy that $K^{*0}\bar{K}^0$ is produced more copiously than $K^{*+}K^-$ (by a factor of at least 20) at *both* energies. Whether these two phenomena are related remains an open question.

BES [19] has observed a $\psi(3770)$-lineshape anomaly in measurement of inclusive cross sections measured in the range $\sqrt{s} = 3.70$–3.87 GeV (see also Sect. 2.1.1). It was suggested by BES and by Dubynskiy and Voloshin [596] that a second structure near 3765 MeV could be responsible. If such a structure exists, and it decays into some of the low-multiplicity LH states discussed above, yet another amplitude comes into play which could interfere and cause observed cross sections to be smaller near the $\psi(3770)$ than on the continuum. A very recent preliminary analysis by KEDR [20] of its e^+e^- scan data near $\psi(3770)$ finds inconsistency with this lineshape anomaly.

The conflict about the fraction of non-$D\bar{D}$ decays from $\psi(3770)$ is not restricted to being between the BESII and CLEO experiments; an inclusive-exclusive rift also remains to be bridged. Inclusive-hadronic measurements alone have difficulty supporting a $\psi(3770) \to$ non-$D\bar{D}$ branching fraction of more than several percent without some confirmation in exclusive mode measurements, which currently show a large number of modes, including the leading candidate low-multiplicity transitions and decays, to be quite small. The large datasets expected at BESIII offer an opportunity for such a multifaceted approach to $\psi(3770) \to$ non-$D\bar{D}$.

3.4.5 *Observation of* $\Upsilon(1S) \to$ antideuteron + X

The appearance of deuterons in fragmentation has been addressed theoretically in the framework of a coalescence model [597, 598], via the binding of a nearby neutron and proton. Experimental constraints on the process from measured deuteron production are limited. ARGUS [599] found evidence in $\Upsilon(1S, 2S)$ decays for such production but with very low statistics. These results have been accommodated in a string-model calculation [600]. Further experimental information is essential.

CLEO [334] addressed this issue with a measurement of antideuteron production in samples of 22, 3.7, and 0.45 million $\Upsilon(1S, 2S, 4S)$ decays, respectively. Only antideuterons were sought due to the presence of a large deuteron background. This background arises from nuclear interactions with matter (such as gas, vacuum chambers, and beam collimators) initiated by either particles created in the e^+e^- annihilation or errant $e^\pm$ from the colliding beams. These interactions result in the appearance of neutrons, protons, and deuterons in the detector. Antideuterons are identified primarily by their distinctive energy loss (dE/dx) as a function of momentum, as measured in tracking chambers, but residual pion and proton backgrounds are additionally suppressed by requiring deuteron-appropriate response in the RICH detectors. This identification is relatively background-free over the momentum range of 0.45–1.45 GeV/c, and is found to correspond to a branching fraction

$$\mathcal{B}\big(\Upsilon(1S) \to \bar{\mathrm{d}}X\big) = (2.86 \pm 0.19 \pm 0.21) \times 10^{-5}. \qquad (151)$$

How often is the baryon-number conservation for the $\bar{\mathrm{d}}$ accomplished with a deuteron? Deuteron background becomes tolerable in $\bar{\mathrm{d}}$-tagged events, and three $\mathrm{d}\bar{\mathrm{d}}$ candidate events are found in the CLEO $\Upsilon(1S)$ data sample, meaning that baryon-number compensation occurs with a deuteron about 1% of the time. By counting events with zero, one, or two protons accompanying an identified $\bar{\mathrm{d}}$, CLEO finds that compensation by each of pn, np, nn, or pp occurs at roughly the same rate, about a quarter of the time.

CLEO also measures antideuteron fractions in the data samples of $\Upsilon(2S)$ and $\Upsilon(4S)$. The $\Upsilon(2S)$ result can be used to calculate the rate of $\chi_{bJ} \to \bar{\mathrm{d}}X$ by subtracting a scaled $\Upsilon(1S)$ $\bar{\mathrm{d}}$-fraction to account for $\Upsilon(2S) \to X\Upsilon(1S)$ transitions, and again, with a different scaling factor, to account for $\Upsilon(2S) \to ggg$, γgg decays, assuming they have the same $\bar{\mathrm{d}}$-fraction as $\Upsilon(1S)$; the balance are attributed to appearance through $\Upsilon(2S) \to \gamma\chi_{bJ}$, $\chi_{bJ} \to \bar{\mathrm{d}}X$. No significant excess is observed, either here nor in $\Upsilon(4S)$ decays, so CLEO reports the upper limits $\mathcal{B}(\chi_{bJ} \to \bar{\mathrm{d}}X) < 1.1 \times 10^{-4}$, averaged over $J = 0, 1, 2$, and $\mathcal{B}(\Upsilon(4S) \to \bar{\mathrm{d}}X) < 1.3 \times 10^{-5}$, both at 90% CL.

Artoisenet and Braaten [333] find this CLEO result useful in tuning parameters of event generators in a study of production of loosely bound hadronic molecules (see Sect. 2.9.3 for a discussion of the relevance of this measurement to the nature of the $X(3872)$). Brodsky [601] comments on how measurements such as these should be extended in order to probe the hidden-color structure of the antideuteron wave function.

3.4.6 *Observation of* Υ, $\chi_{bJ} \to$ *open charm*

Very little is known about the heavy-flavor (i.e., open charm) content of bottomonium hadronic decays, which can be used as a tool to probe of the post-$b\bar{b}$-annihilation fragmentation processes. $\Upsilon(nS)$ hadronic decays are dominated by those materializing through three gluons (ggg), $\chi_{b0,2}$ through gg, and χ_{b1} through $q\bar{q}g$ [495]; each of these processes is expected to have its own characteristic open-charm content.

CLEO [602] first selected events from its $\Upsilon(2S, 3S)$ data samples to have a D^0 (or $\bar{D}^0$) meson exclusively reconstructed, and with momentum >2.5 GeV/c; the momentum cut is required to suppress backgrounds. The single-photon energy spectra for these events were then fit for the presence of narrow peaks corresponding to the radiative transitions $\Upsilon(mS) \to \gamma\chi_{bJ}(nP)$. Significant signals for D^0 production only from $\chi_{b1}(1P, 2P)$ are observed with branching fractions (*not* correcting for the D^0 momentum cut) of about 10%, with upper limits at 90% CL for the others ranging from 2–10%. CLEO then combines these measured fractions with some assumptions and theoretical input to simultaneously extract ρ_8, an NRQCD nonperturbative parameter, and branching fractions for D^0 mesons produced with *any* momenta; the assumptions and ρ_8 directly affect the spectrum of D^0 mesons. With these assumptions, CLEO obtains $\rho_8 \approx 0.09$ and

$$\mathcal{B}\big(\chi_{b1}(1P, 2P) \to D^0X\big) \approx 25\%, \qquad (152)$$

both in agreement with predictions [603, 604]. Upper limits at 90% CL are set for the remaining four χ_{bJ} that are

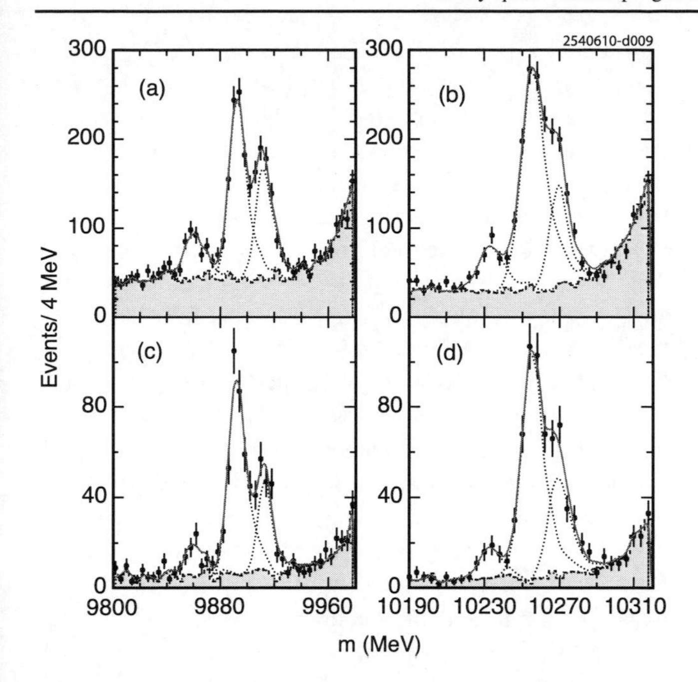

Fig. 49 From CLEO [608], the invariant mass of the sum of exclusive decay modes of the $\chi_{bJ}(1P)$ from (**a**), (**c**) $\Upsilon(2S)$ and (**b**), (**d**) $\Upsilon(3S)$ radiative decays. (**a**) and (**b**) show the sum of 659 exclusive decay modes, while (**c**) and (**d**) show the sum of the 14 decay modes with significant branching fractions, the values of which were measured. *Solid circles* represent data, *shaded histograms* the backgrounds determined from $\Upsilon(1S)$ data, *dotted curves* the contributions of the individual χ_{bJ} signals as determined from fits to the data, and *solid curves* the sum of background and fitted signals. Adapted from [608] with kind permission, copyright (2008) The American Physical Society

also consistent with the predictions, with the exception of $\chi_{b2}(2P)$, which has an upper limit that is half the predicted value of 12%. The largest branching fractions occur for $\chi_{b1}(1P,2P)$, as expected, which are the states expected to decay via $q\bar{q}g$.

BABAR [605] searched its $\Upsilon(1S)$ data sample for the presence of D^{*+} (and D^{*-}) mesons. BABAR measures

$$\mathcal{B}(\Upsilon(1S) \to D^{*+}X) = (2.52 \pm 0.13 \pm 0.15)\%, \tag{153}$$

the first such observation of charm in $\Upsilon(1S)$ decays, and extracts the resulting D^{*+} momentum spectrum. BABAR also finds this rate to be considerably in excess (i.e., roughly double) of that expected from $b\bar{b}$ annihilation into a single photon. The excess is seen to be in agreement with a prediction [606] based on splitting a virtual gluon, but appears to be too small to accommodate an octet-state contribution [607].

3.4.7 Observation of $\chi_{bJ}(1P,2P) \to$ hadrons

Using its full sample of 9 million $\Upsilon(2S)$ and 6 million $\Upsilon(3S)$ decays collected with the CLEO III detector, CLEO made the first measurements of branching fractions of exclusive decays of the $\chi_{bJ}(1P,2P)$ states [608], which were accessed through the allowed E1 radiative transitions $\Upsilon(3S) \to \gamma\chi_{bJ}(2P)$ and $\Upsilon(2S) \to \gamma\chi_{bJ}(1P)$. A comprehensive set of 659 decay exclusive modes was included in the search, where each decay mode was a combination of $\pi^\pm$, π^0, η, $K^\pm$, K_S, and $p^\pm$. Of these 659, 14 were found that have significances of greater than 5σ for both $\chi_{bJ}(1P)$ and $\chi_{bJ}(2P)$ decays (combined for $J = 0, 1, 2$). Figure 49 shows the invariant mass of the exclusive decay modes for all 659 and the selected 14 decay modes of the $\chi_{bJ}(1P)$ and $\chi_{bJ}(2P)$. Branching fractions were measured for the 14 decay modes from both the $\chi_{bJ}(1P)$ and $\chi_{bJ}(2P)$ for $J = 0, 1, 2$. The branching fractions ranged from approximately $(1\text{–}20) \times 10^{-4}$. The largest branching fractions measured were to $6\pi 2\pi^0$ and $8\pi 2\pi^0$. This survey of branching fractions, besides being useful for testing models of bottomonium decays, also gives some indication of which exclusive channels might be most fruitful in searches for new states in the bottomonium region.

4 Production[9]

4.1 Introduction to theoretical concepts

In this subsection, we provide an introduction to some of the theoretical ideas that will appear in subsequent subsections, many of which are based on various factorization formulas for quarkonium production and decay. This subsection also serves to establish notation and nomenclature that is used in subsequent subsections.

4.1.1 Momentum scales and factorization

A heavy quarkonium has at least three intrinsic momentum scales: the heavy-quark mass m_Q; the momentum of the heavy quark or antiquark in the quarkonium rest frame, which is of order $m_Q v$; and the binding energy of the heavy-quark–antiquark ($Q\bar{Q}$) pair, which is of order $m_Q v^2$. Here v is the typical velocity of the heavy quark or antiquark in the quarkonium rest frame. ($v^2 \approx 0.3$ for the J/ψ and $v^2 \approx 0.1$ for the Υ.)

If a heavy quarkonium is produced in a hard-scattering process, then, in addition to the intrinsic scales of the quarkonium, the hard-scattering scale p enters into the description of the production process. The hard-scattering scale p is usually set by a large momentum transfer in the production process. In quarkonium production in hadron–hadron collisions (hadroproduction) and hadron-lepton collisions, p is usually taken to be of order p_T, the transverse

[9] P. Artoisenet, A. Bertolin, G.T. Bodwin†, C.-H. Chang, K.-T. Chao, J.-P. Lansberg, F. Maltoni, A. Meyer†, V. Papadimitriou†, and J.-W. Qiu.

momentum of the quarkonium, while in quarkonium production in e^+e^- collisions, p is usually taken to be of order p^*, the quarkonium momentum in the e^+e^- center-of-mass frame.

One might expect intuitively that the production process could be understood in terms of two distinct steps: the production of the $Q\bar{Q}$ pair, which would occur at the scale p, and the subsequent evolution of the $Q\bar{Q}$ pair into the quarkonium, which would involve the smaller dynamical scales $m_Q v$ and $m_Q v^2$. The first step would be calculable in an expansion in powers of $\alpha_s(p)$, while the second step would typically involve inherently nonperturbative physics. The term "short distance" is often used to refer to the momentum scale p (distance scale $1/p$), while the term "long distance" is often used to refer to typical hadronic momentum scales, such as $m_Q v$, $m_Q v^2$, or $\Lambda_{\rm QCD}$. The term "short distance" is also sometimes used to refer to the scale m_Q in the context of NRQCD.

In order to establish that this intuitive picture of quarkonium production is actually a property of QCD, one must demonstrate that the short-distance, perturbative effects at the scale p can be separated from the long-distance, nonperturbative dynamics. Such a separation is known as "factorization." In proving a "factorization theorem," one must show that an amplitude or cross section can be expressed as a sum of products of infrared-safe, short-distance coefficients with well defined operator matrix elements. Such short-distance coefficients are perturbatively calculable. The operator matrix elements would contain all of the long-distance, nonperturbative physics. They might be determined phenomenologically or, possibly, through lattice simulations. If it can be further demonstrated that the long-distance matrix elements are universal, i.e., process independent, then factorization formulas yield much greater predictive power.

The nonperturbative evolution of the $Q\bar{Q}$ pair into a quarkonium has been discussed extensively in terms of models and in terms of the language of effective theories of QCD [1, 134, 138]. Different treatments of this evolution have led to various theoretical models for inclusive quarkonium production. Most notable among these are the color-singlet model (CSM), the color-evaporation model (CEM), the nonrelativistic QCD (NRQCD) factorization approach, and the fragmentation-function approach.

4.1.2 The color-singlet model

The CSM was first proposed shortly after the discovery of the J/ψ [609–616]. In this model, it is assumed that the $Q\bar{Q}$ pair that evolves into the quarkonium is in a color-singlet state and that it has the same spin and angular-momentum quantum numbers as the quarkonium. In the CSM, the production rate for each quarkonium state is related to the absolute values of the color-singlet $Q\bar{Q}$ wave function and its derivatives, evaluated at zero $Q\bar{Q}$ separation. These quantities can be extracted by comparing theoretical expressions for quarkonium decay rates in the CSM with experimental measurements. Once this extraction has been carried out, the CSM has no free parameters. The CSM was successful in predicting quarkonium production rates at relatively low energy [617]. Recently, it has been found that, at high energies, very large corrections to the CSM appear at next-to-leading order (NLO) and next-to-next-to-leading order (NNLO) in α_s [618–620]. (See Sect. 4.2.1.) Consequently, the possibility that the CSM might embody an important production mechanism at high energies has re-emerged. However, given the very large corrections at NLO and NNLO, it is not clear that the perturbative expansion in α_s is convergent. Furthermore, in the production and decay of P-wave and higher-orbital-angular-momentum quarkonium states, the CSM is known to be inconsistent because it leads to uncanceled infrared divergences. (See [1] and references therein.) As we will describe below, the NRQCD factorization approach encompasses the color-singlet model, but goes beyond it.

4.1.3 The color-evaporation model

The CEM [621–626] is motivated by the principle of quark-hadron duality. In the CEM, it is assumed that every produced $Q\bar{Q}$ pair evolves into a quarkonium if it has an invariant mass that is less than the threshold for producing a pair of open-flavor heavy mesons. It is further assumed that the nonperturbative probability for the $Q\bar{Q}$ pair to evolve into a quarkonium state H is given by a constant F_H that is energy-momentum and process independent. Once F_H has been fixed by comparison with the measured total cross section for the production of the quarkonium H, the CEM can predict, with no additional free parameters, the momentum distribution of the quarkonium production rate. The CEM predictions provide rough descriptions of the CDF data for J/ψ, $\psi(2S)$, and χ_{cJ} production at $\sqrt{s} = 1.8$ TeV [626]. In [627], the CEM predictions are fit to the CDF data for J/ψ, $\psi(2S)$, and χ_{cJ} production at $\sqrt{s} = 1.8$ TeV [628]. The quality of these fits is generally poor, with χ^2/d.o.f. for the J/ψ fits of about 7–8 without initial-state k_T smearing and 2–4.5 with initial-state k_T smearing. In contrast, the NRQCD factorization approach, which we are about to describe, yields fits to the CDF J/ψ data with χ^2/d.o.f. of about 1.

4.1.4 The NRQCD factorization approach

The NRQCD factorization approach [138] to heavy-quarkonium production is by far the most sound theoretically and most successful phenomenologically. NRQCD is an effective theory of QCD and reproduces full QCD dynamics accurately at momentum scales of order $m_Q v$ and smaller. Dynamics involving momentum scales of order m_Q or larger are taken into account through the short-distance coefficients of the operators that appear in the NRQCD action. The NRQCD factorization approach expresses the probability for a $Q\bar{Q}$ pair to evolve into a quarkonium in terms of matrix elements of NRQCD operators. These matrix elements can be characterized in terms of their scaling with the heavy-quark velocity v [138]. In the NRQCD factorization approach, the inclusive cross section for the direct production of a quarkonium state H is written as a sum of products of these NRQCD matrix elements with the corresponding $Q\bar{Q}$ production cross sections:

$$\sigma(H) = \sum_n \sigma_n(\Lambda)\langle \mathcal{O}_n^H(\Lambda)\rangle. \qquad (154)$$

Here Λ is the ultraviolet cut-off of the effective theory, the σ_n are expansions in powers of v of the cross sections to produce a $Q\bar{Q}$ pair in the color, spin, and orbital-angular momentum state n. The σ_n are convolutions of parton-level cross sections at the scale p with parton distribution functions. (The former are short-distance quantities, while the latter are long-distance quantities that depend on the non-perturbative dynamics of the initial hadrons.) The matrix elements $\langle \mathcal{O}_n^H(\Lambda)\rangle$ are vacuum-expectation values of four-fermion operators in NRQCD. We emphasize that (154) represents both processes in which the $Q\bar{Q}$ pair is produced in a color-singlet state and processes in which the $Q\bar{Q}$ pair is produced in a color-octet state. It is conjectured that the NRQCD factorization expression in (154) holds when the momentum transfer p in the hard-scattering production process is of order m_Q or larger.

Unlike the CSM and the CEM expressions for the production cross section, the NRQCD factorization formula for heavy-quarkonium production depends on an infinite number of unknown matrix elements. However, the sum in (154) can be organized as an expansion in powers of v. Hence, the NRQCD factorization formula is a double expansion in powers of v and powers of α_s. In phenomenological applications, the sum in (154) is truncated at a fixed order in v, and only a few matrix elements typically enter into the phenomenology. The predictive power of the NRQCD factorization approach is based on the validity of such a truncation, as well as on perturbative calculability of the $Q\bar{Q}$ cross sections and the universality of the long-distance matrix elements.

If one retains in (154) only the color-singlet contributions of leading order in v for each quarkonium state, then one obtains the CSM. As we have mentioned, such a truncation leads to inconsistencies because the omission of color-octet contributions results in uncanceled infrared divergences in the production rates of P-wave and higher-orbital-angular-momentum quarkonium states.

The CEM implies that certain relationships must hold between the NRQCD long-distance matrix elements [627]. These relationships are generally inconsistent with the scaling of the matrix elements with v that is predicted by NRQCD. The shortcomings of the CEM in describing the Fermilab Tevatron data can be traced, at least in part, to these inconsistencies [627].

As we will explain in more detail below, in the case of inclusive quarkonium production, a compelling proof of NRQCD factorization is still lacking. A further difficulty with the NRQCD factorization formula (154) is that a straightforward perturbative expansion of the short-distance coefficients may not yield an optimal organization of the expression for the cross section. The difficulty with such a straightforward expansion is that it ignores the fact that different orders in α_s in the perturbative expansion may have different dependences on m_Q/p. Consequently, at large p/m_Q, higher orders in the perturbation expansion may be more important than lower orders. Therefore, it may be useful to organize the production cross section in powers of p/m_Q before expanding the short-distance coefficients in powers of α_s [629–631].

Although the application of NRQCD factorization to heavy-quarkonium production processes has had many successes, there remain a number of discrepancies between its predictions and experimental measurements. The most important of these successes and discrepancies are discussed in the remainder of Sect. 4.

4.1.5 The fragmentation-function approach

In the fragmentation-function approach to factorization for inclusive quarkonium production [629–631], one writes the production cross section in terms of convolutions of parton production cross sections with light-cone fragmentation functions. This procedure provides a convenient way to organize the contributions to the cross section in terms of powers of m_Q/p. As we will explain below, it might also represent the first step in proving NRQCD factorization [629–631]. In the second step, one would establish that the light-cone fragmentation functions could be expanded in terms of NRQCD matrix elements.

We now describe the fragmentation-function approach for the specific case of single-inclusive heavy-quarkonium production at transverse momentum $p_T \gg m_Q$. The contribution to the cross section at the leading power in m_Q/p_T is given by the production of a single parton (e.g., a gluon), at a distance scale of order $1/p_T$, which subsequently fragments into a heavy quarkonium [632]. The contribution to the cross section at the first subleading power in m_Q/p_T is given by the production of a $Q\bar{Q}$ pair in a vector- or axial-vector state, at a distance scale of order $1/p_T$, which then fragments into a heavy quarkonium [631]. It was shown in the perturbative-QCD factorization approach [629, 631] that the production cross section can be factorized as

$$
\begin{aligned}
&d\sigma_{A+B\to H+X}(p_T)\\
&\quad=\sum_i d\hat{\sigma}_{A+B\to i+X}(p_T/z,\mu)\otimes D_{i\to H}(z,m_Q,\mu)\\
&\qquad+\sum_{[Q\bar{Q}(\kappa)]} d\hat{\sigma}_{A+B\to[Q\bar{Q}(\kappa)]+X}(P_{[Q\bar{Q}(\kappa)]}=p_T/z,\mu)\\
&\qquad\otimes D_{[Q\bar{Q}(\kappa)]\to H}(z,m_Q,\mu)\\
&\qquad+\mathcal{O}\big(m_Q^4/p_T^4\big),
\end{aligned}
\tag{155}
$$

where the first term in (155) gives the contribution of leading power in m_Q/p, and the second term gives the first contribution of subleading power in m_Q/p. A and B are the initial particles in the hard-scattering process and $\otimes$ represents a convolution in the momentum fraction z. In the first term in (155), the cross section for the inclusive production of a single particle i, $d\hat{\sigma}_{A+B\to i+X}$, contains all of the information about the incoming state and includes convolutions with parton distributions in the cases in which A or B is a hadron. The cross section $d\hat{\sigma}_{A+B\to i+X}$ is evaluated at the factorization scale $\mu\sim p_T$. The quantity $D_{i\to H}$ is the fragmentation function for an off-shell parton of flavor i to fragment into a quarkonium state H [633]. The argument m_Q indicates explicitly the dependence of $D_{i\to H}$ on the heavy-quark mass. Similarly, in the second term in (155), $d\hat{\sigma}_{A+B\to[Q\bar{Q}(\kappa)]+X}$ is the inclusive cross section to produce an on-shell $Q\bar{Q}$ pair with spin and color quantum numbers κ. The cross section $d\hat{\sigma}_{A+B\to[Q\bar{Q}(\kappa)]+X}$ is also evaluated at the factorization scale $\mu\sim p_T$, but it is suppressed by a factor m_Q^2/p_T^2 relative to $d\hat{\sigma}_{A+B\to i+X}$. The quantity $D_{[Q\bar{Q}(\kappa)]\to H}$ is the fragmentation function for an off-shell $Q\bar{Q}$ pair with quantum numbers κ to fragment into a quarkonium state H [631]. The predictive power of the factorization formula in (155) relies on the perturbative calculability of the single-particle inclusive and $Q\bar{Q}$ inclusive cross sections and the universality of the fragmentation functions.

The dependences of the single-parton and the $Q\bar{Q}$-pair fragmentation functions on the factorization scale μ are given by their respective evolution equations. These evolution equations can be used to express the fragmentation functions at the scale $\mu\sim p$ in terms of the fragmentation functions at the scale $\mu_0\sim 2m_Q$, thereby resumming the logarithms of μ/m_Q that are contained in the fragmentation functions.

4.1.6 *Relationship of the fragmentation-function approach to the NRQCD factorization approach*

If the NRQCD factorization formula in (154) is valid for the leading and first subleading power of m_Q^2/p_T^2, then it implies that the fragmentation functions in (155) can be expanded in terms of NRQCD matrix elements [629–631]:

$$
\begin{aligned}
&D_{i\to H}(z,m_Q,\mu_0)=\sum_n d_{i\to n}(z,m_Q,\mu_0)\langle\mathcal{O}_n^H\rangle\\
&D_{[Q\bar{Q}(\kappa)]\to H}(z,m_Q,\mu_0)\\
&\quad=\sum_n d_{[Q\bar{Q}(\kappa)]\to n}(z,m_Q,\mu_0)\langle\mathcal{O}_n^H\rangle.
\end{aligned}
\tag{156}
$$

Here, the short-distance coefficients $d_{i\to n}(z,m_Q,\mu_0)$ and $d_{[Q\bar{Q}(\kappa)]\to n}(z,m_Q,\mu_0)$ describe, respectively, the perturbative evolution at the scale μ_0 of an off-shell parton of flavor i and a $Q\bar{Q}$ pair with quantum numbers κ into a $Q\bar{Q}$ pair in the nonrelativistic state n. Viewed in this way, the factorization formula in (155) is simply a reorganization of the sum over n in (154). (The contributions denoted by $\mathcal{O}(m_Q^4/p_T^4)$ in (155) are the difference between the NRQCD expression in (154) and the first two terms in (155) expanded as a series in α_s [634].) Although (154) and (155) are equivalent if the NRQCD factorization formalism is valid for heavy-quarkonium production, the formula in (155) provides a systematic reorganization of the cross section in term of powers of m_Q/p_T and a systematic method for resumming potentially large logarithms of p_T/m_Q. That reorganization and resummation may make the α_s expansion more convergent.

4.1.7 *Difficulties in establishing NRQCD factorization*

The fragmentation functions $D_{i\to H}$ and $D_{[Q\bar{Q}(\kappa)]\to H}$ include certain contributions whose compatibility with NRQCD factorization is not obvious. These contributions arise from processes which, when viewed in the quarkonium rest frame, involve the emission of a gluon with momentum of order m_Q from the fragmenting parton. That relatively hard gluon can exchange soft gluons with the color-octet $Q\bar{Q}$ pair that evolves into the quarkonium. Such soft interactions can produce logarithmic infrared divergences, which must be absorbed into the NRQCD matrix elements in (156) in order to obtain short-distance coefficients $d_{i\to n}$ and $d_{[Q\bar{Q}(\kappa)]\to n}$ that are infrared safe. The interactions of soft gluons with the gluon that has momentum of order m_Q can be represented by interactions of the soft gluons with a lightlike eikonal line (gauge-field link) [630]. Similar lightlike eikonal lines are required in order to render the color-octet NRQCD long-distance matrix elements gauge invariant [629–631]. If it can be shown that the color-octet NRQCD matrix elements are independent of the directions of these eikonal lines, then it follows that the infrared-divergent soft interactions with gluons that have momenta of order m_Q can be absorbed into universal (i.e., process independent) NRQCD long-distance matrix elements. It has been shown that this is the case through two-loop order and, therefore, that the NRQCD factorization in (156) is valid through two-loop order [629–631]. However, the NRQCD factorization in (156) has not been verified at higher orders and, therefore, it is not known if the NRQCD factorization formula in (154) is valid [629–631]. Note that, because the potential violations of NRQCD factorization at higher loop orders involve gluons with arbitrarily soft momenta, such violations are not suppressed by powers of α_s and, consequently, they could completely invalidate the NRQCD factorization formula in (154).

It is clear that the NRQCD factorization formula cannot apply directly to reactions in which an additional heavy Q or $\bar{Q}$ is produced nearly comoving with the $Q\bar{Q}$ pair that evolves into the heavy quarkonium. That is because the NRQCD factorization formula is designed to take into account only a heavy quark and a heavy antiquark at small relative velocity [635, 636]. If an additional heavy Q or $\bar{Q}$ is nearly comoving with a $Q\bar{Q}$ pair, a color-octet $Q\bar{Q}$ pair could evolve into a color-singlet $Q\bar{Q}$ pair by exchanging soft gluons with the additional Q or $\bar{Q}$. Such nonperturbative color-transfer processes could be taken into account by generalizing the existing NRQCD factorization formalism to include long-distance matrix elements that involve additional heavy quarks or antiquarks. Such color-transfer processes might be identified experimentally by looking for an excess of heavy-flavored mesons near the direction of the quarkonium.

4.1.8 k_T factorization

The k_T-factorization approach is an alternative to standard collinear factorization that has been applied to analyses of inclusive hard-scattering processes. In the case of quarkonium production, the k_T-factorization approach has usually been applied within the CSM [637–642]. In k_T-factorization formulas, the parton distributions for the initial-state hadrons depend on the parton transverse momentum, as well as on the parton longitudinal momentum fraction. The leading-order k_T-factorization expressions for hard-scattering rates contain some contributions that appear in the standard collinear-factorization formulas in higher orders in α_s and Λ_{QCD}/p. In some kinematic situations, these higher-order corrections might be important numerically, and the k_T-factorization predictions in leading order might, in principle, be more accurate than the collinear-factorization predictions in leading order. (An example of such a kinematic situation is the high-energy limit $s \gg \hat{s}$, where s ($\hat{s}$) denotes the square of the total four-momentum of the colliding hadrons (partons).) On the other hand, the k_T-dependent parton distributions are less constrained by phenomenology than are the standard parton distributions, and the uncertainties in the k_T-dependent parton distributions are not yet well quantified in comparison with the uncertainties in the standard parton distributions. Consequently, in practice, the k_T-factorization predictions may be more uncertain than the corresponding collinear-factorization predictions.

4.1.9 Factorization in exclusive quarkonium production

NRQCD factorization has been proven for the amplitudes for two exclusive quarkonium production processes [643, 644]: exclusive production of a quarkonium and a light meson in B-meson decays and exclusive production of two-quarkonium states in e^+e^- annihilation. The proofs begin by factoring nonperturbative processes that involve virtualities of order Λ_{QCD} or smaller from hard processes that involve virtualities of order p. (Here, p is the e^+e^- center-of-mass energy in e^+e^- annihilation, and p is the B-meson mass in B-meson decays.) At this stage, the quarkonia enter through gauge-invariant quarkonium distribution amplitudes. It is then argued that each quarkonium distribution amplitude can be written as a sum of products of perturbatively calculable short-distance coefficients with NRQCD long-distance matrix elements. The difficulties that occur in establishing this step for inclusive quarkonium production do not appear in the case of exclusive quarkonium production because only color-singlet $Q\bar{Q}$ pairs evolve into quarkonia in exclusive production. The proofs of factorization for exclusive quarkonium production reveal that the violations of factorization are generally suppressed by a factor $m_Q v/p$ for each final-state quarkonium and, therefore, vanish in calculations of quarkonium production at order v^0 in the velocity expansion.

The factorization proofs in [643, 644] also establish factorized forms in which the light-cone distribution amplitudes, rather than NRQCD long-distance matrix elements, account for the nonperturbative properties of the quarkonia. In contrast with an NRQCD long-distance matrix element, which is a single number, a light-cone distribution amplitude is a function of the heavy-quark longitudinal momentum fraction. Hence, the light-cone distribution amplitudes are incompletely determined by phenomenology, and predictions that are based on light-cone factorization formulas [645–649] must rely on constrained models for the light-cone distribution amplitudes. Generally, the quantitative effects of the model assumptions on the light-cone factorization predictions are not yet known.

4.1.10 Factorization in quarkonium decays

There are NRQCD factorization formulas for exclusive quarkonium decay amplitudes and for inclusive quarkonium decay rates [138]. As in the NRQCD factorization formula for inclusive quarkonium production in (154), these decay formulas consist of sums of products of NRQCD matrix elements with short-distance coefficients. In the cases of decays, the short-distance coefficients are evaluated at a scale μ of order m_Q and are thought to be calculable as power series in $\alpha_s(\mu)$. It is generally believed that the NRQCD factorization formula for quarkonium decays can be proven by making use of standard methods for establishing perturbative factorization. The color-singlet NRQCD long-distance production matrix elements are proportional, up to corrections of relative order v^4, to the color-singlet NRQCD long-distance decay matrix elements. However, there is no known relationship between the color-octet production and decay matrix elements.

4.1.11 Future opportunities

One of the crucial theoretical issues in quarkonium physics is the validity of the NRQCD factorization formula for inclusive quarkonium production. It is very important either to establish that the NRQCD factorization formula is valid to all orders in perturbation theory or to demonstrate that it breaks down at some fixed order in perturbation theory.

The NRQCD factorization formula is known to break down when an additional heavy quark or antiquark is produced in close proximity to a $Q\bar{Q}$ pair that evolves into a quarkonium. It would help in assessing the numerical importance of such processes if experimental measurements could determine the rate at which heavy-flavored mesons are produced nearby in phase space to a heavy quarkonium. If such processes prove to be important numerically, then it would be useful to extend the NRQCD factorization formalism to include them.

4.2 Production at the Tevatron, RHIC and the LHC

The first measurements by the CDF Collaboration of the *direct* production[10] of the J/ψ and the $\psi(2S)$ at $\sqrt{s} = 1.8$ TeV [628, 650] revealed a striking discrepancy with the existing theoretical calculations: The observed rates were more than an order of magnitude greater than the calculated rates at leading order (LO) in α_s in the CSM. (See Sect. 4.1.2 for a discussion of the CSM.) This discrepancy has triggered many theoretical studies of quarkonium hadroproduction, especially in the framework of NRQCD factorization. (See Sect. 4.1.4 for a discussion of NRQCD factorization.) In the NRQCD factorization approach, mechanisms beyond those in the CSM arise, in which the production of charmonium states proceeds through the creation of a $c\bar{c}$ pair in a color-octet state. For the specific case of the production of the J/ψ or the $\psi(2S)$ (henceforth denoted collectively as "ψ"), these color-octet transitions take place at higher orders in v. Depending on the convergence of the expansions in α_s and v and the validity of the NRQCD factorization formula, the NRQCD factorization approach may provide systematically improvable approximations to the inclusive quarkonium production rates. For some recent reviews, see [1, 651–653]. For some perspectives on quarkonium production at the LHC, see [654].

Despite recent theoretical advances, which we shall detail below, we are still lacking a clear picture of the mechanisms at work in quarkonium hadroproduction. These mechanisms would have to explain, in a consistent way, both the cross section measurements and the polarization measurements for charmonium production at the Tevatron [329, 628, 650, 655–658] and at RHIC [659–664]. For example, the observed p_T spectra in prompt ψ production seem to suggest that a dominant contribution at large p_T arises from a color-octet process in which a gluon fragments into a $Q\bar{Q}$ pair, which then evolves nonrelativistically into a quarkonium. Because of the approximate heavy-quark spin symmetry of NRQCD, the dominance of such a process would lead to a substantial transverse component for the polarization of ψ's produced at large p_T [665–667]. This prediction is clearly challenged by the experimental measurements [658].

A possible interpretation of such a failure of NRQCD factorization is that the charmonium system is too light for relativistic effects to be small and that, in phenomenological analyses, the velocity expansion of NRQCD [138] may have been truncated at too low an order. However, such an explanation would seem to be at odds with other successful predictions of the NRQCD approach to charmonium physics. If the convergence of the velocity expansion of NRQCD is indeed an issue, then one would expect better agreement between theory and the available experimental data on hadroproduction in the case of the bottomonium states and, in particular, in the case of the Υ. Better convergence of the velocity expansion might explain, for example, why a computation that retains only color-singlet contributions [620] (that is, only contributions of leading order in v) seems to be in better agreement with the data for Υ production [668–671] than with the data for ψ production [329]. We will discuss this comparison between theory and experiment in greater detail later.

In efforts to identify the mechanisms that are at work in inclusive ψ or Υ production, it is important to have control of the higher-order perturbative corrections to the short-distance coefficients that appear in the NRQCD factorization formula. Several works have been dedicated to the study of the corrections of higher-order in α_s and their phenomenological implications for the differential production rates. We summarize these results in Sects. 4.2.1 and 4.2.2. An important observable that has been reanalyzed in the context of higher-order perturbative corrections is the polarization of the quarkonium. We review the analyses of the quarkonium polarization in Sect. 4.2.3. In addition to the rates and polarizations in inclusive quarkonium production, other observables have been shown to yield valuable information about

[10]"Prompt production" excludes quarkonium production from weak decays of more massive states, such as the B meson. "Direct production" further excludes quarkonium production from feeddown, via the electromagnetic and strong interactions, from more massive states, such as higher-mass quarkonium states.

the production mechanisms. We discuss them in Sect. 4.2.4. Finally, we summarize the future opportunities for theory and experiment in inclusive quarkonium hadroproduction in Sect. 4.2.5.

4.2.1 Channels at higher-order in α_s

At the LHC, the Tevatron, and RHIC, quarkonium production proceeds predominantly via gluon-fusion processes. The production cross sections differential in p_T for S-wave quarkonium states have been calculated only recently at NLO in α_s [618, 619, 672–674]. These NLO calculations also provide predictions for the quarkonium polarization differential in p_T. In the case of the production of spin-triplet P-wave quarkonium states, the NLO corrections to the production rate differential in p_T have been calculated even more recently [675].

One common outcome of these calculations is that the total cross sections are not much affected by corrections of higher order in α_s [619, 676]. That is, the perturbation series for the total cross sections seem to exhibit normal convergence. However, in the case of ψ, Υ, or χ_{cJ} production via color-singlet channels, very large corrections appear in the cross sections differential in p_T^2 at large p_T. This behavior, which has also been seen in photoproduction [677], is well understood. QCD corrections to the color-singlet parton cross section open new production channels, whose contributions fall more slowly with p_T than do the LO contributions. Hence, the contributions from the new channels increase substantially the cross section in the large-p_T region.

We now discuss this phenomenon briefly. If only the LO (order-α_s^3) contribution to the production of a color-singlet ${}^3S_1$ $Q\bar{Q}$ state is taken into account, then the partonic cross section differential in p_T^2 scales as p_T^{-8} [612–614, 616, 678]. This behavior comes from the contributions that are associated with "box" graphs, such as the one in Fig. 50(a). At NLO in α_s (order α_s^4), several contributions with distinct kinematic properties arise. The loop corrections, illustrated in Fig. 50(b), are not expected to have substantially different p_T scaling than the Born contributions. However, the t-channel gluon-exchange diagrams, such as the one depicted in Fig. 50(c), yield contributions that scale as p_T^{-6} and are, therefore, kinematically enhanced in comparison with the Born contribution. At sufficiently large p_T, this contribution has a kinematic enhancement that compensates for its α_s suppression, and it is expected to dominate over the Born contribution. At NLO, there is also a contribution that arises from the process in which a second $Q\bar{Q}$ pair is produced, in addition to the $Q\bar{Q}$ pair that evolves into a quarkonium. In this "associated-production" process, both $Q\bar{Q}$ pairs have the same flavor. In the limit $p_T \gg m_Q$, the associated-production mechanism reduces to heavy-quark fragmentation. Therefore, in this limit, the corresponding

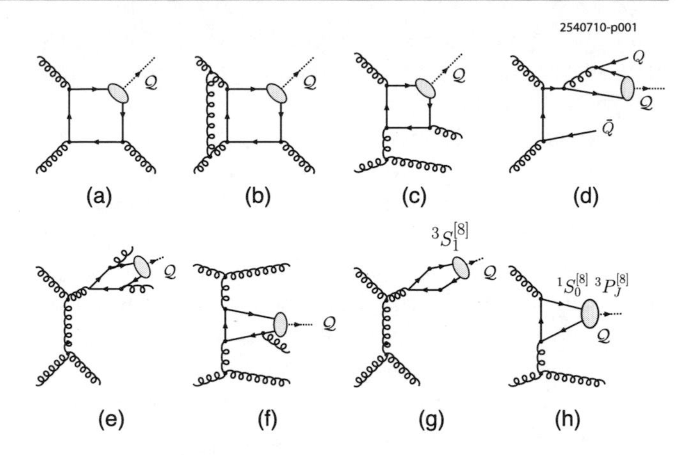

Fig. 50 Representative diagrams that contribute to the hadroproduction of ${}^3S_1$ quarkonium states via color-singlet channels at orders α_s^3 (**a**), α_s^4 (**b, c, d**), and α_s^5 (**e, f**), and via color-octet channels at order α_s^3 (**g, h**). The quark and antiquark that are attached to the ellipses are taken to be on shell, and their relative velocity is set to zero. (**a**)–(**f**), (**g**)–(**h**) from [620, 652] with kind permission, copyrights (2008, 2009), The American Physical Society and Springer, respectively

partonic cross section differential in p_T^2 scales as p_T^{-4}. The associated-production contribution eventually provides the bulk of the color-singlet parton cross section at sufficiently large p_T. In the p_T region that is covered by the current experiments, though, this contribution is relatively small, owing to phase-space suppression [618].

Similar arguments can be used to understand the impact of QCD corrections on the rates differential in p_T^2 for the production of a $Q\bar{Q}$ pair in a color-octet state. At LO in α_s, the production of a color-octet ${}^3S_1$ $Q\bar{Q}$ pair proceeds, at large p_T, predominantly through gluon fragmentation [Fig. 50(g)]. Thus, the rate differential in p_T^2 scales as p_T^{-4}. This is the smallest power of $1/p_T$ that is possible in partonic cross sections. Hence, in this case, the NLO correction cannot contain a kinematically enhanced channel and does not affect substantially the shape of the differential rate [674]. The situation is different for the production of a C-even color-octet $Q\bar{Q}$ pair because, in this case, there is no fragmentation process at LO in α_s. [See Fig. 50(h).] At LO in α_s, the rates for these channels, differential in p_T^2, scale as p_T^{-6}. The fragmentation channels appear at NLO in α_s. Consequently the NLO correction to the differential rate is expected to yield a substantial enhancement at large p_T. This feature has been checked explicitly in [674] in the specific case of the production of a color-octet ${}^1S_0$ $Q\bar{Q}$ state.

In view of the strong impact of the correction at NLO in α_s on the color-singlet differential rate at large transverse momentum, it is natural to examine the QCD corrections that appear at even higher orders in the α_s expansion. At NNLO (order α_s^5), new, important channels with specific kinematic properties continue to appear. Some of these channels have actually been studied for some time in specific kinematic limits in which one can take advantage of large separations between different perturbative energy scales. The color-singlet gluon-fragmentation channel [Fig. 50(e)] has been investigated in the framework of the fragmentation approximation [679], which is relevant in the limit $p_T \gg m_Q$. The processes in which the $Q\bar{Q}$ pair is produced by the exchange of two gluons in the t channel [Fig. 50(f)] were investigated in the k_T-factorization approach [639–642]. This approach is relevant in the high-energy limit $s \gg \hat{s}$, where s ($\hat{s}$) denotes the square of the total four-momentum of the colliding hadrons (partons). In that kinematic limit, other enhanced processes, which are initiated by a symmetric two-gluon color-octet state and an additional gluon, have been investigated more recently in [680]. These processes correspond to higher-order contributions in the framework of the (standard) collinear approximation of perturbative QCD. The advantage in considering either of these kinematic limits is that the perturbation expansion can be reorganized in such a way as to simplify the evaluation of the dominant contribution. Furthermore, the convergence of the perturbation expansion is improved because large logarithms of the ratio of the disparate energy scales are resummed. Away from the asymptotic regime, the corrections to each of these approaches may be important. The impacts of these corrections in the kinematic region that is covered at the current hadron colliders is not known accurately. There is an alternative method that has been proposed for estimating the NNLO corrections to the color-singlet differential rate that is known as the NNLO* method [620]. The NNLO* method does not attempt to separate the various energy scales. Instead, it considers only NNLO corrections involving real gluon emission and imposes an infrared cut-off to control soft and collinear divergences. The NNLO* estimates suffer from large uncertainties, which arise primarily from the sensitivities of these estimates to the infrared cut-off and to the choice of renormalization scale.

A specific higher-order process that has been investigated is the so-called "s-channel $Q\bar{Q}$-cut" process [681, 682]. In this process, an on-shell $Q\bar{Q}$ pair is produced. That pair then rescatters into a quarkonium state. The contribution of LO in α_s to the amplitude for this process is given by the imaginary part of a specific set of one-loop diagrams [683], and the square of this amplitude contributes to the cross section at order α_s^5.

In addition to these new results for the QCD corrections to inclusive quarkonium production, there have also been new results for the QED and relativistic corrections to inclusive quarkonium production. The QED correction to the inclusive J/ψ production rate in hadron–hadron collisions has been computed [684, 685] and turns out to be small in the region of p_T that is covered by the current experiments. Relativistic corrections to the color-singlet rate for inclusive J/ψ hadroproduction have been shown to be negligible over the entire p_T range that is currently accessible [686].

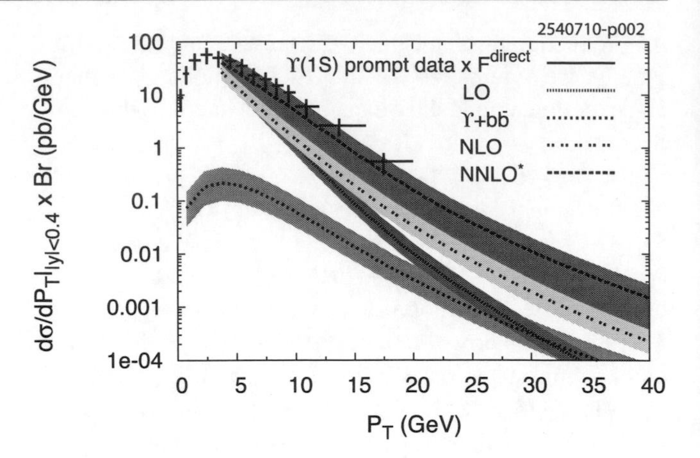

Fig. 51 Comparison between the CSM predictions at NLO and NNLO* accuracy for the Υ cross section as a function of the Υ transverse momentum at the Tevatron at $\sqrt{s} = 1.80$ TeV [620] and the CDF data for $\Upsilon(1S)$ production [669]. The crosses are the CDF data for prompt $\Upsilon(1S)$ production, multiplied by F^{direct}, the fraction of direct $\Upsilon(1S)$'s in prompt $\Upsilon(1S)$ events, as measured by the CDF Collaboration using an older event sample [668]. The *lines* show the central values of the theoretical predictions, and the *bands* depict the theoretical uncertainties. The theoretical uncertainty bands for the LO, NLO, and $\Upsilon + b\bar{b}$ contributions were obtained by combining the uncertainty from m_b with the uncertainties that are obtained by varying the renormalization scale μ_f and the factorization scale μ_r between $2m_T$ and $m_T/2$, where $m_T = \sqrt{4m_b^2 + p_T^2}$. The error band for the NNLO* contribution is obtained by varying the infrared cut-off $s_{ij}^{\min}$ between $2m_b^2$ and $m_b^2/2$ and by varying μ_f and μ_r between $2m_T$ and $m_T/2$. From [620] with kind permission, copyright (2008) The American Physical Society

4.2.2 Phenomenology, including QCD corrections

ψ, Υ, and χ_c production at the Tevatron In the case of Υ production, the contributions of the NLO corrections to the color-singlet channels reduce the discrepancy between the color-singlet contribution to the inclusive cross section and the data collected by the CDF Collaboration, as is illustrated in Fig. 51.[11] However, the predicted NLO rate drops too rapidly at large p_T, indicating that another production mechanism is at work in that phase-space region. A recent study [620] has shown that contributions from channels that open at NNLO (order α_s^5) may fill the remaining gap between the color-singlet contribution at NLO and the data. The estimate of the NNLO contribution from this study, called the "NNLO* contribution," is shown in the (red) band labeled

[11] We note that no phenomenological analysis of χ_{bJ} production is available yet at NLO accuracy. Such an analysis would be necessary in order to predict the prompt Υ cross section to NLO accuracy. As a makeshift, one could multiply the available data by the measurement of the fraction of direct Υ's in the total rate (integrated over p_T) [668]. Note, however, that the NLO calculation of the χ_{cJ} production rate [675] indicates that the fraction of direct J/ψ's may depend rather strongly on p_T.

NNLO⋆ in Fig. 51. Owing to the large theoretical uncertainties, this improved prediction for the color-singlet contribution does not imply any severe constraint on other possible contributions, such as a color-octet contribution. However, in contrast with previous LO analyses, in which color-octet contributions were *required* in order to describe the data, the NNLO⋆ estimate of the color-singlet contribution shows that color-octet contributions are now merely *allowed* by the rather large theoretical uncertainties in the NNLO⋆ estimate.

The impact of the QCD corrections on the color-singlet contribution has also been studied in the case of ψ hadroproduction [652]. The comparison with the data is simpler in the case of the $\psi(2S)$ than in the case of the J/ψ, owing to the absence of significant feeddown from excited charmonium states to the $\psi(2S)$. In a recent paper, the CDF Collaboration has reported a new measurement of the inclusive $\psi(2S)$ cross section [329]. The rates for the prompt production of the $\psi(2S)$ and for the production of the $\psi(2S)$ in B-meson decays were also extracted in that analysis. The reconstructed differential rate for the prompt component is compared to the prediction for the color-singlet rate at LO, NLO and NNLO⋆ accuracy in Fig. 52. At medium values of p_T, the upper limit of the NNLO⋆ rate is compatible with the CDF results. At larger values of p_T, a gap appears between the color-singlet rate and the data [652]. The J/ψ differential production rate has the same qualitative features as the $\psi(2S)$ differential production rate [687]. It is worth emphasizing that the current discrepancy between the color-singlet

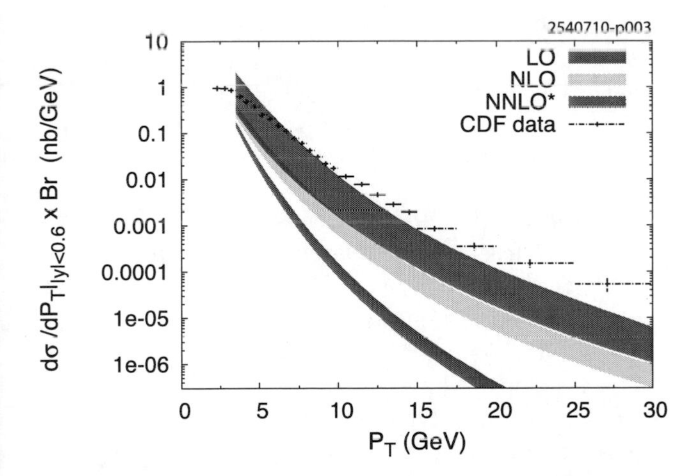

Fig. 52 Comparison between the CSM predictions for the $\psi(2S)$ cross sections at LO, NLO, and NNLO⋆ accuracy as a function of the $\psi(2S)$ p_T at the Tevatron at $\sqrt{s} = 1.96$ TeV [652] and the CDF prompt $\psi(2S)$ data [329]. The theoretical uncertainty bands for the LO and NLO contributions were obtained by combining the uncertainty from m_c with the uncertainties that are obtained by varying the renormalization scale μ_f and the factorization scale μ_r between $2m_T$ and $m_T/2$, where $m_T = \sqrt{4m_c^2 + p_T^2}$. The theoretical uncertainty band for the NNLO⋆ contribution was obtained by varying the infrared cut-off $s_{ij}^{\min}$ between $4m_c^2$ and m_c^2 and by varying μ_f and μ_r between $2m_T$ and $m_T/2$. From [652] with kind permission, copyright (2009) Springer

rate and the Tevatron data has been dramatically reduced by the inclusion of higher-order QCD corrections.

NLO QCD corrections to the color-octet production channels $^3S_1$ and $^1S_0$ have been analyzed for J/ψ production [674] and for Υ production [688]. In both cases, these corrections proved to be small when the p_T of the produced quarkonium is less than 20 GeV. In [674], values of the NRQCD long-distance matrix elements $\langle O^{J/\psi}(^3S_1^{[8]})\rangle$ and $\langle O^{J/\psi}(^1S_0^{[8]})\rangle$ were obtained by fitting the theoretical prediction to the prompt production rate that was measured by the CDF Collaboration [657]. The values of the NRQCD matrix elements that were extracted in this analysis are compatible with the values that were extracted in LO analyses. In the analysis of [674], feeddown contributions were ignored and the P-wave color-octet long-distance matrix elements were set to zero. A satisfactory fit could not be obtained for the experimental data points that have $p_T < 6$ GeV, and so these points were not included in the fit. In this regard, it should be kept in mind that resummation of large logarithms may be needed at small p_T and that NRQCD factorization may break down at small p_T.

The s-channel $c\bar{c}$-cut (sCC) contributions to ψ hadroproduction have been investigated in [681, 682] in the framework of a phenomenological model. A first analysis of this model [682], which was based on the data of [628, 657] and incorporated constraints for the small- and large-p_T regions, supported rates that are significantly larger than those of the CSM prediction. This analysis, which did not include resummation of initial-gluon contributions, yielded a good fit to the p_T dependence of the RHIC data. However, it has been shown recently, by evaluating the leading-order contribution of the sCC amplitude in the framework of NRQCD, that the sCC contributions can account only for a negligible fraction of the J/ψ production rate that is measured by the CDF Collaboration [683].

Some specific NNLO contributions in the CSM were considered in [680]. These contributions are referred to in the literature as the "gluon-tower model." They can be viewed as LO BFKL contributions, and, thus, they are expected to be enhanced in comparison to other NNLO contributions by a factor $\log s/\hat{s}$. At large s, this logarithm may compensate for the α_s^2 suppression of these contributions. At $\sqrt{s} = 1.96$ TeV and $|y| < 0.6$, the gluon-tower model predicts the J/ψ cross section, integrated over p_T, to be $\sigma(|y| < 0.6) = 2.7$ μb [680], in near agreement with the CDF measurement $\sigma(|y| < 0.6) = 4.1^{+0.6}_{-0.5}$ μb [689].[12] However, the theoretical prediction is somewhat sensitive to an effective gluon mass that is introduced as an infrared cut-off in the model. The comparison does not take into account

[12]We note that the LO CSM result for the p_T-integrated direct J/ψ cross section $d\sigma/dy$, evaluated at $y = 0$ [690], is compatible with the CDF [657] and PHENIX [659, 663] measurements.

feedown from P-wave states in the measured cross section, and the model cannot, at present, predict the p_T dependence of the cross section.

At LO in α_s, NRQCD factorization predicts that the ratio of production cross sections, $R_{\chi_c} = \sigma_{\chi_{c2}}/\sigma_{\chi_{c1}}$, is dominated by the color-octet contribution at large p_T and approaches the value $R_{\chi_c} = 5/3$ as p_T increases. This LO prediction is in sharp disagreement with the CDF measurement [691], which finds that $R_{\chi_c} \approx 0.75$ at large p_T. Recently, NLO corrections to χ_{cJ} production have been calculated in [675]. It is found that the NLO corrections are large at large p_T. They make the contributions of the color-singlet ^3P_J channels negative and comparable to the color-octet contribution for large p_T. They also cause the $^3P_1$ color-singlet contribution to fall at a slower rate than the $^3P_2$ color-singlet contribution as p_T increases. Taking into account the large NLO correction, the authors of [675] were able to fit the measured p_T distribution of R_{χ_c}, using a plausible value for the ratio of the relevant NRQCD long-distance matrix elements. Hence, there may now be a resolution of this outstanding conflict between theory and experiment. One interesting prediction of the fit to R_{χ_c} in [675] is that the feeddown from the χ_{cJ} states to the J/ψ state may be quite large—perhaps 30% of the prompt J/ψ rate at $p_T = 20$ GeV. Such a large proportion of feeddown events in the prompt J/ψ rate could have an important effect on the prompt J/ψ polarization.

New NLO hadroproduction results As this article was nearing completion, two papers [692, 693] appeared that give complete calculations of the corrections of NLO in α_s for the color-octet production channels through relative order v^4, that is, for the color-octet $^3S_1$, $^1S_0$, and ^3P_J channels We now give a brief account of these results.

The calculations in [692, 693] are in numerical agreement for the short-distance coefficients for J/ψ production at the Tevatron. The calculations confirm that the corrections to the color-octet $^3S_1$ and $^1S_0$ channels are small, but also show that there is a large, negative K factor in the color-octet ^3P_J channel.

In [692], the NLO calculation was fit to the CDF data for the prompt production of the J/ψ and the $\psi(2S)$ [329, 657], and values were obtained for two linear combinations of NRQCD long-distance matrix elements. In the case of the J/ψ, these fits took into account feeddown from the $\psi(2S)$ state and the χ_{cJ} states, where the latter was obtained from the NLO calculation of χ_c production [675] that was described above. Satisfactory fits could not be obtained to the experimental data points for the J/ψ and the $\psi(2S)$ with $p_T < 7$ GeV, and so these points were excluded from the fits. The fitted values of the linear combinations of matrix elements were used to predict the cross section for J/ψ production at CMS, and good agreement with the CMS data [694] was obtained. This analysis suggests the possibility that the cross section is dominated by the color-octet $^1S_0$ contribution, rather than by the color-octet $^3S_1$ contribution, in contrast with conclusions that had been drawn on the basis of LO fits to the Tevatron data.

In [693], values of the NRQCD long-distance matrix elements were extracted by using the NLO calculation of J/ψ hadroproduction of [693] and an NLO calculation of J/ψ photoproduction from [695] to make a combined fit to the CDF Run II data for prompt J/ψ production [657] and to the HERA I and HERA II H1 data for prompt J/ψ photoproduction [696, 697]. In this fit, only CDF data with $p_T > 3$ GeV were used, as the flattening of the cross section at smaller values of p_T cannot be described by fixed-order perturbation theory. Feeddown of the $\psi(2S)$ and χ_{cJ} states to the J/ψ was not taken into account in the fits. This is the first multiprocess fit of NRQCD long-distance matrix elements for quarkonium production. The values of the NRQCD long-distance matrix elements that were obtained in this fit do not differ greatly from those that were obtained in LO fits. They were used to predict the cross sections for prompt J/ψ production at PHENIX [662] and CMS [694], and good agreement with the data was achieved in both cases.

The values of the linear combinations of J/ψ NRQCD long-distance matrix elements that were obtained in [692] are not consistent with the values of the NRQCD long-distance matrix elements that were obtained in [693]. Since the calculations of [692, 693] are in agreement on the short-distance cross sections, any discrepancies in the extracted NRQCD long-distance matrix elements must be due to differences in the fitting procedures. Clearly, it is necessary to understand the significance of the various choices that have been made in the fitting procedures before any definite conclusions can be drawn about the sizes of the NRQCD long-distance matrix elements.

J/ψ production at RHIC Recently, the STAR Collaboration at RHIC has reported an analysis of prompt J/ψ production for values of p_T up to 12 GeV [664]. In [664], the measured production rate as a function of p_T is compared with predictions based on NRQCD factorization at LO [698] and the CSM up to NNLO* accuracy [620]. The calculations do not include feeddown from the $\psi(2S)$ and the χ_c states. The data clearly favor the NRQCD factorization prediction over the CSM prediction. However, no definite conclusions can be drawn because the effects of feeddown have not been taken into account.

A calculation of prompt J/ψ production at RHIC, including feeddown from the $\psi(2S)$ and χ_c states, has been carried out in [699] in the CSM and the NRQCD factorization formalism at LO. In Fig. 53, we show a comparison between the predictions of [699] for the prompt J/ψ cross section as a function of p_T and data from the PHENIX Collaboration [662, 663]. Again, the NRQCD predictions are

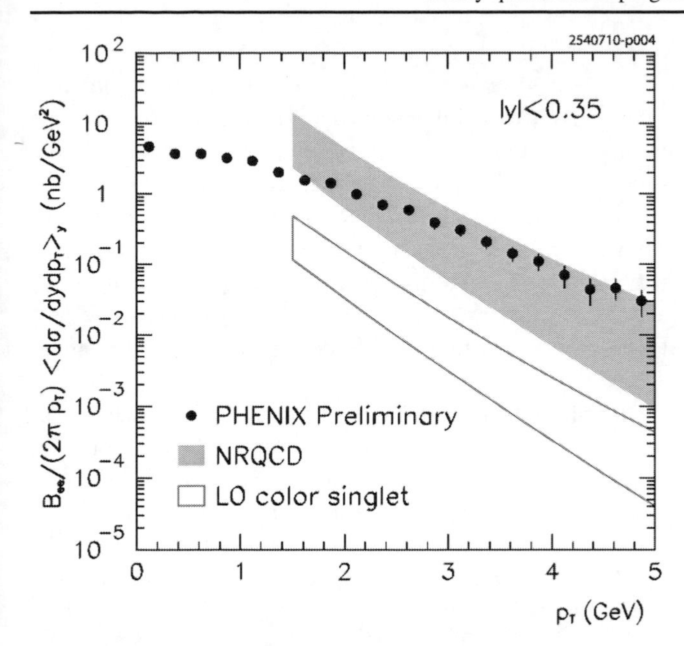

Fig. 53 Comparison of the LO NRQCD and the LO CSM predictions for the J/ψ cross section as a function of the J/ψ transverse momentum [699] with the data from the PHENIX Collaboration [662, 663]. The theoretical uncertainty bands were obtained by combining the uncertainties from m_c and the NRQCD long-distance matrix elements with the uncertainties that are obtained by varying the renormalization scale μ_r and the factorization scale μ_f between $2m_T$ and $m_T/2$. Here $m_T = \sqrt{4m_c^2 + p_T^2}$. From [699] with kind permission, copyright (2010) The American Physical Society

favored over the CSM predictions. However, in this case, the small values of p_T involved may call into question the validity of perturbation theory, and the omission of higher-order corrections to the CSM, which are known to be large, also undermines the comparison.

Higher-order corrections to the color-singlet contribution to J/ψ production at RHIC have been considered in [700] and were found to be large. A comparison between the predictions of [700] for the cross section differential in p_T and the PHENIX and STAR prompt J/ψ data is shown in Fig. 54. The color-singlet contributions through NLO agree with the PHENIX prompt J/ψ data for p_T in the range 1–2 GeV, but fall substantially below the PHENIX and STAR prompt J/ψ data for larger values of p_T. The NNLO* color-singlet contribution can be computed reliably only for $p_T > 5$ GeV. The upper limit of the theoretical uncertainty band for the NNLO* contribution is compatible with the PHENIX and STAR data, although the theoretical uncertainties are very large.

As we have mentioned above, the NLO analysis of [693], which includes the ${}^3S_1$, ${}^1S_0$, and ${}^3P_J$ color-octet channels, uses matrix elements that are extracted from a combined fit to CDF data [657] and H1 data [696, 697] to predict the cross sections for J/ψ production at HERA. This prediction agrees well with PHENIX data [662].

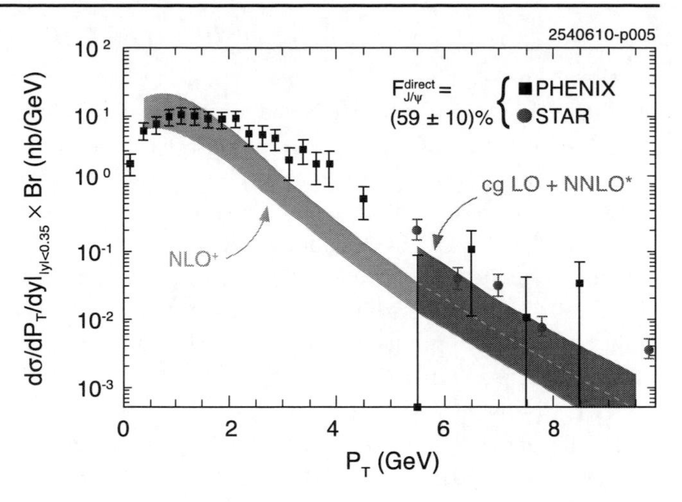

Fig. 54 Comparison between CSM prediction for the J/ψ cross section at NLO accuracy as a function of the J/ψ transverse momentum in pp collisions at RHIC at $\sqrt{s_{NN}} = 200$ GeV and $|y| < 0.35$ [700] and the PHENIX [659] and STAR [664] prompt J/ψ data. The NLO$^+$ contribution contains the contributions from gg and gq fusion at NLO accuracy, where q is a light quark, plus the contribution from cg fusion at LO accuracy. The theoretical uncertainty band for the NLO$^+$ contribution was obtained by combining the uncertainties that are obtained by varying m_c in the range 1.4 GeV $< m_c <$ 1.6 GeV and by varying the factorization scale μ_f and the renormalization scale μ_r through the values $((0.75, 0.75); (1, 1); (1, 2); (2, 1); (2, 2)) \times m_T$. Here $m_T = \sqrt{4m_c^2 + p_T^2}$. The theoretical uncertainty band for the NNLO* contribution was obtained by combining the uncertainties that are obtained by varying m_c in the range 1.4 GeV $< m_c <$ 1.6 GeV, by varying μ_f and μ_r in the range $0.5m_T < \mu_f = \mu_r < 2m_T$, and by varying the infrared cut-off $s_{ij}^{\min}$ in the range 2.25 GeV$^2 < s_{ij}^{\min} <$ 9.00 GeV^2. From [700]

Exclusive production of charmonia The *exclusive* production of charmonium states (plus beam particles) has also been observed at hadron–hadron colliders. In most current theoretical models, the exclusive production of states with charge parity −1, such as the J/ψ or the $\psi(2S)$, is dominated by the process of photon–Pomeron fusion (photoproduction), while the exclusive production of states with charge parity +1, such as the χ_{c0}, is dominated by the process of Pomeron–Pomeron fusion. In perturbative model calculations, the Pomeron is represented as an exchange of two or more gluons in a color-singlet state. Exclusive quarkonium production could provide an important tool with which to probe these mechanisms.

The CDF Collaboration has measured exclusive J/ψ, $\psi(2S)$, and χ_{c0} production in $p\bar{p}$ collisions at $\sqrt{s} =$ 1.96 TeV [701]. The CDF measurements are in agreement with theoretical predictions that are based on models for the Pomeron [702–710]. The PHENIX Collaboration has measured exclusive J/ψ production in Au+Au collisions at $\sqrt{s_{NN}} = 200$ GeV [711] and also finds agreement with theoretical predictions [712–719]. See Sect. 5.7 for more on photoproduction in nuclear collisions.

$X(3872)$ production at the Tevatron and the LHC Since the discovery of the $X(3872)$ by the Belle Collaboration in 2003 [85], this state has attracted a large interest in the particle-physics community. The $X(3872)$ is the exotic state for which the largest experimental data set is available. The production of the $X(3872)$ at the Tevatron has been analyzed by both the CDF Collaboration [88–90] and the DØ Collaboration [91] in the $J/\psi\pi^+\pi^-$ decay channel. The CDF Collaboration has shown that most of the $X(3872)$'s at the Tevatron are produced promptly, rather than through b-hadron decays [720].

As is discussed in Sects. 2.3.1, 2.9.3, and 2.9.4 of this article, the exact nature of the $X(3872)$ is still subject to debate. Nevertheless, it is plausible that the prompt production rate of the $X(3872)$ can be predicted correctly in the factorization framework of NRQCD. The reason for this is that, in all the viable hypotheses as to the nature of the $X(3872)$, the particle content of the state includes a charm-quark pair with a relative momentum $q \ll m_c$. The small size of q suggests that one can make use of the NRQCD expansion of the production rate in powers of q/m_c. It follows that the expression for the cross section is given by the NRQCD factorization formula (154). It was argued in [332, 721] that it is reasonable to truncate the NRQCD series so that it includes only contributions to the $X(3872)$ production rate from $c\bar{c}$ pairs that are created in an S-wave configuration. Moreover, in the case of hadroproduction, it was argued that a truncation that retains only the color-octet $^3S_1$ channel would provide a reliable prediction at large transverse momentum. It follows that the corresponding NRQCD long-distance matrix element can be extracted from the measured production rate at the Tevatron in the $J/\psi\pi^+\pi^-$ decay channel. In [332], the aforementioned simplifying assumptions are used in the NRQCD factorization framework to predict the prompt production rate for $X(3872) \to J/\psi\pi^+\pi^-$ as a function of p_T for various LHC experiments. In the same work, the production of the $X(3872)$ from b-hadron decays is discussed. The data samples at the LHC are predicted to be large, suggesting that the $X(3872)$ can be studied very effectively at the LHC. Measurements of the prompt production rate at the LHC as a function of p_T would provide a key test of the NRQCD factorization approach to $X(3872)$ hadroproduction.

4.2.3 Quarkonium polarization: a key observable

Measurements of quarkonium polarization observables may yield information about quarkonium production mechanisms that is not available from the study of unpolarized cross sections alone.

The three polarization states of a $J = 1$ quarkonium can be specified in terms of a particular coordinate system in the rest frame of the quarkonium. This coordinate system is often called the "spin-quantization frame." In a hadron collider, the J/ψ, $\psi(2S)$ and Υ resonances are reconstructed through their electromagnetic decays into a lepton pair. The information about the polarization of the quarkonium state is encoded in the angular distribution of the leptons. This angular distribution is usually described in the quarkonium rest frame with respect to a particular spin-quantization frame. In that case, the angular distribution of the quarkonium can be expressed in terms of three real parameters that are related to the spin-polarization amplitudes of the $J = 1$ quarkonium state.

In hadron–hadron collisions, polarization analyses are often restricted to the measurement of the distribution as a function of the polar angle with respect to the chosen spin-quantization axis. This distribution is parametrized as

$$1 + \alpha \cos^2 \theta_{\ell\ell}. \tag{157}$$

The parameter α in (157) is directly related to the fraction of the cross section that is longitudinal (or transverse) with respect to the chosen spin-quantization axis: $\alpha = 1$ corresponds to 100% transverse polarization; $\alpha = -1$ corresponds to 100% longitudinal polarization.

In experimental analyses, knowledge of the angular distribution of dileptons from quarkonium decay is important because, typically, detector acceptances fall as dileptons are emitted more along the direction of the quarkonium momentum—especially at small p_T. This effect is included in the corrections to the experimental acceptance. However, it induces systematic experimental uncertainties.

In theoretical calculations, polarization parameters, such as the polar asymmetry α, can be expressed in terms of ratios of polarized quarkonium cross sections. In some cases, these ratios are less sensitive than the production cross sections to the theoretical uncertainties from quantities such as the factorization scale, the renormalization scale, the heavy-quark mass, and the NRQCD long-distance matrix elements. For example, in the cases of the production of the J/ψ or the $\psi(2S)$ in the NRQCD factorization formalism, the polarization parameter α depends, to good approximation, on ratios of color-octet long-distance matrix elements, but not on their magnitudes.

One should keep in mind that a measurement of the polar asymmetry parameter α alone does not give complete information about the polarization state of the produced quarkonium. The importance of measuring all of the parameters of the dilepton angular distribution for a variety of choices of the spin-quantization frame has been emphasized in [722, 723]. The significance of the information that is obtained in measuring α alone depends very much on the orientation of the spin-quantization axis. So far, most of the theoretical studies of polarization in quarkonium production have been carried out for the case in which the spin-quantization axis is taken to be along the direction of the quarkonium momentum in the laboratory frame [618, 620, 639–642, 652, 665–667, 672, 674, 680–682, 724, 725]. That choice of spin-quantization axis [665] is often referred to as the "helicity frame." In [726], it is shown that one can make more sophisticated choices of the spin-quantization axis, which involve not only the kinematics of the quarkonium state, but also the kinematics of other produced particles. These alternative choices of spin-quantization axis can increase the significance of the measurement of α. However, their optimization requires knowledge of the dominant quarkonium production mechanism.

Experimental measurements of quarkonium polarization have been made for a variety of spin-quantization frames. Measurements by the CDF [656, 658, 669], DØ [671], and PHENIX [661–663] Collaborations were carried out in the helicity frame, while some measurements at fixed-target experiments [727, 728] were carried out in the Collins–Soper frame [729]. Recently, the Hera-B Collaboration has analyzed quarkonium polarizations [730] not only in the helicity and Collins–Soper frames, but also in the Gottfried–Jackson frame [731], in which the spin-quantization axis is along the direction of the incident beam. In [722], a global analysis was made of polarization measurements that were carried out in the Collins–Soper and helicity frames. That analysis shows that the results that were obtained in these two spin-quantization frames are plausibly compatible when the experimental rapidity ranges are taken into account. However, it is clear that additional analyses in different spin-quantization frames would be very informative.

According to the CDF Run II measurement of the ψ polarization in the helicity frame [658], the prompt ψ yield becomes increasingly longitudinal as p_T increases. The disagreement of this result with a previous CDF polarization measurement that was based on Run I data [656] has not been resolved. In Fig. 55 the CDF measurement of the polarization parameter α for the prompt J/ψ production at the Tevatron in Run II is compared with the NRQCD factorization prediction at LO in α_s [667]. This prediction ignores possible violations of the heavy-quark spin symmetry, which appear at relative order v^3. The effects of feeddown from the $\psi(2S)$ and the χ_{cJ} states are taken into account in the NRQCD factorization prediction. However, it should be kept in mind that the corrections at NLO in α_s to the χ_c production rate are large [675] and are not taken into account in the NRQCD prediction in Fig. 55. The solid line in Fig. 55 is the prediction from the k_T factorization approach [640], which includes only color-singlet contributions.

At LO accuracy in α_s, the NRQCD factorization prediction for the J/ψ polarization clearly disagrees with the observation of a very small polar asymmetry in the helicity frame. One obvious issue is the effect of corrections of higher order in α_s on the NRQCD factorization prediction.

Corrections of higher order in α_s to J/ψ production via the color-singlet channel dramatically affect the polarization

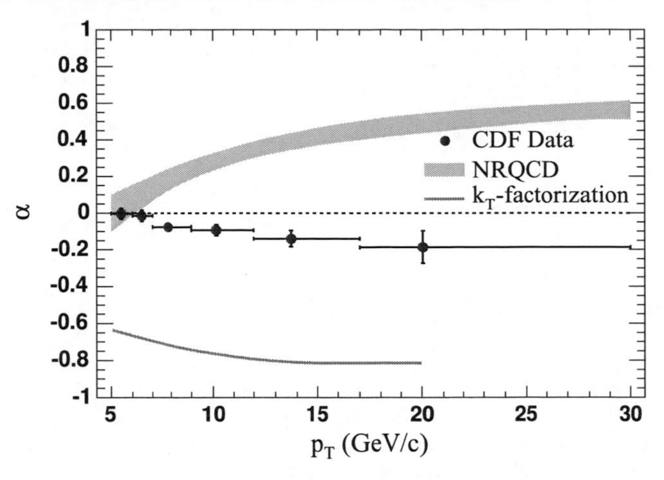

Fig. 55 The polarization parameter α for prompt J/ψ production in $p\bar{p}$ collisions at $\sqrt{s} = 1.96$ TeV as a function of p_T. The *points* are the CDF data [658], the *band* is the prediction from LO NRQCD factorization [667], and the *line* is the prediction from k_T factorization [640]. The theoretical uncertainty in the LO NRQCD factorization prediction was obtained by combining the uncertainties from the parton distributions (estimated by comparing the MRST98LO [732] and the CTEQ5L [733] distributions), the uncertainties from the color-octet NRQCD long-distance matrix elements, the uncertainties that are obtained by varying m_c in the range 1.45 GeV $< m_c <$ 1.55 GeV, and the uncertainties that are obtained by varying the factorization and renormalization scales in the range $0.5m_T < \mu_f = \mu_r < 2m_T$. Here $m_T = \sqrt{4m_c^2 + p_T^2}$. From [658] with kind permission, copyright (2007) The American Physical Society

in that channel. While the prediction at LO in α_s for the helicity of the J/ψ in the color-singlet channel is mainly transverse at medium and large p_T, calculations at NLO or NNLO* accuracy for the color-singlet channel reveal a polarization that is increasingly longitudinal as p_T increases, as can be seen in Fig. 56. A similar trend for the polarization as a function of p_T is found in some other analyses of the color-singlet channel, such those in the k_T factorization approach [639–642] (see Fig. 55), the gluon-tower approach [680], and the s-channel-$Q\bar{Q}$-cut approach [681, 682].

In the case of the color-octet $^3S_1$ channel, the NLO correction to the helicity of the J/ψ is very small [674]. This NLO correction would not change substantially the comparison between the NRQCD factorization prediction and the experimental data that is shown in Fig. 55. As we have explained in Sect. 4.2.2, the NLO analysis of [692] suggests the possibility that the J/ψ direct-production cross section is dominated by the color-octet $^1S_0$ contribution, rather than by the color-octet $^3S_1$ contribution, even at the largest values of p_T that are accessed in the Tevatron measurements. However, the NLO analysis in [693] concludes that the color-octet $^3S_1$ contribution at NLO is not very different from that at LO. A complete NLO analysis of the direct J/ψ polarization, including the contribution of the color-octet ^3P_J channel is still lacking and is an important theoretical goal. Further progress in determining the relevant production mech-

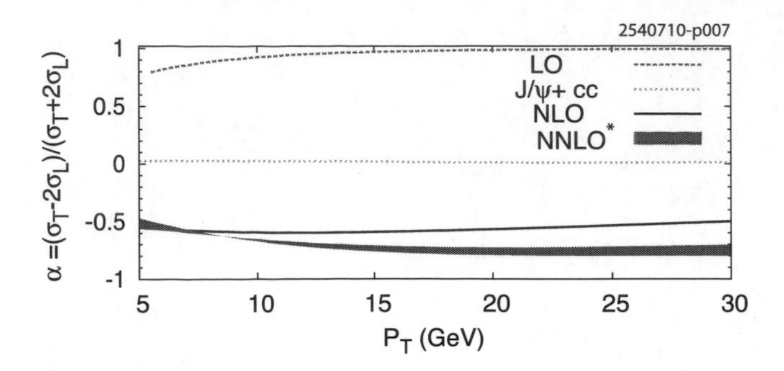

Fig. 56 Predictions for the polarization parameter α for direct J/ψ production in the color-singlet channel in $p\bar{p}$ collisions at the Tevatron at $\sqrt{s} = 1.96$ TeV at LO, NLO and NNLO* accuracy [652]. Most of the uncertainties in α for the LO, $J/\psi + c\bar{c}$, and NLO cases cancel. The theoretical uncertainty band for the NNLO* case was obtained by varying the infrared cut-off $s_{ij}^{\min}$ between $2m_c^2$ and $m_c^2/2$

anisms would be aided significantly by high-statistics measurements of the polarizations of the J/ψ, the χ_{cJ}, and the $\psi(2S)$ in direct production.

The polarization of the J/ψ has also been measured in hadronic collisions at $\sqrt{s} = 200$ GeV. Data from the PHENIX Collaboration for prompt J/ψ polarization as a function of p_T [662, 663] exist in the range $0 < p_T < 3$ GeV and indicate a polarization that is compatible with zero, with a trend toward longitudinal polarization as p_T increases. Comparisons of the data with LO calculations in the CSM and the NRQCD factorization formalism [699] are given in [662, 699]. As can be seen from Fig. 57, the data favor the NRQCD factorization prediction and are in agreement with it. However, the small values of p_T involved may call into question the validity of the NRQCD factorization formula. The NLO color-singlet contribution to the J/ψ polarization at RHIC has been computed in [700]. A comparison of this prediction with the PHENIX prompt J/ψ data differential in p_T [661, 662] is shown in Fig. 58. As can be seen, the CSM contributions to the polarization through NLO are in agreement with the PHENIX prompt J/ψ data.

The polarization of the $\Upsilon(1S)$ in prompt production has been measured by both the CDF Collaboration (Run I) [669] and by the DØ Collaboration (Run II) [671]. The DØ measurement has substantially larger experimental uncertainties than the CDF measurement. The CDF Collaboration has recently reported a new preliminary measurement of the polarization of prompt Υ's that is based on a larger data set from Run II [734]. This Run II measurement is consistent with the CDF Run I measurement. The results of the CDF Run II and DØ measurements are shown in Fig. 59, along with the NRQCD factorization prediction at LO in α_s [735]. The origin of the large discrepancy between the CDF and DØ data is unclear. However, we note that the CDF measurement was made over the rapidity interval $|y| < 0.6$, while the DØ measurement was made over the rapidity interval $|y| < 1.8$. The NRQCD factorization prediction in [735] was

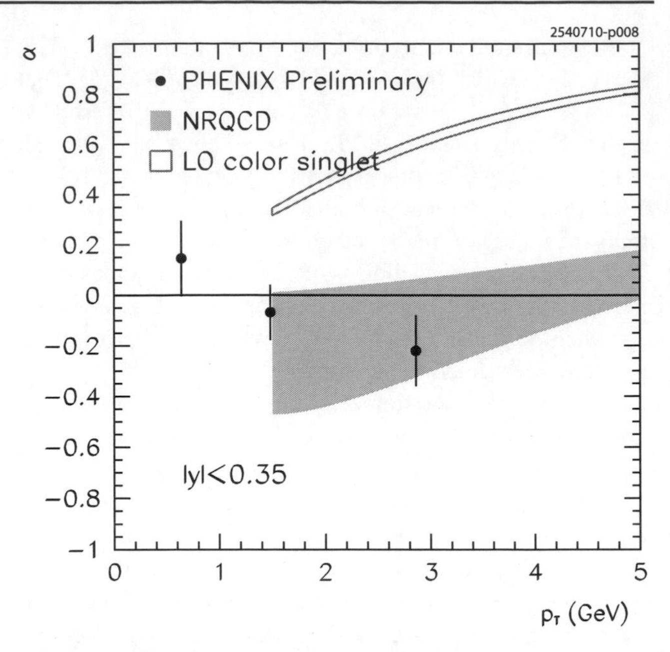

Fig. 57 Comparison of the LO NRQCD and LO CSM predictions for the prompt J/ψ polarization in pp collisions at RHIC at $\sqrt{S_{nn}} = 200$ GeV and $|y| < 0.35$ [699] with the data from the PHENIX Collaboration [662, 663]. The theoretical uncertainty bands were obtained by combining the uncertainties from m_c and the NRQCD long-distance matrix elements with the uncertainties that are obtained by varying the renormalization scale μ_f and the factorization scale μ_r between $2m_T$ and $m_T/2$. Here $m_T = \sqrt{4m_c^2 + p_T^2}$. From [699] with kind permission, copyright (2010) The American Physical Society

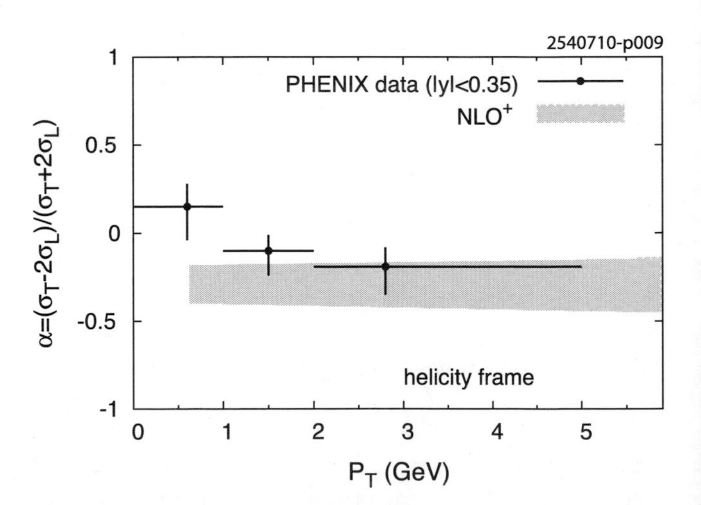

Fig. 58 Comparison of the NLO CSM calculation of J/ψ polarization in pp collisions at RHIC at $\sqrt{S_{NN}} = 200$ GeV and $|y| < 0.35$ [700] with the prompt J/ψ polarization data from the PHENIX Collaboration [661, 662]. The theoretical uncertainty band was obtained by combining the uncertainties that are obtained by varying m_c in the range 1.4 GeV $< m_c <$ 1.6 GeV, by varying the factorization scale μ_f and the renormalization scale μ_r over the values $((0.75, 0.75); (1, 1); (1, 2); (2, 1); (2, 2)) \times m_T$. Here $m_T = \sqrt{4m_c^2 + p_T^2}$. From [700]

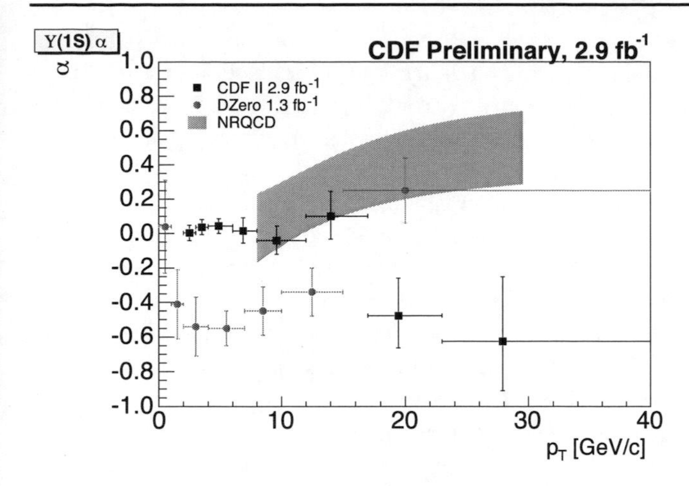

Fig. 59 The polarization parameter α in the helicity frame for prompt $\Upsilon(1S)$ production in $p\bar{p}$ collisions at $\sqrt{s} = 1.96$ TeV. The NRQCD factorization prediction at LO in α_s [735] is compared with the data of the CDF Collaboration [734] and the DØ Collaboration [671]. The CDF measurement was made over the rapidity interval $|y| < 0.6$, while the DØ measurement was made over the rapidity interval $|y| < 1.8$. The NRQCD factorization prediction in [735] was integrated over the range $|y| < 0.4$ [736] and includes feeddown from the $\Upsilon(2S)$, $\Upsilon(3S)$, $\chi_b(1P)$, and $\chi_b(2P)$ states. The theoretical uncertainty band was obtained by combining the uncertainties from the NRQCD long-distance color-singlet and color-octet matrix elements, m_b, the parton distributions, and the quarkonium branching fractions with uncertainties that are obtained by varying the renormalization and factorization scales from $\mu_T/2$ to $2\mu_T$. Here $\mu_T = \sqrt{m_b^2 + p_T^2}$. Figure provided by Hee Sok Chung, using [737], which is based on the analysis of [734]

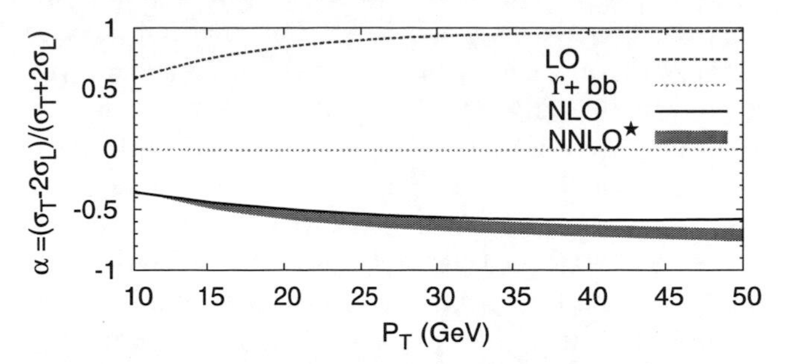

Fig. 60 Polarization parameter α for direct Υ production in the color-singlet channel in $p\bar{p}$ collisions at $\sqrt{s} = 1.96$ TeV at LO, NLO and NNLO* accuracy. Most of the uncertainties in α for the LO, $\Upsilon + b\bar{b}$, and NLO cases cancel. The theoretical uncertainty band for the NNLO* case was obtained by varying the infrared cut-off $s_{ij}^{\min}$ between $2m_b^2$ and $m_b^2/2$ and by varying μ_f and μ_r between $m_T/2$ and $2m_T$. Here $m_T = \sqrt{4m_c^2 + p_T^2}$. The variations with respect to μ_f and μ_r are negligible in comparison with the variation with respect to $s_{ij}^{\min}$. From [620] with kind permission, copyright (2008) The American Physical Society

integrated over the range $|y| < 0.4$ [736]. The LO NRQCD factorization prediction is marginally compatible with the CDF data at medium p_T and incompatible with the CDF data at large p_T, while the LO NRQCD factorization prediction is incompatible with the DØ data at medium p_T and compatible with the DØ data at large p_T. The effects of higher-order QCD corrections on the polarization of direct Υ's produced via the color-singlet channel are shown in Fig. 60. The higher-order corrections in this case have the same qualitative features as in the case of ψ: The higher-order corrections change the polarization from nearly 100% transverse to substantially longitudinal. A complete computation of the prompt Υ polarization at NLO in the NRQCD factorization, including feeddown from χ_{bJ} states, is not yet available.

4.2.4 New observables in hadroproduction

It may be useful, in order to progress in the understanding of the mechanisms that are responsible for heavy-quarkonium production, to identify, to compute, and to measure new observables.

As we have seen from previous discussions, the ψ and Υ production rates in hadron–hadron collisions, differential in p_T, are complicated to calculate because the dominant color-singlet channels at large p_T arise beyond LO in α_s. In the case of the Υ, higher-order corrections bring the color-singlet contribution close to the Tevatron data. However, in the case of ψ, there is a gap, which increases with increasing p_T, between the higher-order color-singlet contributions and the data. In both cases, the uncertainties are very large and, in the Υ case, they are too large to make a definite statement about the relative proportions of color-singlet and color-octet contributions. However, it is worth noting that the uncertainties in the higher-order color-singlet contributions affect the normalization of the differential rates much more than the shape. This suggests the use of observables that do not depend on the total rate. One interesting proposal is to compare ratios of differential cross sections at different values of $\sqrt{s}$, the center-of-mass energy of the colliding hadrons [738]. For a given color channel, this ratio can be predicted quite accurately. If the ratios that are associated with the color-singlet and color-octet channels are sufficiently well separated, then the relative sizes of the color-singlet and color-octet contributions could be extracted by comparing the measured production rate at the Tevatron with the measured production rate at the LHC.

One interesting observable is the hadronic activity near the quarkonium direction [739] or, more generally, the J/ψ-hadron azimuthal correlation. The UA1 Collaboration compared their charged-track distributions with Monte Carlo simulations for a J/ψ produced in the decay of a b hadron and a J/ψ produced in feeddown from a χ_{cJ} state [740, 741]. At the time of the UA1 study, χ_c feeddown was still expected to be the dominant source of prompt J/ψ's. In either the NRQCD factorization formalism or the CSM at higher orders in α_s, it is expected that the production process for prompt J/ψ's is more complex than the χ_{cJ}-feeddown process alone. Recently, the STAR Collaboration reported the first measurement of the J/ψ-hadron azimuthal correlation at RHIC [664, 742]. The STAR Collaboration compared its measurement with up-to-date LO PYTHIA predictions[13] and found no significant hadronic yield in the direction of the J/ψ beyond that which is expected from the LO PYTHIA predictions. Observation of the hadronic activity around the quarkonium might help to disentangle color-octet contributions from color-singlet contributions to the inclusive production process. One expects additional hadronic activity around the quarkonium in color-octet production. However, in practice, it may be difficult to identify this hadronic activity [739] because there are competing effects, such as the suppression of collinear gluon radiation from massive particles (dead-cone effect) and the suppression of soft gluon radiation from color-singlet objects, both of which would be difficult to compute reliably.

It would be useful to identify additional observables that can be computed reliably in order to test the many production models that are available [1, 651]. One such observable could be the rate of production of heavy-flavor mesons in association with a quarkonium. Final states that could be studied include $\psi + c\bar{c}$ and $\Upsilon + b\bar{b}$. Associated production could be investigated first in pp collisions and subsequently in pA and AA collisions. The study of associated production in hadron collisions is motivated by measurements that were carried out at the B factories that show that, in a surprisingly large fraction of J/ψ events, a second $c\bar{c}$ pair is produced. (See Sect. 4.6.) It is not yet known whether such a large fraction of $J/\psi + c\bar{c}$ events occurs in hadroproduction. Analyses at the Tevatron and at RHIC are already possible. The LO prediction for associated production at the Tevatron at $\sqrt{s} = 1.96$ TeV has been computed in [618] and shows that the integrated cross sections are significant:

$$\begin{aligned} &\sigma(J/\psi + c\bar{c}) \times \mathcal{B}\big(J/\psi \to \mu^+\mu^-\big) \approx 1 \text{ nb}; \\ &\sigma(\Upsilon + b\bar{b}) \times \mathcal{B}\big(\Upsilon \to \mu^+\mu^-\big) \approx 1 \text{ pb}. \end{aligned} \tag{158}$$

In order to illustrate the measurement potential at RHIC, the author of [652] computed the differential cross sections as a function of p_T and found them to be on the order of 1 pb/GeV at $p_T = 5$ GeV for the STAR kinematics. Measurements of such processes would provide tests of the NRQCD factorization formalism. They would also provide information about the color-transfer mechanism [635, 636] (see Sect. 4.1.7), which involves soft-gluon exchanges between comoving heavy particles and is known to violate standard NRQCD factorization. In this case, it would be useful to compare heavy-flavor activity near the quarkonium direction and away from the quarkonium direction.

[13] Note that one expects such predictions to be affected by channels that appear beyond LO in α_s.

A new observable that could be measured in existing and future experiments is the rate of production of a photon in association with a J/ψ or an Υ. The QCD corrections to the rates for these processes have been computed recently at NLO [724] and NNLO* [725] accuracy. As is argued in [725], a measurement of such processes would provide information on the quarkonium production mechanisms that is complementary to that which is provided by measurements of inclusive quarkonium production.

In order to facilitate phenomenological studies, an automated tree-level amplitude generator MadOnia [743] has been developed for processes involving quarkonium production or decay. It is now embedded in the online version of MadGraph/MadEvent [744] and, thus, is publicly available. A number of studies [618, 620, 676, 725, 745, 746] have already taken advantage of the flexibility of MadOnia and of the possibility to interface it with showering and hadronization programs.

4.2.5 Future opportunities

The results of the past few years, on both the theoretical and experimental fronts, have yielded important clues as to the mechanisms that are at work in inclusive quarkonium hadroproduction. In general, however, theoretical uncertainties remain too large to draw any definite conclusions about the production mechanisms.

Regarding these uncertainties, one of the key issues is that, in color-singlet production channels, the mechanisms that are dominant at large p_T appear only at higher orders in α_s. (See Sects. 4.2.1 and 4.2.2.) In addition to the new complete calculations of NLO contributions, estimates have been made of the NNLO contributions to ψ and Υ production in the color-singlet channels in several different frameworks (the fragmentation approximation, the k_T factorization approach, the gluon-tower model, and the NNLO* approach). A more accurate treatment of higher-order corrections to the color-singlet contributions at the Tevatron and the LHC is urgently needed. Here, the reorganization of the perturbation series that is provided by the fragmentation-function approach (Sect. 4.1.5) may be an important tool.

Furthermore, the current theoretical predictions suffer from uncertainties that are related to the long-distance dynamics that is involved in quarkonium production. For example, the prediction of the ψ or Υ polarization relies on the approximate heavy-quark spin symmetry of NRQCD. This approximate symmetry is based on the application of the velocity-scaling rules of NRQCD to evaluate the order of suppression of the spin-flip contribution. In the case of inclusive quarkonium decays, calculations of the NRQCD long-distance matrix elements on the lattice [233] have constrained the size of the spin-flip contribution. A similar constraint in case of inclusive quarkonium production would obviously be very valuable.

More generally, lattice determinations of the NRQCD long-distance production matrix elements would provide very useful constraints on the theoretical predictions and would also serve to check the phenomenological determinations of the long-distance matrix elements. An outstanding theoretical challenge is the development of methods for carrying out such lattice calculations, which are, at present, stymied by fundamental issues regarding the correct lattice formulation of single-particle inclusive rates in Euclidean space.

Further light could be shed on the NRQCD velocity expansion and its implications for low-energy dynamics by comparing charmonium and bottomonium production. The heavy-quark velocity v is much smaller in bottomonium systems than in charmonium systems. Hence, the velocity expansion is expected to converge more rapidly for bottomonium systems than for charmonium systems. In particular, spin-flip effects and color-octet contributions are expected to be smaller in bottomonium systems than in charmonium systems. The NRQCD factorization formula for inclusive quarkonium production, if it is correct, becomes more accurate as p_T increases and probably holds only for values of p_T that are greater than the heavy-quark mass. Therefore, the high-p_T reach of the LHC may be crucial in studying bottomonium production.

There are many unresolved theoretical issues at present that bear on the reliability of predictions for prompt J/ψ and Υ production. These issues may affect predictions for both the direct production of the J/ψ and the Υ and the production of the higher-mass quarkonium states that feed down into the J/ψ and the Υ. Therefore, it would be of considerable help in disentangling the theoretical issues in J/ψ and Υ production if experimental measurements could separate direct production of the J/ψ and the Υ from production via feeddown from higher-mass charmonium and bottomonium states. Ideally, the direct-production cross sections and polarizations would both be measured differentially in p_T. Measurement of the direct J/ψ cross section and polarization might be particularly important at large p_T, given the large proportion of χ_{cJ} feeddown events in the prompt J/ψ rate at large p_T that is predicted in [675].

Although it would be ideal to have measurements of direct quarkonium production rates and polarizations, it is, of course, very important to resolve the existing discrepancy between the CDF and DØ measurements of the prompt Υ polarization. The CDF measurement of the $\Upsilon(1S)$ polarization is for the rapidity range $|y| < 0.6$, while the DØ measurement is for the rapidity range $|y| < 1.8$. It would be very useful for the two experiments to provide polarization measurements that cover the same rapidity range.

It might also be useful to formulate new measurements and observables that would provide information that is complementary to that which is provided by the differential rates and polarization observables. The large rates for J/ψ and Υ production that are expected at the LHC open the door to new analyses. As we have mentioned in Sect. 4.2.4, the possibilities include studies of quarkonium production at different values of $\sqrt{s}$, studies of hadronic energy near and away from the quarkonium direction, and studies of the production of heavy-flavor mesons in association with a quarkonium. It is important in all of these studies to identify observables that are accessible under realistic experimental conditions and that can be calculated accurately enough to allow meaningful comparisons with experimental measurements. In this endeavor, communication between the experimental and theoretical experts in these areas will be crucial.

4.3 *ep* collisions

Inelastic production of charmonia in ep collisions at HERA proceeds via photon-gluon fusion: A photon emitted from the incoming electron or positron interacts with a gluon from the proton to produce a $c\bar{c}$ pair that evolves into a color-neutral charmonium state by the radiation of soft and/or hard gluons.

The elasticity observable z is defined as the fraction of energy of the incoming photon, in the proton rest frame, that is carried by the final-state charmonium. The kinematic region of inelastic charmonium production is $0.05 \lesssim z < 0.9$. In the so-called "photoproduction regime," at low photon virtuality Q^2, the incoming electron is scattered through a small angle, and the incoming photon is quasi real. The invariant mass of the γp system $W_{\gamma p}$ depends on the energy of the incoming photon. In photoproduction, photons can interact directly with the charm quark (direct processes), or via their hadronic component (resolved processes). Resolved processes are relevant at low elasticities ($z \lesssim 0.3$). HERA has been a unique laboratory for the observation of photoproduction in the photon–proton center-of-mass range $20 < W_{\gamma p} < 320$ GeV.

In ep scattering at HERA, toward high values of elasticity ($z > 0.95$), another production mechanism that is distinctly different from boson-gluon fusion becomes dominant. In this mechanism, which applies both to exclusive and diffractive charmonium production, the incoming photon fluctuates into a $c\bar{c}$ QCD dipole state which, subsequently interacts with the proton by the exchange of two or more gluons in a colorless state. This colorless interaction transfers momentum that allows the $c\bar{c}$ pair to form a bound quarkonium state. Experiments distinguish between two categories of diffractive processes. In elastic processes, the proton stays intact, i.e., $\gamma p \to J/\psi p$ ($z \approx 1$). In proton-dissociative processes, the proton breaks up into a low-mass final state, i.e., $\gamma p \to J/\psi Y$. In proton-dissociative processes, m_Y, mass of the state Y, is less than about 2 GeV or z lies in the $0.95 \lesssim z \lesssim 1$. Many measurements of the diffractive production of the ρ^0, ω, ϕ, J/ψ, $\psi(2S)$ and Υ states have been performed at HERA [747–758].[14] These measurements were crucial in reaching a new understanding of the partonic structure of hard diffraction, the distributions of partons in the proton, the validity of evolution equations in the low-x limit, and the interplay between soft and hard QCD scales. For a detailed report on diffractive quarkonium production at HERA, we refer the reader to [759]. Possible future opportunities involving measurements of diffractive quarkonium production are discussed in Sect. 6.11.

The ZEUS and H1 Collaborations have published several measurements of inelastic J/ψ and $\psi(2S)$ production that are based on data from HERA Run I [697, 760]. A new measurement, making use of the full Run II data sample, was published recently by the H1 Collaboration [696]. The ZEUS Collaboration has published a new measurement of the J/ψ decay angular distributions in inelastic photoproduction, making use of the collaboration's full data sample [761].

The data samples of J/ψ events that result from the experimental selection cuts are dominated by inelastic production processes in which the J/ψ's do not originate from the decay of a heavier resonance. Subdominant diffractive backgrounds, as well as feeddown contributions from the $\psi(2S)$, the χ_{cJ}, and b-flavored hadrons are estimated to contribute between 15% and 25% of the total J/ψ events, depending on the kinematic region. These backgrounds are usually neglected in theoretical predictions of direct J/ψ production rates.

The measurements of J/ψ cross sections and polarization parameters reported by the ZEUS and H1 Collaborations have been compared extensively to NRQCD factorization predictions at LO in α_s. In these studies, a truncation of the NRQCD velocity expansion is used, in which the independent long-distance matrix elements are $\langle \mathcal{O}_1^{J/\psi}(^3S_1)\rangle$, $\langle \mathcal{O}_8^{J/\psi}(^3S_8)\rangle$, $\langle \mathcal{O}_8^{J/\psi}(^1S_0)\rangle$ and $\langle \mathcal{O}_8^{J/\psi}(^3P_0)\rangle$. In the CSM, all of these matrix elements, except for the first one, are, in effect, set to zero. Usually the values of the long-distance color-octet matrix elements are extracted from the Tevatron data, in which case the comparisons of the resulting predictions for J/ψ production at HERA with the data offer the opportunity to assess the universality of the long-distance matrix elements. The comparisons between the predictions at LO in α_s and the data are summarized in [1].

In calculations in the NRQCD factorization formalism, large uncertainties arise from the sensitivity to the input parameters: the mass of the charm quark, the factorization and renormalization scales, and the values of the color-octet matrix elements, which are obtained from fits to the Tevatron data. One also expects sizable uncertainties owing to the omission of corrections of higher-order in both α_s and v. As we shall see, because of these theoretical uncertainties, the relative sizes of the color-singlet and color-octet contributions in charmonium photoproduction are still unclear.

In fixed-order, tree-level predictions of the color-octet contribution to photoproduction, a large peak appears in the z distribution near the kinematic endpoint $z = 1$. This feature, which was first interpreted as a failure of the universality of the long-distance matrix elements, has since been attributed to the breakdown of the NRQCD velocity expansion and the perturbation expansion in α_s near $z = 1$. In this region, in order to obtain a reliable theoretical prediction, one must resum large perturbative corrections to all orders in α_s and large nonperturbative corrections to all orders in v. It is known that the resummation of the color-octet contribution leads to a significant broadening of the peak at large z [762]. However, the effects of the nonperturbative resummation of the velocity expansion cannot be determined precisely without further information about the so-called "shape function." In principle, that information could be extracted from data on charmonium production in e^+e^- collisions.

J/ψ photoproduction at HERA has also been studied in the k_T-factorization scheme [640, 763]. In this framework, it has been argued that the color-singlet contribution alone can explain the HERA data. A prediction for the color-singlet yield at LO in α_s in the k_T-factorization approach reproduces the measured shapes of the p_T and z distributions reasonably well. However, it should be kept in mind that these predictions rely on unintegrated parton distributions, which are not well known at present. This uncertainty might be reduced as more accurate unintegrated parton distributions sets become available [764].

The pioneering calculation of the correction of NLO in α_s to the color-singlet contribution to the direct J/ψ cross section was presented in [765]. The NLO correction affects not only the normalization of the photoproduction rate, but also the shape of the p_T distribution. Both of these effects bring the color-singlet contribution into better agreement with the data. The large impact of the NLO correction at large transverse momentum can be understood in terms of the kinematic enhancement of NLO color-singlet production processes relative to the LO color-singlet production processes. This effect is similar to the one that appears in hadroproduction in the color-singlet channel. (See Sect. 4.2.1.)

The NLO calculation of [765] suggests that production in the color-singlet channel might be the main mechanism at work in J/ψ photoproduction. However, in more recent work [695, 766–769], which confirms the calculation of [765], it has been emphasized that the factorization and renormalization scales in [765] have been set a value $(m_c/\sqrt{2})$ that is generally considered to be too low in the

[14]Here, only the most recent measurements are cited.

region of large p_T. As we shall see, a more physical choice of scale, such as $\sqrt{4m_c^2 + p_T^2}$, leads to predictions for cross sections differential in p_T^2 or in z that lie considerably below the H1 and ZEUS measurements. Hence, the size of the NLO color-singlet contribution does not exclude the possibility that other contributions, such as those from the color-octet channel, are at least as large as the color-singlet contribution. In the region of low transverse momentum, which is the dominant region for J/ψ production at HERA, the sensitivity to the factorization and renormalization scales is very large and complicates the identification of the dominant production mechanism at HERA.

4.3.1 Phenomenology of the cross section, including NLO corrections

In Fig. 61 we show a comparison between data from the H1 Collaboration [696] for the J/ψ photoproduction cross sections differential in p_T^2 and in z and calculations in the NRQCD factorization formalism from [695, 768, 769].[15] The dashed line depicts the central values of the complete NRQCD factorization prediction (including color-singlet and color-octet contributions) at NLO in α_s, and the band shows the uncertainty in that prediction that arises from the uncertainties in the color-octet long-distance NRQCD matrix elements. Note that the contributions from resolved photoproduction, which are important in the low-z region, and the contributions from diffractive production, which are important near $z = 1$, are not included in the NRQCD factorization prediction. The uncertainties that are shown arise from the uncertainties in the NRQCD color-octet long-distance matrix elements. These matrix elements were obtained through a fit to the Tevatron hadroproduction data that used the NRQCD prediction at LO in α_s, augmented by an approximate calculation of some higher-order corrections from multiple-gluon radiation [770].

The NRQCD factorization prediction at NLO accuracy in α_s is in better agreement with the H1 data than the color-singlet contribution alone. However, it should be kept in mind that the mass and scale uncertainties of the color-singlet contribution have not been displayed here. As we have already mentioned, these uncertainties are large, even at NLO in α_s.

The NRQCD factorization prediction for the cross section differential in z shows a rise near $z = 1$ that is characteristic of the color-octet contributions. As we have mentioned, resummations of the series in α_s and in v are needed in order to obtain a reliable theoretical prediction in this region. In the low-z region, the NRQCD factorization prediction undershoots the data. In this region, the corrections to resolved photoproduction through NLO in α_s may be needed in order to bring the theory into agreement with the data.

[15] A more detailed comparison of the H1 data with theory predictions can be found in [696].

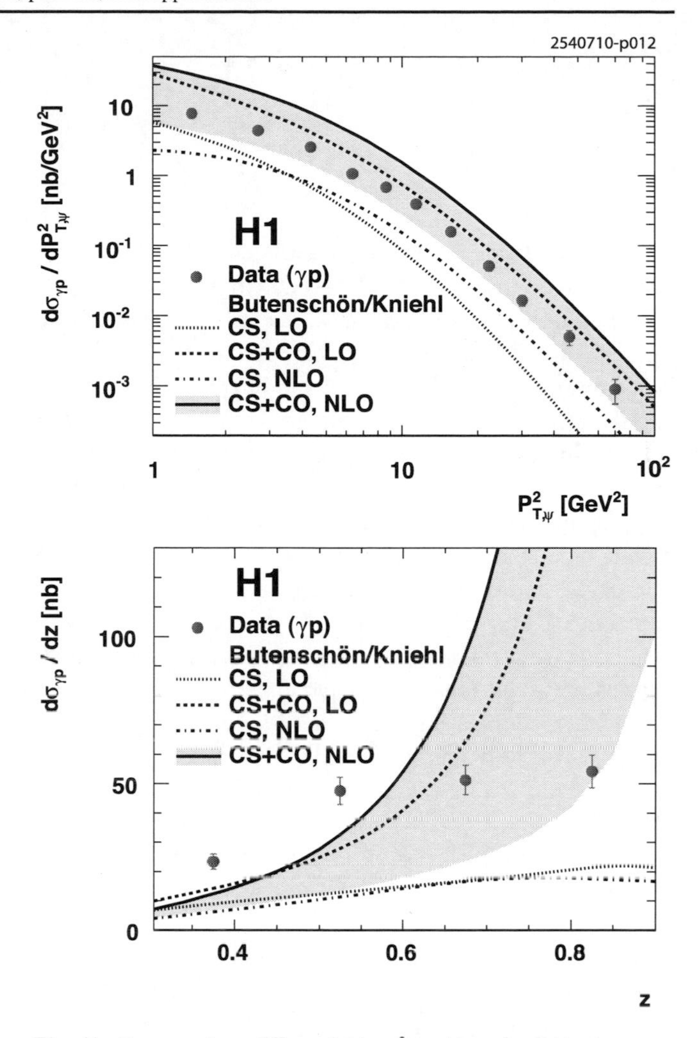

Fig. 61 Cross sections differential in p_T^2 and in z for J/ψ photoproduction at HERA. The measurement by the H1 Collaboration [696] is compared to the CSM and NRQCD predictions at LO and NLO in α_s from [695, 768]. "CS" and "CO" denote the color-singlet and color-octet contributions, respectively. The *dashed line* depicts the central values of the complete NRQCD factorization prediction (including color-singlet and color-octet contributions) at NLO in α_s, and the *band* shows the uncertainty in that prediction that arises from the uncertainties in the color-octet long-distance NRQCD matrix elements. From [696]

4.3.2 Polarization

In addition to the studies of J/ψ differential cross sections that we have mentioned, there have also been recent analyses of polarization in J/ψ photoproduction. The polarization observables may provide additional information about the production mechanisms.

Experimentally, the J/ψ polarization is extracted from the angular distribution of the leptons that originate in J/ψ decays. In the J/ψ rest frame, the distribution takes the general form

$$\frac{d\Gamma(J/\psi \to l^+l^-)}{d\Omega} \propto 1 + \lambda \cos^2\theta + \mu \sin 2\theta \cos\phi + \frac{\nu}{2} \sin^2\theta \cos 2\phi, \quad (159)$$

where θ and ϕ are the polar and azimuthal angles of the l^+ three-momentum with respect to a particular spin-quantization frame. (See Sect. 4.2.3.) The ZEUS Collaboration has published a new measurement of the parameters λ and ν in the target spin-quantization frame [771] that is based on an integrated luminosity of 468 pb^{-1} [761]. The H1 Collaboration has published new measurements of the parameters λ and ν in both the helicity and the Collins–Soper spin-quantization frames that are based on an integrated luminosity of 165 pb^{-1} [696]. The H1 Collaboration uses a more restricted range in the energy fraction z, namely, $0.3 < z < 0.9$, in order to suppress possible contributions from diffractive or feeddown processes. In both experiments, the polarization parameters in each bin are extracted by comparing the data with Monte Carlo distributions for different values of the polarization parameters, using a χ^2 criterion to assess the probability of each distribution.

We show comparisons of several theoretical predictions with the ZEUS data in Fig. 62 and with the H1 data in Fig. 63.

The curves in Fig. 62 that are labeled "LO CS+CO" are the complete NRQCD factorization predictions (including color-singlet and color-octet contributions) at LO in α_s [771]. These agree reasonably well with the ZEUS data in the target frame, except for the value of ν in the lowest-p_T bin. However, at such a low value of p_T, the NRQCD factorization formula is not expected to be valid.

The curves that are labeled "LO k_T" and "CSM k_T" are predictions in the k_T-factorization scheme [772]. The set of unintegrated parton distribution functions that is used is indicated in parentheses. The k_T-factorization predictions bracket the ZEUS data in the target frame for both λ and ν and are in reasonable agreement with the H1 data for λ and ν in the helicity frame, if one takes the difference between the two k_T-factorization predictions to be a measure of the theoretical uncertainty.

The bands that are labeled "NLO CS" and "CSM NLO" correspond to predictions in the color-singlet model at NLO in α_s [766]. (Similar results for the polarization at NLO in α_s in the color-singlet model were obtained in [767].) The uncertainties that are shown in these bands arise from the sensitivity of the polarization to the factorization and renormalization scales and are much larger than the uncertainties that arise from the uncertainty in the value of m_c. A comparison of the LO and NLO predictions of the CSM shows the large impact of the NLO correction on the polarization

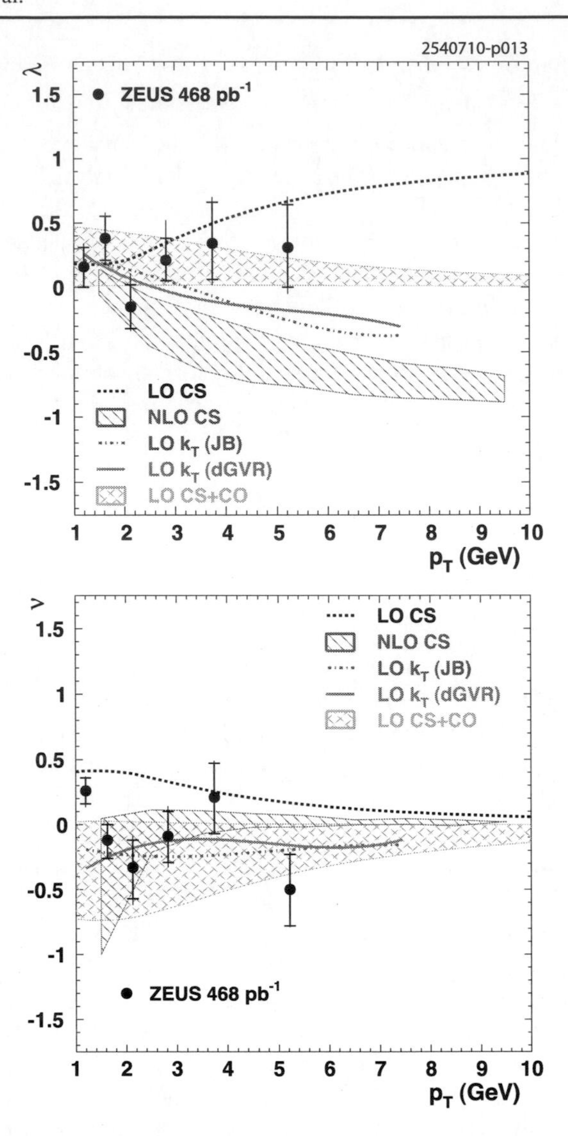

Fig. 62 Polarization parameters λ and ν in the target frame as a function of p_T for J/ψ photoproduction at HERA. The measurement by the ZEUS Collaboration [761] of the polarization parameters in the target frame is compared with the color-singlet contribution at LO in α_s (labeled LO CS) [771] and at NLO in α_s (labeled NLO CS) [766], with predictions in the k_T-factorization approach for two different sets of unintegrated parton-distribution functions (labeled LO k_T (JB) and LO k_T (dGRV)) [772], and with the complete NRQCD factorization predictions (including color-singlet and color-octet contributions) at LO in α_s (labeled LO CS+CO) [771]. The theoretical uncertainty bands labeled NLO CS were obtained by varying the factorization scale μ_f and the renormalization scale μ_r in the range defined by $0.5\mu_0 < \mu_f, \mu_r < 2\mu_0$, and $0.5 < \mu_r/\mu_f < 2$, where $\mu_0 = 4m_c$. The theoretical uncertainty bands labeled LO CS+CO were obtained by considering uncertainties in the values of the color-octet NRQCD long-distance matrix elements. From [761] with kind permission, copyright (2009) Springer

parameters. At NLO, the parameter λ is predicted to decrease with increasing p_T, in both the target and the helicity frames. (This trend is also observed in the LO prediction of the k_T factorization formalism, which effectively accounts for some topologies that occur at higher orders in α_s in the

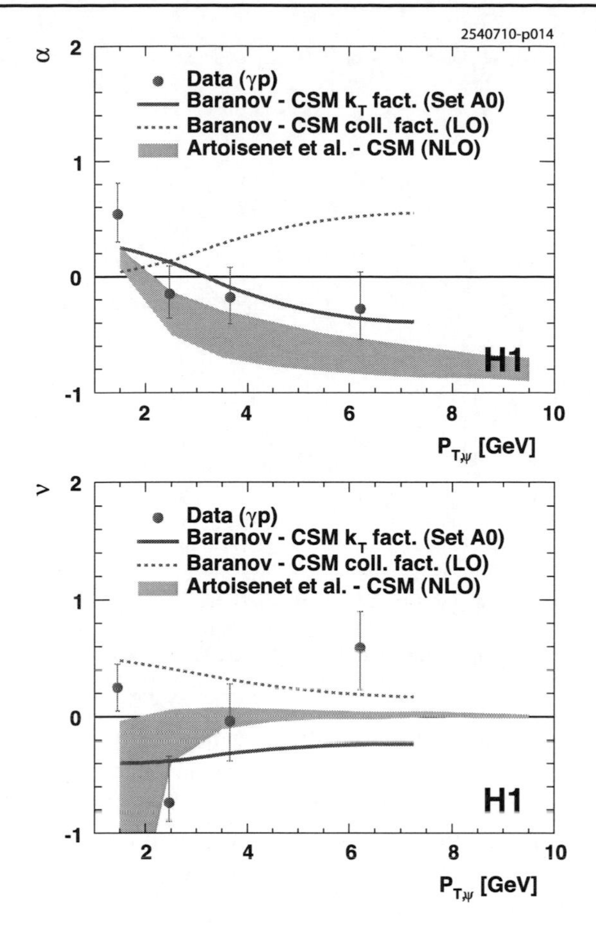

Fig. 63 Polarization parameters $\alpha = \lambda$ and ν in the helicity frame as a function of p_T for J/ψ photoproduction at HERA. The measurement by the H1 Collaboration [696] of the polarization parameters in the helicity frame is compared with the predictions in the k_T-factorization approach (*solid line*) [772] and with the predictions in color-singlet model (CSM) at LO in α_s (*dashed line*) [772] and at NLO α_s (*filled band*) [766, 773]. The theoretical uncertainty bands were obtained by varying the factorization scale μ_f and the renormalization scale μ_r in the range defined by $0.5\mu_0 < \mu_f, \mu_r < 2\mu_0$, and $0.5 < \mu_r/\mu_f < 2$, where $\mu_0 = 4m_c$. From [696]

collinear-factorization scheme.) The NLO color-singlet predictions for λ are compatible with the H1 data in the helicity frame, but differ significantly from the ZEUS data in the target frame. In contrast, the NLO color-singlet prediction for the parameter ν is in reasonable agreement with the ZEUS data in the target frame, as well as with the H1 data in the helicity frame.

A more complete presentation of the comparison between theory and experiment for the polarization parameters λ and ν, including the distributions of these parameters as functions of z, can be found in [696, 761].

4.3.3 Future opportunities

In spite of the recent advances that we have described, it is still unclear which mechanisms are at work in J/ψ photoproduction at HERA. The new computations of NLO corrections to the differential cross sections and the polarization parameters in the collinear-factorization scheme show that there is room for a color-octet contribution. Indeed, the NLO NRQCD factorization prediction for the cross section fits the data reasonably, with central values that are closer to the data points than are the central values of the NLO color-singlet contribution. Recent analyses of the color-singlet contribution in the k_T factorization scheme also show reasonable agreement with the data.

In both k_T factorization and in collinear factorization, theoretical uncertainties remain substantial. Improvement of the situation for k_T factorization will require better knowledge of the k_T-dependent parton distributions. In collinear factorization, the large sensitivity of the NLO color-singlet rates to the renormalization scale signals that QCD corrections beyond NLO might be relevant, especially for the description of the polarization parameters. Here, an analysis in the fragmentation-function approach (see Sect. 4.1.5) might help to bring the perturbation series under better control. Theoretical uncertainties in the NRQCD factorization prediction for the J/ψ polarization could be reduced by computing the color-octet contributions to the polarization parameters at NLO accuracy in α_s. The large uncertainties in the color-octet long-distance matrix elements dominate the uncertainties in the NRQCD factorization prediction for the cross section. Since these matrix elements are obtained by fitting NRQCD factorization predictions to the Tevatron data, improvements in the theoretical uncertainties in the Tevatron (or LHC) predictions are necessary in order to reduce the uncertainties in the matrix elements. Finally, theoretical uncertainties in the region near the kinematic endpoint $z = 1$ might be reduced through a systematic study of resummations of the perturbative and velocity expansions in both ep and e^+e^- quarkonium production.

4.4 Fixed-target production

4.4.1 Phenomenology of fixed-target production

The NRQCD factorization approach has also been tested against the charmonium production data that have been obtained from fixed-target experiments and pp experiments at low energies. Owing to the limited statistics, quarkonium observables measured in these experiments have, in general, been restricted to total cross sections for J/ψ and $\psi(2S)$ production. Also, in the case of J/ψ production, feeddown contributions have not been subtracted.

In the most recent analysis of the fixed-target and low-energy pp quarkonium production data [774], experimental results for the inclusive production rates of the J/ψ and the $\psi(2S)$ have been examined, along with the experimental results for the ratios of these cross sections. A total of 29 experimental results have been analyzed in order to extract the color-octet contribution to the observed rates. By comparing the values of the color-octet matrix elements that are extracted from the fixed-target data with the values that are extracted from the Tevatron data, one can test the universality of the NRQCD matrix elements.

The analysis of [774] made use of the NRQCD short-distance coefficients through order α_s^3 for the P-wave channels and for all of the color-octet channels that are of leading order in v [466]. The LO short-distance coefficients for the color-singlet ${}^3S_1$ channel were employed, as the computation of the QCD correction to these short-distance coefficients in pp collisions had not yet been completed at the time of the analysis of [774].

In extracting the color-octet contribution from the production rates that are measured in fixed-target experiments, the authors of [774] treated the NRQCD long-distance matrix elements as follows: Heavy-quark spin symmetry was employed; the color-singlet matrix elements were taken from the potential-model calculation of [775]; the color-octet matrix element for P-wave charmonium states was set equal to a value that was extracted in [776] from the CDF data [650]. Regarding the color-octet matrix elements for the J/ψ and $\psi(2S)$, it was assumed that $\langle\mathcal{O}_8^H({}^1S_0)\rangle = \langle\mathcal{O}_8^H({}^3P_0)\rangle/m_c^2$ and that the ratios of these matrix elements to $\langle\mathcal{O}_8^H({}^3S_1)\rangle$ are given by the values that were extracted from the Tevatron data. The color-octet contributions to the direct production of the J/ψ and the $\psi(2S)$ were multiplied by rescaling parameters $\lambda_{J/\psi}$ and $\lambda_{\psi'}$, respectively, which were varied in order to fit the fixed-target data. In addition to these two parameters, the factorization and renormalization scales were varied in the fit, while the mass of the charm quark was held fixed at 1.5 GeV.

The χ^2 of the fit favors a nonzero value for the color-octet contribution to direct J/ψ and $\psi(2S)$ production. This can be seen in Fig. 64 for the case of the J/ψ: The color-singlet contribution alone (dotted-dashed line) systematically undershoots the data, while the fitted complete NRQCD contribution (solid line) is in good agreement with the data. However, the fit also indicates that the values of the color-octet matrix elements that are needed to explain the fixed-target data are only about 10% of the values that were extracted from fits to the Tevatron data. A similar result was obtained in the case of the color-octet matrix elements for the $\psi(2S)$. In both cases, the stability of the fitting procedure was checked by changing the selection of measurements used in the fits and by changing the set of parton distribution functions that was used in computing the cross sections. In any of these scenarios, the values of the color-octet matrix elements that were extracted from the fixed-target data are smaller than the values that were extracted from the Tevatron data.

The analysis of fixed-target data that we have described could be updated by making use of the recent results for the

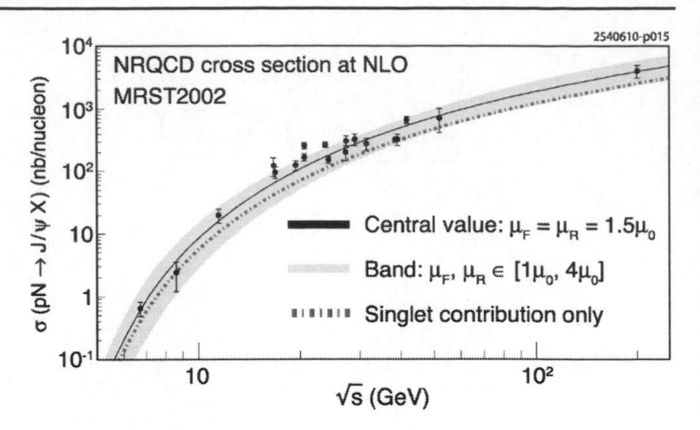

Fig. 64 Fit of the NRQCD factorization cross section for the production of the J/ψ as a function of the center-of-mass energy $\sqrt{s}$ to data from fixed-target experiments and pp experiments at low energies. The *curves* are the result of the fitting procedure that is described in the text. The color-singlet component is shown as a *dotted-dashed line*. The theoretical uncertainty *band* was obtained by varying the factorization scale μ_F and the renormalization scale μ_R in the ranges $\mu_0 < \mu_F < 4\mu_0$ and $\mu_0 < \mu_R < 4\mu_0$, where $\mu_0 = 2m_c$. From [774] with kind permission, copyright (2006) Elsevier

NLO corrections to the hadroproduction of S-wave states [619, 674]. Furthermore, a more accurate value for the J/ψ color-singlet matrix element has been derived recently in [777]. However, it is expected, even with these improvements, that the qualitative conclusions of [774] would still hold.

It is interesting to consider this fixed-target result in light of two recent developments: (1) the possible impact of higher-order corrections to the color-singlet yield on the phenomenology of J/ψ production at the Tevatron (Sect. 4.2.2) and (2) the latest J/ψ polarization measurement by the CDF Collaboration [658] (Sect. 4.2.3). Both of these developments suggest that the values of the color-octet NRQCD matrix elements may be smaller than had previously been supposed. However, in assessing the significance of the fixed-target result, it should be remembered that the fixed-target data, integrated over p_T, are dominated by data from the lowest values of p_T. At low p_T, one would not expect NRQCD factorization to hold.

4.4.2 Future opportunities

Fixed-target experiments provide another means by which to test the various theories of inclusive quarkonium production. In order to test definitively theoretical hypotheses, such as NRQCD factorization, that are based on hard-scattering factorization, it is necessary to make measurements at values of p_T that are much greater than the heavy-quark mass. If fixed-target experiments with such a high reach in p_T could be devised, then it would be very useful to measure both the cross section and the quarkonium polarization as functions of p_T.

Table 43 Experimental measurements and NRQCD predictions for $e^+e^- \to J/\psi + H$, where H is η_c, χ_{c0}, or $\eta_c(2S)$. Cross sections are in units of fb. The quantity $\mathcal{B}_{>2}$ is the branching fraction of the charmonium state that is recoiling against the J/ψ into more than two charged tracks

Quantity	$\eta_c(1S)$	$\chi_{c0}(1P)$	$\eta_c(2S)$
$\sigma \times \mathcal{B}_{>2}$ (Belle [782])	$25.6 \pm 2.8 \pm 3.4$	$6.4 \pm 1.7 \pm 1.0$	$16.5 \pm 3.0 \pm 2.4$
$\sigma \times \mathcal{B}_{>2}$ (*BABAR* [53])	$17.6 \pm 2.8^{+1.5}_{-2.1}$	$10.3 \pm 2.5^{+1.4}_{-1.8}$	$16.4 \pm 3.7^{+2.4}_{-3.0}$
σ (Liu, He, Chao [779])	5.5	6.9	3.7
σ (Braaten, Lee [780])	3.78 ± 1.26	2.40 ± 1.02	1.57 ± 0.52
σ (Hagiwara, Kou, Qiao [781])	2.3		
σ (Bodwin et al. [783])	17.5 ± 5.7		
σ (He, Fan, Chao [784])	20.4		
σ (Bodwin, Lee, Yu [785])	$17.6^{+8.1}_{-6.7}$		

4.5 Exclusive production in e^+e^- collisions

4.5.1 Theory vs. experiment

As recently as three years ago, large discrepancies existed between experimental measurements and theoretical predictions for the process $e^+e^- \to J/\psi + \eta_c$ at the B-factory energy of $\sqrt{s} = 10.58$ GeV. In its initial measurement of this process, the Belle Collaboration obtained $\sigma(e^+e^- \to J/\psi + \eta_c) \times \mathcal{B}_{\geq 4} = 33^{+7}_{-6} + 9$ fb [778], where $\sigma(e^+e^- \to J/\psi + \eta_c)$ is the cross section and $\mathcal{B}_{\geq 4}$ is the branching fraction of the η_c into four or more charged tracks. The first theoretical predictions for $\sigma(e^+e^- \to J/\psi + \eta_c)$ were based on calculations in the NRQCD factorization approach [138]. These initial calculations were carried out at LO in α_s and v. They gave the following predictions for $\sigma(e^+e^- \to J/\psi + \eta_c)$: 5.5 fb [779], 3.78 ± 1.26 fb [780], and 2.3 fb [781].[16] The differences between the calculations of [779–781] arise from QED effects, which are included only in [780]; from contributions from an intermediate Z boson, which are included only in [781]; and from different choices of m_c, the NRQCD long-distance matrix elements, and α_s. The sensitivities of the calculations to the values of these parameters are important sources of theoretical uncertainties, which we will discuss later.

The Belle Collaboration has, more recently, measured the quantity $\sigma(e^+e^- \to J/\psi + \eta_c) \times \mathcal{B}_{>2} = 25.6 \pm 2.8 \pm 3.4$ fb [782], where $\mathcal{B}_{>2}$ is the branching fraction of the charmonium state that is recoiling against the J/ψ (in this case the η_c) into more than two charged tracks. This cross section times branching fraction has also been measured by the *BABAR* Collaboration, which obtains $\sigma(e^+e^- \to J/\psi + \eta_c) \times \mathcal{B}_{>2} = 17.6 \pm 2.8 \pm 2.1$ fb [53]. These more recent experimental results reduced the discrepancy between theory and experiment, but did not eliminate it. In assessing the size of the discrepancy, it is important to recognize that the experimental results are cross sections times branching fractions. Hence, they are lower bounds on the cross sections, which are the quantities that appear in the theoretical predictions.

Table 43 contains a summary of experimental measurements and NRQCD predictions for the process $e^+e^- \to J/\psi + H$, where H is η_c, χ_{c0}, or $\eta_c(2S)$. As can be seen from Table 43, significant discrepancies exist between LO NRQCD predictions and experiment, not only for exclusive production of $J/\psi + \eta_c$, but also for exclusive production of $J/\psi + \chi_{c0}$ and $J/\psi + \eta_c(2S)$. An important step toward resolving the discrepancy between theory and experiment for $\sigma(e^+e^- \to J/\psi + \eta_c)$ was the calculation in [786] of the corrections of NLO in α_s. These corrections yield a K factor of about 1.96. This result has been confirmed in [787]. While this K factor is substantial, it does not, by itself, eliminate the discrepancy between theory and experiment.

In the NRQCD factorization formalism there are, in addition to corrections of higher order in α_s, corrections of higher order in v, i.e., relativistic corrections. In $\sigma(e^+e^- \to J/\psi + \eta_c)$, relativistic corrections can arise in two ways. First, they can appear directly as corrections to the process $e^+e^- \to J/\psi + \eta_c$ itself. Second, they can arise indirectly through the NRQCD long-distance matrix elements that appear in the expression for $\sigma(e^+e^- \to J/\psi + \eta_c)$. For example, the matrix element of leading order in v that appears in J/ψ production can be determined phenomenologically from the experimental value for the width for $J/\psi \to e^+e^-$ and the theoretical expression for that width. There are relativistic corrections to the theoretical expression for the width, which affect the value of the long-distance matrix element that one obtains.

The first relativistic correction appears at relative order v^2, where $v^2 \approx 0.3$ for charmonium. It has been known for some time that this correction is potentially large: In [780], it was found that the order-v^2 correction is $1.95\langle v^2\rangle_{J/\psi} +$

[16] In [780] a cross section of 2.31 ± 1.09 fb was reported initially. Later, a sign error in the QED interference term was corrected and the value cited above was obtained.

$2.37\langle v^2\rangle_{\eta_c}$. Here, $\langle v^2\rangle_H$ is the ratio of an order-v^2 NRQCD long-distance matrix element to the LO matrix element in the quarkonium state H. The authors of [780] estimated the matrix elements of order v^2 by making use of the Gremm–Kapustin relation [788], which follows from the NRQCD equations of motion. On the basis of these estimates, they found that the K factor for the relativistic corrections is about $2.0^{+10.9}_{-1.1}$. The very large uncertainties in this K factor reflect the large uncertainties in the Gremm–Kapustin-relation estimates of the matrix elements of order v^2.

In [789], significant progress was made in reducing the uncertainties in the order-v^2 NRQCD matrix elements. The approach in this work was to make use of a static-potential model to calculate the quarkonium wave functions and, from those wave functions, to compute the dimensionally regulated NRQCD matrix elements. If the static potential is known accurately, for example, from lattice calculations, then the corrections to the static-potential model are of relative order v^2 [140, 179]. Hence, one can regard the potential-model calculation as a first-principles calculation with controlled uncertainties.

Making use of the results of [789], the authors of [783] computed the relativistic corrections to $\sigma(e^+e^- \to J/\psi + \eta_c)$. Taking into account the corrections of NLO in α_s from [786], they obtained $\sigma(e^+e^- \to J/\psi + \eta_c) = 17.5 \pm 5.7$ fb, where the quoted uncertainty reflects only the uncertainties in m_c and the order-v^2 NRQCD long-distance matrix elements. This result includes the effects of a resummation of a class of relativistic corrections that arise from the quarkonium wave function. One might worry that the large relativistic corrections that appear in this calculation, which result in a K factor of about 2.6, are an indication that the v expansion of NRQCD is out of control. However, this large correction is the result of several corrections of a more modest size: a direct correction of about 40% and two indirect corrections (for the J/ψ and the η_c) of about 37% each. Furthermore, higher-order terms in the resummation of wave-function corrections change the direct correction by only about 13%, suggesting that the v expansion is indeed converging well.

The authors of [784] took a different approach to calculating relativistic corrections, determining the NRQCD long-distance matrix elements of LO in v and of relative order v^2 by using $\Gamma(J/\psi \to e^+e^-)$, $\Gamma(\eta_c \to \gamma\gamma)$, and $\Gamma(J/\psi \to \text{light hadrons})$ as inputs. Their result, $\sigma(e^+e^- \to J/\psi + \eta_c) = 20.04$ fb, is in agreement with the result of [783]. However, the values of the NRQCD matrix elements that are given in [784] differ significantly from those that were used in [783]. This difference probably arises mainly because of the very large relativistic corrections to $\Gamma(J/\psi \to \text{light hadrons})$, which may not be under good control.

The results of [783, 784] greatly reduced the difference between the experimental and theoretical central values for $\sigma(e^+e^- \to J/\psi + \eta_c)$. However, in order to assess the significance of these results, it is essential to have a reliable estimate of the theoretical uncertainties. Such an estimate was provided in [785]. In this work, which was based on the method of [783], uncertainties from various input parameters, such as m_c and the electromagnetic widths of the J/ψ and the η_c, were taken into account, as well as uncertainties from the truncations of the α_s and v expansions. Correlations between uncertainties in various components of the calculation were also taken into account. In addition, various refinements were included in the calculation, such as the use of the vector-meson-dominance method to reduce uncertainties in the QED contribution and the inclusion of the effects of interference between relativistic corrections and corrections of NLO in α_s. The conclusion of [785] is that $\sigma(e^+e^- \to J/\psi + \eta_c) = 17.6^{+8.1}_{-6.7}$ fb. This result is in agreement, within uncertainties, with the BABAR result, even if one allows for the fact that the branching fraction $\mathcal{B}_{>2}$ could be as small as 0.5–0.6.

An alternative approach to theoretical calculations of exclusive quarkonium production in e^+e^- annihilation is the light-cone method [645–649]. Generally, the light-cone-method predictions for exclusive quarkonium production cross sections are in agreement with the experimental results. The light-cone approach to quarkonium production can be derived from QCD.[17] In principle, the light-cone approach is as valid as the NRQCD factorization approach. In practice, it is, at present, necessary to model the light-cone wave functions of the quarkonia, possibly making use of constraints from QCD sum rules [790, 791]. Consequently, the existing light-cone calculations are not first-principles calculations, and it is not known how to estimate their uncertainties reliably. The light-cone approach automatically includes relativistic corrections that arise from the quarkonium wave function. As has been pointed out in [792], the light-cone calculations contain contributions from regions in which the quarkonia wave-function momenta are of order m_c or greater. In NRQCD, such contributions are contained in corrections to the short-distance coefficients of higher order in α_s. Therefore, in order to avoid double counting, one should refrain from combining light-cone results with NRQCD corrections of higher order in α_s. Finally, we note that, in [792], it was suggested that resummations of logarithms of $\sqrt{s}/m_c$ have not been carried out correctly in some light-cone calculations.

4.5.2 Future opportunities

Clearly, it would be desirable to reduce both the theoretical and experimental uncertainties in the rates for the exclusive production of quarkonia in e^+e^- annihilation and to

[17] Light-cone factorization formulas are derived in [643] in the course of proving NRQCD factorization formulas.

extend theory and experiment to processes involving additional quarkonium states.

On the experimental side, the central values of the Belle and BABAR measurements of $\sigma(e^+e^- \to J/\psi + \eta_c) \times \mathcal{B}_{>2}$ differ by about twice the uncertainty of either measurement. Although those uncertainties are small in comparison with the theoretical uncertainties, the rather large difference in central values suggests that further experimental work would be useful. Furthermore, it would be very useful, for comparisons with theory, to eliminate the uncertainty in the cross section that arises from the unmeasured branching fraction $\mathcal{B}_{>2}$.

On the theoretical side, the largest uncertainty in $\sigma(e^+e^- \to J/\psi + \eta_c)$ arises from the uncertainty in m_c. One could take advantage of the recent reductions in the uncertainty in m_c [228, 248] to reduce the theoretical uncertainty in $\sigma(e^+e^- \to J/\psi + \eta_c)$ from this source. The next largest source of theoretical uncertainty arises from the omission of the correction to the J/ψ electromagnetic width at NNLO in α_s [793, 794]. Unfortunately, the large scale dependence of this correction is a serious impediment to progress on this issue. The uncalculated correction to $\sigma(e^+e^- \to J/\psi + \eta_c)$ of relative order $\alpha_s v^2$ is potentially large, as is the uncalculated correction of relative order α_s^4. While the calculation of the former correction may be feasible, the calculation of the latter correction is probably beyond the current state of the art. However, one might be able to identify large contributions that could be resummed to all orders in α_s.

Finally, it would be desirable to extend the theoretical calculations to include P-wave and higher S-wave states. In the NRQCD approach, a serious obstacle to such calculations is the fact that the relativistic corrections become much larger for excited states, possibly spoiling the convergence of the NRQCD velocity expansion.

4.6 Inclusive production in e^+e^- collisions

4.6.1 Experiments and LO theoretical expectations

In 2001, the prompt J/ψ inclusive production cross section $\sigma(e^+e^- \to J/\psi + X)$ was measured to be $\sigma_{\rm tot} = 2.52 \pm 0.21 \pm 0.21$ pb by the BABAR Collaboration [795]. A smaller value, $\sigma_{\rm tot} = 1.47 \pm 0.10 \pm 0.13$ pb was found by the Belle Collaboration [796]. The color-singlet contribution to the prompt J/ψ inclusive production cross section at LO in α_s, including contributions from the processes $e^+e^- \to J/\psi + c\bar{c}$, $e^+e^- \to J/\psi + gg$, and $e^+e^- \to J/\psi + q\bar{q} + gg$ ($q = u, d, s$), was estimated to be only about 0.3–0.5 pb [797–802], which is much smaller than the measured value. This would suggest that the color-octet contribution might play an important role in inclusive J/ψ production [797–802]. However, it was found by the Belle Collaboration [778] that the associated-production cross section

$$\sigma\big(e^+e^- \to J/\psi + c\bar{c}\big) = \big(0.87^{+0.21}_{-0.19} \pm 0.17\big)\ \text{pb}, \qquad (160)$$

is larger, by at least a factor of 5, than the LO NRQCD factorization prediction, which includes both the color-singlet contribution [798–803] and the color-octet contribution [803]. The ratio of the $J/\psi + c\bar{c}$ cross section to the J/ψ inclusive cross section was found by the Belle Collaboration [778] to be

$$R_{c\bar{c}} = \frac{\sigma(e^+e^- \to J/\psi + c\bar{c})}{\sigma(e^+e^- \to J/\psi + X)} = 0.59^{+0.15}_{-0.13} \pm 0.12, \qquad (161)$$

which is also much larger than LO NRQCD factorization prediction. If one includes only the color-singlet contribution at LO in α_s, then the ratio is predicted to be $R_{c\bar{c}} =$ 0.1–0.3 [797–802, 804], depending on the values of input parameters, such as α_s, m_c, and, especially, the color-singlet matrix elements. A large color-octet contribution could enhance substantially the J/ψ inclusive cross section (the denominator) but could enhance only slightly the $J/\psi + c\bar{c}$ cross section (the numerator) [803] and, therefore, would have the effect of decreasing the prediction for $R_{c\bar{c}}$. Thus, the LO theoretical results and the experimental results in (160) and (161) presented a serious challenge to the NRQCD factorization picture.[18]

Several theoretical studies were made with the aim of resolving this puzzle in J/ψ production. The authors of [807] used soft-collinear effective theory (SCET) to resum the color-octet contribution to the J/ψ inclusive cross section, the authors of [808] used SCET to analyze the color-singlet contribution to $e^+e^- \to J/\psi + gg$, and the authors of [809] resummed the LO and NLO logarithms in the color-singlet contribution to the J/ψ inclusive cross section. These resummation calculations, while potentially useful, did not resolve the puzzle.

Very recently, the Belle Collaboration reported new measurements [810]:

$$\sigma(J/\psi + X) = (1.17 \pm 0.02 \pm 0.07)\ \text{pb}, \qquad (162)$$

$$\sigma(J/\psi + c\bar{c}) = \big(0.74 \pm 0.08^{+0.09}_{-0.08}\big)\ \text{pb}, \qquad (163)$$

$$\sigma(J/\psi + X_{\text{non}\ c\bar{c}}) = (0.43 \pm 0.09 \pm 0.09)\ \text{pb}. \qquad (164)$$

The value of the inclusive J/ψ cross section in (162) is significantly smaller than the values that were obtained previously by the BABAR Collaboration [795] and by the Belle Collaboration [796], but it is still much larger than the LO NRQCD prediction. The cross section $\sigma(e^+e^- \to$

[18] The light-cone perturbative-QCD approach [805] gives a prediction that $R_{c\bar{c}} = 0.1$–0.3, while the CEM gives a prediction that $R_{c\bar{c}} = 0.06$ [806], both of which are far below the measured value of $R_{c\bar{c}}$.

$J/\psi + c\bar{c}$) in (163) is also much larger than the LO color-singlet and color-octet predictions.

4.6.2 $e^+e^- \to J/\psi + c\bar{c}$ at NLO.

An important step toward resolving the puzzle of the J/ψ production cross section was taken in [811], where it was found that the correction of NLO in α_s to $e^+e^- \to J/\psi + c\bar{c}$ gives a large enhancement. In this work, a value for the square of the J/ψ wave function at the origin was obtained by comparing the observed J/ψ leptonic decay width ($5.55 \pm 0.14 \pm 0.02$ keV) with the theoretical expression for that width, including corrections of NLO in α_s. (The square of the wave function at the origin is proportional to the color-singlet NRQCD long-distance matrix element of leading order in v.) The value $|R_S(0)|^2 = 1.01$ GeV^3 that was obtained in this work is a factor of 1.25 larger than the value $|R_S(0)|^2 = 0.810$ GeV^3 that was obtained in potential-model calculations and used in [803] to calculate the $e^+e^- \to J/\psi + c\bar{c}$ cross section. Taking $m_c = 1.5$ GeV, the renormalization scale $\mu_R = 2m_c$, and $\Lambda^{(4)}_{\overline{\mathrm{MS}}} = 0.338$ GeV, the authors of [811] found the direct-production cross section at NLO in α_s to be

$$\sigma^{\mathrm{NLO}}\big(e^+e^- \to J/\psi + c\bar{c} + X\big) = 0.33 \text{ pb}, \tag{165}$$

which is a factor of 1.8 larger than the LO result (0.18 pb) that is obtained with the same set of input parameters. Results for the NLO cross section for other values of the input parameters can be found in Table 44. From Table 44 and Fig. 65, it can be seen that the renormalization-scale dependence of the NLO cross section for $e^+e^- \to J/\psi + c\bar{c}$ is quite strong. This strong μ_R dependence is related to the large size of the NLO contribution relative to the LO contribution. The cross section is also sensitive to the value of the charm-quark mass. It is larger for smaller values of m_c.

In [811] the QED contribution at order $\alpha_s\alpha^3$ and the contribution from $e^+e^- \to 2\gamma^* \to J/\psi + c\bar{c}$ were found to increase the $J/\psi + c\bar{c}$ cross section by only a small amount. The feeddown contributions $e^+e^- \to \psi(2S) + c\bar{c} \to J/\psi + c\bar{c} + X$ and $e^+e^- \to \chi_{cJ} + c\bar{c} \to J/\psi + c\bar{c} + X$ were also estimated in [811]. The primary feeddown contribution comes from the $\psi(2S)$ and produces an enhancement factor of 1.355 for the prompt $J/\psi c\bar{c}$ cross section.

Taking into account all of the aforementioned contributions, we obtain the following estimate for the prompt cross section:

$$\sigma^{\mathrm{NLO}}_{\mathrm{prompt}}\big(e^+ + e^- \to J/\psi + c\bar{c} + X\big) = 0.51 \text{ pb}, \tag{166}$$

where the input values $m_c = 1.5$ GeV and $\mu_R = 2m_c$ have been used. As is shown in Table 44, despite the uncertainties in the input parameters, the NLO correction to the color-singlet contribution to $e^+e^- \to J/\psi + c\bar{c}$ substantially increases the cross section and largely reduces the discrepancy between experiment and theory. Furthermore, the NLO relativistic correction to this process is found to be negligible [784], in contrast with the NLO relativistic correction to the process $e^+ + e^- \to J/\psi + \eta_c$ (Sect. 4.5).

Table 44 Color-singlet contributions to cross sections for J/ψ production in e^+e^- annihilation at the B-factory energy, $\sqrt{s} = 10.58$ GeV. The table shows cross sections for the production of $J/\psi + c\bar{c}$ [811] and $J/\psi + gg$ [812], as well as the ratio $R_{c\bar{c}}$, which is computed from the expression in (161) by summing, in the denominator, over only the $J/\psi + c\bar{c}$ and $J/\psi + gg$ cross sections. The cross sections, in units of pb, are shown for different values of m_c (1.4 and 1.5 GeV) and μ_R ($2m_c$ and $\sqrt{s}/2$), along with the corresponding value of $\alpha_s(\mu_R)$. The prompt $J/\psi + c\bar{c}$ cross sections include feeddown from the $\psi(2S)$ and the χ_{cJ} states, while the prompt $J/\psi + gg$ cross sections include feeddown from the $\psi(2S)$ state

μ_R (GeV)	2.8	3.0	5.3	5.3
$\alpha_s(\mu_R)$	0.267	0.259	0.211	0.211
m_c (GeV)	1.4	1.5	1.4	1.5
$\sigma^{\mathrm{LO}}(gg)$	0.42	0.32	0.26	0.22
$\sigma^{\mathrm{NLO}}(gg)$	0.50	0.40	0.39	0.32
$\sigma^{\mathrm{NLO}}_{\mathrm{prompt}}(gg)$	0.67	0.54	0.53	0.44
$\sigma^{\mathrm{LO}}(c\bar{c})$	0.27	0.18	0.17	0.12
$\sigma^{\mathrm{NLO}}(c\bar{c})$	0.47	0.33	0.34	0.24
$\sigma^{\mathrm{NLO}}_{\mathrm{prompt}}(c\bar{c})$	0.71	0.51	0.53	0.39
$R^{\mathrm{LO}}_{c\bar{c}}$	0.39	0.36	0.40	0.35
$R^{\mathrm{NLO}}_{c\bar{c}}$	0.51	0.49	0.50	0.47

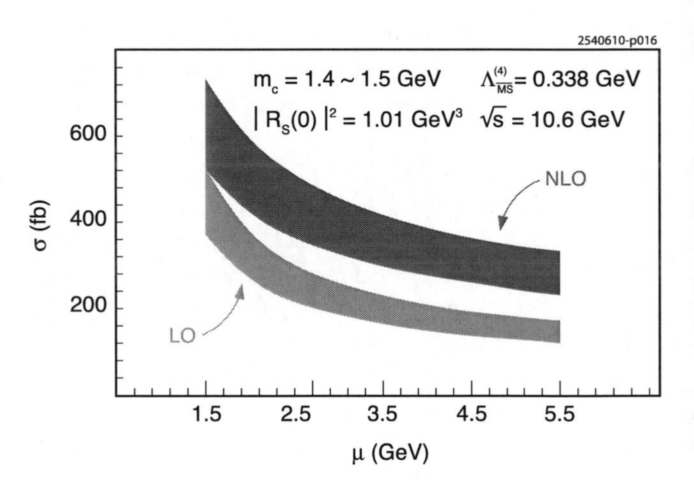

Fig. 65 Color-singlet contributions to the direct-production cross sections for $e^+e^- \to J/\psi + c\bar{c}$ as functions of the renormalization scale μ_R. The following values of the input parameters were taken: $|R_S(0)|^2 = 1.01$ GeV^3, $\Lambda^{(4)}_{\overline{\mathrm{MS}}} = 0.338$ GeV, $\sqrt{s} = 10.6$ GeV. Results at NLO in α_s are represented by *upper band*, and results at LO in α_s are represented by *lower band*. In each case, the *upper border* corresponds to the input value $m_c = 1.4$ GeV and the *lower border* corresponds to the input value $m_c = 1.5$ GeV. From [811] with kind permission, copyright (2007) The American Physical Society

Recently, the authors of [813] confirmed the results of [811] and presented a more detailed analysis of the J/ψ angular distributions and polarization parameters, using slightly different input parameters than those in [811].

4.6.3 $e^+e^- \to J/\psi + gg$ at NLO

In NRQCD factorization, the production cross section for the J/ψ in association with light hadrons, $\sigma(e^+e^- \to J/\psi + X_{\text{non-}c\bar{c}})$, includes the color-singlet contribution $\sigma(J/\psi + gg)$, and the color-octet contributions $\sigma(J/\psi({}^3P_J^{[8]}, {}^1S_0^{[8]}) + g)$. Contributions from other Fock states are suppressed by powers of α_s or v. The corrections of NLO in α_s to $\sigma(J/\psi + gg)$ were calculated in [812, 814] and found to enhance the LO cross section by about 20–30%. The prompt production cross section $\sigma(e^+e^- \to J/\psi + gg)$ at NLO in α_s, including the feed-down contribution from the $\psi(2S)$, can be found in Table 44. From Fig. 66, it can be seen that the renormalization-scale dependence at NLO is moderate and much improved in comparison with the renormalization-scale dependence at LO. It can also be seen that the NLO result is consistent with the latest Belle measurement of $\sigma(e^+e^- \to J/\psi + X_{\text{non-}c\bar{c}})$ [810], given the experimental uncertainties. Resummation of the leading logarithms near the kinematic endpoint of the J/ψ momentum distribution is found to change the endpoint momentum distribution, but to have only a small effect on the total $J/\psi + gg$ cross section [812].

Results for the color-singlet contributions to $\sigma(e^+e^- \to J/\psi + c\bar{c})$ [811], $\sigma(J/\psi + gg)$ [812], and the corresponding ratio $R_{c\bar{c}}$ are summarized in Table 44. In Table 44, the color-octet contribution $\sigma(J/\psi({}^3P_J^{[8]}, {}^1S_0^{[8]}) + g)$ is ignored. It can be seen that the NLO results significantly reduce the discrepancies between theory and experiment.

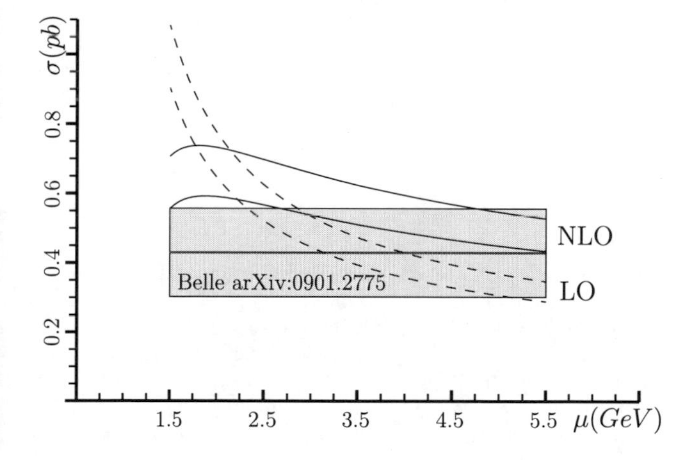

Fig. 66 Prompt cross sections for the process $e^+e^- \to J/\psi + gg$. The *curves* are the predictions of [812], plotted as functions of the renormalization scale μ_R at LO (*solid curves*) and NLO (*dashed curves*) in α_s. In each case, the *upper curves* correspond to $m_c = 1.4$ GeV, and the *lower curves* correspond to $m_c = 1.5$ GeV. The experimental datum (*shaded band*) is for the process $e^+e^- \to J/\psi + X_{\text{non-}c\bar{c}}$ [810]. From [812] with kind permission, copyright (2009) The American Physical Society

In Fig. 67, the measured distributions in the J/ψ momentum p^* are shown for the processes $e^+e^- \to J/\psi + c\bar{c}$ and $e^+e^- \to J/\psi + X_{\text{non-}c\bar{c}}$ [810]. The calculated J/ψ momentum distributions for the process $e^+e^- \to J/\psi + gg$ at NLO in α_s [812, 814] are roughly compatible with the $e^+e^- \to J/\psi + X_{\text{non-}c\bar{c}}$ data. We note that the NLO J/ψ momentum distribution is much softer than the LO J/ψ momentum distribution (see Fig. 6 of [812]), resulting in better agreement with data.

Finally, it was found in [816] that the relative-order-v^2 relativistic correction can enhance the $e^+e^- \to J/\psi + gg$ cross section by 20–30%, which is comparable to the enhancement that arises from the corrections of NLO in α_s [812, 814]. This relativistic correction has been confirmed in [817]. If one includes both the correction of NLO in α_s and the relative-order-v^2 relativistic correction, then the color-singlet contribution to $e^+e^- \to J/\psi + gg$ saturates the latest observed cross section for $e^+e^- \to J/\psi + X_{\text{non-}c\bar{c}}$ in (164), even if a significantly smaller color-singlet matrix element is chosen. This leaves little room for the color-octet contribution $\sigma(J/\psi({}^3P_J^{[8]}, {}^1S_0^{[8]}) + g)$ and may imply that the true values of the color-octet matrix elements are much smaller than those that have been extracted in LO fits to the

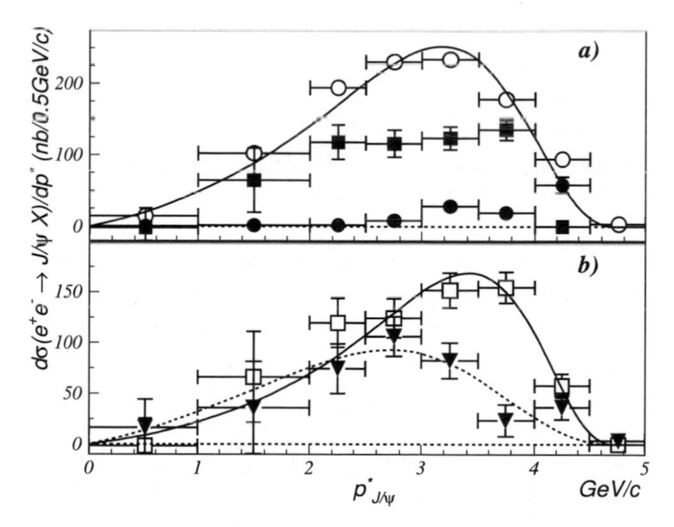

Fig. 67 J/ψ momentum distributions: (**a**) the Belle measurements [810] for the inclusive distribution (*open circles*), the distribution from $e^+e^- \to J/\psi + H_c + X$ (*filled squares*), where H_c is a charmed hadron, and the distribution from double-charmonium production (*filled circles*); (**b**) the Belle measurements [810] for the distribution for the sum of all $e^+e^- \to J/\psi + c\bar{c}$ processes (*open squares*) and the distribution from the $e^+e^- \to J/\psi + X_{\text{non-}c\bar{c}}$ processes (*filled triangles*). The *curves* are the results of using the Peterson function [815] to fit the inclusive distribution [*solid curve in* (**a**)], the $e^+e^- \to J/\psi + c\bar{c}$ distribution [*solid curve in* (**b**)], and the $e^+e^- \to J/\psi + X_{\text{non-}c\bar{c}}$ distribution [*dashed curve in* (**b**)]. From [810] with kind permission, copyright (2009) The American Physical Society

Tevatron data or those that would be expected from a naive application of the NRQCD velocity-scaling rules.

4.6.4 Color-octet process $e^+e^- \to J/\psi(^3P_J^{[8]}, {}^1S_0^{[8]}) + g$

The color-octet contribution to the cross section at LO in α_s is given at LO in v by $\sigma(J/\psi(^3P_J^{[8]}, {}^1S_0^{[8]}) + g)$. This contribution was calculated in [818], and an enhancement near the kinematic endpoint, $z = 1$, was predicted. Here, $z = E_{c\bar{c}}/E_{c\bar{c}}^{\max}$ is the energy of the $c\bar{c}$ pair divided by the maximum possible energy of the $c\bar{c}$ pair. As can be seen from Fig. 67, measurements of the J/ψ momentum distribution do not show any enhancement near the kinematic endpoint. In [807], resummations of the NRQCD velocity expansion near the endpoint and resummations of logarithms of $1-z$ were considered. These resummations smear out the peak near $z = 1$ and shift it to smaller values of z, making the theory more compatible with the data. However, the resummation results rely heavily on a nonperturbative shape function that is not well known, and so it is not clear if they can reconcile the theoretical and experimental results. The nonobservation of an enhancement near $z = 1$ might also point to a possibility that we mentioned in Sect. 4.6.3, namely, that the ^3P_J and $^1S_0$ color-octet matrix elements are much smaller than would be expected from the LO fits to the Tevatron data or the NRQCD velocity-scaling rules.

Very recently, the corrections at NLO in α_s to the color-octet contribution to inclusive J/ψ production have been calculated [819]. In comparison with the LO result, the NLO contributions are found to enhance the short-distance coefficients in the color-octet contributions $\sigma(e^+e^- \to c\bar{c}(^1S_0^{(8)}) + g)$ and $\sigma(e^+e^- \to c\bar{c}(^3P_J^{(8)}) + g)$ (with $J = 0$, 1, 2) by a factor of about 1.9. Moreover, the NLO corrections smear the peak at the endpoint in the J/ψ energy distribution, although the bulk of the color-octet contribution still comes from the region of large J/ψ energy. One can obtain an upper bound on the sizes of the color-octet matrix elements by setting the color-singlet contribution to be zero in $\sigma(e^+e^- \to J/\psi + X_{\text{non-}c\bar{c}})$. The result, at NLO in α_s, is

$$\begin{aligned} &\langle 0|\mathcal{O}^{J/\psi}(^1S_0^{(8)})|0\rangle + 4.0\langle 0|\mathcal{O}^{J/\psi}(^3P_0^{(8)})|0\rangle/m_c^2 \\ &\quad < (2.0 \pm 0.6) \times 10^{-2}\ \text{GeV}^3. \end{aligned} \tag{167}$$

This bound is smaller by about a factor of 2 than the values that were extracted in LO fits to the Tevatron data for the combination of matrix elements in (167).

4.6.5 J/ψ production in $\gamma\gamma$ collisions

Photon–photon ($\gamma\gamma$) collisions can be can be studied at e^+e^- colliders by observing processes in which the incoming e^+ and e^- each emit a virtual γ that is very close to its mass shell. The inclusive cross section differential in p_T for the production of J/ψ in $\gamma\gamma$ collisions at LEP has been measured by the DELPHI Collaboration [820, 821].

The cross section differential in p_T has been calculated in the CSM and the NRQCD factorization approach at LO in α_s [822–826]. The computations include three processes: the direct-γ process $\gamma\gamma \to (c\bar{c}) + g$, which is of order $\alpha^2\alpha_s$; the single-resolved-γ process $i\gamma \to (c\bar{c}) + i$, which is of order $\alpha\alpha_s^2$; and the double-resolved-γ process $ij \to (c\bar{c}) + k$, which is of order α_s^3. Here, $ij = gg$, gq, $g\bar{q}$, or $q\bar{q}$, where q is a light quark. Because the leading contribution to the distribution of a parton in a γ is of order α/α_s, all of the processes that we have mentioned contribute to the $\gamma\gamma$ production rate in order $\alpha^2\alpha_s$. The corrections of NLO in α_s to the direct-γ process have been computed in [827].

A comparison of the LO CSM and NRQCD factorization predictions with the DELPHI data is shown in Fig. 68. The data clearly favor the NRQCD factorization prediction. However, it should be kept in mind that there may be large NLO corrections to the color-singlet contribution, as is the

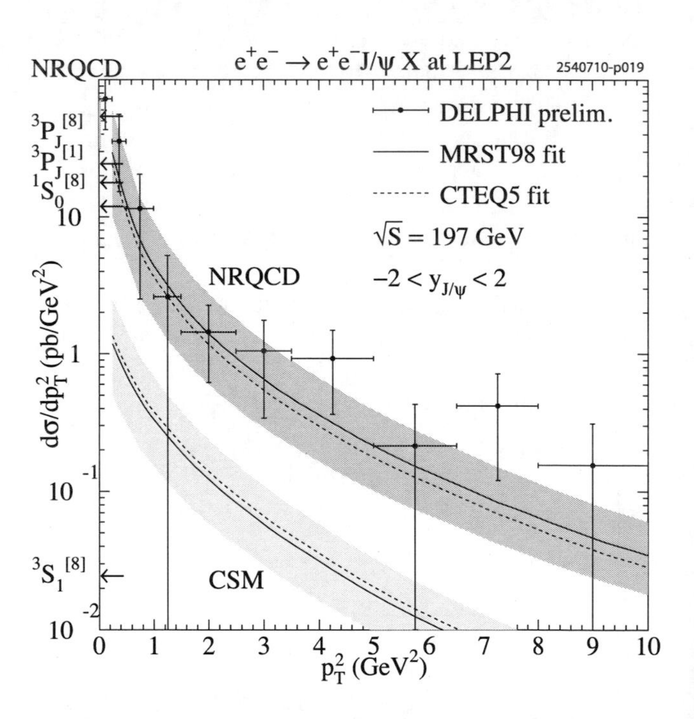

Fig. 68 Comparison of the inclusive cross section differential in p_T for $\gamma\gamma$ production of the J/ψ [820, 821] with the predictions at LO in α_s of the CSM and the NRQCD factorization approach. The *upper two curves* are the NRQCD factorization predictions, and the *lower two curves* are the CSM predictions. The *solid and dashed curves* correspond to the MRST98LO [732] and the CTEQ5L [733] parton distributions, respectively. The *arrows* indicate the NRQCD factorization predictions at $p_T = 0$ for the $^3P_J^{[1]}$, $^1S_0^{[8]}$, $^3S_1^{[8]}$, and $^3P_J^{[8]}$ contributions. The theoretical uncertainty bands were obtained by combining the uncertainties from $m_c = 1.5 \pm 0.1$ GeV, the decay branching fractions of the $\psi(2S)$ and the χ_{cJ} states, the parton distributions, and the NRQCD long-distance matrix elements with the uncertainties that are obtained by varying the renormalization factorization scales between $2m_T$ and $m_T/2$. Here $m_T = \sqrt{4m_c^2 + p_T^2}$. From [826] with kind permission, copyright (2002) The American Physical Society

case in pp and ep charmonium production. Therefore, no firm conclusions can be drawn until a complete NLO calculation of the production cross section is available.

4.6.6 Future opportunities

The central values for the prompt J/ψ inclusive production cross section that were obtained by the BABAR Collaboration and the Belle Collaboration differ more than a factor of 2. It would certainly be desirable to clear up this discrepancy. Furthermore, the BABAR Collaboration has not presented results for $\sigma(e^+e^- \to J/\psi + c\bar{c})$ and $\sigma(e^+e^- \to J/\psi + X_{\text{non}-c\bar{c}})$. It is very important that the BABAR Collaboration check the Belle results for these cross sections, which play a central role in efforts to understand the mechanisms of inclusive quarkonium production. Additionally, measurements of greater accuracy of the charmonium angular distributions and polarization parameters would be useful in understanding the mechanisms of quarkonium production. Measurements of inclusive cross sections for the production of charmonium states other than the J/ψ might also yield important clues regarding the production mechanisms.

In the theoretical prediction for the color-singlet contribution to $\sigma(e^+e^- \to J/\psi + c\bar{c})$, large uncertainties arise from the uncertainties in the renormalization scale μ_R and m_c. (The uncertainties in the theoretical prediction for the color-singlet contribution to $\sigma(e^+e^- \to J/\psi + gg)$ seem to be under rather good control.) The uncertainties in the color-singlet contribution to $\sigma(e^+e^- \to J/\psi + c\bar{c})$ might be reduced by understanding the origins of the large corrections at NLO in α_s and by taking advantage of recent progress in determining m_c [227]. It is important to examine further whether the observed value of $\sigma(e^+e^- \to J/\psi + X_{\text{non}-c\bar{c}})$ is actually saturated by $\sigma(e^+e^- \to J/\psi + gg)$, which would imply that the color-octet $({}^3P_J^{[8]}, {}^1S_0^{[8]})$ contributions are negligible. In this regard, one might obtain additional information by examining the J/ψ angular distribution and polarization parameters, in addition to the total cross section and momentum distribution. Finally, we mention, as we did in Sect. 4.3.3, that theoretical uncertainties in the region near the kinematic endpoint $z = 1$ might be reduced through a systematic study of resummations of the perturbative and velocity expansions in both ep and e^+e^- quarkonium production.

Quarkonium production in $\gamma\gamma$ collisions provides yet another opportunity to understand quarkonium production mechanisms. In order to make a definitive comparison of the DELPHI data for the J/ψ cross section differential in p_T with the NRQCD factorization prediction, it is necessary to have a complete calculation of all of the direct and resolved contributions at least through NLO in α_s. Further measurements of quarkonium production in $\gamma\gamma$ collisions should be carried out at the next opportunity at an e^+e^- or ep collider. It may also be possible to measure quarkonium production in $\gamma\gamma$ and γp collisions at the LHC [828].

4.7 B_c production

4.7.1 Experimental progress

The first observation of the B_c was reported by the CDF Collaboration in [73]. Subsequently, this unique double-heavy-flavored meson has been observed by both the CDF and DØ Collaborations at the Tevatron via two decay channels: $B_c \to J/\psi + \bar{l}\nu_l$ and $B_c \to J/\psi + \pi^+$ [57, 58, 75, 77, 173]. Using an event sample corresponding to an integrated luminosity of 360 pb^{-1} at $\sqrt{s} = 1.96$ TeV, the CDF Collaboration has measured the B_c lifetime in the decay $B_c^+ \to J/\psi e^+\nu_e$ [75] and obtained

$$\tau_{B_c} = 0.463^{+0.073}_{-0.065} \pm 0.036 \text{ ps}. \tag{168}$$

The B_c lifetime has also been measured by the CDF Collaboration in the decay $B_c \to J/\psi + l^\pm + X$ [829]. The result of this measurement, which is based on an event sample corresponding to an integrated luminosity of 1 fb^{-1} at $\sqrt{s} = 1.96$ TeV, is

$$\tau_{B_c} = 0.475^{+0.053}_{-0.049} \pm 0.018 \text{ ps}. \tag{169}$$

The CDF Collaboration has measured the B_c mass in the decay $B_c^\pm \to J/\psi\pi^\pm$ [57, 173], obtaining in its most recent measurement [57], which is based on an integrated luminosity of 2.4 fb^{-1} at $\sqrt{s} = 1.96$ TeV, the value

$$m_{B_c} = 6275.6 \pm 2.9 \pm 2.5 \text{ MeV}. \tag{170}$$

The DØ Collaboration, making use of an event sample based on an integrated luminosity of 1.3 fb^{-1} at $\sqrt{s} = 1.96$ TeV, has also provided measurements of the B_c lifetime [58],

$$\tau_{B_c} = 0.448^{+0.038}_{-0.036} \pm 0.032 \text{ ps}, \tag{171}$$

and the B_c mass [77],

$$m_{B_c} = 6300 \pm 14 \pm 5 \text{ MeV}. \tag{172}$$

The results obtained by the two collaborations are consistent with each other. (See Sect. 2.2.4 of this article for a further discussion of the B_c mass and lifetime.)

Recently, the CDF Collaboration has updated a previous measurement [830] of the ratio

$$R_{B_c} = \frac{\sigma(B_c^+)\mathcal{B}(B_c \to J/\psi + \mu^+ + \nu_\mu)}{\sigma(B^+)\mathcal{B}(B^+ \to J/\psi + K^+)}. \tag{173}$$

The new analysis [831], which used a data set that corresponds an integrated luminosity of 1 fb^{-1}, yielded the results

$$R_{B_c} = 0.295 \pm 0.040(\text{stat})^{+0.033}_{-0.026}(\text{sys}) \pm 0.036(p_T), \tag{174}$$

for $p_T > 4$ GeV, and

$$R_{B_c} = 0.227 \pm 0.040(\text{stat})^{+0.024}_{-0.017}(\text{sys}) \pm 0.014(p_T), \quad (175)$$

for $p_T > 6$ GeV, where p_T is the B_c transverse momentum. The measurements in [830, 831] provided the first, indirect, experimental information on the production cross section.

4.7.2 Calculational schemes

Experimental studies of B_c production could help to further theoretical progress in understanding the production mechanisms for heavy-quark bound states. On the other hand, experimental observations of the B_c are very challenging and might benefit from theoretical predictions of B_c production rates, which could be of use in devising efficient observational strategies, for example, in selecting decay channels to study.

So far, two theoretical approaches have been used to obtain predictions for B_c hadroproduction. Both of them are based on the NRQCD factorization approach.

The simplest approach conceptually is to calculate the contributions from all of the hard subprocesses, through a fixed order in α_s, that produce a $c\bar{c}$ pair and a $b\bar{b}$ pair. The c quark (antiquark) is required to be nearly comoving with the b antiquark (quark) in order to produce a B_c. We call this approach the "fixed-order approach." A typical Feynman diagram that appears in this approach is shown in Fig. 69. If the initial-state partons are light (i.e., gluons, or u, d, or s quarks or antiquarks), then the leading order for B_c production in the fixed-order approach is α_s^4. LO computations in this approach can be found in [832–840].

An alternative approach is the fragmentation approximation, in which the production process is factorized into convolutions of fragmentation functions with a simple perturbative-QCD hard-scattering sub-process [841, 842]. An example of such a factorized contribution is one in which $pp \to b\bar{b}$, with the final-state b or $\bar{b}$ fragmenting to $B_c + \bar{c}$. This process is illustrated in Fig. 70. The fragmentation approximation drastically simplifies the calculation and also provides a formalism with which to resum large final-state logarithms of p_T/m_b. However, it has been shown in [843] that the fragmentation diagrams are not dominant at the Tevatron unless the B_c is produced at very large (experimentally inaccessible) values of p_T. A similar behavior has been found in [618] in the case of charmonium or bottomonium production at the Tevatron and the LHC in association with a heavy-quark pair. Therefore, we can conclude that there is no need to resum logarithms of p_T/m_b at the values of p_T that are accessible at the current hadron colliders. That is, a fixed-order calculation is sufficient, and the fragmentation approach is not relevant.

Within either the fixed-order approach or the fragmentation approach, various choices of factorization scheme are

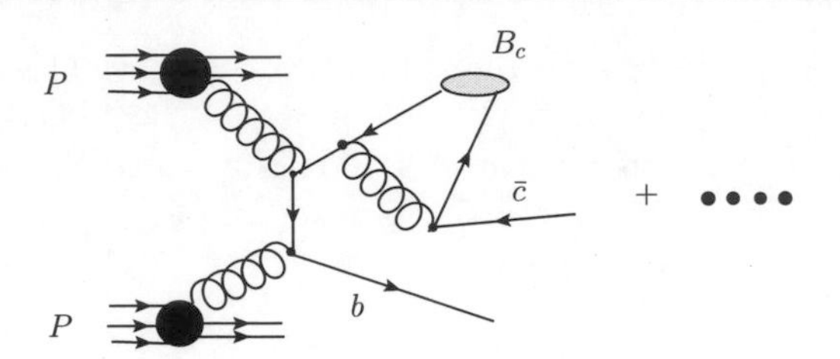

Fig. 69 Typical Feynman diagram for B_c production via gluon-gluon fusion in the fixed-order approach at order α_s^4

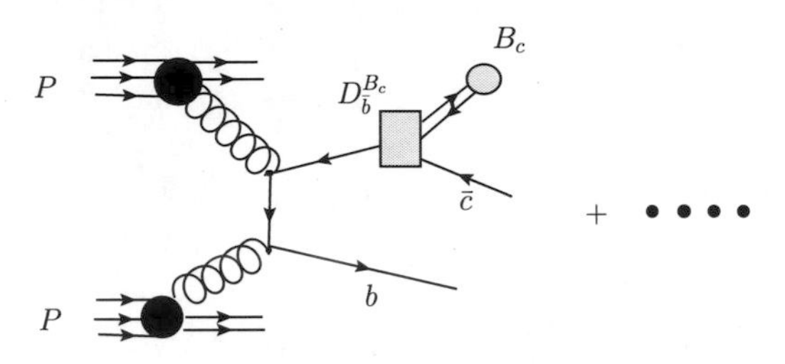

Fig. 70 Typical Feynman diagram for B_c production via gluon-gluon fusion in the fragmentation approach at order α_s^2. $D_{\bar{b}}^{B_c}$ is the fragmentation function for a b antiquark into a B_c

possible. In the simplest scheme, which is known as the fixed-flavor-number (FFN) scheme, the flavor content of the proton is held fixed and includes only the light flavors. Specifically, it is assumed that only the light quarks are "active" flavors, which means that there are contributions involving light-quark (and antiquark) parton distributions, but there are no contributions involving c- and b-quark (and antiquark) parton distributions.

Alternatives to the FFN scheme have been proposed with the aim of improving upon FFN calculations by resumming large initial-state collinear logarithms of Q/m_c and Q/m_b, where Q is a kinematic scale in the production process (e.g., p_T). Such logarithms arise, for example, in a process in which an initial-state gluon splits into a quark–antiquark pair. The quark or antiquark then participates in a hard scattering, giving rise to a B_c. (An example of such a process is $g \to c\bar{c}$, followed by $g\bar{c} \to B_c\bar{b}$.) It has been suggested that one can resum these initial-state collinear logarithms by making use of heavy-flavor parton distributions. The basic idea is that one can absorb logarithms of Q/m_c or Q/m_b into massless c-quark (or antiquark) and b-quark (or antiquark) parton distributions, in which they can be computed through DGLAP evolution. In such methods, there are contributions in which the B_c is produced in a $2 \to 2$ scattering that involves a c or b initial-state parton. Because the contributions involving massless c and b parton distributions are good approximations to the physical process only at large Q, the formalism must suppress these contributions when Q is of the order of the heavy-quark mass or less. A factorization scheme in which the number of active heavy quarks varies with Q in this way is called a general-mass variable-flavor-number (GM-VFN) scheme. Typical Feynman dia-

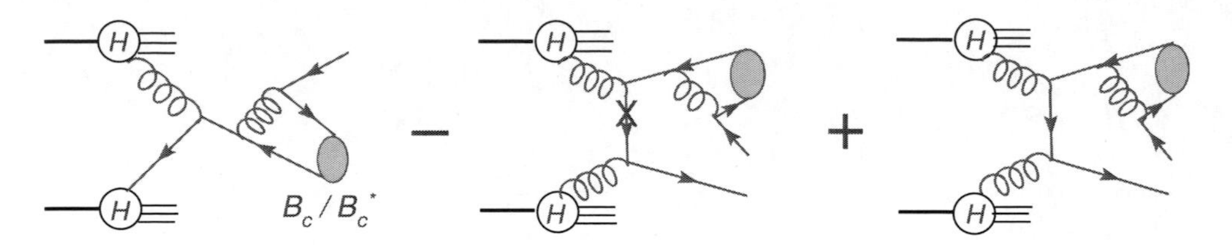

Fig. 71 Typical Feynman diagrams for B_c production via gluon-gluon fusion in a GM-VFN scheme. The middle diagram represents a subtraction that removes double-counting of the contribution in the right-hand diagram in which the lower initial-state gluon produces a heavy-quark–antiquark pair that is collinear to that gluon. From [844] with kind permission, copyright (2006) The American Physical Society

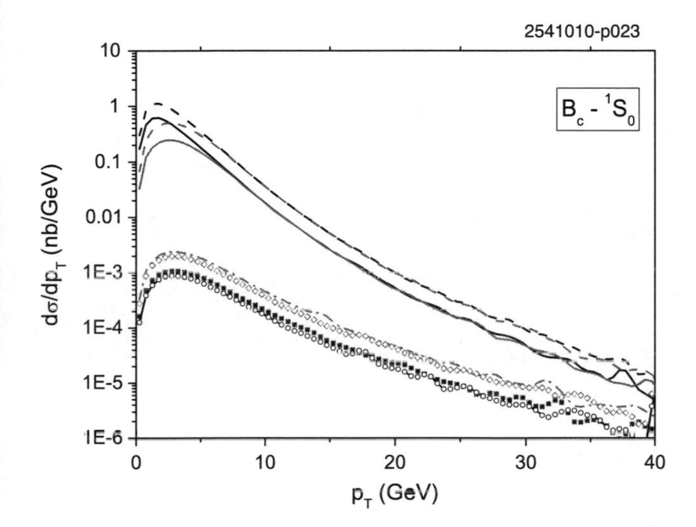

Fig. 72 Predictions for p_T distributions of the $B_c(^1S_0)$ in production at the Tevatron [845]. Four pairs of *curves* are shown, corresponding, from top to bottom, to the following contributions: gg-fusion with the cut $|y| < 1.3$, gg-fusion with the cut $|y| < 0.6$, $q\bar{q}$-annihilation with the cut $|y| < 1.3$, and $q\bar{q}$-annihilation with the cut $|y| < 0.6$. Here, q is a light quark. In each pair of *curves*, the *upper curve* is the GM-VFN prediction, and the *lower curve* is the FFN prediction

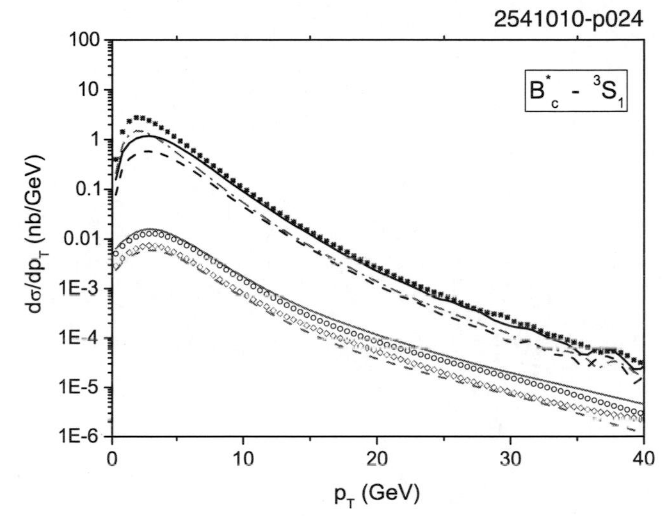

Fig. 73 Predictions for p_T distributions of the $B_c^*(^3S_1)$ in production at the Tevatron [845]. Four pairs of curves are shown, corresponding, from top to bottom, to the following contributions: gg-fusion with the cut $|y| < 1.3$, gg-fusion with the cut $|y| < 0.6$, $q\bar{q}$-annihilation with the cut $|y| < 1.3$, and $q\bar{q}$-annihilation with the cut $|y| < 0.6$. Here, q is a light quark. In each pair of *curves*, the *upper curve* is the GM-VFN prediction, and the *lower curve* is the FFN prediction

grams that enter into the calculation of B_c production in a GM-VFN scheme are shown in Fig. 71.

The GM-VFN scheme was first applied to B_c production in [844, 845]. (Earlier calculations [846–849] in the GM-VFN scheme did *not* address B_c production.) Comparisons [845] of results from both FFN and GM-VFN schemes for $B_c(^1S_0)$ and $B_c(^3S_1)$ production at the Tevatron are shown in Figs. 72 and 73, respectively. The FFN and GM-VFN approaches yield different results at small p_T, possibly because, in the implementation of the GM-VFN scheme in [845], the factorization scale $\mu_F = \sqrt{p_T^2 + m_{B_c}^2}$ was chosen, which implies that there are contributions from initial c and b quarks down to $p_T = 0$. Outside the small-p_T region, the FFN and GM-VFN approaches give very similar results. This implies that the resummations that are contained in the GM-VFN scheme have only a small effect and that there is no compelling need to resum initial-state logarithms.

In conclusion, there is no evidence of a need to improve the FFN order-α_s^4 computation of B_c production by resumming large final-state or initial-state logarithms. On the other hand, a complete computation of the NLO corrections in the FFN approach would be very welcome in order to allow one to improve the dependence of the predictions on the renormalization scale and to reduce the theoretical uncertainties.

4.7.3 Phenomenology

The study of B_c production and decays will be possible at the LHC, as well as at the Tevatron. The LHCb experiment, in particular, has been designed especially for the study of b hadrons (e.g., $B^{\pm}$, B^0, $\bar{B}^0$, B_s, B_c, Λ_b). Thus, many new results for the B_c meson are expected. New data will be useful in understanding the production mechanism itself and also in determining the decay branching ratios.

Because the excitations of the B_c carry both b and c flavor quantum numbers, they must decay into the B_c ground

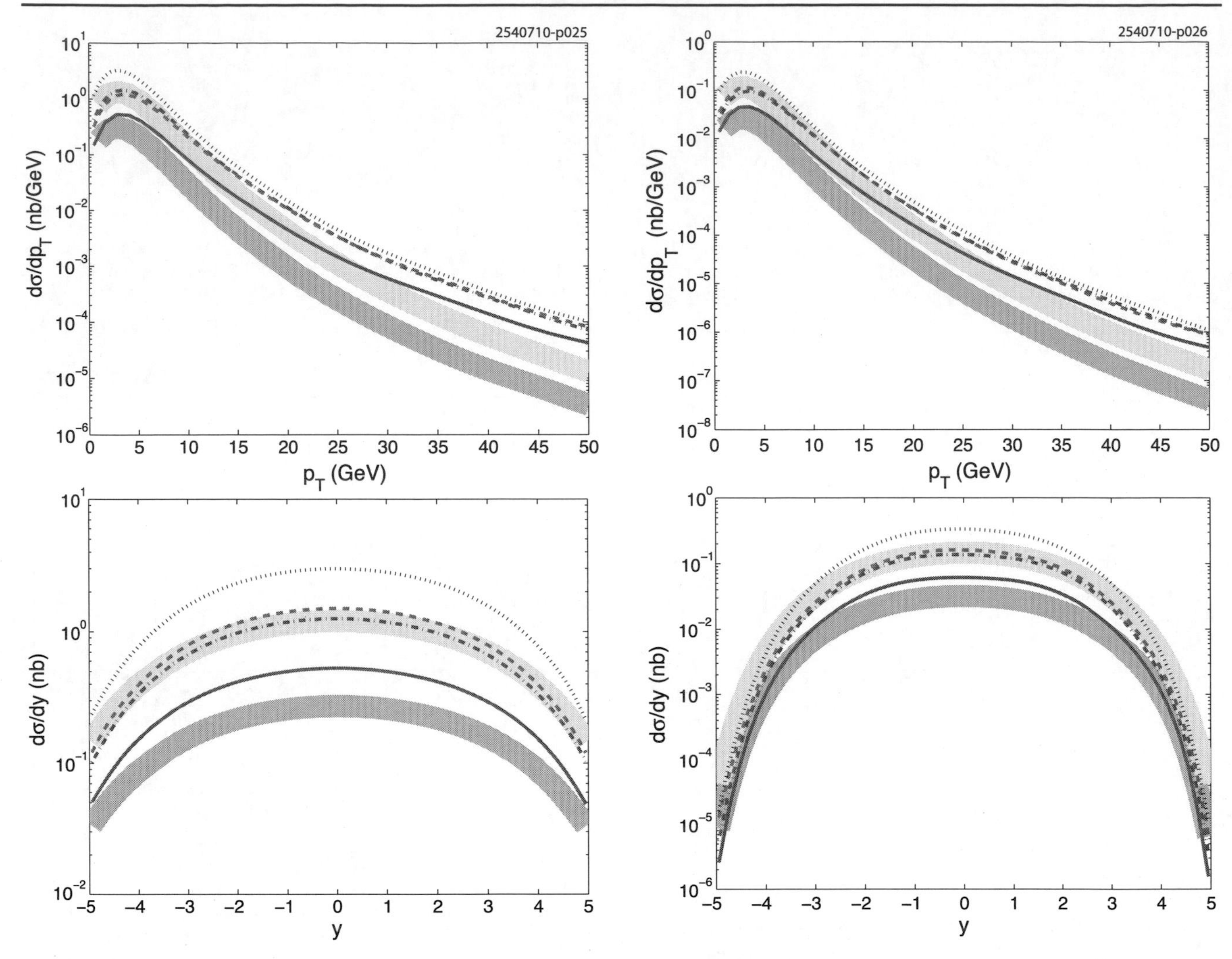

Fig. 74 Distributions in p_T and y for production of $c\bar{b}$ mesons at the LHC. The *dashed*, *solid*, *dash-dotted*, and *dotted lines* represent the color-singlet ${}^1P_1$, ${}^3P_0$, ${}^3P_1$, and ${}^3P_2$ contributions, respectively. The *lower* and *upper shaded bands* represent the color-octet ${}^1S_0$ and ${}^3S_1$ contributions, respectively. From [851] with kind permission, copyright (2005) The American Physical Society

Fig. 75 Distributions in p_T and y for production of $c\bar{b}$ mesons at the Tevatron. The *dashed*, *solid*, *dash-dotted*, and *dotted lines* represent the color-singlet ${}^1P_1$, ${}^3P_0$, ${}^3P_1$, and ${}^3P_2$ contributions, respectively. The *lower* and *upper shaded bands* represent the color-octet ${}^1S_0$ and ${}^3S_1$ contributions, respectively. From [851] with kind permission, copyright (2005) The American Physical Society

state with almost 100% probability, either directly or via strong and/or electromagnetic cascades. No experimental distinction among direct or indirect production is possible at the moment. It is therefore important that predictions for B_c production include feeddown from the excitations of the B_c, such as the $B_c^*({}^3S_1)$, $B^*_{cJ,L=1}$, $B^{**}_{cJ,L=2}$, etc., and the radially excited states. Predictions in LO perturbative QCD for the production cross sections of the B_c^* and the low-lying P-wave excited states $B^*_{cJ,L=1}$ are available in analytic and numerical form in [850, 851] and in numerical form in [743]. These predictions show that the contributions of the color-octet channels, $[(c\bar{b})_{\mathbf{8}},{}^1S_0]$ and $[(c\bar{b})_{\mathbf{8}},{}^3S_1]$, to the production of low-lying P-wave excited states, such as the $B^*_{cJ,L=1}$, are comparable to the contributions of the leading color-singlet channels, provided that the values of the color-octet matrix elements are consistent with NRQCD velocity scaling.

Simulations of the production of $c\bar{b}$ and $\bar{c}b$ mesons are now implemented in the programs BCVEGPY [852] and MadOnia [743]. Both programs are interfaced to PYTHIA [853].

In Figs. 74 and 75, we show typical distributions in y and p_T for the production of excited $(c\bar{b})$ mesons at the LHC and the Tevatron [851]. The contributions from the various parton-level production channels are shown separately.

4.7.4 Future opportunities

The observation and study of the B_c mesons and their excitations are new and exciting components of the quarkonium-physics plans for both the Tevatron and the LHC. Much has yet to be learned about the production and decay of these states.

A number of improvements in the theoretical predictions for B_c mesons are needed, such as computations of the NLO corrections to the production cross sections. In addition, new mechanisms for B_c production at high p_T should be explored. These include production via Z^0 or top-quark decays [854].

5 In medium[19]

5.1 Quarkonia as a probe of hot and dense matter

It is expected that strongly interacting matter shows qualitatively new behavior at temperatures and/or densities which are comparable to or larger than the typical hadronic scale. It has been argued that under such extreme conditions deconfinement of quarks and gluons should set in and the thermodynamics of strongly interacting matter could then be understood in terms of these elementary degrees of freedom. This new form of matter is called *quark–gluon plasma* [855], or QGP. The existence of such a transition has indeed been demonstrated from first principles using Monte Carlo simulations of lattice QCD. The properties of this new state of matter have also been studied [856–859].

In addition to theoretical efforts, the deconfinement transition and the properties of hot, strongly interacting matter are also studied experimentally in heavy-ion collisions [860, 861]. A significant part of the extensive experimental heavy-ion program is dedicated to measuring quarkonium yields since Matsui and Satz suggested that quarkonium suppression could be a signature of deconfinement [862]. In fact, the observation of anomalous suppression was considered to be a key signature of deconfinement at SPS energies [863].

However, not all of the observed quarkonium suppression in nucleus–nucleus (AB) collisions relative to scaled proton–proton (pp) collisions is due to quark–gluon plasma formation. In fact, quarkonium suppression was also observed in proton–nucleus (pA) collisions, so that part of the nucleus–nucleus suppression is due to cold-nuclear-matter effects. Therefore it is necessary to disentangle hot- and cold-medium effects. We first discuss cold-nuclear-matter effects at different center-of-mass energies. Then we discuss what is known about the properties of heavy $Q\overline{Q}$ states in hot, deconfined media. Finally, we review recent experimental results on quarkonium production from pA collisions at the SPS and from pp, d+Au, and AA collisions at RHIC.

[19]Contributing authors: A.D. Frawley†, P. Petreczky†, R. Vogt†, R. Arnaldi, N. Brambilla, P. Cortese, S.R. Klein, C. Lourenço, A. Mocsy, E. Scomparin, and H.K. Wöhri.

5.2 Cold-nuclear-matter effects

The baseline for quarkonium production and suppression in heavy-ion collisions should be determined from studies of cold-nuclear-matter (CNM) effects. The name cold matter arises because these effects are observed in hadron–nucleus interactions where no hot, dense matter effects are expected. There are several CNM effects. Modifications of the parton distribution functions in the nucleus, relative to the nucleon, (i.e., *shadowing*) and energy loss of the parton traversing the nucleus before the hard scattering are both assumed to be initial-state effects, intrinsic to the nuclear target. Another CNM effect is absorption (i.e., destruction) of the quarkonium state as it passes through the nucleus. Since the latter occurs after the $Q\overline{Q}$ pair has been produced and while it is traversing the nuclear medium, this absorption is typically referred to as a final-state effect. In order to disentangle the mechanisms affecting the produced $Q\overline{Q}$, data from a variety of center-of-mass energies and different phase-space windows need to be studied. In addition, the inclusive J/ψ yield includes contributions from χ_c and $\psi(2S)$ decays to J/ψ at the 30–35% level [864]. While there is some information on the A dependence of $\psi(2S)$ production, that on χ_c is largely unknown [865].

Even though the contributions to CNM effects may seem rather straightforward, there are a number of associated uncertainties. First, while nuclear modifications of the quark densities are relatively well measured in nuclear deep-inelastic scattering (nDIS), the modifications of the gluon density are not directly measured. The nDIS measurements probe only the quark and antiquark distributions directly. The scaling violations in nDIS can be used to constrain the nuclear gluon density. Overall momentum conservation provides another constraint. However, more direct probes of the gluon density are needed. Current shadowing parametrizations are derived from global fits to the nuclear parton densities and give wide variations in the nuclear gluon density, from almost no effect to very large shadowing at low-x, compensated by strong antishadowing around $x \sim 0.1$. The range of the possible shadowing effects is illustrated in Fig. 76 by the new EPS09 [866] parametrization and its associated uncertainties, employing the scale values used to fix the J/ψ and Υ cross sections below the open heavy-flavor threshold [867].

The color-glass condensate (CGC) is expected to play an important role in quarkonium production at RHIC and the LHC since the saturation scale $Q_{S,A}(x)$ is comparable to the charm quark mass [868]. In this picture, collinear factorization of J/ψ production is assumed to break down and forward J/ψ production is suppressed. Indeed, CGC suppression of J/ψ formation may mask some QGP effects [869].

The nuclear absorption survival probability depends on the quarkonium absorption cross section. There are more

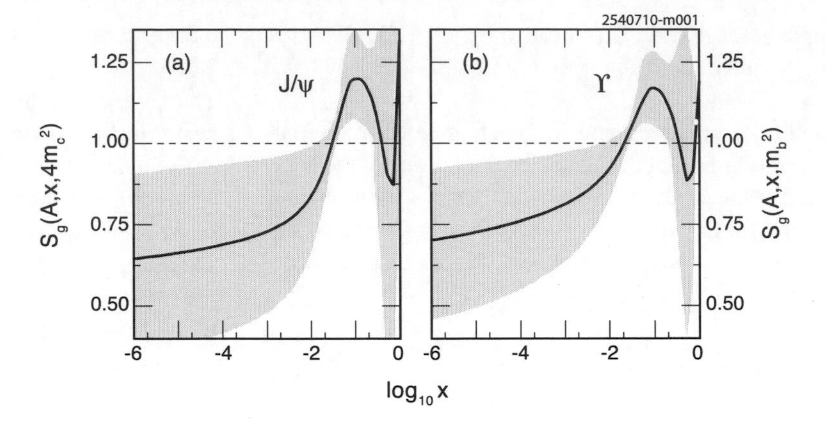

Fig. 76 The EPS09 gluon-shadowing parametrization [866] at $Q = 2m_c$ and m_b. The central value (*solid curves*) and the associated uncertainty (*shaded band*) are shown

inherent uncertainties in absorption than in the shadowing parametrization, which is obtained from data on other processes and is independent of the final state. Typically an absorption cross section is fit to the A dependence of J/ψ and/or $\psi(2S)$ production at a given energy. This is rather simplistic since it is unknown whether the object traversing the nucleus is a precursor color-octet state or a fully formed color-singlet quarkonium state. If it is an octet state, it is assumed to immediately interact with a large, finite cross section since it is a colored object [870]. In this case, it has often been assumed that all precursor quarkonium states will interact with the same cross section. If it is produced as a small color singlet, the absorption cross section immediately after the production of the $Q\overline{Q}$ pair should be small and increasing with proper time until, at the formation time, it reaches its final-state size [871]. High-momentum color-singlet quarkonium states will experience negligible nuclear absorption effects since they will be formed well outside the target. See [865] for a discussion of the A dependence of absorption for all the quarkonium states.

Fixed-target data taken in the range $400 \leq E_{\text{lab}} \leq 800$ GeV have shown that the J/ψ and $\psi(2S)$ absorption cross sections are not identical, as the basic color-octet absorption mechanism would suggest [872–874]. The difference between the effective A dependence of J/ψ and $\psi(2S)$ seems to decrease with beam energy. The J/ψ absorption cross section at $y \sim 0$ is seen to decrease with energy, regardless of the chosen shadowing parametrization [875], as shown in Fig. 77.

Recent analyses of J/ψ production in fixed-target interactions [875] show that the effective absorption cross section depends on the energy of the initial beam and the rapidity or x_F of the observed J/ψ. One possible interpretation is that low-momentum color-singlet states can hadronize in the target, resulting in larger effective absorption cross sections at lower center-of-mass energies and backward x_F (or center-of-mass rapidity). At higher energies, the states traverse the target more rapidly so that the x_F values at which they can hadronize in the target move back from midrapidity toward

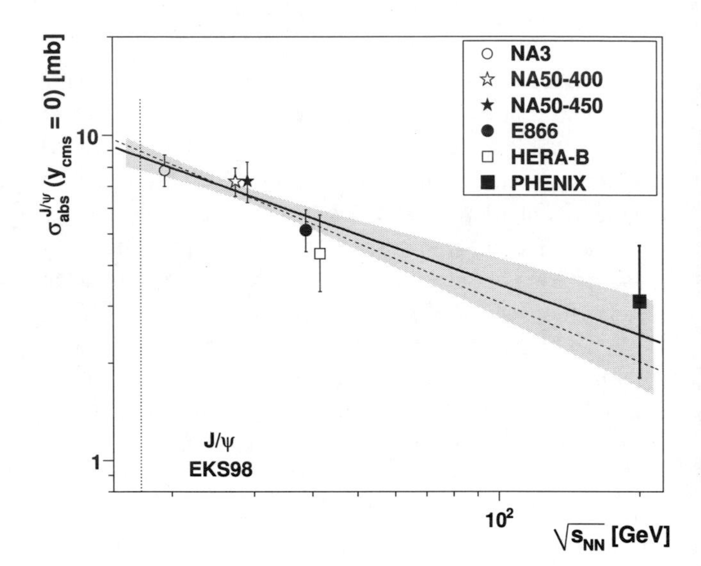

Fig. 77 The extracted energy dependence of $\sigma_{\text{abs}}^{J/\psi}$ at midrapidity. The *solid line* is a power-law approximation to $\sigma_{\text{abs}}^{J/\psi}(y = 0, \sqrt{s_{NN}})$ using the EKS98 [876, 877] shadowing parametrization with the CTEQ61L parton densities [878, 879]. The *band* indicates the uncertainty in the extracted cross sections. The *dashed curve* shows an exponential fit for comparison. The data at $y_{\text{cms}} \sim 0$ from NA3 [880], NA50 at 400 GeV [872] and 450 GeV [873], E866 [874], HERA-B [881], and PHENIX [663] are also shown. The *vertical dotted line* indicates the energy of the Pb+Pb and In+In collisions at the CERN SPS. Adapted from [875] with kind permission, copyright (2009) Springer

more negative x_F. Finally, at sufficiently high energies, the quarkonium states pass through the target before hadronizing, resulting in negligible absorption effects. Thus the *effective* absorption cross section decreases with increasing center-of-mass energy because faster states are less likely to hadronize inside the target.

At higher x_F, away from midrapidity, the effective absorption cross section becomes very large, as shown in the top panel of Fig. 78. The increase in $\sigma_{\text{abs}}^{J/\psi}$ begins closer to midrapidity for lower incident energies.

There appears to be some saturation of the effect since the 800 GeV fixed-target data exhibit the same trend as the most recent (preliminary) PHENIX data [884] as a function

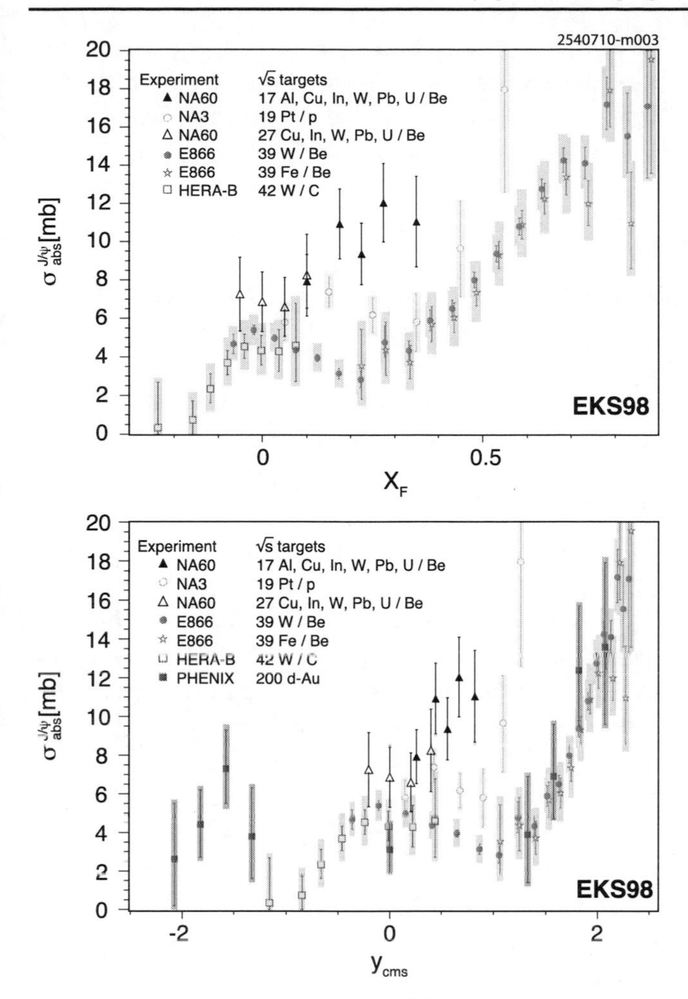

Fig. 78 *Top*: The x_F dependence of $\sigma_{\mathrm{abs}}^{J/\psi}$ for incident fixed-target energies from 158 GeV [882], 200 GeV [880], 400 GeV [872], 450 GeV [873], 800 GeV [874], and 920 GeV [881] obtained using the EKS98 shadowing parametrization [876, 877]. The E866 [874] and HERA-B [881] results were previously shown in [875]. *Bottom*: The same results as above but as a function of center-of-mass rapidity y_{CMS}. The absorption cross sections extracted from the preliminary PHENIX [663] results at $|y_{\mathrm{CMS}}| > 0$ and the central rapidity result [883] are also included. The *boxes* surrounding the data *points* represent the statistical and total systematic uncertainties added in quadrature, except for PHENIX where the global term ($^{+5.0}_{-4.1}$ mb, common to all rapidity bins) is missing

of center-of-mass rapidity, y_{cms}, as seen in the bottom panel of Fig. 78. Model calculations including CGC effects can reproduce the general trend of the high-x_F behavior of J/ψ production at 800 GeV without invoking energy loss [869]. However, the fact that the NA3 data at $\sqrt{s_{NN}} = 19$ GeV exhibit the same trend in x_F as E866 calls the CGC explanation into question.

As previously discussed, such an increase in the apparent absorption cannot be due to interactions with nucleons. In addition, since the large-x_F dependence seems to be independent of the quarkonium state (i.e., the same for J/ψ and $\psi(2S)$ [874], and also for $\Upsilon(1S)$ and $\Upsilon(2S) + \Upsilon(3S)$) [885], it likely cannot be attributed to the size of the final state and should thus be an initial-state effect, possibly energy loss. (See [886] for a discussion of several types of energy-loss models and their effect on J/ψ production.) Work is in progress to incorporate this effect using a new approach, based on the number of soft collisions the projectile parton undergoes before the hard scattering to produce the $Q\overline{Q}$ pair.

It is also well known that feeddown from P and higher-S states through radiative and hadronic transitions, respectively, accounts for almost half of the observed J/ψ and $\Upsilon(1S)$ yields. The excited quarkonium states have very different sizes and formation times and should thus have different absorption cross sections. For example, the absorption cross section of quarkonium state C may be proportional to its area, $\sigma_C \propto r_C^2$ [887].

It should be noted, however, that the fitted absorption cross sections used for extracting the "normal absorption" baseline for Pb+Pb collisions at the SPS have treated J/ψ and $\psi(2S)$ absorption independently, ignoring feeddown and formation times, and have not taken initial-state shadowing into account [872, 873]. As discussed above, more detailed analyses show that the quarkonium absorption cross section decreases with increasing energy [875, 888]. More recent fixed-target analyses [882, 889], comparing measurements at 158 and 400 GeV, have begun to address these issues (see Sect. 5.4). Indeed, the extracted absorption cross section is found to be larger at 158 GeV than at 400 GeV, contrary to previous analyses, which assumed a universal, constant absorption cross section [872, 873]. When these latest results are extrapolated to nucleus–nucleus collisions at the same energy, the anomalous suppression is significantly decreased relative to the new baseline [882].

The cold-nuclear-matter effects suggested (initial-state energy loss, shadowing, final-state breakup, etc.) depend differently on the quarkonium kinematic variables and the collision energy. It is clearly unsatisfactory to combine all these mechanisms into an *effective* absorption cross section, as employed in the Glauber formalism, that only evaluates final-state absorption. Simply taking the σ_{abs} obtained from the analysis of the pA data and using it to define the Pb+Pb baseline may not be sufficient.

A better understanding of absorption requires more detailed knowledge of the production mechanism. Most calculations of the A dependence use the color-evaporation model (CEM), in which all quarkonia are assumed to be produced with the same underlying kinematic distributions [890]. This model works well for fixed-target energies and for RHIC [867], as does the LO color-singlet model (CSM) [676]. In the latter case, but contrary to the CEM at LO, J/ψ production is necessarily accompanied by the emission of a perturbative final-state gluon which can be seen as an extrinsic source of transverse momentum. This induces modifications in the relations between the initial-state gluon momentum fractions and the momentum of the

J/ψ. In turn, this modifies [891] the gluon-shadowing corrections relative to those expected from the LO CEM where the transverse momentum of the J/ψ is intrinsic to the initial-state gluons. Further studies are being carried out, including the impact of feeddown, the extraction of absorption cross sections for each of the charmonium states, and the dependence on the partonic J/ψ production mechanism. A high-precision measurement of the $\psi(2S)$ and $\Upsilon(3S)$ production ratios in pA interactions as a function of rapidity would be desirable since they are not affected by feeddown contributions. In addition, measurements of the feeddown contributions to J/ψ and $\Upsilon(1S)$ as a function of rapidity and p_T would be very useful.

On the other hand, the higher-p_T Tevatron predictions have been calculated within the nonrelativistic QCD (NRQCD) approach [138], which includes both singlet and octet matrix elements. These high p_T calculations can be tuned to agree with the high p_T data but cannot reproduce the measured quarkonium polarization at the same energy [1]. If some fraction of the final-state quarkonium yields can be attributed to color-singlet production, then absorption need not be solely due to either singlet or octet states but rather some mixture of the two, as dictated by NRQCD [865, 892, 893]. A measurement of the A dependence of χ_c production would be particularly helpful to ensure significant progress toward understanding the production mechanism.

5.3 Quarkonium in hot medium

5.3.1 Spectral properties at high temperature

There has been considerable interest in studying quarkonia in hot media since publication of the famous Matsui and Satz paper [862]. It has been argued that color screening in a deconfined QCD medium will destroy all $Q\overline{Q}$ bound states at sufficiently high temperatures. Although this idea was proposed long ago, first principle QCD calculations, which go beyond qualitative arguments, have been performed only recently. Such calculations include lattice QCD determinations of quarkonium correlators [894–898], potential model calculations of the quarkonium spectral functions with potentials based on lattice QCD [899–906], as well as effective field theory approaches that justify potential models and reveal new medium effects [907–910]. Furthermore, better modeling of quarkonium production in the medium created by heavy-ion collisions has been achieved. These advancements make it possible to disentangle the cold- and hot-medium effects on the quarkonium states, crucial for the interpretation of heavy-ion data.

5.3.2 Color screening and deconfinement

At high temperatures, strongly interacting matter undergoes a deconfining phase transition to a quark–gluon plasma. This transition is triggered by a rapid increase of the energy and entropy densities as well as the disappearance of hadronic states. (For a recent review, see [858].) According to current lattice calculations [911–920] at zero net-baryon density, deconfinement occurs at $T \sim 165$–195 MeV. The QGP is characterized by color screening: the range of interaction between heavy quarks becomes inversely proportional to the temperature. Thus at sufficiently high temperatures, it is impossible to produce a bound state between a heavy quark (c or b) and its antiquark.

Color screening is studied on the lattice by calculating the spatial correlation function of a static quark and antiquark in a color-singlet state which propagates in Euclidean time from $\tau = 0$ to $\tau = 1/T$, where T is the temperature (see [921, 922] for reviews). Lattice calculations of this quantity with dynamical quarks have been reported [923–928]. The logarithm of the singlet correlation function, also called the singlet free energy, is shown in Fig. 79. As expected, in the zero-temperature limit the singlet free energy coincides with the zero-temperature potential. Figure 79 also illustrates that, at sufficiently short distances, the singlet free energy is temperature independent and equal to the zero-temperature potential. The range of interaction decreases with increasing temperature. For temperatures above the transition temperature, T_c, the heavy-quark interaction range becomes comparable to the charmonium radius. Based on this general observation, one would expect that the charmonium states, as well as the excited bottomonium states, do not remain bound at temperatures just above the deconfinement transition, often referred to as *dissociation* or *melting*.

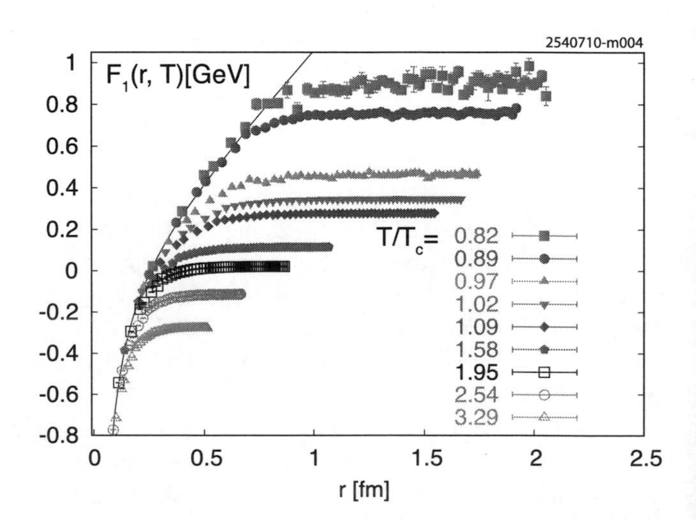

Fig. 79 Heavy-quark-singlet free energy versus quark separation calculated in 2 + 1 flavor QCD on $16^3 \times 4$ lattices at different temperatures [928, 929]

5.3.3 Quarkonium spectral functions and quarkonium potential

In-medium quarkonium properties are encoded in the corresponding spectral functions, as is quarkonium dissolution at high temperatures. Spectral functions are defined as the imaginary part of the retarded correlation function of quarkonium operators. Bound states appear as peaks in the spectral functions. The peaks broaden and eventually disappear with increasing temperature. The disappearance of a peak signals the melting of the given quarkonium state.

In lattice QCD, the meson correlation functions, $G(\tau, T)$, are calculated in Euclidean time. These correlation functions are related to the spectral functions $\sigma(\omega, T)$ by

$$G(\tau, T) = \int_0^\infty d\omega \sigma(\omega, T) \frac{\cosh(\omega(\tau - 1/(2T)))}{\sinh(\omega/(2T))}. \quad (176)$$

Detailed information on $G(\tau, T)$ would allow reconstruction of the spectral function from the lattice data. In practice, however, this turns out to be a very difficult task because the time extent is limited to $1/T$ (see the discussion in [897] and references therein). Lattice artifacts in the spectral functions at high energies ω are also a problem when analyzing the correlation functions calculated on the lattice [930].

The quarkonium spectral functions can be calculated in potential models using the singlet free energy from Fig. 79 or with different lattice-based potentials obtained using the singlet free energy as an input [905, 906] (see also [931] for a review). The results for quenched QCD calculations are shown in Fig. 80 for S-wave charmonium (top) and bottomonium (bottom) spectral functions [905]. All charmonium states are dissolved in the deconfined phase while the bottomonium $1S$ state may persist up to $T \sim 2T_c$. The temperature dependence of the Euclidean correlators can be predicted using (176). Somewhat surprisingly, the Euclidean correlation functions in the pseudoscalar channel show very little temperature dependence, irrespective of whether a state remains bound (η_b) or not (η_c). Note also that correlators from potential models are in accord with the lattice calculations (see insets in Fig. 80). Initially, the weak temperature dependence of the pseudoscalar correlators was considered to be evidence for the survival of $1S$ quarkonium states [896]. It is now clear that this conclusion was premature. In other channels one sees significant temperature dependence of the Euclidean correlation functions, especially in the scalar and axial-vector channels where it has been interpreted as evidence for dissolution of the quarkonium $1P$ states. However, this temperature dependence is due to the zero-mode contribution, i.e., a peak in the finite-temperature spectral functions at $\omega \simeq 0$ [932, 933]. After subtracting the zero-mode contribution, the Euclidean correlation functions show no temperature dependence within the uncertainties [934]. Thus melting of the quarkonium states is not visible in the Euclidean correlation functions.

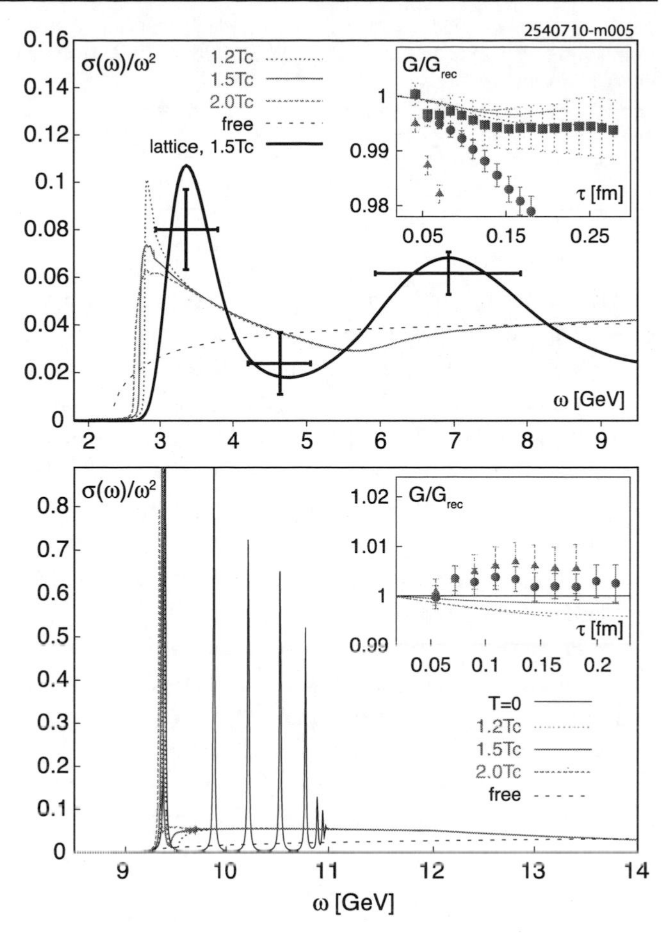

Fig. 80 The S-wave charmonium (*upper*) and bottomonium (*lower*) spectral functions calculated in potential models. Insets: correlators compared to lattice data. The *dotted curves* are the free spectral functions. Adapted from [905] with kind permission, copyright (2008) The American Physical Society

There is a large enhancement in the threshold region of the spectral functions relative to the free spectral function, as shown in Fig. 80. This threshold enhancement compensates for the absence of bound states and leads to Euclidean correlation functions with very weak temperature dependencies [905]. It further indicates strong residual correlations between the quark and antiquark, even in the absence of bound states. Similar analyses were done for the P-wave charmonium and bottomonium spectral functions [905, 906]. An upper bound on the dissociation temperature (the temperatures above which no bound states peaks can be seen in the spectral function and bound state formation is suppressed) can be obtained from the analysis of the spectral functions. Conservative upper limits on the dissociation temperatures for the different quarkonium states obtained from a full QCD calculation [906] are given in Table 45.

Table 45 Upper bounds on the dissociation temperatures. Adapted from [906] with kind permission, copyright (2007) The American Physical Society

State	$\chi_{cJ}(1P)$	$\psi(2S)$	J/ψ	$\Upsilon(2S)$	$\chi_{bJ}(1P)$	Υ
$T_{\rm diss}$	$\le T_c$	$\le T_c$	$1.2T_c$	$1.2T_c$	$1.3T_c$	$2T_c$

The problems with the potential model approach described above are how to relate it to first principles QCD and how to define the in-medium quark–antiquark potential. These problems can be addressed using an EFT approach to heavy-quark bound states in a real-time formalism. The first steps in this direction were taken in [907, 935]. The static potential (for infinitely heavy $Q\overline{Q}$ pairs) was calculated in the regime $T \gg 1/r \gtrsim m_D$, where m_D is the Debye mass and r the $Q\overline{Q}$ separation, by analytically continuing the Euclidean Wilson loop to real time. The calculation was done in weak-coupling resummed perturbation theory. The imaginary part of the gluon self-energy gives an imaginary part to the static potential and thus introduces a thermal width in the $Q\overline{Q}$ bound state. The heavy-quarkonium contribution to the spectral function of the electromagnetic current was calculated in the same framework [908, 936]. Static $Q\overline{Q}$ pairs in a real-time formalism are considered in [937], where the potential for distances $1/r \sim m_D$ in a hot QED plasma is derived. The real part of the static potential was in agreement with the singlet free energy while the damping factor (imaginary part of the potential) at large distances agreed with the one found in [935]. This damping can be thought of as quarkonia scattering with light partons in a thermal bath where the collisional width increases with temperature. The real part of the static potential was found to agree with the singlet free energy and the damping factor with the one found in [935]. In [938], a study of bound states in a hot QED plasma was performed in a nonrelativistic EFT framework. In particular, the hydrogen atom was studied for temperatures ranging from $T \ll m\alpha^2$ to $T \sim m$, where the imaginary part of the potential becomes larger than the real part and the hydrogen ceases to exist. The same study has been extended to muonic hydrogen in [939].

The EFT framework quarkonium at finite temperature for different distance regimes was developed in [910] using the real-time formalism and weak-coupling techniques. In a zero-temperature medium, the behavior of quarkonia is characterized by different energy and momentum scales related to nonrelativistic bound states with heavy-quark velocity v and mass m. The scale related to the inverse distance between the Q and $\overline{Q}$ is mv while mv^2 is the scale related to the binding energy of the state. Finally, $\Lambda_{\rm QCD}$ is related to the nonperturbative features of the QCD vacuum. In the weak-coupling regime, $v \sim \alpha_{\rm s}$ and the hierarchy of scales follows $m \gg mv \gg mv^2 \gg \Lambda_{\rm QCD}$.[20] In addition, there are thermodynamical scales: the temperature T; the inverse of the screening length of chromo-electric interactions, the Debye mass ($m_D \sim gT$); and the static magnetic scale, g^2T. In the weak-coupling regime at finite temperature, the ordering of these scales follows $T \gg gT \gg g^2T$.

If there exists such a hierarchy of scales, any quantity of interest maybe expanded in some ratio of the scales. If the contributions from the different scales are separated explicitly at the Lagrangian level, this amounts to replacing QCD with a hierarchy of EFTs equivalent to QCD order by order in the expansion parameters. As described in Sect. 2.5, at $T = 0$ the EFTs that follow from QCD by integrating over the scales m and mv are called nonrelativistic QCD (NRQCD) and potential NRQCD (pNRQCD) respectively. This procedure for constructing EFTs can be generalized to finite temperatures. The construction of EFTs for heavy-quark bound states and their resulting forms depend on how the bound-state scales are related to the thermal scales. If all the bound state scales are larger than T, the relevant EFT is zero temperature pNRQCD. In this case, there are no corrections to the heavy $Q\overline{Q}$ potential even though there may be thermal corrections to the quarkonium binding energies and widths. If $mv > T > mv^2$ the relevant EFT can be constructed similarly to the $T = 0$ case but now the temperature scale is integrated over. Now there are thermal corrections to the potential which have both real and imaginary parts [910, 940]. Finally, when $T > mv$ the temperature scale must be integrated out after the scale m is integrated over but before the scale mv. This procedure leads to an EFT very similar to NRQCD but a modification of the Lagrangian corresponding to gluon and light quark fields is necessary. This part of the Lagrangian is replaced by the hard thermal loop (HTL) Lagrangian [941–944]. The resulting effective theory is referred to as $\mathrm{NRQCD_{HTL}}$ [910, 940]. Subsequent integration over the scale mv leads to a new EFT called $\mathrm{pNRQCD_{HTL}}$ similar to pNRQCD. Now the heavy-quark potential receives both real and imaginary thermal corrections. Furthermore, for $r \sim 1/m_D$, the real part of the potential is exponentially screened and the imaginary part is much larger than the real part [910, 940]. Below we summarize what has been learned from the EFT approach to heavy-quark bound states at finite temperature.

The thermal part of the potential has both a real and an imaginary part. The imaginary part of the potential smears out the bound state peaks of the quarkonium spectral function, leading to their dissolution prior to the onset of Debye screening in the real part of the potential (see, e.g., the discussion in [909]). Therefore, quarkonium dissociation appears to be a consequence of the thermal decay width rather

[20] A hierarchy of bound state scales may exist beyond the weak-coupling regime, e.g., $mv \simeq \Lambda_{\rm QCD}$, but has not yet been considered at finite temperature.

than color screening of the real part of the potential. This conclusion follows from the observation that the thermal decay width becomes comparable to the binding energy at temperatures below that required for the onset of color screening.

Two mechanisms contribute to the thermal decay width: the imaginary part of the gluon self-energy induced by Landau damping, as also observed in QED, see [935] and the quark–antiquark color-singlet to color-octet thermal breakup (a new effect, specific to QCD) [910]. Parametrically, the first mechanism dominates at temperatures where the Debye mass m_D is larger than the binding energy, while the latter effect dominates for temperatures where m_D is smaller than the binding energy. The dissociation temperature is related to the coupling by $\pi T_{\text{dissoc}} \sim mg^{\frac{4}{3}}$ [909, 938].

The derived color-singlet thermal potential, V, is neither the color-singlet $Q\overline{Q}$ free energy nor its internal energy. Instead it has an imaginary part and may contain divergences that eventually cancel in physical observables [910].

Finally, there may be other finite-temperature effects other than screening. These typically may take the form of power-law corrections or have a logarithmic temperature dependence [910, 938].

The EFT framework thus provides a clear definition of the potential and a coherent and systematic approach for calculating quarkonium masses and widths at finite temperature. In [945], heavy-quarkonium energy levels and decay widths in a quark–gluon plasma, below the quarkonium dissociation temperature where the temperature and screening mass satisfy the hierarchy $m\alpha_s \gg \pi T \gg m\alpha_s^2 \gg m_D$, have been calculated to order $m\alpha_s^5$, relevant for bottomonium $1S$ states ($\Upsilon(1S)$, η_b) at the LHC. At leading order the quarkonium masses increase quadratically with T, the same functional increase with energy as dileptons produced in electromagnetic decays [945]. A thermal correction proportional to T^2 appears in the quarkonium electromagnetic decay rates. The leading-order decay width grows linearly with temperature, implying that quarkonium dissociates by decaying to the color-octet continuum.

This EFT approach was derived assuming weak coupling and neglecting nonperturbative effects. In particular, the role of the color-octet degrees of freedom in pNRQCD beyond perturbation theory needs to be better understood (see e.g., [946]). Comparison of certain static $Q\overline{Q}$ correlators calculated in the EFT framework with results from lattice QCD, partly discussed in Sect. 5.3.2, could prove useful in this respect. The correlation function of two Polyakov loops, which is gauge invariant and corresponds to the free energy of a static $Q\overline{Q}$ pair could be particularly suitable for this purpose. Therefore, in [947] the Polyakov loop and the correlator of two Polyakov loops have been calculated to next-to-next-to-leading order at finite temperature in the weak-coupling regime and at $Q\overline{Q}$ separations shorter than the inverse of the temperature and for Debye masses larger than the Coulomb potential. The relationship between the Polyakov loop correlator and the singlet and octet $Q\overline{Q}$ correlator has been established in the EFT framework. A related study of cyclic Wilson loops at finite temperature in perturbation theory was reported in [948]. A further attempt to relate static $Q\overline{Q}$ correlation functions to the real-time potential was discussed in [949]. Very recently the first lattice NRQCD calculations of bottomonium at finite temperature have appeared [950]. The initial discussion on the possibility of calculating quarkonium properties at finite temperature using NRQCD goes back to the first meeting of the QWG at CERN in 2002.

In addition, the effects of medium anisotropies on the quarkonium states have been considered, both on the real [951] and imaginary [952–954] parts of the potential, as well as on bound-state production. Polarization of the P states has been predicted to arise from the medium anisotropies, resulting in a significant (∼30%) effect on the χ_b states [955]. A weak medium anisotropy may also be related to the shear viscosity [956]. Thus the polarization can directly probe the properties of the medium produced in heavy-ion collisions.

5.3.4 Dynamical production models

While it is necessary to understand the quarkonium spectral functions in equilibrium QCD, this knowledge is insufficient for predicting effects on quarkonium production in heavy-ion collisions because, unlike the light degrees of freedom, heavy quarks are not fully thermalized in heavy-ion collisions. Therefore it is nontrivial to relate the finite-temperature quarkonium spectral functions to quarkonium production rates in heavy-ion collisions without further model assumptions. The bridge between the two is provided by dynamical models of the matter produced in heavy-ion collisions. Some of the simple models currently available are based on statistical recombination [957], statistical recombination and dissociation rates [958], or sequential melting [959]. Here we highlight a more recent model, which makes closer contact with both QCD and experimental observations [960].

The bulk evolution of the matter produced in heavy-ion collisions is well described by hydrodynamics (see [961] for a recent review). The large heavy-quark mass makes it possible to model its interaction with the medium by Langevin dynamics [962]. Such an approach successfully describes the anisotropic flow of charm quarks observed at RHIC [962, 963] (see also the review [964] and references therein). Potential models have shown that, in the absence of bound states, the $Q\overline{Q}$ pairs are correlated in space [905, 906]. This correlation can be modeled classically using Langevin dynamics, including a drag force and a random force between the Q (or $\overline{Q}$) and the medium as well as the forces between the Q and $\overline{Q}$ described by the potential. It was recently shown that a model combining an ideal hydrodynamic expansion of the medium with a description of the correlated $Q\overline{Q}$ pair dynamics by the Langevin equation can describe charmonium suppression at RHIC quite well [960]. In particular, this model can explain why, despite the fact that a deconfined medium is created at RHIC, there is only a 40–50% suppression in the charmonium yield. The attractive potential and the finite lifetime of the system prevents the complete decorrelation of some of the $Q\overline{Q}$ pairs [960]. Once the matter has cooled sufficiently, these residual correlations make it possible for the Q and $\overline{Q}$ to form a bound state. Charmonium production by recombination can also be calculated in this approach [965]. Although recombination was found to be significant for the most central collisions, it is still subdominant [965].

The above approach, which neglects quantum effects, is applicable only if there are no bound states, as is likely to be the case for the J/ψ. If heavy-quark bound states are present, as is probable for the $\Upsilon(1S)$, the thermal dissociation rate will be most relevant for understanding the quarkonium yield. It is expected that the interaction of a color-singlet quarkonium state with the medium is much smaller than that of heavy quarks. Thus, to first approximation, medium effects will only lead to quarkonium dissociation.

5.3.5 Summary of hot-medium effects

Potential model calculations based on lattice QCD, as well as resummed perturbative QCD calculations, indicate that all charmonium states and the excited bottomonium states dissolve in the deconfined medium. This leads to the reduction of the quarkonium yields in heavy-ion collisions compared to the binary scaling of pp collisions. Recombination and edge effects, however, guarantee a nonzero yield.

One of the great opportunities of the LHC and RHIC-II heavy-ion programs is the ability to study bottomonium yields. From a theoretical perspective, bottomonium is an important and clean probe for at least two reasons. First, the effective field theory approach, which provides a link to first principles QCD, is more applicable for bottomonium due to better separation of scales and higher dissociation temperatures. Second, the heavier bottom quark mass reduces the importance of statistical recombination effects, making bottomonium a good probe of dynamical models.

5.4 Recent results at SPS energies

Studies of charmonium production and suppression in cold and hot nuclear matter have been carried out by the NA60 Collaboration [882, 966, 967]. In particular, data have been taken for In+In collisions at 158 GeV/nucleon and for pA collisions at 158 and 400 GeV. In the following, the primary NA60 results and their impact on the understanding of the anomalous J/ψ suppression, first observed by the NA50 Collaboration in Pb+Pb collisions [968], are summarized. A preliminary comparison between the suppression patterns observed at the SPS and RHIC is discussed in Sect. 5.6.

5.4.1 J/ψ production in pA collisions

One of the main results of the SPS heavy-ion program was the observation of anomalous J/ψ suppression. Results obtained in Pb+Pb collisions at 158 GeV/nucleon by the NA50 Collaboration showed that the J/ψ yield was suppressed with respect to estimates that include only cold-nuclear-matter effects [968]. The magnitude of the cold-nuclear-matter effects has typically been extracted by extrapolating the J/ψ production data obtained in pA collisions. Until recently the reference SPS pA data were based on samples collected at 400/450 GeV by the NA50 Collaboration, at higher energy than the nuclear collisions and in a slightly different rapidity domain [872, 873, 969].

The need for reference pA data taken under the same conditions as the AA data was a major motivation for the NA60 run with an SPS primary proton beam at 158 GeV in 2004. Seven nuclear targets (Be, Al, Cu, In, W, Pb, and U) were simultaneously exposed to the beam. The sophisticated NA60 experimental setup [970], based on a high-resolution vertex spectrometer coupled to the muon spectrometer inherited from NA50, made it possible to unambiguously identify the target in which the J/ψ was produced as well as measure muon pairs from its decay with a ~70 MeV invariant-mass resolution. During the same period, a 400 GeV pA data sample was taken with the same experimental setup.

Cold-nuclear-matter effects were evaluated comparing the cross section ratio

$$\frac{\sigma_{pA}^{J/\psi}}{\sigma_{p\mathrm{Be}}^{J/\psi}}, \tag{177}$$

for each nucleus with mass number A, relative to the lightest target (Be). The beam luminosity factors cancel out in the ratio, apart from a small beam-attenuation factor. However, since the sub-targets see the vertex telescope from slightly different angles, the track reconstruction efficiencies do not completely cancel out. Therefore an accurate evaluation of the time evolution of such quantities was performed target-by-target, with high granularity and on a run-by-run basis. The results [882, 967], shown in Fig. 81, are integrated over p_T and are given in the rapidity region covered by all the sub-targets, $0.28 < y_{\mathrm{CMS}} < 0.78$ for the 158 GeV sample and $-0.17 < y_{\mathrm{CMS}} < 0.33$ for the 400 GeV sample. Systematic errors include uncertainties in the target thickness,

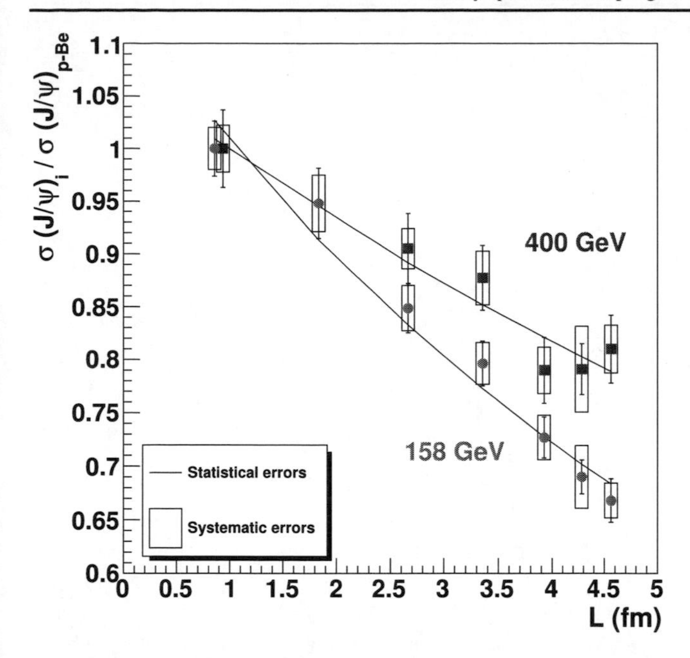

Fig. 81 The J/ψ cross section ratios for pA collisions at 158 GeV (*circles*) and 400 GeV (*squares*), as a function of L, the mean thickness of nuclear matter traversed by the J/ψ. From [882] with kind permission, copyright (2009) Elsevier

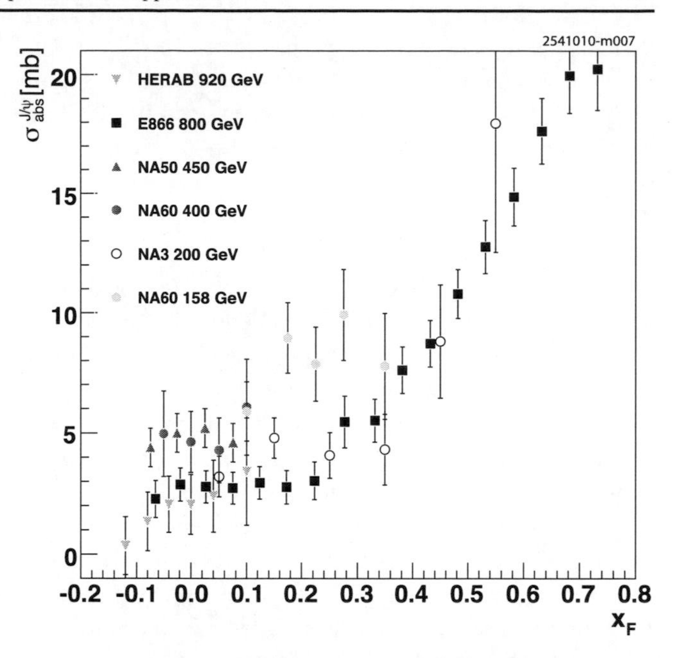

Fig. 82 Compilation of $\sigma_{\rm abs}^{J/\psi}$ as a function of x_F with no additional cold-matter effects included. From [967] with kind permission, copyright (2009) Elsevier

the rapidity distribution used in the acceptance calculation, and the reconstruction efficiency. Only the fraction of systematic errors not common to all the points is shown since it affects the evaluation of nuclear effects.

Nuclear effects have usually been parametrized by fitting the A dependence of the J/ψ production cross section using the expression

$$\sigma_{pA}^{J/\psi} = \sigma_{pp}^{J/\psi} A^{\alpha}. \tag{178}$$

Alternatively, the effective absorption cross section, $\sigma_{\rm abs}^{J/\psi}$, can be extracted from the data using the Glauber model. Both α and $\sigma_{\rm abs}^{J/\psi}$ are effective quantities since they represent the strength of the cold-nuclear-matter effects that reduce the J/ψ yield. However, they cannot distinguish among the different effects, e.g., shadowing and nuclear absorption, contributing to this reduction. The results in Fig. 81 were used to extract

$$\begin{aligned} &\sigma_{\rm abs}^{J/\psi} = 7.6 \pm 0.7(\text{stat.}) \pm 0.6(\text{syst.})\ \text{mb}; \\ &\alpha = 0.882 \pm 0.009 \pm 0.008 \end{aligned} \tag{179}$$

at 158 GeV and

$$\begin{aligned} &\sigma_{\rm abs}^{J/\psi} = 4.3 \pm 0.8(\text{stat.}) \pm 0.6(\text{syst.})\ \text{mb}; \\ &\alpha = 0.927 \pm 0.013 \pm 0.009 \end{aligned} \tag{180}$$

at 400 GeV. Thus $\sigma_{\rm abs}^{J/\psi}$ is larger at 158 GeV than at 400 GeV by three standard deviations. The 400 GeV result is, on the other hand, in excellent agreement with the previous NA50 result obtained at the same energy [872].

The study of cold-nuclear-matter effects at fixed-target energies is a subject which has attracted considerable interest. In Fig. 82, a compilation of previous results for $\sigma_{\rm abs}^{J/\psi}$ as a function of x_F [873, 874, 880, 881] is presented, together with the new NA60 results [967]. Contrary to Fig. 78, the values of $\sigma_{\rm abs}^{J/\psi}$ in Fig. 82 do not include any shadowing contribution, only absorption. There is a systematic increase in the nuclear effects going from low to high x_F as well as when from high to low incident proton energies. As shown in Fig. 82, the new NA60 results at 400 GeV confirm the NA50 values obtained at a similar energy. On the other hand, the NA60 158 GeV data suggest higher values of $\sigma_{\rm abs}^{J/\psi}$ and hint at increased absorption over the x_F range. Note also that the older NA3 J/ψ results are in partial contradiction with these observations, giving lower values of $\sigma_{\rm abs}^{J/\psi}$, similar to those obtained from the higher energy data samples. Such a complex pattern of nuclear effects results from a delicate interplay of various nuclear effects (final-state absorption, shadowing, initial-state energy loss, etc.) and has so far not been satisfactorily explained by theoretical models [886]. A first attempt to disentangle the contribution of shadowing from $\sigma_{\rm abs}^{J/\psi}$ (as extracted from the NA60 results) has been carried out using the EKS98 [877] parametrization of the nuclear PDFs. It was found that a larger $\sigma_{\rm abs}^{J/\psi}$ is needed to describe the measured data:

$$\begin{aligned} &\sigma_{\rm abs}^{J/\psi}(158\ \text{GeV}) = 9.3 \pm 0.7 \pm 0.7\ \text{mb}; \\ &\sigma_{\rm abs}^{J/\psi}(400\ \text{GeV}) = 6.0 \pm 0.9 \pm 0.7\ \text{mb}. \end{aligned} \tag{181}$$

The results thus depend on the parametrization of the nuclear modifications of the PDFs. For example, slightly higher (5–10%) values of $\sigma_{\rm abs}^{J/\psi}$ are obtained if the EPS08 [971] parametrization is used.

5.4.2 Anomalous J/ψ suppression

The pA results at 158 GeV shown in the previous section have been collected at the same energy and in the same x_F range as the SPS AA data. It is therefore natural to use these results to calculate the expected magnitude of cold-nuclear-matter effects on J/ψ production in nuclear collisions. In order to do so, the expected shape of the J/ψ distribution as a function of the forward energy in the zero degree calorimeter, $dN_{J/\psi}^{\rm expect}/dE_{\rm ZDC}$, has been determined using the Glauber model. The J/ψ yield is assumed to scale with the number of NN collisions. The effective J/ψ absorption cross section in nuclear matter is assumed to be the same as the value at 158 GeV deduced in the previous section.

The measured J/ψ yield, $dN_{J/\psi}/dE_{\rm ZDC}$, is normalized to $dN_{J/\psi}^{\rm expect}/dE_{\rm ZDC}$ using the procedure detailed in [966]. This procedure previously did not take shadowing effects into account when extrapolating from pA to AA interactions. In pA collisions, only the target partons are affected by shadowing, while in AA collisions, effects on both the projectile and target must be taken into account. If shadowing is neglected in the pA to AA extrapolation, a small bias is introduced, resulting in an artificial $\sim$5% suppression of the J/ψ yield with the EKS98 parametrization [889]. Therefore, if shadowing is properly accounted for in the pA to AA extrapolation, the amount of the anomalous J/ψ suppression is reduced. Figure 83 presents the new results for the anomalous J/ψ suppression in In+In and Pb+Pb collisions [882, 967] as a function of $N_{\rm part}$, the number of participant nucleons. Up to $N_{\rm part} \sim 200$ the J/ψ yield is, within errors, compatible with the extrapolation of cold-nuclear-matter effects. When $N_{\rm part} > 200$, there is an anomalous suppression of up to $\sim$20–30% in the most central Pb+Pb collisions. This new, smaller anomalous suppression is primarily due to the larger $\sigma_{\rm abs}^{J/\psi}$ extracted from the evaluation of cold-nuclear-matter effects.

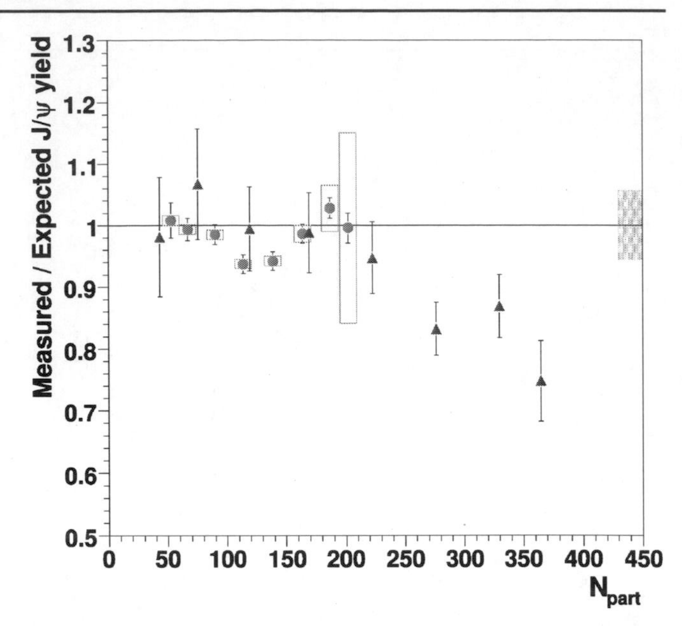

Fig. 83 Anomalous J/ψ suppression in In+In (*circles*) and Pb+Pb collisions (*triangles*) as a function of $N_{\rm part}$. The *boxes* around the In+In *points* represent correlated systematic errors. The *filled box* on the right corresponds to the uncertainty in the absolute normalization of the In+In *points*. A 12% global error, due to the uncertainty on $\sigma_{\rm abs}^{J/\psi}$ at 158 GeV is not shown. From [967] with kind permission, copyright (2009) Elsevier

5.5 Recent hadroproduction results from RHIC

The strategy of the RHIC J/ψ program has been to measure production cross sections in $\sqrt{s_{NN}} = 200$ GeV collisions for pp, d+Au, Au+Au and Cu+Cu collisions. RHIC has also studied J/ψ production in Cu+Cu collisions at $\sqrt{s_{NN}} = 62$ GeV and will also study J/ψ production in pp collisions at $\sqrt{s} = 500$ GeV. The pp collisions are studied both to learn about the J/ψ production mechanism and to provide baseline production cross sections needed for understanding the d+Au and AA data. Similarly, the d+Au measurements are inherently interesting because they study the physical processes that modify J/ψ production cross sections in nuclear targets and also provide the crucial cold-nuclear-matter baseline for understanding J/ψ production in AA collisions. Note that d+Au collisions are studied at RHIC instead of p+Au collisions for convenience—p+Au collisions are possible at RHIC, but would require a dedicated p+Au run.

The last few years of the RHIC program have produced J/ψ data from PHENIX for pp, d+Au and Au+Au collisions with sufficient statistical precision to establish the centrality dependence of both hot and cold-nuclear-matter effects at $\sqrt{s_{NN}} = 200$ GeV. The data cover the rapidity range $|y| < 2.4$.

We introduce some quantities that have been applied to d+Au and AA collisions at RHIC to describe the impact parameter, b (also called *centrality*) , dependence of the quarkonium results. While most of the data taken are at large impact parameter (peripheral collisions), the small impact-parameter (central) collisions are more likely to produce a quark–gluon plasma. Therefore it is important to study quantities over a range of centralities, using impact-parameter dependent variables such as the number of participant nucleons, $N_{\rm part}$, and the number of collisions, $N_{\rm coll}$.

The number of participants depends on b as

$$N_{\text{part}}(b) = \int d^2s \times \big[T_A(s)\big(1 - \exp\big[-\sigma_{\text{inel}}(s_{NN})T_B\big(|\vec{b} - \vec{s}|\big)\big]\big) + T_B\big(|\vec{b} - \vec{s}|\big) \times \big(1 - \exp\big[-\sigma_{\text{inel}}(s_{NN})T_A(s)\big]\big)\big]. \qquad (182)$$

Here σ_{inel} is the inelastic nucleon–nucleon cross section, 42 mb at RHIC, and $T_{A/B}(s) = \int dz\rho_{A/B}(s,z)$, the line integral of the nuclear density, ρ, in the beam direction, is the nuclear profile function. Large values of N_{part} are obtained for small impact parameters with $N_{\text{part}}(b=0) = 2A$ for spherical nuclei. Small values of N_{part} occur in very peripheral collisions. The number of collisions, $N_{\text{coll}}(s_{NN}; b) = \sigma_{\text{inel}}(s_{NN})T_{AB}(b)$, depends on the nuclear overlap integral,

$$T_{AB}(b) = \int d^2s\,dz\,dz'\rho_A(s,z)\rho_B\big(|\vec{b} - \vec{s}|, z'\big). \qquad (183)$$

In pA collisions, we assume that the proton has a negligible size, $\rho_A(s,z) = \delta(s)\delta(z)$ so that $T_{AB}(b)$ collapses to the nuclear profile function. The deuteron cannot be treated as a point particle since it is large and diffuse. Thus the Hülthen wave function [972, 973] is used to calculate the deuteron density distribution. No shadowing effects are included on the deuteron.

The nuclear suppression factor, R_{AB}, for dA, and AA collisions is defined as the ratio

$$R_{AB}(N_{\text{part}}; b) = \frac{d\sigma_{AB}/dy}{T_{AB}(b)\,d\sigma_{pp}/dy} \qquad (184)$$

where $d\sigma_{AB}/dy$ and $d\sigma_{pp}/dy$ are the quarkonium rapidity distributions in AB and pp collisions and T_{AB} is the nuclear overlap function, defined in (183). In AA collisions, R_{AA} is sometimes shown relative to the extracted cold-nuclear-matter baseline, R_{AA}^{CNM}. PHENIX has also shown both d+Au and AA data as a function of R_{CP}, the ratio of AB cross sections in central relative to peripheral collisions,

$$R_{CP}(y) = \frac{T_{AB}(b_P)}{T_{AB}(b_C)} \frac{d\sigma_{AB}(b_C)/dy}{d\sigma_{AB}(b_P)/dy}, \qquad (185)$$

where b_C and b_P correspond to the central and peripheral values of the impact parameter, since systematic uncertainties cancel in the ratio. Another quantity of interest is v_2, the second harmonic of the azimuthal Fourier decomposition of the momentum distribution, $dN/dp_T \propto 1 + 2v_2\cos(2(\phi - \phi_r))$ where ϕ is the particle emission angle and ϕ_r is the reaction plane angle, known as the elliptic flow. It gives some indication of the particle response to the thermalization of the medium. A finite J/ψ v_2 would give some indication of whether the J/ψ distribution becomes thermal. The strength of v_2 depends on the proportion of J/ψ produced by coalescence.

In the next few years the increased RHIC luminosity and the commissioning of upgraded detectors and triggers for PHENIX and STAR will enable a next generation of RHIC measurements, extending the program to the Υ family, excited charmonium states, and J/ψ v_2 and high-p_T suppression measurements. There have already been low-precision, essentially proof-of-principle, measurements of most of those signals. Very importantly, upgraded silicon vertex detectors for both PHENIX and STAR are expected to produce qualitatively better open charm measurements that will provide important inputs to models of J/ψ production in heavy-ion collisions.

In addition to the results discussed here, there have been PHENIX results on J/ψ photoproduction in peripheral Au+Au collisions [974] and a proof-of-principle measurement of the J/ψ v_2 in Au+Au collisions by PHENIX [975] with insufficient precision for physics conclusions.

5.5.1 Charmonium from pp collisions

PHENIX [662] has reported measurements of the inclusive J/ψ polarization in 200 GeV pp collisions at midrapidity. Results for the polarization parameter λ, defined in the Helicity frame, are shown in Fig. 84 and compared to COM [699] and s-channel-cut CSM [682] predictions. The latter has been shown to describe the rapidity and p_T dependence of the PHENIX 200 GeV pp J/ψ data [976] using a two-parameter fit to CDF data at $\sqrt{s} = 1.8$ TeV.

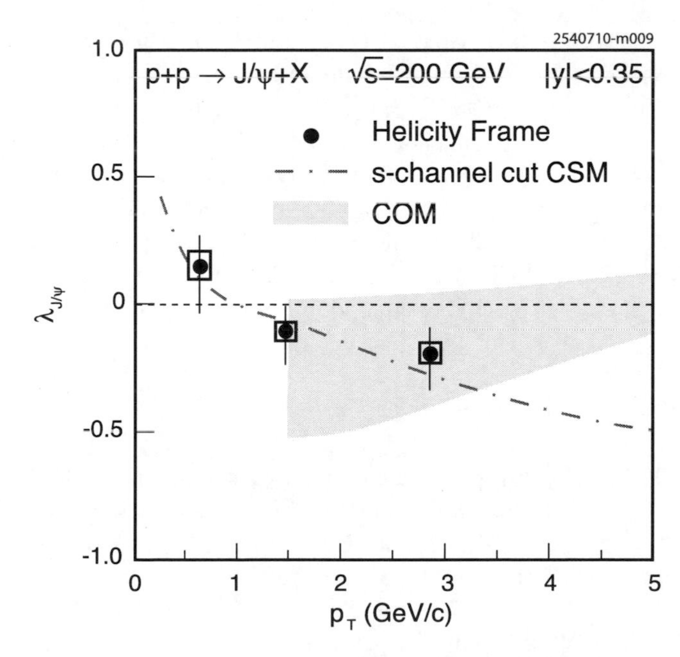

Fig. 84 The polarization extracted from 200 GeV PHENIX pp data at midrapidity as a function of p_T. The data are compared with the s-channel-cut CSM [682] and a COM [699] prediction. Adapted from [662] with kind permission, copyright (2010) The American Physical Society

At Quark Matter 2009, PHENIX [663] showed preliminary measurements of the p_T dependence of the $\psi(2S)$ cross section at 200 GeV. This is the first measurement of the p_T-dependence of an excited charmonium state at RHIC. PHENIX measured the feeddown contribution of the $\psi(2S)$ to the J/ψ to be $(8.6 \pm 2.3)\%$, in good agreement with the world average.

STAR [664] has published measurements of the J/ψ cross section in 200 GeV pp collisions for $5 < p_T < 13$ GeV/c. This greatly extends the p_T range over which J/ψ data are available at RHIC. Although PHENIX can trigger at all p_T, it has so far been limited to p_T below about 9 GeV/c [663] because of its much smaller acceptance.

5.5.2 Charmonium from Cu+Cu collisions

Measured quarkonium production rates from heavy-ion collisions are commonly presented in terms of a nuclear modification factor, R_{AA}, defined in (184). PHENIX [977] results on the rapidity and p_T dependence of R_{AA} values for J/ψ from 200 GeV Cu+Cu collisions were published some time ago. However, those results were limited to $p_T < 5$ GeV/c, and do not address the high-p_T behavior of the measurements very well. STAR [664] has now published Cu+Cu R_{AA} data for J/ψ at 5.5 and 7 GeV/c that yield an average $\langle R_{AA}\rangle = 1.4 \pm 0.4(\text{stat}) \pm 0.2(\text{sys})$ above 5 GeV/c for the 0–20% most central collisions. The R_{AA} data for the 0–60% most central collisions have very similar values, in contrast to the PHENIX data below 5 GeV/c that yield $\langle R_{AA}\rangle \approx 0.52$ for central Cu+Cu collisions.

PHENIX [975] has also released preliminary data on the R_{AA} for J/ψ from minimum bias (0–94% centrality) Cu+Cu data at 7 and 9 GeV/c. The minimum bias PHENIX data should be comparable to the STAR 0–60% data, but the PHENIX results are more consistent with a nearly p_T-independent R_{AA}. However, both measurements have large statistical uncertainties and a direct comparison [978] of the STAR and PHENIX Cu+Cu R_{AA} data at high p_T suggests that more data will be required to definitively determine the high-p_T behavior of R_{AA} in central collisions.

5.5.3 Bottomonium production

PHENIX [979] showed a preliminary result for the $\Upsilon(1S) + \Upsilon(2S) + \Upsilon(3S)$ cross section at forward and backward rapidity ($1.2 < |y| < 2.4$) at Quark Matter 2006. More recently, PHENIX [663] showed a preliminary result at Quark Matter 2009 for $\Upsilon(1S) + \Upsilon(2S) + \Upsilon(3S)$ production in 200 GeV pp collisions at midrapidity ($|y| < 0.35$). The measured cross section is $\mathcal{B}\,d\sigma/dy = 114^{+46}_{-45}$ pb at $y = 0$, where the presence of the $\mathcal{B}$ reflects that the results have not been separated by individual $\Upsilon(nS)$ resonance nor corrected for the dilepton branching fractions $\mathcal{B}(\Upsilon(nS) \to e^+e^-)$.

STAR [980] published a measurement of the $\Upsilon(1S) + \Upsilon(2S) + \Upsilon(3S) \to e^+e^-$ cross section at $|y| < 0.5$ for 200 GeV pp collisions. The measured value is $\mathcal{B}\,d\sigma/dy = 114 \pm 38(\text{stat})^{+23}_{-24}(\text{syst})$ pb at $y = 0$. STAR [981] also has a preliminary result for the $\Upsilon(1S) + \Upsilon(2S) + \Upsilon(3S) \to e^+e^-$ cross section at midrapidity in d+Au collisions at 200 GeV/c. The cross section was found to be $\mathcal{B}\,d\sigma/dy = 35 \pm 4(\text{stat}) \pm 5(\text{syst})$ nb. The midrapidity value of R_{dAu} was found to be $0.98 \pm 0.32(\text{stat}) \pm 0.28(\text{syst})$, consistent with binary scaling.

PHENIX has made a preliminary measurement of the dielectron yield in the $\Upsilon(1S) + \Upsilon(2S) + \Upsilon(3S)$ mass range at midrapidity in Au+Au collisions [975]. In combination with the PHENIX $\Upsilon(1S) + \Upsilon(2S) + \Upsilon(3S)$ pp result at midrapity, a 90% CL upper limit on R_{dAu} of 0.64 was found for the $\Upsilon(1S) + \Upsilon(2S) + \Upsilon(3S)$ mass region. The significance of this result is not yet very clear since the measurement is for all three Υ states combined.

5.5.4 J/ψ production from d+Au collisions

As discussed previously, modification of the J/ψ production cross section due to the presence of a nuclear target is expected to be caused by shadowing, breakup of the precursor J/ψ state by collisions with nucleons, initial-state energy loss, and other possible effects. Parametrizing these effects by employing a Glauber model with a fitted effective J/ψ-absorption cross section, $\sigma_{\text{abs}}^{J/\psi}$, results in an effective cross section with strong rapidity and $\sqrt{s_{NN}}$ dependencies [875] that are not well understood. A large increase in the effective absorption cross section is observed by E866/NuSea [874] at forward rapidity. This increase cannot be explained by shadowing models alone, suggesting that there are important physics effects omitted from the Glauber absorption-plus-shadowing model.

The extraction of hot-matter effects in the Au+Au J/ψ data at RHIC has been seriously hampered by the poor understanding of J/ψ production in nuclear targets, including the underlying J/ψ production mechanism. Thus the cold-nuclear-matter baseline has to be obtained experimentally.

The PHENIX J/ψ data obtained in the 2003 RHIC d+Au run did not have sufficient statistical precision either for studies of cold-nuclear-matter effects or for setting a cold-nuclear-matter baseline for the Au+Au data [883]. This low-statistics measurement has been augmented by the large J/ψ data set obtained in the 2008 d+Au run. PHENIX [663] has released d+Au R_{CP} data for J/ψ production in nine rapidity bins over $|y| < 2.4$. Systematic uncertainties associated with the beam luminosity, detector acceptance, trigger efficiency, and tracking efficiency cancel in R_{CP}, defined in (185). There is a remaining systematic uncertainty due to the centrality dependence of the tracking and particle identification efficiencies.

The use of a Glauber model also gives rise to significant systematic uncertainties in the centrality dependence of R_{CP}. The model is used to calculate the average number of nucleon–nucleon collisions as a means of estimating the relative normalization between different centrality bins. The systematic uncertainty due to this effect is independent of rapidity.

The PHENIX d+Au R_{CP} data have been independently fitted at each of the nine rapidities [884] employing a model including shadowing and J/ψ absorption. The model calculations [888] use the EKS98 and nDSg shadowing parametrizations with $0 \leq \sigma_{\text{abs}} \leq 15$ mb. The best fit absorption cross section was determined at each rapidity, along with the $\pm 1\sigma$ uncertainties associated with both rapidity-dependent and rapidity-independent systematic effects. The results are shown in Fig. 85. The most notable feature is the stronger effective absorption cross section at forward rapidity, similar to the behavior observed at lower energies [874]. In fact, it is striking that the extracted cross sections at forward rapidity are very similar for PHENIX ($\sqrt{s_{NN}} = 200$ GeV) and E866 [875] ($\sqrt{s_{NN}} = 38.8$ GeV) (see the lower panel of Fig. 78), despite the large difference in center-of-mass energies.

Note the large global systematic uncertainty in σ_{abs} extracted from the PHENIX R_{CP} data, dominated by the uncertainty in the Glauber estimate of the average number of collisions at each centrality. Although it does not affect the shape of the rapidity dependence of $\sigma_{\text{abs}}^{J/\psi}$, it results in considerable uncertainty in the magnitude of the effective absorption cross section.

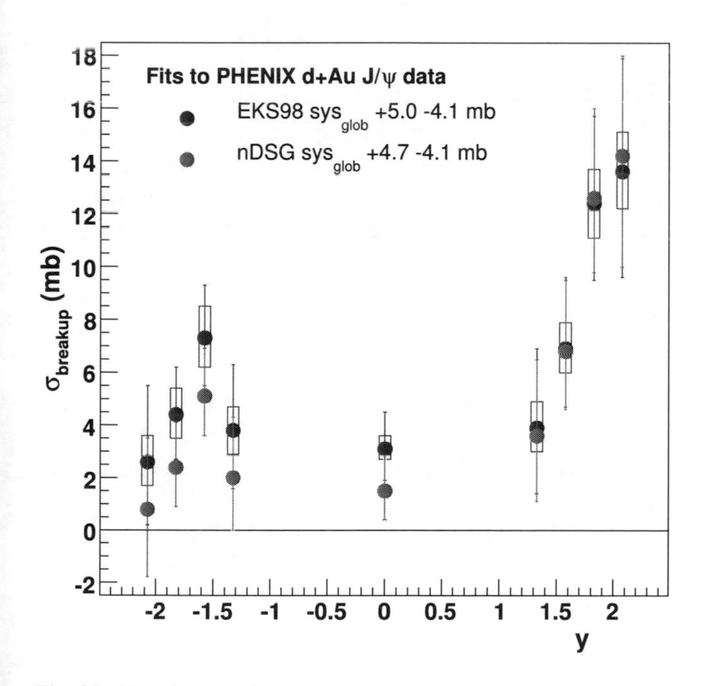

Fig. 85 The effective absorption cross section as a function of rapidity extracted from PHENIX d+Au R_{CP} data using the EKS98 and nDSg shadowing parametrizations. The *vertical bars* show uncorrelated point-to-point uncertainties, the *boxes* show correlated uncertainties, and the global uncertainties are given in the legend

It has been suggested [891] that the large increase in effective absorption cross section at forward rapidity obtained from a CEM calculation [884] may be moderated significantly if the $2 \to 2$ kinematics of the leading-order CSM is used. This difference emphasizes the importance of understanding the underlying production mechanism.

PHENIX has very recently released [982] final R_{dAu} and R_{CP} data from the 2008 d+Au RHIC run. The final R_{CP} data are in good agreement with the preliminary data, discussed earlier, as well as in the next section. A comparison [982] of the R_{dAu} data, which has not been shown before, with the R_{CP} data shows that a simultaneous description of the two observables will require a stronger than linear dependence of the J/ψ suppression on the nuclear thickness function at forward rapidity. The dependence of the suppression on nuclear thickness is at least quadratic, and is likely higher. The result has important implications for the understanding of forward-rapidity d+Au physics. Since the calculations of the cold matter contributions to R_{AA} assumed that shadowing depends linearly on the nuclear thickness, the calculations of R_{AA} shown in the next section should be revisited.

5.5.5 J/ψ production from Au+Au collisions

PHENIX [983] has published the centrality dependence of R_{AA} for Au+Au collisions using Au+Au data from the 2004 RHIC run and *pp* data from the 2005 run. The data are shown in Fig. 86. The suppression is considerably stronger at forward rapidity than at midrapidity. The significance of this difference with respect to hot-matter effects will not be clear, however, until the suppression due to cold-nuclear-matter effects is more accurately known.

To estimate the cold-nuclear-matter contribution to the Au+Au J/ψ $R_{AA}$the d+Au J/ψ R_{CP} data were extracted using the EKS98 and nDSg shadowing parametrizations, as described earlier, except that, in this case, the $\sigma_{\text{abs}}^{J/\psi}$ values in d+Au collisions were fitted independently in three rapidity intervals: $-2.2 < y < -1.2$, $|y| < 0.35$, and $1.2 < y < 2.2$. In effect, this tunes the calculations to reproduce the d+Au R_{CP} independently in each of the three rapidity windows in which the Au+Au R_{AA} data were measured. The cold-nuclear-matter R_{AA} for Au+Au collisions was then estimated in a Glauber calculation using the fitted absorption cross sections and the centrality-dependent $R_{p\text{Au}}$ values calculated using EKS98 and nDSg shadowing parametrizations [984]. Each nucleon–nucleon collision contributes differently to the R_{AA} in each rapidity window. To more directly simulate nucleon–nucleus interactions, the analysis assumes that R_{AA} can be treated as a convolution of p+Au and Au+p collisions in the three rapidity windows. The

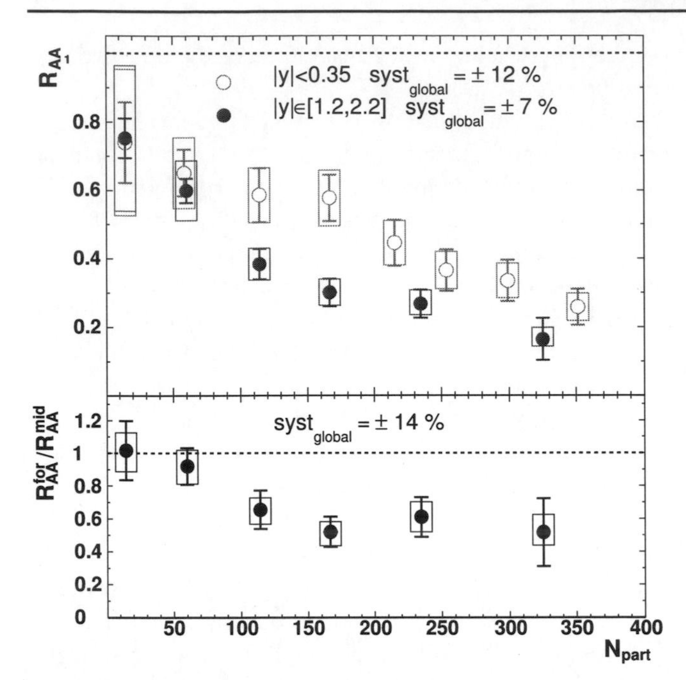

Fig. 86 The PHENIX Au+Au R_{AA} as a function of centrality for $|y| < 0.35$ and $1.2 < |y| < 2.2$

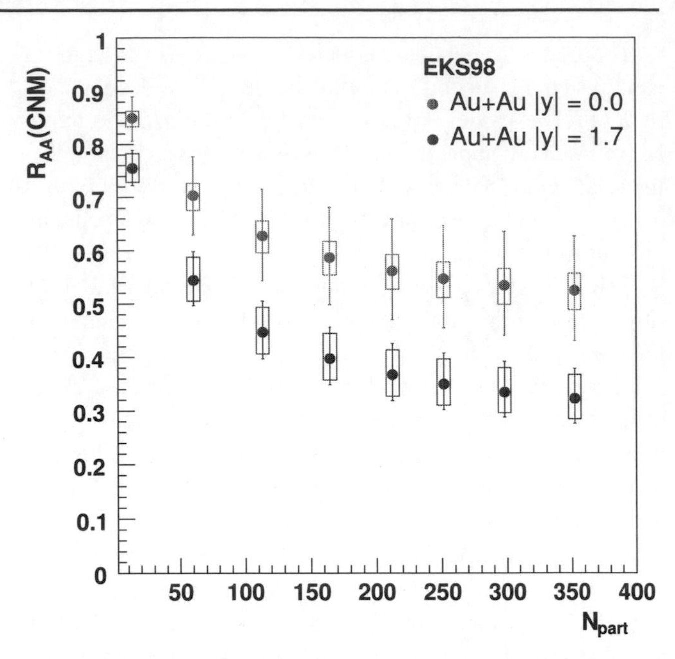

Fig. 87 The estimated Au+Au cold-nuclear-matter R_{AA} as a function of centrality for $|y| < 0.35$ and $1.2 < |y| < 2.2$. The *vertical bar* represents the rapidity-dependent systematic uncertainty in the fitted $\sigma_{abs}^{J/\psi}$

impact-parameter dependence of R_{pAu} is determined separately to infer the R_{AA} centrality dependence for a rapidity-dependent absorption cross section. Thus the value of R_{pAu} at the impact parameter of nucleon 1 in the projectile is convoluted with the value of R_{Aup} at the impact parameter of nucleon 2 in the target. Effectively, this means that to obtain R_{AA} for $1.2 < |y| < 2.2$, R_{pAu} for the forward-moving nucleon ($1.2 < y < 2.2$) is multiplied by R_{pAu} for the backward-moving nucleon ($-2.2 < y < -1.2$). When $|y| < 0.35$, the R_{pAu} calculations at midrapidity are used. The number of participants, obtained from a Glauber calculation, is used to bin the collisions in centrality with a cut on peripheral events to mimic the effect of the PHENIX trigger efficiency at large impact parameter. The uncertainty in the calculated CNM R_{AA} was estimated by repeating the calculation with $\sigma_{abs}^{J/\psi}$ varied away from best-fit values. This variation ranged over the rapidity-dependent systematic uncertainty determined when fitting the d+Au R_{CP}.

The global systematic uncertainty in $\sigma_{abs}^{J/\psi}$ was neglected in the calculation of the CNM R_{AA}. This was done because the same Glauber model was used to obtain both the number of nucleon–nucleon collisions, N_{coll}, in d+Au and Au+Au interactions and the fitted $\sigma_{abs}^{J/\psi}$ values. Therefore, if, for example, N_{coll} is underestimated for the d+Au R_{CP}, the fitted absorption cross section will be overestimated. However, this would be compensated in the calculated CNM R_{AA} by the underestimated N_{coll} value. Any possible differences in the details of the d+Au and Au+Au Glauber calculations would result in an imprecise cancelation of the uncertainties. This effect has not yet been studied.

Note that there is a significant difference between the impact-parameter dependence of the R_{pAu} and R_{dAu} calculations [884], primarily for peripheral collisions, due to the smearing caused by the finite size of the deuteron. Since R_{dAu} and R_{pAu} are calculated using the same basic model, this smearing does not present a problem in the present analysis. However, if the measured R_{dAu} was used directly in a Glauber model, as was done with the RHIC 2003 data [883], a correction would be necessary.

The resulting Glauber calculations of the cold-nuclear-matter R_{AA} using the EKS98 shadowing parametrization are shown in Fig. 87. The values obtained with nDSg are almost identical, as they should be since both methods parametrize the same data.

We emphasize that the kinematic-dependent differences in the effective absorption cross sections noted in the previous section do not affect the cold-nuclear-matter R_{AA} derived from the data. As long as the method of fitting the d+Au data is consistent with the estimate of the cold nuclear matter R_{AA}, the result should be model independent.

The J/ψ suppression beyond CNM effects in Au+Au collisions can be estimated by dividing the measured R_{AA} by the estimates of the CNM R_{AA}. The result for EKS98 is shown in Fig. 88. The result for nDSg is nearly identical.

Assuming that the final PHENIX R_{dAu} confirms the strong suppression at forward rapidity seen in R_{CP}, it would suggest that the stronger suppression seen at forward/backward rapidity in the PHENIX Au+Au R_{AA} data

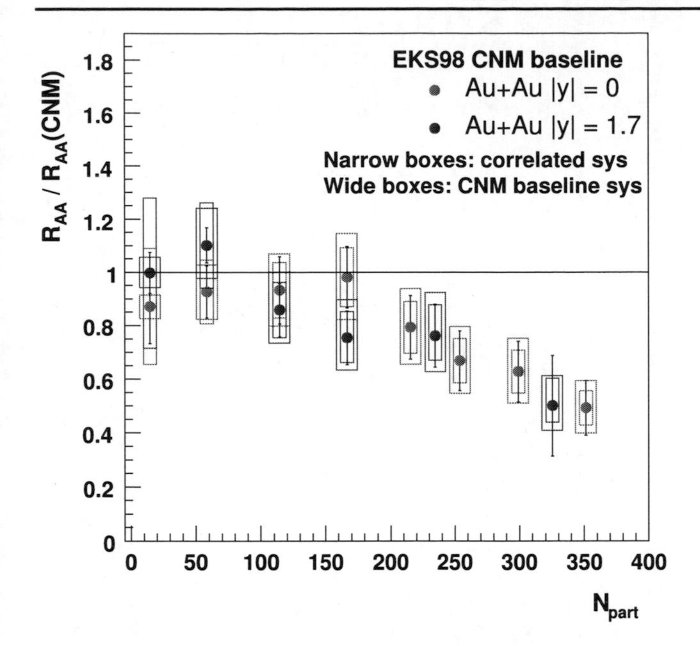

Fig. 88 The estimated Au+Au suppression relative to the cold-nuclear-matter R_{AA} as a function of centrality for $|y| < 0.35$ and $1.2 < |y| < 2.2$. The systematic uncertainty of the baseline cold-nuclear-matter R_{AA} is depicted by the *wide box* around each *point*. The *narrow box* is the systematic uncertainty in the Au+Au R_{AA}

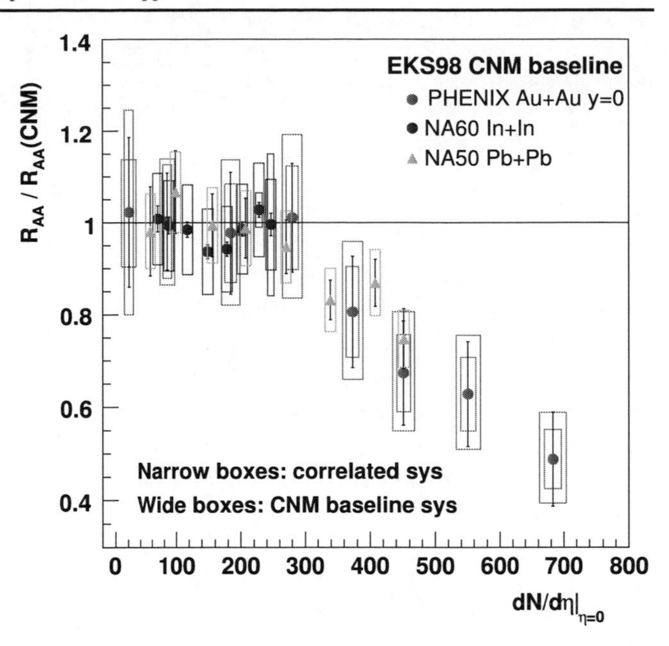

Fig. 89 Comparison of the anomalous suppression at the SPS and RHIC as a function of $dN_{\rm ch}/d\eta$ at $\eta = 0$

is primarily due to cold-nuclear-matter effects. The suppression due to hot-matter effects seems to be comparable at midrapidity and at forward/backward rapidity.

Finally, it is possible to use the effective absorption cross sections obtained from the d+Au J/ψ R_{CP} data in a similar Glauber calculation of $R_{p\rm Cu}$ to estimate the cold-nuclear-matter R_{AA} for Cu+Cu collisions. However, the resulting CNM R_{AA} for Cu+Cu is significantly different for EKS98 and nDSg [884], most likely due to the different A dependences of EKS98 and nDSg. Measurements of J/ψ production in p+Cu or d+Cu collisions would be needed to reduce the model dependence of the estimated CNM R_{AA} for Cu+Cu collisions.

5.6 Anomalous suppression: SPS vs. RHIC

The preliminary PHENIX d+Au results at $\sqrt{s} = 200$ GeV are, for the first time, based on a high-statistics sample [884]. Comparing these results with the previous Au+Au data gives an estimate of the magnitude of the anomalous J/ψ suppression at RHIC. The newly available NA60 pA results at 158 GeV, described in Sect. 5.4, allow significant comparisons of the centrality dependence of the anomalous suppression at the SPS and that obtained at RHIC. Work is in progress to make such a comparison as a function of several variables of interest, such as the charged particle multiplicity, $dN_{\rm ch}/d\eta$, and the Bjorken energy density reached in the collision. The anomalous suppression patterns in In+In and Pb+Pb collisions at the SPS and the midrapidity Au+Au results at RHIC are presented as a function of $dN_{\rm ch}/d\eta$ in Fig. 89 [985]. Note that the magnitude of the anomalous J/ψ suppression is practically system- and $\sqrt{s}$-independent when expressed as a function of $dN_{\rm ch}/d\eta|_{\eta=0}$.

5.7 Photoproduction in nuclear collisions

In addition to in-medium hadroproduction, photoproduction of quarkonium may also occur in nucleus–nucleus collisions. In this case, one nucleus acts as a photon source (the photon flux is given by the Weizsäcker–Williams formalism). The photons fluctuate to virtual quark–antiquark pairs which interact with the opposite (target) nucleus [986, 987] and emerge as heavy quarkonia (e.g., J/ψ and Υ) or other, light, vector mesons. Such J/ψ photoproduction has been observed in Au+Au collisions with PHENIX [974] and in $\overline{p}p$ collisions at the Tevatron [701]. CDF [701] has also observed $\psi(2S)$ photoproduction.

At the LHC, photoproduction can be studied at far higher energies than available at fixed-target facilities or at HERA. At the maximum pp energy of the LHC, γp collisions with center-of-mass energies up to $\sqrt{s_{\gamma p}} = 8.4$ TeV are accessible, forty times the energy reached at HERA. With Pb beams at maximum energy, the per-nucleon center-of-mass energy can reach $\sqrt{s_{\gamma N}} = 950$ GeV [988], equivalent to a 480 TeV photon beam on a fixed target.

Photoproduction is of interest because it is sensitive to the gluon distribution in the target nucleus. The cross section for $\gamma p \to Vp$ scales as [989] $[xg(x, Q^2)]^2$, where x is the gluon momentum fraction, $Q^2 = m_V^2/4$ is the photon virtuality, and m_V is the vector-meson mass. For low p_T vector mesons, the gluon momentum fraction, x, may be related to the final-state rapidity, y, by

$$y = -\frac{1}{2}\ln\left(\frac{2x\gamma m_p}{m_V}\right), \quad (186)$$

where γ is the Lorentz boost of the nuclear beam and m_p is the proton mass. The higher energies available at the LHC allow studies at much lower x than previously available, possibly down to 10^{-6} [988, 990].

Photoproduction cross sections The cross section for vector-meson production may be calculated by integrating over photon momentum k (equivalent to integrating over rapidity y):

$$\sigma(AA \to AAV) = 2\int dk \frac{dN_\gamma}{dk}\sigma(\gamma A \to VA), \quad (187)$$

where dN_γ/dk is the photon flux, determined from the Weizsäcker–Williams method, and $\sigma(\gamma A \to VA)$ is the photoproduction cross section. This cross section may be extrapolated from HERA data. A Glauber calculation is used to determine the cross sections for nuclear targets. Two Glauber calculations of J/ψ and Υ photoproduction are available [704, 712, 713, 991, 992]; a third uses a color-glass condensate/saturation approach to describe the nuclear target [706].

The Glauber calculations successfully predict the rapidity distribution and cross section for ρ^0 photoproduction in Au+Au collisions [993, 994], while the saturation calculation predicts a somewhat higher cross section. Calculations have also provided a reasonable estimate of the cross sections of excited meson production, such as ρ^* states [995–997]. The Tevatron J/ψ and $\psi(2S)$ cross sections are compatible with expectations [701]. The J/ψ photoproduction cross section in Au+Au collisions is sensitive to nuclear shadowing. The uncertainty of the PHENIX [974] measurement is still large, but, as Fig. 90 shows, the central point indicates that shadowing is not large. (At RHIC, midrapidity J/ψ photoproduction corresponds to $x \approx 0.015$.)

Transverse momentum spectra The p_T spectrum of quarkonium photoproduction is the sum of the photon and Pomeron p_T-dependent contributions. Since the photon p_T is small, the spectrum is dominated by the momentum transfer from the target nucleus. In pp collisions, the typical momentum scale is ~300 MeV, set by the size of the nucleon, while for heavy-ion collisions, the momentum scale is $\sim\hbar c/R_A$, where R_A is the nuclear radius.

Photoproduction has a unique feature [998]: either nucleus can emit the photon while the other serves as the target. Because the two possibilities are indistinguishable, their

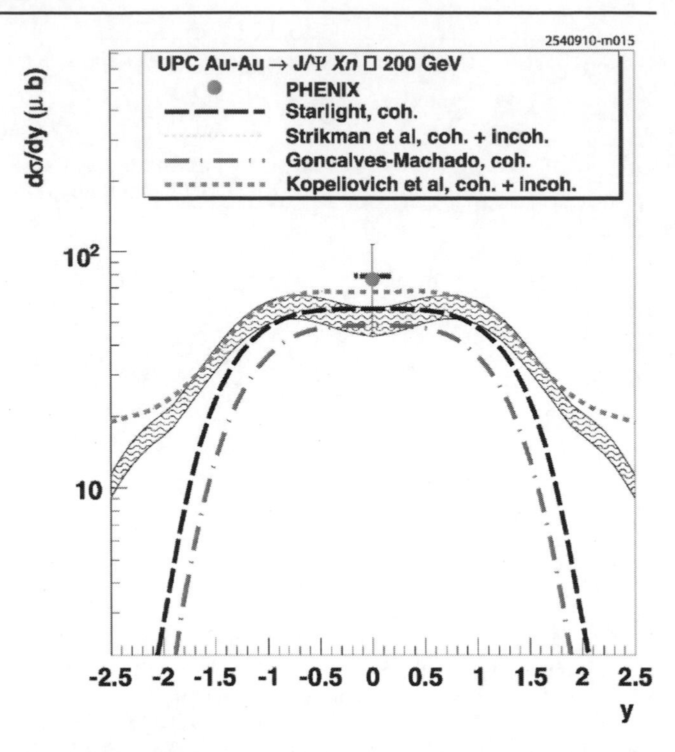

Fig. 90 The rapidity distribution, $d\sigma/dy$, of J/ψ photoproduction measured by PHENIX, compared with three calculations. Coherent and incoherent predictions are summed. Coherent production dominates until $|y| > 2$, where the coherent cross section is kinematically suppressed and incoherent production becomes important. From [974] with kind permission, copyright (2009) Elsevier

amplitudes add. In pp and AA collisions, the possibilities are related by a parity transformation. Since vector mesons have negative parity, the two amplitude subtract, leading to a net amplitude $A \approx A_1 - A_2 \exp(ip_T \cdot b)$ where b is the impact parameter. The two amplitudes, A_1 and A_2, are equal at midrapidity, but may differ for $y \neq 0$ because the photon energies differ, depending on which proton or nucleus emits the photon. The exponential is a propagator from one nucleus to the other. The cross section is suppressed for $p_T < \langle b\rangle$ with a suppression factor proportional to p_T^2. Such suppression has been observed by STAR [999].

The bulk of the cross section from a nuclear target is due to coherent production since the virtual $q\overline{q}$ pair interacts in phase with the entire nucleus. The p_T transfer from the nucleus is small with a p_T scale on the order of a few times $\hbar c/R_A$. The cross section for coherent photoproduction scales as Z^2 (from the photon flux) times A^δ, where $4/3 < \delta < 2$. Here, $\delta = 2$ corresponds to small interaction probabilities, as expected for heavy quarkonia. Larger interaction probabilities lead to smaller values of δ. Studies of ρ^0 photoproduction at RHIC suggest $\delta \approx 5/3$.

At larger p_T, the $q\overline{q}$ pair interactions are out of phase so that the pair effectively interacts with a single nucleon. This contribution thus gives a harder slope in momentum transfer, t, corresponding to the size of a single nucleon, as can

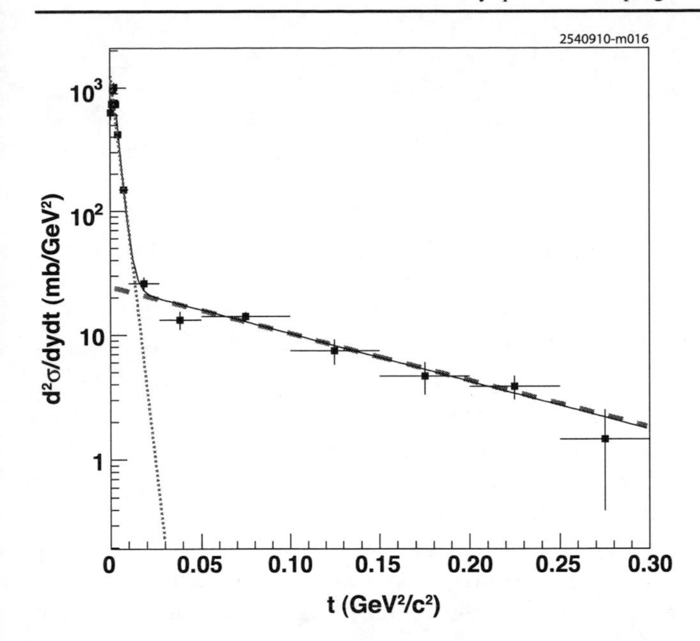

Fig. 91 The $t = p_T^2$ spectrum for ρ^0 photoproduction observed by STAR in 200 GeV Au+Au collisions, averaged over $|y| < 1$. The data are shown by *points with error bars*, and the *solid curve* is a fit of the data to a sum of two components, each exponential in t, representing that is coherent (*dashed curve*) at low t and incoherent (*dotted*) at high t. Adapted from [993] with kind permission, copyright (2008) The American Physical Society

be seen in the STAR data in Fig. 91. At these higher p_T, the struck nucleon may be ejected from the nucleus, resulting in nuclear dissociation, making it possible to probe the dynamics of hard Pomerons [996, 997]. Expected at higher p_T are additional components which probe the nucleon substructure. In this regime, the cross sections become sensitive to the quark distributions [1000]. Because of the higher momentum transfer from the target, incoherent interactions contribute at larger rapidities than coherent interactions, explaining the 'wings' at large $|y|$ in Fig. 90.

Plans for the LHC ALICE, CMS, and ATLAS are all planning to study quarkonium photoproduction [988]. These events have a very clean topology: two nearly back-to-back electrons or muons, with almost nothing else in the detector. At RHIC, STAR and PHENIX found that such an analysis is relatively straightforward. The most difficult part of the study is devising a trigger to select these events. However, the LHC experiments will benefit from vastly more sophisticated triggers than are available at RHIC. Indeed CMS, ATLAS, and the ALICE forward muon spectrometer have triggers primitives that can be employed for this purpose. CMS, in particular, may be able to separately measure $\Upsilon(1S)$, $\Upsilon(2S)$ and $\Upsilon(3S)$ photoproduction.

The LHC energy and luminosity are sufficient for copious J/ψ and significant $\psi(2S)$ and Υ signals. The greatest physics interest may be in probing the gluon distributions, and, in particular, measuring nuclear shadowing. At midrapidity, quarkonium production probes x values between 2×10^{-4} (J/ψ in pp collisions) and 1.7×10^{-3} (Υ in Pb+Pb collisions). Away from midrapidity it is possible to probe x values as low as 10^{-6} [988].

In the case of AA collisions, since the photon can be emitted from either nucleus, ambiguities arise because the photon energies, and hence the x values probed, are different for the two possibilities. We mention two possible ways to resolve this two-fold ambiguity. Conceptually, the easiest is to study pA collisions (or dA at RHIC). Here, the ion is usually the photon emitter. In addition, it is possible to employ the difference in the p_T spectra for photons scattering on protons and ions to separate the two possibilities. Unfortunately, pA runs at the LHC are some years off. A second possibility is to use bootstrapping, usually by comparing results at different beam energies. At each beam energy, the midrapidity cross section can be unambiguously determined, giving the cross section at a specific photon energy. At a different beam energy, the same photon energy corresponds to a different rapidity. By measuring the cross section at this rapidity and subtracting the known cross section determined previously, one obtains the cross section at the new photon energy. Unfortunately, the uncertainties add each time the cross sections are subtracted, increasing the relative error. A similar procedure may also employed by using data taken under different running conditions, such as exclusive J/ψ production relative to J/ψ production accompanied by mutual Coulomb excitation. At a given rapidity, these two processes contribute differently to the cross section, depending on the direction.

The LHC measurements allow for a relatively clean measurement of nuclear shadowing by taking the ratio of the AA and pp cross sections. In this approach, many of the theoretical and experimental uncertainties cancel in the ratio, resulting in a relatively clean determination. Additional pA data would help this study by allowing cross checks between pA and pp interactions as well as between AA and pA interactions. Another possibility, for pp collisions, is to use Roman pots or other small-angle detectors to tag the outgoing protons [1001]. The proton that emitted the photon will usually have lower p_T. Of course, some of these techniques are also applicable at RHIC, where the experiments are collecting large data sets with improved triggers and particle identification.

Such measurements of the nuclear gluon distributions will be important for understanding the properties of cold nuclear matter, which, in turn, clarifies the interpretation of the quarkonium signals in central heavy-ion collisions.

6 Experimental outlook[21]

Moving beyond the present status of heavy-quarkonium physics described in the previous sections poses major challenges to the next generation of accelerators and experiments. In this section the future "players" in the field will be described with special emphasis on the potential to resolve the important open questions. Here we will present the rationale for and status of the newer facilities and experiments, from those already running (BESIII at BEPCII; ALICE, ATLAS, CMS, and LHCb at the LHC) to those under construction or only planned ($\overline{\text{P}}$ANDA and CBM at FAIR, SuperB and tau–charm factories, lepton–hadron colliders, and high-energy linear e^+e^- colliders).

6.1 BESIII

For BESIII, the future is now. The Beijing Electron–Positron Collider (BEPC) and the Beijing Spectrometer (BES) operated in the tau–charm center-of-mass energy region between 2 and 5 GeV from 1990 until 2003. Now, BEPC has been upgraded to a two-ring collider (BEPCII), and a brand new detector (BESIII) has been constructed. Commissioning of the upgraded accelerator and new detector began in spring 2008, and the first event was obtained on July 20, 2008. Approximately 13×10^6 $\psi(2S)$ events were accumulated in fall 2008, which provided data for studies of the new detector and for calibration. In spring 2009 after running for about one month, 106×10^6 $\psi(2S)$ events were obtained, and in summer 2009 after running for six weeks, about 226×10^6 J/ψ events were accumulated. These are the world's largest such data sets and are approximately four times larger than the CLEO-c $\psi(2S)$ sample and the BESII J/ψ sample, respectively. The new data will allow more detailed studies of detector performance, and offers many physics opportunities.

The peak design luminosity of BEPCII is 10^{33} cm^{-2} s^{-1} (1 nb^{-1}s^{-1}) at a beam energy of 1.89 GeV, an improvement of a factor of 100 with respect to the BEPC. It will operate at a center-of-mass energy between 2 and 4.6 GeV, which allows production of almost all known charmonium and charmonium-like states. The detector performance is also greatly improved compared to BESII. BESIII [1002] is a new, general-purpose detector. It features a beryllium beam pipe; a small-cell, helium-based drift chamber (MDC); a Time-of-Flight (TOF) system; a CsI(Tl) electromagnetic calorimeter; a 1T superconducting solenoidal magnet; and a muon identifier using the magnet yoke interleaved with Resistive Plate Chambers.

Running at design luminosity, BESIII will be able to accumulate 10×10^9 J/ψ events or 3×10^9 $\psi(2S)$ events in one year's running. It will take around 20 fb^{-1} of data each at 3.77 GeV and 4.17 GeV for charm physics. There is also the possibility of a high-statistics fine scan between 2 and 4.6 GeV, allowing the direct study of states with $J^{PC} = 1^{--}$. States with even charge parity may be studied using radiative decays of high mass excited ψ states, such as $\psi(2S)$, $\psi(3770)$, $\psi(4040)$, $\psi(4160)$, and $\psi(4415)$. All these data samples allow detailed studies of charmonium physics, including the spectroscopy of conventional charmonium (see Sects. 2.1–2.2) and charmonium-like (see Sect. 2.3) states, charmonium transitions (see Sects. 3.1 and 3.3), and charmonium decays (see Sects. 3.2 and 3.4). Charmonium hadronic decay dynamics are especially interesting because of the $\rho\pi$ puzzle (see Sect. 3.4.1). The new datasets should also enable a better understanding of the physics of the strong interaction in the transition region between perturbative and nonperturbative QCD.

6.1.1 *Spin singlets: $h_c(1P)$, $\eta_c(1S,2S)$*

Below open charm threshold there are three spin-singlet states, the S-wave spin singlet, $\eta_c(1S)$, its radially excited state, $\eta_c(2S)$, and the P-wave spin singlet, $h_c(1P)$. All these may be reached from $\psi(2S)$ transitions. The $\eta_c(1S)$ can also be studied in J/ψ radiative decays. Their properties are less well measured because of their low production rates in previous e^+e^- experiments.

BESIII will measure the $h_c(1P)$ mass, width, spin-parity, production rate via $\psi(2S) \to \pi^0 h_c(1P)$, and its $E1$ transition rate $h_c(1P) \to \gamma\eta_c(1S)$ (see Sects. 2.2.1 and 3.1.3). BESIII will also search for its hadronic decays (see Sect. 3.4.2), which are expected to be about 50% of the total decay width, and search for other transitions.

Extraction of the $\eta_c(1S)$ mass and width from radiative J/ψ or $\psi(2S)$ radiative transitions is not straightforward due to the unexpected lineshape observed in such transitions (see Sect. 3.1.2). With theoretical guidance and more data, these transitions may become competitive with other $\eta_c(1S)$ production sources in determination of its mass and width. In addition to increased statistics, more decay modes will be found and their branching fractions measured.

Despite the passage of eight years since the observation of the $\eta_c(2S)$ (see Sect. 2.2.2), the discovery mode $\eta_c(2S) \to K\bar{K}\pi$ remained the only mode observed until the summer of 2010, at which time Belle [70] reported preliminary observation of several hadronic $\eta_c(2S)$ decay modes in two-photon production of $\eta_c(2S)$. BESIII will search for $\eta_c(2S)$ in $\psi(2S)$ radiative decays. With much less data, CLEO-c sought 11 exclusive hadronic decay modes in radiative transitions but saw none (see Sect. 3.1.4), even in

[21] Contributing authors: S. Eidelman†, P. Robbe†, A. Andronic, D. Bettoni, J. Brodzicka, G.E. Bruno, A. Caldwell, J. Catmore, E. Chudakov, P. Crochet, P. Faccioli, A.D. Frawley, C. Hanhart, F.A. Harris, D.M. Kaplan, H. Kowalski, E. Levichev, V. Lombardo, C. Lourenço, M. Negrini, K. Peters, W. Qian, E. Scomparin, P. Senger, F. Simon, S. Stracka, Y. Sumino, C. Weiss, H.K. Wöhri, and C.-Z. Yuan.

the discovery mode and in the three new modes found by Belle. With more data, BESIII will have a better chance with exclusive decay modes. However, it will be a challenge to isolate the low-energy ($\simeq$50 MeV) radiative photon due to the many background photon candidates, both genuine and fake. Observation of a signal in the inclusive photon spectrum is even more challenging, but is the only way to get the absolute $\psi(2S) \to \gamma\eta_c(2S)$ transition rate. This task will require a good understanding of both backgrounds and the electromagnetic calorimeter performance. If the $\eta_c(2S)$ is found in radiative $\psi(2S)$ decays, the photon energy lineshape can then be studied and compared to that of the $\eta_c(1S)$ (see Sect. 3.1.2).

6.1.2 Vectors above $\psi(3770)$: ψ's and Y's

There are many structures between 3.9 GeV and 4.7 GeV, including the excited ψ (see Sect. 2.1.1) and the Y (see Sect. 2.3.2) states [33, 82, 1003]. By doing a fine scan in this energy range, BESIII may study the inclusive cross section, as well as the cross sections of many exclusive modes, such as $D\bar{D}$, $D^*\bar{D}+c.c.$, $D^*\bar{D}^*+c.c.$, $D\bar{D}\pi$, etc. (see Sect. 2.1.1). This will help in understanding the structures, for instance, whether they are really resonances, due to coupled-channel effects, final-state interactions, or even threshold effects. BESIII will also measure the hadronic and radiative transitions of these excited ψ and the Y states. Other XYZ particles can also be sought in these transitions.

6.1.3 Hadronic decays

As discussed in Sect. 3.4.1, the 12% rule is expected to hold for exclusive and inclusive decays, but is violated by many such modes, including the namesake mode of the $\rho\pi$ puzzle. A plethora of experimental results exists (see Sect. 3.4.1 and references therein). With much larger datasets, a variety of theoretical explanations can be tested by BESIII at higher accuracy [571]. Moreover, studies should be made not only of ratios of $\psi(2S)$ to J/ψ decays, but also of other ratios such as those between $\eta_c(2S)$ and $\eta_c(1S)$ [1004], between $\psi(3770)$ and J/ψ [565], and other ratios between different resonances for the same channel or between different channels from the same resonance [1005] (e.g., $\gamma\eta$ and $\gamma\eta'$, as discussed in Sect. 3.2.4). All such studies are important to our understanding of charmonium decays.

BESIII also has an opportunity to measure the direct photon spectrum in both J/ψ and $\psi(2S)$ decays and values of R_γ (see (22)) for both resonances (see discussion in Sects. 2.8.1 and 3.2.2 and in [1002]), building on the work of CLEO for J/ψ [235] and $\psi(2S)$ [236].

6.1.4 Excited C-even charmonium states

Above open charm threshold, there are still many C-even charmonium states not yet observed, especially the excited P-wave spin-triplet and the S-wave spin singlet [31, 1006]. In principle, these states can be produced in the E1 or M1 transitions from excited ψ states. As BESIII will accumulate much data at 4.17 GeV for the study of charm physics, the sample can be used for such a search.

6.1.5 Decays of $\chi_{cJ}(1P)$

Approximately 30% of $\psi(2S)$ events decay radiatively to χ_{cJ}, which decay hadronically via two or more gluons (see Sect. 3.4.3). These events and radiative J/ψ decays are thought to be important processes for the production of glueball, hybrid, and other non-$q\bar{q}$ states. BESIII will study these processes and also search for charmonium rare decays. The decay $\chi_{c1} \to \eta\pi\pi$ is a golden channel for the study of states with exotic quantum numbers $I^G(J^{PC}) = 1^-(1^{-+})$, that is, the π_1 states [18], since these states, can be produced in χ_{c1} S-wave decays. A detailed partial wave analysis with a large χ_{c1} sample will shed light on these exotic states.

6.1.6 Prospects

The present and future large BESIII data sets and excellent new detector will allow extensive studies of charmonium states and their decays.

6.2 ALICE

ALICE [1007] is the experiment dedicated to the study of nucleus–nucleus collisions at the Large Hadron Collider (LHC). The study of heavy-quarkonium production in nuclear collisions is one of the most important sources of information on the characteristics of the hadronic/partonic medium. (For a discussion of quarkonium physics in this medium, see Sect. 5.) ALICE will study Pb+Pb collisions at top LHC Pb energy ($\sqrt{s_{NN}} = 5.5$ TeV), at a nominal luminosity $L = 5 \times 10^{26}$ cm^{-2} s^{-1}.

In the ALICE physics program [1008, 1009], the study of pp collisions is also essential, in order to provide reference data for the interpretation of nuclear collision results. In addition, many aspects of genuine pp physics can be addressed. The pp luminosity in ALICE will be restricted so as to not exceed $L = 3 \times 10^{30}$ cm^{-2} s^{-1}. Despite this luminosity limitation, most physics topics related to charmonium and bottomonium production remain accessible.

Heavy quarkonia will be measured in the central barrel, covering the pseudorapidity range $-0.9 < \eta < 0.9$, and in the forward muon arm, which has a coverage $2.5 < \eta < 4$. In the central barrel, heavy quarkonia will be detected through the e^+e^- decay. ALICE can push its transverse momentum (p_T) reach for charmonium down to $p_T \sim 0$. Electron identification is performed jointly in the TPC through the

dE/dx measurement and in the Transition Radiation Detector (TRD). In the forward region, quarkonia will be studied via their decay into muon pairs. Muons with momenta larger than 4 GeV/c are detected by means of a spectrometer which includes a 3 Tm dipole magnet, a front absorber, a muon filter, tracking (Cathode Pad Chambers, CPC) and triggering (Resistive Plate Chambers, RPC) devices.

In the following sections we will review the ALICE physics capabilities for heavy-quarkonium measurements at the top LHC energy within the running conditions specified above. A short overview of the measurements that could be performed in the first high-energy run of the LHC will also be presented. For the ALICE physics run in 2010, the forward muon spectrometer and most of the central barrel detectors have been installed and commissioned, including seven TRD supermodules (out of 18).

In the central barrel, the geometrical acceptance for J/ψ produced at rapidity $|y| < 0.9$ (with no p_T cut on either the J/ψ or the decay electrons) is 29% for the complete TRD setup. The electron reconstruction and identification efficiency in the TRD is between 80 and 90% for $p_T > 0.5$ GeV/c, while the probability of misidentifying a pion as an electron is $\sim$1%. Below a few GeV/c, particle identification in the TPC [1010] contributes substantially to hadron rejection, with an overall TPC+TRD electron reconstruction efficiency of $\sim$75%.

The acceptance of the forward spectrometer, relative to the rapidity range $2.5 < y < 4$, is $\sim$35% for the J/ψ. Since most of the background is due to low transverse momentum muons, a p_T cut is applied to each muon at the trigger level. With a 1 GeV/c p_T cut, there is a $\sim$20% reduction of the J/ψ acceptance. The combined efficiency for J/ψ detection in the forward spectrometer acceptance, taking into account the efficiency of tracking and triggering detectors, is expected to be about 70%.

6.2.1 *J/ψ production from Pb+Pb collisions*

Heavy quarkonium states probe the medium created in heavy-ion collisions. Color screening in a deconfined state is expected to suppress the charmonium and bottomonium yields. In addition, at the LHC, a large multiplicity of heavy quarks (in particular, charm) may lead to significant regeneration of bound states in the dense medium during the hadronization phase. ALICE will investigate these topics through a study of the yields and differential distributions of various quarkonium states, performed as a function of the centrality of the collision.

A simulation has been performed [1011] for J/ψ production in the forward muon arm, using as an input a Color-Evaporation Model (CEM) calculation, based on the MRST HO set of Parton Distribution Functions (PDF), with $m_c = 1.2$ GeV and $\mu = 2m_c$ [1012]. (For a discussion of the CEM, see Sect. 4.1.3.) With such a choice of parameters, the total pp J/ψ cross section at $\sqrt{s_{NN}} = 5.5$ TeV, including the feeddown from higher resonances, amounts to 31 μb. The pp cross section has been scaled to Pb+Pb assuming binary collision scaling and taking into account nuclear shadowing through the EKS98 [877] parametrization. The differential p_T and y shapes have been obtained via an extrapolation of the CDF measurements and via CEM predictions, respectively, and assuming that the J/ψ are produced unpolarized. The hadronic background was simulated using a parametrized HIJING generator tuned to $dN_{ch}/dy = 8000$ for central events at midrapidity (such a high value, 3–4 times that realistically expected, represents a rather extreme evaluation of this source). Open heavy-quark production was simulated using PYTHIA, tuned to reproduce the single-particle results of NLO pQCD calculations.

At nominal luminosity, the expected J/ψ Pb+Pb statistics for a 10^6 s run, corresponding to the yearly running time with the Pb beam, are of the order of 7×10^5 events. The mass resolution will be $\sim$70 MeV [1008, 1009]. A simulation of the various background sources to the muon-pair invariant mass spectrum in the J/ψ region (including combinatorial π and K decays, as well as semileptonic decays of open heavy flavors) shows that the signal-to-background ratio, S/B, ranges from 0.13 to $\approx$7 when moving from central to peripheral collisions. With such statistics and S/B values it will be possible to study the proposed theoretical scenarios for the modification of the J/ψ yield in the hot medium.

The transverse momentum distributions can be addressed with reasonable statistics even for the relatively less populated peripheral Pb+Pb collisions. In particular, for collisions with an impact parameter $b > 12$ fm, we expect having more than 1000 events with $p_T > 8$ GeV/c.

Finally, a study of the J/ψ polarization will be performed by measuring the angular distribution of the decay products. With the expected statistics, the polarization parameter λ extracted from the fit $d\sigma/d\cos\theta = \sigma_0(1 + \lambda\cos^2\theta)$ can be measured, defining five impact-parameter bins, with a statistical error <0.05 for each bin.

Another simulation of J/ψ production at central rapidity has been carried out using as input the rates obtained from the CEM calculation described above. For high-mass electron pairs, the main background sources are misidentified pions and electrons from semileptonic B and D decays. The value $dN_{ch}/dy = 3000$ for central events at $y = 0$ was used for the simulation of the hadronic background. PYTHIA was used for open heavy-quark production, with the same tuning used for the forward-rapidity simulations.

For Pb+Pb, the expected J/ψ statistics, measured for a 10^6 s running time at the nominal luminosity, are about 2×10^5 candidates from the 10^8 collisions passing the 10% most central impact-parameter criteria. The mass resolution will be $\sim$30 MeV [1013]. The background under the J/ψ

peak, dominated by misidentified pions, is at a rather comfortable level ($S/B = 1.2$). As for the forward region, it will therefore be possible to test the proposed theoretical models.

The S/B ratio is expected to increase as a function of p_T, reaching a value of ~5 at 10 GeV/c. The expected statistics at that p_T are still a few hundred counts, implying that differential J/ψ spectra can also be studied.

6.2.2 *J/ψ production from pp collisions*

Quarkonium hadroproduction is an issue which is not yet quantitatively understood theoretically. A study of J/ψ production in pp collisions at ALICE aims at a comprehensive measurement of interesting observables (production cross sections, p_T spectra, polarization) useful to test theory in a still unexplored energy regime. Furthermore, the forward-rapidity measurement offers a possibility to access the gluon PDFs at very low x ($<10^{-5}$).

In the forward muon arm, J/ψ production at $\sqrt{s} =$ 14 TeV has been simulated using the CEM, with parameters identical to those listed above for Pb+Pb collisions. The J/ψ total cross section turns out to be 53.9 μb, including the feeddown from higher-mass resonances. A typical data-taking period of one year (assuming 10^7 s running time) at $L = 3 \times 10^{30}$ cm^{-2} s^{-1} gives an integrated luminosity of 30 pb^{-1}. The corresponding dimuon invariant-mass spectrum, for opposite-sign pairs, is shown in Fig. 92.

The expected J/ψ statistics are $\sim 2.8 \times 10^6$ events [1014]. The background under the J/ψ peak is dominated by correlated decays of heavy flavors but is anyway expected to be quite small ($S/B = 12$). It will be possible to study the transverse momentum distribution of the J/ψ with negligible statistical errors up to at least $p_T = 20$ GeV/c. By studying the shape of the J/ψ rapidity distribution in the region $2.5 < y < 4$ it will be possible to put strong constraints on the gluon PDFs and, in particular, to discriminate between the currently available extrapolations in the region around $x = 10^{-5}$. With the expected statistics it will also be possible to carry out a detailed analysis of the p_T dependence of the J/ψ polarization.

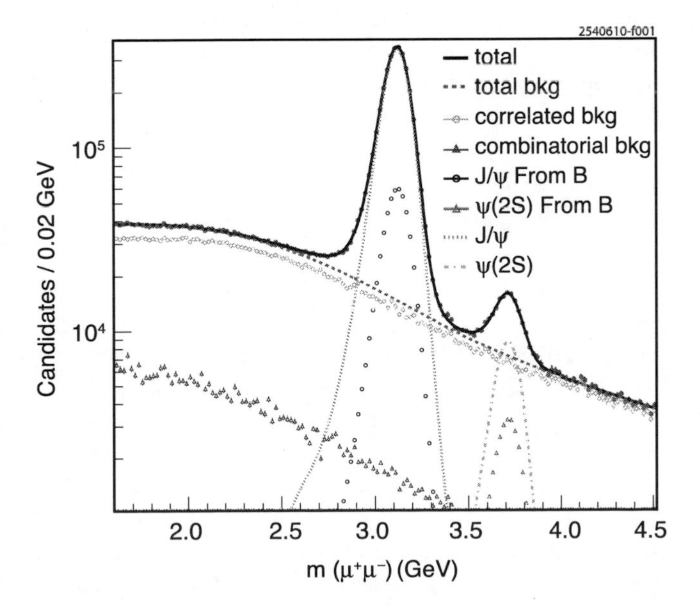

Fig. 92 Opposite-sign dimuon mass spectrum in pp collisions at $\sqrt{s} = 14$ TeV for a 10^7 s running time at $L = 3 \times 10^{30}$ cm^{-2} s^{-1}

In a pp run, a sample of a few thousand J/ψ events is expected to be acquired in minimum bias collisions. With such statistics it will be possible to measure $d\sigma/dy$ at midrapidity. In order to improve these statistics, the implementation of a Level-1 trigger for electrons is foreseen. Assuming a conservative trigger efficiency of 10%, one would get about 7×10^5 J/ψ. Such a yield would open up the possibility of measuring differential spectra up to high p_T and polarization.

6.2.3 *J/ψ production from b-hadron decays*

When measuring J/ψ production at the LHC, a significant fraction of the measured yield comes from b-hadron decays. This J/ψ source is a very interesting physics signal for the evaluation of the open-bottom production cross section, nicely complementing measurements performed via single leptons. It is also an important component to be disentangled when one wants to identify prompt J/ψ production, as it is the case for studies of yield modifications in nuclear collisions.

At midrapidity, thanks to the excellent vertexing capabilities of ALICE, the secondary b-decay vertex can be separated. A good measure of the separation from the main vertex is L_{xy}, the signed projection of the J/ψ flight distance, $\vec{L}$, on its transverse momentum, $\vec{p}_T$, defined as

$$L_{xy} \equiv \frac{\vec{L} \times \vec{p}_T}{p_T}. \tag{188}$$

To reduce the dependence on the J/ψ transverse momentum distribution, the variable x is used instead of L_{xy},

$$x \equiv L_{xy} \times \frac{m(J/\psi)}{p_T}, \tag{189}$$

where $m(J/\psi)$ is the known J/ψ mass. Studies based on Monte Carlo simulation have shown that the fractions of secondary J/ψ as a function of p_T can be extracted by a likelihood fit to the dielectron invariant mass and the x variable defined above with uncertainties smaller than 10%. This approach will also provide a measurement of the open-bottom p_T-differential cross section down to $p_T \approx 0$.

The situation is more difficult at forward rapidity. Due to the presence of a thick hadron absorber in the path of the muons, the accuracy on the position of the J/ψ production vertex is not sufficient. Work is in progress in order to evaluate the secondary J/ψ yield starting from the study of events with three muons detected in the muon spectrometer in pp collisions. Finally, the option of introducing a Si vertex tracker covering the $2.5 < \eta < 4$ rapidity domain is currently under study. It should be noted that open-bottom production at forward rapidity will be estimated from the study of the single-μ p_T distributions and from the contribution to the dimuon continuum of correlated semileptonic decays of b-hadrons. These measurements will allow us to estimate the fraction of the J/ψ yield coming from b decays.

6.2.4 *Production of $\chi_{cJ}(1P)$ and $\psi(2S)$*

It is well known that a significant fraction (up to $\approx$40%) of the measured J/ψ yield comes from χ_{cJ} and $\psi(2S)$ decays. An accurate measurement of the yield of these resonances is therefore an important ingredient in the interpretation of the J/ψ production data. At the same time, these higher-mass resonances suffer from a much smaller feeddown contribution than the J/ψ and may represent cleaner signals for theoretical calculations.

The dilepton yield from $\psi(2S)$ is much smaller than that of J/ψ. At the nominal LHC energy and luminosity described above, one expects about 7.5×10^4 events in the forward muon arm for a standard pp run, with $S/B \approx 0.6$ [1014]. In Pb+Pb collisions, the situation is not so favorable, due to the much larger combinatorial background. The expected statistics are about 1.5×10^4 events, but with a S/B ratio ranging from 18% to only 1% from peripheral to central collisions. The background levels at midrapidity are prohibitive for Pb+Pb collisions; a measurement in pp collisions also appears to be problematic.

Concerning χ_c, a feasibility study has been performed on the detection of the radiative decay $\chi_c \to J/\psi\gamma$ at midrapidity in pp collisions [1015]. The J/ψ has been reconstructed via its e^+e^- decay, while the photon conversion has been reconstructed from opposite-sign tracks with opening angle <0.1rad and mass <0.175 GeV. The χ_{c1} and χ_{c2} states can be separated in the $\Delta m = m(e^+e^-\gamma) - m(e^+e^-)$ spectrum. The mean reconstruction efficiency is 0.9%. As for J/ψ production at midrapidity, triggering is crucial also for this signal. With a 10% trigger efficiency, several thousand events could be collected in a pp run.

6.2.5 *Υ production*

In nucleus–nucleus collisions, the yield of $\Upsilon(1S, 2S, 3S)$ states should exhibit various degree of suppression due to the screening of the color force in a Quark–gluon Plasma. Results from pp collisions will be essential as a normalization for Pb+Pb results and extremely interesting in order to understand the related QCD topics (see Sect. 5).

In the forward-rapidity region, where the muon-pair invariant-mass resolution is $\sim$100 MeV, the Υ states can be clearly separated. The expected yields are of the order of 7×10^3 events for the $\Upsilon(1S)$ in Pb+Pb collisions, and factors $\approx$4 and $\approx$6.5 smaller for the higher-mass resonances $\Upsilon(2S)$ and $\Upsilon(3S)$, respectively [1011]. The S/B ratios will be more favorable than for the J/ψ ($\approx$1.7 for the $\Upsilon(1S)$ in central collisions). In pp collisions, about 2.7×10^4 $\Upsilon(1S)$ events are expected for one run [1014]. These statistics will allow, in addition to the integrated cross section measurement, a study of p_T distributions and polarization.

At midrapidity, a possibility of measuring the Υ states is closely related to the implementation of a Level-1 trigger on electrons [1013]. Assuming a conservative 10% trigger efficiency, about 7000 Υ events could be collected in a pp run. In a Pb+Pb run, a significant $\Upsilon(1S)$ sample (several thousand events) can be collected with a comfortable $S/B \approx 1$. The statistics for the higher-mass resonances depend crucially on the production mechanism. Assuming binary scaling, as for $\Upsilon(1S)$, a measurement of $\Upsilon(2S)$ looks very promising ($\approx$1000 events with $S/B = 0.35$).

6.2.6 *First LHC high-energy running*

In 2010 the LHC has begun to deliver proton beams at $\sqrt{s} = 7$ TeV. Under the present running conditions, during 2010 it is expected that a few 10^4 $J/\psi \to \mu^+\mu^-$ will be collected in the forward spectrometer using a single-muon trigger. With these statistics a measurement of the p_T distribution and a p_T-integrated polarization estimate could be within reach. Several hundred $\psi(2S)$ and $\Upsilon(1S)$ events could be collected, enough for an estimate of the p_T-integrated cross sections.

Assuming a sample of 10^9 minimum bias events, the expected J/ψ statistics in the central barrel are a few hundred events (due to the reduced coverage provided by the presently installed TRD supermodules). Employing the TRD trigger would enhance this sample significantly and would enable measurements of other charmonium states, as well as of the Υ.

6.3 ATLAS

ATLAS [1016, 1017] is a general-purpose 4π detector at the LHC. Although primarily designed for the discovery of physics beyond the Standard Model through the direct observation of new particles, indirect constraints through precise measurements of known phenomena are also an important avenue of activity. The quarkonium program of ATLAS falls into this category. Of particular importance to these studies are the tracking detector and muon spectrometer. The silicon pixels and strips close to the interaction point allow primary- and secondary-vertex reconstruction

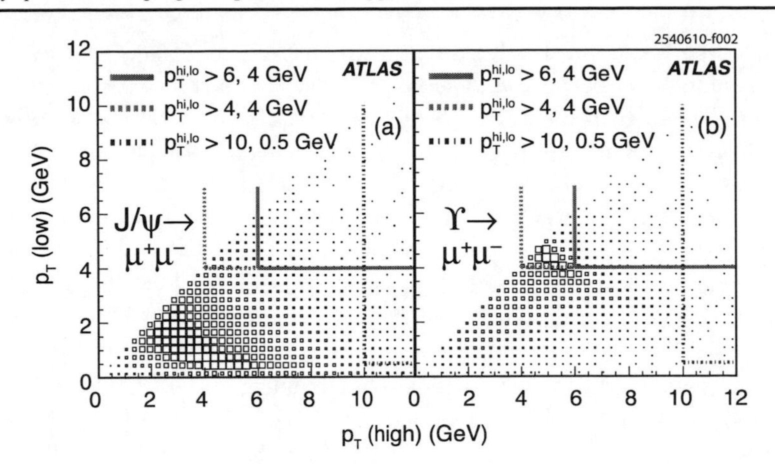

Fig. 93 Density of production cross section for J/ψ and Υ

with good resolution. The vast muon spectrometer in the outer parts of the machine provides a flexible muon trigger scheme that can fire on pairs of low-momentum (4 GeV/c) muons as well as efficient muon identification and reconstruction. Together these factors have allowed ATLAS to assemble a strong quarkonium physics program.

We review here the ATLAS capability relevant for prompt quarkonium production at the LHC, in particular the methods of separating promptly produced $J/\psi \to \mu^+\mu^-$ and $\Upsilon \to \mu^+\mu^-$ decays from the various backgrounds. The outlook for the first measurements at 7 TeV is also discussed. All of the results shown here are taken from the Computing Services Commissioning [1018, pp. 1083–1110] exercises on Monte Carlo carried out in 2008.

6.3.1 Trigger considerations

A detailed account of the ATLAS trigger can be found in [1019], and the full details of the trigger scheme to be used in the ATLAS bottom and quarkonia program are available in [1018, pp. 1044–1082]. The quarkonium program relies on two trigger methods in particular. The first requires the lowest-level trigger to fire on two over-threshold muons independently, forming two conical "Regions of Interest" (RoIs) around the muon candidate. Full track reconstruction on hits within the RoIs is then performed by higher-level trigger algorithms to confirm and refine the low-level signature. The second method requires only one muon at the lowest level; the RoI in this case encompasses a larger volume and the second muon is sought only in the higher-level algorithms. These methods allow thresholds in p_T down to 4 GeV/c. An alternative approach requires only one muon, with a higher p_T threshold of 10 GeV/c; in this case the other muon is sought offline.

Any determination of the quarkonia cross sections requires a detailed understanding of the trigger efficiencies. With around 10 pb^{-1} of data it will be possible to measure the efficiency maps directly from the data, using the narrow J/ψ resonance in the so-called "Tag and Probe" method [1018, pp. 1069–1081]. With fewer data such maps will have to be made from Monte Carlo.

Table 46 Predicted cross sections for various prompt quarkonia production and decay into dimuons for three trigger scenarios

State	Cross section, nb			
	$\mu4\mu4$	$\mu6\mu4$	$\mu10$	$\mu6\mu4 \cap \mu10$
J/ψ	28	23	23	5
$\psi(2S)$	1.0	0.8	0.8	0.2
$\Upsilon(1S)$	48	5.2	2.8	0.8
$\Upsilon(2S)$	16	1.7	0.9	0.3
$\Upsilon(3S)$	9.0	1.0	0.6	0.2

Figure 93 shows the density of the production cross section for $pp \to J/\psi \to \mu^+\mu^-$ and $pp \to \Upsilon \to \mu^+\mu^-$ as a function of the p_T of the two muons, with cut lines representing dimuon triggers of (4, 4) and (6, 4) GeV/c and the single-muon trigger threshold of 10 GeV/c. It can be seen immediately that the situation for the two states is very different. In the case of J/ψ, most of the decays produce muons with p_T well below the (4, 4) GeV/c threshold, which is as low as the ATLAS muon triggers can go. Furthermore, it is clear that increasing the thresholds to (6, 4) GeV/c does not lose many additional events. On the other hand, the Υ, which is three times as massive as the J/ψ, decays into muons with significantly higher p_T. In this case the difference between thresholds of (4, 4) and (6, 4) GeV/c is critical, with the lower cut capturing many more events and resulting in an order-of-magnitude increase in the accessible cross section. See Table 46 for expected cross sections from a variety of quarkonium states for different trigger configurations. Although excited Υ states are included in the table, it is unlikely that ATLAS will have good enough mass resolution to be able to separate them. It should also be noted that the muon trigger configuration used early-on will have a nonzero efficiency below the (4, 4) GeV/c threshold, which will allow ATLAS to collect more events than suggested by Table 46, which assumes hard cuts. Finally, the opening angle between the muons in Υ decay is typically much larger than for J/ψ, which presents a difficulty for RoI-guided triggers because the RoI is generally too narrow. Physics studies of Υ will therefore benefit from the "full scan" dimuon triggers, which allow the whole tracking volume to be accessed by the higher-level trigger algorithms rather than just hits in the RoI. Full-scan triggers are CPU-intensive and will only be available at low luminosity.

The angle $\cos\theta^*$, used in quarkonium spin-alignment analyses, is defined (by convention) as the angle in the quarkonium rest frame between the positive muon from the quarkonium decay and the flight direction of the quarkonium itself in the laboratory frame. The distribution of this angle may depend on the relative contributions of the different quarkonium production mechanisms that are not fully understood. Different angular distributions can have different trigger acceptances: until the spin alignment is properly understood, a proper determination of the trigger acceptance will not be possible. For quarkonium decays in which the two muons have roughly equal p_T, $\cos\theta^* \approx 0$; such decays will have a high chance of being accepted by the trigger. Conversely, quarkonia decays with $|\cos\theta^*| \approx 1$ will have muons with very different p_T, and as the lower p_T muon is likely to fall below the trigger threshold, such events have a greater chance of being rejected. Figure 94 shows the $\cos\theta^*$ distributions for J/ψ and Υ after trigger cuts of $p_T > (6, 4)$ GeV (*solid line*) and a single-muon trigger cut of $p_T > 10$ GeV (*dashed line*). The samples were generated with zero spin alignment, so without trigger selection the $\cos\theta^*$ distribution would be flat across the range -1 to $+1$. The figures show, first, that a narrow acceptance in $\cos\theta^*$ would impair the spin-alignment measurements, and second, that the single-muon trigger has much better acceptance at the extreme ends of the $\cos\theta^*$ distribution, since it has a much better chance of picking up events with one low-p_T and one high-p_T muon. At low luminosity such a trigger will have an acceptable rate, and, used in conjunction with the dimuon triggers, will provide excellent coverage across the whole $\cos\theta^*$ range.

6.3.2 Event selection

Events passing the triggers are processed offline. Oppositely charged pairs of tracks identified as muons by the offline reconstruction are fit to a common vertex, after which the invariant mass is calculated from the refitted track parameters. Candidates whose refitted mass is within 300 MeV (1 GeV) of the J/ψ (Υ) table mass of 3097 (9460) MeV are regarded as quarkonia candidates and are accepted for further analysis. Table 47 shows the mass resolution for J/ψ and Υ candidates for three cases: both muon tracks reconstructed in the barrel ($|\eta| < 1.05$), both in the endcaps ($|\eta| > 1.05$), and one each in the barrel and an endcap.

For prompt quarkonia candidates accepted by a dimuon trigger there are five major sources of background:

- $J/\psi \to \mu^+\mu^-$ candidates from $b\bar{b}$ events
- nonresonant $\mu^+\mu^-$ from $b\bar{b}$ events
- nonresonant $\mu^+\mu^-$ from charm decays
- nonresonant $\mu^+\mu^-$ from the Drell–Yan mechanism
- nonresonant $\mu^+\mu^-$ from π and K decays-in-flight

The first two in the list are the largest: decays of the form $b \to J/\psi(\to \mu^+\mu^-)X$ and dimuons from $b\bar{b}$ events. While the charm background may be higher in aggregate, the p_T spectrum of the muons falls off sharply and the probability of a dimuon having an invariant mass close to either of the quarkonia is much lower than for the $b\bar{b}$. Monte Carlo studies indicate that the background from Drell–Yan is negligible because only a tiny fraction passes the trigger thresholds. Muons from decays-in-flight also have a very sharply falling p_T spectrum and also need to be in coincidence with another muon, such that the two form an accepted quarkonium candidate, and hence are not dominant background contributors.

Table 47 Mass peak positions and resolutions for prompt quarkonia production in various pseudorapidity ranges

State	$m_{rec} - m_{PDG}$ (MeV)	Resolution σ (MeV)			
		Average	Barrel	Mixed	Endcap
J/ψ	$+4 \pm 1$	53	42	54	75
Υ	$+15 \pm 1$	161	129	170	225

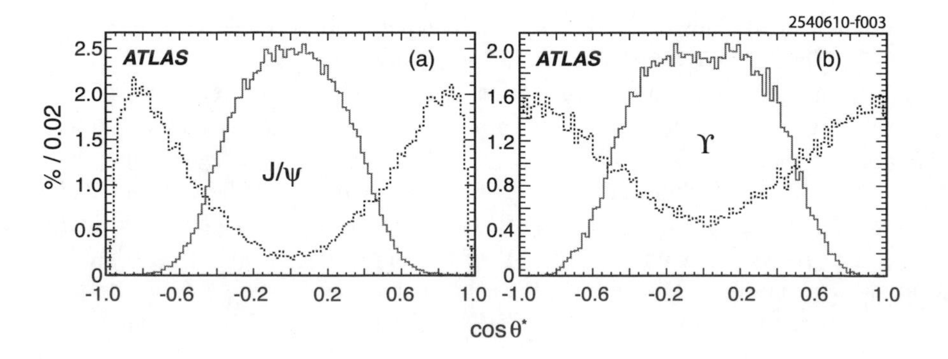

Fig. 94 Polarization angle $\cos\theta^*$ distributions for J/ψ and Υ dimuon decays

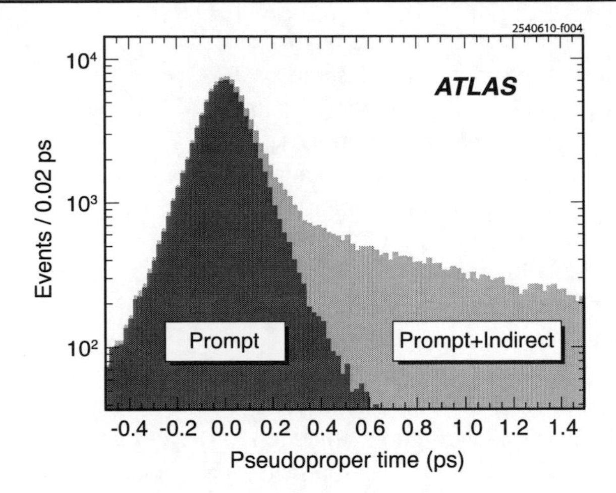

Fig. 95 Pseudoproper decay-time distribution for reconstructed prompt J/ψ (*dark shading*) and the sum of direct and indirect contributions (*lighter shading*)

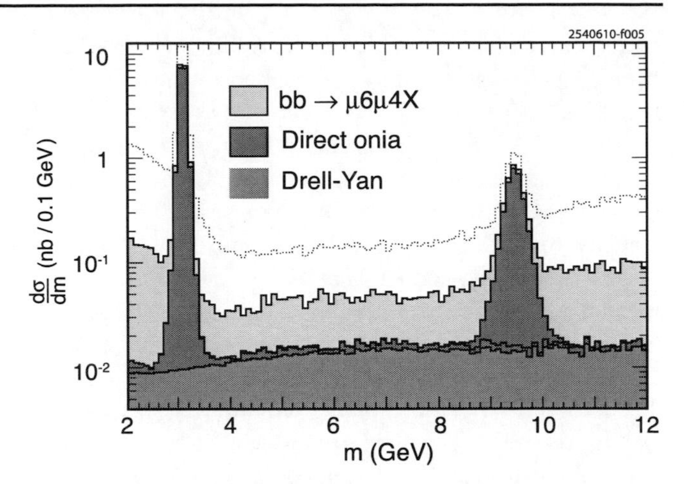

Fig. 96 Cumulative plot of the invariant mass of dimuons from various sources, reconstructed with a dimuon trigger with thresholds of $p_T > (6, 4)$ GeV, with the requirements that both muons are identified as coming from the primary vertex and with a pseudoproper-time cut of 0.2 ps. The *dotted line* shows the cumulative distribution without vertex and pseudoproper-time cuts

Since all of the sources above (aside from Drell–Yan) produce muons which emerge from a secondary vertex, it is possible to suppress them by means of a secondary-vertex cut based on the pseudoproper time, defined as

$$\tau = \frac{L_{xy} \times m}{p_T \times c}, \tag{190}$$

where m and p_T are the invariant mass and transverse momentum of the quarkonium candidate, respectively, and L_{xy} is the measured radial displacement of the two-track vertex from the beamline as in (188). A collection of prompt quarkonia will have a pseudoproper-time distribution around zero, while distributions for nonprompt candidates will have an exponentially decaying tail on the positive side due to the nonzero lifetime of the parent, as shown in Fig. 95. By making a cut on τ, it is possible to exclude the nonprompt component by, for instance, removing all prompt J/ψ candidates with $\tau > 0.2$ ps, thereby obtaining a sample with an efficiency of 93% and a purity of 92%. In the case of Υ there is no background from $b \to J/\psi(\mu^+\mu^-)X$ to address. But $b\bar{b} \to \mu^+\mu^-$ is more problematic in this higher-mass region: the two muons must have come from different decays, rendering the use of pseudoproper time less effective. However, it is possible, e.g., to insist that both muon tracks in the candidate are used to build the same primary vertex: in this case the $b\bar{b} \to \mu^+\mu^-$ background under the Υ can be reduced by a factor of three or more while losing about 5% of the signal.

Figure 96 shows the quarkonia signals and the principal backgrounds for the dimuon trigger with thresholds of $p_T > (6, 4)$ GeV. The higher resonances of the Υ were not included in the simulation; hence their absence from the plot. A pseudoproper-time cut of 0.2 ps has been applied as described above, and both muon tracks in a candidate are required to have been fitted to the same primary vertex. Table 48 summarizes the reconstruction efficiencies of all of the cuts described above for the different trigger schemes.

Table 48 For prompt quarkonia with various selection and background suppression cuts, predicted and observed cross sections, and efficiencies relative to generator-level Monte Carlo

	Quantity	J/ψ	J/ψ	Υ	Υ
	Trigger type	$\mu 6 \mu 4$	$\mu 10$	$\mu 6 \mu 4$	$\mu 10$
	MC cross sections	23 nb	23 nb	5.2 nb	2.8 nb
ϵ_1	Trigger, reconstruction, and vertexing	75%	90%	51%	90%
ϵ_2	Offline cuts	90%	76%	95%	75%
ϵ	Overall efficiency $\epsilon_1 \times \epsilon_2$	67%	69%	49%	68%
	Observed signal σ	15 nb	16 nb	2.5 nb	2.0 nb
	N_s (10 pb^{-1})	150 K	160 K	25 K	20 K
	N_b (10 pb^{-1})	7 K	700 K	16 K	2000 K
	Signal/bgd at peak	60	1.2	10	0.05

6.3.3 Prompt quarkonium polarization

The Color Octet Model (COM) predicts that prompt quarkonia are transversely polarized, with the degree of polarization increasing as a function of the transverse momentum of the quarkonium. Other models predict different p_T-behaviors of the polarization, so this measurable quantity serves as an important discriminator of the various quarkonia production models. As discussed above, the polarization can be accessed via the distribution of the angle $\cos\theta^*$. This measurement is challenging due to reduced acceptance at high $|\cos\theta^*|$ and the difficulty of disentangling acceptance corrections from the spin alignment. Additionally, feeddown from χ_c and B mesons may act to reduce polarization in the final quarkonia sample.

The ATLAS program, in this respect, seeks to measure the polarization of prompt quarkonia states up to transverse momenta of ~50 GeV/c, with the coverage in $\cos\theta^*$ extended through the use of both single- and double-muon triggers. The high quarkonia rate at the LHC will allow ATLAS to obtain a high-purity prompt quarkonia sample through the use of the pseudoproper-time cut, which reduces the depolarization due to contamination from nonprompt quarkonia. Taken together, these techniques will allow ATLAS to control the systematics of the polarization measurement. Of the two main quarkonium states, the J/ψ is easier to deal with than the Υ due to a higher production cross section and much lower backgrounds with the single-muon trigger. Indeed, the background to the single-muon trigger for Υ renders the sample available at 10 pb^{-1} essentially unusable: the reduced acceptance in the high $|\cos\theta^*|$ part of the angular distribution cannot be offset with use of a single-muon trigger in the same way as for J/ψ. For this reason the uncertainties on the spin alignment for Υ are much higher than J/ψ.

The uncertainties from both integrated luminosity and spin alignment need to be factored into measurement errors for prompt quarkonia production cross sections. Both are expected to be high in the early running of the new machine. However, the relative magnitudes of the cross sections measured in separate p_T slices will be unaffected by both luminosity and spin-alignment uncertainties.

Summarizing the main conclusion of [1018, pp. 1083–1110]: after 10 pb^{-1} it should be possible to measure the spin alignment, α, of prompt J/ψ with a precision of $\Delta\alpha = \pm 0.02$–0.06 for $p_T > 12$ GeV/c, depending on the level of polarization. For the reasons discussed above, in the case of Υ, the precision is about ten times worse—of order 0.2. With an integrated luminosity increased by a factor of 10, the uncertainties on Υ polarization could drop by a factor of around 5 because the sample obtained with the $\mu10$ trigger will become more useful.

6.3.4 Early 7 TeV LHC running

The quarkonium program in ATLAS has begun with the first runs of the LHC. The first task is to observe the resonances in the data, using the peaks as calibration points to assess the performance of the muon- and inner-detector track reconstruction and the muon triggers. These studies are being carried out in a rapidly changing luminosity and trigger environment as the LHC itself is commissioned.

After about 1 pb^{-1} ATLAS should have collected some 15K J/ψ and 2.5K Υ candidates decaying to pairs of muons passing the dimuon trigger requiring both muons to have a p_T of 4 GeV/c and one having at 6 GeV/c. The single-muon trigger with a threshold of 10 GeV/c will provide largely independent additional samples of 16K J/ψ and 2K Υ decays. Separately from these, some 7K $J/\psi \to \mu^+\mu^-$ events are expected from b-hadron decays. All of these decays can be used for detector performance studies. Furthermore, a measurement of the fraction of J/ψ arising from B decays will be possible at this level, although the muon trigger and reconstruction efficiencies will have to be estimated with Monte Carlo at this stage.

After about 10 pb^{-1} there will be sufficient statistics to use the data-driven tag-and-probe method to calculate the efficiencies, leading to a reduction in the systematic uncertainties on the ratio measurement. The p_T-dependence of the production cross section for both J/ψ and Υ should be fairly well measured by then, over a wide range of transverse momenta ($10 \le p_T \le 50$ GeV/c).

After around 100 pb^{-1} the J/ψ and Υ differential cross sections will be measured up to transverse momenta around 100 GeV/c. With several million J/ψ and around 500 K Υ, and a good understanding of the efficiency and acceptance, polarization measurements should reach precisions of a few percent. Additional luminosity may allow the observation of resonant pairs of J/ψ in the Υ mass region from the decays of η_b and χ_b states.

6.4 CMS

The primary goal of the Compact Muon Solenoid (CMS) experiment [1020] is to explore particle physics at the TeV energy scale exploiting the proton–proton collisions delivered by the LHC. The central feature of the CMS apparatus is a superconducting solenoid of 6 m internal diameter which provides an axial magnetic field of 3.8 T. Within the field volume are the silicon tracker, the crystal electromagnetic calorimeter and the brass/scintillator hadronic calorimeter in barrel and endcap configurations. CMS also has extensive forward calorimetry, including a steel/quartz-fiber forward calorimeter covering the $2.9 < |\eta| < 5.2$ region. Four stations of muon detectors are embedded in the steel return yoke, covering the $|\eta| < 2.4$ window. Each station consists of several layers of drift tubes in the barrel region and cathode strip chambers in the endcap regions, both complemented by resistive plate chambers.

Having a high-quality muon measurement was one of the basic pillars in the design of CMS. Around 44% of the J/ψ mesons produced in pp collisions are emitted within the almost 5 units of pseudorapidity covered by the muon stations, which cover an even larger phase-space fraction for dimuons from Υ decays. These detectors are crucial for triggering and for muon identification purposes; CMS can easily select collisions which produced one or more muons for writing on permanent storage. The good quality of the muon measurement, however, is mostly due to the granularity of the silicon tracker (1440 silicon-pixel and 15 148 silicon-strip modules) and to the very strong bending power of the magnetic field [1021]. The silicon tracker also provides the vertex position with ~15 μm accuracy [1022].

The performance of muon reconstruction in CMS has been evaluated using a large data sample of cosmic-ray muons recorded in 2008 [1023]. Various efficiencies, measured for a broad range of muon momenta, were found to be in good agreement with expectations from Monte Carlo simulation studies. The relative momentum resolution for muons crossing the barrel part of the detector is better than 1% at 10 GeV/c.

The CMS experiment, thanks to its good performance for the measurement of dimuons, including the capability of distinguishing prompt dimuons from dimuons produced in a displaced vertex, should be ideally placed to study the production of several quarkonia, including the J/ψ, $\psi(2S)$ and $\Upsilon(1S, 2S, 3S)$ states. Complementing the dimuon measurements with the photon information provided by the electromagnetic calorimeter should also allow reconstruction of the χ_c and χ_b states. Such measurements will lead to several studies of quarkonium production. Some will be simple analyses that will lead to the first CMS physics publications. Other rather complex ones will come later, such as the measurement of the polarization of the directly produced J/ψ mesons as a function of their p_T, after subtraction of feed-down contributions from χ_c and B-meson decays.

Here we do not describe an exhaustive description of all the many interesting quarkonium physics analyses that can, in principle, be performed by CMS. Instead, we focus on only a few representative studies. We only mention measurements with dimuons in proton–proton collisions, despite the fact that similar studies could also be made with electron pairs, and/or in heavy-ion collisions, at least to some extent.

6.4.1 Quarkonium production

At midrapidity, the strong magnetic field imposes a minimum transverse momentum of around 3 GeV/c for muons to reach the muon stations. At forward angles, the material thickness imposes a minimum energy on the detected muons, rather than a minimum p_T. In general, for a muon to trigger it needs to cross at least two muon stations. This requirement rejects a significant fraction of the low p_T J/ψ dimuons which could be reconstructed from a data sample collected with a "minimum bias" trigger. In the first few months of LHC operation, while the instantaneous luminosity will be low enough, less selective triggers can be used. For instance, it is possible to combine (in the "high-level trigger" online farm) a single-muon trigger with a silicon track, such that their pair mass is in a mass window surrounding the J/ψ peak. In this way, sizeable samples of low p_T J/ψ dimuons can be collected before the trigger rates become too large.

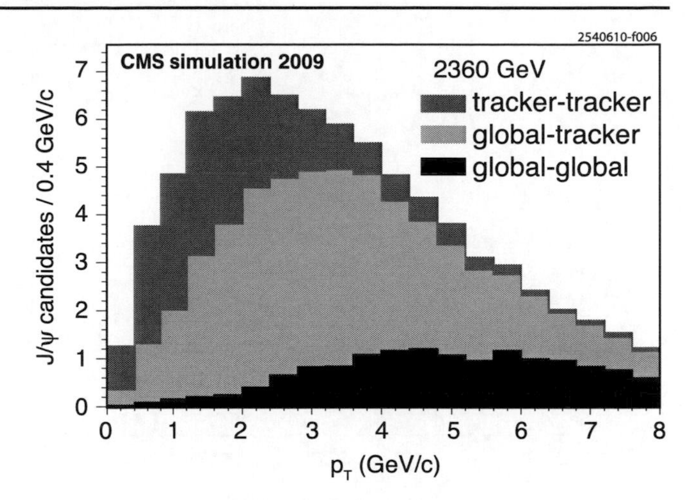

Fig. 97 Transverse momentum distribution of dimuons from J/ψ, reconstructed by CMS in three different muon-pair categories, which depend upon the number of stations crossed by the muons (MC study) [1024]

Figure 97 shows a J/ψ p_T distribution resulting from a Monte Carlo simulation study (based on a tuned [1025] version of the PYTHIA [1026] event generator). This study [1024] was made for pp collisions at 2360 GeV and corresponds to a minimum-bias event sample, collected without any trigger selection of muon-station signals. We see that CMS should have the capability of measuring very low p_T J/ψ dimuons, especially if one of the two muons (or both) is reconstructed as a "tracker muon," meaning that it only traverses one muon station. In fact, most of the yield that could be reconstructed by CMS is contained in the muon-pair category where only one of the muons crosses two or more muon stations (the "global-tracker" pairs).

By accepting events with one of the muons measured only in one station, the signal-to-background ratio in the J/ψ dimuon mass region becomes smaller than in the "global-global" category. However, it remains rather good, as illustrated in Fig. 98, where we see the J/ψ peak reconstructed from pp collisions at 7 TeV after applying certain selection cuts on the muons and requiring a minimum dimuon vertex quality.

With ~10 pb^{-1} of integrated luminosity for pp collisions at 7 TeV, CMS should collect a few hundred thousand J/ψ dimuons and a few tens of thousands of $\Upsilon(1S)$ dimuons. It is important to note that CMS can measure muon pairs resulting from decays of zero p_T Υs. Indeed, the high mass of the Υ states (~10 GeV) gives single muons enough energy to reach the muon stations even when the Υ is produced at rest.

Given its very good muon momentum resolution, better than 1% (2%) for the barrel (endcap) region for muon momenta up to 100 GeV/c [1028], CMS will reconstruct the J/ψ and Υ peaks with a dimuon mass resolution of around

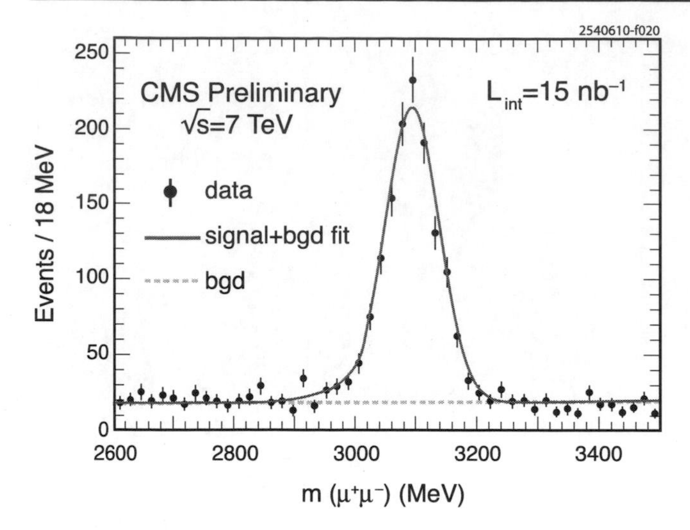

Fig. 98 Dimuon mass distribution reconstructed for pp collisions at 7 TeV, for an integrated luminosity of 15 nb^{-1} [1027]

30 and 80 MeV, respectively, better at midrapidity than at forward rapidity [1029].

The very good electromagnetic calorimeter of CMS, covering the range $|\eta| < 3.0$, enables the study of χ_c and χ_b production through the measurement of their radiative decays. Such measurements are crucial to evaluate non-negligible feeddown contributions to prompt J/ψ and Υ production, a mandatory ingredient to fully understand the physics of quarkonium production from measurements of differential cross sections and polarization. The contribution of $\psi(2S)$ and χ_c decays to prompt J/ψ production has recently been evaluated to be $8.1 \pm 0.3\%$ and $25 \pm 5\%$, respectively [864], while around half of the $\Upsilon(1S)$ yield is due to decays of heavier bottomonium states, at least for $p_{\mathrm{T}}(\Upsilon) > 8\ \mathrm{GeV}/c$ [668]. The decays of b-hadrons also contribute to the observed J/ψ yield. This further complication can be kept under control through the measurement of nonprompt J/ψ production, which CMS can do efficiently thanks to very good vertexing and b-tagging capabilities, and profiting from the long b-hadron lifetimes. In the Υ sector there are no feeddown decays from nonprompt sources.

Given the performance capabilities of the CMS detector, which include a good dimuon mass resolution, a broad rapidity coverage, acceptance down to zero p_{T} for the Υ states (and also to relatively low p_{T} for the J/ψ), CMS is in a very good position to do detailed studies of the quarkonium production mechanisms, hopefully answering some of the questions left open by the lower energy experiments. Naturally, the CMS quarkonium physics program foresees measurements of the differential production cross sections, versus p_{T} and rapidity, of many quarkonium states. Given the large charm and bottom production cross sections at LHC energies, CMS should collect large J/ψ and Υ event samples in only a few months of LHC operation, leading to physics publications on, in particular, their p_{T} distributions, very competitive with respect to the presently available Tevatron results.

6.4.2 Quarkonium polarization

Polarization studies, particularly challenging because of their multidimensional character, will exploit the full capabilities of the CMS detector and the ongoing optimization of dedicated trigger selections. CMS will study the complete dilepton decay distributions, including polar and azimuthal anisotropies and as functions of p_{T} and rapidity, in the Collins-Soper (CS) and the helicity (HX) frames. These analyses will require considerably larger event samples than the cross section measurements. The acceptance in the lepton decay angles is drastically limited by the minimum-p_{T} requirements on the accepted leptons (rather than reflecting geometrical detector constraints). Polarization measurements will therefore profit crucially from looser muon triggers. Moreover, such trigger-specific acceptance limitations determine a significant dependence of the global acceptances (in different degrees for different quarkonium states) on the knowledge of the polarization. The systematic contributions of the as-yet unknown polarizations to early cross section measurements will be estimated, adopting the same multidimensional approach of the polarization analyses. Plans for high-statistics runs include separate determinations of the polarizations of quarkonia produced directly and of those coming from the decays of heavier states. Current studies indicate that CMS should be able to measure the polarization of the J/ψ's that result from χ_c decays, together with the p_{T}-dependent J/ψ feeddown contribution from χ_c decays, from very low to very high p_{T}.

All measurements will also be reported in terms of frame-invariant quantities, which will be determined, for cross-checking purposes, in more than one reference frame. These plans reflect our conviction that robust measurements of quarkonium polarization can only be provided by fully taking into account the intrinsic multidimensionality of the problem. As emphasized in [722, 1030], the measurements should report the full decay distribution in possibly more than one frame and avoid kinematic averages (for example, over the whole rapidity acceptance range) as much as possible.

Figure 99 shows, as a simple example, how a hypothetical Drell–Yan-like polarization (fully transverse and purely polar in the CS frame) in the Υ mass region would translate into different p_{T}-dependent polarizations measured in the HX frame by experiments with different rapidity acceptances. The anisotropy parameters λ_ϑ, λ_φ and $\lambda_{\vartheta\varphi}$ are defined as in [722]. This example illustrates the following general concepts:

- The polarization depends very strongly on the reference frame. The very concepts of "transverse" and "longitudinal" are frame-dependent.

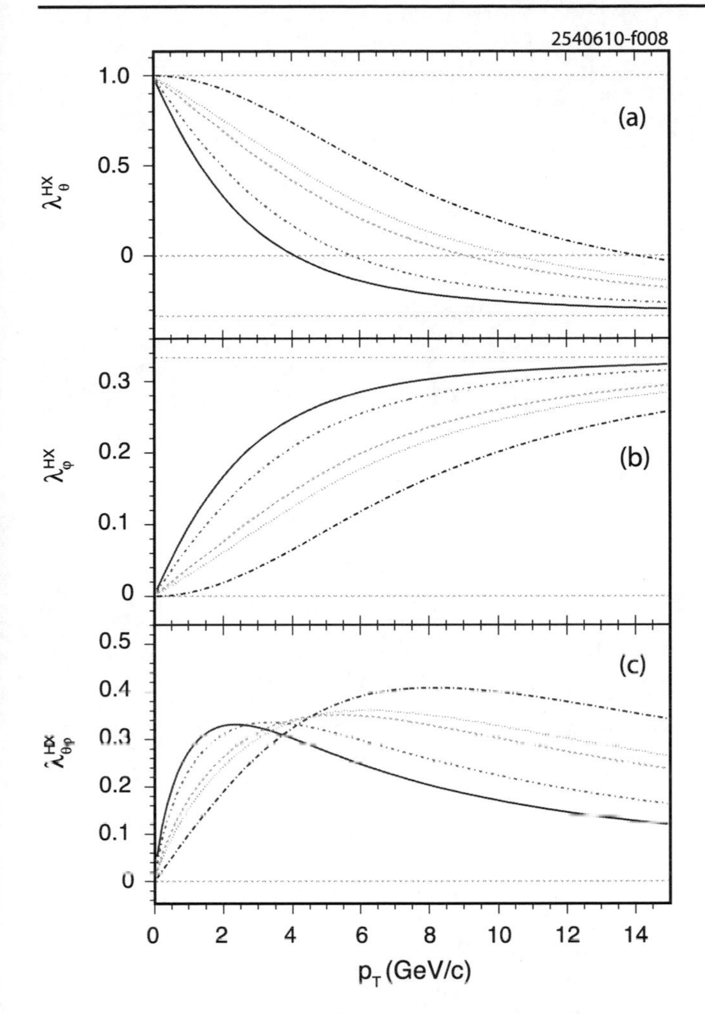

Fig. 99 Anisotropy parameters in (**a**) polar and (**b**) azimuthal angle vs. transverse momentum for $\Upsilon \to \ell^+\ell^-$ decays in the HX frame for a natural polarization $\lambda_\vartheta = +1$ in the CS frame. The *curves* in each plot correspond to different rapidity intervals representative of different experiments. Starting from the *solid line*: $|y| < 0.6$ (CDF), $|y| < 0.9$ (ALICE, e^+e^- channel), $|y| < 1.8$ (DØ), $|y| < 2.5$ (ATLAS and CMS), $2 < |y| < 5$ (LHCb). For simplicity, the event populations have been assumed to be flat in rapidity. The vertical axis of the polarization frame is here defined as $\mathrm{sign}(p_L)(\vec{P}_1' \times \vec{P}_2')/|\vec{P}_1' \times \vec{P}_2'|$, where $\vec{P}_1'$ and $\vec{P}_2'$ are the momenta of the colliding protons in the quarkonium rest frame (the sign of $\lambda_{\vartheta\varphi}$ depends on this definition)

- The fundamental nature of the polarization observed in one chosen frame can be correctly interpreted (without relying on assumptions) only when the azimuthal anisotropy is measured together with the polar anisotropy.
- The measured polarization may be affected by "extrinsic" kinematic dependencies due to a nonoptimal choice of the observation frame. Such extrinsic dependencies can introduce artificial differences among the results obtained by experiments performed in different acceptance windows and give a misleading view of the polarization scenario.

On the other hand, these spurious effects cannot always be eliminated by a suitable frame choice, as is shown by the further illustrative case represented in Fig. 100. Here it is

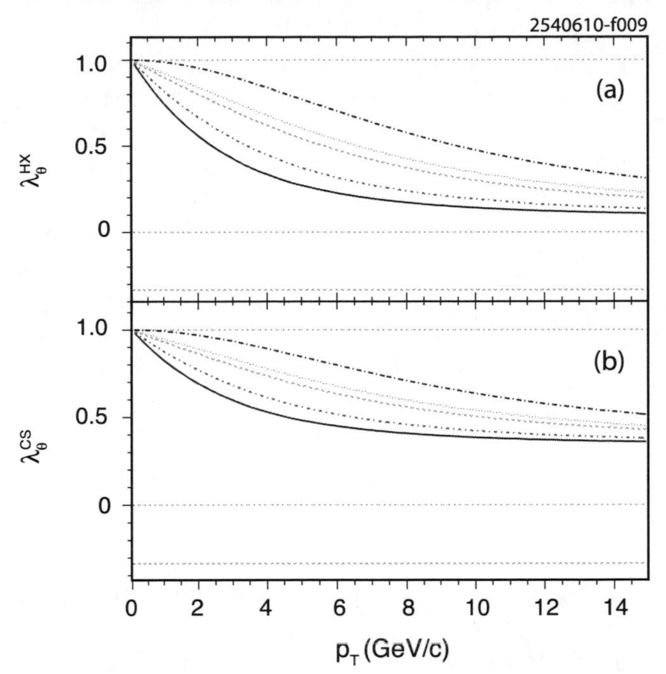

Fig. 100 Polar angle anisotropy parameter vs. transverse momentum for $\Upsilon \to \ell^+\ell^-$ decay, observed in the (**a**) HX and (**b**) CS frames when 40% (60%) of the events have full transverse polarization in the HX (CS) frame. The *curves* in each plot represent measurements performed in different experimental acceptance ranges, as detailed in Fig. 99

assumed that 60% of the Υ events has natural polarization $\lambda_\vartheta = +1$ in the CS frame and the remaining fraction has natural polarization $\lambda_\vartheta = +1$ in the HX frame. While the polarizations of the two event subsamples are intrinsically independent of the production kinematics, in neither frame will measurements performed in different transverse and longitudinal momentum windows find identical results for λ_ϑ (the same is true for the other two anisotropy parameters, not shown here). This example provides a first motivation for the complementary use of a frame-invariant approach [1030], consisting of the measurement of intrinsically rotation-invariant polarization parameters like

$$K = \frac{1 + \lambda_\theta + 2\lambda_\phi}{3 + \lambda_\theta}. \tag{191}$$

In the example of Fig. 100, all experiments would measure a constant, frame-independent value $K = 1/2$. This method facilitates the comparison between different experiments, as well as between measurements and theory. Furthermore, since the acceptance distributions for the polar and azimuthal decay angles can be very different in different frames, checking whether quantities like K are, as they should be, numerically independent of the reference frame provides a nontrivial systematic test of the experimental analyses.

6.5 LHCb

LHCb [1031] is a dedicated experiment for b-physics at the LHC. Since b production is peaked in the forward region at LHC energies, the LHCb detector has a forward spectrometer geometry covering an angle between 15 mrad and 300 mrad with respect to the beam axis. This corresponds to an η range between 2 and 5, which will allow LHCb to have a unique acceptance coverage among the LHC experiments. Good vertex resolution and particle identification over a wide momentum range are key characteristics of LHCb. The trigger system retains muons with moderate p_T as well as purely hadronic final states.

6.5.1 Charmonium physics

The J/ψ selection studies and in general all studies presented here have been performed using the full LHCb Monte Carlo simulation based on PYTHIA [1026], EvtGen [1032], and GEANT4 [1033]. At the generation level, color-octet J/ψ production models in PYTHIA have been tuned to reproduce the cross section and p_T spectrum observed at Tevatron energies [1025]. A full-event reconstruction is applied to the simulated events [1034]. J/ψ candidates are selected using track and vertex quality requirements, and also muon identification information. Since the first-level trigger (L0) requires at least one muon with a p_T larger than 1 GeV/c, a tighter selection is applied at reconstruction level to keep only candidates formed with at least one muon with a p_T larger than 1.5 GeV/c. The J/ψ selection yields an expected number of reconstructed events equal to 3.2×10^6 at $\sqrt{s} = 14$ TeV, with $S/B = 4$, for an integrated luminosity equal to 5 pb^{-1}. This number is obtained assuming a J/ψ production cross section equal to 290 μb. This amount of data could be collected in a few days under nominal LHC running conditions. The mass resolution is 11.4 ± 0.4 MeV.

One of the first goals of the LHCb experiment will be to measure the differential J/ψ cross section in bins of p_T and η in the range $0 < p_T < 7$ GeV/c and $2 < \eta < 5$. Both the prompt-J/ψ and the $b \to J/\psi X$ production cross sections will be accessible, thereby measuring the total $b\bar{b}$-production cross section. The two contributions will be separated using a variable which approximates the b-hadron proper time along the beam axis

$$t \equiv \frac{dz \times m(J/\psi)}{p_z^{J/\psi} \times c}, \tag{192}$$

where dz is the distance between the J/ψ decay vertex and the primary vertex of the event projected along the beam (z) axis, $p_z^{J/\psi}$ is the signed projection of the J/ψ momentum along the z axis, and $m(J/\psi)$ is the known J/ψ mass. (Note that this is analogous to the ATLAS pseudoproper-time definition in (190), which uses the transverse decay length and momentum instead of the longitudinal component employed here.) The expected distribution of the t variable is shown in Fig. 101. The distribution can be described by

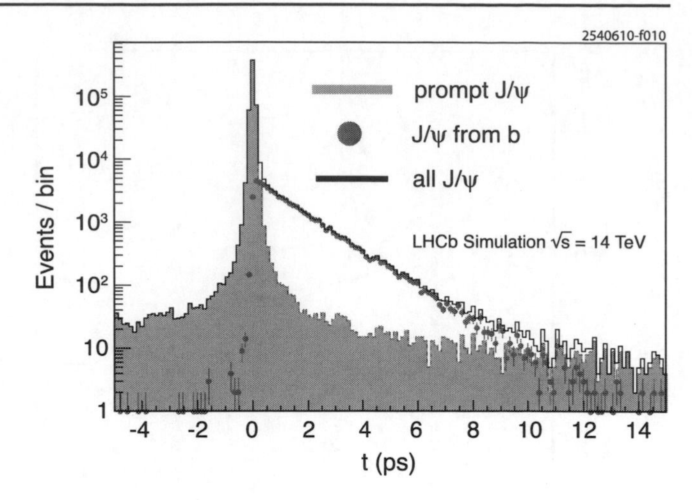

Fig. 101 Time distribution of J/ψ candidates obtained with LHCb Monte Carlo simulation

- a prompt J/ψ component produced at the primary vertex of the event, represented by a Gaussian distribution to account for vertex resolution;
- an exponential J/ψ component coming from b-hadron decays, convoluted with a Gaussian resolution function;
- a combinatorial background component due to random combinations of tracks coming from the primary vertex (the form of this component will be extracted using events in the sidebands of the dilepton mass distribution);
- a tail due to a wrong association of primary vertex when computing the t variable (the shape of this component will be determined from data, associating the J/ψ vertex with a different event's primary vertex).

A combined fit of the mass and t distributions will extract the number of reconstructed J/ψ in each p_T and η bin [1035]. The absolute cross section in each of the bins will be obtained from this measured yield, efficiencies that will be computed from Monte Carlo simulations, and integrated luminosity. The measurement uncertainty will be dominated by systematic errors in the integrated luminosity, the resolution model, and the reconstruction and trigger efficiencies.

The unknown polarization at production of the J/ψ will complicate the measurement. The acceptance of the LHCb detector is not uniform as a function of the J/ψ polarization angle, θ, defined as the angle between the μ^+ direction in the center-of-mass frame of the J/ψ and the direction of the J/ψ in the laboratory. Ignoring this effect adds a 25% uncertainty to the cross section measurement. Performing the measurement in bins of θ will allow determination of the J/ψ production polarization.

Measurements of the rates and polarization of the J/ψ, and more generally of the charmonium and bottomonium

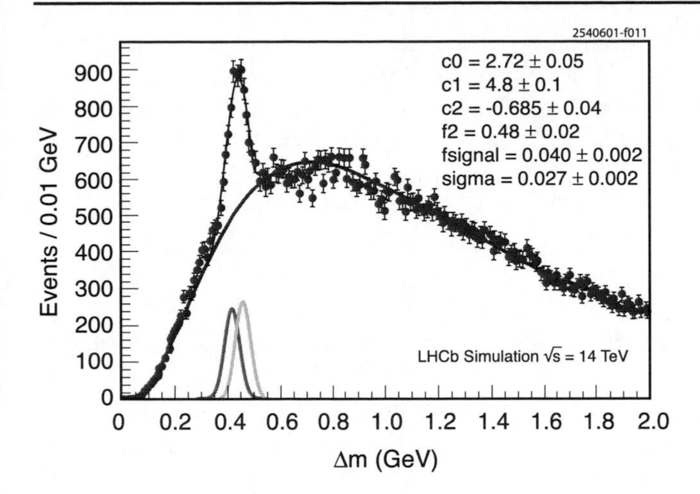

Fig. 102 Δm distribution of χ_c candidates obtained with LHCb Monte Carlo simulation

states, will be compared to predictions of different theoretical models.

Reconstruction of χ_{c1} and χ_{c2} has also been studied [1034] using the decay modes $\chi_{c1,2} \to J/\psi\gamma$. A photon candidate reconstructed in the electromagnetic calorimeter with $p_T > 500$ MeV is associated with a J/ψ candidate. Figure 102 shows the $\Delta m = m(\mu^+\mu^-\gamma) - m(\mu^+\mu^-)$ distribution obtained from a Monte Carlo simulation. A clear signal peak can be observed. The Δm resolution of 27 MeV is dominated by the uncertainty in photon energy. Since the mass difference between χ_{c1} and χ_{c2} is known (55 MeV), imposing this constraint on the analysis should allow separation of the χ_{c1} and χ_{c2} contributions.

A large number of $\psi(2S)$ mesons will also be collected at LHCb. It is expected that the number of reconstructed $\psi(2S) \to \mu^+\mu^-$ decays will be equal to 2 to 4% of the number of reconstructed $J/\psi \to \mu^+\mu^-$, with S/B between 1 and 2 [1036]. Because the masses of the J/ψ and $\psi(2S)$ are close, trigger and reconstruction efficiencies for the two states are similar. A measurement of

$$\frac{\sigma(\text{prompt}\, J/\psi)}{\sigma(\text{prompt}\, \psi(2S))}, \tag{193}$$

where a number of systematic errors cancel, will be possible.

A large sample of $X(3872) \to J/\psi\pi^+\pi^-$ decays will be reconstructed at LHCb, either prompt $X(3872)$, or $X(3872)$ from b-hadron decays. In particular, the decay channel[22] $B^+ \to X(3872)(\to J/\psi\rho^0)K^+$ will be studied because an angular analysis of the decay products can lead to the determination of the now-ambiguous quantum numbers of the $X(3872)$ (see Sect. 2.3.1), allowing separation of the 1^{++} and the 2^{-+} hypotheses. 1850 reconstructed events are expected for 2 fb^{-1} of data at $\sqrt{s} =$ 14 TeV, with B/S between 0.3 and 3.4 [1037]. $Z(4430)$ will be sought in the decay $B^0 \to Z(4430)^{\mp}K^{\pm}$, with $Z(4430)^{\mp} \to \psi(2S)(\to \mu^+\mu^-)\pi^{\mp}$. 6200 events are expected with B/S between 2.7 and 5.3, for 2 fb^{-1} of data at $\sqrt{s} = 14$ TeV [1037].

[22] A charge-conjugate decay mode is implied in the rest of the text.

6.5.2 B_c physics

The expected B_c^+ cross section at the LHC is at the level of 1 μb, so a very large number of B_c will be produced and recorded at LHCb. (For the present status of B_c measurements, see Sects. 2.2.4 and 4.7.) First studies will use the reconstruction of the mode with a large branching fraction $B_c^+ \to J/\psi\mu^+\nu$, and most promising results are expected using the clean $B_c^+ \to J/\psi\pi^+$ decay mode. But the large number of B_c^+ produced will allow a systematic study of the B_c family at LHCb.

The selection of the decay channel $B_c^+ \to J/\psi\pi^+$, with $J/\psi \to \mu^+\mu^-$ has been studied using full Monte Carlo simulation of events reconstructed by the LHCb detector [1038–1040]. A specific generator, BCVEGPY [1041] has been used to generate B_c events. Since the B_c vertex is displaced with respect to the primary vertex, impact-parameter selections are imposed to the π and J/ψ candidates. Particle identification, quality of track and vertex fits, and minimum p_T requirements are applied to B_c candidates in order to reduce the large background due to other b-hadron decays with a J/ψ in the final state. The total reconstruction efficiency is estimated to be $(1.01 \pm 0.02)\%$, with $1 < B/S < 2$ at 90% CL. Assuming a B_c production cross section of $\sigma(B_c^+) = 0.4$ μb for $\sqrt{s} = 14$ TeV and a branching fraction $\mathcal{B}(B_c^+ \to J/\psi\pi^+) = 1.3 \times 10^{-3}$, 310 signal events are expected with 1 fb^{-1} of data.

The potential of a mass measurement has been studied using an unbinned maximum-likelihood method to extract the B_c^+ mass from the reconstructed sample of $B_c^+ \to J/\psi\pi^+$ candidates. Describing the invariant-mass distribution of the signal by a single Gaussian and the combinatorial background by a first-order polynomial, the fit procedure gives a B_c^+ mass of $m(B_c^+) = 6399.6 \pm 1.7$ MeV, where the error is statistical only, consistent with the input value of 6400 MeV. The size of the sample used corresponds to the expected yield for 1 fb^{-1} of data. The result of the fit and the mass distribution are shown in Fig. 103. The mass resolution is $\sigma = 17.0 \pm 1.6$ MeV.

The reconstructed B_c^+ candidates will also be used to measure the lifetime of the B_c^+. A combined mass-lifetime fit is performed. The proper-time distribution is described by an exponential function convoluted with a resolution function and multiplied by an acceptance function $\epsilon(t)$ which describes the distortion of the proper-time distribution due to the trigger and offline event selections through impact-parameter requirements. The form of these functions is determined from the full Monte Carlo simulation. Since the

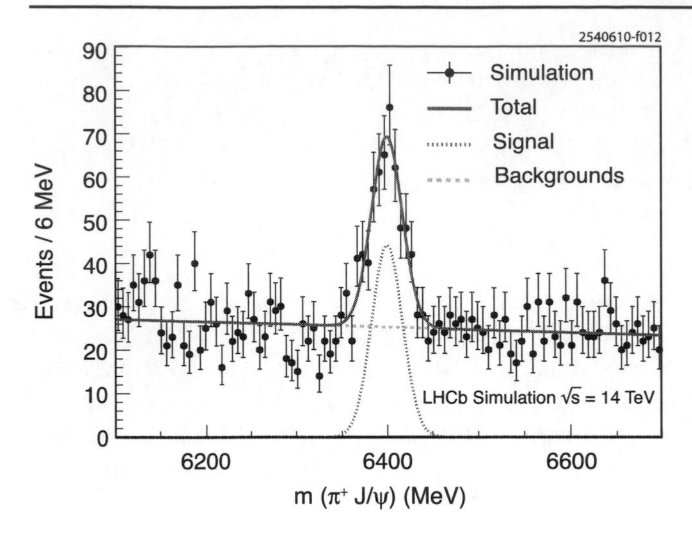

Fig. 103 B_c^+ candidate mass distribution obtained with LHCb Monte Carlo simulation

resolution of the impact parameter depends on the transverse momentum of the tracks, the proper-time acceptance function $\epsilon(t)$ depends on the p_T distribution of the B_c^+, and then on the generation model used for the B_c^+ when determining $\epsilon(t)$. In order to evaluate the systematics associated with this effect, a fit was performed on a B_c^+ sample generated with a p_T spectrum identical to the B^+ spectrum observed in the simulation. A bias of 0.023 ps is then observed in the lifetime determination. In order to reduce this bias, the lifetime fit is performed simultaneously on two samples with different p_T ranges, $5 < p_T < 12$ GeV/c and $p_T > 12$ GeV/c. The resulting bias is then reduced to 0.004 ps. The fit procedure applied to a sample corresponding to 1 fb^{-1} of data gives a B_c^+ lifetime of $\tau(B_c^+) = 0.438 \pm 0.027$ ps, where the error is statistical only, consistent with the input value of 0.46 ps [1042].

LHCb capabilities should allow comprehensive studies of many other aspects of B_c physics [1043]. Spectroscopy of the B_c excited states, both below and above the $m(B^0) + m(D^0)$ threshold will be performed. For example, searches for $B_c^{**} \to B_c^+\pi^+\pi^-$ have been envisaged [1044]. Searches for new decay modes of the B_c will also be made, e.g., modes with a weak decay of the c quark, such as $B_c^+ \to B_s^0\pi^+$ or those with a $\bar{b}c$ annihilation [1045], such as $B_c^+ \to \bar{K}^{*0}K^+$.

6.5.3 Bottomonium physics

Analyses of bottomonium in LHCb have begun. First results show that the reconstruction of $\Upsilon(1S) \to \mu^+\mu^-$ is possible with a mass resolution of 37 MeV. Other Υ states will also be observed in the detector. Using these candidates, measurements of the production cross section and of the production polarization will be performed as a function of p_T and η. Similarly, the decays $\chi_{b1,2} \to \Upsilon\gamma$ are also possible to reconstruct in the LHCb detector, due to low-p_T-photon reconstruction provided by the electromagnetic calorimeter [1034]. In addition, the equivalent for the b family of the exotic X, Y, and Z states (see Sect. 2.3) will be sought at LHCb, for example, in the decay mode $Y_b \to \Upsilon(1S)\pi^+\pi^-$ (for status of the Y_b, see Sects. 2.3.2 and 3.3.11).

6.6 RHIC

The Relativistic Heavy-Ion Collider, RHIC, provides d+Au, Au+Au, and Cu+Cu collisions at $\sqrt{s_{NN}} = 200$ GeV, and polarized pp collisions at both 200 and 500 GeV. Polarized pp collisions are used for spin studies, with those from 200 GeV serving as reference data for the heavy-ion program. The d+Au collisions are used for studies of forward physics and to establish the cold-nuclear-matter baseline for heavy-ion collisions. The primary focus of the heavy-ion program is to quantify the differences between the hot, dense final state and scaled pp and d+Au reference data.

In 200 GeV pp collisions, PHENIX [663, 979] has measured cross sections for production of the (unresolved) Υ states and the $\psi(2S)$. A χ_{cJ} measurement will be published soon. There is also a preliminary PHENIX [975] result showing that the Υ R_{AA}, defined in (184)), measured in Au+Au collisions is below 0.64 at 90% CL. All of these measurements involve low yields and will benefit greatly from additional luminosity. PHENIX [976] has also measured the polarization of the J/ψ in 200 GeV pp collisions and will also do so with existing 500 GeV data.

So far, the high-statistics heavy-quarkonium data available from RHIC are the J/ψ data sets measured by PHENIX in $\sqrt{s_{NN}} = 200$ GeV pp [976], d+Au [663, 883], Cu+Cu [977], and Au+Au [983] collisions. The data were measured in three rapidity ranges: $-2.2 < y < -1.2$; $-0.5 < y < 0.5$; and $1.2 < y < 2.2$. There are also more recent data sets with much higher yields in d+Au (2008) and Au+Au (2007 and 2010) that are still being analyzed. While preliminary J/ψ R_{CP} (see (185)) results were presented by PHENIX [663] at Quark Matter 2009, the final data have not yet been released. Section 5.5 reviews the status of the J/ψ program at RHIC.

STAR [1046] has measured J/ψ-hadron azimuthal angular correlations which have been used to infer the B-meson feeddown contribution to the J/ψ. STAR has also measured $R_{\rm dAu}$ for the combined Υ states using 2008 d+Au [981] data and 2006 pp [980] data, and made measurements of the high p_T J/ψ suppression factor, R_{AA}, from Cu+Cu [664] collisions.

The ongoing RHIC luminosity upgrades will be completed by the 2013 run. The introduction in 2011 and 2012 of silicon vertex detectors into PHENIX at mid- and forward rapidity, respectively, and of the STAR Heavy Flavor Tracker in 2014, will enable open charm and open bottom to be measured independently with greatly improved precision. The detector upgrades will also allow improved measurements of quarkonium states due to improved mass resolution and background rejection.

The RHIC run plan for the next five years or so will be centered on exploiting the capabilities of the new silicon vertex detectors and other upgrades, combined with the increased RHIC luminosity. Of greatest interest to in-medium heavy-flavor physics, there will likely be long pp, d+Au, and Au+Au runs at 200 GeV, plus shorter runs with the same species at 62 GeV to explore the energy dependence of open and hidden heavy-flavor production. The luminosity increase and the PHENIX detector upgrades will enhance the PHENIX heavy-quarkonium program in several ways: increased p_T reach for the J/ψ; new studies of J/ψ suppression with respect to the reaction plane; a first J/ψ v_2 measurement; better understanding of cold-nuclear-matter effects on J/ψ production; and low-statistics measurements of the modification of the combined Υ states. The increased luminosity will enable the large-acceptance STAR detector to extend its p_T reach to considerably higher values for Υ and J/ψ measurements than previously possible.

The RHIC schedule for the period beyond about 5 years is still under development. RHIC experiments are presently engaged in preparing a decadal plan that will lay out their proposed science goals and detector upgrades for 2011 to 2020. PHENIX is considering a conceptual plan that would keep the new silicon vertex detectors but completely replace the central magnet and the outer central arm detectors. The new magnet would be a 2T solenoid with an inner radius of 70 cm. Two new silicon tracking layers would be placed inside the solenoid at 40 and 60 cm, followed by a compact electromagnetic calorimeter of 8 cm depth and a preshower layer. A hadronic calorimeter would be added outside the magnet with an acceptance of $|\eta| < 1$ and 2π in azimuth. The conceptual design is still being evaluated, but it promises to allow powerful measurements of light-quark and gluon jets; dijets and γ+jet coincidences; charm and bottom jets; the J/ψ modification factor over a range of energies; simultaneous studies of the modification of the three bound Υ states; and direct γ^* flow. Removing the south muon spectrometer and replacing it with an electron/photon endcap spectrometer has also been discussed. This replacement would be aimed at addressing spin physics questions and possibly providing electron-ion capabilities in PHENIX. The north muon spectrometer would be retained to provide forward-rapidity heavy-flavor measurements.

In addition to the Heavy Flavor Tracker, STAR will add a new detector that is of major importance to their quarkonium program. The Muon Telescope Detector will be a large-acceptance muon detector located outside the STAR magnet at midrapidity, covering $|y| < 0.5$. It will provide a good signal-to-background ratio for measurements of the three $\Upsilon(nS)$ states and add the capability of measuring J/ψ elliptic flow and suppression at high p_T. The upgrades being pursued by PHENIX and STAR, combined with very high RHIC luminosity, will provide the opportunity to compare, between RHIC and the LHC, in-medium quarkonium modification at energies of 62, 200, and 5500 GeV. This energy regime spans a wide range of medium temperatures and lifetimes, and also provides measurements with very different underlying heavy-quark rapidity densities and thus very different contributions to quarkonia from processes involving coalescence of heavy quarks from different hard collisions.

6.7 Super flavor factories

At e^+e^- machines, quarkonium can be produced through several processes: directly, i.e., during energy scans for $J^{PC} = 1^{--}$ states in the $c\bar{c}$ and $b\bar{b}$ regions; through ISR (for $J^{PC} = 1^{--}$ states below the e^+e^- center-of-mass energy) or two-photon (for $C = +1$ states) processes; in B-meson decays through color-suppressed $b \to c$ transitions. All these have been successfully employed for quarkonium studies in the CLEO, BABAR, Belle, and/or BESIII experiments, the first two of which permanently ceased taking data in early 2008. Belle acquired data until the middle of 2010, and then shut down for a significant upgrade. CLEO-c collected 48 pb^{-1} at the $\psi(2S)$ (about 27 million $\psi(2S)$ produced) and 1485 pb^{-1} in the CM energy range 3.67–4.26 GeV; BESIII has already quadrupled the CLEO-c $\psi(2S)$ sample, acquired 200M J/ψ decays, and will, in time, exceed the CLEO-c samples above open-charm threshold as well (Sect. 6.1). What will happen to e^+e^- quarkonium physics after BESIII and Belle programs are complete?

A new generation of Super Flavor Factories has recently been proposed in order to perform precision measurements in the flavor sector and complement New Physics (NP) searches at hadronic machines [1047, 1048]. An increase in statistics by a factor of 50–100 with respect to the current generation of still-running experiments is essential to such physics program.

Two different approaches have been devised to reach a design peak luminosity of 10^{36} $\text{cm}^{-2}\,\text{s}^{-1}$. In the original SuperKEKB [1047] design this was to be achieved by increasing the beam currents, and introducing crab crossing to maintain large beam-beam parameters [1049]. In the SuperB design [1048] a similar luminosity goal is pursued through the reduction of the interaction point size using very small emittance beams, and with a "crab" of the focal plane to compensate for a large crossing angle and maintain optimal collisions [1050]. After KEK revised their design in favor of the nanobeam collision option [1050], the machine parameters for the SuperB factory and the KEKB upgrade are very similar [1048, 1049].

An NP-oriented program suggests that the machine be operated primarily at the $\Upsilon(4S)$ resonance [1051], with integrated luminosity of order 25–75 ab^{-1}. However, the ability to run at other Υ resonances or at energies in the $c\bar{c}$ region would substantially enhance the physics potential of the machine. In one month at design luminosities it would be possible to collect about 150 fb^{-1} at $D\bar{D}$ threshold [1048]. In about the same time, an $\Upsilon(5S)$ run would integrate about 1 ab^{-1}, corresponding to a "short run" scenario. In a "long run" scenario at $\Upsilon(5S)$, about 30 ab^{-1} could be collected [1048].

The detector design will primarily address the requirements of an NP search program at the $\Upsilon(4S)$. Other constraints are posed by the possibility to re-use components from the BABAR and Belle detectors. Improvements in the vertex resolution, to compensate a reduced beam asymmetry, as well as increased hermeticity of the detector are foreseen. Assuming the same magnetic field, a similar momentum resolution is expected. In a simplified approach, one can assume similar backgrounds and detector performances as at BABAR [1052] and Belle [1053].

A heavy-quarkonium to-do list in a Super Flavor Factory physics program can be found in the Spectroscopy (Sect. 2), Decays (Sect. 3), and Summary (Sect. 7) sections of this review. Precision measurements or simply observation of some conventional, expected processes is warranted; the as-yet-unexplained phenomena also demand attention and offer great reward. Highlights of such a program would include the following:

- Precision measurements of $\eta_c(1S)$ and $\eta_c(2S)$ masses and widths (probably in $\gamma\gamma$-fusion), understanding of their observed lineshapes in radiative J/ψ and $\psi(2S)$ decays, and a comprehensive inventory of radiative and hadronic decay modes and respective branching fractions (Sects. 2.2.2, 3.1.2 and Table 4).
- First observation and study of $\eta_b(2S)$, $h_b(^1P_1)$, and $\Upsilon(1^3D_J)$ $(J = 1, 3)$ (Sect. 2.2.7 and Tables 4 and 8).
- For $X(3872)$ (see Sect. 2.3.1 and Tables 9, 10, 11, 12, and 13), detailed study of the $\pi^+\pi^- J/\psi$ and $D^{*0}\bar{D}^0$ lineshapes, high-statistics measurements of $\gamma J/\psi$ and $\gamma\psi(2S)$ branching fractions, and a search for decay to $\pi^0\pi^0 J/\psi$. The fruit of such an effort could be an answer to how much tetraquark, molecular, and/or conventional charmonium content this state contains.
- For the $Y(4260)$ and nearby states (see Sect. 2.3.2 and Tables 9, 14, 15, and 16), a comprehensive lineshape measurement accompanied by precision branching fractions could shed light on the molecular, tetraquark, or hybrid hypotheses.
- Confirmation and study of the exotic charged Z^+ states (Sect. 2.3.4 and Table 9).
- Confirmation and study of the Y_b (Sect. 2.3.2 and Table 9). A high-statistics scan of 20fb^{-1} per point, necessary to reduce the relative error to the 10^{-3} level, might be needed [1054]. If the Y_b is below 10800 MeV, it could also be produced by ISR with $\Upsilon(5S)$ data collected by a Super B-factory. If it is above that energy, a direct scan will be necessary.

6.8 $\overline{\text{P}}$ANDA

$\overline{\text{P}}$ANDA is one of the major projects at the future Facility for Antiproton and Ion Research (FAIR) [1055] at GSI in Darmstadt, which is expected to start its operations in 2014. $\overline{\text{P}}$ANDA will use the antiprotons circulating in the High Energy Storage Ring (HESR), to study their interactions with protons or nuclei on a fixed target. The antiproton momentum in the range 1.5 to 15 GeV/c corresponds to a center-of-mass energy in $p\bar{p}$ collisions in the range 2.5–5.5 GeV. The purpose of the $\overline{\text{P}}$ANDA experiment is to investigate QCD in the nonperturbative regime. This is achieved through the study of several topics, like QCD bound states, nonperturbative QCD dynamics, study of hadrons in nuclear matter, hypernuclear physics, electromagnetic processes, and electroweak physics. In particular, $\overline{\text{P}}$ANDA has an extensive research program in charmonium physics. Experimentally, charmonium has been studied mainly in e^+e^- and $p\bar{p}$ experiments. While in e^+e^- collisions, only the $J^{PC} = 1^{--}$ states can be directly formed, in $p\bar{p}$ interactions, direct formation is possible for all the states with different quantum numbers, through coherent annihilation of the three quarks of the protons with the three antiquarks of the antiproton. An additional advantage of this technique is that the interaction energy can be precisely determined from the beam parameters and it is not limited by the detector resolution, allowing fine energy scans of narrow resonances. Historically, this experimental technique was successfully used at CERN and at Fermilab. With a higher luminosity and a better beam energy resolution with respect to previous $p\bar{p}$ experiments, $\overline{\text{P}}$ANDA will be able to obtain high-precision data on charmonia and measure the masses, widths, and excitation curves for the recently observed states with unprecedented precision.

6.8.1 *Experimental technique*

Quarkonium physics is one of the main research fields for $\overline{\text{P}}$ANDA. The precision study of resonance parameters and excitation curves is an area where the close interplay between machine and detector is fundamental. For the $\overline{\text{P}}$ANDA charmonium program, the antiproton beam collides with a fixed hydrogen target. In order to achieve the design luminosity (2×10^{32} cm^{-2} s^{-1}), a high-density target thickness is needed. Two solutions are under study: the cluster-jet target and the pellet target, with different implications for the beam quality and the definition of the interaction point.

Thanks to the cooling of the antiproton beam, at the HESR the energy spread of the beam will be approximately

30 keV, which is comparable with the width of the narrowest charmonia states. The knowledge of the total interaction energy from the beam parameters and the narrowness of the beam energy distribution allow the direct measurement of the mass and the width of narrow states through scans of the excitation curve.

The resonance parameters are determined from a maximum-likelihood fit to the number of observed events in a specific channel N_i, where the subscript i refers to different center-of-mass energies:

$$N_i = \epsilon_i \mathcal{L} \int \sigma_{\mathrm{BW}}(E') B_i(E')\, dE' \qquad (194)$$

where σ_{BW} is the Breit–Wigner resonance cross section to be measured and B_i is the center-of-mass energy distribution from beam parameters.

The energy scan of a resonance is schematically represented in Fig. 104: the energy of the interaction, shown on the horizontal axis, is obtained from the beam parameters, and is varied in the energy region to be explored. The detector is used as a simple event counter and the cross section values (vertical axis) can be obtained from the number of events observed at each energy point. Using (194), it is then possible to determine the resonance parameters. By means of fine scans it will be possible to measure masses with accuracies of the order of 100 keV and widths at the 10% level. The entire region below and above the open charm threshold (from 2.5 GeV to 5.5 GeV in the center-of-mass energy) will be explored.

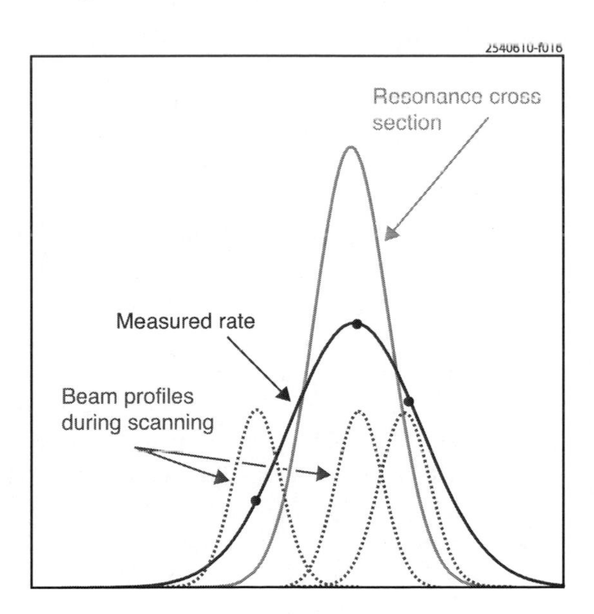

Fig. 104 Schematic representation of the scan of a resonance (*lighter solid line*): the center-of-mass energy is varied with the HESR to several values across the resonance (the energy distributions are shown by (*dotted lines*). The *solid circles* represent the observed cross section for each energy point, defining the measured excitation curve (*darker solid line*), which is the convolution of the resonance curve and the beam profile

As described above, antiproton–proton formation will become an important tool for decisive studies of the natural widths and lineshapes of open and hidden charm states. It will also have a powerful role in measuring decay modes of very narrow states. Unfortunately, charm production cross sections in proton–antiproton annihilation in this energy regime are either unknown or, at best, poorly measured. Thus predictions for sensitivities are not reliable. In order to provide experimental input to event generators, and thereby allow credible sensitivity studies, it would be worthwhile to measure inclusive open charm production at the existing antiproton facility at Fermilab with a relatively simple and small test apparatus equipped with a microvertex detector and a tracking device. Such an experiment would likely require less than a month of data taking to acquire a sufficient statistics.

6.8.2 Detector

The $\overline{\text{P}}$ANDA detector is a multipurpose detector, designed to obtain the highest possible acceptance, good tracking resolution, particle identification and calorimetry. It is composed of two parts: a target spectrometer with an axial field generated by a superconducting solenoid for the measurement of the particles emitted at large angles with respect to the beam direction, and a forward spectrometer with a dipole magnet in the forward direction for the measurement of the particles leaving the interaction region at small angles with respect to the beam.

The target spectrometer is arranged in a barrel part, for angles larger than 22°, and an endcap part, covering the forward region down to 5° in the vertical plane and 10° in the horizontal plane. A silicon tracker is located close to the interaction region. It is composed of two inner layers of hybrid pixel detectors and two outer layers with silicon-strip detectors, with cylindrical symmetry around the beam pipe. In addition, six silicon discs with both pixel and strip modules are located in the forward direction. The tracking system (solutions with straw tubes detector or TPC are under study) is required to provide momentum measurement with resolution $\delta p/p$ at the percent level, handling the high particle fluxes that are foreseen at the maximum luminosity. In the forward part of the target spectrometer, particles exiting the interaction region at polar angles below 22° are tracked with three GEM detectors placed downstream. Charged particle identification is obtained by collecting complementary information from different detectors. Most charged particles with momentum exceeding 0.8 GeV$/c$ are identified using a Cherenkov detector. A detector of internally reflected Cherenkov light based on the BABAR design [1056] will provide particle identification in the region 22° to 140°. Time-of-flight will be also used in $\overline{\text{P}}$ANDA for the identification of low-momentum particles. To obtain muon–pion separation, the yoke of the superconducting solenoid is segmented in 12 layers and the gaps are instrumented for the measurement of the interaction length in iron. Electromagnetic calorimetry is required over a wide energy range, from the MeV to the GeV scale. A lead-tungstate ($PbWO_4$) calorimeter, with crystal length corresponding to $22X_0$, will measure energy with a resolution better than 2% at 2 GeV and good time resolution.

A dipole magnet with 1 m gap and 2 m aperture is located between the target and the forward spectrometer. The measurement of the deflection of charged tracks is obtained by using a set of wire chambers placed before, within, and behind the dipole magnet. The expected $\delta p/p \sim 3\%$ for 3 GeV/c protons is limited by the scattering on gas and wires in the chambers. In the forward spectrometer, RICH or time-of-flight detectors are proposed as particle identification systems. Electromagnetic calorimetry in the forward region will be performed by a shashlyk-type calorimeter with high resolution and efficiency. An energy resolution of $4\%/\sqrt{E}$ is foreseen. In the very forward part, a muon detector similar to the one employed in the target region will provide muon–pion separation.

6.8.3 Charmonium and open charm physics

The study of charmonium and open charm are among the main topics in the $\overline{\text{P}}$ANDA physics program. Compared to Fermilab E760/E835, a multipurpose detector with magnetic field, $\overline{\text{P}}$ANDA has better momentum resolution and higher machine luminosity, which will allow the realization of an extensive research program in these fields.

The existence of a large hadronic background in $p\bar{p}$ annihilation represents a challenge for the study of many final states and the capability of observing a particular final state depends on the signal-to-background ratio after the selection. A full detector simulation has been developed and used to test the separation of signal from background sources, with cross sections that are orders of magnitude larger than the channels of interest, and to prove the capability of background reduction at the level needed to perform charmonium studies in $\overline{\text{P}}$ANDA. Here we will summarize the results, which appear in more detail in the $\overline{\text{P}}$ANDA Physics Performance Report [1057], of using the full simulation on some channels of interest. In all the following cases, a simple selection is adopted for the channel of interest: particles are reconstructed starting from detected tracks, using associated PID criteria, and a kinematic constraint to the energy and momentum of the beam is imposed in order to improve on detector resolutions.

The identification of charmonium states decaying into J/ψ is relatively clean due to the presence of a pair of leptons $\ell^+\ell^-$ in the final state. These channels can be used to study decays from $\psi(2S)$, χ_{cJ}, $X(3872)$, and $Y(4260)$ that contain a J/ψ in the final state. These analyses rely on the positive identification of the two leptons in the final state for the reconstruction of $J/\psi \to \ell^+\ell^-$, where the main background is represented by pairs of tracks, like $\pi^+\pi^-$, associated with large energy deposition in the electromagnetic calorimeter. The simulation shows that the resolution of the e^+e^- invariant mass for a reconstructed J/ψ in the final state is in the range 4–8 MeV, depending on the total center-of-mass energy of the interaction.

The $J/\psi\pi^+\pi^-$ final state is a key decay channel in charmonium studies. New states in the charmonia mass region, like the $X(3872)$ and the $Y(4260)$, have been discovered at the B-factories through this decay mode (see Sect. 2.3). The $\overline{\text{P}}$ANDA performance on $J/\psi\pi^+\pi^-$ has been analyzed in detail through a complete simulation of this final state in the detector. A simple selection has been performed: adding two charged pions to a reconstructed J/ψ and performing a kinematic fit with vertex constraint, the efficiency of the complete selection is approximately 30% over the energy region of interest (3.5 to 5.0 GeV). The main source of hadronic background for this channel comes from $p\bar{p} \to \pi^+\pi^-\pi^+\pi^-$, where two pions may be erroneously identified as electrons and contaminate the signal. At a center-of-mass energy around 4.26 GeV, the cross section for this process is a few tens of μb [1058], which is 10^6 times larger than the expected signal, estimated from previous results of Fermilab E835. With the selection described, the rejection power for this background source is of the order of 10^6 and the signal-to-background ratio is about 2, which should provide well-identified and relatively clean $J/\psi\pi^+\pi^-$final states in $\overline{\text{P}}$ANDA.

The discovery mode (see Sect. 2.2.1) for $h_c(1P)$ is the electromagnetic transition to the ground state charmonia $h_c(1P) \to \eta_c(1S)\gamma$, where the $\eta_c(1S)$ can then be detected through many decay modes. The decay $\eta_c(1S) \to \gamma\gamma$ is characterized by a reasonably clean signature, due to the presence of two energetic photons in the detector with the $\eta_c(1S)$ mass, albeit with a small branching fraction (see Sect. 3.2.5). A study has been performed to assess the $\overline{\text{P}}$ANDA capability to detect the $h_c(1P) \to \eta_c(1S)\gamma \to 3\gamma$ decay, in presence of background sources due to final states such as $p\bar{p} \to \pi^0\pi^0$, $\pi^0\eta$, and $\eta\eta$. Each presents hard photons in the final state and no charged tracks, a signature that could mimic the channel of interest. (See Sect. 2.2.1 for a description of a previous $p\bar{p} \to h_c(1P) \to \gamma\eta_c(2S) \to 3\gamma$ analysis and Sect. 3.2.3 for a description of a $J/\psi \to 3\gamma$ observation in $\psi(2S) \to \pi^+\pi^-J/\psi$ decays.) In order to improve the background rejection, additional cuts are applied after the event reconstruction:

- a cut on the center-of-mass energy of the energy of the γ coming from the radiative transition: $0.4 < E_\gamma < 0.6$ GeV;

- an angular cut $|\cos\theta_{CM}| < 0.6$ allows rejection of a large fraction of backgrounds (like $\pi^0\pi^0$) that are strongly peaked in the forward direction;
- to suppress η' decays, the invariant mass of the radiative γ paired with either photon coming from the decay of the $\eta_c(1S)$ candidate is required to be larger than 1 GeV.

After these cuts, the efficiency on the signal is about 8% and the background suppression of the order of 10^{-6} or larger on many background channels. The production cross section observed by E835 [48], although with large uncertainties, can be combined with the present background suppression, to obtain an estimate of the order of 90 or more for the signal-to-background ratio.

As a benchmark channel of hadronic decays, we consider $h_c(1P) \to \eta_c(1S)\gamma \to \phi\phi\gamma$, with $\phi \to K^+K^-$. Three reactions are considered to be dominant contributions to the background: $p\bar{p} \to K^+K^-K^+K^-\pi^0$, $p\bar{p} \to \phi K^+K^-\pi^0$, and $p\bar{p} \to \phi\phi\pi^0$, with one photon from the π^0 undetected. To suppress such background, it is additionally required that no π^0 candidates (photon pair with invariant mass in the 0.115–0.150 GeV region) be present in the event. The overall efficiency for signal events is ~25%. Since no experimental data is available for the three background cross sections, the only way to estimate the background contribution is to use the dual parton model (DPM); none out of 2×10^7 simulated events pass the selection. The three main background channels have been also simulated separately. With a total $p\bar{p}$ cross section of 60 mb, we estimate that $\sigma(p\bar{p} \to K^+K^-K^+K^-\pi^0) = 345$ nb, $\sigma(p\bar{p} \to \phi K^+K^-\pi^0) = 60$ nb, and $\sigma(p\bar{p} \to \phi\phi\pi^0) = 3$ nb. Using these values, a signal-to-background ratio of ≥ 8 for each of the background channels is obtained. Using these signal-to-background values, it is possible to estimate the $\overline{\text{P}}$ANDA sensitivity in the $h_c(1P)$ width measurement. A few scans of the $h_c(1P)$ have been simulated for different values of $\Gamma(h_c(1P))$. The expected shape of the measured cross section is obtained from the convolution of the Breit–Wigner resonance curve with the normalised beam energy distribution plus a background term. The simulated $h_c(1P)$ resonance shape for the case $\Gamma(h_c(1P)) = 0.5$ MeV, assuming five days of data taking per point in high-resolution mode, is shown in Fig. 105. The accuracy on the width measurement is of the order of 0.2 MeV for $\Gamma(h_c(1P))$ values in the range 0.5–1.0 MeV.

The ability to study charmonium states above $D\bar{D}$ threshold is important for the major part of the $\overline{\text{P}}$ANDA physics program, in topics like the study of open charm spectroscopy, the search for hybrids, and CP violation studies. The study of $p\bar{p} \to D\bar{D}$ as a benchmark channel will also assess the capability to separate a hadronic decay channel from a large source of hadronic background. Two benchmark channels are studied in detail:

- $p\bar{p} \to D^+D^-$ (with $D^+ \to K^-\pi^+\pi^+$)

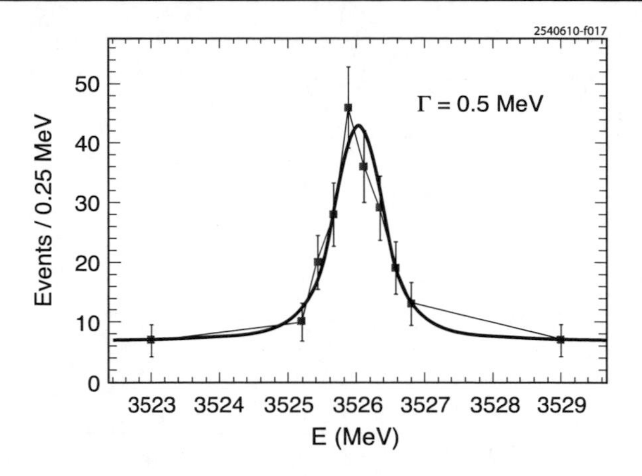

Fig. 105 Simulation of the scan of the $h_c(1P)$ (with $\Gamma(h_c(1P)) = 0.5$ MeV) for the measurement of the resonance width. Each point corresponds to five days of data taking with $\overline{\text{P}}$ANDA

- $p\bar{p} \to D^{*+}D^{*-}$ (with $D^{*+} \to D^0\pi^+$, $D^0 \to K^-\pi^+$)

The first one is simulated at the $\psi(3770)$ and the second at the $\psi(4040)$ mass energies. We assume a conservative estimate for the charmonium production cross section above the open-charm threshold, on the order of 3 nb for D^+D^- and 0.9 nb for $D^{*+}D^{*-}$ production. The background is simulated using the DPM to produce inelastic reaction in $p\bar{p}$ annihilations. A background suppression of the order of 10^7 is achieved with the previous selection. A detailed study of specific background reactions is also performed. In particular, nonresonant production of $K^+K^-2\pi^+2\pi^-$ has a cross section which is 10^6 times larger than the D^+D^- signal. A cut on the longitudinal and transverse momentum of the $D^\pm$ can reduce the background by a factor ~26, and the remaining events leave a nonpeaking background in the loose mass region defined in the preselection. The reconstructed decay vertex location will further improve the background rejection, reaching a signal-to-background ratio near unity with an efficiency for signal events of ≈8%. Under these assumptions, a conservative estimate of the number of reconstructed events per year of $\overline{\text{P}}$ANDA operation is of the order of 10^4 and 10^3 for D^+D^- and $D^{*+}D^{*-}$, respectively.

Performing fine energy scans, $\overline{\text{P}}$ANDA will be able to observe the energy dependence of a cross section in proximity to a threshold. Here we report the result obtained in the simulation of $p\bar{p} \to D_s^\pm D_{s0}^*(2317)^\mp$, to test the sensitivity to resonance parameters measurements with this technique. The assumptions used in this study are:

- a 12-point scan in a 4 MeV-wide region;
- 14 days of data with a total integrated luminosity of 9 pb^{-1}/day;
- a signal-to-background ratio of 1:3;
- 1 MeV total width for the $D_{s0}^*(2317)$.

The results of a fit to the scan simulation are presented in Fig. 106, where the mass and the width of the $D_{s0}^*(2317)$

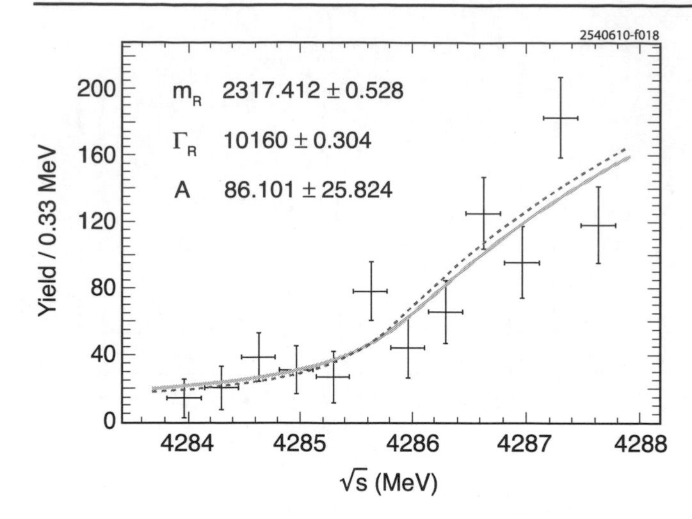

Fig. 106 Fit of the simulated excitation function for near-threshold $p\bar{p} \to D_s^{\pm} D_{s0}^*(2317)^{\mp}$. The *dashed line* corresponds to the simulated function and the *solid line* is the reconstructed curve

are free parameters. The study yields the results:

$$m = 2317.41 \pm 0.53 \text{ MeV},$$
$$\Gamma = 1.16 \pm 0.30 \text{ MeV}, \tag{195}$$

to be compared with the input values of the simulation: $m = 2317.30$ MeV and $\Gamma = 1.00$ MeV, demonstrating that such a scan would yield accurate values of mass and width with the precisions shown.

In conventional charmonia, the quantum numbers are derived directly from the excitation of the $c\bar{c}$ pair. The glue tube adds degrees of freedom that manifest themselves in unconventional quantum numbers; in the simplest $c\bar{c}g$ scenario this corresponds to the addition of a single gluon quantum number ($J^{PC} = 1^-$ or 1^+ for color-electric or color-magnetic excitation). This would result in charmonium hybrids with non-exotic and exotic quantum numbers which are expected in the 3–5 GeV mass region. Here we will sketch out the strategy for hybrids studies in P̄ANDA, with more details available in [1057]. Formation experiments would generate non-exotic charmonium hybrids with high cross sections while production experiments would yield a hybrid together with another particle like π or η. In P̄ANDA both processes are possible; the strategy would be to start searching for hybrids in production processes, fixing the $p\bar{p}$ center-of-mass energy at the highest possible value ($\sqrt{s} \simeq 5.5$ GeV) and studying all the production channels. Then hybrid formation could be studied through energy scans over the regions where possible signals have been observed in production measurements.

Aside from the benchmark channels used for the detection of conventional charmonia, hybrids can be identified through reactions like:

- $p\bar{p} \to \widetilde{\eta}_{c0,1,2}\eta \to \chi_{c1}\pi^0\pi^0\eta$;
- $p\bar{p} \to \widetilde{h}_{c0,1,2}\eta \to \chi_{c1}\pi^0\pi^0\eta$;
- $p\bar{p} \to \widetilde{\psi}\eta \to J/\psi\omega\eta$;
- $p\bar{p} \to [\widetilde{\eta}_{c0,1,2}, \widetilde{h}_{c0,1,2}, \widetilde{\chi_{c1}}]\eta \to DD^*\eta$;

namely final states with charmonia accompanied by light hadrons or final states with a DD^* pair. As a case study, we will present the results obtained for the benchmark channel $p\bar{p} \to \widetilde{\eta}_{c1}\eta \to \chi_{c1}\pi^0\pi^0\eta$. It can be assumed that the $p\bar{p} \to \widetilde{\eta}_{c1}\eta$ production cross section is of the same order of $p\bar{p} \to \psi(2S)\eta$, which is estimated to be 33 ± 8 pb [1059]. As possible sources of background, several reactions with similar topology have been considered:

- $p\bar{p} \to \chi_{c0}(1P)\pi^0\pi^0\eta$;
- $p\bar{p} \to \chi_{c1}(1P)\pi^0\eta\eta$;
- $p\bar{p} \to \chi_{c0}(1P)\pi^0\pi^0\pi^0\eta$;
- $p\bar{p} \to J/\psi\pi^0\pi^0\pi^0\eta$.

A simple analysis is carried out. Two photons are accepted as π^0 or η candidates if their invariant mass is in the range 115–150 MeV or 470–610 MeV, respectively. The χ_{c1} is formed adding a radiative photon to a J/ψ candidate, with total invariant mass within 3.3–3.7 GeV. From these, $\chi_{c1}\pi^0\pi^0\eta$ candidates are created and kinematically fit to the original beam energy-momentum, with an additional constraint for the J/ψ mass. Additional cuts on the kinematic fit CL and on the invariant masses of the intermediate decay products are applied, obtaining a total efficiency around 7% for this channel. The $\widetilde{\eta}_{c1}$ peak reconstructed in this way has a FWHM of 30 MeV. The background suppression is estimated applying the same analysis to background events. The results are summarized in Table 49.

The $J/\psi N$ dissociation cross section is as yet experimentally unknown, except for indirect information deduced from high-energy J/ψ production from nuclear targets. Apart from being a quantity of its own interest, this cross section is closely related to the attempt of identifying quark–gluon plasma (QGP) formation in ultrarelativistic nucleus–nucleus collisions: the interpretation of the J/ψ suppression observed at the CERN SPS [966, 968, 1060] as a signal for QGP formation relies on the knowledge of the "normal" suppression effect due to J/ψ dissociation in a hadronic environment. Nuclear J/ψ absorption can only be deduced from models, since the available data do not cover the kinematic regime relevant for the interpretation of the SPS re-

Table 49 Background suppression (η) for the individual background reactions

Background channel	Suppression (10^3)
$\chi_{c0}(1P)\pi^0\pi^0\eta$	5.3
$\chi_{c1}(1P)\pi^0\eta\eta$	26
$\chi_{c0}(1P)\pi^0\pi^0\pi^0\eta$	>80
$J/\psi\pi^0\pi^0\pi^0\eta$	10

sults. In antiproton–nucleus collisions the $J/\psi N$ dissociation cross section can be determined for momenta around 4 GeV/c with very little model dependence. The determination of the $J/\psi N$ dissociation cross section is in principle straightforward: the J/ψ production cross section is measured for different target nuclei of mass number ranging from light (d) to heavy (Xe or Au) by scanning the $\bar{p}$ momentum across the J/ψ yield profile whose width is essentially given by the known internal target-nucleon momentum distribution. The J/ψ is identified by its decay to e^+e^- or $\mu^+\mu^-$. The attenuation of the J/ψ yield per effective target proton is a direct measure of the $J/\psi N$ dissociation cross section, which can be deduced by a Glauber-type analysis. These studies may be extended to higher charmonium states like the $\psi(2S)$, which would allow determination of the cross section for the inelastic process $\psi(2S)N \to J/\psi N$, which is also relevant for the interpretation of the ultrarelativistic heavy-ion data. The benchmark channel studied in this context is the reaction:

$$\bar{p}^{40}\text{Ca} \to J/\psi X \to e^+e^- X. \quad (196)$$

The cross section for this process is estimated to be nine orders of magnitude smaller than the total antiproton–nucleus cross section. The results of the simulations show that it is possible to identify the channel of interest with good efficiency and acceptable signal-to-background ratio [1057].

6.9 CBM at FAIR

The Compressed Baryonic Matter (CBM) experiment will be one of the major scientific activities at FAIR [1055]. The goal of the CBM research program is to explore the QCD phase diagram in the region of high baryon densities using high-energy nucleus–nucleus collisions. This includes a study of the equation-of-state of nuclear matter at high densities, and a search for the deconfinement and chiral phase transitions. The CBM research program comprises a comprehensive scan of observables, beam energies, and collision systems. The observables include low mass dileptons, charmonia and open charm, but also collective flow of rare and bulk particles, correlations and fluctuations.

Particles with open and hidden charm are expected to provide valuable information about the conditions inside the dense fireball. For example, the excitation function of the charm particle ratios such as the $\psi(2S)/(J/\psi)$ ratio and the $(J/\psi)/D$ ratio may vary when passing the deconfinement phase transition. In addition, the initial pressure of the partonic phase influences the elliptic flow of charmonium. The transport properties of open charm mesons in dense matter, which depend on the interaction with the medium and hence on the structure of the medium, can be studied via the yield, the elliptic flow, and the momentum distributions of charmed particles. In a baryon-dominated medium, these observables are expected to differ for D and $\bar{D}$ mesons.

The experimental goal is to measure these rare probes with unprecedented precision. In order to compensate for the low yields, the measurements will be performed at exceptionally high reaction rates (up to 10 MHz for certain observables). These conditions require the development of ultrafast and extremely radiation-hard detectors and electronics. A particular challenge for the detectors, the front-end electronics, and the data-acquisition system is the online selection of displaced vertices with the extraordinarily high speed and precision needed for open-charm measurements.

A schematic view of the proposed CBM experimental facility is shown in Fig. 107. The core of the setup is a Silicon Tracking and Vertexing System located inside a large aperture dipole magnet. The vertex detector consists of 2 stations of Monolithic Active Pixel Sensors, and the tracker comprises 8 stations of double-sided microstrip sensors. Particle identification will be performed using the momentum information from the silicon tracker and the time-of-flight measured with a large area Resistive Plate Chamber (RPC) wall. Figure 107(a) depicts the setup with the Ring Imaging Cherenkov (RICH) detector for the identification of electrons from low-mass vector-

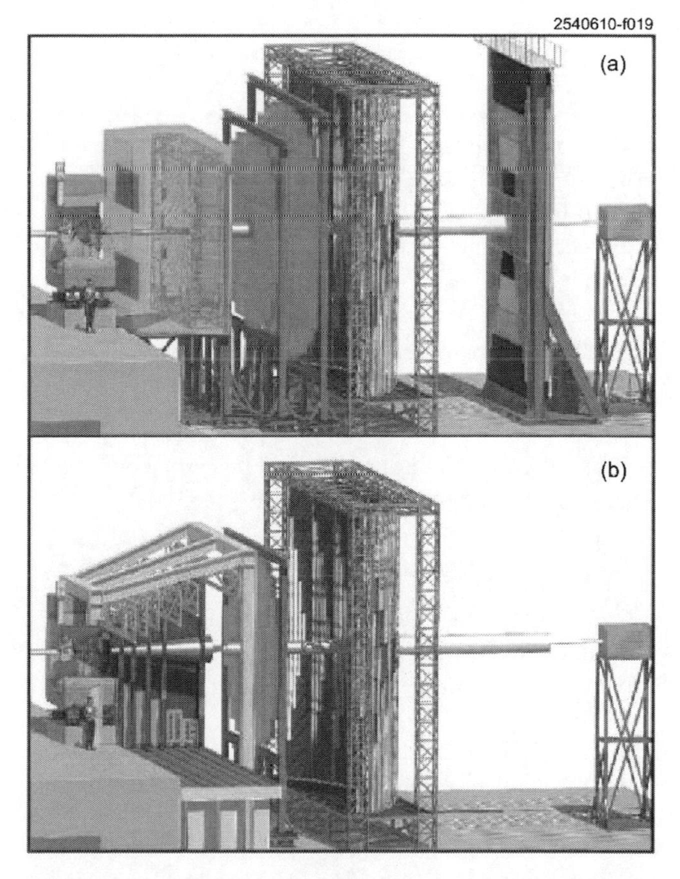

Fig. 107 The CBM experimental facility with (**a**) the RICH and TRD electron detectors, or (**b**) the muon detection system

meson decays. The Transition Radiation Detector (TRD) will provide charged particle tracking and the identification of high-energy electrons and positrons. The Electromagnetic Calorimeter (ECAL) will be used for the identification of electrons and photons. The muon detection/hadron absorber system is shown in Fig. 107(b). It consists of 5 double- or triple-stations of highly granulated gaseous micropattern chambers, e.g., Gas–Electron Multiplier (GEM) detectors, sandwiched by iron plates with a total thickness equivalent to 13 absorption lengths. The status of detector R&D and recent results of detailed simulations are documented in [1061]. The CBM Collaboration consists of more than 450 scientists from 55 institutions and 14 countries.

6.10 Tau–charm factory in Novosibirsk

A tau–charm factory can address various issues concerning τ-leptons, charmonia, open-charm particles, and light-quark spectroscopy in a unique manner. Indeed, the B-factories have inadequate sensitivity for some of these physics topics, leaving a tau–charm factory as the only practical avenue for substantial progress. A next-generation tau–charm factory is now under consideration in Novosibirsk. A novel approach of the Crab Waist collision scheme [1050, 1062] allows reaching luminosity of (1–2) $\times$ 10^{35} cm^{-2} s^{-1}. Suggested priorities include

- $D\bar{D}$ mixing;
- search for CP violation in charm decays;
- study of rare and forbidden decays of open charm mesons;
- high-precision study of regular charmonia and charmonium-like states;
- tests of the Standard Model in τ-lepton decays;
- search for lepton flavor violation;
- search for CP/T violation in τ-lepton decays;
- extensive study of light-quark (u, d, s) states between 1 and 3 GeV using ISR;
- production of polarized antinucleons.

This experimental program would be carried out at a facility with the following basic features [1063]:

- collision energy varying from 2–5 GeV;
- luminosity of $\sim$5 $\times$ 10^{34} cm^{-2} s^{-1} at 2 GeV and more than 10^{35} cm^{-2} s^{-1} at the τ-production threshold;
- a longitudinally polarized-electron beam at the interaction point (IP) extending the experimental possibilities of the facility;
- extensive use of superconducting wigglers allowing control of damping parameters and tuning for optimal luminosity in the whole energy range;
- e^+e^- center-of-mass energy calibration with relative accuracy of $\approx$5 $\times$ 10^{-4}, achieved with the Compton backscattering technique.

The planned factory has separate rings for electrons and positrons and one interaction region. Each ring features $\sim$800 m circumference and a racetrack shape, with two arcs and two long ($\sim$100 m) straight sections to accommodate the injection and radio-frequency (RF) equipment for the machine and the interaction region for the experiment. The injection facility includes a full-energy (2.5 GeV) linear accelerator equipped with the polarized-electron source and a positron complex with a 500 MeV linac, converter, and accumulating ring. The injection facility operates at 50 Hz and can produce $\geq 10^{11}$ positrons per second. High luminosity is provided by the Crab Waist collision approach, which assumes the beam intersection at large Piwinski angle and local focusing of the beams at the IP by means of two Crab Sextupoles with properly matched betatron phase advance in between. Local focusing rotates the vertical waist at IP according to the horizontal displacement of each individual particle, decouples the betatron oscillations, and therefore effectively reduces the beam-beam coupling betatron resonances.

6.11 Charmonium photoproduction facilities

The mechanism of charmonia photoproduction and their interactions with hadrons and nuclei have been the subject of much interest since their discovery (see Sect. 4.3). Generally, because of the small size of these heavy mesons on the hadronic scale of $\sim$1 fm, it is expected that one can apply QCD to describe their interactions with hadronic matter. Heavy quarkonium production thus probes the local color fields in the target and can reveal properties such as their response to momentum transfer and their spatial distribution, which are of fundamental interest for understanding nucleon structure in QCD. While this interpretation is valid at all energies, the details (what mechanism produces the relevant color fields, which configurations in the target are their main source) vary considerably between high energies and the near-threshold region, calling for detailed experimental and theoretical study of this fascinating landscape.

6.11.1 J/ψ photoproduction at high energy

The mechanism of exclusive J/ψ photoproduction is well-understood at high energies ($W > 10$ GeV) and $|t| < 1$ GeV, where the coherence length is large compared to the nucleon size, $l_{\text{coh}} \gg 1$ fm: the process takes the form of the scattering of a small-size color dipole off the target (Fig. 108(a)). The leading interaction in the small-size expansion is via two-gluon exchange with the target. The nucleon structure probed in this case is the gluon generalized parton distribution (or GPD), which describes the two-gluon form factor of the target and is normalized to the usual gluon density in the zero momentum-transfer limit. In the process shown in

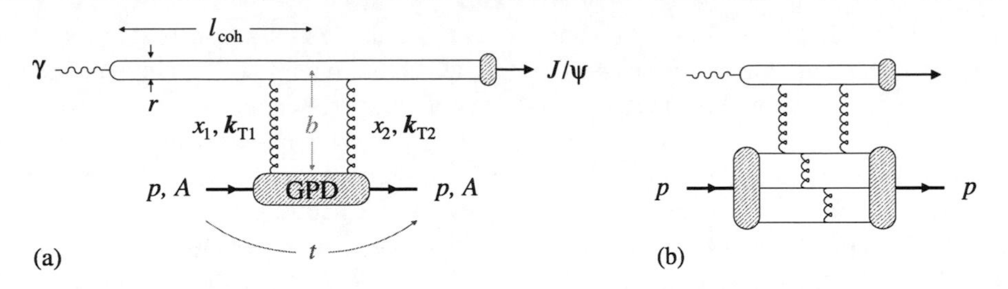

Fig. 108 Dynamical mechanisms of J/ψ photoproduction. (**a**) Two-gluon exchange mechanism at high and intermediate energies, $W - W_{th} >$ few GeV, based on QCD factorization for hard exclusive meson production. The space-time evolution in the target rest frame corresponds to the dipole picture of high-energy scattering [1064, 1065], where l_{coh} is the coherence length, r the transverse size of the $c\bar{c}$ dipole and b its impact parameter with the target. Here $x_{1,2}$ and $\boldsymbol{k}_{T1,2}$ denote the longitudinal momentum fractions and transverse momenta of the exchanged gluons. The invariant momentum transfer to the target proton or nucleus is small and of the order of the inverse target size, $|t| \sim |\boldsymbol{k}_{T1} - \boldsymbol{k}_{T2}|^2 \sim R_{target}^{-2}$. (**b**) Coherent multi-gluon exchange proposed for near-threshold production on the proton, analogous to the hard scattering mechanism for high-t elastic form factors. Near threshold the minimum value of the invariant momentum transfer $|t_{min}|$ is large

Fig. 108(a), the transverse momenta of the exchanged gluons are large, $|\boldsymbol{k}_{T1,2}| \sim m_{J/\psi}$, but their difference can be small, resulting in a small invariant momentum transfer to the target $|t| \sim |\boldsymbol{k}_{T1} - \boldsymbol{k}_{T2}|^2$. Hence the reaction can leave a proton or a nuclear target in its ground state or a slightly excited state. Experiments in exclusive J/ψ photo- and electroproduction at HERA [751, 752] have confirmed this picture through detailed measurements of the Q^2-independence of t-slopes in electroproduction, energy dependence of the cross section, and comparison with other exclusive vector-meson channels (universality of the gluon GPD); see [1066] for a review. They have also measured the t-slope of the differential cross section and its change with energy, which allows one to infer the average transverse radius of gluons in the nucleon and its change with x (Fig. 109). This information represents an essential input to small-x physics (initial condition of evolution equations) and the phenomenology of high-energy pp collisions with hard processes [1067].

Of particular interest is elastic or quasi-elastic scattering of a charmed dipole on nuclei (Fig. 108(a)), with a subsequent transformation of the $c\bar{c}$ into a J/ψ. The momentum transfer is measurable in this reaction because it is the difference between the transverse momentum of the incoming virtual photon and the final meson. Caldwell and Kowalski [1068] propose investigation of the properties of nuclear matter by measuring the elastic scattering of J/ψ on nuclei with high precision. The J/ψ mesons are produced from photons emitted in high-energy electron–proton or electron–nucleus scattering in the low-x region. Over the next few years, some relevant data should be available from RHIC, where STAR is collecting data on J/ψ photoproduction. Based on the recent RHIC Au+Au luminosity, a sample of order 100 events might be expected. Such measurements could be performed at the future ENC (Electron–nucleon Collider, GSI), Electron–Ion Collider (EIC, Brookhaven National Lab, or Jefferson Lab (JLab)), or LHeC (Large Hadron-electron Collider, CERN) facilities. The advantage of J/ψ photoproduction compared to the electroproduction of light vector mesons is its high cross section and small dipole size, even at $Q^2 = 0$. In addition, the momenta of the decay products of $J/\psi \to \mu^+\mu^-$ or e^+e^- can be precisely measured. The smallness of the dipole in low-x reactions

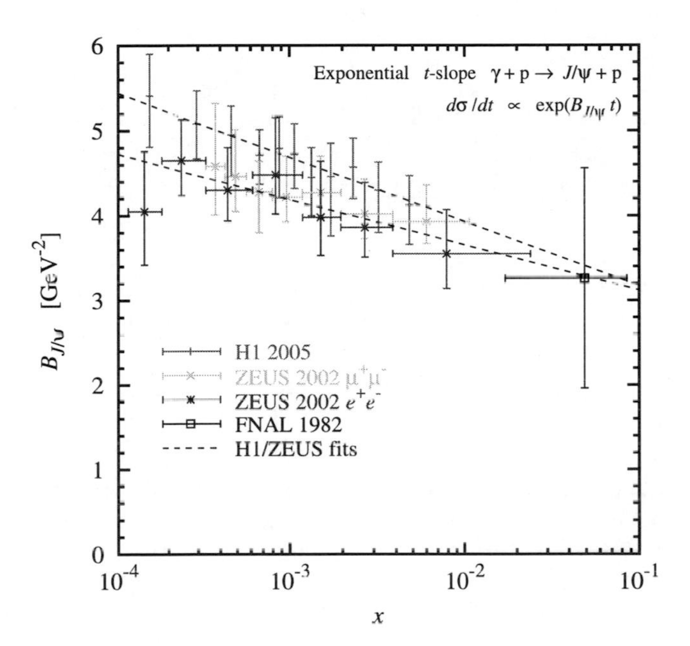

Fig. 109 The exponential t-slope, $B_{J/\psi}$, of the differential cross section of exclusive J/ψ photoproduction measured in the FNAL E401/E458 [1070], HERA H1 [751], and ZEUS [753] experiments, as a function of $x = m_{J/\psi}^2/W^2$ [1109]. (In the H1 and ZEUS results, the quoted statistical and systematic uncertainties were added linearly.) The *dashed lines* represent the published two-dimensional fits to the H1 and ZEUS data. The average squared transverse radius of gluons with momentum fraction x in the nucleon can be inferred from the measured slope as $\langle b^2 \rangle_g = 2(B_{J/\psi} - \Delta B)$, where $\Delta B \approx 0.3$–0.6 GeV^{-2} accounts for the finite transverse size of the $c\bar{c}$ pair in the production amplitude [1064, 1065]. The data show that the nucleon's gluonic transverse radius at $x \sim 10^{-1}$ is smaller than the transverse charge (Dirac) radius and increases slowly toward small x; see [1065, 1109] for details

assures that the interaction is mediated by gluon exchange only. Thus the deflection of the J/ψ directly measures the intensity and the spatial distribution of the nuclear gluon field.

The measurement of J/ψ scattering on nuclei could become an important source of information on nuclear structure and high-density QCD. In the absence of nuclear shadowing, the interaction of a dipole with a nucleus can be viewed as a sum of dipole scatterings of the nucleons forming the nucleus. The size of the $c\bar{c}$ dipole in elastic J/ψ scattering is around 0.15 fm, i.e., it is much smaller than the nucleon radius. It is therefore possible that dipoles interact with smaller objects than nucleons; e.g., with constituent quarks or hot spots. The conventional assumption is that the nucleus consists of nucleons and that dipoles scatter on an ensemble of nucleons according to the Woods–Saxon [1069] distribution,

$$\rho_{WS}(r) = \frac{N}{1+\exp(\frac{r-R_A}{\delta})},$$
$$R_A \equiv 1.12A^{\frac{1}{3}} - 0.86A^{-\frac{1}{3}} \text{ fm}, \tag{197}$$

where R_A is the nuclear radius for atomic number A, $\delta = 0.54$ fm is the skin depth, and N is chosen to normalize $\int d^3\vec{r}\rho_{WS}(r) = 1$. Under the foregoing assumption (or slight variations thereon), the dipole model predicts the coherent and incoherent nuclear cross sections shown in Fig. 110. Deviations from the predicted $|t|$-distribution reveal the effects of nuclear shadowing, which depend on the effective thickness of the target and thus change with the dipole impact parameter, b. In the coherent process, the nucleus remains in its ground state. In the incoherent process, the nucleus gets excited and frequently breaks into nucleons or nucleonic fragments. Experimentally we expect to be able to distinguish cases where the nucleus remains intact and cases where the nucleus breaks up. In the nuclear breakup process, there are several free neutrons and protons in the final state, as well as other fragments. The number of free nucleons could depend on the value of the momentum transfer. The free nucleons and fragments have high momenta and different charge-to-mass ratios than the nuclear beam and should therefore be measurable in specialized detectors.

The transverse momenta of the J/ψ can be determined, in a TPC detector with 2 m radius and 3.5 T magnetic field, with a precision of $O(1)$ MeV. The momenta of the breakup protons can be precisely measured in the forward detector. Therefore a measurement of t-distributions together with a measurement of nuclear debris could become a source of invaluable information about the inner structure of the gluonic fields of nuclei [1068].

6.11.2 J/ψ photoproduction at low energy

Measurements of exclusive J/ψ photoproduction at lower energies were performed in the Fermilab broadband beam experiment [1070], which detected the recoiling proton, as well as several other experiments [1071, 1072]. The few existing data suggest that the two-gluon exchange mechanism of Fig. 108(a) continues to work in this region, and the gluonic size of the nucleon observed in these experiments consistently extrapolates to the HERA values (Fig. 109) [1073]. However, no precise differential measurements are available for detailed tests of the reaction mechanism. New data in

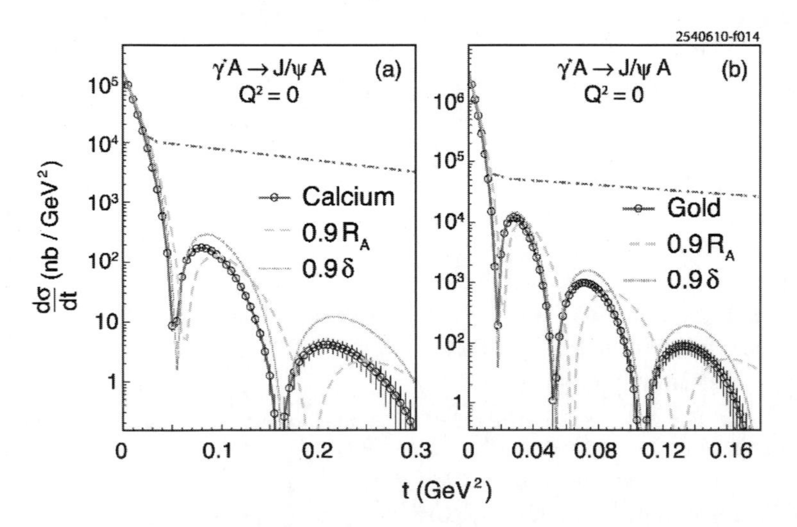

Fig. 110 The prediction [1068] of the dipole model for the t-distribution of coherent J/ψ photoproduction on (**a**) calcium and (**b**) gold nuclei, assuming that the single-nucleon distribution can be identified with the Woods–Saxon distribution, parametrized by the nuclear radius R_A and skin depth δ as in (197). The simulated measurements with nominal R_A and δ (*open circles and solid curve*) have statistical error bars based on an assumed collected sample of 10^6 events. Somewhat smaller values of R_A or δ (by 10% or 20%, respectively) result in the *dashed* and *dotted* curves, respectively. The *dot-dashed curve* shows the sum of the coherent and incoherent processes in the case of no correlations

this region are expected from the COMPASS experiment at CERN and a future Electron-Ion Collider (EIC).

Charmonium photoproduction near threshold will be studied in experiments at JLab. The Continuous Electron Beam Accelerator Facility (CEBAF) at JLab delivers 500 MHz electron beams with energies up to 6 GeV to three experimental halls. The ongoing upgrade will increase the maximum energy to 12 GeV. The first beam delivery is planned for 2014. Halls A and C are equipped with small-acceptance, high-resolution spectrometers and able to receive beam currents up to 100 μA. Hall B is equipped with a large-acceptance toroidal spectrometer (CLAS) and is able to receive up to 0.1 μA. A new Hall D, being built as a part of the upgrade, will use a tagged-photon beam with a 12 GeV endpoint and intensity up to 100 MHz/GeV.

In exclusive J/ψ photoproduction near threshold ($E_{\gamma,\text{thr}} = 8.21$ GeV for the free nucleon), to be studied with JLab 12 GeV, the reaction mechanism is expected to change in several important aspects. First, the minimum invariant momentum transfer to the nucleon becomes large: $|t_{\text{min}}| = 2.23\ \text{GeV}^2$ at threshold, and $|t_{\text{min}}| = 1.3\text{–}0.3\ \text{GeV}^2$ at $E_\gamma = 8.5\text{–}12$ GeV. This requires more exceptional high-momentum configurations in the target to bring about an exclusive transition, similar to elastic eN scattering at large t. It is expected that, in this situation, color correlations in the wave functions play an essential role, suggesting a new avenue for their experimental study. Second, while at high energies the production amplitude is mostly imaginary (absorptive), near threshold the real part plays an essential role, rendering the partonic interpretation of the production process more complex.

Two possible scenarios for near-threshold exclusive production on the nucleon have been proposed. One scenario assumes that the two-gluon exchange mechanism of Fig. 108(a) continues to apply, justified by the small size of the J/ψ, and extends this description to the near-threshold region [1073]. Here the challenge lies in modeling the gluon GPD in the "extreme" near-threshold region, characterized by large t_{min} and large "skewness" (difference in momentum gluon fractions $x_1 \neq x_2$). Some support for this picture comes from the fact that the exclusive ϕ electroproduction data at JLab with 6 GeV beam energy [1074, 1075] are well described by a dynamical model based on gluon GPDs [1076]. The other scenario is based on analogy with the hard scattering mechanism for high-t elastic form factors. It assumes that the production process happens predominantly in the valence (3-quark) configuration of the nucleon and that the momentum transfer is balanced via hard-gluon exchange [1077] as illustrated in Fig. 108(b). The two basic pictures make different predictions for the energy dependence of the cross section and t-slope near threshold, and can be tested with the expected JLab 12 GeV data. The quantitative implementation of the above scenarios, and a possible unified description, are the subjects of ongoing theoretical research. Independently of the details, the expected JLab data will greatly advance our knowledge of color correlations and the gluonic response of the nucleon at low energies.

The cross section for exclusive J/ψ photoproduction at 11 GeV is expected to be 0.2–0.5 nb, rapidly falling toward the threshold. The JLab experiment plans to map out the J/ψ differential cross section in the region from 9 to 12 GeV.

Another objective of J/ψ production experiments is to study the interaction of the produced system with nuclear matter at low energies. The small coherence and formation lengths allow extraction of the J/ψ-nucleon cross sections from the A-dependence of the J/ψ photoproduction cross section with minimal corrections to the color-transparency effects. The only such experiment done at sufficiently low energies [1078] obtained 3.5 ± 0.9 mb; the signal was extracted from a single-muon transverse momentum spectrum and the background level was not well understood. A new experiment at JLab can reduce the statistical and systematic errors by a factor of three.

A proposal to study J/ψ photoproduction close to threshold in Hall C has been conditionally approved by the JLab Program Advisory Committee. The lepton decay modes of J/ψ will be detected. In spite of a small acceptance of $\sim 3 \times 10^{-4}$ to J/ψ decay products, the expected rate of detected J/ψ is 150–200 per hour in a 2.2% radiation-length-thick liquid hydrogen target. The effective photon flux of the 50 μA electron beam will be increased due to a radiator in front of the target which has a thickness of 7% of a radiation length. The high resolution of the Hall C spectrometers will provide strong background suppression.

6.12 Proposed $\bar{p}p$ project at Fermilab

A uniquely capable and cost-effective multipurpose experiment could be mounted by adding a magnetic spectrometer to the existing Fermilab E760 lead-glass calorimeter [1079] using an available BESS solenoid [1080], fine-pitch scintillating fibers (SciFi), the DØ SciFi readout system [1081], and hadron ID via fast timing [1082]. If the relevant cross sections are as large as expected, this apparatus could produce world-leading measurements of $X(3872)$ properties, along with those of other charmonium and nearby states, as is now proposed[23] to Fermilab [1083, 1084]. The Fermilab Antiproton Accumulator's 8 GeV maximum kinetic energy and ability to decelerate down to $\approx$3.5 GeV suit it well for studies in this mass region. If approved, the experiment could start about a year after completion of the Tevatron Collider run.

[23] The experiment would also address nonquarkonium topics, such as charm mixing and CP violation.

Antiproton Accumulator experiments E760 and E835 made the world's most precise ($\lesssim$100 keV) measurements of charmonium masses and widths [1085, 1086], thanks to the precisely known collision energy of the stochastically cooled $\bar{p}$ beam (with its $\approx$0.02% energy spread) with a hydrogen cluster-jet target [1087]. Significant charmonium-related questions remain, most notably the nature of the mysterious $X(3872)$ state [32] and improved measurements of the $h_c(1P)$ and $\eta_c(2S)$ [1]. The width of the X may well be $\ll$1 MeV [122]. This unique $\bar{p}p$ precision would have a crucial role in establishing whether the $X(3872)$ is a $D^{*0}\bar{D}^0$ molecule [1088], a tetraquark state [285], or something else entirely.

The $\bar{p}p \to X(3872)$ formation cross section may be similar to that of the χ_c states [721, 1089]. The E760 χ_{c1} and χ_{c2} detection rates of 1 event/nb^{-1} at the mass peak [1090] and the lower limit $\mathcal{B}(X(3872) \to \pi^+\pi^- J/\psi) > 0.042$ at 90% CL [124] imply that, at the peak of the $X(3872)$, about 500 events/day can be observed. (Although CDF and DØ could also amass $\sim$10^4 $X(3872)$ decays, backgrounds and energy resolution limit their incisiveness.) Large samples will also be obtained in other modes besides $\pi^+\pi^- J/\psi$, increasing the statistics and improving knowledge of $X(3872)$ branching ratios.

While the above may be an under- or overestimate, perhaps by as much as an order of magnitude, it is likely that a new experiment at the Antiproton Accumulator could obtain the world's largest clean samples of $X(3872)$, in perhaps as little as a month of running. The high statistics, event cleanliness, and unique precision available in the $\bar{p}p$ formation technique could enable the world's smallest systematics. Such an experiment could provide a definitive test of the nature of the $X(3872)$.

6.13 Future linear collider

A high-energy e^+e^- linear collider provides excellent possibilities for precision and discovery, within the Standard Model as well as for new physics. The International Linear Collider ILC [1091] is a proposed machine based on superconducting RF cavities that will provide center-of-mass energies of up to 500 GeV, with the possibility for an upgrade to 1 TeV. At the design energy of 500 GeV, it will deliver a luminosity of 2×10^{34} cm^{-2} s^{-1}, providing high-statistics datasets for precision studies. Two mature ILC detector concepts, ILD [1092] and SID [1093] exist, both sophisticated general-purpose detectors with excellent tracking and vertexing capabilities and unprecedented jet energy resolution achieved with highly granular calorimeter systems and particle flow reconstruction algorithms. In parallel, the technology for the Compact Linear Collider CLIC [1094], using a two-beam acceleration scheme to reach center-of-mass energies up to 3 TeV, is being developed. In a staged construction, such a machine would initially run at energies comparable to the design goals for the ILC before reaching the multi-TeV regime. The detector concepts for CLIC are based on the already-mature ILC detectors, with some modifications to account for the higher final collision energy. These planned colliders are excellent tools for top physics. The precise control of the beam energy at a linear e^+e^- collider allows a scan of the $t\bar{t}$ production threshold to determine the top mass, width, and production cross section [1095].

There are several interesting observables in the $t\bar{t}$ threshold region that can be used for these measurements. These are the total cross section, top momentum distribution, top forward–backward asymmetry and the lepton angular distribution in the decay of the top quark. The total cross section as a function of $\sqrt{s}$ rises sharply below the threshold and peaks roughly at the position of the would-be $1S$ $t\bar{t}$ resonance mass [1096]. The normalization at the peak is proportional to the square of the resonance wave function at the origin and inversely proportional to the top-quark decay width. Theoretically, the resonance mass (hence the peak position) can be predicted very accurately as a function of m_t and α_s, so that the top-quark mass can be determined by measuring the peak position accurately. We may determine the top-quark width and the top-quark Yukawa coupling from the normalization, since the exchange of a light Higgs boson between t and $\bar{t}$ induces a Yukawa potential, which affects the resonance wave function at the origin [1097].

The top-quark momentum distribution is proportional to the square of the resonance wave function in momentum space [1098, 1099]. The wave function of the would-be $1S$ toponium resonance can be measured, since the top quark momentum can be experimentally reconstructed from the final state, unlike in the bottomonium or charmonium cases. The wave function is also predicted theoretically and is determined by α_s and m_t. The top-quark forward–backward asymmetry below threshold is generated as a result of interference between the S-wave and P-wave resonance states [1100]. This interference is sizable, since the top-quark width is not very different from the level splitting between the $1S$ and $1P$ states, and since the width is much larger than the S-P splittings of the excited states. Thus, by measuring the forward–backward asymmetry, one can obtain information on the level structure of the S and P-wave resonances in the threshold region. Both the top momentum distribution and forward–backward asymmetry provide information on $t\bar{t}$ dynamics, which are otherwise concealed because of the smearing due to the large top decay width.

It is known that the charged lepton angular distribution is sensitive to the top-quark spin. By measuring the top-quark spin in the threshold region [1101], it is possible to extract the chromoelectric and electric dipole moments of the top quark, which are sensitive to CP-violations originating from beyond-the-standard-model physics [1102].

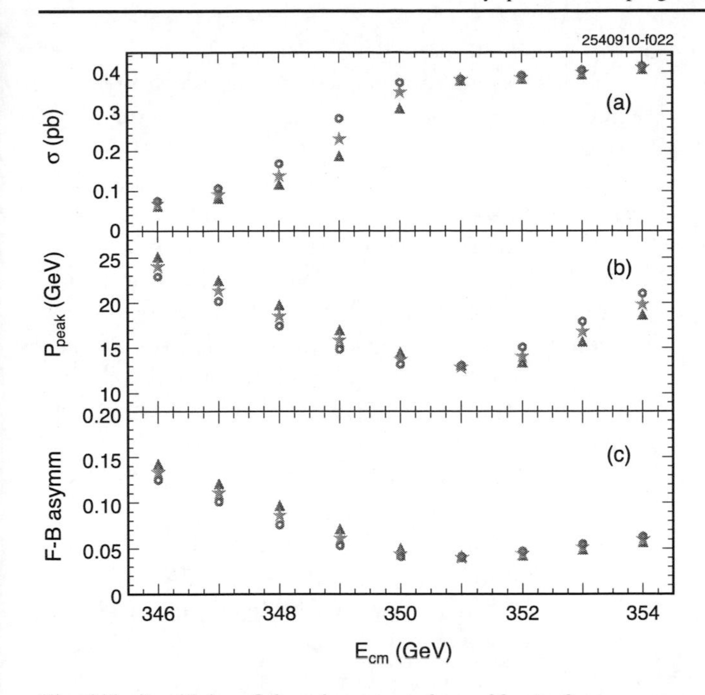

Fig. 111 Sensitivity of three important observables to the top mass from a simulated threshold scan at a linear collider: (**a**) the top production cross section, (**b**) the peak of the top momentum distribution P_{peak}, and (**c**) the forward–backward asymmetry. The *solid triangles* (*open circles*) correspond to input MC generated with top mass that is 200 MeV higher (lower) than that used for the MC sample with nominal top mass, represented by *stars*. Adapted from [1103] with kind permission, copyright (2003) Springer

By combining the collision energy dependence of these observables, precise measurements of the top parameters are possible. Figure 111 shows simulations of the sensitivity of the cross section, the peak of the top momentum distribution and of the forward–backward asymmetry to the top mass in a threshold scan at a linear collider [1103]. (See [1104] for an earlier, similar analysis.) To illustrate the sensitivity, three different input masses, spaced by 200 MeV, are shown. From a simultaneous fit to these three observables, a statistical uncertainty of 19 MeV on m_t was obtained, neglecting sources of systematic errors such as the determination of the luminosity spectrum of the collider or theoretical uncertainties. To achieve the highest possible precision, detailed understanding of the beam energy spectrum, of beamstrahlung and of initial-state radiation is crucial. The measured cross section in such a scan is theoretically well described, as discussed in detail in Sect. 6 of [1] and references therein. Using NNLO QCD calculations, an overall experimental and theoretical error of 100 MeV on the top mass is reachable with a threshold scan at a linear collider [1105]. Further important improvements in precision arise from NNNLO QCD calculations, the summation of large logarithms at NNLL order and a systematic treatment of electroweak and top-quark finite-lifetime effects. Also, the determination of corrections beyond NLO for differential observables such as the momentum distribution and the forward–backward asymmetry

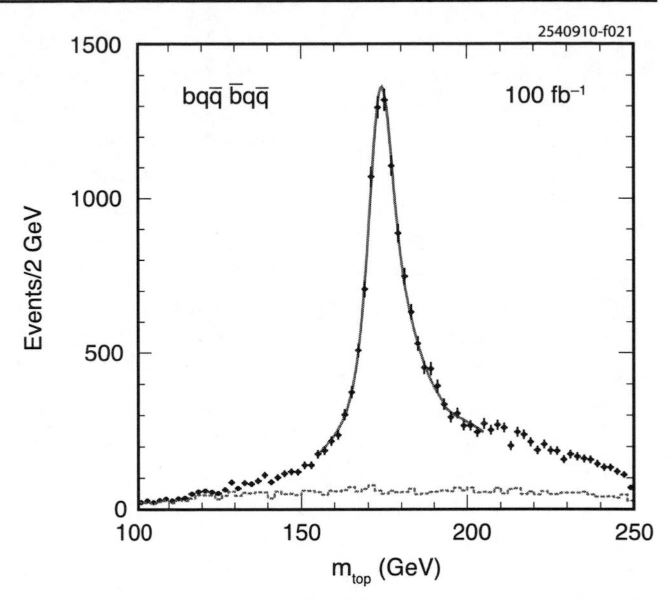

Fig. 112 Distribution of the reconstructed top-quark invariant mass for the fully hadronic $t\bar{t} \to bq\bar{q}\bar{b}q\bar{q}$ decay channel in the ILD detector concept for an integrated luminosity of 100 fb^{-1}, including the full non-top background, indicated in the *dotted histogram*. The *solid curve* represents a fit for the top mass. From [1092]

used in the study described above would be required to fully exploit the precision of the expected data.

A method has been developed to improve accuracies of kinematic variables in reconstructed top-quark events in the threshold region [1106]. This method would improve measurements of the top-quark observables mentioned above.

An alternative to the threshold-region measurements described above is reconstruction of decay products in tt pair production far above threshold. This approach has the advantage that the top measurement is possible at full collider energy in parallel with other measurements, resulting in much higher integrated luminosity. The top quarks decay essentially exclusively into a b quark and a W boson, giving rise to two b jets and additional jets or leptons from the W decays in the final state. The expected performance of the ILC detectors leads to very high precision in the invariant-mass determination. Due to the clean environment in e^+e^- collisions, a measurement in the full-hadronic six-jet mode, which has the highest branching fraction, is possible. Figure 112 shows the reconstructed top-quark invariant mass in the fully hadronic decay mode obtained from a simulation for the ILD detector concept [1092] at an energy of 500 GeV and corresponding to an integrated luminosity of 100 fb^{-1}. The simulation uses a complete detector modeling and standard model backgrounds. The statistical error on the mass reconstruction is 90 MeV in this channel alone. With a higher integrated luminosity of 500 fb^{-1} and by combining measurements in fully hadronic and in semileptonic top decays, a statistical precision of 30 MeV seems reachable, comparable to the experimental precision of a threshold scan. However, the theoretical interpretation of the invariant mass is considerably more complicated than in the case of threshold measurements. Recently, a systematic formalism to relate the measured invariant mass to theoretically meaningful parameters has been established [1107, 1108]. This formalism has the potential to compete with the threshold-scan calculations in terms of achievable precision. The computations were carried out to NLL. Further higher-order calculations are needed to reduce the present theoretical uncertainties and fully exploit the experimental precision achievable at a future linear collider. See also Sect. 2.8.3.

Beyond precision measurements of the top mass and width, a Linear Collider would also provide excellent conditions for a wealth of other top-related studies, such as the measurement of the top Yukawa coupling in the process $e^+e^- \to t\bar{t}H$ and the search for anomalous couplings in decays.

7 Conclusions and priorities

Below we present a summary of the most crucial developments in each of the major topics and suggested directions for further advancement.

Spectroscopy An overview of the last decade's progress in heavy-quarkonium spectroscopy was given in Sect. 2.

With regard to experimental progress, we conclude:

1. New measurements of inclusive hadronic cross sections (i.e., R) for e^+e^- collisions just above open $c\bar{c}$ and $b\bar{b}$ flavor thresholds have enabled improved determinations of some resonance parameters but more precision and fine-grained studies are needed to resolve puzzles and ambiguities. Likewise, progress has been made studying exclusive open-flavor two-body and multibody composition in these regions, but further data are needed to clarify the details. Theory has not yet been able to explain the measured exclusive two-body cross sections.
2. Successful observations were made (Table 4) of six new conventional heavy-quarkonium states (four $c\bar{c}$, two $b\bar{b}$); of these, only the $\eta_b(1S)$ lacks a second, independent 5σ confirmation. Improved measurement of $\eta_c(1S)$ and $\eta_c(2S)$ masses and widths would be quite valuable. Unambiguous observations and precise mass and width measurements are needed for $\eta_b(2S)$, $h_b(^1P_1)$, $\Upsilon(1^3D_1)$, and $\Upsilon(1^3D_3)$ in order to constrain theoretical descriptions.
3. Experimental evidence has been gathered (Table 9) for up to 17 unconventional heavy-quarkonium-like states. All but $Y_b(10888)$ are in the charmonium mass region, and all but five remain unconfirmed at the 5σ level. Confirmation or refutation of the remaining 12 is a high priority.
4. Theoretical interpretations for the unconventional states (Table 20) range from coupled-channel effects to quark–gluon hybrids, mesonic molecules, and tetraquarks. More measurements and theoretical investigations are necessary to narrow the possibilities. In particular, high-resolution measurements of lineshapes promise deeper insights into the nature of various of those states.
5. The $X(3872)$ was the first unexpected state to be observed and has generated the most experimental and theoretical attention. Its sub-MeV proximity to $D^{*0}\bar{D}^0$-threshold (Tables 10–11) and dominant $D^0\bar{D}^0\pi^0$ branching fraction suggest a $D^{*0}\bar{D}^0$-molecular component, although this interpretation is not universally shared. The $X(3872)$ has been confirmed in four decay modes (Table 12). The discovery mode, $\pi^+\pi^-J/\psi$, is still the best measured, and has a branching fraction comparable in size to that of $\omega J/\psi$; $D^0\bar{D}^0\pi^0$ is ten times more common and $\gamma J/\psi$ three times less. The $X(3872)$ quantum numbers have been narrowed to 1^{++} or 2^{-+}.
6. The charged Z states observed in $Z^- \to \pi^-\psi(2S)$ and $\pi^-\chi_{c1}$ would be, if confirmed, manifestly exotic. Hence their confirmation or refutation is of the utmost importance.

With regard to lattice QCD calculations:

7. Lattice QCD technology has progressed to the point that it may provide accurate calculations of the energies of quarkonium states below the open flavor threshold, and also provide information about higher states.
8. Precise and definitive calculations of the $c\bar{c}$ and $b\bar{b}$ meson spectra below threshold are needed. Unquenching effects, valence quark annihilation channels and spin contributions should be fully included.
9. Unquenched calculations of states above the open-flavor thresholds are needed. These would provide invaluable clues to the nature of these states.
10. The complete set of Wilson loop field strength averages entering the definition of the nonperturbative $Q\bar{Q}$ potentials must be calculated on the lattice.
11. Calculations of local and nonlocal gluon condensates on the lattice are needed as inputs to weakly coupled pNRQCD spectra and decay calculations.
12. NRQCD matching coefficients in the lattice scheme at one loop (or more) are needed.
13. Higher-order calculations of all the relevant quantities due to the lattice-to-$\overline{MS}$ scheme change are required in order to relate lattice and continuum results in the EFT.
14. Lattice calculations of the overlap between quarkonia and heavy-light states in the threshold region, as well as with hybrids or exotic states, should be performed.
15. A better determination of the r_0 lattice scale and a nonperturbative determination of $\Lambda_{\overline{\rm MS}}$ with $2+1$ or $2+1+1$ sea quarks is needed.

With regard to effective field theories (EFTs),

16. Higher-order perturbative EFT calculations of static energies, static potentials, and relativistic corrections to the potentials and energy levels have appeared recently for different heavy-quark/antiquark configurations. Further efforts in this direction are needed, and the emerging patterns of renormalons should be studied in relation to the behavior of the bound states perturbative series.
17. We have described simulation studies for a future linear collider which demonstrate that precise determinations of the top mass and the top Yukawa coupling can be obtained from a $t\bar{t}$ production scan near threshold. To at least match this expected precision, a complete NNLL computation is necessary to obtain a firmer grasp of the theoretical uncertainties. The complete NNNLO computation is also desirable to assess the importance of the resummation of logarithms. The electroweak and non-factorizable corrections may be calculated by developing an effective theory description for unstable particles.
18. Better experimental data for $b\bar{b}$ production above threshold would diminish the impact of the experimental error in nonrelativistic sum rules determinations of the bottom quark mass.
19. The 2σ discrepancy between the EFT calculation and measurements of the $\eta_b(1S)$ mass needs to be resolved.

Decay Section 3 described the enormous progress on heavy-quarkonium decays, showing that many branching fraction, width, and spectra measurements have attained high precision. Some of these results provide crucial anchors for theoretical approaches, while others have just scratched the surface of what may be attainable with more data and improved techniques. Accomplishments and priorities are:

20. Precise measurements of the dileptonic and total widths in charmonium (Tables 28–31) and bottomonium (Tables 32–35) below the respective open-flavor thresholds have been performed.
21. A novel and empirical measurement of the nontrivial radiative photon lineshape in J/ψ and $\psi(2S)$ decays to $\eta_c(1S)$, which in turn enabled determination of much-improved branching fractions, has appeared. This has stimulated theoretical activity to explain the photon spectral shape and raised the importance of measuring the corresponding spectra for $\psi(2S) \to \gamma\eta_c(2S)$, $\psi(2S) \to \gamma\chi_{cJ}(1P)$, $\Upsilon(nS) \to \gamma\eta_b(1S, 2S)$, and $\Upsilon(nS) \to \gamma\chi_{bJ}(nP)$. Precise measurement of the $\psi(2S) \to \gamma\chi_{cJ}(1P)$ lineshapes will be essential in quantifying the branching fraction for the nonresonant $\psi(2S) \to \gamma\gamma J/\psi$ decay.
22. There are new branching-fraction measurements for decays of J/ψ, $\psi(2S)$, and Υ to $\gamma\eta(')$ that present a puzzling and unexplained pattern (Table 26).
23. Measurements of the γgg spectra and branching fractions (Table 25) for $\Upsilon(1S, 2S, 3S)$, J/ψ, and $\psi(2S)$ all have been published. Considerably more attention to measurements, in which background-subtraction uncertainties have limited the precision of charmonium results much more severely than for bottomonium, and predictions, which have had mixed success on predicting the radiative photon spectra, is needed.
24. First measurements of the two- and three-photon partial widths of $\chi_{c0,2}$ (Table 27) and J/ψ, respectively, have been performed. Although the experimental relative uncertainties on these widths are not yet below the 10% and 30%, respectively, they already present challenges to the theory. Better experimental precision for the two-photon couplings of $\eta_c(1S)$ and $\eta_c(2S)$ would be very useful to constrain cross-particle branching-fraction measurements as well as relevant theoretical descriptions.
25. Detailed and provocative measurements of dipion and η transitions for the ψ (Table 37) and Υ (Tables 38–39) systems below the respective open-flavor thresholds challenge theoretical rate and dipion-mass-spectra predictions, while the surprisingly high rates observed above these thresholds remain a mystery.
26. Measurements of $\psi(3770) \to$ non-$D\bar{D}$ (Table 40) conflict with one another. Definitive measurements of *exclusive decays* can best supply confidence in such a rate being more than a few percent, because indirect or aggregate (Table 41) comparisons appear to be quite challenging.
27. A multitude of measurements has been accumulated in the realm of exclusive hadronic decays of heavy quarkonia. These have deepened theoretical mysteries (e.g., $\rho\pi$ puzzle, conflicting measurements of non-$D\bar{D}$ decays of $\psi(3770)$) but whetted the community's appetite for more information on such decays of χ_{cJ}, χ_{bJ}, $\eta_c(1S)$, $\eta_c(2S)$, $\eta_b(1S)$, $h_c(^1P_1)$, and $h_b(^1P_1)$.
28. Initial but nevertheless intriguing measurements of the rate for deuterons to appear in Υ decays and the mechanism of baryon-number compensation therein have been reported. Further experimental and theoretical attention in Υ decays, the LHC, and future facilities are warranted. This information is useful for tuning MC generators and may be relevant for the molecular interpretation of $X(3872)$ (Sect. 2.9.3) and other loosely bound states.
29. A new measurement (Table 22) has resolved the long-standing discrepancy between experiment and theory on multipole amplitudes in $\psi(2S) \to \gamma\chi_{cJ}$, $\chi_{cJ} \to \gamma J/\psi$.

30. A theoretical understanding of the photon energy spectrum from $J/\psi \to \gamma\eta_c(1S)$ and $\psi(2S) \to \gamma\eta_c(1S)$ (see item #7) is required (Sects. 3.1.1 and 3.1.2).
31. It would be important to have a coherent EFT treatment for all magnetic and electric transitions. In particular, a rigorous treatment of the relativistic corrections contributing to the E1 transitions and a nonperturbative analysis of the M1 transitions is missing. The first is relevant for transitions involving P states, the second for any transition from above the ground state.
32. New resummation schemes for the perturbative expressions of the quarkonium decay widths should be developed. At the moment, this is the major obstacle to precise theoretical determinations of the $\Upsilon(1S)$ and $\eta_b(1S)$ inclusive and electromagnetic decays (Sect. 3.2.1).
33. More rigorous techniques to describe above-threshold quarkonium decays and transitions, whose descriptions still rely upon models, should be developed (Sects. 3.3.1 and 3.4).

Production The theoretical and experimental status of production of heavy quarkonia was given in Sect. 4. Conclusions and priorities are as follows:

34. It is very important either to establish that the NRQCD factorization formula is valid to all orders in perturbation theory or to demonstrate that it breaks down at some fixed order.
35. A more accurate treatment of higher-order corrections to the color-singlet contributions at the Tevatron and the LHC is urgently needed. The reorganization of the perturbation series that is provided by the fragmentation-function approach (Sect. 4.1.5) may be an important tool.
36. An outstanding theoretical challenge is the development of methods to compute color-octet long-distance NRQCD production matrix elements on the lattice.
37. If NRQCD factorization is valid, it likely holds only for values of p_T that are much greater than the heavy-quark mass. Therefore, it is important for experiments to make measurements of quarkonium production, differentially in p_T, at the highest possible values of p_T.
38. Further light could be shed on the NRQCD velocity expansion and its implications for low-energy dynamics by comparing studies of charmonium production and bottomonium production. The higher p_T reach of the LHC may be particularly important for studying bottomonium production at values of p_T that are much greater than the bottomonium mass.
39. It would be of considerable help in disentangling the theoretical issues in production of the J/ψ and Υ if experimental measurements could separately quantify direct and feeddown contributions. Ideally, the direct-production cross sections and polarizations would both be measured differentially in p_T.
40. It is important to resolve the apparent discrepancy between the CDF and DØ measurements of the Υ polarization, which were performed for different rapidity ranges, $|y| < 0.6$ (CDF) and $|y| < 1.8$ (DØ). A useful first step would be for the two experiments to provide polarization measurements that cover the same rapidity range.
41. It would be advantageous to measure complete quarkonium polarization information in a variety of spin-quantization frames and to make use of frame-invariant quantities to cross-check measurements in different frames [722, 723, 1030]. Care should be taken in comparing different polarization measurements to ensure that dependences on the choices of frame and the kinematic ranges of the experiments have been taken into account.
42. Measurements of inclusive cross sections, charmonium angular distributions, and polarization parameters for P-wave charmonium states would provide further important information about quarkonium production mechanisms.
43. Studies of quarkonium production at different values of $\sqrt{s}$ at the Tevatron and the LHC, studies of hadronic energy near to and away from the quarkonium direction at the Tevatron and the LHC, and studies of the production of heavy-flavor mesons in association with a quarkonium at e^+e^-, ep, $p\bar{p}$, and pp machines could give information that is complementary to that provided by traditional observations of quarkonium production rates and polarizations.
44. Theoretical uncertainties in the region near the kinematic endpoint of maximum quarkonium energy might be reduced through a systematic study of resummations of the perturbative and velocity expansions in both ep and e^+e^- quarkonium production.
45. In predictions for exclusive and inclusive quarkonium production in e^+e^- annihilation, large corrections appear at NLO. An important step would be to identify the origins of these large corrections. It might then be possible to improve the convergence of perturbation series by resumming specific large contributions to all orders in α_s.
46. The central values of the Belle and BABAR measurements of $\sigma(e^+e^- \to J/\psi + \eta_c(1S)) \times \mathcal{B}_{>2}$, where $\mathcal{B}_{>2}$ is the branching fraction for the $\eta_c(1S)$ to decay into a final state with more than two charged particles, differ by about twice the uncertainty of either measurement, suggesting that further experimental attention would be valuable. Comparisons with theory would be more informative if the uncertainty from the unmeasured branching fraction $\mathcal{B}_{>2}$ were reduced or eliminated.
47. The central values for the prompt J/ψ inclusive production cross section that were obtained by BABAR and

Belle differ by more than a factor of two. It would be very desirable to clear up this discrepancy.

48. Belle has presented results for $\sigma(e^+e^- \to J/\psi + c\bar{c})$ and $\sigma(e^+e^- \to J/\psi + X_{\text{non-}c\bar{c}})$. It would be beneficial to have similar results from BABAR.
49. Measurements of quarkonium production in γp and $\gamma\gamma$ collisions could provide additional information about production mechanisms and should be carried out at the next opportunity at an e^+e^- or ep collider. It may also be possible to measure quarkonium production in $\gamma\gamma$ and γp collisions at the LHC [828].
50. The observation and study of the B_c mesons and their excitations are new and exciting components of the quarkonium-physics plans for both the Tevatron and the LHC.
51. Theoretical progress should always be mirrored by the development of simulation tools for experiment. A particularly important goal would be to develop experiment-friendly simulation tools that incorporate the state-of-the-art theory for inclusive and associative quarkonium production.

In Medium The status of heavy-quarkonium production in cold and hot matter was presented in Sect. 5. Conclusions are, for cold matter:

52. Studies are now attempting to place a limit on the allowed level of quark energy loss in next-to-leading order Drell–Yan dilepton production. These will determine the maximum amount of gluon energy loss that can be applied to J/ψ production and suppression models of pA interactions.
53. A limit on the level of energy loss apparent in J/ψ production as a function of nuclear mass, A, and longitudinal momentum, expressed either as a function of x_F or rapidity, y, will put constraints on the nuclear absorption cross section. These constraints will help determine the importance of formation-time effects and feeddown on the quarkonium absorption cross section. Ultimately, the cold-nuclear-matter baseline should include different asymptotic absorption cross sections based on their final-state radii and formation times.

For hot matter:

54. It is important to calculate the quarkonium spectral functions at non zero temperature using an effective field theory approach with different hierarchies of relevant scales amended with lattice QCD calculations of the relevant correlation functions.
55. We would like to be able to compare the hot-matter effects more directly to heavy-ion data. To do this, more realistic dynamical models of quarkonium production in heavy-ion collisions must be developed that rely on state-of-the-art calculations of the quarkonium spectral functions.

Important future RHIC measurements include:

56. The open charm and open bottom cross sections as a function of rapidity and centrality in d+Au and Au+Au collisions, as well as a measurement of the rapidity distribution in pp collisions. Such a measurement is complementary to quarkonium production and necessary to establish whether the reduced J/ψ production cross section at forward rapidity, manifested in the larger effective absorption cross shown in Figs. 78, 82 and 85, is associated with reduced charm production or is particular to bound-state formation. This should be feasible with the upgraded vertex detectors being installed in 2011 and 2012, in conjunction with another d+Au run.
57. The J/ψ elliptic flow, v_2, in Au+Au collisions. Statistical recombination models predict strong secondary J/ψ production in heavy-ion collisions. This measurement will be an important test of this recombination picture.
58. Higher-statistics measurements of ψ and Υ production in d+Au and Au+Au collisions, which can be expected with increasing RHIC luminosity. These will provide further tests of the quarkonium production and suppression mechanisms in cold and hot matter.

Experimental outlook Section 6 gives an overview of newly commissioned, under-construction, and only-planned experimental facilities and what their activities relating to heavy quarkonium will be:

59. BESIII operating at BEPCII will continue where BESII and CLEO-c left off, with a robust program of charmonium spectroscopy and decay investigations. Initial datasets at the J/ψ and $\psi(2S)$ already exceed previous accumulations, with additional data acquisition at $\psi(3770)$ and the $D\bar{D}^*$ peak at $\sqrt{s} =$ 4170 MeV planned for the near future. Fine scans above open-charm threshold are likely. At its highest energy, BEPCII will directly produce $Y(4260)$ for much-needed further study.
60. ALICE, ATLAS, CMS, and LHCb are being commissioned along with the LHC, and have potent heavy-quarkonium programs underway, planning important production and polarization measurements in both pp and heavy-ion collisions. The four experiments have distinct and complementary experimental strengths and specialties, but also will have significant overlaps in many measurements for cross-checking of results.
61. The $\overline{\text{P}}$ANDA ($p\bar{p}$) and CBM (nucleus–nucleus) experiments at the FAIR facility at GSI will complement the activities at other laboratories.
62. The knowledge of open-charm cross section in proton–antiproton annihilations is extremely important to shape

the initial physics program of $\overline{\textsf{P}}$ANDA at FAIR. A collaboration should be formed to prepare a proposal to perform inclusive measurements at Fermilab with $\bar{p}p$ collisions in the charmonium energy region using existing detector elements.
63. Lepton–hadron colliders have significant role in advancing the heavy-quarkonium-physics agenda, with the energy, intensity, and experimental upgrades at JLab likely to contribute before other similar proposed facilities.
64. A tau–charm factory and/or a more flexible super flavor factory have been proposed to continue the giant strides in heavy-quarkonium physics taken at e^+e^- machines, from Mark-I at SPEAR to the very recent landmark results from both B-factories, CLEO-c, and BESIII. The factory-level luminosities combined with sophisticated detectors and well-defined initial-state energy-momentum and quantum numbers give e^+e^- collisions many important advantages.
65. A future linear collider (CLIC or ILC) would offer important opportunities to measure top-quark properties in the $t\bar{t}$ threshold region (see item #7 above).

Acknowledgements We dedicate this review to the memories of three friends of the Quarkonium Working Group. Richard Galik, building on his leadership role in quarkonium physics at CLEO, supported and guided our efforts from the start. His relentless focus, unyielding objectivity, and insistent collegiality continue to inspire those who knew him. Beate Naroska pioneered quarkonium measurements at HERA and enthusiastically advocated these for the 2007 QWG meeting at DESY as a member of the local organizing committee. A visionary scientist and dedicated teacher, she made invaluable and enduring contributions to our field. Andrzej Zieminski had a longstanding devotion to the study of quarkonium production in hadron collisions. He worked energetically to ensure the success of the inaugural QWG meeting in 2002 as a convener of the QCD Tests Working Group, and continued his sustaining commitment as liaison to the DØ Collaboration. We remember Rich, Beate, and Andrzej with fondness and gratitude.

The authors appreciate and acknowledge support for work on this document provided, in part or in whole, by

- the US Department of Energy (DOE), under contracts DE-FG02-91-ER40690 (P. Artoisenet),
  DE-AC02-06-CH11357 (G.T. Bodwin),
  DE-AC05-06-OR23177 (E. Chudakov and C. Weiss)
  DE-AC02-07-CH11359, through FNAL, which is operated for DOE by the Fermi Research Alliance, LLC, under Grant No. DE-FG02-91-ER40676 (E. Eichten and V. Papadimitriou),
  DE-AC02-76-SF00515 (A. Gabareen Mokhtar and J.P. Lansberg),
  DE-AC02-05-CH11231 (S.R. Klein),
  DE-AC02-98-CH10886 (P. Petreczky and J.W. Qiu),
  DE-FG02-96-ER41005 (A.A. Petrov),
  DE-AC52-07-NA27344f (R. Vogt), and
  DE-FG02-94-ER40823 (M. Voloshin);
- the German Research Foundation (DFG) Collaborative Research Center 55 (SFB) and the European Union Research Executive Agency (REA) Marie Curie Initial Training Network (www.physik.uni-regensburg.de/STRONGnet), under Grant Agreement PITN-GA-2009-238353 (G. Bali);
- the European Union Marie Curie Research Training Network (RTN) Flavianet, under Contract MRTN-CT-2006-035482, and the German Research Foundation (DFG) Cluster of Excellence *Origin and Structure of the Universe* (www.universe-cluster.de) (N. Brambilla and A. Vairo);
- the Polish Ministry of Science and Higher Education (J. Brodzicka);
- the National Natural Science Foundation of China (NSFC) under Grants
  10875155 and 10847001 (C.-H. Chang),
  10721063 (K.-T. Chao),
  10920101072 and 10845003 (W. Qian),
  and 10775412, 10825524, and 10935008 (C.-Z. Yuan);
- the Ministry of Science and Technology of China, under Grant 2009CB825200 (K.-T. Chao);
- The German Research Foundation (DFG) under grant GZ 436 RUS 113/769/0-3 and the Russian Foundation for Basic Research (RFBR) under grants 08-02-13516 and 08-02-91969 (S. Eidelman);
- the US National Science Foundation (NSF), under contracts PHY-07-56474 (A.D. Frawley),
  PHY-07-58312 and PHY-09-70024 (B.K. Heltsley),
  CAREER Award PHY-05-47794 (A. Petrov), and
  PHY-05-55660 (R. Vogt);
- Science and Engineering Research Canada (NSERC) (X. Garcia i Tormo);
- the Helmholtz Association, through funds provided to the virtual institute *Spin and strong QCD* (VH-VI-231), the German Research Foundation (DFG) (under grants SFB/TR 16 and 436 RUS 113/991/0-1) and the European Community-Research Infrastructure Integrating Activity *Study of Strongly Interacting Matter* (acronym HadronPhysics2, Grant Agreement 227431) under the European Union Seventh Framework Programme (C. Hanhart);
- the Belgian American Educational Foundation and the Francqui Foundation (J.P. Lansberg);
- the Belgian Federal Science Policy (IAP 6/11)
  (F. Maltoni);
- the Brazil National Council for Scientific and Technological Development (CNPq) and Foundation for Research Support of the State of São Paulo (FAPESP) (F.S. Navarra and M. Nielson);
- the World Class University (WCU) project of the National Research Foundation of Korea, under contract R32-2008-000-10155-0 (S. Olsen);
- the Ministry of Education and Science of the Russian Federation and the State Atomic Energy Corporation "Rosatom" (P. Pakhlov and G. Pakhlova);
- the France–China Particle Physics Laboratory (FCPPL) (W. Qian);
- the French National Research Agency (ANR) under Contract "BcLHCb ANR-07-JCJC-0146" (P. Robbe);
- the Spanish Ministry of Science and Innovation (MICNN), under grant FPA2008-02878 and Generalitat Valenciana under grant GVPROMETEO2010-056 (M.A. Sanchis-Lozano);
- the Portuguese Foundation for Science and Technology (FCT), under contracts SFRH/BPD/42343/2007 and SFRH/BPD/42138/2007 (P. Faccioli and H.K. Wöhri)

N. Brambilla, A.D. Frawley, C. Lourenço, A. Mocsy, P. Petreczky, H.K. Wöhri, A. Vairo, and R. Vogt acknowledge the hospitality of Institute for Nuclear Theory at the University of Washington and the US Department of Energy for partial support during their attendance at the CATHIE-INT mini-program on heavy quarkonium in Seattle.

We thank Hee Sok Chung for preparing Fig. 59, and Roman Mizuk for his perspective on the Z^+ analyses. We thank Eric Braaten for his feedback on the manuscript, and Dave Besson, Mathias Butenschön, and Bernd Kniehl for useful discussions. We thank Jeanne Butler for assistance with preparation of the figures.

References

1. N. Brambilla et al., CERN-2005-005 (CERN, Geneva, 2005). arXiv:hep-ph/0412158
2. D.C. Hom et al., Phys. Rev. Lett. **36**, 1236 (1976)
3. H.D. Snyder et al., Phys. Rev. Lett. **36**, 1415 (1976)
4. S.W. Herb et al., Phys. Rev. Lett. **39**, 252 (1977)
5. C. Edwards et al., Phys. Rev. Lett. **48**, 70 (1982)
6. S.B. Athar et al. (CLEO Collaboration), Phys. Rev. D **70**, 112002 (2004). arXiv:hep-ex/0408133
7. J. Siegrist et al. (Mark I Collaboration), Phys. Rev. Lett. **36**, 700 (1976)
8. P.A. Rapidis et al. (MARK I Collaboration), Phys. Rev. Lett. **39**, 526 (1977). Erratum-ibid. **39**, 974 (1977)
9. W. Bacino et al. (DELCO Collaboration), Phys. Rev. Lett. **40**, 671 (1978)
10. R. Brandelik et al. (DASP Collaboration), Phys. Lett. B **76**, 361 (1978)
11. R.H. Schindler et al. (MARK II Collaboration), Phys. Rev. D **21**, 2716 (1980)
12. A. Osterheld et al. (Crystal Ball Collaboration), SLAC-PUB-4160, 1986
13. J.Z. Bai et al. (BES Collaboration), Phys. Rev. Lett. **88**, 101802 (2002). arXiv:hep-ex/0102003
14. K.K. Seth, Phys. Rev. D **72**, 017501 (2005)
15. M. Ablikim et al. (BES Collaboration), Phys. Lett. B **660**, 315 (2008). arXiv:0705.4500 [hep-ex]
16. D. Cronin-Hennessy et al. (CLEO Collaboration), Phys. Rev. D **80**, 072001 (2009). arXiv:0801.3418 [hep-ex]
17. S. Eidelman et al. (Particle Data Group), Phys. Lett. B **592**, 1 (2004)
18. C. Amsler et al. (Particle Data Group), Phys. Lett. B **667**, 1 (2008)
19. M. Ablikim et al., Phys. Rev. Lett. **101**, 102004 (2008)
20. K. Todyshev (KEDR Collaboration), Presentation at ICHEP2010, the *35th International Conference on High Energy, July 22–28, 2010, Paris, France*
21. G. Pakhlova et al. (Belle Collaboration), Phys. Rev. D **77**, 011103 (2008). arXiv:0708.0082 [hep-ex]
22. G. Pakhlova et al. (Belle Collaboration), Phys. Rev. Lett. **98**, 092001 (2007). arXiv:hep-ex/0608018
23. G. Pakhlova et al. (Belle Collaboration), Phys. Rev. Lett. **100**, 062001 (2008). arXiv:0708.3313 [hep-ex]
24. G. Pakhlova et al. (Belle Collaboration), Phys. Rev. D **80**, 091101 (2009). arXiv:0908.0231 [hep-ex]
25. G. Pakhlova et al. (Belle Collaboration), Phys. Rev. Lett. **101**, 172001 (2008). arXiv:0807.4458 [hep-ex]
26. G. Pakhlova et al. (Belle Collaboration), Phys. Rev. D **83**, 011101(R) (2011). arXiv:1011.4397 [hep-ex]
27. B. Aubert et al. (BABAR Collaboration), Phys. Rev. D **76**, 111105 (2007). arXiv:hep-ex/0607083
28. B. Aubert et al. (BABAR Collaboration), Phys. Rev. D **79**, 092001 (2009). arXiv:0903.1597 [hep-ex]
29. P. del Amo Sanchez et al. (BABAR Collaboration), Phys. Rev. D **82**, 052004 (2010). arXiv:1008.0338 [hep-ex]
30. S. Dobbs et al. (CLEO Collaboration), Phys. Rev. D **76**, 112001 (2007). arXiv:0709.3783 [hep-ex]
31. T. Barnes, S. Godfrey, E.S. Swanson, Phys. Rev. D **72**, 054026 (2005). arXiv:hep-ph/0505002
32. E.J. Eichten, K. Lane, C. Quigg, Phys. Rev. D **73**, 014014 (2006). Erratum-ibid. D **73**, 079903 (2006). arXiv: hep-ph/0511179
33. E.S. Swanson, Phys. Rep. **429**, 243 (2006). arXiv:hep-ph/0601110
34. E. Eichten, K. Gottfried, T. Kinoshita, K.D. Lane, T.M. Yan, Phys. Rev. D **21**, 203 (1980)
35. W.M. Yao et al. (Particle Data Group), J. Phys. G **33**, 1 (2006)
36. B. Aubert et al. (BABAR Collaboration), Phys. Rev. Lett. **102**, 012001 (2009). arXiv:0809.4120 [hep-ex]
37. K.-F. Chen et al. (Belle Collaboration), Phys. Rev. D **82**, 091106(R) (2010). arXiv:0808.2445 [hep-ex]
38. D. Besson et al. (CLEO Collaboration), Phys. Rev. Lett. **54**, 381 (1985)
39. D.M.J. Lovelock et al. Phys. Rev. Lett. **54**, 377 (1985)
40. N.A. Tornqvist, Phys. Rev. Lett. **53**, 878 (1984)
41. A. Drutskoy et al. (Belle Collaboration), Phys. Rev. D **81**, 112003 (2010). arXiv:1003.5885 [hep-ex]
42. Yu.A. Simonov, A.I. Veselov, JETP Lett. **88**, 5 (2008). arXiv:0805.4518 [hep-ph]
43. L. Lellouch, L. Randall, E. Sather, Nucl. Phys. B **405**, 55 (1993). arXiv:hep-ph/9301223
44. P. Rubin et al. (CLEO Collaboration), Phys. Rev. D **72**, 092004 (2005). arXiv:hep-ex/0508037
45. J.L. Rosner et al. (CLEO Collaboration), Phys. Rev. Lett. **95**, 102003 (2005). arXiv:hep-ex/0505073
46. S. Dobbs et al. (CLEO Collaboration), Phys. Rev. Lett. **101**, 182003 (2008). arXiv:0805.4599 [hep-ex]
47. M. Ablikim et al. (BESIII Collaboration), Phys. Rev. Lett. **104**, 132002 (2010). arXiv:1002.0501 [hep-ex]
48. M. Andreotti et al. (E835 Collaboration), Phys. Rev. D **72**, 032001 (2005)
49. S.K. Choi et al. (Belle Collaboration), Phys. Rev. Lett. **89**, 102001 (2002). Erratum-ibid. **89**, 129901 (2002), arXiv: hep-ex/0206002
50. B. Aubert et al. (BABAR Collaboration), Phys. Rev. Lett. **92**, 142002 (2004). arXiv:hep-ex/0311038
51. D.M. Asner et al. (CLEO Collaboration), Phys. Rev. Lett. **92**, 142001 (2004). arXiv:hep-ex/0312058
52. H. Nakazawa (Belle Collaboration), Nucl. Phys. Proc. Suppl. **184**, 220 (2008)
53. B. Aubert et al. (BABAR Collaboration), Phys. Rev. D **72**, 031101 (2005). arXiv:hep-ex/0506062
54. K. Abe et al. (Belle Collaboration), Phys. Rev. Lett. **98**, 082001 (2007). arXiv:hep-ex/0507019
55. S. Uehara et al. (Belle Collaboration), Phys. Rev. Lett. **96**, 082003 (2006). arXiv:hep-ex/0512035
56. B. Aubert et al. (BABAR Collaboration), Phys. Rev. D **81**, 092003 (2010). arXiv:1002.0281 [hep-ex]
57. T. Aaltonen et al. (CDF Collaboration), Phys. Rev. Lett. **100**, 182002 (2008). arXiv:0712.1506 [hep-ex]
58. V.M. Abazov et al. (DØ Collaboration), Phys. Rev. Lett. **101**, 012001 (2008). arXiv:0802.4258 [hep-ex]
59. B. Aubert et al. (BABAR Collaboration), Phys. Rev. Lett. **101**, 071801 (2008). Erratum-ibid. **102**, 029901 (2009). arXiv: 0807.1086 [hep-ex]
60. G. Bonvicini et al. (CLEO Collaboration), Phys. Rev. D **81**, 031104 (2010). arXiv:0909.5474 [hep-ex]
61. B. Aubert et al. (BABAR Collaboration), Phys. Rev. Lett. **103**, 161801 (2009). arXiv:0903.1124 [hep-ex]
62. G. Bonvicini et al. (CLEO Collaboration), Phys. Rev. D **70**, 032001 (2004). arXiv:hep-ex/0404021
63. P. del Amo Sanchez et al. (BABAR Collaboration), Phys. Rev. D **82**, 111102 (2010). arXiv:1004.0175 [hep-ex]
64. J.S. Whitaker et al. (Mark I Collaboration), Phys. Rev. Lett. **37**, 1596 (1976)
65. C.J. Biddick et al. Phys. Rev. Lett. **38**, 1324 (1977)
66. W.M. Tanenbaum et al. (Mark I Collaboration), Phys. Rev. D **17**, 1731 (1978)
67. W. Bartel et al., Phys. Lett. B **79**, 492 (1978)
68. D. Cronin-Hennessy et al. (CLEO Collaboration), Phys. Rev. D **81**, 052002 (2010). arXiv:0910.1324 [hep-ex]
69. R.E. Mitchell et al. (CLEO Collaboration), Phys. Rev. Lett. **102**, 011801 (2009). arXiv:0805.0252 [hep-ex]

70. H. Nakazawa (Belle Collaboration), Presentation at ICHEP-2010, the *35th International Conference on High Energy Physics, July 22–28, 2010, Paris, France*
71. J.J. Dudek, R. Edwards, C.E. Thomas, Phys. Rev. D **79**, 094504 (2009). arXiv:0902.2241 [hep-ph]
72. K. Ackerstaff et al. (OPAL Collaboration), Phys. Lett. B **420**, 157 (1998). arXiv:hep-ex/9801026
73. F. Abe et al. (CDF Collaboration), Phys. Rev. Lett. **81**, 2432 (1998). arXiv:hep-ex/9805034
74. I.F. Allison et al. (HPQCD, Fermilab Lattice, and UKQCD Collaborations), Phys. Rev. Lett. **94**, 172001 (2005). arXiv: hep-lat/0411027
75. A. Abulencia et al. (CDF Collaboration), Phys. Rev. Lett. **97**, 012002 (2006). arXiv:hep-ex/0603027
76. T.S. Nigmanov, K.R. Gibson, M.P. Hartz, P.F. Shepard (CDF Collaboration), arXiv:0910.3013 [hep-ex]
77. V.M. Abazov et al. (DØ Collaboration), Phys. Rev. Lett. **102**, 092001 (2009). arXiv:0805.2614 [hep-ex]
78. S. Brandt, C. Peyrou, R. Sosnowski, A. Wroblewski, Phys. Lett. **12**, 57 (1964)
79. M. Artuso et al. (CLEO Collaboration), Phys. Rev. Lett. **94**, 032001 (2005). arXiv:hep-ex/0411068
80. B. Fulsom (BABAR Collaboration), Presentation at ICHEP2010, the *35th International Conference on High Energy, Physics, July 22–28, 2010, Paris, France*
81. E. Eichten, S. Godfrey, H. Mahlke, J.L. Rosner, Rev. Mod. Phys. **80**, 1161 (2008). arXiv:hep-ph/0701208
82. S. Godfrey, S.L. Olsen, Annu. Rev. Nucl. Part. Sci. **58**, 51 (2008). arXiv:0801.3867 [hep-ph]
83. T. Barnes, S.L. Olsen, Int. J. Mod. Phys. A **24**, 305 (2009)
84. G.V. Pakhlova, P.N. Pakhlov, S.I. Eidelman, Phys. Usp. **53** (2010) 219 [Usp. Fiz. Nauk **180** (2010) 225]
85. S.K. Choi et al. (Belle Collaboration), Phys. Rev. Lett. **91**, 262001 (2003). arXiv:hep-ex/0309032
86. I. Adachi et al. (Belle Collaboration), arXiv:0809.1224 [hep-ex]
87. B. Aubert et al. (BABAR Collaboration), Phys. Rev. D **77**, 111101 (2008). arXiv:0803.2838 [hep-ex]
88. D.E. Acosta et al. (CDF II Collaboration), Phys. Rev. Lett. **93**, 072001 (2004). arXiv:hep-ex/0312021
89. A. Abulencia et al. (CDF Collaboration), Phys. Rev. Lett. **98**, 132002 (2007). arXiv:hep-ex/0612053
90. T. Aaltonen et al. (CDF Collaboration), Phys. Rev. Lett. **103**, 152001 (2009). arXiv:0906.5218 [hep-ex]
91. V.M. Abazov et al. (DØ Collaboration), Phys. Rev. Lett. **93**, 162002 (2004). arXiv:hep-ex/0405004
92. K. Abe et al. (Belle Collaboration), arXiv:hep-ex/0505037
93. P. del Amo Sanchez et al. (BABAR Collaboration), Phys. Rev. D **82**, 011101 (2010). arXiv:1005.5190 [hep-ex]
94. G. Gokhroo et al. (Belle Collaboration), Phys. Rev. Lett. **97**, 162002 (2006). arXiv:hep-ex/0606055
95. T. Aushev et al., Phys. Rev. D **81**, 031103 (2010). arXiv:0810.0358 [hep-ex]
96. B. Aubert et al. (BABAR Collaboration), Phys. Rev. D **77**, 011102 (2008). arXiv:0708.1565 [hep-ex]
97. B. Aubert et al. (BABR Collaboration), Phys. Rev. D **74**, 071101 (2006). arXiv:hep-ex/0607050
98. B. Aubert et al. (BABAR Collaboration), Phys. Rev. Lett. **102**, 132001 (2009). arXiv:0809.0042 [hep-ex]
99. V. Bhardwaj (Belle Collaboration), Presentation at the *International Workshop on Heavy Quarkonium* May 18–21, 2010, Fermilab, Batavia, IL
100. K. Abe et al. (Belle Collaboration), Phys. Rev. Lett. **94**, 182002 (2005). arXiv:hep-ex/0408126
101. B. Aubert et al. (BABAR Collaboration), Phys. Rev. Lett. **101**, 082001 (2008). arXiv:0711.2047 [hep-ex]
102. S. Uehara et al. (Belle Collaboration), Phys. Rev. Lett. **104**, 092001 (2010). arXiv:0912.4451 [hep-ex]
103. P. Pakhlov et al. (Belle Collaboration), Phys. Rev. Lett. **100**, 202001 (2008). arXiv:0708.3812 [hep-ex]
104. C.Z. Yuan et al. (Belle Collaboration), Phys. Rev. Lett. **99**, 182004 (2007). arXiv:0707.2541 [hep-ex]
105. R. Mizuk et al. (Belle Collaboration), Phys. Rev. D **78**, 072004 (2008). arXiv:0806.4098 [hep-ex]
106. T. Aaltonen et al. (CDF Collaboration), Phys. Rev. Lett. **102**, 242002 (2009). arXiv:0903.2229 [hep-ex]
107. Kai Yi (for the CDF Collaboration), arXiv:1010.3470 [hep-ex]
108. B. Aubert et al. (BABAR Collaboration), Phys. Rev. Lett. **95**, 142001 (2005). arXiv:hep-ex/0506081
109. B. Aubert et al. (BABAR Collaboration), arXiv:0808.1543v2 [hep-ex]
110. Q. He et al. (CLEO Collaboration), Phys. Rev. D **74**, 091104 (2006). arXiv:hep-ex/0611021
111. T.E. Coan et al. (CLEO Collaboration), Phys. Rev. Lett. **96**, 162003 (2006). arXiv:hep-ex/0602034
112. C.P. Shen et al. (Belle Collaboration), Phys. Rev. Lett. **104**, 112004 (2010). arXiv:0912.2383 [hep-ex]
113. B. Aubert et al. (BABAR Collaboration), Phys. Rev. Lett. **98**, 212001 (2007). arXiv:hep-ex/0610057
114. X.L. Wang et al. (Belle Collaboration), Phys. Rev. Lett. **99**, 142002 (2007). arXiv:0707.3699 [hep-ex]
115. S.K. Choi et al. (Belle Collaboration), Phys. Rev. Lett. **100**, 142001 (2008). arXiv:0708.1790 [hep-ex]
116. R. Mizuk et al. (Belle Collaboration), Phys. Rev. D **80**, 031104 (2009). arXiv:0905.2869 [hep-ex]
117. K.F. Chen et al. (Belle Collaboration), Phys. Rev. Lett. **100**, 112001 (2008). arXiv:0710.2577 [hep-ex]
118. S. Barlag et al. (ACCMOR Collaboration), Z. Phys. C **46**, 563 (1990)
119. C. Cawlfield et al. (CLEO Collaboration), Phys. Rev. Lett. **98**, 092002 (2007). arXiv:hep-ex/0701016
120. V.V. Anashin et al. (KEDR Collaboration), Phys. Lett. B **686**, 84 (2010). arXiv:0909.5545 [hep-ex]
121. E. Braaten, M. Lu, Phys. Rev. D **76**, 094028 (2007). arXiv:0709.2697 [hep-ph]
122. E. Braaten, J. Stapleton, Phys. Rev. D **81**, 014019 (2010). arXiv:0907.3167 [hep-ph]
123. M.L. Goldberger, K. Watson, *Collision Theory* (Wiley, New York, 1964)
124. B. Aubert et al. (BABAR Collaboration), Phys. Rev. Lett. **96**, 052002 (2006). arXiv:hep-ex/0510070
125. B. Aubert et al. (BABAR Collaboration), Phys. Rev. D **71**, 031501 (2005). arXiv:hep-ex/0412051
126. S. Godfrey, N. Isgur, Phys. Rev. D **32**, 189 (1985)
127. Yu.S. Kalashnikova, A.V. Nefediev, Phys. Rev. D **82**, 097502 (2010). arXiv:1008.2895 [hep-ph]
128. Z.Q. Liu, X.S. Qin, C.Z. Yuan, Phys. Rev. D **78**, 014032 (2008). arXiv:0805.3560 [hep-ex]
129. X.H. Mo et al., Phys. Lett. B **640**, 182 (2006). arXiv: hep-ex/0603024
130. K. Abe et al. (Belle Collaboration), Phys. Rev. Lett. **93**, 051803 (2004). arXiv:hep-ex/0307061
131. B. Aubert et al. (BABAR Collaboration), Phys. Rev. D **79**, 112001 (2009). arXiv:0811.0564 [hep-ex]
132. M. Neubert, Phys. Rep. **245**, 259 (1994). arXiv:hep-ph/9306320
133. A.V. Manohar, M.B. Wise, Camb. Monogr. Part. Phys. Nucl. Phys. Cosmol. **10**, 1 (2000)
134. N. Brambilla, A. Pineda, J. Soto, A. Vairo, Rev. Mod. Phys. **77**, 1423 (2005). arXiv:hep-ph/0410047
135. N. Brambilla, D. Gromes, A. Vairo, Phys. Lett. B **576**, 314 (2003). arXiv:hep-ph/0306107

136. N. Brambilla, D. Gromes, A. Vairo, Phys. Rev. D **64**, 076010 (2001). arXiv:hep-ph/0104068
137. W.E. Caswell, G.P. Lepage, Phys. Lett. B **167**, 437 (1986)
138. G.T. Bodwin, E. Braaten, G.P. Lepage, Phys. Rev. D **51**, 1125 (1995). Erratum-ibid. D **55**, 5853 (1997). arXiv: hep-ph/9407339
139. A. Pineda, J. Soto, Nucl. Phys. Proc. Suppl. **64**, 428 (1998). arXiv:hep-ph/9707481
140. N. Brambilla, A. Pineda, J. Soto, A. Vairo, Nucl. Phys. B **566**, 275 (2000). arXiv:hep-ph/9907240
141. N. Brambilla, X. Garcia i Tormo, J. Soto, A. Vairo, Phys. Lett. B **647**, 185 (2007). arXiv:hep-ph/0610143
142. N. Brambilla, A. Vairo, X. Garcia i Tormo, J. Soto, Phys. Rev. D **80**, 034016 (2009). arXiv:0906.1390 [hep-ph]
143. A.V. Smirnov, V.A. Smirnov, M. Steinhauser, Phys. Lett. B **668**, 293 (2008). arXiv:0809.1927 [hep-ph]
144. C. Anzai, Y. Kiyo, Y. Sumino, Phys. Rev. Lett. **104**, 112003 (2010). arXiv:0911.4335 [hep-ph]
145. A.V. Smirnov, V.A. Smirnov, M. Steinhauser, Phys. Rev. Lett. **104**, 112002 (2010). arXiv:0911.4742 [hep-ph]
146. N. Brambilla, X.G.i. Tormo, J. Soto, A. Vairo, Phys. Rev. Lett. **105**, 212001 (2010). arXiv:1006.2066 [hep-ph]
147. Y. Sumino, Phys. Rev. D **76**, 114009 (2007). arXiv:hep-ph/0505034
148. C. Anzai, Y. Kiyo, Y. Sumino, Nucl. Phys. B **838**, 28 (2010). arXiv:1004.1562 [hep-ph]
149. B.A. Kniehl, A.A. Penin, Y. Schroder, V.A. Smirnov, M. Steinhauser, Phys. Lett. B **607**, 96 (2005). arXiv:hep-ph/0412083
150. N. Brambilla, J. Ghiglieri, A. Vairo, Phys. Rev. D **81**, 054031 (2010). arXiv:0911.3541 [hep-ph]
151. N. Brambilla, A. Vairo, T. Rosch, Phys. Rev. D **72**, 034021 (2005). arXiv:hep-ph/0506065
152. S. Fleming, T. Mehen, Phys. Rev. D **73**, 034502 (2006). arXiv:hep-ph/0509313
153. H. Suganuma, A. Yamamoto, N. Sakumichi, T.T. Takahashi, H. Iida, F. Okiharu, Mod. Phys. Lett. A **23**, 2331 (2008). arXiv:0802.3500 [hep-ph]
154. A. Yamamoto, H. Suganuma, H. Iida, Prog. Theor. Phys. Suppl. **174**, 270 (2008). arXiv:0805.4735 [hep-ph]
155. J. Najjar, G. Bali, PoS **LAT2009**, 089 (2009). arXiv:0910.2824 [hep-lat]
156. N. Brambilla, A. Pineda, J. Soto, A. Vairo, Phys. Lett. B **470**, 215 (1999). arXiv:hep-ph/9910238
157. B.A. Kniehl, A.A. Penin, Nucl. Phys. B **563**, 200 (1999). arXiv:hep-ph/9907489
158. M.B. Voloshin, Nucl. Phys. B **154**, 365 (1979)
159. H. Leutwyler, Phys. Lett. B **98**, 447 (1981)
160. A. Pineda, Phys. Rev. D **65**, 074007 (2002). arXiv:hep-ph/0109117
161. A. Pineda, J. Soto, Phys. Lett. B **495**, 323 (2000). arXiv: hep-ph/0007197
162. A.H. Hoang, I.W. Stewart, Phys. Rev. D **67**, 114020 (2003). arXiv:hep-ph/0209340
163. A.V. Manohar, I.W. Stewart, Phys. Rev. D **62**, 014033 (2000). arXiv:hep-ph/9912226
164. X. Garcia i Tormo, J. Soto, Phys. Rev. Lett. **96**, 111801 (2006). arXiv:hep-ph/0511167
165. J.L. Domenech-Garret, M.A. Sanchis-Lozano, Phys. Lett. B **669**, 52 (2008). arXiv:0805.2916 [hep-ph]
166. M. Beneke, Y. Kiyo, K. Schuller, Nucl. Phys. B **714**, 67 (2005). arXiv:hep-ph/0501289
167. A.A. Penin, V.A. Smirnov, M. Steinhauser, Nucl. Phys. B **716**, 303 (2005). arXiv:hep-ph/0501042
168. A.A. Penin, M. Steinhauser, Phys. Lett. B **538**, 335 (2002). arXiv:hep-ph/0204290
169. B.A. Kniehl, A.A. Penin, V.A. Smirnov, M. Steinhauser, Nucl. Phys. B **635**, 357 (2002). arXiv:hep-ph/0203166
170. N. Brambilla, A. Vairo, Phys. Rev. D **62**, 094019 (2000). arXiv:hep-ph/0002075
171. N. Brambilla, Y. Sumino, A. Vairo, Phys. Rev. D **65**, 034001 (2002). arXiv:hep-ph/0108084
172. N. Brambilla, Y. Sumino, A. Vairo, Phys. Lett. B **513**, 381 (2001). arXiv:hep-ph/0101305
173. A. Abulencia et al. (CDF Collaboration), Phys. Rev. Lett. **96**, 082002 (2006). arXiv:hep-ex/0505076
174. N. Brambilla, A. Vairo, Phys. Rev. D **71**, 034020 (2005). arXiv:hep-ph/0411156
175. B.A. Kniehl, A.A. Penin, A. Pineda, V.A. Smirnov, M. Steinhauser, Phys. Rev. Lett. **92**, 242001 (2004). Erratum-ibid. **104**, 199901 (2010). arXiv:hep-ph/0312086
176. S. Recksiegel, Y. Sumino, Phys. Lett. B **578**, 369 (2004). arXiv:hep-ph/0305178
177. A.A. Penin, A. Pineda, V.A. Smirnov, M. Steinhauser, Phys. Lett. B **593**, 124 (2004). Erratum-ibid. **677**, 343 (2009). Erratum-ibid. **683**, 358 (2010). arXiv:hep-ph/0403080
178. N. Brambilla, A. Pineda, J. Soto, A. Vairo, Phys. Rev. D **63**, 014023 (2001). arXiv:hep-ph/0002250
179. A. Pineda, A. Vairo, Phys. Rev. D **63**, 054007 (2001). Erratum-ibid. D **64**, 039902 (2001). arXiv:hep-ph/0009145
180. A. Laschka, N. Kaiser, W. Weise, PoS **CONFINEMENT8**, 168 (2008). arXiv:0901.2260 [hep-ph]
181. N. Brambilla, D. Eiras, A. Pineda, J. Soto, A. Vairo, Phys. Rev. D **67**, 034018 (2003). arXiv:hep-ph/0208019
182. C.T.H. Davies, K. Hornbostel, I.D. Kendall, G.P. Lepage, C. McNeile, J. Shigemitsu, H. Trottier (HPQCD Collaboration), Phys. Rev. D **78**, 114507 (2008). arXiv:0807.1687 [hep-lat]
183. Y. Namekawa et al. (PACS-CS Collaboration), PoS **LATTICE2008**, 121 (2008). arXiv:0810.2364 [hep-lat]
184. Y. Taniguchi (PACS-CS Collaboration), PoS **LAT2009**, 208 (2009). arXiv:0910.5105 [hep-lat]
185. M. Göckeler et al. (QCDSF Collaboration), PoS **LAT2009**, 125 (2009). arXiv:0912.0167 [hep-lat]
186. M. Creutz, Phys. Lett. B **649**, 230 (2007). arXiv: hep-lat/0701018
187. M. Creutz, PoS **CONFINEMENT8**, 016 (2008). arXiv: 0810.4526 [hep-lat]
188. C. Bernard, M. Golterman, Y. Shamir, S.R. Sharpe, Phys. Lett. B **649**, 235 (2007). arXiv:hep-lat/0603027
189. M. Golterman, PoS **CONFINEMENT8**, 014 (2008). arXiv: 0812.3110 [hep-ph]
190. G.P. Lepage, L. Magnea, C. Nakhleh, U. Magnea, K. Hornbostel, Phys. Rev. D **46**, 4052 (1992). arXiv:hep-lat/9205007
191. A.X. El-Khadra, A.S. Kronfeld, P.B. Mackenzie, Phys. Rev. D **55**, 3933 (1997). arXiv:hep-lat/9604004
192. A.S. Kronfeld, Phys. Rev. D **62**, 014505 (2000). arXiv:hep-lat/0002008
193. X. Liao, T. Manke, arXiv:hep-lat/0210030
194. M. Okamoto et al. (CP-PACS Collaboration), Phys. Rev. D **65**, 094508 (2002). arXiv:hep-lat/0112020
195. J. Harada, A.S. Kronfeld, H. Matsufuru, N. Nakajima, T. Onogi, Phys. Rev. D **64**, 074501 (2001). arXiv:hep-lat/0103026
196. H.W. Lin et al. (Hadron Spectrum Collaboration), Phys. Rev. D **79**, 034502 (2009). arXiv:0810.3588 [hep-lat]
197. J.J. Dudek, R.G. Edwards, N. Mathur, D.G. Richards, Phys. Rev. D **77**, 034501 (2008). arXiv:0707.4162 [hep-lat]
198. J.J. Dudek, E. Rrapaj, Phys. Rev. D **78**, 094504 (2008). arXiv:0809.2582 [hep-ph]
199. M. Göckeler et al., Phys. Rev. D **82**, 114511 (2010). arXiv:1003.5756 [hep-lat]
200. C. Ehmann, G. Bali, PoS **LAT2007**, 094 (2007). arXiv: 0710.0256 [hep-lat]
201. C. Ehmann, G.S. Bali, PoS **LATTICE2008**, 114 (2008). arXiv:0903.2947 [hep-lat]

202. L. Levkova, C.E. DeTar, PoS **LATTICE2008**, 133 (2008). arXiv:0809.5086 [hep-lat]
203. G. Bali, C. Ehmann, PoS **LAT2009**, 113 (2009). arXiv:0911.1238 [hep-lat]
204. G.Z. Meng et al. (CLQCD Collaboration), Phys. Rev. D **80**, 034503 (2009). arXiv:0905.0752 [hep-lat]
205. T.W. Chiu, T.H. Hsieh (TWQCD Collaboration), Phys. Lett. B **646**, 95 (2007). arXiv:hep-ph/0603207
206. S. Prelovsek, Acta Phys. Pol. Suppl. **3**, 975 (2010). arXiv:1004.3636 [hep-lat]
207. C.W. Bernard et al. Phys. Rev. D **64**, 054506 (2001). arXiv:hep-lat/0104002
208. G.P. Lepage, Phys. Rev. D **59**, 074502 (1999). hep-lat/9809157
209. T. Burch et al. (Fermilab Lattice and MILC Collaborations), Phys. Rev. D **81**, 034508 (2010). arXiv:0912.2701 [hep-lat]
210. L. Liu, H.W. Lin, K. Orginos, A. Walker-Loud, Phys. Rev. D **81**, 094505 (2010). arXiv:0909.3294 [hep-lat]
211. A. Gray et al. (HPQCD and UKQCD Collaborations), Phys. Rev. D **72**, 094507 (2005). arXiv:hep-lat/0507013
212. S. Meinel, Phys. Rev. D **79**, 094501 (2009). arXiv:0903.3224 [hep-lat]
213. C. Allton et al. (RBC-UKQCD Collaboration), Phys. Rev. D **78**, 114509 (2008). arXiv:0804.0473 [hep-lat]
214. S. Meinel, W. Detmold, C.J. Lin, M. Wingate, PoS **LAT2009**, 105 (2009). arXiv:0909.3837 [hep-lat]
215. E.B. Gregory et al., Phys. Rev. Lett. **104**, 022001 (2010). arXiv:0909.4462 [hep-lat]
216. E.B. Gregory et al. PoS **LAT2009**, 092 (2009). arXiv:0911.2133 [hep-lat]
217. S. Meinel, Phys. Rev. D **82**, 114514 (2010). arXiv:1008.3154 [hep-lat]
218. G.S. Bali, H. Neff, T. Düssel, T. Lippert, K. Schilling (SESAM Collaboration), Phys. Rev. D **71**, 114513 (2005). arXiv:hep-lat/0505012
219. A. Yamamoto, H. Suganuma, H. Iida, Phys. Rev. D **78**, 014513 (2008). arXiv:0806.3554 [hep-lat]
220. W. Detmold, K. Orginos, M.J. Savage, Phys. Rev. D **76**, 114503 (2007). arXiv:hep-lat/0703009
221. Y. Koma, M. Koma, H. Wittig, Phys. Rev. Lett. **97**, 122003 (2006). arXiv:hep-lat/0607009
222. Y. Koma, M. Koma, Nucl. Phys. B **769**, 79 (2007). arXiv:hep-lat/0609078
223. Y. Koma, M. Koma, PoS **LAT2009**, 122 (2009). arXiv:0911.3204 [hep-lat]
224. A.A. Penin, arXiv:0905.4296 [hep-ph]
225. S. Meinel, Phys. Rev. D **82**, 114502 (2010). arXiv:1007.3966 [hep-lat]
226. K. Maltman, D. Leinweber, P. Moran, A. Sternbeck, Phys. Rev. D **78**, 114504 (2008). arXiv:0807.2020 [hep-lat]
227. C. McNeile, C.T.H. Davies, E. Follana, K. Hornbostel, G.P. Lepage, Phys. Rev. D **82**, 034512 (2010). arXiv:1004.4285 [hep-lat]
228. I. Allison et al. (HPQCD Collaboration), Phys. Rev. D **78**, 054513 (2008). arXiv:0805.2999 [hep-lat]
229. S. Bethke, Eur. Phys. J. C **64**, 689 (2009). arXiv:0908.1135 [hep-ph]
230. D. Besson et al. (CLEO Collaboration), Phys. Rev. D **74**, 012003 (2006). arXiv:hep-ex/0512061
231. X. Garcia i Tormo, J. Soto, Phys. Rev. D **72**, 054014 (2005). arXiv:hep-ph/0507107
232. N. Brambilla, X. Garcia i Tormo, J. Soto, A. Vairo, Phys. Rev. D **75**, 074014 (2007). arXiv:hep-ph/0702079
233. G.T. Bodwin, J. Lee, D.K. Sinclair, Phys. Rev. D **72**, 014009 (2005). arXiv:hep-lat/0503032
234. X. Garcia i Tormo, J. Soto, Phys. Rev. D **69**, 114006 (2004). arXiv:hep-ph/0401233
235. D. Besson et al. (CLEO Collaboration), Phys. Rev. D **78**, 032012 (2008). arXiv:0806.0315 [hep-ex]
236. J. Libby et al. (CLEO Collaboration), Phys. Rev. D **80**, 072002 (2009). arXiv:0909.0193 [hep-ex]
237. X. Garcia i Tormo, J. Soto, arXiv:hep-ph/0701030
238. P. Colangelo, A. Khodjamirian, arXiv:hep-ph/0010175
239. P. Colangelo, P. Santorelli, E. Scrimieri, arXiv:0912.1081 [hep-ph]
240. K.G. Chetyrkin, J.H. Kuhn, C. Sturm, Eur. Phys. J. C **48**, 107 (2006). arXiv:hep-ph/0604234
241. R. Boughezal, M. Czakon, T. Schutzmeier, Phys. Rev. D **74**, 074006 (2006). arXiv:hep-ph/0605023
242. B.A. Kniehl, A.V. Kotikov, Phys. Lett. B **642**, 68 (2006). arXiv:hep-ph/0607201
243. A. Maier, P. Maierhofer, P. Marqaurd, Phys. Lett. B **669**, 88 (2008). arXiv:0806.3405 [hep-ph]
244. A. Maier, P. Maierhofer, P. Marquard, A.V. Smirnov, Nucl. Phys. B **824**, 1 (2010). arXiv:0907.2117 [hep-ph]
245. A.H. Hoang, V. Mateu, S. Mohammad Zebarjad, Nucl. Phys. B **813**, 349 (2009). arXiv:0807.4173 [hep-ph]
246. Y. Kiyo, A. Maier, P. Maierhofer, P. Marquard, Nucl. Phys. B **823**, 269 (2009). arXiv:0907.2120 [hep-ph]
247. J.H. Kuhn, M. Steinhauser, C. Sturm, Nucl. Phys. B **778**, 192 (2007). arXiv:hep-ph/0702103
248. K.G. Chetyrkin, J.H. Kuhn, A. Maier, P. Maierhofer, P. Marquard, M. Steinhauser, C. Sturm, Phys. Rev. D **80**, 074010 (2009). arXiv:0907.2110 [hep-ph]
249. A. Hoang, Presentation at Euroflavour2010, the *Fifth Workshop of the European Flavour Physics Network FLAVIAnet*
250. A. Pineda, A. Signer, Phys. Rev. D **73**, 111501 (2006). arXiv:hep-ph/0601185
251. A. Signer, Phys. Lett. B **672**, 333 (2009). arXiv:0810.1152 [hep-ph]
252. S. Narison, Phys. Lett. B **693**, 559 (2010). arXiv:1004.5333 [hep-ph]
253. M. Beneke, Y. Kiyo, K. Schuller, Phys. Lett. B **658**, 222 (2008). arXiv:0705.4518 [hep-ph]
254. M. Beneke, Y. Kiyo, A.A. Penin, Phys. Lett. B **653**, 53 (2007). arXiv:0706.2733 [hep-ph]
255. M. Beneke, Y. Kiyo, Phys. Lett. B **668**, 143 (2008). arXiv:0804.4004 [hep-ph]
256. P. Marquard, J.H. Piclum, D. Seidel, M. Steinhauser, Nucl. Phys. B **758**, 144 (2006). arXiv:hep-ph/0607168
257. P. Marquard, J.H. Piclum, D. Seidel, M. Steinhauser, Phys. Lett. B **678**, 269 (2009). arXiv:0904.0920 [hep-ph]
258. D. Eiras, M. Steinhauser, Nucl. Phys. B **757**, 197 (2006). arXiv:hep-ph/0605227
259. Y. Kiyo, D. Seidel, M. Steinhauser, J. High Energy Phys. **0901**, 038 (2009). arXiv:0810.1597 [hep-ph]
260. A. Pineda, A. Signer, Nucl. Phys. B **762**, 67 (2007). arXiv:hep-ph/0607239
261. A.H. Hoang, M. Stahlhofen, Phys. Rev. D **75**, 054025 (2007). arXiv:hep-ph/0611292
262. A.H. Hoang, C.J. Reisser, Phys. Rev. D **71**, 074022 (2005). arXiv:hep-ph/0412258
263. A.H. Hoang, C.J. Reisser, P. Ruiz-Femenia, Phys. Rev. D **82**, 014005 (2010). arXiv:1002.3223 [hep-ph]
264. M. Beneke, B. Jantzen, P. Ruiz-Femenia, Nucl. Phys. B **840**, 186 (2010). arXiv:1004.2188 [hep-ph]
265. Tevatron Electroweak Working Group and CDF and DØ Collaborations, arXiv:0903.2503 [hep-ex]
266. V.S. Fadin, V.A. Khoze, T. Sjostrand, Z. Phys. C **48**, 613 (1990)
267. K. Hagiwara, Y. Sumino, H. Yokoya, Phys. Lett. B **666**, 71 (2008). arXiv:0804.1014 [hep-ph]
268. Y. Kiyo, J.H. Kuhn, S. Moch, M. Steinhauser, P. Uwer, Eur. Phys. J. C **60**, 375 (2009). arXiv:0812.0919 [hep-ph]

269. Y. Sumino, H. Yokoya, J. High Energy Phys. **1009**, 034 (2010). arXiv:1007.0075 [hep-ph]
270. A. Vairo, Int. J. Mod. Phys. A **22**, 5481 (2007). arXiv: hep-ph/0611310
271. N. Brambilla, A. Vairo, A. Polosa, J. Soto, Nucl. Phys. Proc. Suppl. **185**, 107 (2008)
272. D. Horn, J. Mandula, Phys. Rev. D **17**, 898 (1978)
273. P. Hasenfratz, R.R. Horgan, J. Kuti, J.M. Richard, Phys. Lett. B **95**, 299 (1980)
274. K.J. Juge, J. Kuti, C. Morningstar, Phys. Rev. Lett. **90**, 161601 (2003). arXiv:hep-lat/0207004
275. G.S. Bali, A. Pineda, Phys. Rev. D **69**, 094001 (2004). arXiv:hep-ph/0310130
276. S.L. Zhu, Phys. Lett. B **625**, 212 (2005). arXiv:hep-ph/0507025
277. E. Kou, O. Pene, Phys. Lett. B **631**, 164 (2005). arXiv: hep-ph/0507119
278. F.E. Close, P.R. Page, Phys. Lett. B **628**, 215 (2005). arXiv:hep-ph/0507199
279. G. Chiladze, A.F. Falk, A.A. Petrov, Phys. Rev. D **58**, 034013 (1998). arXiv:hep-ph/9804248
280. A. Pineda, J. High Energy Phys. **0106**, 022 (2001). arXiv:hep-ph/0105008
281. N.A. Tornqvist, Phys. Rev. Lett. **67**, 556 (1991)
282. S. Dubynskiy, M.B. Voloshin, Phys. Lett. B **666**, 344 (2008). arXiv:0803.2224 [hep-ph]
283. C.F. Qiao, Phys. Lett. B **639**, 263 (2006). arXiv:hep-ph/0510228
284. R.L. Jaffe, Phys. Rev. D **15**, 267 (1977)
285. L. Maiani, F. Piccinini, A.D. Polosa, V. Riquer, Phys. Rev. D **71**, 014028 (2005). arXiv:hep-ph/0412098
286. D. Ebert, R.N. Faustov, V.O. Galkin, Phys. Lett. B **634**, 214 (2006). arXiv:hep-ph/0512230
287. Yu.S. Kalashnikova, Phys. Rev. D **72**, 034010 (2005). arXiv:hep-ph/0506270
288. E. Braaten, M. Kusunoki, Phys. Rev. D **69**, 074005 (2004). arXiv:hep-ph/0311147
289. H. Høgaasen, J.M. Richard, P. Sorba, Phys. Rev. D **73**, 054013 (2006). arXiv:hep-ph/0511039
290. F. Buccella, H. Hogaasen, J.M. Richard, P. Sorba, Eur. Phys. J. C **49**, 743 (2007). arXiv:hep-ph/0608001
291. N.A. Törnqvist, Z. Phys. C **61**, 525 (1994). arXiv:hep-ph/9310247
292. N.A. Tornqvist, arXiv:hep-ph/0308277
293. E.S. Swanson, Phys. Lett. B **588**, 189 (2004). arXiv:hep-ph/0311229
294. E.S. Swanson, Phys. Lett. B **598**, 197 (2004). arXiv:hep-ph/0406080
295. S. Pakvasa, M. Suzuki, Phys. Lett. B **579**, 67 (2004). arXiv:hep-ph/0309294
296. M.B. Voloshin, Phys. Lett. B **579**, 316 (2004). arXiv:hep-ph/0309307
297. M.B. Voloshin, Phys. Lett. B **604**, 69 (2004). arXiv:hep-ph/0408321
298. C. Quigg, Nucl. Phys. Proc. Suppl. **142**, 87 (2005). arXiv: hep-ph/0407124
299. S. Weinberg, Phys. Rev. **130**, 776 (1963)
300. S. Weinberg, Phys. Rev. **131**, 440 (1963)
301. S. Weinberg, Phys. Rev. B **137**, 672 (1965)
302. V. Baru, J. Haidenbauer, C. Hanhart, Yu. Kalashnikova, A.E. Kudryavtsev, Phys. Lett. B **586**, 53 (2004). arXiv: hep-ph/0308129
303. V. Baru, J. Haidenbauer, C. Hanhart, A.E. Kudryavtsev, U.G. Meissner, Eur. Phys. J. A **23**, 523 (2005). arXiv: nucl-th/0410099
304. D. Gamermann, J. Nieves, E. Oset, E. Ruiz Arriola, Phys. Rev. D **81**, 014029 (2010). arXiv:0911.4407 [hep-ph]
305. D. Morgan, Nucl. Phys. A **543**, 632 (1992)
306. F.K. Guo, C. Hanhart, U.G. Meissner, Phys. Lett. B **665**, 26 (2008). arXiv:0803.1392 [hep-ph]
307. A. De Rujula, H. Georgi, S.L. Glashow, Phys. Rev. Lett. **38**, 317 (1977)
308. M.B. Voloshin, L.B. Okun, JETP Lett. **23**, 333 (1976). Pisma Z. Eksp. Teor. Fiz. **23**, 369 (1976)
309. E. van Beveren, G. Rupp, Phys. Rev. Lett. **91**, 012003 (2003). arXiv:hep-ph/0305035
310. E.E. Kolomeitsev, M.F.M. Lutz, Phys. Lett. B **582**, 39 (2004). arXiv:hep-ph/0307133
311. T.E. Browder, S. Pakvasa, A.A. Petrov, Phys. Lett. B **578**, 365 (2004). arXiv:hep-ph/0307054
312. F.K. Guo, C. Hanhart, S. Krewald, U.G. Meissner, Phys. Lett. B **666**, 251 (2008). arXiv:0806.3374 [hep-ph]
313. A. Martinez Torres, K.P. Khemchandani, D. Gamermann, E. Oset, Phys. Rev. D **80**, 094012 (2009). arXiv:0906.5333 [nucl-th]
314. M.T. AlFiky, F. Gabbiani, A.A. Petrov, Phys. Lett. B **640**, 238 (2006). arXiv:hep-ph/0506141
315. S. Fleming, M. Kusunoki, T. Mehen, U. van Kolck, Phys. Rev. D **76**, 034006 (2007). arXiv:hep-ph/0703168
316. S. Fleming, T. Mehen, Phys. Rev. D **78**, 094019 (2008). arXiv:0807.2674 [hep ph]
317. S. Fleming, T. Mehen, AIP Conf. Proc. **1182**, 491 (2009). arXiv:0907.4142 [hep-ph]
318. E. Braaten, M. Lu, Phys. Rev. D **74**, 054020 (2006). arXiv: hep-ph/0606115
319. E. Braaten, M. Kusunoki, Phys. Rev. D **72**, 014012 (2005). arXiv:hep-ph/0506087
320. E. Braaten, M. Lu, Phys. Rev. D **79**, 051503 (2009). arXiv: 0712.3885 [hep-ph]
321. G. Cotugno, R. Faccini, A.D. Polosa, C. Sabelli, Phys. Rev. Lett. **104**, 132005 (2010). arXiv:0911.2178 [hep-ph]
322. F.K. Guo, C. Hanhart, U.G. Meissner, Phys. Rev. Lett. **102**, 242004 (2009). arXiv:0904.3338 [hep-ph]
323. C. Hanhart, Yu.S. Kalashnikova, A.V. Nefediev, Phys. Rev. D **81**, 094028 (2010). arXiv:1002.4097 [hep-ph]
324. F.K. Guo, J. Haidenbauer, C. Hanhart, U.G. Meissner, Phys. Rev. D **82**, 094008 (2010). arXiv:1005.2055 [hep-ph]
325. C. Hanhart, Yu.S. Kalashnikova, A.E. Kudryavtsev, A.V. Nefediev, Phys. Rev. D **76**, 034007 (2007). arXiv:0704.0605 [hep-ph]
326. Yu.S. Kalashnikova, A.V. Nefediev, Phys. Rev. D **80**, 074004 (2009). arXiv:0907.4901 [hep-ph]
327. D.L. Canham, H.W. Hammer, R.P. Springer, Phys. Rev. D **80**, 014009 (2009). arXiv:0906.1263 [hep-ph]
328. C. Bignamini, B. Grinstein, F. Piccinini, A.D. Polosa, C. Sabelli, Phys. Rev. Lett. **103**, 162001 (2009). arXiv: 0906.0882 [hep-ph]
329. T. Aaltonen et al. (CDF Collaboration), Phys. Rev. D **80**, 031103 (2009). arXiv:0905.1982 [hep-ex]
330. G. Corcella et al., J. High Energy Phys. **0101**, 010 (2001). arXiv:hep-ph/0011363
331. T. Sjostrand, P. Eden, C. Friberg, L. Lonnblad, G. Miu, S. Mrenna, E. Norrbin, Comput. Phys. Commun. **135**, 238 (2001). arXiv:hep-ph/0010017
332. P. Artoisenet, E. Braaten, Phys. Rev. D **81**, 114018 (2010). arXiv:0911.2016 [hep-ph]
333. P. Artoisenet, E. Braaten, arXiv:1007.2868 [hep-ph]
334. D.M. Asner et al. (CLEO Collaboration), Phys. Rev. D **75**, 012009 (2007). arXiv:hep-ex/0612019
335. T.J. Burns, F. Piccinini, A.D. Polosa, C. Sabelli, Phys. Rev. D **82**, 074003 (2010). arXiv:1008.0018 [hep-ph]
336. T. Sjostrand, Private communication to C. Sabelli
337. P.L. Cho, M.B. Wise, Phys. Rev. D **51**, 3352 (1995). arXiv: hep-ph/9410214

338. R.L. Jaffe, F. Wilczek, Phys. Rev. Lett. **91**, 232003 (2003). arXiv:hep-ph/0307341
339. L. Maiani, F. Piccinini, A.D. Polosa, V. Riquer, Phys. Rev. Lett. **93**, 212002 (2004). arXiv:hep-ph/0407017
340. G. 't Hooft, G. Isidori, L. Maiani, A.D. Polosa, V. Riquer, Phys. Lett. B **662**, 424 (2008). arXiv:0801.2288 [hep-ph]
341. C. Amsler, T. Gutsche, S. Spanier, N.A. Tornqvist, Note on Scalar Mesons, in [447]
342. G. 't Hooft, arXiv:hep-th/0408148
343. C. Alexandrou, Ph. de Forcrand, B. Lucini, Phys. Rev. Lett. **97**, 222002 (2006). arXiv:hep-lat/0609004
344. N.V. Drenska, R. Faccini, A.D. Polosa, Phys. Rev. D **79**, 077502 (2009). arXiv:0902.2803 [hep-ph]
345. L. Maiani, V. Riquer, F. Piccinini, A.D. Polosa, Phys. Rev. D **72**, 031502 (2005). arXiv:hep-ph/0507062
346. B. Aubert et al. (BABAR Collaboration), Phys. Rev. D **74**, 091103 (2006). arXiv:hep-ex/0610018
347. M. Ablikim et al. (BES Collaboration), Phys. Rev. Lett. **100**, 102003 (2008). arXiv:0712.1143 [hep-ex]
348. C.P. Shen et al. (Belle Collaboration), Phys. Rev. D **80**, 031101 (2009). arXiv:0808.0006 [hep-ex]
349. N.V. Drenska, R. Faccini, A.D. Polosa, Phys. Lett. B **669**, 160 (2008). arXiv:0807.0593 [hep-ph]
350. A.D. Polosa, Private communication
351. L. Maiani, A.D. Polosa, V. Riquer, C.A. Salgado, Phys. Lett. B **645**, 138 (2007). arXiv:hep-ph/0606217
352. A. Ali, C. Hambrock, M.J. Aslam, Phys. Rev. Lett. **104**, 162001 (2010). arXiv:0912.5016 [hep-ph]
353. A. Ali, C. Hambrock, I. Ahmed, M.J. Aslam, Phys. Lett. B **684**, 28 (2010). arXiv:0911.2787 [hep-ph]
354. D. Ebert, R.N. Faustov, V.O. Galkin, Eur. Phys. J. C **58**, 399 (2008). arXiv:0808.3912 [hep-ph]
355. M.B. Voloshin, Prog. Part. Nucl. Phys. **61**, 455 (2008). arXiv:0711.4556 [hep-ph]
356. M.E. Peskin, Nucl. Phys. B **156**, 365 (1979)
357. G. Bhanot, M.E. Peskin, Nucl. Phys. B **156**, 391 (1979)
358. A.B. Kaidalov, P.E. Volkovitsky, Phys. Rev. Lett. **69**, 3155 (1992)
359. A. Sibirtsev, M.B. Voloshin, Phys. Rev. D **71**, 076005 (2005). arXiv:hep-ph/0502068
360. M.B. Voloshin, Mod. Phys. Lett. A **19**, 665 (2004). arXiv: hep-ph/0402011
361. J. Erlich, E. Katz, D.T. Son, M.A. Stephanov, Phys. Rev. Lett. **95**, 261602 (2005). arXiv:hep-ph/0501128
362. A. Karch, E. Katz, D.T. Son, M.A. Stephanov, Phys. Rev. D **74**, 015005 (2006). arXiv:hep-ph/0602229
363. S. Dubynskiy, A. Gorsky, M.B. Voloshin, Phys. Lett. B **671**, 82 (2009). arXiv:0804.2244 [hep-th]
364. M.A. Shifman, A.I. Vainshtein, V.I. Zakharov, Nucl. Phys. B **147**, 385 (1979)
365. L.J. Reinders, H. Rubinstein, S. Yazaki, Phys. Rep. **127**, 1 (1985)
366. S. Narison, Camb. Monogr. Part. Phys. Nucl. Phys. Cosmol. **17**, 1 (2002). arXiv:hep-ph/0205006
367. R.D. Matheus, F.S. Navarra, M. Nielsen, R. Rodrigues da Silva, Phys. Rev. D **76**, 056005 (2007). arXiv:0705.1357 [hep-ph]
368. R.D. Matheus, S. Narison, M. Nielsen, J.M. Richard, Phys. Rev. D **75**, 014005 (2007). arXiv:hep-ph/0608297
369. S.H. Lee, M. Nielsen, U. Wiedner, arXiv:0803.1168 [hep-ph]
370. T.V. Brito, F.S. Navarra, M. Nielsen, M.E. Bracco, Phys. Lett. B **608**, 69 (2005). arXiv:hep-ph/0411233
371. F.S. Navarra, M. Nielsen, Phys. Lett. B **639**, 272 (2006). arXiv:hep-ph/0605038
372. R.D. Matheus, F.S. Navarra, M. Nielsen, C.M. Zanetti, Phys. Rev. D **80**, 056002 (2009). arXiv:0907.2683 [hep-ph]
373. J. Sugiyama, T. Nakamura, N. Ishii, T. Nishikawa, M. Oka, Phys. Rev. D **76**, 114010 (2007). arXiv:0707.2533 [hep-ph]
374. M. Nielsen, C.M. Zanetti, Phys. Rev. D **82**, 116002 (2010). arXiv:1006.0467 [hep-ph]
375. R.M. Albuquerque, M. Nielsen, Nucl. Phys. A **815**, 53 (2009). arXiv:0804.4817 [hep-ph]
376. S.H. Lee, K. Morita, M. Nielsen, Nucl. Phys. A **815**, 29 (2009). arXiv:0808.0690 [hep-ph]
377. G.J. Ding, Phys. Rev. D **79**, 014001 (2009). arXiv:0809.4818 [hep-ph]
378. S.H. Lee, A. Mihara, F.S. Navarra, M. Nielsen, Phys. Lett. B **661**, 28 (2008). arXiv:0710.1029 [hep-ph]
379. M.E. Bracco, S.H. Lee, M. Nielsen, R. Rodrigues da Silva, Phys. Lett. B **671**, 240 (2009). arXiv:0807.3275 [hep-ph]
380. K.m. Cheung, W.Y. Keung, T.C. Yuan, Phys. Rev. D **76**, 117501 (2007). arXiv:0709.1312 [hep-ph]
381. M. Nielsen, F.S. Navarra, S.H. Lee, Phys. Rep. **497**, 41 (2010). arXiv:0911.1958 [hep-ph]
382. J.R. Zhang, M.Q. Huang, Phys. Rev. D **80**, 056004 (2009). arXiv:0906.0090 [hep-ph]
383. Y.b. Dong, A. Faessler, T. Gutsche, V.E. Lyubovitskij, Phys. Rev. D **77**, 094013 (2008). arXiv:0802.3610 [hep-ph]
384. Y. Dong, A. Faessler, T. Gutsche, S. Kovalenko, V.E. Lyubovitskij, Phys. Rev. D **79**, 094013 (2009). arXiv:0903.5416 [hep-ph]
385. I.W. Lee, A. Faessler, T. Gutsche, V.E. Lyubovitskij, Phys. Rev. D **80**, 094005 (2009). arXiv:0910.1009 [hep-ph]
386. T. Branz, T. Gutsche, V.E. Lyubovitskij, Phys. Rev. D **80**, 054019 (2009). arXiv:0903.5424 [hep-ph]
387. T. Branz, T. Gutsche, V.E. Lyubovitskij, Phys. Rev. D **82**, 054025 (2010). arXiv:1005.3168 [hep-ph]
388. W.S. Hou, Phys. Rev. D **74**, 017504 (2006). arXiv:hep-ph/0606016
389. M. Drees, K.-i. Hikasa, Phys. Rev. D **41**, 1547 (1990)
390. E. Fullana, M.A. Sanchis-Lozano, Phys. Lett. B **653**, 67 (2007). arXiv:hep-ph/0702190
391. F. Domingo, U. Ellwanger, E. Fullana, C. Hugonie, M.A. Sanchis-Lozano, J. High Energy Phys. **0901**, 061 (2009). arXiv:0810.4736 [hep-ph]
392. U. Ellwanger, C. Hugonie, A.M. Teixeira, Phys. Rep. **496**, 1 (2010). arXiv:0910.1785 [hep-ph]
393. S. Schael et al. (ALEPH, DELPHI, L3, and OPAL Collaborations), Eur. Phys. J. C **47**, 547 (2006). arXiv:hep-ex/0602042
394. R. Dermisek, J.F. Gunion, Phys. Rev. D **73**, 111701 (2006). arXiv:hep-ph/0510322
395. R. Dermisek, J.F. Gunion, Phys. Rev. D **75**, 075019 (2007). arXiv:hep-ph/0611142
396. F. Domingo, U. Ellwanger, M.A. Sanchis-Lozano, Phys. Rev. Lett. **103**, 111802 (2009). arXiv:0907.0348 [hep-ph]
397. B. Aubert et al. (BABAR Collaboration), Phys. Rev. Lett. **103**, 181801 (2009). arXiv:0906.2219 [hep-ex]
398. M.A. Sanchis-Lozano, Mod. Phys. Lett. A **17**, 2265 (2002). arXiv:hep-ph/0206156
399. C. Balazs, M.S. Carena, A. Menon, D.E. Morrissey, C.E.M. Wagner, Phys. Rev. D **71**, 075002 (2005). arXiv: hep-ph/0412264
400. M. Drees, M.M. Nojiri, Phys. Rev. D **49**, 4595 (1994). arXiv:hep-ph/9312213
401. K. Hagiwara, K. Kato, A.D. Martin, C.K. Ng, Nucl. Phys. B **344**, 1 (1990)
402. S.P. Martin, Phys. Rev. D **77**, 075002 (2008). arXiv:0801.0237 [hep-ph]
403. S.P. Martin, J.E. Younkin, Phys. Rev. D **80**, 035026 (2009). arXiv:0901.4318 [hep-ph]
404. J.E. Younkin, S.P. Martin, Phys. Rev. D **81**, 055006 (2010). arXiv:0912.4813 [hep-ph]
405. B. McElrath, Phys. Rev. D **72**, 103508 (2005). arXiv:hep-ph/0506151

406. L.N. Chang, O. Lebedev, J.N. Ng, Phys. Lett. B **441**, 419 (1998). arXiv:hep-ph/9806487
407. P. Fayet, Phys. Lett. B **84**, 421 (1979)
408. P. Fayet, J. Kaplan, Phys. Lett. B **269**, 213 (1991)
409. P. Fayet, Phys. Rev. D **74**, 054034 (2006). arXiv:hep-ph/0607318
410. G.K. Yeghiyan, Phys. Rev. D **80**, 115019 (2009). arXiv:0909.4919 [hep-ph]
411. A.L. Fitzpatrick, D. Hooper, K.M. Zurek, Phys. Rev. D **81**, 115005 (2010)
412. D.E. Kaplan, M.A. Luty, K.M. Zurek, Phys. Rev. D **79**, 115016 (2009). arXiv:0901.4117 [hep-ph]
413. T. Cohen, D.J. Phalen, A. Pierce, K.M. Zurek, Phys. Rev. D **82**, 056001 (2010). arXiv:1005.1655 [hep-ph]
414. H. Albrecht et al. (ARGUS Collaboration), Phys. Lett. B **179**, 403 (1986)
415. P. Rubin et al. (CLEO Collaboration), Phys. Rev. D **75**, 031104 (2007). arXiv:hep-ex/0612051
416. O. Tajima et al. (Belle Collaboration), Phys. Rev. Lett. **98**, 132001 (2007). arXiv:hep-ex/0611041
417. B. Aubert et al. (BABAR Collaboration), Phys. Rev. Lett. **103**, 251801 (2009). arXiv:0908.2840 [hep-ex]
418. M. Ablikim et al. (BES Collaboration), Phys. Rev. Lett. **100**, 192001 (2008). arXiv:0710.0039 [hep-ex]
419. P. Fayet, Phys. Rev. D **81**, 054025 (2010). arXiv:0910.2587 [hep-ph]
420. B. Aubert et al. (BABAR Collaboration), arXiv:0808.0017 [hep-ex]
421. J. Insler et al. (CLEO Collaboration), Phys. Rev. D **81**, 091101 (2010). arXiv:1003.0417 [hep-ex]
422. N. Brambilla, Y. Jia, A. Vairo, Phys. Rev. D **73**, 054005 (2006). arXiv:hep-ph/0512369
423. G. Feinberg, J. Sucher, Phys. Rev. Lett. **35**, 1740 (1975)
424. J. Sucher, Rep. Prog. Phys. **41**, 1781 (1978)
425. E. Eichten, K. Gottfried, T. Kinoshita, K.D. Lane, T.M. Yan, Phys. Rev. D **17**, 3090 (1978). Erratum-ibid. D **21**, 313 (1980)
426. J.S. Kang, J. Sucher, Phys. Rev. D **18**, 2698 (1978)
427. K.J. Sebastian, Phys. Rev. D **26**, 2295 (1982)
428. G. Karl, S. Meshkov, J.L. Rosner, Phys. Rev. Lett. **45**, 215 (1980)
429. H. Grotch, K.J. Sebastian, Phys. Rev. D **25**, 2944 (1982)
430. P. Moxhay, J.L. Rosner, Phys. Rev. D **28**, 1132 (1983)
431. R. McClary, N. Byers, Phys. Rev. D **28**, 1692 (1983)
432. H. Grotch, D.A. Owen, K.J. Sebastian, Phys. Rev. D **30**, 1924 (1984)
433. Fayyazuddin, O.H. Mobarek, Phys. Rev. D **48**, 1220 (1993)
434. T.A. Lahde, Nucl. Phys. A **714**, 183 (2003). arXiv:hep-ph/0208110
435. D. Ebert, R.N. Faustov, V.O. Galkin, Phys. Rev. D **67**, 014027 (2003). arXiv:hep-ph/0210381
436. A.Y. Khodjamirian, Phys. Lett. B **90**, 460 (1980)
437. A. Le Yaouanc, L. Oliver, O. Pene, J.C. Raynal, *Hadron Transitions in the Quark Model* (Gordon and Breach, New York, 1988)
438. J.J. Dudek, R.G. Edwards, D.G. Richards, Phys. Rev. D **73**, 074507 (2006). arXiv:hep-ph/0601137
439. N. Brambilla, P. Roig, A. Vairo, TUM-EFT 9/10 (in preparation). arXiv:1012.0773
440. G. Li, Q. Zhao, Phys. Lett. B **670**, 55 (2008). arXiv:0709.4639 [hep-ph]
441. V.V. Anashin et al., arXiv:1002.2071 [hep-ex]
442. S. Uehara et al. (Belle Collaboration), Eur. Phys. J. C **53**, 1 (2008). arXiv:0706.3955 [hep-ex]
443. J.P. Lees et al. (BABAR Collaboration), Phys. Rev. D **81**, 052010 (2010). arXiv:1002.3000 [hep-ex]
444. B. Aubert et al. (BABAR Collaboration), Phys. Rev. D **78**, 012006 (2008). arXiv:0804.1208 [hep-ex]
445. N.E. Adam et al. (CLEO Collaboration), Phys. Rev. Lett. **94**, 232002 (2005). arXiv:hep-ex/0503028
446. H. Mendez et al. (CLEO Collaboration), Phys. Rev. D **78**, 011102 (2008). arXiv:0804.4432 [hep-ex]
447. K. Nakamura et al. (Particle Data Group), J. Phys. G **37**, 075021 (2010)
448. G. Li (BESIII Collaboration), Presentation at ICHEP2010, *the 35th International Conference on High Energy Physics, July 22–28, 2010, Paris, France*
449. X.-R. Lu (BESIII Collaboration), Presentation at MESON2010, *the 11th International Workshop on Meson Production, Properties and Interaction Kraków, Poland, 10–15 June 2010*
450. M. Artuso et al. (CLEO Collaboration), Phys. Rev. D **80**, 112003 (2009). arXiv:0910.0046 [hep-ex]
451. J.L. Rosner, Phys. Rev. D **64**, 094002 (2001). arXiv:hep-ph/0105327
452. E.J. Eichten, K. Lane, C. Quigg, Phys. Rev. D **69**, 094019 (2004). arXiv:hep-ph/0401210
453. T.E. Coan et al. (CLEO Collaboration), Phys. Rev. Lett. **96**, 182002 (2006). arXiv:hep-ex/0509030
454. R.A. Briere et al. (CLEO Collaboration), Phys. Rev. D **74**, 031106 (2006). arXiv:hep-ex/0605070
455. S. Godfrey, J.L. Rosner, Phys. Rev. D **64**, 074011 (2001). Erratum-ibid. D **65**, 039901 (2002). arXiv:hep-ph/0104253
456. G.T. Bodwin, E. Braaten, G.P. Lepage, Phys. Rev. D **46**, 1914 (1992). arXiv:hep-lat/9205006
457. N. Brambilla, E. Mereghetti, A. Vairo, J. High Energy Phys. **0608**, 039 (2006). arXiv:hep-ph/0604190
458. N. Brambilla, E. Mereghetti, A. Vairo, Phys. Rev. D **79**, 074002 (2009). arXiv:0810.2259 [hep-ph]
459. A. Vairo, Eur. Phys. J. A **31**, 728 (2007). arXiv:hep-ph/0610251
460. A.A. Penin, A. Pineda, V.A. Smirnov, M. Steinhauser, Nucl. Phys. B **699**, 183 (2004). arXiv:hep-ph/0406175
461. G.T. Bodwin, Y.Q. Chen, Phys. Rev. D **64**, 114008 (2001). arXiv:hep-ph/0106095
462. Y. Kiyo, A. Pineda, A. Signer, arXiv:1006.2685 [hep-ph]
463. R.D. Field, Phys. Lett. B **133**, 248 (1983)
464. S.G. Karshenboim, Phys. Rep. **422**, 1 (2005). arXiv:hep-ph/0509010
465. W. Kwong, P.B. Mackenzie, R. Rosenfeld, J.L. Rosner, Phys. Rev. D **37**, 3210 (1988)
466. A. Petrelli, M. Cacciari, M. Greco, F. Maltoni, M.L. Mangano, Nucl. Phys. B **514**, 245 (1998). arXiv:hep-ph/9707223
467. A. Abele et al. (Crystal Barrel Collaboration), Phys. Lett. B **411**, 361 (1997)
468. R. Partridge et al., Phys. Rev. Lett. **44**, 712 (1980)
469. G.S. Adams et al. (CLEO Collaboration), Phys. Rev. Lett. **101**, 101801 (2008). arXiv:0806.0671 [hep-ex]
470. M. Ablikim et al. (BES Collaboration), Phys. Rev. D **73**, 052008 (2006). arXiv:hep-ex/0510066
471. T.K. Pedlar et al. (CLEO Collaboration), Phys. Rev. D **79**, 111101 (2009). arXiv:0904.1394 [hep-ex]
472. S.B. Athar et al. (CLEO Collaboration), Phys. Rev. D **76**, 072003 (2007). arXiv:0704.3063 [hep-ex]
473. K.T. Chao, Nucl. Phys. B **335**, 101 (1990)
474. J.P. Ma, Phys. Rev. D **65**, 097506 (2002). arXiv:hep-ph/0202256
475. B.A. Li, Phys. Rev. D **77**, 097502 (2008). arXiv:0712.4246 [hep-ph]
476. J.L. Rosner, Phys. Rev. D **27**, 1101 (1983)
477. J.L. Rosner, in *Proceedings of the 1985 Int. Symp. on Lepton and Photon Interactions at High Energies*, Kyoto, Japan, Aug. 19–24, 1985, ed. by M. Konuma, K. Takahashi (Kyoto Univ., Research Inst. Fund. Phys., Kyoto, 1986), p. 448
478. F.J. Gilman, R. Kauffman, Phys. Rev. D **36**, 2761 (1987). Erratum-ibid. D **37**, 3348 (1988)

479. J.Z. Bai et al. (BES Collaboration), Phys. Rev. D **58**, 097101 (1998). arXiv:hep-ex/9806002
480. F. Ambrosino et al. (KLOE Collaboration), Phys. Lett. B **648**, 267 (2007). arXiv:hep-ex/0612029
481. R. Escribano, J. Nadal, J. High Energy Phys. **0705**, 006 (2007). arXiv:hep-ph/0703187
482. J. Libby et al. (CLEO Collaboration), Phys. Rev. Lett. **101**, 182002 (2008). arXiv:0806.2344 [hep-ex]
483. B. Aubert et al. (BABAR Collaboration), Phys. Rev. D **74**, 012002 (2006). arXiv:hep-ex/0605018
484. J.L. Rosner, Phys. Rev. D **79**, 097301 (2009). arXiv:0903.1796 [hep-ph]
485. H. Nakazawa et al. (Belle Collaboration), Phys. Lett. B **615**, 39 (2005). arXiv:hep-ex/0412058
486. C.C. Kuo et al. Phys. Lett. B **621**, 41 (2005). arXiv:hep-ex/0503006
487. W.T. Chen et al., Phys. Lett. B **651**, 15 (2007). arXiv:hep-ex/0609042
488. S. Uehara et al. (Belle Collaboration), Phys. Rev. D **78**, 052004 (2008). arXiv:0810.0655 [hep-ex]
489. S. Uehara et al. (Belle Collaboration), Phys. Rev. D **79**, 052009 (2009). arXiv:0903.3697 [hep-ex]
490. S. Dobbs et al. (CLEO Collaboration), Phys. Rev. D **73**, 071101 (2006). arXiv:hep-ex/0510033
491. S. Uehara et al. (Belle Collaboration), Phys. Rev. D **82**, 114031 (2010). arXiv:1007.3779 [hep-ex]
492. K.M. Ecklund et al. (CLEO Collaboration), Phys. Rev. D **78**, 091501 (2008). arXiv:0803.2869 [hep-ex]
493. L. Landau, Phys. Abstr. A **52**, 125 (1949)
494. C.N. Yang, Phys. Rev. **77**, 242 (1950)
495. R. Barbieri, R. Gatto, R. Kogerler, Phys. Lett. B **60**, 183 (1976)
496. J. Wicht et al. (Belle Collaboration), Phys. Lett. B **662**, 323 (2008). arXiv:hep-ex/0608037
497. B. Aubert et al. (BABAR Collaboration), Phys. Rev. D **69**, 011103 (2004). arXiv:hep-ex/0310027
498. G.S. Adams et al. (CLEO Collaboration), Phys. Rev. D **73**, 051103 (2006). arXiv:hep-ex/0512046
499. Z. Li et al. (CLEO Collaboration), Phys. Rev. D **71**, 111103 (2005). arXiv:hep-ex/0503027
500. N.E. Adam et al. (CLEO Collaboration), Phys. Rev. Lett. **96**, 082004 (2006). arXiv:hep-ex/0508023
501. V.V. Anashin et al. (KEDR Collaboration), Phys. Lett. B **685**, 134 (2010). arXiv:0912.1082 [hep-ex]
502. M. Ablikim et al. (BES Collaboration), Phys. Rev. Lett. **97**, 121801 (2006). arXiv:hep-ex/0605107
503. M. Ablikim et al. (BES Collaboration), Phys. Lett. B **659**, 74 (2008)
504. M. Ablikim et al., Phys. Rev. D **74**, 112003 (2006)
505. R. Brandelik et al. (DASP Collaboration), Phys. Lett. B **73**, 109 (1978)
506. J.Z. Bai et al. (BES Collaboration), Phys. Rev. D **65**, 052004 (2002). arXiv:hep-ex/0010072
507. V.V. Anashin et al. JETP Lett. **85**, 347 (2007)
508. D. Besson et al. (CLEO Collaboration), Phys. Rev. Lett. **96**, 092002 (2006). Erratum-ibid. **104** (2010) 159901. arXiv:hep-ex/0512038
509. M. Ablikim et al. (BES Collaboration), Phys. Lett. B **652**, 238 (2007). arXiv:hep-ex/0612056
510. G.S. Adams et al. (CLEO Collaboration), Phys. Rev. Lett. **94**, 012001 (2005). arXiv:hep-ex/0409027
511. J.L. Rosner et al. (CLEO Collaboration), Phys. Rev. Lett. **96**, 092003 (2006). arXiv:hep-ex/0512056
512. D. Besson et al. (CLEO Collaboration), Phys. Rev. Lett. **98**, 052002 (2007). arXiv:hep-ex/0607019
513. B. Aubert et al. (BABAR Collaboration), Phys. Rev. D **72**, 032005 (2005). arXiv:hep-ex/0405025
514. K. Gottfried, Phys. Rev. Lett. **40**, 598 (1978)
515. G. Bhanot, W. Fischler, S. Rudaz, Nucl. Phys. B **155**, 208 (1979)
516. T.M. Yan, Phys. Rev. D **22**, 1652 (1980)
517. Y.P. Kuang, T.M. Yan, Phys. Rev. D **24**, 2874 (1981)
518. W. Buchmuller, S.H.H. Tye, Phys. Rev. D **24**, 132 (1981)
519. L.S. Brown, R.N. Cahn, Phys. Rev. Lett. **35**, 1 (1975)
520. S. Dubynskiy, M.B. Voloshin, Phys. Rev. D **76**, 094004 (2007). arXiv:0707.1272 [hep-ph]
521. Y.P. Kuang, Front. Phys. China **1**, 19 (2006). arXiv:hep-ph/0601044
522. B.L. Ioffe, M.A. Shifman, Phys. Lett. B **95**, 99 (1980)
523. C. Cawlfield et al. (CLEO Collaboration), Phys. Rev. D **73**, 012003 (2006). arXiv:hep-ex/0511019
524. W. Kwong, J.L. Rosner, Phys. Rev. D **38**, 279 (1988)
525. Z.G. He, Y. Fan, K.T. Chao, Phys. Rev. D **81**, 074032 (2010). arXiv:0910.3939 [hep-ph]
526. Y.P. Kuang, T.M. Yan, Phys. Rev. D **41**, 155 (1990)
527. Y.P. Kuang, Phys. Rev. D **65**, 094024 (2002). arXiv:hep-ph/0201210
528. H.W. Ke, J. Tang, X.Q. Hao, X.Q. Li, Phys. Rev. D **76**, 074035 (2007). arXiv:0706.2074 [hep-ph]
529. P. Moxhay, Phys. Rev. D **37**, 2557 (1988)
530. Y.P. Kuang, S.F. Tuan, T.M. Yan, Phys. Rev. D **37**, 1210 (1988)
531. Yu.A. Simonov, A.I. Veselov, Phys. Rev. D **79**, 034024 (2009). arXiv:0804.4635 [hep-ph]
532. P. Moxhay, Phys. Rev. D **39**, 3497 (1989)
533. H.Y. Zhou, Y.P. Kuang, Phys. Rev. D **44**, 756 (1991)
534. F.K. Guo, C. Hanhart, U.G. Meissner, Phys. Rev. Lett. **103**, 082003 (2009). arXiv:0907.0521 [hep-ph]
535. F.K. Guo, C. Hanhart, G. Li, U.G. Meissner, Q. Zhao, Phys. Rev. D **82**, 034025 (2010). arXiv:1002.2712 [hep-ph]
536. G.S. Adams et al. (CLEO Collaboration), Phys. Rev. D **80**, 051106 (2009). arXiv:0906.4470 [hep-ex]
537. J.Z. Bai et al. (BES Collaboration), Phys. Lett. B **605**, 63 (2005). arXiv:hep-ex/0307028
538. M.B. Voloshin, Phys. Rev. D **71**, 114003 (2005). arXiv:hep-ph/0504197
539. Q. He et al. (CLEO Collaboration), Phys. Rev. Lett. **101**, 192001 (2008). arXiv:0806.3027 [hep-ex]
540. D. Cronin-Hennessy et al. (CLEO Collaboration), Phys. Rev. Lett. **92**, 222002 (2004). arXiv:hep-ex/0311043
541. M.B. Voloshin, Mod. Phys. Lett. A **18**, 1067 (2003). arXiv:hep-ph/0304165
542. D. Cronin-Hennessy et al. (CLEO Collaboration), Phys. Rev. D **76**, 072001 (2007). arXiv:0706.2317 [hep-ex]
543. M.B. Voloshin, V.I. Zakharov, Phys. Rev. Lett. **45**, 688 (1980)
544. S.R. Bhari et al. (CLEO Collaboration), Phys. Rev. D **79**, 011103 (2009). arXiv:0809.1110 [hep-ex]
545. B. Aubert et al. (BABAR Collaboration), Phys. Rev. D **78**, 112002 (2008). arXiv:0807.2014 [hep-ex]
546. B. Aubert et al. (BABAR Collaboration), Phys. Rev. Lett. **96**, 232001 (2006). arXiv:hep-ex/0604031
547. A. Sokolov et al. (Belle Collaboration), Phys. Rev. D **79**, 051103 (2009). arXiv:0901.1431 [hep-ex]
548. Yu.A. Simonov, A.I. Veselov, Phys. Lett. B **671**, 55 (2009). arXiv:0805.4499 [hep-ph]
549. C. Meng, K.T. Chao, Phys. Rev. D **77**, 074003 (2008). arXiv:0712.3595 [hep-ph]
550. C. Meng, K.T. Chao, Phys. Rev. D **78**, 034022 (2008). arXiv:0805.0143 [hep-ph]
551. X. Liu, B. Zhang, X.Q. Li, Phys. Lett. B **675**, 441 (2009). arXiv:0902.0480 [hep-ph]
552. Y.J. Zhang, G. Li, Q. Zhao, Phys. Rev. Lett. **102**, 172001 (2009). arXiv:0902.1300 [hep-ph]
553. D.Y. Chen, Y.B. Dong, X. Liu, Eur. Phys. J. C **70**, 177 (2010). arXiv:1005.0066 [hep-ph]

554. T. Appelquist, H.D. Politzer, Phys. Rev. Lett. **34**, 43 (1975)
555. A. De Rujula, S.L. Glashow, Phys. Rev. Lett. **34**, 46 (1975)
556. M.E.B. Franklin et al., Phys. Rev. Lett. **51**, 963 (1983)
557. J.Z. Bai et al. (BES Collaboration), Phys. Rev. D **69**, 072001 (2004). arXiv:hep-ex/0312016
558. J.Z. Bai et al. (BES Collaboration), Phys. Rev. Lett. **92**, 052001 (2004). arXiv:hep-ex/0310024
559. M. Ablikim et al. (BES Collaboration), Phys. Rev. D **70**, 112007 (2004). Erratum-ibid. D **71**, 019901 (2005). arXiv: hep-ex/0410031
560. M. Ablikim et al. (BES Collaboration), Phys. Rev. D **70**, 112003 (2004). arXiv:hep-ex/0408118
561. M. Ablikim et al. (BES Collaboration), Phys. Lett. B **614**, 37 (2005). arXiv:hep-ex/0407037
562. M. Ablikim et al. (BES Collaboration), Phys. Lett. B **648**, 149 (2007). arXiv:hep-ex/0610079
563. M. Ablikim et al. (BES Collaboration), Phys. Rev. D **74**, 012004 (2006). arXiv:hep-ex/0605031
564. N.E. Adam et al. (CLEO Collaboration), Phys. Rev. Lett. **94**, 012005 (2005). arXiv:hep-ex/0407028
565. C.Z. Yuan, AIP Conf. Proc. **814**, 65 (2006). arXiv:hep-ex/0510062
566. L. Chen, W.M. Dunwoodie (MARK-III Collaboration), SLAC-PUB-5674 (1991)
567. M. Ablikim et al. (BES Collaboration), Phys. Lett. B **619**, 247 (2005)
568. M. Ablikim et al. (BES Collaboration), Phys. Rev. D **71**, 072006 (2005). arXiv:hep-ex/0503030
569. R.A. Briere et al. (CLEO Collaboration), Phys. Rev. Lett. **95**, 062001 (2005). arXiv:hep-ex/0505101
570. T.K. Pedlar et al. (CLEO Collaboration), Phys. Rev. D **72**, 051108 (2005). arXiv:hep-ex/0505057
571. X.H. Mo, C.Z. Yuan, P. Wang, arXiv:hep-ph/0611214
572. M. Ablikim et al. (BES Collaboration), Phys. Lett. B **630**, 7 (2005). arXiv:hep-ex/0506045
573. D.M. Asner et al. (CLEO Collaboration), Phys. Rev. D **79**, 072007 (2009). arXiv:0811.0586 [hep-ex]
574. P. Naik et al. (CLEO Collaboration), Phys. Rev. D **78**, 031101 (2008). arXiv:0806.1715 [hep-ex]
575. S.B. Athar et al. (CLEO Collaboration), Phys. Rev. D **75**, 032002 (2007). arXiv:hep-ex/0607072
576. M. Ablikim et al. Phys. Rev. D **74**, 072001 (2006). arXiv: hep-ex/0607023
577. M. Ablikim et al. (BES Collaboration), Phys. Rev. D **70**, 092002 (2004). arXiv:hep-ex/0406079
578. M. Ablikim et al. (BES Collaboration), Phys. Rev. D **72**, 092002 (2005). arXiv:hep-ex/0508050
579. M. Ablikim et al. (BES Collaboration), Phys. Lett. B **630**, 21 (2005). arXiv:hep-ex/0410028
580. M. Ablikim et al. (BES Collaboration), Phys. Lett. B **642**, 197 (2006). arXiv:hep-ex/0607025
581. Q. He et al. (CLEO Collaboration), Phys. Rev. D **78**, 092004 (2008). arXiv:0806.1227 [hep-ex]
582. M. Ablikim et al., Phys. Rev. D **73**, 052006 (2006). arXiv: hep-ex/0602033
583. Q. Zhao, Phys. Lett. B **659**, 221 (2008). arXiv:0705.0101 [hep-ph]
584. M. Ablikim et al. (BES Collaboration), Phys. Lett. B **641**, 145 (2006). arXiv:hep-ex/0605105
585. M. Ablikim et al., Phys. Rev. D **76**, 122002 (2007)
586. M. Ablikim et al. (BES Collaboration), Phys. Lett. B **650**, 111 (2007). arXiv:0705.2276 [hep-ex]
587. M. Ablikim et al. (BES Collaboration), Phys. Lett. B **656**, 30 (2007). arXiv:0710.0786 [hep-ex]
588. M. Ablikim et al. (BES Collaboration), Eur. Phys. J. C **52**, 805 (2007). arXiv:0710.2176 [hep-ex]
589. M. Ablikim et al. (BES Collaboration), Phys. Lett. B **670**, 179 (2008). arXiv:0810.5608 [hep-ex]
590. M. Ablikim et al. (BES Collaboration), Phys. Lett. B **670**, 184 (2008). arXiv:0810.5611 [hep-ex]
591. M. Ablikim et al. (BES Collaboration), Eur. Phys. J. C **64**, 243 (2009)
592. M. Ablikim et al. (BES Collaboration), Eur. Phys. J. C **66**, 11 (2010)
593. G.S. Huang et al. (CLEO Collaboration), Phys. Rev. Lett. **96**, 032003 (2006). arXiv:hep-ex/0509046
594. D. Cronin-Hennessy et al. (CLEO Collaboration), Phys. Rev. D **74**, 012005 (2006). Erratum-ibid. D **75**, 119903 (2007). arXiv:hep-ex/0603026
595. G.S. Adams et al. (CLEO Collaboration), Phys. Rev. D **73**, 012002 (2006). arXiv:hep-ex/0509011
596. S. Dubynskiy, M.B. Voloshin, Phys. Rev. D **78**, 116014 (2008). arXiv:0809.3780 [hep-ph]
597. H.H. Gutbrod et al., Phys. Rev. Lett. **37**, 667 (1976)
598. H. Sato, K. Yazaki, Phys. Lett. B **98**, 153 (1981)
599. H. Albrecht et al. (ARGUS Collaboration), Phys. Lett. B **236**, 102 (1990)
600. G. Gustafson, J. Hakkinen, Z. Phys. C **61**, 683 (1994)
601. S.J. Brodsky, arXiv:0904.3037 [hep-ph]
602. R.A. Briere et al. (CLEO Collaboration), Phys. Rev. D **78**, 092007 (2008). arXiv:0807.3757 [hep-ex]
603. R. Barbieri, M. Caffo, E. Remiddi, Phys. Lett. B **83**, 345 (1979)
604. G.T. Bodwin, E. Braaten, D. Kang, J. Lee, Phys. Rev. D **76**, 054001 (2007). arXiv:0704.2599 [hep-ph]
605. B. Aubert et al. (BABAR Collaboration), Phys. Rev. D **81**, 011102 (2010). arXiv:0911.2024 [hep-ex]
606. D. Kang, T. Kim, J. Lee, C. Yu, Phys. Rev. D **76**, 114018 (2007). arXiv:0707.4056 [hep-ph]
607. Y.J. Zhang, K.T. Chao, Phys. Rev. D **78**, 094017 (2008). arXiv:0808.2985 [hep-ph]
608. D.M. Asner et al. (CLEO Collaboration), Phys. Rev. D **78**, 091103 (2008). arXiv:0808.0933 [hep-ex]
609. M.B. Einhorn, S.D. Ellis, Phys. Rev. D **12**, 2007 (1975)
610. S.D. Ellis, M.B. Einhorn, C. Quigg, Phys. Rev. Lett. **36**, 1263 (1976)
611. C.E. Carlson, R. Suaya, Phys. Rev. D **14**, 3115 (1976)
612. C.H. Chang, Nucl. Phys. B **172**, 425 (1980)
613. E.L. Berger, D.L. Jones, Phys. Rev. D **23**, 1521 (1981)
614. R. Baier, R. Ruckl, Phys. Lett. B **102**, 364 (1981)
615. R. Baier, R. Ruckl, Nucl. Phys. B **201**, 1 (1982)
616. R. Baier, R. Ruckl, Z. Phys. C **19**, 251 (1983)
617. G.A. Schuler, arXiv:hep-ph/9403387
618. P. Artoisenet, J.P. Lansberg, F. Maltoni, Phys. Lett. B **653**, 60 (2007). arXiv:hep-ph/0703129
619. J.M. Campbell, F. Maltoni, F. Tramontano, Phys. Rev. Lett. **98**, 252002 (2007). arXiv:hep-ph/0703113
620. P. Artoisenet, J.M. Campbell, J.P. Lansberg, F. Maltoni, F. Tramontano, Phys. Rev. Lett. **101**, 152001 (2008). arXiv: 0806.3282 [hep-ph]
621. H. Fritzsch, Phys. Lett. B **67**, 217 (1977)
622. F. Halzen, Phys. Lett. B **69**, 105 (1977)
623. M. Gluck, J.F. Owens, E. Reya, Phys. Rev. D **17**, 2324 (1978)
624. V.D. Barger, W.Y. Keung, R.J.N. Phillips, Phys. Lett. B **91**, 253 (1980)
625. J.F. Amundson, O.J.P. Eboli, E.M. Gregores, F. Halzen, Phys. Lett. B **372**, 127 (1996). arXiv:hep-ph/9512248
626. J.F. Amundson, O.J.P. Eboli, E.M. Gregores, F. Halzen, Phys. Lett. B **390**, 323 (1997). arXiv:hep-ph/9605295
627. G.T. Bodwin, E. Braaten, J. Lee, Phys. Rev. D **72**, 014004 (2005). arXiv:hep-ph/0504014
628. F. Abe et al. (CDF Collaboration), Phys. Rev. Lett. **79**, 578 (1997)

629. G.C. Nayak, J.W. Qiu, G. Sterman, Phys. Lett. B **613**, 45 (2005). arXiv:hep-ph/0501235
630. G.C. Nayak, J.W. Qiu, G. Sterman, Phys. Rev. D **72**, 114012 (2005). arXiv:hep-ph/0509021
631. Z.B. Kang, J.W. Qiu, G. Sterman (in preparation)
632. E. Braaten, S. Fleming, T.C. Yuan, Annu. Rev. Nucl. Part. Sci. **46**, 197 (1996). arXiv:hep-ph/9602374
633. J.C. Collins, D.E. Soper, Nucl. Phys. B **194**, 445 (1982)
634. E.L. Berger, J.W. Qiu, X.f. Zhang, Phys. Rev. D **65**, 034006 (2002). arXiv:hep-ph/0107309
635. G.C. Nayak, J.W. Qiu, G. Sterman, Phys. Rev. Lett. **99**, 212001 (2007). arXiv:0707.2973 [hep-ph]
636. G.C. Nayak, J.W. Qiu, G. Sterman, Phys. Rev. D **77**, 034022 (2008). arXiv:0711.3476 [hep-ph]
637. P. Hagler, R. Kirschner, A. Schafer, L. Szymanowski, O.V. Teryaev, Phys. Rev. Lett. **86**, 1446 (2001). arXiv:hep-ph/0004263
638. Ph. Hagler, R. Kirschner, A. Schafer, L. Szymanowski, O.V. Teryaev, Phys. Rev. D **63**, 077501 (2001). arXiv:hep-ph/0008316
639. F. Yuan, K.T. Chao, Phys. Rev. Lett. **87**, 022002 (2001). arXiv:hep-ph/0009224
640. S.P. Baranov, Phys. Rev. D **66**, 114003 (2002)
641. S.P. Baranov, N.P. Zotov, JETP Lett. **86**, 435 (2007). arXiv:0707.0253 [hep-ph]
642. S.P. Baranov, A. Szczurek, Phys. Rev. D **77**, 054016 (2008). arXiv:0710.1792 [hep-ph]
643. G.T. Bodwin, X. Garcia i Tormo, J. Lee, Phys. Rev. Lett. **101**, 102002 (2008). arXiv:0805.3876 [hep-ph]
644. G.T. Bodwin, X. Garcia i Tormo, J. Lee, Phys. Rev. D **81**, 114014 (2010). arXiv:1003.0061 [hep-ph]
645. J.P. Ma, Z.G. Si, Phys. Rev. D **70**, 074007 (2004). arXiv:hep-ph/0405111
646. A.E. Bondar, V.L. Chernyak, Phys. Lett. B **612**, 215 (2005). arXiv:hep-ph/0412335
647. V.V. Braguta, A.K. Likhoded, A.V. Luchinsky, Phys. Rev. D **72**, 074019 (2005). arXiv:hep-ph/0507275
648. V.V. Braguta, A.K. Likhoded, A.V. Luchinsky, Phys. Lett. B **635**, 299 (2006). arXiv:hep-ph/0602047
649. V.V. Braguta, Phys. Rev. D **78**, 054025 (2008). arXiv:0712.1475 [hep-ph]
650. F. Abe et al. (CDF Collaboration), Phys. Rev. Lett. **79**, 572 (1997)
651. J.P. Lansberg, Int. J. Mod. Phys. A **21**, 3857 (2006). arXiv:hep-ph/0602091
652. J.P. Lansberg, Eur. Phys. J. C **61**, 693 (2009). arXiv:0811.4005 [hep-ph]
653. M. Kramer, Prog. Part. Nucl. Phys. **47**, 141 (2001). arXiv:hep-ph/0106120
654. J.P. Lansberg et al., AIP Conf. Proc. **1038**, 15 (2008). arXiv:0807.3666 [hep-ph]
655. S. Abachi et al. (DØ Collaboration), Phys. Lett. B **370**, 239 (1996)
656. A.A. Affolder et al. (CDF Collaboration), Phys. Rev. Lett. **85**, 2886 (2000). arXiv:hep-ex/0004027
657. D.E. Acosta et al. (CDF Collaboration), Phys. Rev. D **71**, 032001 (2005). arXiv:hep-ex/0412071
658. A. Abulencia et al. (CDF Collaboration), Phys. Rev. Lett. **99**, 132001 (2007). arXiv:0704.0638 [hep-ex]
659. A. Adare et al. (PHENIX Collaboration), Phys. Rev. Lett. **98**, 232002 (2007). arXiv:hep-ex/0611020
660. S.S. Adler et al. (PHENIX Collaboration), Phys. Rev. Lett. **92**, 051802 (2004). arXiv:hep-ex/0307019
661. E.T. Atomssa (PHENIX Collaboration), Eur. Phys. J. C **61**, 683 (2009). arXiv:0805.4562 [nucl-ex]
662. A. Adare et al. (PHENIX Collaboration), Phys. Rev. D **82**, 012001 (2010). arXiv:0912.2082 [hep-ex]
663. C.L. da Silva (PHENIX Collaboration), Nucl. Phys. A **830**, 227C (2009). arXiv:0907.4696 [nucl-ex]
664. B.I. Abelev et al. (STAR Collaboration), Phys. Rev. C **80**, 041902 (2009). arXiv:0904.0439 [nucl-ex]
665. P.L. Cho, M.B. Wise, Phys. Lett. B **346**, 129 (1995). arXiv:hep-ph/9411303
666. A.K. Leibovich, Phys. Rev. D **56**, 4412 (1997). arXiv:hep-ph/9610381
667. E. Braaten, B.A. Kniehl, J. Lee, Phys. Rev. D **62**, 094005 (2000). arXiv:hep-ph/9911436
668. A.A. Affolder et al. (CDF Collaboration), Phys. Rev. Lett. **84**, 2094 (2000). arXiv:hep-ex/9910025
669. D.E. Acosta et al. (CDF Collaboration), Phys. Rev. Lett. **88**, 161802 (2002)
670. V.M. Abazov et al. (DØ Collaboration), Phys. Rev. Lett. **94**, 232001 (2005). Erratum-ibid. **100**, 049902 (2008). arXiv:hep-ex/0502030
671. V.M. Abazov et al. (DØ Collaboration), Phys. Rev. Lett. **101**, 182004 (2008). arXiv:0804.2799 [hep-ex]
672. B. Gong, J.X. Wang, Phys. Rev. Lett. **100**, 232001 (2008). arXiv:0802.3727 [hep-ph]
673. B. Gong, J.X. Wang, Phys. Rev. D **78**, 074011 (2008). arXiv:0805.2469 [hep-ph]
674. B. Gong, X.Q. Li, J.X. Wang, Phys. Lett. B **673**, 197 (2009). arXiv:0805.4751 [hep-ph]
675. Y.Q. Ma, K. Wang, K.T. Chao, arXiv:1002.3987 [hep-ph]
676. S.J. Brodsky, J.P. Lansberg, Phys. Rev. D **81**, 051502 (2010). arXiv:0908.0754 [hep-ph]
677. M. Kramer, Nucl. Phys. B **459**, 3 (1996). arXiv:hep-ph/9508409
678. V.G. Kartvelishvili, A.K. Likhoded, S.R. Slabospitsky, Sov. J. Nucl. Phys. **28**, 678 (1978). Yad. Fiz. **28**, 1315 (1978)
679. E. Braaten, T.C. Yuan, Phys. Rev. D **52**, 6627 (1995). arXiv:hep-ph/9507398
680. V.A. Khoze, A.D. Martin, M.G. Ryskin, W.J. Stirling, Eur. Phys. J. C **39**, 163 (2005). arXiv:hep-ph/0410020
681. J.P. Lansberg, J.R. Cudell, Yu.L. Kalinovsky, Phys. Lett. B **633**, 301 (2006). arXiv:hep-ph/0507060
682. H. Haberzettl, J.P. Lansberg, Phys. Rev. Lett. **100**, 032006 (2008). arXiv:0709.3471 [hep-ph]
683. P. Artoisenet, E. Braaten, Phys. Rev. D **80**, 034018 (2009). arXiv:0907.0025 [hep-ph]
684. Z.G. He, R. Li, J.X. Wang, Phys. Rev. D **79**, 094003 (2009). arXiv:0904.2069 [hep-ph]
685. Z.G. He, R. Li, J.X. Wang, arXiv:0904.1477 [hep-ph]
686. Y. Fan, Y.Q. Ma, K.T. Chao, Phys. Rev. D **79**, 114009 (2009). arXiv:0904.4025 [hep-ph]
687. P. Artoisenet, AIP Conf. Proc. **1038**, 55 (2008)
688. B. Gong, J.X. Wang, H.F. Zhang, arXiv:1009.3839 [hep-ph]
689. T. Shears (on behalf of the CDF Collaboration), Eur. Phys. J. C **33**, S475 (2004)
690. J.P. Lansberg, arXiv:1006.2750 [hep-ph]
691. A. Abulencia et al. (CDF Collaboration), Phys. Rev. Lett. **98**, 232001 (2007). arXiv:hep-ex/0703028
692. Y.Q. Ma, K. Wang, K.T. Chao, arXiv:1009.3655 [hep-ph]
693. M. Butenschoen, B.A. Kniehl, arXiv:1009.5662 [hep-ph]
694. (CMS Collaboration) CMS Physics Analysis Summary, CMS PAS BPH-10-002 (2010)
695. M. Butenschoen, B.A. Kniehl, Phys. Rev. Lett. **104**, 072001 (2010). arXiv:0909.2798 [hep-ph]
696. F.D. Aaron et al. (H1 Collaboration), Eur. Phys. J. C **68**, 401 (2010). arXiv:1002.0234 [hep-ex]
697. C. Adloff et al. (H1 Collaboration), Eur. Phys. J. C **25**, 25 (2002). arXiv:hep-ex/0205064
698. G.C. Nayak, M.X. Liu, F. Cooper, Phys. Rev. D **68**, 034003 (2003). arXiv:hep-ph/0302095

699. H.S. Chung, S. Kim, J. Lee, C. Yu, Phys. Rev. D **81**, 014020 (2010). arXiv:0911.2113 [hep-ph]
700. J.P. Lansberg, Phys. Lett. B **695**, 149 (2010). arXiv:1003.4319 [hep-ph]
701. T. Aaltonen et al. (CDF Collaboration), Phys. Rev. Lett. **102**, 242001 (2009). arXiv:0902.1271 [hep-ex]
702. V.A. Khoze, A.D. Martin, M.G. Ryskin, Eur. Phys. J. C **19**, 477 (2001). Erratum-ibid. C **20**, 599 (2001), arXiv:hep-ph/0011393
703. F. Yuan, Phys. Lett. B **510**, 155 (2001). arXiv:hep-ph/0103213
704. S.R. Klein, J. Nystrand, Phys. Rev. Lett. **92**, 142003 (2004). arXiv:hep-ph/0311164
705. V.A. Khoze, A.D. Martin, M.G. Ryskin, W.J. Stirling, Eur. Phys. J. C **35**, 211 (2004). arXiv:hep-ph/0403218
706. V.P. Goncalves, M.V.T. Machado, Eur. Phys. J. C **40**, 519 (2005). arXiv:hep-ph/0501099
707. A. Bzdak, Phys. Lett. B **619**, 288 (2005). arXiv:hep-ph/0506101
708. A. Bzdak, L. Motyka, L. Szymanowski, J.R. Cudell, Phys. Rev. D **75**, 094023 (2007). arXiv:hep-ph/0702134
709. W. Schafer, A. Szczurek, Phys. Rev. D **76**, 094014 (2007). arXiv:0705.2887 [hep-ph]
710. L. Motyka, G. Watt, Phys. Rev. D **78**, 014023 (2008). arXiv:0805.2113 [hep-ph]
711. Z. Conesa del Valle (PHENIX Collaboration), Nucl. Phys. A **830**, 511C (2009). arXiv:0907.4452 [nucl-ex]
712. S. Klein, J. Nystrand, Phys. Rev. C **60**, 014903 (1999). arXiv:hep-ph/9902259
713. A.J. Baltz, S.R. Klein, J. Nystrand, Phys. Rev. Lett. **89**, 012301 (2002). arXiv:nucl-th/0205031
714. J. Nystrand, Nucl. Phys. A **752**, 470 (2005). arXiv:hep-ph/0412096
715. M. Strikman, M. Tverskoy, M. Zhalov, Phys. Lett. B **626**, 72 (2005). arXiv:hep-ph/0505023
716. V.P. Goncalves, M.V.T. Machado, J. Phys. G **32**, 295 (2006). arXiv:hep-ph/0506331
717. V.P. Goncalves, M.V.T. Machado, arXiv:0706.2810 [hep-ph]
718. Yu.P. Ivanov, B.Z. Kopeliovich, I. Schmidt, arXiv:0706.1532 [hep-ph]
719. A.L. Ayala Filho, V.P. Goncalves, M.T. Griep, Phys. Rev. C **78**, 044904 (2008). arXiv:0808.0366 [hep-ph]
720. G. Bauer (CDF II Collaboration), Int. J. Mod. Phys. A **20**, 3765 (2005). arXiv:hep-ex/0409052
721. E. Braaten, Phys. Rev. D **73**, 011501 (2006). arXiv:hep-ph/0408230
722. P. Faccioli, C. Lourenço, J. Seixas, H.K. Wohri, Phys. Rev. Lett. **102**, 151802 (2009). arXiv:0902.4462 [hep-ph]
723. P. Faccioli, C. Lourenço, J. Seixas, Phys. Rev. D **81**, 111502 (2010). arXiv:1005.2855 [hep-ph]
724. R. Li, J.X. Wang, Phys. Lett. B **672**, 51 (2009). arXiv:0811.0963 [hep-ph]
725. J.P. Lansberg, Phys. Lett. B **679**, 340 (2009). arXiv:0901.4777 [hep-ph]
726. E. Braaten, D. Kang, J. Lee, C. Yu, Phys. Rev. D **79**, 054013 (2009). arXiv:0812.3727 [hep-ph]
727. C.N. Brown et al. (FNAL E866 Collaboration and NuSea Collaboration), Phys. Rev. Lett. **86**, 2529 (2001). arXiv:hep-ex/0011030
728. T.H. Chang et al. (FNAL E866/NuSea Collaboration), Phys. Rev. Lett. **91**, 211801 (2003). arXiv:hep-ex/0308001
729. J.C. Collins, D.E. Soper, Phys. Rev. D **16**, 2219 (1977)
730. I. Abt et al. (HERA-B Collaboration), Eur. Phys. J. C **60**, 517 (2009). arXiv:0901.1015 [hep-ex]
731. K. Gottfried, J.D. Jackson, Nuovo Cimento **33**, 309 (1964)
732. A.D. Martin, R.G. Roberts, W.J. Stirling, R.S. Thorne, Eur. Phys. J. C **4**, 463 (1998). arXiv:hep-ph/9803445
733. H.L. Lai et al. (CTEQ Collaboration), Eur. Phys. J. C **12**, 375 (2000). arXiv:hep-ph/9903282
734. CDF Collaboration, CDF public note 9966
735. E. Braaten, J. Lee, Phys. Rev. D **63**, 071501 (2001). arXiv:hep-ph/0012244
736. Rapidity interval communicated privately by J. Lee
737. CDF Collaboration, CDF public plot
738. Berkeley Workshop on Physics Opportunities with Early LHC Data
739. A.C. Kraan, AIP Conf. Proc. **1038**, 45 (2008). arXiv:0807.3123 [hep-ex]
740. C. Albajar et al. (UA1 Collaboration), Phys. Lett. B **200**, 380 (1988)
741. C. Albajar et al. (UA1 Collaboration), Phys. Lett. B **256**, 112 (1991)
742. Z. Tang (STAR Collaboration), Nucl. Phys. A **834**, 282C (2010)
743. P. Artoisenet, F. Maltoni, T. Stelzer, J. High Energy Phys. **0802**, 102 (2008). arXiv:0712.2770 [hep-ph]
744. J. Alwall et al., J. High Energy Phys. **0709**, 028 (2007). arXiv:0706.2334 [hep-ph]
745. P. Artoisenet, in *The Proceedings of 9th Workshop on Non-Perturbative Quantum Chromodynamics, Paris, France, 4–8 June 2007, p. 21*, arXiv:0804.2975 [hep-ph]
746. M. Klasen, J.P. Lansberg, Nucl. Phys. Proc. Suppl. **179–180**, 226 (2008). arXiv:0806.3662 [hep-ph]
747. F.D. Aaron et al. (H1 Collaboration), J. High Energy Phys. **1005**, 032 (2010). arXiv:0910.5831 [hep-ex]
748. S. Kananov (ZEUS Collaboration), Nucl. Phys. Proc. Suppl. **184**, 252 (2008)
749. S. Chekanov et al. (ZEUS Collaboration), Nucl. Phys. B **718**, 3 (2005). arXiv:hep-ex/0504010
750. J. Breitweg et al. (ZEUS Collaboration), Phys. Lett. B **487**, 273 (2000). arXiv:hep-ex/0006013
751. A. Aktas et al. (H1 Collaboration), Eur. Phys. J. C **46**, 585 (2006). arXiv:hep-ex/0510016
752. S. Chekanov et al. (ZEUS Collaboration), Nucl. Phys. B **695**, 3 (2004). arXiv:hep-ex/0404008
753. S. Chekanov et al. (ZEUS Collaboration), Eur. Phys. J. C **24**, 345 (2002). arXiv:hep-ex/0201043
754. A. Aktas et al. (H1 Collaboration), Phys. Lett. B **568**, 205 (2003). arXiv:hep-ex/0306013
755. S. Chekanov et al. (ZEUS Collaboration), J. High Energy Phys. **1005**, 085 (2010). arXiv:0910.1235 [hep-ex]
756. C. Adloff et al. (H1 Collaboration), Phys. Lett. B **541**, 251 (2002). arXiv:hep-ex/0205107
757. C. Adloff et al. (H1 Collaboration), Phys. Lett. B **483**, 23 (2000). arXiv:hep-ex/0003020
758. S. Chekanov et al. (ZEUS Collaboration), Phys. Lett. B **680**, 4 (2009). arXiv:0903.4205 [hep-ex]
759. I.P. Ivanov, N.N. Nikolaev, A.A. Savin, Phys. Part. Nucl. **37**, 1 (2006). arXiv:hep-ph/0501034
760. S. Chekanov et al. (ZEUS Collaboration), Eur. Phys. J. C **27**, 173 (2003). arXiv:hep-ex/0211011
761. S. Chekanov et al. (ZEUS Collaboration), J. High Energy Phys. **0912**, 007 (2009). arXiv:0906.1424 [hep-ex]
762. S. Fleming, A.K. Leibovich, T. Mehen, Phys. Rev. D **74**, 114004 (2006). arXiv:hep-ph/0607121
763. B.A. Kniehl, D.V. Vasin, V.A. Saleev, Phys. Rev. D **73**, 074022 (2006). arXiv:hep-ph/0602179
764. A.D. Martin, M.G. Ryskin, G. Watt, Eur. Phys. J. C **66**, 163 (2010). arXiv:0909.5529 [hep-ph]
765. M. Kramer, J. Zunft, J. Steegborn, P.M. Zerwas, Phys. Lett. B **348**, 657 (1995). arXiv:hep-ph/9411372
766. P. Artoisenet, J.M. Campbell, F. Maltoni, F. Tramontano, Phys. Rev. Lett. **102**, 142001 (2009). arXiv:0901.4352 [hep-ph]
767. C.H. Chang, R. Li, J.X. Wang, Phys. Rev. D **80**, 034020 (2009). arXiv:0901.4749 [hep-ph]

768. M. Butenschoen, Nucl. Phys. Proc. Suppl. **191**, 193 (2009)
769. M. Butenschoen, B.A. Kniehl, PoS D **IS2010**, 157 (2010). arXiv:1006.1776 [hep-ph]
770. B.A. Kniehl, G. Kramer, Eur. Phys. J. C **6**, 493 (1999). arXiv: hep-ph/9803256
771. M. Beneke, M. Kramer, M. Vanttinen, Phys. Rev. D **57**, 4258 (1998). arXiv:hep-ph/9709376
772. S.P. Baranov, JETP Lett. **88**, 471 (2008)
773. P. Artoisenet, Private communication
774. F. Maltoni et al., Phys. Lett. B **638**, 202 (2006). arXiv:hep-ph/0601203
775. E.J. Eichten, C. Quigg, Phys. Rev. D **52**, 1726 (1995). arXiv: hep-ph/9503356
776. P. Nason et al., arXiv:hep-ph/0003142
777. G.T. Bodwin, H.S. Chung, D. Kang, J. Lee, C. Yu, Phys. Rev. D **77**, 094017 (2008). arXiv:0710.0994 [hep-ph]
778. K. Abe et al. (Belle Collaboration), Phys. Rev. Lett. **89**, 142001 (2002). arXiv:hep-ex/0205104
779. K.Y. Liu, Z.G. He, K.T. Chao, Phys. Lett. B **557**, 45 (2003). arXiv:hep-ph/0211181
780. E. Braaten, J. Lee, Phys. Rev. D **67**, 054007 (2003). Erratum-ibid. D **72**, 099901 (2005). arXiv:hep-ph/0211085
781. K. Hagiwara, E. Kou, C.F. Qiao, Phys. Lett. B **570**, 39 (2003). arXiv:hep-ph/0305102
782. K. Abe et al. (Belle Collaboration), Phys. Rev. D **70**, 071102 (2004). arXiv:hep-ex/0407009
783. G.T. Bodwin, D. Kang, T. Kim, J. Lee, C. Yu, AIP Conf. Proc. **892**, 315 (2007). arXiv:hep-ph/0611002
784. Z.G. He, Y. Fan, K.T. Chao, Phys. Rev. D **75**, 074011 (2007). arXiv:hep-ph/0702239
785. G.T. Bodwin, J. Lee, C. Yu, Phys. Rev. D **77**, 094018 (2008). arXiv:0710.0995 [hep-ph]
786. Y.J. Zhang, Y.j. Gao, K.T. Chao, Phys. Rev. Lett. **96**, 092001 (2006). arXiv:hep-ph/0506076
787. B. Gong, J.X. Wang, Phys. Rev. D **77**, 054028 (2008). arXiv: 0712.4220 [hep-ph]
788. M. Gremm, A. Kapustin, Phys. Lett. B **407**, 323 (1997). arXiv:hep-ph/9701353
789. G.T. Bodwin, D. Kang, J. Lee, Phys. Rev. D **74**, 014014 (2006). arXiv:hep-ph/0603186
790. V.V. Braguta, A.K. Likhoded, A.V. Luchinsky, Phys. Lett. B **646**, 80 (2007). arXiv:hep-ph/0611021
791. V.V. Braguta, A.K. Likhoded, A.V. Luchinsky, Phys. Rev. D **79**, 074004 (2009). arXiv:0810.3607 [hep-ph]
792. G.T. Bodwin, D. Kang, J. Lee, Phys. Rev. D **74**, 114028 (2006). arXiv:hep-ph/0603185
793. M. Beneke, A. Signer, V.A. Smirnov, Phys. Rev. Lett. **80**, 2535 (1998). arXiv:hep-ph/9712302
794. A. Czarnecki, K. Melnikov, Phys. Lett. B **519**, 212 (2001). arXiv:hep-ph/0109054
795. B. Aubert et al. (BABAR Collaboration), Phys. Rev. Lett. **87**, 162002 (2001). arXiv:hep-ex/0106044
796. K. Abe et al. (BELLE Collaboration), Phys. Rev. Lett. **88**, 052001 (2002). arXiv:hep-ex/0110012
797. F. Yuan, C.F. Qiao, K.T. Chao, Phys. Rev. D **56**, 321 (1997). arXiv:hep-ph/9703438
798. P.L. Cho, A.K. Leibovich, Phys. Rev. D **54**, 6690 (1996). arXiv:hep-ph/9606229
799. S. Baek, P. Ko, J. Lee, H.S. Song, J. Korean Phys. Soc. **33**, 97 (1998). arXiv:hep-ph/9804455
800. G.A. Schuler, Eur. Phys. J. C **8**, 273 (1999). arXiv:hep-ph/9804349
801. V.V. Kiselev, A.K. Likhoded, M.V. Shevlyagin, Phys. Lett. B **332**, 411 (1994). arXiv:hep-ph/9408407
802. K.Y. Liu, Z.G. He, K.T. Chao, Phys. Rev. D **68**, 031501 (2003). arXiv:hep-ph/0305084
803. K.Y. Liu, Z.G. He, K.T. Chao, Phys. Rev. D **69**, 094027 (2004). arXiv:hep-ph/0301218
804. K. Hagiwara, E. Kou, Z.H. Lin, C.F. Qiao, G.H. Zhu, Phys. Rev. D **70**, 034013 (2004). arXiv:hep-ph/0401246
805. A.V. Berezhnoy, A.K. Likhoded, Phys. At. Nucl. **67**, 757 (2004). Yad. Fiz. **67**, 778 (2004), arXiv:hep-ph/0303145
806. D. Kang, J.W. Lee, J. Lee, T. Kim, P. Ko, Phys. Rev. D **71**, 094019 (2005). arXiv:hep-ph/0412381
807. S. Fleming, A.K. Leibovich, T. Mehen, Phys. Rev. D **68**, 094011 (2003). arXiv:hep-ph/0306139
808. Z.H. Lin, G.h. Zhu, Phys. Lett. B **597**, 382 (2004). arXiv: hep-ph/0406121
809. A.K. Leibovich, X. Liu, Phys. Rev. D **76**, 034005 (2007). arXiv:0705.3230 [hep-ph]
810. P. Pakhlov et al. (Belle Collaboration), Phys. Rev. D **79**, 071101 (2009). arXiv:0901.2775 [hep-ex]
811. Y.J. Zhang, K.T. Chao, Phys. Rev. Lett. **98**, 092003 (2007). arXiv:hep-ph/0611086
812. Y.Q. Ma, Y.J. Zhang, K.T. Chao, Phys. Rev. Lett. **102**, 162002 (2009). arXiv:0812.5106 [hep-ph]
813. B. Gong, J.X. Wang, Phys. Rev. D **80**, 054015 (2009). arXiv:0904.1103 [hep-ph]
814. B. Gong, J.X. Wang, Phys. Rev. Lett. **102**, 162003 (2009). arXiv:0901.0117 [hep-ph]
815. C. Peterson, D. Schlatter, I. Schmitt, P.M. Zerwas, Phys. Rev. D **27**, 105 (1983)
816. Z.G. He, Y. Fan, K.T. Chao, Phys. Rev. D **81**, 054036 (2010). arXiv:0910.3636 [hep-ph]
817. Y. Jia, Phys. Rev. D **82**, 034017 (2010). arXiv:0912.5498 [hep-ph]
818. E. Braaten, Y.Q. Chen, Phys. Rev. Lett. **76**, 730 (1996). arXiv: hep-ph/9508373
819. Y.J. Zhang, Y.Q. Ma, K. Wang, K.T. Chao, Phys. Rev. D **81**, 034015 (2010). arXiv:0911.2166 [hep-ph]
820. S. Todorova-Nova, arXiv:hep-ph/0112050
821. J. Abdallah et al. (DELPHI Collaboration), Phys. Lett. B **565**, 76 (2003). arXiv:hep-ex/0307049
822. J.P. Ma, B.H.J. McKellar, C.B. Paranavitane, Phys. Rev. D **57**, 606 (1998). arXiv:hep-ph/9707480
823. G. Japaridze, A. Tkabladze, Phys. Lett. B **433**, 139 (1998). arXiv:hep-ph/9803447
824. R.M. Godbole, D. Indumathi, M. Krämer, Phys. Rev. D **65**, 074003 (2002). arXiv:hep-ph/0101333
825. M. Klasen, B.A. Kniehl, L. Mihaila, M. Steinhauser, Nucl. Phys. B **609**, 518 (2001). arXiv:hep-ph/0104044
826. M. Klasen, B.A. Kniehl, L.N. Mihaila, M. Steinhauser, Phys. Rev. Lett. **89**, 032001 (2002). arXiv:hep-ph/0112259
827. M. Klasen, B.A. Kniehl, L.N. Mihaila, M. Steinhauser, Nucl. Phys. B **713**, 487 (2005). arXiv:hep-ph/0407014
828. J. de Favereau de Jeneret et al., arXiv:0908.2020 [hep-ph]
829. CDF Collaboration, CDF public note 9294
830. CDF Collaboration, CDF public note 7649
831. CDF Collaboration, CDF public note 9740
832. W. Beenakker, H. Kuijf, W.L. van Neerven, J. Smith, Phys. Rev. D **40**, 54 (1989)
833. P. Nason, S. Dawson, R.K. Ellis, Nucl. Phys. B **327**, 49 (1989). Erratum-ibid. B **335**, 260 (1990)
834. C.H. Chang, Y.Q. Chen, Phys. Rev. D **48**, 4086 (1993)
835. C.H. Chang, Y.Q. Chen, G.P. Han, H.T. Jiang, Phys. Lett. B **364**, 78 (1995). arXiv:hep-ph/9408242
836. K. Kolodziej, A. Leike, R. Ruckl, Phys. Lett. B **355**, 337 (1995). arXiv:hep-ph/9505298
837. A.V. Berezhnoy, V.V. Kiselev, A.K. Likhoded, Z. Phys. A **356**, 89 (1996)
838. S.P. Baranov, Phys. Rev. D **56**, 3046 (1997)
839. A.V. Berezhnoi, V.V. Kiselev, A.K. Likhoded, Phys. At. Nucl. **60**, 100 (1997). Yad. Fiz. **60**, 108 (1997)

840. S. Frixione, M.L. Mangano, P. Nason, G. Ridolfi, Adv. Ser. Dir. High Energy Phys. **15**, 609 (1998). arXiv:hep-ph/9702287
841. E. Braaten, K.m. Cheung, T.C. Yuan, Phys. Rev. D **48**, 5049 (1993). arXiv:hep-ph/9305206
842. K.m. Cheung, Phys. Lett. B **472**, 408 (2000). arXiv:hep-ph/9908405
843. C.H. Chang, Y.Q. Chen, R.J. Oakes, Phys. Rev. D **54**, 4344 (1996). arXiv:hep-ph/9602411
844. C.H. Chang, C.F. Qiao, J.X. Wang, X.G. Wu, Phys. Rev. D **73**, 094022 (2006). arXiv:hep-ph/0601032
845. C.H. Chang, C.F. Qiao, J.X. Wang, X.G. Wu, Phys. Rev. D **72**, 114009 (2005). arXiv:hep-ph/0509040
846. M.A.G. Aivazis, J.C. Collins, F.I. Olness, W.K. Tung, Phys. Rev. D **50**, 3102 (1994). arXiv:hep-ph/9312319
847. M.A.G. Aivazis, F.I. Olness, W.K. Tung, Phys. Rev. D **50**, 3085 (1994). arXiv:hep-ph/9312318
848. F.I. Olness, R.J. Scalise, W.K. Tung, Phys. Rev. D **59**, 014506 (1999). arXiv:hep-ph/9712494
849. J. Amundson, C. Schmidt, W.K. Tung, X. Wang, J. High Energy Phys. **0010**, 031 (2000). arXiv:hep-ph/0005221
850. C.H. Chang, J.X. Wang, X.G. Wu, Phys. Rev. D **70**, 114019 (2004). arXiv:hep-ph/0409280
851. C.H. Chang, C.F. Qiao, J.X. Wang, X.G. Wu, Phys. Rev. D **71**, 074012 (2005). arXiv:hep-ph/0502155
852. C.H. Chang, J.X. Wang, X.G. Wu, Comput. Phys. Commun. **175**, 624 (2006). arXiv:hep-ph/0604238
853. T. Sjostrand, S. Mrenna, P.Z. Skands, J. High Energy Phys. **0605**, 026 (2006). arXiv:hep-ph/0603175
854. C.H. Chang, J.X. Wang, X.G. Wu, Phys. Rev. D **77**, 014022 (2008). arXiv:0711.1898 [hep-ph]
855. E.V. Shuryak, Phys. Rep. **61**, 71 (1980)
856. P. Petreczky, Nucl. Phys. A **830**, 11C (2009). arXiv:0908.1917 [hep-ph]
857. Z. Fodor, S.D. Katz, arXiv:0908.3341 [hep-ph]
858. C. DeTar, U.M. Heller, Eur. Phys. J. A **41**, 405 (2009). arXiv:0905.2949 [hep-lat]
859. P. Petreczky, Nucl. Phys. Proc. Suppl. **140**, 78 (2005). arXiv:hep-lat/0409139
860. H. Satz, Rep. Prog. Phys. **63**, 1511 (2000). arXiv:hep-ph/0007069
861. B. Muller, J.L. Nagle, Annu. Rev. Nucl. Part. Sci. **56**, 93 (2006). arXiv:nucl-th/0602029
862. T. Matsui, H. Satz, Phys. Lett. B **178**, 416 (1986)
863. CERN press release, *New State of Matter created at CERN*, Feb. 10, 2000
864. P. Faccioli, C. Lourenço, J. Seixas, H.K. Woehri, J. High Energy Phys. **0810**, 004 (2008). arXiv:0809.2153 [hep-ph]
865. R. Vogt, Nucl. Phys. A **700**, 539 (2002). arXiv:hep-ph/0107045
866. K.J. Eskola, H. Paukkunen, C.A. Salgado, J. High Energy Phys. **0904**, 065 (2009). arXiv:0902.4154 [hep-ph]
867. A.D. Frawley, T. Ullrich, R. Vogt, Phys. Rep. **462**, 125 (2008). arXiv:0806.1013 [nucl-ex]
868. D. Kharzeev, K. Tuchin, Nucl. Phys. A **735**, 248 (2004). arXiv:hep-ph/0310358
869. D. Kharzeev, E. Levin, M. Nardi, K. Tuchin, Phys. Rev. Lett. **102**, 152301 (2009). arXiv:0808.2954 [hep-ph]
870. D. Kharzeev, H. Satz, Phys. Lett. B **366**, 316 (1996). arXiv:hep-ph/9508276
871. J.P. Blaizot, J.Y. Ollitrault, Phys. Lett. B **217**, 386 (1989)
872. B. Alessandro et al. (NA50 Collaboration), Eur. Phys. J. C **48**, 329 (2006). arXiv:nucl-ex/0612012
873. B. Alessandro et al. (NA50 Collaboration), Eur. Phys. J. C **33**, 31 (2004)
874. M.J. Leitch et al. (FNAL E866/NuSea Collaboration), Phys. Rev. Lett. **84**, 3256 (2000). arXiv:nucl-ex/9909007
875. C. Lourenço, R. Vogt, H.K. Woehri, J. High Energy Phys. **0902**, 014 (2009). arXiv:0901.3054 [hep-ph]
876. K.J. Eskola, V.J. Kolhinen, P.V. Ruuskanen, Nucl. Phys. B **535**, 351 (1998). arXiv:hep-ph/9802350
877. K.J. Eskola, V.J. Kolhinen, C.A. Salgado, Eur. Phys. J. C **9**, 61 (1999). arXiv:hep-ph/9807297
878. J. Pumplin, D.R. Stump, J. Huston, H.L. Lai, P.M. Nadolsky, W.K. Tung, J. High Energy Phys. **0207**, 012 (2002). arXiv:hep-ph/0201195
879. D. Stump, J. Huston, J. Pumplin, W.K. Tung, H.L. Lai, S. Kuhlmann, J.F. Owens, J. High Energy Phys. **0310**, 046 (2003). arXiv:hep-ph/0303013
880. J. Badier et al. (NA3 Collaboration), Z. Phys. C **20**, 101 (1983)
881. I. Abt et al. (HERA-B Collaboration), Eur. Phys. J. C **60**, 525 (2009). arXiv:0812.0734 [hep-ex]
882. E. Scomparin (NA60 Collaboration), Nucl. Phys. A **830**, 239c (2009). arXiv:0907.3682 [nucl-ex]
883. A. Adare et al. (PHENIX Collaboration), Phys. Rev. C **77**, 024912 (2008). Erratum-ibid. C **79**, 059901 (2009). arXiv:0711.3917 [nucl-ex]
884. A.D. Frawley, Presentation at the *ECT* workshop on Quarkonium Production in Heavy-Ion Collisions, Trento (Italy), May 25–29, 2009* and at the *Joint CATHIE-INT mini-program "Quarkonia in Hot QCD", June 16–26, 2009*
885. D.M. Alde et al., Phys. Rev. Lett. **66**, 2285 (1991)
886. R. Vogt, Phys. Rev. C **61**, 035203 (2000). arXiv:hep-ph/9907317
887. B. Povh, J. Hufner, Phys. Rev. Lett. **58**, 1612 (1987)
888. R. Vogt, Phys. Rev. C **71**, 054902 (2005). arXiv:hep-ph/0411378
889. R. Arnaldi, P. Cortese, E. Scomparin, Phys. Rev. C **81**, 014903 (2010). arXiv:0909.2199 [hep-ph]
890. R. Gavai, D. Kharzeev, H. Satz, G.A. Schuler, K. Sridhar, R. Vogt, Int. J. Mod. Phys. A **10**, 3043 (1995). arXiv:hep-ph/9502270
891. E.G. Ferreiro, F. Fleuret, J.P. Lansberg, A. Rakotozafindrabe, Phys. Lett. B **680**, 50 (2009). arXiv:0809.4684 [hep-ph]
892. M. Beneke, I.Z. Rothstein, Phys. Rev. D **54**, 2005 (1996). Erratum-ibid. D **54**, 7082 (1996). arXiv:hep-ph/9603400
893. X.F. Zhang, C.F. Qiao, X.X. Yao, W.Q. Chao, arXiv:hep-ph/9711237
894. T. Umeda, K. Nomura, H. Matsufuru, Eur. Phys. J. C **39S1**, 9 (2005). arXiv:hep-lat/0211003
895. M. Asakawa, T. Hatsuda, Phys. Rev. Lett. **92**, 012001 (2004). arXiv:hep-lat/0308034
896. S. Datta, F. Karsch, P. Petreczky, I. Wetzorke, Phys. Rev. D **69**, 094507 (2004). arXiv:hep-lat/0312037
897. A. Jakovác, P. Petreczky, K. Petrov, A. Velytsky, Phys. Rev. D **75**, 014506 (2007). arXiv:hep-lat/0611017
898. G. Aarts et al., Phys. Rev. D **76**, 094513 (2007). arXiv:0705.2198 [hep-lat]
899. S. Digal, P. Petreczky, H. Satz, Phys. Rev. D **64**, 094015 (2001). arXiv:hep-ph/0106017
900. C.Y. Wong, Phys. Rev. C **72**, 034906 (2005)
901. Á. Mócsy, P. Petreczky, Phys. Rev. D **73**, 074007 (2006). arXiv:hep-ph/0512156
902. A. Mocsy, P. Petreczky, Eur. Phys. J. C **43**, 77 (2005). arXiv:hep-ph/0411262
903. W.M. Alberico, A. Beraudo, A. De Pace, A. Molinari, Phys. Rev. D **75**, 074009 (2007)
904. D. Cabrera, R. Rapp, Phys. Rev. D **76**, 114506 (2007). arXiv:hep-ph/0611134
905. Á. Mócsy, P. Petreczky, Phys. Rev. D **77**, 014501 (2008). arXiv:0705.2559 [hep-ph]
906. Á. Mócsy, P. Petreczky, Phys. Rev. Lett. **99**, 211602 (2007). arXiv:0706.2183 [hep-ph]
907. M. Laine, O. Philipsen, M. Tassler, J. High Energy Phys. **0709**, 066 (2007). arXiv:0707.2458 [hep-lat]

908. M. Laine, J. High Energy Phys. **0705**, 028 (2007). arXiv:0704.1720 [hep-ph]
909. M. Laine, Nucl. Phys. A **820**, 25C (2009). arXiv:0810.1112 [hep-ph]
910. N. Brambilla, J. Ghiglieri, A. Vairo, P. Petreczky, Phys. Rev. D **78**, 014017 (2008). arXiv:0804.0993 [hep-ph]
911. Y. Aoki, Z. Fodor, S.D. Katz, K.K. Szabo, Phys. Lett. B **643**, 46 (2006). arXiv:hep-lat/0609068
912. M. Cheng et al., Phys. Rev. D **77**, 014511 (2008). arXiv: 0710.0354 [hep-lat]
913. Y. Aoki, S. Borsanyi, S. Durr, Z. Fodor, S.D. Katz, S. Krieg, K.K. Szabo, J. High Energy Phys. **0906**, 088 (2009). arXiv: 0903.4155 [hep-lat]
914. A. Bazavov et al. Phys. Rev. D **80**, 014504 (2009). arXiv: 0903.4379 [hep-lat]
915. M. Cheng et al., Phys. Rev. D **81**, 054504 (2010). arXiv: 0911.2215 [hep-lat]
916. A. Bazavov, P. Petreczky, PoS **LAT2009**, 163 (2009). arXiv: 0912.5421 [hep-lat]
917. A. Bazavov, P. Petreczky (HotQCD Collaboration), J. Phys. Conf. Ser. **230**, 012014 (2010). arXiv:1005.1131 [hep-lat]
918. A. Bazavov, P. Petreczky, arXiv:1009.4914 [hep-lat]
919. S. Borsanyiet al. (Wuppertal-Budapest Collaboration), J. High Energy Phys. **1009**, 073 (2010). arXiv:1005.3508 [hep-lat]
920. S. Borsanyi et al., J. High Energy Phys. **1011**, 077 (2010). arXiv:1007.2580 [hep-lat]
921. A. Bazavov, P. Petreczky, A. Velytsky, arXiv:0904.1748 [hep-ph]
922. P. Petreczky, Eur. Phys. J. C **43**, 51 (2005). arXiv:hep-lat/0502008
923. O. Kaczmarek, F. Karsch, P. Petreczky, F. Zantow, Phys. Lett. B **543**, 41 (2002). arXiv:hep-lat/0207002
924. S. Digal, S. Fortunato, P. Petreczky, Phys. Rev. D **68**, 034008 (2003). arXiv:hep-lat/0304017
925. P. Petreczky, K. Petrov, Phys. Rev. D **70**, 054503 (2004). arXiv:hep-lat/0405009
926. O. Kaczmarek, F. Zantow, Phys. Rev. D **71**, 114510 (2005). arXiv:hep-lat/0503017
927. O. Kaczmarek, PoS C **POD07**, 043 (2007). arXiv:0710.0498 [hep-lat]
928. P. Petreczky, J. Phys. G **37**, 094009 (2010). arXiv:1001.5284 [hep-ph]
929. P. Petreczky, arXiv:0906.0502 [nucl-th]
930. F. Karsch, E. Laermann, P. Petreczky, S. Stickan, Phys. Rev. D **68**, 014504 (2003). arXiv:hep-lat/0303017
931. A. Mocsy, Eur. Phys. J. C **61**, 705 (2009). arXiv:0811.0337 [hep-ph]
932. P. Petreczky, D. Teaney, Phys. Rev. D **73**, 014508 (2006). arXiv:hep-ph/0507318
933. T. Umeda, Phys. Rev. D **75**, 094502 (2007). arXiv: hep-lat/0701005
934. P. Petreczky, Eur. Phys. J. C **62**, 85 (2009). arXiv:0810.0258 [hep-lat]
935. M. Laine, O. Philipsen, P. Romatschke, M. Tassler, J. High Energy Phys. **0703**, 054 (2007). arXiv:hep-ph/0611300
936. Y. Burnier, M. Laine, M. Vepsalainen, J. High Energy Phys. **0801**, 043 (2008). arXiv:0711.1743 [hep-ph]
937. A. Beraudo, J.P. Blaizot, C. Ratti, Nucl. Phys. A **806**, 312 (2008). arXiv:0712.4394 [nucl-th]
938. M.A. Escobedo, J. Soto, Phys. Rev. A **78**, 032520 (2008). arXiv:0804.0691 [hep-ph]
939. M.A. Escobedo, J. Soto, Phys. Rev. A **82**, 042506 (2010). arXiv:1008.0254 [hep-ph]
940. A. Vairo, PoS C **CONFINEMENT8**, 002 (2008). arXiv: 0901.3495 [hep-ph]
941. E. Braaten, R.D. Pisarski, Nucl. Phys. B **337**, 569 (1990)
942. E. Braaten, R.D. Pisarski, Nucl. Phys. B **339**, 310 (1990)
943. E. Braaten, R.D. Pisarski, Phys. Rev. D **45**, 1827 (1992)
944. J. Frenkel, J.C. Taylor, Nucl. Phys. B **334**, 199 (1990)
945. N. Brambilla, M.A. Escobedo, J. Ghiglieri, J. Soto, A. Vairo, J. High Energy Phys. **1009**, 038 (2010). arXiv:1007.4156 [hep-ph]
946. O. Jahn, O. Philipsen, Phys. Rev. D **70**, 074504 (2004). arXiv:hep-lat/0407042
947. N. Brambilla, J. Ghiglieri, P. Petreczky, A. Vairo, Phys. Rev. D **82**, 074019 (2010). arXiv:1007.5172 [hep-ph]
948. Y. Burnier, M. Laine, M. Vepsalainen, J. High Energy Phys. **1001**, 054 (2010). arXiv:0911.3480 [hep-ph]
949. A. Rothkopf, T. Hatsuda, S. Sasaki, PoS **LAT2009**, 162 (2009). arXiv:0910.2321 [hep-lat]
950. G. Aarts, S. Kim, M.P. Lombardo, M.B. Oktay, S.M. Ryan, D.K. Sinclair, J.I. Skullerud, arXiv:1010.3725 [hep-lat]
951. A. Dumitru, Y. Guo, M. Strickland, Phys. Lett. B **662**, 37 (2008). arXiv:0711.4722 [hep-ph]
952. O. Philipsen, M. Tassler, arXiv:0908.1746 [hep-ph]
953. A. Dumitru, Y. Guo, M. Strickland, Phys. Rev. D **79**, 114003 (2009). arXiv:0903.4703 [hep-ph]
954. Y. Burnier, M. Laine, M. Vepsalainen, Phys. Lett. B **678**, 86 (2009). arXiv:0903.3467 [hep-ph]
955. A. Dumitru, Y. Guo, A. Mocsy, M. Strickland, Phys. Rev. D **79**, 054019 (2009). arXiv:0901.1998 [hep-ph]
956. M. Asakawa, S.A. Bass, B. Muller, Phys. Rev. Lett. **96**, 252301 (2006). arXiv:hep-ph/0603092
957. A. Andronic, P. Braun-Munzinger, K. Redlich, J. Stachel, Nucl. Phys. A **789**, 334 (2007). arXiv:nucl-th/0611023
958. X. Zhao, R. Rapp, Phys. Lett. B **664**, 253 (2008). arXiv: 0712.2407 [hep-ph]
959. F. Karsch, D. Kharzeev, H. Satz, Phys. Lett. B **637**, 75 (2006). arXiv:hep-ph/0512239
960. C. Young, E. Shuryak, Phys. Rev. C **79**, 034907 (2009). arXiv:0803.2866 [nucl-th]
961. D. Teaney, arXiv:0905.2433 [nucl-th]
962. G.D. Moore, D. Teaney, Phys. Rev. C **71**, 064904 (2005). arXiv:hep-ph/0412346
963. H. van Hees, R. Rapp, Phys. Rev. C **71**, 034907 (2005). arXiv:nucl-th/0412015
964. R. Rapp, H. van Hees, arXiv:0903.1096 [hep-ph]
965. C. Young, E. Shuryak, Phys. Rev. C **81**, 034905 (2010). arXiv:0911.3080 [nucl-th]
966. R. Arnaldi et al. (NA60 Collaboration), Phys. Rev. Lett. **99**, 132302 (2007)
967. R. Arnaldi (NA60 Collaboration), Nucl. Phys. A **830**, 345c (2009). arXiv:0907.5004 [nucl-ex]
968. B. Alessandro et al. (NA50 Collaboration), Eur. Phys. J. C **39**, 335 (2005). arXiv:hep-ex/0412036
969. B. Alessandro et al. (NA50 Collaboration), Phys. Lett. B **553**, 167 (2003)
970. R. Arnaldi et al. (NA60 Collaboration), Eur. Phys. J. C **59**, 607 (2009). arXiv:0810.3204 [nucl-ex]
971. K.J. Eskola, H. Paukkunen, C.A. Salgado, J. High Energy Phys. **0807**, 102 (2008). arXiv:0802.0139 [hep-ph]
972. D. Kharzeev, E. Levin, M. Nardi, Nucl. Phys. A **730**, 448 (2004). Erratum-ibid. A **743**, 329 (2004). arXiv:hep-ph/0212316
973. L. Hulthen, M. Sagawara, Hand. Phys. **39** (1957)
974. S. Afanasiev et al. (PHENIX Collaboration), Phys. Lett. B **679**, 321 (2009). arXiv:0903.2041 [nucl-ex]
975. E.T. Atomssa (PHENIX Collaboration), Nucl. Phys. A **830**, 331c (2009). arXiv:0907.4787 [nucl-ex]
976. J.P. Lansberg, H. Haberzettl, AIP Conf. Proc. **1038**, 83 (2008). arXiv:0806.4001 [hep-ph]
977. A. Adare et al. (PHENIX Collaboration), Phys. Rev. Lett. **101**, 122301 (2008). arXiv:0801.0220 [nucl-ex]

978. L.A. Linden Levy, Nucl. Phys. A **830**, 353c (2009). arXiv:0908.2361 [nucl-ex]
979. M.J. Leitch, J. Phys. G **34**, S453 (2007). arXiv:nucl-ex/0701021
980. B.I. Abelev et al. (STAR Collaboration), Phys. Rev. D **82**, 012004 (2010). arXiv:1001.2745 [nucl-ex]
981. H. Liu (STAR Collaboration), Nucl. Phys. A **830**, 235c (2009). arXiv:0907.4538 [nucl-ex]
982. A. Adare et al. (PHENIX Collaboration), arXiv:1010.1246 [nucl-ex]
983. A. Adare et al. (PHENIX Collaboration), Phys. Rev. Lett. **98**, 232301 (2007). arXiv:nucl-ex/0611020
984. R. Vogt, Phys. Rev. C **81**, 044903 (2010). arXiv:1003.3497 [hep-ph]
985. R. Arnaldi (NA60 Collaboration), Presentation at the *ECT* Workshop on Quarkonium Production in Heavy-Ion Collisions, Trento (Italy), May 25–29, 2009*
986. C.A. Bertulani, S.R. Klein, J. Nystrand, Annu. Rev. Nucl. Part. Sci. **55**, 271 (2005). arXiv:nucl-ex/0502005
987. G. Baur, K. Hencken, D. Trautmann, S. Sadovsky, Y. Kharlov, Phys. Rep. **364**, 359 (2002). arXiv:hep-ph/0112211
988. K. Hencken et al., Phys. Rep. **458**, 1 (2008). arXiv:0706.3356 [nucl-ex]
989. M.G. Ryskin, Z. Phys. C **57**, 89 (1993)
990. V. Rebyakova, M. Strikman, M. Zhalov, Phys. Rev. D **81**, 031501 (2010). arXiv:0911.5169 [hep-ph]
991. L. Frankfurt, M. Strikman, M. Zhalov, Phys. Rev. C **67**, 034901 (2003). arXiv:hep-ph/0210303
992. L. Frankfurt, M. Strikman, M. Zhalov, Phys. Lett. B **537**, 51 (2002). arXiv:hep-ph/0204175
993. B.I. Abelev et al. (STAR Collaboration), Phys. Rev. C **77**, 034910 (2008). arXiv:0712.3320 [nucl-ex]
994. C. Adler et al. (STAR Collaboration), Phys. Rev. Lett. **89**, 272302 (2002). arXiv:nucl-ex/0206004
995. B.I. Abelev et al. (STAR Collaboration), Phys. Rev. C **81**, 044901 (2010). arXiv:0912.0604 [nucl-ex]
996. L. Frankfurt, M. Strikman, M. Zhalov, Phys. Lett. B **670**, 32 (2008). arXiv:0807.2208 [hep-ph]
997. L. Frankfurt, M. Strikman, M. Zhalov, Phys. Lett. B **640**, 162 (2006). arXiv:hep-ph/0605160
998. S.R. Klein, J. Nystrand, Phys. Rev. Lett. **84**, 2330 (2000). arXiv:hep-ph/9909237
999. B.I. Abelev et al. (STAR Collaboration), Phys. Rev. Lett. **102**, 112301 (2009). arXiv:0812.1063 [nucl-ex]
1000. V.P. Goncalves, W.K. Sauter, Phys. Rev. D **81**, 074028 (2010). arXiv:0911.5638 [hep-ph]
1001. B.E. Cox, J.R. Forshaw, R. Sandapen, J. High Energy Phys. **0906**, 034 (2009). arXiv:0905.0102 [hep-ph]
1002. D.M. Asner et al. Int. J. Mod. Phys. A **24**, 1 (2009). arXiv:0809.1869 [hep-ex]
1003. E. Klempt, A. Zaitsev, Phys. Rep. **454**, 1 (2007). arXiv:0708.4016 [hep-ph]
1004. T. Feldmann, P. Kroll, Phys. Rev. D **62**, 074006 (2000). arXiv:hep-ph/0003096
1005. V.L. Chernyak, A.R. Zhitnitsky, Phys. Rep. **112**, 173 (1984)
1006. B.Q. Li, K.T. Chao, Phys. Rev. D **79**, 094004 (2009). arXiv:0903.5506 [hep-ph]
1007. K. Aamodt et al. (ALICE Collaboration), JINST **3**, S08002 (2008)
1008. F. Carminati et al. (ALICE Collaboration), J. Phys. G **30**, 1517 (2004)
1009. B. Alessandro et al. (ALICE Collaboration), J. Phys. G **32**, 1295 (2006)
1010. J. Alme et al., Nucl. Instrum. Methods A **622**, 316 (2010). arXiv:1001.1950 [physics.ins-det]
1011. S. Grigoryan, A. De Falco, ALICE-INT-2008-01
1012. M. Bedjidian et al., arXiv:hep-ph/0311048
1013. W. Sommer, C. Blume, F. Kramer, J.F. Grosse-Oetringhaus (ALICE Collaboration), Int. J. Mod. Phys. E **16**, 2484 (2007). arXiv:nucl-ex/0702045
1014. D. Stocco et al., ALICE-INT-2006-029
1015. P. Gonzalez, P. Ladron de Guevara, E. Lopez Torres, A. Marin, E. Serradilla (ALICE Collaboration), Eur. Phys. J. C **61**, 899 (2009). Erratum-ibid. C **61**, 915 (2009). arXiv:0811.1592 [hep-ex]
1016. ATLAS Collaboration, CERN-LHCC-99-014
1017. ATLAS Collaboration, CERN-LHCC-99-015
1018. G. Aad et al. (ATLAS Collaboration), arXiv:0901.0512 [hep-ex]
1019. G. Aad et al. (ATLAS Collaboration), JINST **3**, S08003 (2008)
1020. R. Adolphi et al. (CMS Collaboration), JINST **3**, S08004 (2008)
1021. F. Ragusa, L. Rolandi, New J. Phys. **9**, 336 (2007)
1022. W. Adam et al. (CMS Collaboration), JINST **3**, P07009 (2008)
1023. S. Chatrchyan et al. (CMS Collaboration), JINST **5**, T03022 (2010)
1024. CMS Collaboration, *Dimuons in CMS at 900 and 2360 GeV*, CMS performance note (2010), DP-2010/005
1025. M. Bargiotti, V. Vagnoni, CERN-LHCB-2007-042
1026. T. Sjostrand, L. Lonnblad, S. Mrenna, P. Skands, arXiv:hep-ph/0308153
1027. CMS Collaboration, *First two months of data taking at 7 TeV: $J/\psi \to \mu\mu$, $W \to \mu\nu$, $Z \to \mu\mu$ mass plots and displays of $Z \to \mu\mu$ candidates*, CMS performance note (2010), DP-2010/016
1028. D. Acosta et al. (CMS Collaboration), CERN/LHCC 2006-001, CMS-TDR-008-1
1029. CMS Collaboration, *Feasibility study of a J/ψ cross section measurement with early CMS data*, CMS PAS BPH-07-02
1030. P. Faccioli, C. Lourenço, J. Seixas, H.K. Wohri, arXiv:1006.2738 [hep-ph]
1031. A.A. Alves et al. (LHCb Collaboration), JINST **3**, S08005 (2008)
1032. D.J. Lange, Nucl. Instrum. Methods A **462**, 152 (2001)
1033. S. Agostinelli et al. (GEANT4 Collaboration), Nucl. Instrum. Methods A **506**, 250 (2003)
1034. M. Needham (LHCb Collaboration), CERN-LHCB-CONF-2009-011
1035. Y. Gao, W. Qian, P. Robbe, M.-H. Schune, LHCB-PUB-2010-011
1036. G. Sabatino (LHCb Collaboration), HEP 2009, CERN-Poster-2009-121
1037. L. Nicolas, EPFL thesis #4530, Lausanne, October 2009
1038. J. He (LHCb Collaboration), CERN-LHCB-CONF-2009-013
1039. Y. Gao, J. He, Z. Yang, CERN-LHCB-2008-059
1040. O. Yushchenko, CERN-LHCb-2003-113
1041. C.H. Chang, C. Driouichi, P. Eerola, X.G. Wu, Comput. Phys. Commun. **159**, 192 (2004). arXiv:hep-ph/0309120
1042. Y. Gao, J. He, Z. Yang, CERN-LHCB-2008-077
1043. I.P. Gouz, V.V. Kiselev, A.K. Likhoded, V.I. Romanovsky, O.P. Yushchenko, Phys. At. Nucl. **67**, 1559 (2004). Yad. Fiz. **67**, 1581 (2004). arXiv:hep-ph/0211432
1044. H.W. Ke, X.Q. Li, Sci. China **G53**, 2019 (2010). arXiv:0910.1158 [hep-ph]
1045. S. Descotes-Genon, J. He, E. Kou, P. Robbe, Phys. Rev. D **80**, 114031 (2009). arXiv:0907.2256 [hep-ph]
1046. S. Sakai (STAR Collaboration), J. Phys. G **36**, 064056 (2009)
1047. S. Hashimoto et al., KEK-REPORT-2004-4
1048. M. Bona et al., arXiv:0709.0451 [hep-ex]
1049. K. Oide, Prog. Theor. Phys. **122**, 69 (2009)
1050. P. Raimondi, D.N. Shatilov, M. Zobov, arXiv:physics/0702033
1051. T.E. Browder, T. Gershon, D. Pirjol, A. Soni, J. Zupan, Rev. Mod. Phys. **81**, 1887 (2009). arXiv:0802.3201 [hep-ph]

1052. B. Aubert et al. (BABAR Collaboration), Nucl. Instrum. Methods A **479**, 1 (2002). arXiv:hep-ex/0105044
1053. A. Abashian et al. (Belle Collaboration), Nucl. Instrum. Methods A **479**, 117 (2002)
1054. D.G. Hitlin et al., arXiv:0810.1312 [hep-ph]
1055. FAIR Baseline Technical Report (2006)
1056. H. Staengle et al., Nucl. Instrum. Methods A **397**, 261 (1997)
1057. M.F. Lutz, B. Pire, O. Scholten, R. Timmermans (PANDA Collaboration), arXiv:0903.3905 [hep-ex]
1058. V. Flaminio et al., CERN-HERA 70-03 (1970)
1059. A. Lundborg, T. Barnes, U. Wiedner, Phys. Rev. D **73**, 096003 (2006). arXiv:hep-ph/0507166
1060. R. Arnaldi et al. (NA60 Collaboration), Eur. Phys. J. C **43**, 167 (2005)
1061. V. Friese, W.F.J. Müller (Eds.), CBM Progress Report 2008, GSI Report 2009-03, ISBN 978-3-9811298-6-1
1062. P. Raimondi, Presentation at the *2nd SuperB Workshop, Frascati, March 2006*
1063. E. Levichev, Phys. Part. Nucl. Lett. **5**, 554 (2008)
1064. L. Frankfurt, W. Koepf, M. Strikman, Phys. Rev. D **57**, 512 (1998). arXiv:hep-ph/9702216
1065. A. Caldwell, H. Kowalski, Phys. Rev. C **81**, 025203 (2010)
1066. L. Frankfurt, M. Strikman, C. Weiss, Annu. Rev. Nucl. Part. Sci. **55**, 403 (2005). arXiv:hep-ph/0507286
1067. L. Frankfurt, C.E. Hyde, M. Strikman, C. Weiss, Phys. Rev. D **75**, 054009 (2007). arXiv:hep-ph/0608271
1068. A. Caldwell, H. Kowalski, arXiv:0909.1254 [hep-ph]
1069. A. Bohr, B.R. Mottelson, *Nuclear Structure* (Benjamin, New York, 1969)
1070. M.E. Binkley et al., Phys. Rev. Lett. **48**, 73 (1982)
1071. B. Gittelman, K.M. Hanson, D. Larson, E. Loh, A. Silverman, G. Theodosiou, Phys. Rev. Lett. **35**, 1616 (1975)
1072. U. Camerini et al., Phys. Rev. Lett. **35**, 483 (1975)
1073. L. Frankfurt, M. Strikman, Phys. Rev. D **66**, 031502 (2002). arXiv:hep-ph/0205223
1074. K. Lukashin et al. (CLAS Collaboration), Phys. Rev. C **63**, 065205 (2001). arXiv:hep-ex/0101030
1075. J.P. Santoro et al. (CLAS Collaboration), Phys. Rev. C **78**, 025210 (2008). arXiv:0803.3537 [nucl-ex]
1076. S.V. Goloskokov, P. Kroll, Eur. Phys. J. C **50**, 829 (2007). arXiv:hep-ph/0611290
1077. S.J. Brodsky, E. Chudakov, P. Hoyer, J.M. Laget, Phys. Lett. B **498**, 23 (2001). arXiv:hep-ph/0010343
1078. R.L. Anderson et al., Phys. Rev. Lett. **38**, 263 (1977)
1079. L. Bartoszek et al., Nucl. Instrum. Methods A **301**, 47 (1991)
1080. A. Moiseev et al. (BESS Collaboration), Astrophys. J. **474**, 479 (1997)
1081. V.M. Abazov et al. (DØ Collaboration), Nucl. Instrum. Methods A **565**, 463 (2006). arXiv:physics/0507191
1082. See http://psec.uchicago.edu/
1083. D.M. Asner et al., *P-986 Letter of Intent: Medium-Energy Antiproton Physics at Fermilab* (2009)
1084. See http://capp.iit.edu/hep/pbar/
1085. T.A. Armstrong et al. (E760 Collaboration), Phys. Rev. D **47**, 772 (1993)
1086. M. Andreotti et al. (Fermilab E835 Collaboration), Phys. Lett. B **654**, 74 (2007). arXiv:hep-ex/0703012
1087. G. Garzoglio et al., Nucl. Instrum. Methods A **519**, 558 (2004)
1088. N.A. Tornqvist, Phys. Lett. B **590**, 209 (2004). arXiv:hep-ph/0402237
1089. E. Braaten, Phys. Rev. D **77**, 034019 (2008). arXiv:0711.1854 [hep-ph]
1090. T.A. Armstrong et al. (E760 Collaboration), Nucl. Phys. B **373**, 35 (1992)
1091. J. Brau et al. (ILC Collaboration), arXiv:0712.1950 [physics.acc-ph]
1092. T. Abe et al. (ILD Concept Group—Linear Collider Collaboration), arXiv:1006.3396 [hep-ex]
1093. H. Aihara et al., arXiv:0911.0006 [physics.ins-det]
1094. R.W. Assmann et al., CERN-2000-008
1095. G. Aarons et al. (ILC Collaboration), arXiv:0709.1893 [hep-ph]
1096. V.S. Fadin, V.A. Khoze, Sov. J. Nucl. Phys. **48**, 309 (1988). Yad. Fiz. **48**, 487 (1988)
1097. M. Jezabek, J.H. Kuhn, Phys. Lett. B **316**, 360 (1993)
1098. Y. Sumino, K. Fujii, K. Hagiwara, H. Murayama, C.K. Ng, Phys. Rev. D **47**, 56 (1993)
1099. M. Jezabek, J.H. Kuhn, T. Teubner, Z. Phys. C **56**, 653 (1992)
1100. H. Murayama, Y. Sumino, Phys. Rev. D **47**, 82 (1993)
1101. R. Harlander, M. Jezabek, J.H. Kuhn, T. Teubner, Phys. Lett. B **346**, 137 (1995). arXiv:hep-ph/9411395
1102. M. Jezabek, T. Nagano, Y. Sumino, Phys. Rev. D **62**, 014034 (2000). arXiv:hep-ph/0001322
1103. M. Martinez, R. Miquel, Eur. Phys. J. C **27**, 49 (2003). arXiv:hep-ph/0207315
1104. K. Fujii, T. Matsui, Y. Sumino, Phys. Rev. D **50**, 4341 (1994)
1105. A.H. Hoang et al., Eur. Phys. J. C **2**, 1 (2000). arXiv:hep-ph/0001286
1106. K. Ikematsu, K. Fujii, Z. Hioki, Y. Sumino, T. Takahashi, Eur. Phys. J. C **29**, 1 (2003). arXiv:hep-ph/0302214
1107. S. Fleming, A.H. Hoang, S. Mantry, I.W. Stewart, Phys. Rev. D **77**, 074010 (2008). arXiv:hep-ph/0703207
1108. S. Fleming, A.H. Hoang, S. Mantry, I.W. Stewart, Phys. Rev. D **77**, 114003 (2008). arXiv:0711.2079 [hep-ph]
1109. L. Frankfurt, M. Strikman, C. Weiss, arXiv:1009.2559 [hep-ph]

Properties of the top quark

Daniel Wicke[a]

Inst. für Physik, Johannes Gutenberg-Universität, Staudinger Weg 7, 55099 Mainz, Germany
Bergische Universität, Gaußstr. 20, 42097 Wuppertal, Germany

Abstract The top quark was discovered at the CDF and D0 experiments in 1995. As the partner of the bottom quark its properties within the Standard Model are fully defined. Only the mass is a free parameter. The measurement of the top quark mass and the verification of the expected properties have been an important topic of experimental top quark physics since. In this review the recent results on top quark properties obtained by the Tevatron experiments CDF and D0 are summarised. At the advent of the LHC special emphasis is given to the basic measurement methods and the dominating systematic uncertainties.

Contents

1 Introduction

The aim of particle physics is the understanding of elementary particles and their interactions. The current theory of elementary particle physics, the Standard Model, contains twelve different types of fermions which (neglecting gravity) interact through the gauge bosons of three forces [1–4]. In addition a scalar particle, the Higgs boson, is needed for theoretical consistency [5]. These few building blocks explain all experimental results found in the context of particle physics, so far.

Nevertheless, it is believed that the Standard Model is only an approximation to a more complete theory. First of all the fourth known force, gravity, has withstood all attempts to be included until now. Furthermore, the Standard Model describes several features of the elementary particles like the existence of three families of fermions or the quantisation of charges, but does not explain these properties from underlying principles. Finally, the lightness of the Higgs boson needed to explain the symmetry breaking is difficult to maintain in the presence of expected corrections from gravity at high scales. This is the so-called hierarchy problem.

[a] e-mail: Daniel.Wicke@physik.uni-wuppertal.de

In addition astrophysical results indicate that the universe consists only to a very small fraction of matter described by the Standard Model. Large fractions of dark energy and dark matter are needed to describe the observations [6–10]. Both do not have any correspondence in the Standard Model. Also the very small asymmetry between matter and antimatter that results in the observed universe built of matter (and not of antimatter) cannot be explained until now.

It is thus an important task of experimental particle physics to test the predictions of the Standard Model to the best possible accuracy and to search for deviations pointing to necessary extensions or modifications of our current theoretical understanding.

The top quark was predicted to exist by the Standard Model as the partner of the bottom quark. It was first observed in 1995 by the Tevatron experiments CDF and DØ [11, 12] and was the last of the quarks to be discovered. As the partner of the bottom quark the top quark is expected to have quantum numbers identical to that of the other known up-type quarks. Only the mass is a free parameter. We now know that it is more than 30 times heavier than the next heaviest quark, the bottom quark.

Thus, within the Standard Model all production and decay properties are fully defined. Having the complete set of quarks further allows to verify constraints that the Standard Model puts on the sum of all quarks or particles. This alone is reason enough to experimentally study the top quark properties. The high value of the top quark mass and its closeness to the electroweak scale has inspired people to speculate that the top quark could have a special role in the electroweak symmetry breaking, see e.g. [13, 14]. Confirming the expected properties of the top quark experimentally establishes the top quark as we expect it to be. Any deviation from the expectations gives hints to new physics that may help to solve the outstanding questions.

Since the last review on top quark physics in this journal [15] the luminosity at the Tevatron has been increased by more than an order of magnitude. Now measurements of the top quark mass are no longer limited by statistical but by systematic uncertainties. Even discussions about the knowledge of the implicitly used theoretical mass definition become relevant. Moreover measurements of additional top quark properties now reach a viable precision. In this review the recent results on top quark properties obtained by the Tevatron experiments CDF and DØ are summarised. At the advent of the LHC special emphasis is given to the basic measurement methods and the dominating systematic uncertainties. Other reviews with different emphases are available in the literature [16–24].

After a short introduction to the Standard Model and the experimental environment in the remainder of this section, Sect. 2 describes the current status of top quark mass measurements. Then measurements of interaction properties are described in Sect. 3. Finally, Sect. 4 deals with analyses that consider hypothetical particles beyond the Standard Model in the observed events.

1.1 Theory

The physics of elementary particles is described by the Standard Model of particle physics. It describes three of the four known forces that act on elementary particles, the electromagnetic and the weak force are unified in the GSW-theory [1–3], the strong force is described by quantum chromodynamics (QCD) [4]. So far the influence of gravitation could not be unified with the other three forces in a consistent quantum theory. Due to its weakness it is usually safe to neglect its influence in the context of particle physics.

In the following a short description of the Standard Model shall be given to set the stage for later descriptions. For this the natural units where $\hbar = c = 1$ will be used.

1.1.1 The Standard Model

Lagrangian The Standard Model is a quantum field theory with local gauge symmetry under the group $SU(3) \times SU(2) \times U(1)$. Its Lagrangian contains fields corresponding to three types of particles: Gauge or vector bosons, fermions and scalars. Gauge boson are described by vector fields, A_μ, scalars by complex fields, ϕ. Fermions can be described by Weyl spinors, ψ, with left- and right-handed helicity. Using Einstein's summation convention the Lagrangian can be written as

$$\mathcal{L} = -\frac{1}{4} F^A_{\mu\nu} F^{A\mu\nu} + i\overline{\psi}_\alpha \not{D} \psi_\alpha + (D_\mu \phi^a)(D^\mu \phi^a) + y^a_{\alpha\beta} \overline{\psi}_\alpha \psi_\beta \phi^a + V(\phi) + \text{(ghost- and gauge terms)}. \tag{1}$$

Here capital Latin letters run over the gauge bosons, lower case Latin letters over the scalar fields and Greek letters α, β index the fermions of the standard model. μ and ν are Dirac indices. The field tensor is defined as

$$F^A_{\mu\nu} = \partial_\mu A^A_\nu - \partial_\nu A^A_\mu - g C_{ABC} A^B_\mu A^C_\nu, \tag{2}$$

where C_{ABC} are the structure constants of the gauge group. The covariant derivative is defined from the gauge symmetry to be

$$D_\mu = \partial_\mu - i \frac{g_{(B)}}{2} t^B A^B_\mu. \tag{3}$$

Here $g_{(B)}$ is the coupling strength for the gauge boson A^B, i.e. g, g' or g_s. t^B are the generators of the gauge symmetry in the representation that corresponds to the particle field on which the derivative acts.

$$\not{D} = \sigma^\mu D_\mu \quad \text{with } \sigma^\mu = (1, \pm\boldsymbol{\sigma}), \tag{4}$$

where $\boldsymbol{\sigma}$ is the vector of Pauli matrices. The positive sign applies to fermions of right-handed helicity, the negative for left-handed ones.

The Yukawa couplings, $y^a_{\alpha\beta}$, are free parameters that may be non-zero only when the combination of the fermions α and β with the scalars is gauge invariant. $V(\phi)$ is a quartic form.

Particle content The particle content of the Standard Model is specified by defining the representations of the gauge symmetry for each of the fields contained.

The Weyl spinors describing the fermions transform according to fundamental representations for each of the subgroups or may be invariant under a subgroup. Quarks transform under the three-dimensional representation of the $SU(3)$ group, **3**. All left-handed spinors transform under $SU(2)$ according to the two-dimensional representation, **2**, while the right-handed spinors are singlets, i.e. invariant under $SU(2)$ rotations. This reflects the left-handed nature of weak interactions. The transformation properties under $U(1)$ are specified by the hypercharge. In the Standard Model all representations are repeated three times and build the three generation of fermions.

In Table 1 the representations of the fermions of the Standard Model are summarised. The representations of the non-Abelian gauge groups $SU(3)$ and $SU(2)$ are specified by their dimension (in bold-face), the hypercharge corresponding to the $U(1)$ group is given as index. For a few years it is known from the measurement of neutrino oscillations [6, 25–27] that also the neutrinos have mass and thus right-handed neutrinos should be added to Table 1. The necessary extensions of the Standard Model are not unique and thus they are usually not considered part of the Standard Model.

Table 1 Fermions of the Standard Model and their representations

Generation	Quarks			Leptons	
	$(\mathbf{3},\mathbf{2})_{\frac{1}{6}}$	$(\mathbf{3},\mathbf{1})_{\frac{2}{3}}$	$(\mathbf{3},\mathbf{1})_{-\frac{1}{3}}$	$(\mathbf{1},\mathbf{2})_{-\frac{1}{2}}$	$(\mathbf{1},\mathbf{1})_{-1}$
1st	$\begin{pmatrix} u \\ d \end{pmatrix}_L$	u_R	d_R	$\begin{pmatrix} \nu_e \\ e \end{pmatrix}_L$	e_R
2nd	$\begin{pmatrix} c \\ s \end{pmatrix}_L$	c_R	s_R	$\begin{pmatrix} \nu_\mu \\ \mu \end{pmatrix}_L$	μ_R
3rd	$\begin{pmatrix} t \\ b \end{pmatrix}_L$	t_R	b_R	$\begin{pmatrix} \nu_\tau \\ \tau \end{pmatrix}_L$	τ_R

Table 2 Gauge bosons of the Standard Model and their representations

Symbol	Representation	Coupling strength
g	$(\mathbf{8},\mathbf{1})_0$	g_s
(W^1, W^2, W^3)	$(\mathbf{1},\mathbf{3})_0$	$g = g_2$
B	$(\mathbf{1},\mathbf{1})_0$	$g' = \sqrt{\frac{3}{5}} g_1$

In the context of top quark physics neutrino oscillations do not play any role and can thus be ignored for the purpose of this review.

The gauge bosons of a quantum field theory need to transform according to the adjoint representation of their subgroup. Thus we get eight gauge bosons for the $SU(3)$ symmetry, the gluons, three for $SU(2)$ and one for the $U(1)$ symmetry, cf. Table 2. In addition to the bosons and fermions described the Standard Model contains a complex scalar doublet, the Higgs doublet Φ, which transforms according to $(\mathbf{1},\mathbf{2})_{\frac{1}{2}}$. It is needed for symmetry breaking.

Symmetry breaking: Higgs mechanism In nature the symmetry of the Standard Model is broken. The symmetry breaking is implemented by the Higgs mechanism [5], which assumes that the scalar isospin doublet Φ has a vacuum expectation value. This is achieved by proper choice of parameters in the most general potential, $V(\phi)$, for the scalar field in (1). According to the symmetry this can be chosen to exist in the lower component of the doublet:

$$\Phi = \begin{pmatrix} \phi_1 \\ \phi_2 \end{pmatrix} \quad \text{with } \langle \Phi \rangle = \begin{pmatrix} 0 \\ \langle \phi_2 \rangle \end{pmatrix}. \tag{5}$$

By expanding the complex scalar field around the vacuum expectation value first four real scalar fields are specified. Three of these can behave like longitudinal components of the $SU(2)$ gauge bosons, W^i. Usually the theory is thus written in terms of three massive vector bosons W^+, Z, W^-, a massless vector boson, the photon A, and the fourth real scalar field, the Higgs boson H:

$$\begin{aligned} W^\pm &= \frac{1}{\sqrt{2}}(W^1 \mp i W^2) \\ Z &= \frac{g'B - gW^3}{\sqrt{g'^2 + g^2}} = \sin\theta_W\, B - \cos\theta_W\, W^3 \\ A &= \frac{gB + g'W^3}{\sqrt{g'^2 + g^2}} = \cos\theta_W\, B + \sin\theta_W\, W^3. \end{aligned} \tag{6}$$

Here B is the gauge boson of the (hypercharge) $U(1)$ symmetry and the Weinberg angle θ_W is defined by the ratio of coupling constants

$$\tan\theta_W := \frac{g'}{g}. \tag{7}$$

After this rewriting the theory remains $SU(3) \times U(1)$ invariant. The $U(1)$ symmetry now corresponds to the electrical charge. So the Standard Model parts are quantum chromodynamics (QCD) and quantum electrodynamics with their $SU(3)$ and $U(1)$ symmetries, respectively.

The Higgs mechanism not only yields massive vector bosons, it is also responsible for the masses of the fermions. For the specified particle content, the Yukawa terms in (1) may be non-zero for left-handed quark or lepton doublets paired with the corresponding right-handed quark and lepton $SU(2)$-singlets. For the first generation these terms are

$$y_{ee}\overline{(\nu_e, e)}_L \Phi e_R + y_{dd}\overline{(u, d)}_L \Phi d_R + y_{uu}\overline{(u, d)}_L i\sigma^2 \Phi u_R + \text{h.c.} \tag{8}$$

Corresponding terms can be written not only for the other generations, but in general also for fermion pairs between different generations. Unitary rotations in the three-dimensional space of generations are commonly used to redefine the lepton and quark fields such that Yukawa couplings occur only between particles of the same generation. This provides the mass eigenstates of the quark and lepton fields. These rotations cancel in most terms of the Lagrangian. The only observable remainder of this rotation is the Cabibbo–Kobayashi–Maskawa (CKM) matrix [6, 28, 29] which occurs in the coupling of the $W^\pm$ bosons to quarks:

$$V_{\text{CKM}} = \begin{pmatrix} V_{ud} & V_{us} & V_{ub} \\ V_{cd} & V_{cs} & V_{cb} \\ V_{td} & V_{ts} & V_{tb} \end{pmatrix}. \tag{9}$$

At the same time this is the only process in the Standard Model that connects the different generations. Numerically this unitary matrix has diagonal entries close to unity and off-diagonal entries that are around 0.2 between the first and second generation, around 0.04 between the second and third generation and even smaller for the transition of the first to the third generation [6].

1.1.2 Perturbation theory

Predictions of the Standard Model for high energy reactions are so far generally performed in perturbation theory. The reactions are described as one or more point-like interactions between otherwise free particles. The allowed reactions can be read off the Lagrangian and are usually represented by Feynman diagrams. For example the Yukawa term $y\bar{\psi}\phi\psi$ yields an interaction vertex of the strength y with two fermions ψ and the scalar field ϕ.

Each Feynman diagram serves simultaneously as a diagrammatic description of the reaction and as a short hand notation for the corresponding computation of the transition amplitude. The quantum mechanical amplitude of a given process that transforms a set of initial state particles to a set of final state particles is given by the sum of all possible diagrams with the corresponding initial and final state particles as external lines.

Diagrams with few interactions usually give the largest contributions and higher order corrections are suppressed by factors of the additional coupling strengths. Thus calculations are usually performed in a fixed order of the coupling constant(s). In some cases, however, kinematic enhancements of logarithmic type may compensate the suppression by additional powers of the coupling constant. Notably these occur in cases of collinear or soft gluon radiation and for top quark pair production near threshold. In these cases resummation of the leading (or next-to-leading) logarithms to all order of the coupling are performed.

Diagrams of higher orders generally involve loops which require to integrate over all possible momenta of the internal lines. Naively, such integrals diverge. It is necessary to renormalise the theory in order to obtain finite predictions. There are several possible schemes to perform this renormalisation. Most commonly the so-called $\overline{\text{MS}}$ scheme is used, which itself depends on a continuous parameter the renormalisation scale μ. The dependence of results on the choice of this parameter is often used as a measure for theoretical uncertainties of a prediction.

Perturbation theory described so far deals with the particles of the Standard Model named above, i.e. with quarks, leptons, gauge bosons and the Higgs boson. In nature, however, quarks have not been observed as free particles, rather they are confined in bound states of colour neutral hadrons. The dynamics of quarks and gluons inside hadrons cannot be described by perturbation theory.

To describe the collisions of hadrons with perturbation theory the (soft) physics that governs the behaviour of quarks and gluons in the hadron needs to be factorised from the hard process in the collisions. The partons inside an incoming hadron are considered as a number of "free" partons that may enter the hard interaction. The distribution of partons inside the incoming hadron is taken from parton distribution functions (PDFs) that are derived from experiments. With this the cross-section, $\sigma(p\bar{p} \to X; s)$, to produce final state particles X can be described in terms of cross-sections $\hat{\sigma}$ of incoming quarks and gluons to produce X:

$$\sigma(p\bar{p} \to X; s) = \sum_{i,j=g,u,\bar{u},d,\bar{d},\ldots} \int \mathrm{d}x_1\,\mathrm{d}x_2\, f_i(x_1)\bar{f}_j(x_2)\hat{\sigma}(ij \to X; \hat{s}). \tag{10}$$

Here s is the centre-of-mass energy squared of the incoming hadrons. At the Tevatron Run II $\sqrt{s} = 1.96$ TeV. The colliding partons carry momentum fractions x_1 and x_2, and $\hat{s} = x_1 x_2 s$ is their centre-of-mass energy squared. f_i and $\bar{f}_j$ are the parton distribution functions for the partons i and j, i.e. the probabilities to find this parton in the proton and antiproton, respectively. Formally, the parton density functions f, $\bar{f}$ and the partonic cross-section, $\hat{\sigma}$, depend on the factorisation scale, μ_F, which specifies which effects are included in the PDFs and which remain to be described by the hard matrix element. They also depend on the renormalisation scale, μ. These dependencies are not written out in (10).

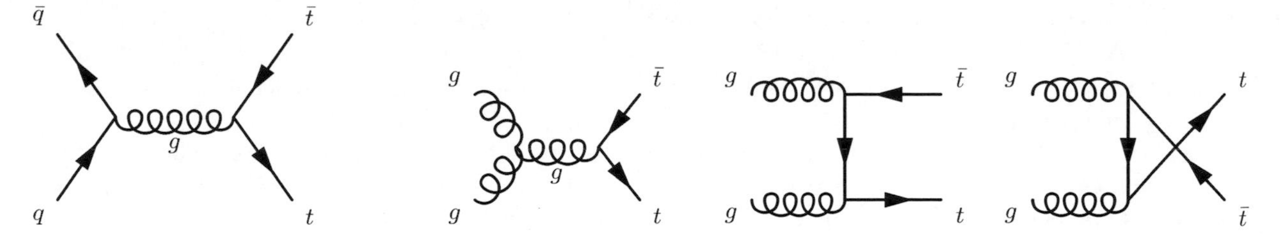

Fig. 1 Born level Feynman diagrams contributing to top quark pair production. The quark annihilation (*leftmost diagram*) is dominating top pair production at the Tevatron. The gluon fusion processes (*three right diagrams*) contribute about 15%, only

1.1.3 Top quark production in hadron collisions

In hadron collisions, like the proton–antiproton collisions at the Tevatron, top quarks can be produced singly or in pairs. The pair production occurs dominantly via the strong interaction. In leading order the quark–antiquark annihilation and gluon fusion processes shown in Fig. 1 contribute. In higher orders also quark–gluon processes exist. The relative contribution of these diagrams depend on the parton distribution functions. At the Tevatron with a centre-of-mass energy of 1.96 TeV next-to-leading order predictions lead to an expectation of 85% contribution from $q\bar{q}$ annihilation and 15% from gluon fusion. The total cross-section of top quark pair production has been computed in perturbation theory using various approximations [30–35]. Its value has a significant dependence on the top quark mass, which near the world average value is about −0.2 pb/GeV. For a top quark (pole) mass of 175 GeV Moch and Uwer [34] find $\sigma_{t\bar{t}} = 6.90^{+0.46}_{-0.64}$ pb, based on the CTEQ6.6 [36] PDF. The uncertainty includes uncertainties of the PDF and the scale uncertainty.

Single top quark production can only take place via the weak interaction. The leading processes are quark annihilation through a W boson, also called s-channel, the quark–gluon fusion with a W boson in a t-channel, cf. Fig. 2, and production of single top quarks in association with a (close to) on-shell W-boson. Charge conjugate diagrams apply for antitop quark production. At the Tevatron Run II the cross-section to produce a single top or antitop quark is 3.4 ± 0.22 pb. The s- and t-channel contribute a little less than 1/3 and 2/3, respectively. The associated production contributes less than 10% [37–39]. These numbers were derived assuming $m_t = 175$ GeV and using the MRST2004 [40] PDFs.

After production top quarks decay very rapidly through the weak interaction into a W boson and a b quark. In the Standard Model contributions from decays to light quarks are suppressed due to the smallness of the corresponding entries of the CKM matrix. The expected decay width of about 1.34 GeV corresponds to a lifetime of the order of 5×10^{-25} s [6]. Thus the top quark decays before it can couple to light quarks and form hadrons. The lifetime of $t\bar{t}$

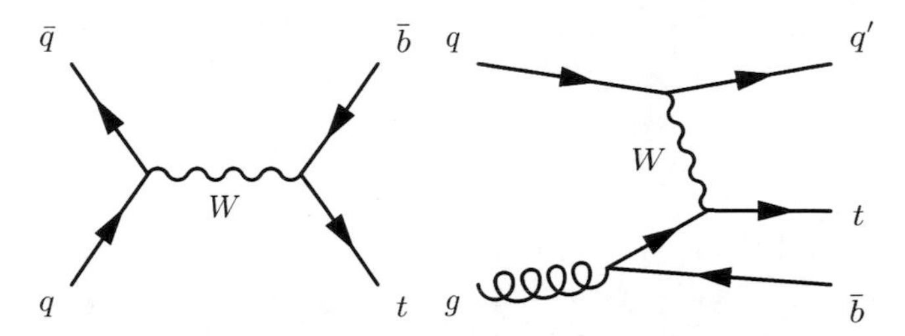

Fig. 2 Leading order Feynman diagrams contributing to single top quark production at the Tevatron. According to the structure of the diagrams *the left* process is called s-channel and *the right* t-channel production

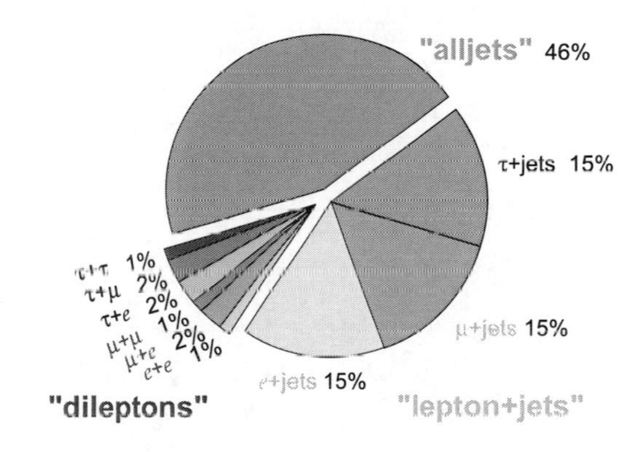

Fig. 3 Top quark pair branching fractions [42]

bound states, toponium, is too small, $\Gamma_{t\bar{t}} \sim 2\,\Gamma_t$, to allow for a proper definition of a bound state as already pointed out in the early 1980s [41].

The decay modes are defined purely by the W boson decays. W bosons may decay to quarks, i.e. hadronically, or leptonically. For top quark pairs this yields three basic decay modes: the all hadronic channel, the semileptonic channel and the dilepton channel. The all hadronic channel has a branching fraction of 46%. Each of the three charged Standard Model leptons contributes 15% in the semileptonic channel. The dileptonic decays have a total branching fraction of 9%. See Fig. 3 for a graphical representation. Decays that involve τ leptons are usually not considered in analyses of the semileptonic and dileptonic decay modes, because τ leptons are difficult to identify. The analyses, however, include the events in which the tau decayed to an electron or muon.

Following this experimental nomenclature the semileptonic and dileptonic channels are considered to include only electrons and muons. If an analysis considers identified tau leptons this fact shall be explicitly stated. For single top quark analyses so far only the leptonic W boson decays (to electrons and muons) are considered.

1.2 Experiments

Up to recently only one collider provides centre-of-mass energies sufficiently high to produce top quarks: the Tevatron at the Fermi National Accelerator Laboratory (FNAL) near Chicago, IL, USA. At the two collision points typical general purpose experiments of present collider physics are positioned, CDF and DØ. Each consists of a cylindrical part that covers particles produced at large angles to the beam and two end-caps that detect particles at smaller angles. Close to the beam tracking and vertex reconstruction components are placed, followed to the outside by calorimetry and muon detection systems. Some more details of the Tevatron and the two detectors shall be described below.

1.2.1 The Tevatron

The Tevatron collides beams of protons and antiprotons. The protons and antiprotons are produced and pre-accelerated in a series of smaller machines and then filled into the Tevatron to circulate in opposite directions. In a first phase of operation between 1992 and 1996 the beams were accelerated to 900 GeV. This phase is commonly called Run I. After an upgrade the Run II started in 2001. The upgrade enables the Tevatron to accelerate the beams to a final energy of 980 GeV to yield a centre-of-mass energy of 1.96 TeV. Also the peak luminosity was gradually improved and now reaches a factor of approximately 10 over the Run I performance [43, 44].

In Run I an integrated luminosity of about 160 pb^{-1} was delivered to both experiments. This was sufficient to discover the top quark in proton–antiproton collisions with a centre-of-mass energy of 1.8 TeV [11, 12]. With the increased centre-of-mass energy a 40% increased cross-section was achieved for top quark pair production. At the time of writing an integrated luminosity of more than 6 fb^{-1} was delivered to both experiments. Preliminary results using up to 3.6 fb^{-1} have been made public by the experiments.

It is currently foreseen to continue running the Tevatron at least until the end of 2010 and an extension into 2011 is being discussed. The total integrated luminosity is expected to increase by about 2.5 fb^{-1} for each year of running. Thus until the end of the Tevatron program the total luminosity will more than double compared to what has been analysed so far [45].

1.2.2 The CDF detector

The Collider Detector at Fermilab (CDF) [46–48] is one of two detectors recording collisions at the Tevatron. The vertexing and tracking components, the calorimetry and muon detection systems as used in the Run II measurements shall be described shortly in turn.

The tracking system of CDF is placed in a 1.4 T solenoidal magnetic field. The heart of the CDF tracking for Run II consists of three separate silicon detectors. The innermost (L00) consists of a single layer of single sided sensors at a radius of about 1.5 cm. The upgraded silicon vertex detector (SVX II) consists of five cylindrical double-sided layers that along the beam axis reach to ±45 cm from the centre of the detector at radii between 2.5 cm and 10.6 cm [49, 50] (Fig. 4). Finally the intermediate silicon layer (ISL) consists of double-sided sensors in one central layer at 23 cm and in two layers at 20 cm and 29 cm for the forward regions [51]. The tracking system is completed by wire drift chambers with an outer radius of 1.32 m. Their length of 3.2 m allows to cover the central region of $|\eta| < 1$. The tracking systems provides full coverage and thus precise tracking for $|\eta| \leq 2.0$ to measure the momentum of charged particles, to find the primary vertex of the collision as well as to find possible secondary vertices from long lived particles like b quark hadrons.

Outside the tracking system a time of flight system (TOF) based on scintillator bars is positioned. This allows particle identification and is used in identifying b hadrons [52]. The solenoid that provides the magnetic field for the tracking system is placed between the TOF and the calorimetry.

The central electromagnetic calorimeter is composed of alternating layers of lead and scintillator. It covers the pseudo-rapidity range of $|\eta| < 3.5$. To detect more forwardly produced particles layers of lead and proportional chambers are installed at pseudo-rapidities $|\eta| < 4.2$. The hadronic calorimeter uses lead as absorber material. The central region and the end cap wall use scintillator as active material. The forward regions also use gas proportional

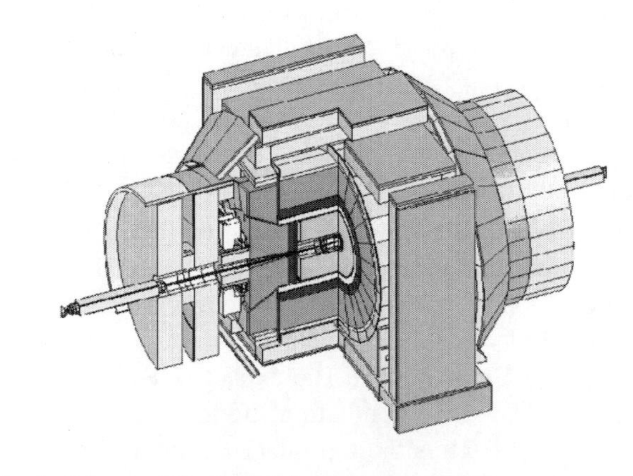

Fig. 4 Schema of the CDF detector at Run II [49]

chambers. In total the calorimeter covers pseudo-rapidities of $|\eta| < 4.2$. The calorimeter allows energy measurements and identification of electrons, photons, jets and missing transverse momentum.

The calorimeter is surrounded by several systems of drift chambers to identify muons. The 'central muon system' is placed at a radius of 347 cm. Behind an additional 3.3 interaction lengths of 60 cm steel of the return joke the central muon upgrade system is located. Both systems cover pseudo-rapidities of $|\eta| < 0.6$. The third system, the central muon extension, extends this coverage at $0.6 < |\eta| < 1.0$. So-called barrel muon chambers extend the coverage for $1.0 < |\eta| < 1.5$. Track segments in these components are used to identify muons.

CDF uses a three-level trigger system. The level 1 system has a pipeline for 42 beam-crossings. After the level 1 trigger the event rate is approximately 10 kHz. At level 2 trigger processors analyse a substantial fraction of the event and further reduce the rate to 200 Hz. The 3rd level consists of a cluster of computers which perform an optimised event reconstruction. With this information the event rate is reduced to about 40 Hz. The events selected by level 3 are stored permanently to tape.

1.2.3 The DØ detector

The DØ (Dzero) detector [53, 54] is the second of the two detectors recording collisions at the Tevatron. It has been significantly upgraded to adapt to the reduced bunch distance and increased luminosity of the Tevatron Run II. The upgraded DØ detector shall be shortly described here (Fig. 5).

The tracking system at the centre of DØ has been fully replaced since Run I and now consists of a silicon micro-strip tracker (SMT) and a scintillating-fibre tracker within a 2 T solenoidal magnet. The SMT consists of a barrel with four layers of single and double-sided silicon micro-strip detectors with a total length of about 70 cm. The barrel is separated into six subsections along the beam pipe. Each subsection is capped with a disk of silicon detectors (F-disks). Three additional F-disks are placed on each side further outside of the barrel. Larger disks (H-disks) are placed at distances of 100 cm and 121 cm from the beam pipe. The SMT barrel provides excellent r–ϕ-information, the disks provide r–z as well as r–ϕ measurements. During a shutdown period in 2006 the DØ silicon system was extended by adding an additional layer of sensors directly on top of the beam pipe. This Layer-0 significantly improves the ability of vertex reconstruction. DØ measurements are usually performed separately for data taken before and after the installation of Layer-0. The two run periods are commonly denoted as Run IIa and Run IIb. Outside the SMT the central fibre tracker (CFT) is placed. It consists of scintillating

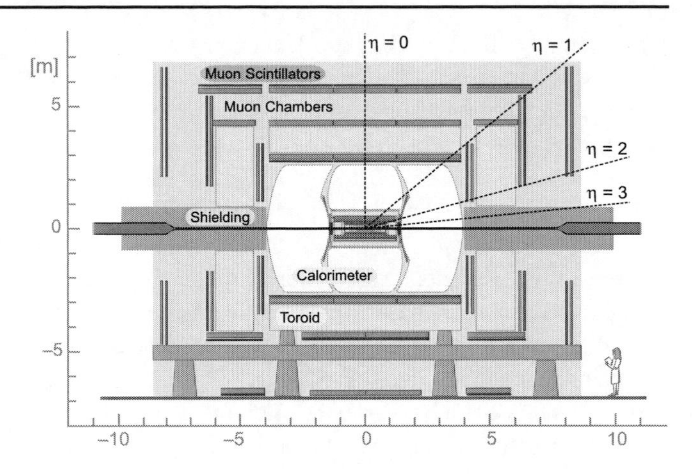

Fig. 5 Schematic view of the Run II DØ detector [55]

fibres mounted on eight concentric support cylinders at radial distances between 20 and 52 cm covering $|\eta|$ to about 1.7. At each distance one layer of fibres is oriented along the beam axis and a second is mounted with stereo angles of $\pm 3°$.

To the outside of the CFT the solenoidal magnet is placed that produces a 2 T magnetic field for the tracking components. It is followed by the calorimetry that consists of a pre-shower detector that is placed in front of the cryostat and the sampling calorimeter based primarily on uranium/liquid-argon inside the cryostat. The end-caps that cover the forward regions have a similar structure, i.e. a pre-shower detector and a calorimeter cryostat. The calorimeters consists of three regions. The innermost is the electromagnetic calorimeter which uses thin (3–4 mm) plates of depleted uranium as absorber material. It is followed by the fine hadronic calorimeter with 6 mm thick plates of uranium-niobium. The outermost subsection, the coarse hadronic calorimeters, uses thick (about 47 mm) absorber plates of copper (in the central) and stainless steel (in the end-caps). The active medium in all three regions is liquid argon. Additionally, inter-cryostat-detectors (ICD) are mounted between the central and the forward cryostat to improve on the incomplete coverage of the calorimeters at $0.8 < |\eta| < 1.4$.

The DØ muon systems outside the calorimeter consist of tree layers of drift chambers, one inside and two outside a toroidal magnetic field. The central muon chambers cover $|\eta| \lesssim 1.0$ with proportional drift chambers and have a magnetic field of 1.9 T. The forward muon chambers use mini drift chambers and a toroidal field of 2.0 T. They extend the coverage to $|\eta| \approx 2.0$.

Also DØ uses a trigger system with three stages. The first level consists of a set of hardware trigger elements that provide a trigger accept rate of 2 kHz. In the second level hardware engines and embedded microprocessors provide information to a global processor that considers individual detector information as well as correlations. It reduces the rate by a factor of about 2. The third level consists of a farm of computers that reduces the rate to 50 Hz based on a limited event reconstruction. The accepted events are stored to tape for offline analysis.

1.3 Basic event selection

The selection of events in general, in particular the selection of top quark events, is based on the reconstruction of a number of different objects: The primary vertex of the collision, electrons, muons, jets and transverse missing energy. In addition, to reduce the background, often methods to identify jets from b-quarks are applied. In the following these objects shall be shortly described in turn.

The reconstruction of the primary vertex is determined by assigning well measured tracks to a common origin in the interaction region. The primary vertex is constructed event-by-event and is used as reference for some of the following objects.

Electrons are reconstructed by a combination of tracking and calorimeter information. Quality cuts on the tracks typically include a p_T threshold of the order of 10 GeV and the matching energy deposit in the calorimeters should be well contained in the electromagnetic subsection. In the absence of bremsstrahlung the energy deposit is expected to have a small radius in the η–ϕ plane. Sophisticated algorithms take into account bremsstrahlung photons. Discriminating observables include the relative sizes of the energy deposit in the hadronic and the electromagnetic calorimeter. DØ uses a fixed ratio of the electromagnetic to the total energy $f_{\text{em}} = E_{\text{em}}/E < 0.9$ [56] while CDF uses an energy dependent cut on $E_{\text{em}}/E_{\text{had}} \leq 0.055 + 0.00045\ \text{GeV}^{-1} E$ [57].

Muons are identified by the presence of signals in the dedicated muon chambers that can be matched to a track found in the tracking system. For this purpose CDF and DØ extrapolate isolated tracks with standardised quality cuts through the calorimeter out to the muon chambers to find matching track segments.

For both lepton objects different quality classes are defined, named "loose", "medium", "tight", etc., with increasingly stricter requirements on the isolation of the track and the calorimeter cluster. These different criteria can be used to obtain an acceptable signal to background ratio. The selection of "loose" leptons excluding those that also have a "tighter" identification is often used to define sideband samples to extract background estimates from data. The triggering and reconstruction efficiencies are usually studied in $Z \to \ell\ell$ events using the tag and probe method.

Jets are reconstructed from all calorimeter towers. They usually show a substantial contribution from the hadronic subsection and are usually broader than signals from electromagnetic particles. At the Tevatron experiments the "improved legacy" cone algorithms [58] with radii of 0.4 and 0.5 are used by CDF and DØ, respectively. Quality cuts typically require that the energy of a jet is not contained to more than 90% in a single tower and deposits from electron and photon candidates are removed. The jet energy reconstructed from the calorimeter cells needs to be corrected for a number of effects. These corrections include imperfections of the calorimeter but also energy offsets due to contributions from the underlying event, multiple hadron interactions and noise in the electronics. This correction, usually called the Jet Energy Scale (JES), is obtained from precisely measured electromagnetic objects by assuming momentum conservation in the transverse plane for γ + jets events.

Jets stemming from b-quarks can be identified due to the long lifetime of about 1.5 ps of the B hadrons in such jets. At the typical energies in top quark events of 50 to 100 GeV the mean decay length is of the order of 5 mm. This fact is exploited by computing the impact parameter for tracks or by explicitly reconstructing a secondary vertex from the tracks that is displaced from the primary vertex. To identify b-jets CDF uses only tracks that fall within the cone radius of the considered jet. A secondary vertex is reconstructed in two passes with different track requirements [59]. The 2d decay length is computed as the distance of primary to the secondary vertex. The significance, i.e. the decay length over its uncertainty, is required to be larger than 3. If the direction from the primary to the secondary vertex is opposite to the direction of the jet, the tag is called a negative tag. These negative tags are useful to determine the purity of mis-tag rates. In DØ first track jets are built with a cone algorithm independent of the calorimeter jets described above. Secondary vertices are reconstructed with tracks from a given track jet. Calorimeter jets are considered as identified b-jets, if an identified track jet falls within a radius of $\Delta R < 0.5$. The most recent DØ tagging algorithm uses the impact parameters of tracks matched to a given jet and information on the secondary vertex mass, the significance of displacement, and the number of participating tracks for any reconstructed secondary vertex within the cone of the given jet. The information is combined in a neural network to obtain the output variable, NN_B, which tends towards one for b-jets and towards zero for light jets [60].

The presence of a neutrino is inferred from the momentum imbalance in the transverse plane, which occurs because neutrinos are invisible to the detectors. If all objects of an event were measured the sum of the transverse momenta should vanish. Thus the sum of the transverse momenta of all neutrinos can be deduced from the missing transverse momentum needed to ensure momentum conservation. It is derived from the calorimetric measurements and their direction with respect to the interaction region as the negative vector sum of the transverse energies and thus usually named Missing Transverse Energy, $\not{E}_T$. The sum is taken over all calorimeter cells that remain after noise suppression.

In general corrections for identified objects with known energies like electrons, muons or jets are applied.

Triggering and preselection of single top quark events and of top quark pair events in the semileptonic and dileptonic channels are based on reconstructed lepton, jet and $\not{E}_T$ objects described above. Typically the leptons and the missing transverse energy are required to have $p_T > 20$ GeV. Signal selection in addition requires the presence of jets also with a typical momentum requirement $p_T > 20$ GeV. The number of jets required of course depends on the channel under consideration. Some analyses here improve the signal to background ratio by requiring one or more identified b-jets. Others construct a likelihood for an event being top quark like based on topological quantities to consciously avoid b-jet identification. Additional cuts may be introduced to enhance the signal to background ratio or to improve the data to Monte Carlo agreement. In the semileptonic selection DØ e.g. recently requires the leading jet to fulfil $p_T > 40$ GeV and avoids events in which missing transverse momentum and the selected lepton are aligned.

Events of the all hadronic channel have to be selected by requiring multijet final states. Due to the overwhelming background from multijet events in these analyses usually stronger cuts are applied on the transverse momentum of the jets.

2 Top quark mass measurements

The mass is the only property of the top quark that is not fixed within the Standard Model of particle physics. The Yukawa coupling responsible for the coupling of the top quark to the Higgs boson and thus for the mass of the top quark is a free parameter of the Standard Model. This already illustrates the importance of measuring the top quark mass.

Moreover Standard Model predictions of electroweak precision observables depend on the value of the top quark mass via radiative corrections. By correlating the W boson mass with the top quark mass the mass range for the yet undiscovered Higgs boson can be constrained. Measurements of the top quark mass are thus an important preparation for discovering the Higgs boson and will serve as a consistency check of the Standard Model after its discovery.

In the following theoretical aspects of fermion mass measurements are discussed, before the experimental methods used in the various decay channels are explained. Then issues of modelling non-perturbative effects in $p\bar{p}$ collisions and their influence on the existing measurements are discussed and the combination of the various measurements to a final result is reviewed. The conclusions contain a critical comparison of current results with expectations and future prospects.

2.1 Theoretical aspects

For free particles the physical mass is usually taken to correspond to the pole of their propagator, i.e. their value of the four-momentum squared, $p^2 = E^2 - \boldsymbol{p}^2$. Because of confinement quarks cannot exist as free particles and this definition becomes ambiguous.

The definition of the pole mass is still possible on an order by order basis in perturbation theory, but is considered to be intrinsically ambiguous on the order of the confinement scale $\mathcal{O}(\Lambda_{\text{QCD}})$ [61]. Mass definitions can also be obtained following other renormalisation schemes such as the $\overline{\text{MS}}$-mass in the $\overline{\text{MS}}$-scheme. Other definitions have been suggested by [62, 63].

The relation between the various mass definitions can usually be computed in perturbation theory. For the top quark the numerical values of the different definitions may differ significantly. In NNLO for example the $\overline{\text{MS}}$-mass of the top quark is about 10 GeV smaller than the pole mass.

This large difference makes it important to understand which definition and order of perturbation theory of the top quark mass is measured by the experiments. As we shall see below, all direct methods to determine the top quark mass use Monte Carlo simulation to either extract the mass or calibrate the procedure. Thus the question really is, which definition of the top quark mass is used as a parameter in these Monte Carlo generators.

Unfortunately, it is not well understood which field theoretic mass definition the currently used generators correspond to. Clearly, as the hard matrix elements of (most) generators is implemented in leading order, they correspondingly use the leading order mass. It is usually argued that this corresponds to the pole mass, because in the pole mass definition any shifts of the position are to be absorbed in the mass definition. Monte Carlo generators do not absorb corrections from the parton shower or the hadronisation into the mass definition. And it is not clear in how far the parton shower and the modelling of hadronisation alter the mass definition, nor which approximation of QCD this mass parameter corresponds to. Partial answers have been given in [63], but at this point a conceptual uncertainty remains that is considered to be of the order of 1 GeV.

Comparisons of the top quark mass from direct measurements with electroweak precision data to determine e.g. the Higgs boson mass currently assume the measured top quark mass values corresponds to the pole mass definition. They thus have to be interpreted with care.

Experimentally precise measurements of the top quark mass are nevertheless useful. On the one hand they are needed to set the mass parameter for simulating top quark events, which will be important backgrounds for some LHC searches. On the other hand the consistency between the various experimental methods and the top quark decay channels gives confidence in these methods. Finally, it seems feasible that a theoretically more precise specification of the top quark mass definition used in the Monte Carlo generators can be derived in the future [64]. Any bias from the interpretation as a pole mass may then be corrected for.

2.2 Lepton plus jets channel

The lepton plus jets channel is considered to be the golden channel in top quark mass measurements. Due to a lepton and a neutrino in the final state it has a good signal to background ratio. In addition one of the top quark quarks is fully measured in the detector.

Several method have been applied by CDF and DØ to measure the top quark mass in this channel. The methods exploit the kinematics of the events to different levels and make different assumptions about details of the production mechanisms.

2.2.1 Template method

The basic template method was already applied to lepton plus jets events in the papers describing the first evidence and discovery of the top quark.

In this method a top quark mass is reconstructed for the each of the selected events using the momenta measured for lepton and jets and the transverse missing energy. The distribution of the reconstructed masses is then compared to template distributions from simulation. These templates are constructed from signal Monte Carlo with varying top quark mass values and contain the expected amount of background events. The method thus relies on good Monte Carlo modelling of signal and background.

CDF Run II template measurement CDF has applied an improved template method in up to 5.6 fb^{-1} of data combining the lepton plus jets channel and the dilepton channel [65–68]. For clarity the two analysis parts will be described separately. The dilepton portion of this analysis can be found in Sect. 2.3.1.

Lepton plus jets events are selected requiring a single high p_T lepton, large missing transverse momentum and at least four jets. Events are separated by the number of jets identified as b-jets based on the transverse decay length of track inside the jet [69]. In case of only a single identified jet only events with exactly four jets are considered. For events with more than one identified b-jet more than four jets are allowed.

In each event a top quark mass m_t^{reco} is fitted using a constrained fit. Besides the top quark mass, the momenta of the top quark decay products (the quarks and leptons) are fitted to the observed transverse jet and lepton momenta, p_T^{obs}, and the unclustered energy, $\boldsymbol{U}_T^{\text{obs}}$. The unclustered energy is the calorimetric energy in the transverse direction not associated with any reconstructed object. Jet momenta are corrected to parton level with CDFs common jet energy scale correction. The invariant masses of the W boson decay products on both sides, $M_{q\bar{q}}$ and $M_{\ell\nu}$, are constrained to be consistent with the nominal W boson mass within the W boson width. The reconstructed top quark mass m_t^{reco} is required to be consistent within the top quark width with the invariant mass of the top quark decay products on both sides, M_{bqq} and $M_{b\ell\nu}$. The fit χ^2 thus is written as

$$\chi^2 = \frac{(p_T^{\ell} - p_T^{\ell,\text{obs}})^2}{\sigma_\ell^2} + \sum_{i=1}^{4} \frac{(p_T^{i,q} - p_T^{i,\text{jet}})^2}{\sigma_i^2} + \frac{(\boldsymbol{U}_T^{\text{fit}} - \boldsymbol{U}_T^{\text{obs}})^2}{\sigma_\ell^2} + \frac{(M_{q\bar{q}} - M_W)^2}{\Gamma_W^2} + \frac{(M_{\ell\nu} - M_W)^2}{\Gamma_W^2} + \frac{(M_{bqq} - m_t^{\text{reco}})^2}{\Gamma_t^2} + \frac{(M_{b\ell\nu} - m_t^{\text{reco}})^2}{\Gamma_t^2}. \quad (11)$$

The fitted unclustered energy, $\boldsymbol{U}_T^{\text{fit}}$, is related to the transverse momentum of the neutrino used in the computation of $M_{\ell\nu}$ and $M_{b\ell\nu}$. This χ^2 is computed for all possible associations of quarks to four jets, allowing b-jets only to be matched with b quarks. Only the top quark mass value of the association with the best fit χ^2 is kept. If this best $\chi^2 > 9.0$ the event is rejected.

To constrain the jet energy scale simultaneously with the top quark mass, the dijet mass, m_{jj}, is measured from the non-b-tagged jets among the four leading jets without applying the above kinematic fit. This is only unique for events with two identified jets. In other events the two jets that yield the dijet mass closest to the W-boson mass are chosen.

Thus for each event one top quark mass value, m_t^{reco}, and one dijet mass, m_{jj}, is entering the following analysis.

Monte Carlo simulations are used to determine the expected behaviour of signal events as well as the background contributions. The dominant $W+$jets background is simulated with ALPGEN + PYTHIA [70, 71] with a normalisation derived from data. The multijet background is modelled with samples containing non-isolated leptons. Smaller background from single-top and diboson events are taken from Monte Carlo with normalisation to theoretical cross-sections. All backgrounds are assumed to have no dependence on the nominal top quark mass, but are allowed to vary with the jet energy scale. Signal samples for various nominal top quark mass values, m_t, are generated using PYTHIA. All simulations are passed through the full CDF detector simulation and reconstruction.

These simulations are used to generate probability density functions, P^{sig}, to find a signal event with measured values m_t^{reco} and m_{jj} given a nominal top quark mass, m_t and jet energy scale shift, Δ_{JES}. Similarly a probability density for background events, P^{bkg}, is computed as function

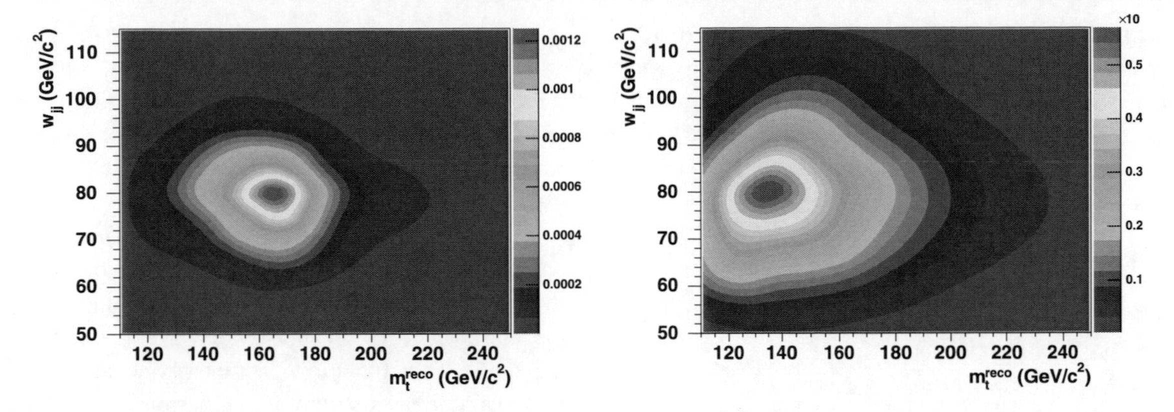

Fig. 6 Probability densities for signal (*left*) and background (*right*) for nominal top quark mass of 170 GeV and no jet energy scale shift as obtained for single b-tag events [66]

of the jet energy scale shift, only. CDF uses a Kernel Density Approach, where each simulated event contributes not only with its reconstructed values m_t^{reco} and m_{jj}, but also in the neighbourhood around these. The size of this neighbourhood is controlled by smoothing parameters. The parameters are chosen dynamically: small near the maximum of a given distribution where the statistics require less smoothing and larger in the tails. Examples of the resulting density functions are shown in Fig. 6.

With these probability densities the likelihood $\mathcal{L}(m_t, \Delta_{\text{JES}})$ is constructed. The number of signal and background events in the single b-tag and the double b-tag sample are used as additional parameters with Gaussian constraints on the background estimations obtained above. Also the jet energy shift, Δ_{JES}, is constrained by a Gaussian term to be consistent with zero within its nominal uncertainty, σ_c.

$$\begin{aligned}
&\mathcal{L}(m_t, \Delta_{\text{JES}}) \\
&\quad = \exp\left(-\frac{\Delta_{\text{JES}}^2}{2\sigma_c}\right) \times \mathcal{L}_1(m_t, \Delta_{\text{JES}}) \times \mathcal{L}_2(m_t, \Delta_{\text{JES}}) \\
&\mathcal{L}_i(m_t, \Delta_{\text{JES}}) \\
&\quad = \exp\left(-\frac{(b_i - b_i^{\text{e}})^2}{2\sigma_{b_i}^2}\right) \prod_{\text{events}} \frac{s_i P_i^{\text{sig}} + b_i P_i^{\text{bkg}}}{s_i + b_i}.
\end{aligned} \tag{12}$$

Here $i = 1, 2$ indicates the subsamples with one or more than one identified b-jets, respectively. s_i and b_i are the number of signal and background events and b_i^{e} the background expectations in the samples.

The described likelihood formulae can only be evaluated at the discrete values of m_t and Δ_{JES} for which simulations were run. To obtain the likelihood for arbitrary values of m_t and Δ_{JES} a quadratic interpolation is used. The central result is obtained by maximising the likelihood and its uncertainty is quoted as the (largest) mass shift corresponding to a likelihood change of $\Delta \log \mathcal{L} = 0.5$.

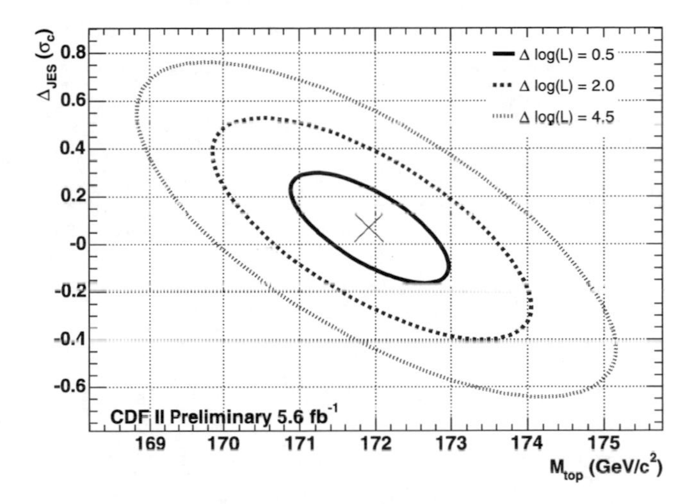

Fig. 7 Likelihood contours for the combined lepton plus jet and dilepton fit of 5.6 fb^{-1} CDF data [68]

Systematic uncertainties are evaluated by modifying several aspects of the analysis described. The dominating uncertainty in top quark mass measurements is the jet energy scale. Through the simultaneous fit its contribution is part of the statistical uncertainty in this measurement. Residual effects remain through uncertainties in the p_T and η dependence. The uncertainty of modelling the top quark pair signal events is evaluated by comparing pseudo-experiments generated with PYTHIA and HERWIG. These two systematic uncertainties contribute with about 0.7 GeV to the uncertainty and are the two largest single contributions to the systematic uncertainty. Additional contributions in order of decreasing importance include the uncertainties on modelling colour reconnection [72–74], the background shape, the parton density functions.

With 5.6 fb^{-1} of data using the lepton plus jets channel only CDF determines a top quark mass of [68]

$$m_t = 172.2 \pm 1.2_{\text{stat}} \pm 0.9_{\text{syst}}\ \text{GeV}. \tag{13}$$

For the jet energy scale shift the fit yields $\Delta_{\text{JES}} = (0.10 \pm 0.26)\sigma_c$, very consistent with the external jet energy scale but with a significantly reduced uncertainty. Combining the results with an analogous measurement in the dilepton channel which uses a consistent jet energy scale shift (cf. Sect. 2.3.1) yields

$$m_t = 172.1 \pm 1.1_{\text{stat}} \pm 0.9_{\text{syst}}\ \text{GeV}. \tag{14}$$

The likelihood contour of the two-dimensional measurement of m_t and Δ_{JES} is shown in Fig. 7.

DØ Run II template measurements DØ has produced preliminary results for top quark mass measurements using the template method [75, 76]. The method differs from the CDF method as the templates are not smoothened and in the use of a binned likelihood. As these results were not updated nor published in the last years they shall not be described in more detail here.

2.2.2 Ideogram method

The ideogram method was transferred to top quark mass measurements from a method used in the DELPHI W boson mass measurement [77, 78].

Also this method relies on an event-by-event reconstruction of top quark mass values. But in contrast to the template method the signal likelihood is constructed event-by-event from the theoretically expected Breit–Wigner distribution smeared with the experimental resolution of each individual event. Events with a configuration that allows a more precise reconstruction thus contribute more than events with a configuration that is difficult to reconstruct.

DØ DØ has published an analysis of 425 pb^{-1} of data based on semileptonic top quark pair events [79]. The event selection requires an isolated lepton, missing transverse energy and four or more jets. To identify b quark jets the decay length significance of a secondary vertex reconstruction is used [80]. No cut is placed on the number of identified b-jets.

The top quark mass in each event is reconstructed using a constrained fit that determines the momenta of the top quark pair decay products ($\ell \nu b\bar{b}q\bar{q}'$) from the measured momenta of the charged lepton, the four leading jets and the missing transverse momentum and their uncertainties. Constraints are placed on the invariant mass of the two light quarks and the two leptons, respectively, which are required to be consistent with the W boson mass. The reconstructed top and antitop quark masses are required to be equal. The 12 possible assignments of jets to quarks that yield different constraints are considered. In addition two possible solutions for the neutrino z momentum are considered, which results in 24 top quark mass values per event. As the common jet energy scale of DØ corresponds to the particle level, i.e. what would be visible in an ideal detector, the fit uses jet–parton mapping functions determined from PYTHIA simulation. These mappings contain an overall scale factor, f_{JES}, for the jet energy scale that may modify the default jet energy scale.

The described fit is repeated for various values of the jet energy scale factor, so that the resulting mass values and the fit χ^2 are functions of f_{JES}. Through the constraints to the nominal W boson mass, the fit χ^2 is expected to be the smallest for the correct jet energy scale (and top quark mass value). Events with no jet–parton combination reaching $\chi^2 < 10$ at the central jet energy scale are rejected at this point.

To determine the expected performance of the selection and the corresponding background contamination, top quark pair signal for various nominal top quark mass values and $W+$jets background events are generated using ALPGEN + PYTHIA. The events are passed through the full detector simulation and reconstruction. Multijet background is modelled with side band data obtained from inverted lepton quality cuts. The sample composition is determined from a likelihood discriminant that combines topological observables, with tracking based jet shape and b tagging information, cf. Fig. 8. The observables are selected to have low correlation to the top quark mass.

An event-by-event likelihood is now constructed to observe the top quark masses reconstructed in the kinematic fit, the discriminant and the number of identified b-jets given the nominal values of m_t and f_{JES}:

$$\begin{aligned}\mathcal{L}_{\text{evt}}(x; m_t, f_{\text{JES}}, f_{\text{top}}) &= f_{\text{top}}\, P_{\text{sig}}(x; m_t, f_{\text{JES}}) \\ &\quad + (1 - f_{\text{top}}) P_{\text{bkg}}(x; f_{\text{JES}}).\end{aligned} \tag{15}$$

Here x represents the measured observables for the event under consideration and f_{top} is the signal fraction of the corresponding sample. The signal and background probabilities, P_{sig} and P_{bkg}, are proportional to the probabilities to find signal or background at the observed discriminant value. Because of the selection of observables in the discriminant, this factor does not depend on the top quark mass and the jet energy scale and is factorised as

$$\begin{aligned}P_{\text{sig}}(x; m_t, f_{\text{JES}}) &= P_{\text{sig}}(D)\tilde{P}_{\text{sig}}(x; m_t, f_{\text{JES}}) \\ P_{\text{bkg}}(x; f_{\text{JES}}) &= P_{\text{bkg}}(D)\tilde{P}_{\text{bkg}}(x; f_{\text{JES}}).\end{aligned} \tag{16}$$

The top quark mass and jet energy scale dependent portion of the signal probability, $\tilde{P}_{\text{sig}}$, is computed as the sum over all 24 jet–parton assignments and neutrino solutions. Their relative weight, w_i, is computed from the χ^2 probability of the kinematic fit and in presence of identified b-jets includes the probability for the quarks associated to the b-jets to pro-

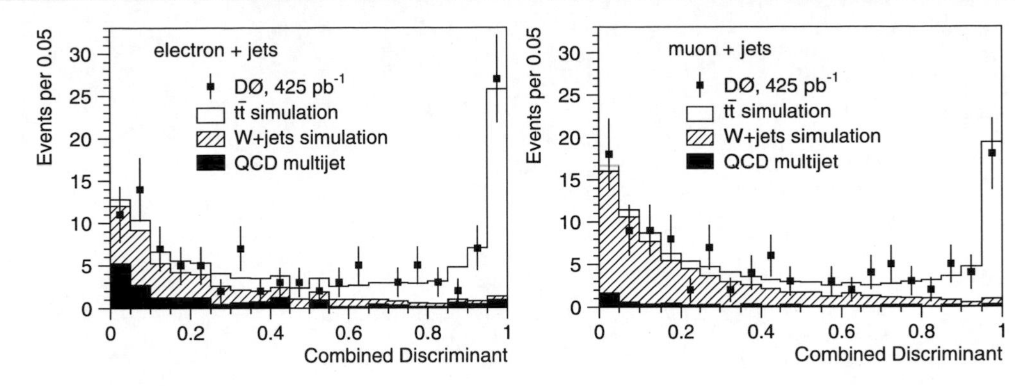

Fig. 8 Combined likelihood discriminant for e + jets (*left*) and μ + jets (*right*). The top quark pair, W + jets and multijet background are normalised according to a fit of their shapes to data [79]

duce a tagged jet. Because the result of the kinematic fit depends on the jet energy scale factor the weights depend on f_{JES}.

$$\tilde{P}_{\mathrm{sig}}(x; m_t, f_{\mathrm{JES}}) = \sum_{i-1}^{24} w_i(f_{\mathrm{JES}}) S(m_i, f_{\mathrm{JES}})$$
$$\tilde{P}_{\mathrm{bkg}}(x; f_{\mathrm{JES}}) = \sum_{i=1}^{24} w_i(f_{\mathrm{JES}}) B(m_i). \tag{17}$$

The background probabilities $B(m_i)$ are obtained from the simulated W + jets events. The signal probabilities $S(m_i, f_{\mathrm{JES}})$ are computed as the convolution of a Breit–Wigner, **BW**, which describes the theoretical distribution of mass values and a Gaussian, **G**, which represents the detector resolution. As this ansatz is valid only for the correct jet–parton assignment and because the wrong pairings do contain information about the top quark mass, a second contribution to describe wrong pairings is added:

$$S(m_i, f_{\mathrm{JES}}) = f_c \int_{m_{\min}}^{m_{\max}} \mathbf{G}(m_i, m', \sigma_i) \mathbf{BW}(m', m_t)\, dm' + (1 - f_c) S_{\mathrm{wrong}}(m_i, m_t; n_{\mathrm{tag}}), \tag{18}$$

with f_c being the fraction of events in which the weight is assigned to the correct jet–parton pairing. For the width of the Gaussian the uncertainty, σ_i, of the mass determination obtained in the constrained kinematic fit is used. Thus events with a well-determined mass from the kinematic fit contribute more than events with a less precise fit result. The expected mass value distribution in wrong pairings, S_{wrong}, also contains information about the true top quark mass. It is determined from simulation as function of the number of identified b-jets. Also the background function, $B(m_i)$, is taken from simulation.

Due to the presence of wrong jet–parton assignments and background events the described likelihood does not yield unbiased results for the jet energy scale factor, f_{JES}. The likelihood is thus corrected with an f_{JES}-dependent but mass-independent correction factor.

The likelihood for the complete sample of observed events is then simply the product of the jet energy scale corrected likelihood, $\mathcal{L}_{\mathrm{evt}}^{\mathrm{corr}}$, for all individual events:

$$\mathcal{L}(m_t, f_{\mathrm{JES}}, f_{\mathrm{top}}) = \prod_{\mathrm{events}} \mathcal{L}_{\mathrm{evt}}^{\mathrm{corr}}(x; m_t, f_{\mathrm{JES}}, f_{\mathrm{top}}). \tag{19}$$

This likelihood is maximised simultaneously with respect to the top quark mass, m_t, the jet energy scale factor, f_{JES} and the signal fraction f_{top}.

Before this procedure is applied to data its performance is determined on large numbers of pseudo-experiments with varying nominal parameter values. The bias is determined as the mean difference between the nominal value and the measured result. The correctness of the fit uncertainty is checked from the distribution of pull values, i.e. the distribution of differences between the measured values and their means normalised to the individual fit error. The bias and the width of the pull distributions for the fitted top quark mass and jet energy scale factor, f_{JES}, are determined as function of the nominal top quark mass and jet energy scale factor simultaneously. Linear corrections are applied to correct for the obtained biases and to correct the uncertainty for any deviation of the pull width from the ideal value of one.

Systematic uncertainties for this measurement are determined from pseudo-experiments with events shifted according to the systematic variation under study. Due to the two-dimensional fit the uncertainty of the overall jet energy scale is contained in the uncertainty obtained from the likelihood fit. Residual discrepancies between the data and Monte Carlo energy scales still affect the result and give the largest contributions to the systematic uncertainties. DØ evaluates uncertainties due to b quark fragmentation modelling [81] and the calorimeter response to b-jets to ± 1.30 and ± 1.15 GeV, respectively. The uncertainty from the p_T of the jet energy scale yields ± 0.45 GeV. Another large contribution of ± 0.73 GeV comes from the signal modelling which is estimated by varying the fraction of high energy gluons produced in addition to a top quark pair. Further uncertainties include the influence of uncertainties on the trigger efficiencies, the background modelling and the calibration.

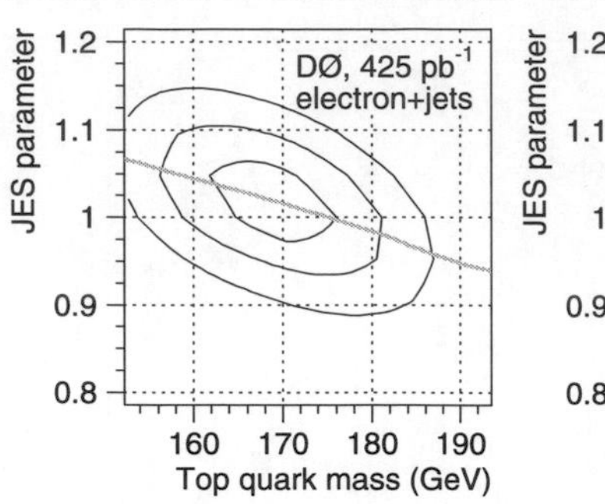

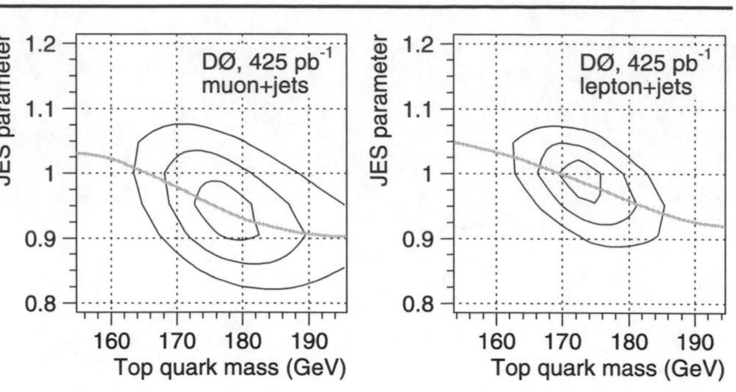

Fig. 9 Contours for the two-dimensional likelihood of determined for $e +$jets (*left*), $\mu +$jets and the combined dataset in 425 pb^{-1} [79]. The contours correspond to log likelihood differences of 0.5, 2.0 and 4.5

The ideogram method with in situ jet energy scale calibration applied in 425 pb^{-1} of data [79] yields a top quark mass of

$$m_t = 173.7 \pm 4.4_{\text{stat+JES}}{}^{+2.1}_{-2.0}{}_{\text{syst}}\ \text{GeV}, \tag{20}$$

the jet energy scale factor is determined to be $f_{\text{JES}} = 0.989 \pm 0.029_{\text{stat}}$ consistent with the nominal value of $f_{\text{JES}} = 1.0$, which corresponds to the calibration obtained in jet-photon events. The fit probability contour lines for the individual lepton channels and the combined results are shown in Fig. 9.

2.2.3 Matrix Element method

The Matrix Element method in the lepton plus jets channel was pioneered by DØ in Run I [82] and is based on ideas described in [83, 84]. In this method the likelihood is constructed according to the expected distribution from the mass dependent top quark pair production matrix element. With respect to the previously described methods this includes additional information from top quark mass dependent kinematics into the measurement.

DØ Results for the top quark mass using the Matrix Element method have been regularly updated with DØ Run II data [85–88] with the latest result including 3.6 fb^{-1} of data. The analyses select events from semileptonic top quark pair decays by requiring a single isolated lepton, missing transverse momentum and exactly four jets. In the recent analyses at least one of the jets needs to be identified using DØ's neural network b jet identification [60]. Vetoes are applied on additional leptons and on events in which the lepton and the missing transverse momentum are close in the azimuthal direction.

The top quark mass and an overall jet energy scale factor are determined simultaneously from an event-by-event probability that the observed event may occur given the assumed values of the top quark mass, m_t, the jet energy scale factor, f_{JES} and the signal fraction f_{top}:

$$\begin{aligned} P_{\text{evt}}(x; m_t, f_{\text{JES}}, f_{\text{top}}) &= A(x) \cdot \big(f_{\text{top}} P_{\text{sig}}(x; m_t, f_{\text{JES}}) \\ &\quad + (1 - f_{\text{top}}) P_{\text{bkg}}(x; f_{\text{JES}})\big). \end{aligned} \tag{21}$$

Here x represents the measured momenta of the lepton, the jets and the missing transverse momentum. $A(x)$ is the probability for the configuration x to be accepted in the analysis. The probabilities, P_{sig} and P_{bkg}, for signal and background events to be observed in the configuration x given the set of parameters, m_t, f_{JES} and f_{top}, are computed from matrix elements of the dominating processes.

The signal probability is computed from the matrix element for top quark pair production and decay through quark antiquark annihilation $\mathcal{M}_{t\bar{t}}(y; m_t)$, convoluted with the transfer function, $W(x, y; f_{\text{JES}})$, which describes the probability to observe a parton configuration, y, as the measured quantities, x.

$$\begin{aligned} P_{\text{sig}}(x; m_t, f_{\text{JES}}) &= \frac{1}{\sigma_{\text{obs}}(m_t, f_{\text{JES}})} \\ &\quad \times \sum_{\text{flavours}} \int \mathrm{d}q_1\, \mathrm{d}q_2\, \mathrm{d}\Phi_6\, f(q_1) f(q_2) \\ &\quad \times \frac{(2\pi)^4 |\mathcal{M}_{t\bar{t}}(y; m_t)|^2}{q_1 q_2 s} W(x, y; f_{\text{JES}}). \end{aligned} \tag{22}$$

The sum is over the possible flavours of the incoming quarks and the integral over their momentum fractions, q_1 and q_2, as well as the 6-body phase space for the outgoing particles, Φ_6. $f(q_i)$ are the parton densities for the incoming quarks and s is the centre-of-mass energy squared. The normalisation factor $\sigma_{\text{obs}}(m_t, f_{\text{JES}})$ is the cross-section observable with the selection used, i.e. it includes effects of efficiencies and geometric acceptance.

For the background probability a corresponding formula is used with the top quark pair matrix element replaced by the $W +$jets matrix element, which of course is independent of m_t. The contribution from multijet events is considered to have a similar shape and is not included separately.

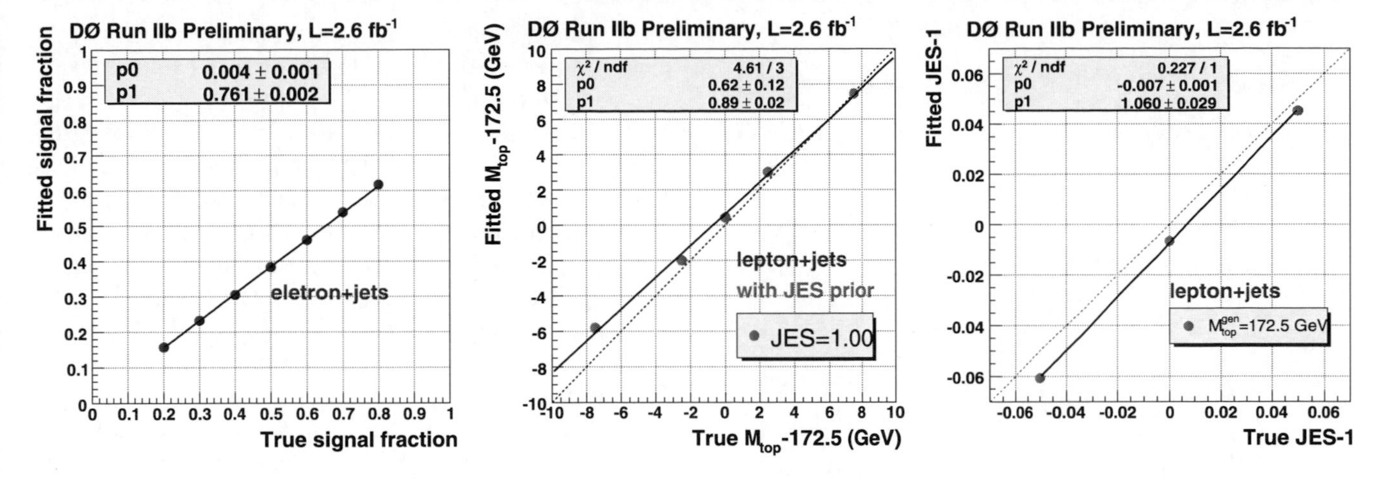

Fig. 10 Calibration curves for the mass measurement with the Matrix Element method obtained for 2.6 fb^{-1} of Run IIb data by DØ. *Left*: Calibration of signal fraction determination in $e+$jets. *Middle*: top quark mass calibration curve for nominal jet energy scale factor. *Right*: Calibration curve for jet energy scale factor [88]

Both, P_{sig} and P_{bkg} contain the same transfer function, $W(x, y; f_{\text{JES}})$. It is derived from full simulation for individual jets and leptons. Only changes of the size of the momenta but not of the directions are considered. Because it is not known which jet stems from which parton, a weighted sum over all 24 possible assignments is made. The weight, w_i, reflects the probability of the event's b-tags to be consistent with the jet–parton assignment under consideration:

$$W(x, y; f_{\text{JES}}) = W_\ell(x_\ell, y_\ell; f_{\text{JES}}) \sum_{i=1}^{24} w_i \prod_{j=1}^{4} W_{\text{j}}(x_j, y_{i,j}; f_{\text{JES}}). \quad (23)$$

The values x_ℓ and y_ℓ are the measured and the assumed momenta of the lepton, x_j is the measured momentum of the jth jet and $y_{i,j}$ the momentum of the matrix element parton associated to the jth jet in the jet–parton association number i. W_ℓ and W_j are the transfer functions for leptons and jets, respectively, which are zero when the directions do not coincide.

The likelihood to observe the measured data is computed as the product of the individual event probabilities, P_{evt}:

$$\mathcal{L}(m_t, f_{\text{JES}}, f_{\text{top}}) = \prod_{\text{events}} P_{\text{evt}}(x; m_t, f_{\text{JES}}, f_{\text{top}}). \quad (24)$$

For each assumed pair of the nominal top quark mass and the jet energy scale factor, m_t and f_{JES}, the likelihood is maximised with respect to the top quark fraction, f_{top}. In [85–87] the top quark mass and jet energy scale factor are then determined by maximising the two-dimensional likelihood $\mathcal{L}(m_t, f_{\text{JES}}, f_{\text{top}}^{\text{best}}((m_t, f_{\text{JES}}))$. In [88] the jet energy scale factor is constrained with a Gaussian probability distribution to its nominal value and its uncertainty as obtained from photon plus jet and dijet events.

Before the method is applied to data, its performance is calibrated in pseudo-experiments. Random events are drawn from a large pool of simulated top quark pair signal and $W+$jets background events with proper fluctuations of the signal and background contribution such that the total number of events corresponds to the number of events observed in data. The simulated events were generated with ALPGEN + PYTHIA and passed through the full DØ simulation and reconstruction. This procedure is repeated 1000 times for several fixed nominal values of m_t, f_{JES} and f_{top}. Thus for each of the nominal values the signal fraction, the top quark mass and the jet energy factor can be measured with the described procedure 1000 times. The mean result for each set of pseudo-experiments with fixed nominal m_t, f_{JES} and f_{top} are compared to the nominal values in the calibration curves in Fig. 10. The final result is corrected for any deviation of these calibration curves from the ideal diagonal. Also the pull width is computed and the statistical uncertainty corrected accordingly.

The leading source of systematic uncertainty contributing ± 0.81 GeV (Run IIb) stems from the modelling of differences in the detector response between light quark and b quark jets. The next important contribution arises from uncertainties modelling hadronisation and underlying event. It is estimated from the difference between PYTHIA and HERWIG and contributes with nearly ± 0.6 GeV. Also the sample dependence of jet energy scale corrections in simulation contributes with this size. For the first time in [88] this measurement includes an estimate of the uncertainty due to colour reconnection effects [72–74], which contributes 0.4 GeV to the uncertainty. The squared sum of the individual contributions yields a total uncertainty of ± 1.4 GeV (Run IIb). DØ has applied this analysis to their 3.6 fb^{-1} dataset separated by run periods. The 1.0 fb^{-1} Run IIa result and the 2.6 fb^{-1} Run IIb result, shown in Fig. 11, are

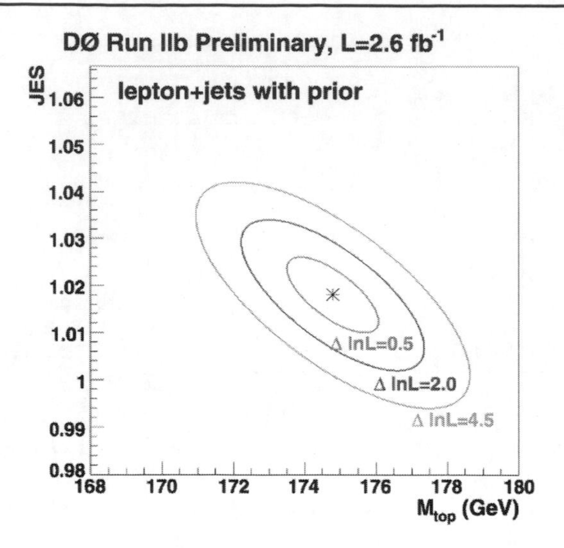

Fig. 11 Result of two-dimensional fit of top quark mass and jet energy scale factor obtained with the Matrix Element method by DØ in 2.6 fb^{-1} of Run IIb data [88]. The ellipses correspond to log likelihood differences of 0.5, 2.0 and 4.5

combined with the BLUE method [89] following the error categories used by the Tevatron Electroweak Working Group [90] (see also Sect. 2.7) and yield

$$m_t = 173.7 \pm 0.8_{\text{stat}} \pm 1.6_{\text{syst}}\ \text{GeV}, \tag{25}$$

where in this combined result the uncertainty due to the overall jet energy scale factor is contained in the systematic uncertainty. This measurement is currently DØ's most precise top quark mass result.

CDF The CDF collaboration is using the concept to measure the top quark mass from a likelihood based on the production matrix element in the lepton plus channel in several variations [91–96].

The recent analyses all base on an event selection that requires a single isolated lepton, missing transverse momentum and exactly four jets, at least one of which is required to be identified as b jet. Then the top quark mass and an overall jet energy scale factor are determined simultaneously from an event-by-event probability that the observed event may occur given the true values of the top quark mass, m_t, the jet energy scale factor, f_{JES} and the signal fraction f_{top}. The various CDF analyses differ in the construction of the likelihood.

The *CDF Matrix Element method (MEM)* [91, 92] follows closely the procedure outlined in Sect. 2.2.3. The matrix element used to describe signal is that of the $q\bar{q} \to t\bar{t}$ process with its decay. For background the $W+4$jets matrix element is employed. Finally, the transfer functions of (23) use only parton-jet assignments consistent with b-tagging information and assumes the lepton measurement to be exact. For the background probability, P_{bkg}, the dependence on the jet energy scale is taken from a parameterisation of the average likelihood response rather than an explicit change of the transfer functions.

The *Dynamical Likelihood method (DLM)* [93, 94] differs from the Matrix Element method in that it bases its likelihood only on the signal contribution, $P_{\text{sig}}(x; m_t, \Delta_{\text{JES}})$. In this signal term the squared matrix element is factorised into a production matrix element, the (anti-)top quark propagators and their decay matrix elements: $|\mathcal{M}(a_1 a_2 \to t\bar{t} \to \ell\nu b\bar{b}qq')|^2 = |\mathcal{M}_{a_1 a_2 \to t\bar{t}}|^2 \mathcal{P}_t \mathcal{P}_{\bar{t}} |\mathcal{D}_t|^2 |\mathcal{D}_{\bar{t}}|^2$, which removes spin correlations. However, in contrast to previously described measurements it includes gluon diagrams, i.e. $a_1 a_2$ can be $q\bar{q}$ or gg. For the treatment of background contributions the method fully relies on results obtained in ensembles of pseudo-datasets similar to the calibration step of the other methods. This correction is computed depending on the jet energy scale correction and the background fraction.

The *Matrix Element Method with Quasi-MC-Integration* (MTM) [97–99] uses an again more complete matrix element to construct the signal likelihood. The applied matrix element [100] includes the $q\bar{q}$ and the gg production channels with full spin correlations. The method treats the background by subtracting the expected contribution of the logarithm of the likelihood. Equation (21) is thus rewritten as

$$\begin{aligned} &\log P_{\text{evt}}(x; m_t, \Delta_{\text{JES}}) \\ &\quad = \log P_{\text{sig}}(x; m_t, \Delta_{\text{JES}}) \\ &\qquad - f_{\text{bkg}}(q) \log P_{\text{bkg}}(m_t, \Delta_{\text{JES}}) \end{aligned} \tag{26}$$

with $\log P_{\text{bkg}}$ being the average $\log P_{\text{sig}}$ obtained in background events. The expected background fraction, f_{bkg}, is computed from simulation and applied event-by-event as function of the output q of a neural network discriminant,

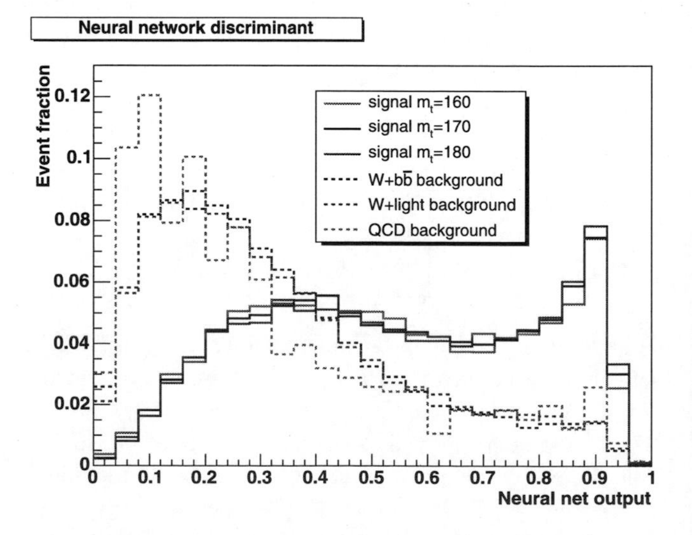

Fig. 12 Expected performance of the neural network discriminant used in the MTM Matrix Element method to determine the event-by-event background fraction [98]

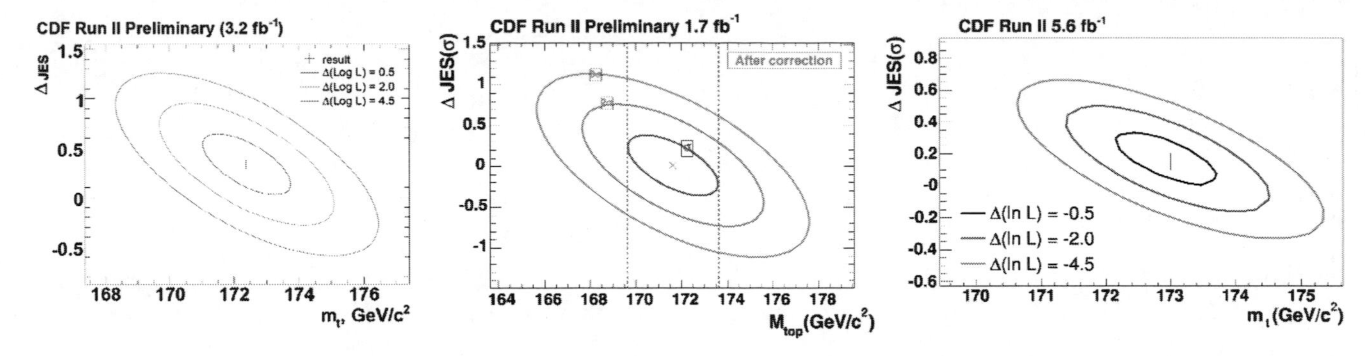

Fig. 13 Results of CDF top quark mass measurements with in situ jet energy calibration obtained in three variations of the Matrix Element method. *Left*: Matrix Element method à la DØ (MEM) [92], *Middle*: Dynamical Likelihood method (DLM) [94], *Right*: Matrix Element method (MTM) [99]. *The vertical axes* show the results obtained for the jet energy shift. In *the left* plot it is a scale factor; *the middle* and *right* plot show it in units of the nominal jet energy uncertainty

cf. Fig. 12. Moreover the analysis removes events which are likely to be mismeasured, e.g. due to extra jets, misidentification etc. For this events are required to have $\log P_{\text{evt}} > 10$ for at least some range of m_t.

In all methods the overall likelihood is computed from the various event likelihoods following (24). It is maximised to find the optimal top quark mass and jet energy scale parameters. For the MEM also the background fraction is fitted. The performance is then checked by applying the top quark mass measurement on ensembles of pseudo-experiments with various nominal values of the top quark mass, m_t, and the jet energy scale shifts, Δ_{JES}. The results are corrected for any observed shifts and the uncertainties scaled to fit the observed spread of results in the ensemble tests. The observed shifts vary between 1.4 GeV for the Dynamical Likelihood method [93], which relies on this calibration to describe the background effects, and 0.09 GeV for the Matrix Element method [92], which has the most complete background term.

As all methods apply an in situ jet energy calibration this important systematic is already covered by the uncertainty of the fit result. The residual effect of the jet energy scale is estimated by varying the p_T and η dependence. The uncertainty of the jet energy scale of b-jets from various sources is considered separately. The uncertainty of signal modelling is determined from comparison of PYTHIA and HERWIG and by varying the amount of initial and final state radiation produced in the parton shower. Further uncertainties include an estimate of the background modelling and treatment in each method. The DLM has the uncertainty due to initial and final state radiation as the most important single contribution, while in the MEM and the MTM the residual jet energy scale and the difference between generators used for signal simulation are the two leading contributions. In MEM and MTM colour reconnection effects [72–74] are estimated and yield significant contributions of 0.56 GeV and 0.37 GeV, respectively.

The most recent results of the various Matrix Element like methods of CDF were obtained at different integrated luminosities. All results yield in situ jet energy scale corrections consistent with the default calibration used in CDF.

$$\begin{aligned} &\text{MEM:} \quad 3.2\ \text{fb}^{-1}\ [92]: && m_t = 172.4 \pm 1.4 \pm 1.3\ \text{GeV} \\ &\text{DLM:} \quad 1.7\ \text{fb}^{-1}\ [94]: && m_t = 171.6 \pm 2.0 \pm 1.3\ \text{GeV} \\ &\text{MTM:} \quad 5.6\ \text{fb}^{-1}\ [98, 99]: && m_t = 173.0 \pm 0.9 \pm 0.9\ \text{GeV}, \end{aligned} \tag{27}$$

where the first uncertainty is the statistical one and includes the overall jet energy scale uncertainty, the second uncertainty is the systematic one. The two-dimensional representation of these results which simultaneously determine the jet energy scale are shown in Fig. 13. These measurements of course use (partially) the same data, thus for the combination only the most precise one is used, until their correlation is determined.

2.2.4 Decay length and lepton momentum methods

The top quark mass measurements described so far, aimed to use the maximal available information to yield the best statistical uncertainty for the given number of events. However, all these measurements have significant uncertainties from the jet energy scale. Alternative observables that have little or no dependence on the jet energy scale are therefore an interesting complement.

In the lepton plus jets channel of top quark pair decays CDF has investigated two observables: the transverse decay length of b tagged jets and the lepton transverse momentum. One analysis combines the mean values of both observables [101, 102] while the other uses the mean and distribution of the lepton transverse momentum alone [103].

Two observable mean value method For the method using both observables the data selection requires one isolated

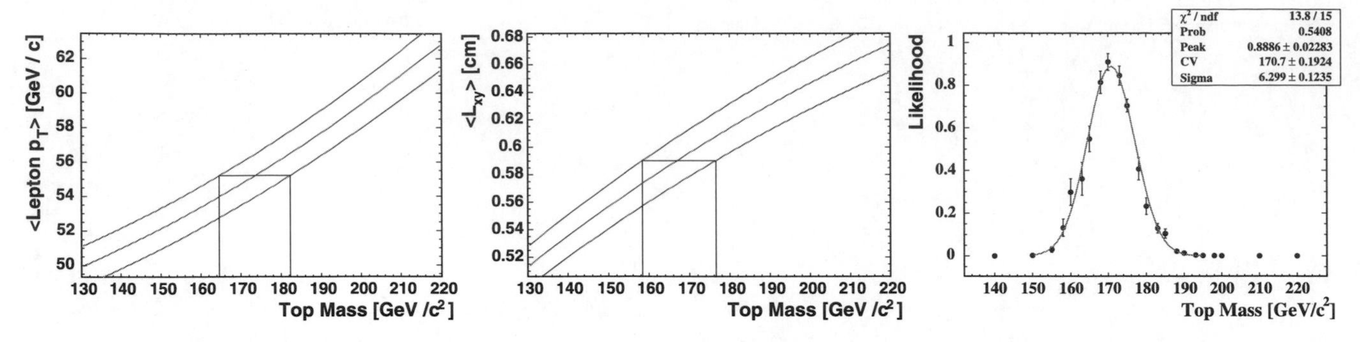

Fig. 14 *Left* and *middle*: Calibration curves for the expected mean decay length, $\langle L_{2D}\rangle$, and mean lepton momentum, $\langle p_T^\ell\rangle$, as function of the nominal top quark mass (*left* and *middle*). *Right*: The likelihood for the combined measurement [102]

lepton, missing transverse momentum and at least three jets. For events with exactly three jets two of the jets need to be identified as b jets with CDFs secondary vertex algorithm [69]. For events with four or more jets only one needs to have such a reconstructed secondary vertex.

The expected sample composition is determined from a combination of data and simulation. Top quark pair signal is simulated using PYTHIA with CTEQ5L parton densities for various nominal top quark mass values. Also single top quark samples are generated with various nominal top quark mass values. The dominating background $W+$jets is simulated with ALPGEN + PYTHIA, the multijet background is modelled from a data sample with modified lepton selection criterion. All simulated events are passed through the full CDF detector simulation and reconstruction.

For the top quark mass determination the transverse decay length L_{2D} of the b tagged jets is measured with the vertex algorithm used for the selection above. The actual measurement is performed using the means of the observed decay lengths, $\langle L_{2D}\rangle$, and the transverse lepton momentum, $\langle p_T^\ell\rangle$. The expected contribution from background is taken from the described background model. Because the observables chosen are sensitive to the details of the event kinematics the signal simulation is corrected to yield parton distributions consistent with CTEQ6M [36]. In addition the total contribution from gluon fusion is corrected as function of the top quark mass. The performance of the simulation for determining the decay length L_{2D} of the b tagged jets is calibrated in a dijet control sample.

With these corrections the expected means of decay lengths and transverse lepton momentum and their expected statistical spread are determined as function of the nominal top quark mass using ensembles of pseudo-data. The curves obtained are fitted by quadratic polynomials to obtain smooth curves. Figure 14 (left, middle) shows the mean values $\langle L_{2D}\rangle$ and $\langle p_T^\ell\rangle$ and the resulting top quark mass ranges from the individual observables. The combined result is obtained from the likelihood to find the observed mean values at various nominal top quark masses, cf. Fig. 14 (right).

The dominant systematic uncertainties for the measurement from $\langle p_T^\ell\rangle$ arises from the uncertainty on the modelling of initial and final state radiation in the signal simulation, from the lepton momentum scale and the shape of the background description. In the measurement from $\langle L_{2D}\rangle$ the uncertainty of the data to Monte Carlo correction for the decay length dominates the uncertainties. The jet energy scale gives a non-negligible contribution to the measurement from the average decay length through effects on the event selection.

In 1.9 fb^{-1} of data CDF measures the top quark mass from the mean decay length, $\langle L_{2D}\rangle$, and the mean lepton momentum, $\langle p_T^\ell\rangle$ to be [102]

$$m_t = 170.7 \pm 6.3_{\text{stat}} \pm 2.6_{\text{syst}}\ \text{GeV}, \tag{28}$$

where the jet energy scale uncertainty is included (as a small) contribution to the systematic uncertainty. The larger statistical uncertainty (compared to the Matrix Element methods) prevents a significant weight to the current combination of top quark masses [90].

Lepton momentum shape method The analysis that concentrates on the transverse lepton momentum alone [103] is based on events with a single isolated lepton, missing transverse energy and at least four jets. At least one of the jets is required to be identified as b-jet.

Top quark pair signal events are simulated for various assumed top quark mass values. Backgrounds due to $W+$jets, $Z+$jets diboson and single top quark production are described with simulation. Estimates for multijet background with fake leptons and the normalisation for $W+$light flavour jets are derived from data. These signal and background models are used to determine parametrised templates for the distribution of the lepton p_T spectrum as function of the top quark mass. To extract the top quark mass these templates are compared to the observed data using an unbinned maximum likelihood fit. Leptonic Z boson decays are used to calibrate the lepton p_T.

The uncertainty of the final result is dominated by the statistical uncertainty. The systematic uncertainties are dominated by effects related to signal and background modelling. The uncertainty on the fake lepton description contributes with ± 1.8 GeV; differences between generators for the signal modelling contribute with ± 1.8 GeV to the systematic uncertainty.

In 2.7 fb^{-1} of CDF data the top quark mass determined from the shape of the lepton momentum distribution is [103]

$$m_t = 176.9 \pm 8.0_{\text{stat}} \pm 2.7_{\text{syst}}\ \text{GeV}. \tag{29}$$

This result has been combined with a corresponding study of dilepton events [104], cf. Sect. 2.3.3, to yield [105]

$$m_t = 172.8 \pm 7.2_{\text{stat}} \pm 2.3_{\text{syst}}\ \text{GeV}. \tag{30}$$

The large statistical uncertainty of these methods prevents a significant contribution to the world average from such measurements at the Tevatron. Its independent and low systematics will make them more relevant at the LHC.

2.3 Dilepton channel

Due to the small branching fraction the dilepton channel has much fewer events, but the two charged leptons also yield a much cleaner signature. For a measurement of the top quark mass the dilepton channel has the additional complication that the kinematics are under-constrained due to the two unmeasured neutrinos. Two of the six missing numbers can be recovered from the transverse momentum balance, two more from requiring that the invariant mass of the charged lepton and its neutrino should be consistent with the W boson mass in the top and the antitop quark decay, a fifth constraint can be obtained by forcing the top and the antitop quark masses to be equal. Thus for a full reconstruction of an event including the top quark mass one constraint is missing.

2.3.1 Weighting methods

To completely recover the event kinematics additional assumptions can be made on a statistical basis, i.e. by assuming the distribution of one or more kinematic quantities. For a given event top quark masses corresponding to certain values of these kinematic quantities are weighted by the probability that these kinematic values occur. The top quark mass reconstructed in the given event is then chosen such that it corresponds to the largest weight. This basic idea leads to various so-called weighting methods for measuring the top quark mass in dilepton events that mainly differ in the distribution that is assumed a priori. In the following the relevant methods recently used by the two collaborations are described.

CDF neutrino weighting One of these weighting methods is the neutrino weighting method that is used in CDFs combined lepton plus jets and dilepton analysis [65–68]. The lepton plus jets part of this analysis is described in Sect. 2.3.1.

Dilepton events are selected by requiring two oppositely charged leptons, missing transverse energy and exactly two energetic jets. In addition the scalar sum of the transverse momenta, H_T, is required to be greater than 200 GeV. Events in which at least one of the jets was identified as b jet are kept separately throughout the analysis.

To reconstruct one top quark mass for each selected event the distribution of neutrino pseudo-rapidities is assumed a priori. The distribution is taken from top quark pair simulation and found to be Gaussian with an approximate width of one. The weight is computed in two steps for each assumed top quark mass.

First, for fixed values of the neutrino pseudo-rapidities, η_ν and $\eta_{\bar{\nu}}$, the event kinematics are reconstructed using constraints on the assumed top quark mass and the nominal W boson mass (ignoring the measurement of the transverse momentum). The weight of these η choices is constructed as the χ^2 probability that the sum of the reconstructed transverse neutrino momenta agree with the measurement of the missing transverse momentum, $\not{\boldsymbol{p}}_T$:

$$w(m_t, \eta_\nu, \eta_{\bar{\nu}}) = \sum_{i=1}^{8} \exp\left(-\frac{(\not{\boldsymbol{p}}_T - \boldsymbol{p}_T^{\nu}(i) - \boldsymbol{p}_T^{\bar{\nu}}(i))^2}{2\sigma_T^2}\right). \tag{31}$$

Here, the sum adds the four possible sign choices that occur in solving the above constraints and the two possibilities to assign the two measured jets to the b or $\bar{b}$ quark. $\boldsymbol{p}_T^{\nu}(i)$ and $\boldsymbol{p}_T^{\bar{\nu}}(i)$ are the transverse neutrino momenta for the choice i and σ_T is the experimental resolution of the measurement of the missing transverse momentum.

The second step now folds the weights obtained for fixed pseudo-rapidities with the a priori probability, $P(\eta_\nu, \eta_{\bar{\nu}})$, that these pseudo-rapidities occur:

$$W(m_t) = \sum_{\eta_\nu, \eta_{\bar{\nu}}} P(\eta_\nu, \eta_{\bar{\nu}}) w(m_t, \eta_\nu, \eta_{\bar{\nu}}). \tag{32}$$

The top quark mass that yields the highest $W(m_t)$ is the reconstructed value, m_t^{reco}, for the event under consideration. In addition to this top quark mass the variable m_{T2} is used for further analysis. $m_{T2} = \min[\max(m_T^{(1)}, m_T^{(2)})]$, where $m_T^{(i)}$ are the transverse mass of the top and the antitop quark, respectively. The minimisation is performed over all possible neutrino momenta consistent with the missing transverse energy. Previous versions of the analysis instead used the scalar sum of the jets and lepton transverse momenta and the missing transverse momentum, H_T.

Simulation is now used to determine the expected distribution of m_t^{reco} and m_{T2}. The backgrounds in the dilepton sample stem from fake events with a jet misidentified as

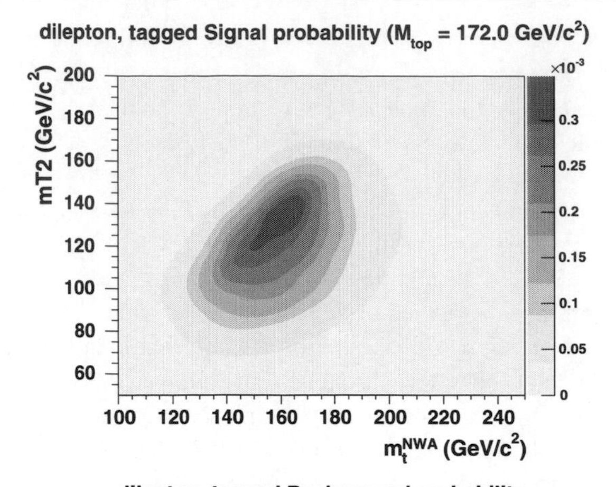

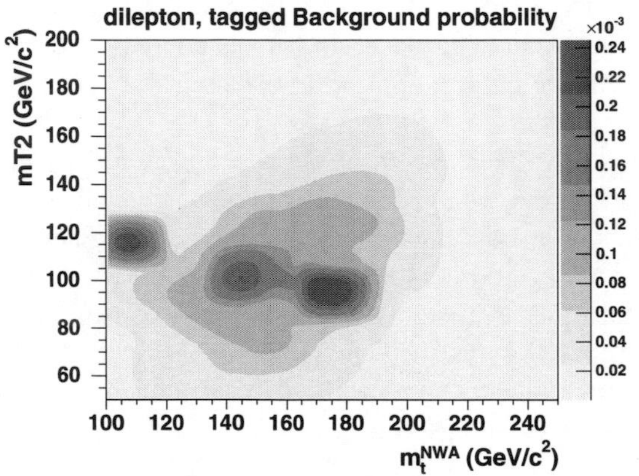

Fig. 15 Probability density in the m_t^{reco}–m_{T2} plane for a dilepton signal at $m_t = 172$ GeV (*top*) and background (*bottom*) at nominal jet energy scale [67]

lepton, from Drell–Yan and from diboson production. The dominating fake background is modelled from data. Drell–Yan events are simulated with ALPGEN + PYTHIA, diboson events with PYTHIA. The simulated events are passed through the CDF detector simulation and reconstruction.

These simulations are used to generate probability density functions for the two observables, m_t^{reco} and m_{T2}, as function of the nominal top quark mass, m_t, and the jet energy scale shift, Δ_{JES}, see Fig. 15. From the probability densities a likelihood is constructed that is maximised to find the final result. The construction of the probability densities and the likelihood and the determination of the resulting measured top quark mass follows the lepton plus jets analysis described in Sect. 2.3.1 The dependence on the jet energy shift is only used for simultaneously fitting the lepton plus jets and dilepton events. Systematic uncertainties for the pure dilepton measurement are dominated by jet energy scale uncertainties, followed by the uncertainty of modelling colour reconnection effects.

In 5.6 fb^{-1} of data CDF determines the top quark mass [68] in dilepton events using the neutrino weighting method and m_{T2} to be:

$$m_t = 170.3 \pm 2.0_{\text{stat}} \pm 3.1_{\text{syst}}\ \text{GeV}. \tag{33}$$

A combined fit including lepton plus jet events and in situ jet energy scale determination the dataset yields

$$m_t = 172.1 \pm 1.1_{\text{stat}} \pm 0.9_{\text{syst}}\ \text{GeV}. \tag{34}$$

With the combined fit the dilepton result profits from the constraint of the jet energy scale. The overall precision is still dominated by the lepton plus jets result.

CDF neutrino ϕ weighting methods The neutrino ϕ weighting method uses the distribution of neutrino azimuthal directions, ϕ, as an a priori distribution. CDF has analysed up to 2.9 fb^{-1} of dilepton events to measure the top quark mass [106] with this method.

Events are selected by requiring one isolated well-identified lepton, one oppositely charged isolated track, missing transverse momentum and at least two jets. Vetoes are applied on events where the missing transverse momentum is close to a jet, on cosmic events, conversions and Z boson events.

For each value pair $(\phi_\nu, \phi_{\bar{\nu}})$ on a grid of azimuthal directions the top quark mass is reconstructed using a constrained fit that determines the top quark mass, m_t, the lepton and (anti-)b quark momenta using the measured lepton, track and jet momenta, the missing transverse momentum and the constraints on the W boson mass and the equality of the top and antitop quark mass. The χ^2 is defined similarly to (11), but uses Breit–Wigner distributions rather than Gaussians for the mass constraints. For this fit two possibilities exist to assign the two measured jets to the b and anti-b quarks. In addition the quadratic nature of the W boson mass constraints gives a fourfold ambiguity. Of the corresponding eight top quark mass results at each $(\phi_\nu, \phi_{\bar{\nu}})$ pair only the one with the lowest χ^2 is kept. To arrive at a single mass value for the event under consideration the masses obtained at the various azimuthal directions are averaged weighted with their χ^2 probability. Only events with a weight of at least 30% of the maximum are considered in the average.

The top quark mass is finally measured by comparing the distribution of mass values reconstructed for each event in data to templates in a likelihood fit. The templates are derived from simulation. For the dilepton signal HERWIG [107] is used at many nominal top quark mass values. The background is simulated with PYTHIA for the Drell–Yan background and ALPGEN + HERWIG for the fakes from W + jets production. The diboson background was simulated with PYTHIA and ALPGEN + HERWIG [65]. Signal and background templates were parametrised to obtain smooth templates. The signal templates yield a smooth dependence on the nominal top quark mass.

The likelihood for N events is built as the product over the probabilities that each event agrees with the sum of the signal and background templates, P_{sig} and P_{bkg}, weighted according to the signal and background contributions, s and b. The likelihood contains a Poissonian term on the total number of expected events and a Gaussian constraint on the background contribution, b, to be consistent with the a priori expectation, b^e:

$$\mathcal{L}(m_t) = e^{-\frac{(b-b^e)^2}{2\sigma_b^2}} \left(\frac{e^{-(s+b)}(s+b)^N}{N!} \right) \times \prod_{i=1}^{N} \frac{s P_{\text{sig}}(m_i; m_t) + b P_{\text{bkg}}(m_i)}{s+b} . \tag{35}$$

In addition a term, $\mathcal{L}_{\text{param}}$, that describes the parametrisation uncertainties of the template curves is included into the likelihood. Minimising the total likelihood yields the top quark mass results, m_t, and estimates for the signal and background contributions, s and b. The statistical uncertainty is obtained by finding the top quark masses corresponding to a log likelihood decrease of 0.5. Due to the inclusion of $\mathcal{L}_{\text{param}}$ in the likelihood this includes the uncertainties from the template parametrisation.

The systematic uncertainties are evaluated by measuring the top quark mass in ensembles of pseudo-data generated from simulation with modifications that reflect a one sigma change of the systematic under consideration. The systematics are dominated by the jet energy scale that has to be taken from the external calibration of the CDF calorimeter. It contributes ±2.9 GeV. The next-to-leading contributions from the background composition and the b jet energy scale contribute less than 20% of this uncertainty.

In 2.9 fb^{-1} of CDF data with the described lepton plus track selection the method yields [106]

$$m_t = 165.5^{+3.4}_{-3.3\,\text{stat}} \pm 3.1_{\text{syst}}\ \text{GeV}. \tag{36}$$

The method shows similar statistical precision as the neutrino weighting method and similar sensitivity to the jet energy scale systematics.

DØ Neutrino Weighting method DØ has applied the Neutrino Weighting method in up to 5.3 fb^{-1} of data [108–110]. Events are selected by requiring two isolated, oppositely charged, identified leptons or an isolated, identified lepton and an isolated track of opposite charge. In addition the events are required to have two energetic jets and missing transverse energy. For lepton plus track events at least one of the jets needs to be identified as b-jet. Vetoes on the lepton pair (or lepton plus track) invariant mass and the scalar sum of transverse momenta are applied to reject Z + jets and other backgrounds depending on the identified lepton types.

The event kinematics is reconstructed by assuming the neutrino rapidity distribution which according to simulations is expected to be Gaussian. Scanning through the range of possible top quark masses an event weight is computed as function of the top quark mass.

First, for given values of the neutrino rapidities and the top quark mass a constrained fit is performed to determine the momenta of the top quark decay products, $b\bar{b}\ell^+\ell^-\nu\bar{\nu}$. As constraints the W boson mass and the assumed top quark mass are used, ignoring the measured values of the missing transverse momentum. An individual weight, $w(m_t, \eta_\nu, \eta_{\bar{\nu}})$, is computed from a χ^2 term that compares the sum of neutrino momenta in the transverse plane with the measured missing transverse momentum, cf. (31). Then the a priori Gaussian distribution of the neutrino rapidities are folded into a total weight, $W_i(m_t)$, by adding the $w(m_t, \eta_\nu, \eta_{\bar{\nu}})$ at 10 values of the neutrino rapidity with appropriate unequal distance.

To take detector resolution for jet and lepton energies into account the determination of $W_i(m_t)$ is repeated for a number of jet and lepton momenta fluctuated according to their experimental resolution. The weight averaged over these fluctuations, $W(m_t) = \langle W_i(m_t) \rangle$, shows a much smoother distribution and yields fit results for a wider range of top quark masses, see Fig. 16 (left).

To determine the top quark mass the mean of the weight distribution and its variance are computed for each individual event:

$$\mu_w = \int m_t W(m_t)\, \mathrm{d}m_t, \qquad \sigma_w^2 = \int m_t^2 W(m_t)\, \mathrm{d}m_t - \mu_w^2. \tag{37}$$

Compared to using μ_w alone, including σ_w to the following extraction of the top quark mass yields an 16% improvement in the statistical uncertainty. Previous analyses of DØ used even more detailed information about $W(m_t)$ and thus the statistical information was exploited slightly better but at the cost of higher complexity [109, 110].

The distribution of μ_w and σ_w values in data is now compared to templates derived from simulation. Top quark pair signal events were generated with ALPGEN + PYTHIA for various nominal top quark masses. Background contributions from Z/γ + jets are simulated using ALPGEN + PYTHIA, diboson events with PYTHIA. All simulated events are passed through the full DØ detector simulation and reconstruction. For Z/γ + jets events with $Z/\gamma \to e^+e^-$ or $\mu^+\mu^-$ the amount of fake missing transverse momenta and of fake isolated leptons or tracks is derived from control samples in data and used in the normalisation of the samples. Template histograms are obtained from the above signal and background estimates. The observed data distributions are compared directly to the template histograms or are compared to parameterised smooth functions fitted to the templates, see Fig. 16 (right) for an example. This yields two methods, a histogram based and a function based method.

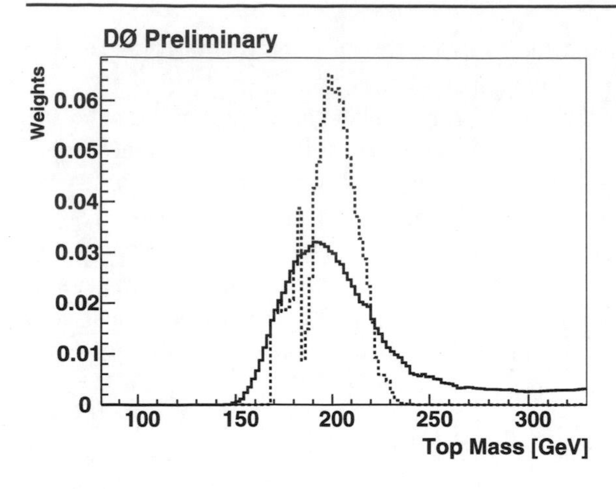

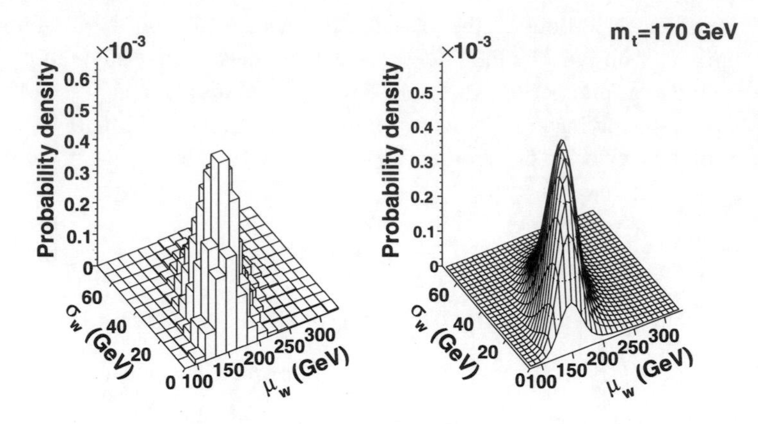

Fig. 16 Normalised weight distribution for an example simulated event with $m_t = 175$ GeV before (*dashed*) and after (*full line*) smearing lepton and jet momenta [108] (*left*). Example of the template histograms (*middle*) and functions (*right*) of μ_w vs. σ_w evaluated at $m_t = 170$ GeV for the $e\mu$ channel [109]

The final top quark mass result is extracted by maximising a likelihood as function of the number of signal and background events as well as the top quark mass. The likelihood describes the agreement of the data with the templates, a Gaussian constraint on the expected amount of background and a Poisson term for the total number of observed events as in (35).

The performance and precision of the method is tested on a large number of pseudo-data with known nominal top quark mass composed from the signal and background models. Calibration curves of the nominal vs. the measured top quark mass are obtained by comparing the average of the measured top quark masses on many such pseudo-data with their nominal top quark mass. The observed offsets of around 1 GeV and deviations of the pull distribution from the normal distribution are corrected for.

As for the other dilepton measurements the dominating systematic uncertainty stems from the energy scale uncertainties for the two b jets in the event. In this analysis it contributes about ± 1.5 GeV. Sub-leading uncertainties arise from limited template statistics and from the difference between modelling the top quark pairs with pure PYTHIA and using ALPGEN + PYTHIA.

The histogram based method and the method using parametrised templates show a correlation of only 85%. They are thus averaged using the BLUE method. Using dilepton events selected in 1 fb^{-1} of data DØ determines the top quark mass to be [109]

$$m_t = 176.2 \pm 4.8_{\text{stat}} \pm 2.1_{\text{syst}}\ \text{GeV}. \tag{38}$$

An updated result including up to 5.3 fb^{-1} of data and combining with the Matrix Weighting method is described at the end of the next section. Despite the large systematic uncertainty from the jet energy scale the measurement is statistically limited at this luminosity.

DØ Matrix Weighting method Another method to measure the top quark mass in dilepton events is the Matrix Weighting method. This has been applied by DØ in Run II to 5.3 fb^{-1} of data [109]. The event selection requires two identified oppositely charged leptons, large missing transverse momentum and at least two jets. As above, vetoes on the lepton pair invariant mass and the scalar sum of transverse momenta are applied to reject Z + jets and other backgrounds depending on the identified lepton types.

To reconstruct the top quark mass in each event this method assumes the distribution of lepton energies in the top quark rest frame [111] and the parton density functions. First the kinematics is reconstructed as function of the top quark mass. At each assumed top quark mass a kinematic fit is applied to reconstruct the top quark decay products from the measured quantities. The two leading jets are identified with the b quarks, the missing transverse momentum is required to be consistent with the sum of neutrino momenta in the transverse plane. In addition kinematic constraints from the W boson mass and the assumed top quark mass are applied. With the reconstructed kinematics a weight as function of the top quark mass is computed:

$$w_j(m_t) = f(x_1)f(x_2)P\big(E^*_{\ell^+}|m_t\big)P\big(E^*_{\ell^-}|m_t\big). \tag{39}$$

Here the quark parton densities, $f(x)$, at the quark and antiquark momentum fractions, x_1 and x_2, are explicitly included in the weight. For the neutrino weighting described above they are included implicitly in the expected η distribution. $P(E^*_{\ell^\pm}|m_t)$ are the probabilities that (given the hypothetical top quark mass, m_t) the reconstructed energy of the lepton $\ell^\pm$ in the rest frame of the corresponding top quark is $E^*_{\ell^\pm}$. The distribution of these energies is taken from the top quark pair production and decay matrix element [111].

In each event there are up to four possibilities to solve the constraint from the W boson mass and two possibilities to assign the two jets to the two b quarks, thus up to eight values of $w_i(m_t)$ may exist. The total event weight is computed as the sum of these.

$$W_i(m_t) = \sum_j w_j(m_t). \quad (40)$$

As for the neutrino weighting method resolution effects are included by recomputing the W_i with the measured quantities fluctuated according to their resolutions and averaging the results: $W(m_t) = \langle W_i(m_t) \rangle$. In each event the top quark mass which maximises this smeared weight, $W(m_t)$, is used as the estimator of the top quark mass for this event.

The distribution of the top quark mass estimators observed in data is compared to templates obtained from simulation for a range of nominal top quark mass values. The simulations use PYTHIA to generate top quark pair signal and diboson background events. Backgrounds from $Z+$jets are generated with ALPGEN + PYTHIA. The generated events are passed through full DØ detector simulation and reconstruction. For the comparison and to extract the final top quark mass a binned likelihood is used.

The method is calibrated by using ensembles of pseudo-data constructed from the simulated events for various nominal top quark masses. The obtained bias in the determination of the central result and its statistical error are corrected for. Ensemble tests are also used to determine the effect of systematic uncertainties. As for the other dilepton mass measurements the systematic uncertainty is dominated by uncertainties related to the jet energy scale. They account for ± 1.2 GeV.

In 1.0 fb^{-1} DØ measures the top quark mass using the Matrix Weighting method in dilepton events to be [109, 112].

$$m_t = 173.2 \pm 4.9_{\text{stat}} \pm 2.0_{\text{syst}}\ \text{GeV}. \quad (41)$$

With respect to the DØ neutrino weighting this Matrix Weighting method yields a slightly worse statistical precision, but has a smaller dependence on the jet energy scale. DØ has averaged the two weighting methods described taking correlations into account using the BLUE method [109]. The combined result updated to 5.3 fb^{-1} yields [113]:

$$m_t = 173.3 \pm 2.4_{\text{stat}} \pm 2.1_{\text{syst}}\ \text{GeV}. \quad (42)$$

As different events contribute to the two methods to a different amount, this allows a significant improvement of the statistical uncertainty over the individual results.

2.3.2 Matrix Element methods

An alternative to the weighting methods of measuring the top quark mass in dilepton events is the Matrix Element method described above already for the lepton plus jets channel, see Sect. 2.2.3. Instead of assuming individual distributions to add effective constraints, the full knowledge from the matrix element is used to check the agreement of the measured events with different top quark mass assumptions.

CDF Matrix Element method CDF has applied the Matrix Element method to the dilepton channel in several analyses [114–116]. The analyses require two oppositely charged leptons, missing transverse energy and two or more energetic jets.

The probability to observe a given event depends on the top quark mass and is computed from folding the matrix element with parton densities and detector resolutions along (21) and (22). In a dilepton sample of course the integration needs to include the unmeasured momenta of *two* neutrinos. The transfer functions that connect the measured quantities with the parton momenta correspondingly contains terms for only two jets. The momenta of the charged leptons are kept unsmeared.

In the 1 fb^{-1} analysis [115] an additional contribution to the transfer function was used to connect unclustered energy and energy from additional sub-leading jets with the transverse momentum of the top quark pair system. This more accurate treatment was found to yield better statistical power, but also larger systematics. In the recent update to 1.8 fb^{-1} [116] such a term is no longer used, instead events with more than two jets are rejected.

The matrix element used for top quark pair events corresponds to the quark annihilation process. Background contributions are computed using the matrix elements for $Z/\gamma+$jets, $WW+$jets and $W+3$jets. The contributions are added according to their expected fractions in the selected dataset, depending on the number of identified b-jets in the event.

The probability for the full sample, $P(m_t)$, is the product of the event-by-event probabilities. As the jet energy scale cannot be constrained in dilepton events the external nominal CDF energy scale is used. The mean, $m_t^{\text{raw}} = \int m_t P(m_t)\, dm_t$, is used as the raw measured mass. The result is calibrated using ensembles of pseudo-data with varying nominal top quark masses. The pseudo-data are modelled using HERWIG for signal events, ALPGEN + PYTHIA for $Z/\gamma+$jets and pure PYTHIA for diboson samples. Misidentified leptons are modelled from data. The observed bias and pull width observed by applying the above procedure on the ensembles are used to correct the raw mass measurement and its statistical uncertainty.

The systematic uncertainties are again dominated by the jet energy scale uncertainty. In the 1.0 fb^{-1} analysis [115] the additional inclusion of additional jets is found to increase the dependence on the jet energy scale by 15%. In the 1.8 fb^{-1} analysis [116] the jet energy scale accounts for a ±2.6 GeV mass uncertainty. The next leading uncertainties in this analysis stem from simulation statistics for the background description, which of course can be addressed, and the uncertainty in the calibration. Differences of signal simulation with PYTHIA vs. HERWIG give only a marginally smaller contribution.

In 1.8 fb^{-1} of dilepton events CDF measured the top quark mass with the Matrix Element method as [116]

$$m_t = 170.4 \pm 3.1_{\text{stat}} \pm 3.0_{\text{syst}}\ \text{GeV}. \tag{43}$$

With the increased statistics the method of [115] that yields a statistical improvement at the cost of increased systematics is no longer applied.

DØ Matrix Element method DØ has applied the Matrix Element method to measure the top quark mass in dilepton events to 3.6 fb^{-1} [117]. The event selection requires one isolated electron and one isolated muon of opposite charge and at least two jets.

The probability to observe a given event is computed as function of the top quark mass by folding the matrix element with parton densities and detector resolutions along (21) and (22). The integration includes the additional unknowns due to the second unseen neutrino and the unknown transverse momentum of the top quark pair system. The weight function, $W(x, y)$, contains terms to describe the detector resolution for two leptons and two jets. The two possible jet to (anti-)b-quark assignments are summed. For the signal description DØ uses the matrix element of quark annihilation to top quark pairs and the subsequent decay. The backgrounds are all described with the matrix element for $Z+$jets process with $Z \to \tau^+\tau^-$. Special transfer functions were used to relate the measured electron and muon momenta from a tau decay to the tau momentum.

The likelihood for the observed data sample as function of the assumed top quark mass, $\mathcal{L}(m_t)$, is the product of the event probabilities, cf. (24). As dilepton events cannot constrain the jet energy scale, the externally determined nominal jet energy scale is used. The measured top quark mass is the one that maximises the sample likelihood $\mathcal{L}(m_t)$. Its statistical uncertainty is taken from the masses that yield an $\mathcal{L}(m_t)$ decreased by half a unit from the maximum.

The performance of the method is evaluated using ensembles of pseudo-data that are created from simulation. Top quark pair signal and $Z+$jets background events were generated with ALPGEN + PYTHIA, diboson events are generated with pure PYTHIA. The samples were normalised to theoretical cross-sections. All simulated events are passed through the full DØ detector simulation and reconstruction. The bias and widths of the pull distribution observed in the ensemble tests are used to correct the above measured numbers.

Systematics are dominated by uncertainties related to the jet energy scale of the two b jets in the event. They account to about ±2 GeV of the total systematic uncertainty.

DØ has applied the Matrix Element method to determine the top quark mass to the Run IIa (1.1 fb^{-1}) and Run IIb (2.5 fb^{-1}) run periods separately and combined the individual results using the BLUE method [89] with the error categories used by the Tevatron Electroweak Working Group [90] (see also Sect. 2.7 below):

$$m_t = 174.8 \pm 3.3_{\text{stat}} \pm 2.6_{\text{syst}}\ \text{GeV}. \tag{44}$$

Combinations with the other DØ results in the dilepton are available in [117] also.

2.3.3 Lepton momentum method

The so far described methods of determining the top quark mass in the dilepton channel, suffer from uncertainties of the jet energy scale. This uncertainty dominates here the uncertainties much more than in the lepton plus jets channel, where the in situ calibration allows to constrain the jet energy scale. Measurements with lepton based quantities thus yield complementary results.

CDF has performed an analysis of 2.8 fb^{-1} of data in the dilepton channel using the transverse momentum of the leptons to extract the top quark mass [104]. Events are selected by requiring two isolated leptons of opposite charge and large missing transverse momentum. For the signal sample at least two jets are required, one of which needs to be identified as b jet.

The background in this selection is dominated by fake signals from multijet events, which is estimated from events with same charge leptons. Contributions from diboson and Z/γ events are simulated with PYTHIA passed through full detector simulation and reconstruction. Events of $W + b\bar{b} +$ jets are generated with ALPGEN + PYTHIA.

To extract the top quark mass from the selected events the distribution of the lepton transverse momenta is employed. The expected distribution of the leptons' transverse momenta is parametrised as function of the top quark mass. A binned likelihood is used to compare the observed data to parametrised templates. The observed distribution and the best fit templates are shown in Fig. 17. Systematic uncertainties of this measurement are dominated by effects related to signal simulation. The choice of the generator and the uncertainty in modelling initial and final state radiation contribute with 1.5 GeV and 1.3 GeV, respectively. The momentum scale contributes with only 0.7 GeV.

In this preliminary result based on 2.7 fb^{-1} of data the top quark mass determined from the shape of the lepton p_T distribution is [104]

$$m_t = 154.6 \pm 13.3_{\text{stat}} \pm 2.3_{\text{syst}}\ \text{GeV}. \tag{45}$$

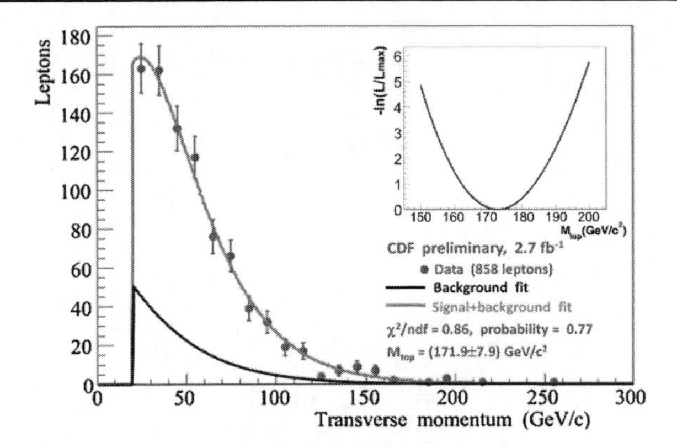

Fig. 17 Distribution and fit of lepton transverse momenta observed in $2.7\ \mathrm{fb}^{-1}$ of CDF data using dileptonic events. For fitted contribution of the background is shown in *black*. The inset shows the corresponding likelihood as function of the top quark mass [104]

The combination with the corresponding study of semileptonic events [103], cf. Sect. 2.2.4, yields [105]

$$m_t = 172.8 \pm 7.2_{\text{stat}} \pm 2.3_{\text{syst}}\ \text{GeV}. \tag{46}$$

The statistical power of these methods is not sufficient to yield significant contributions to the average from measurements at the Tevatron. Due to its independent and low systematics it will become more relevant at the LHC, though.

2.4 All hadronic channel

The full hadronic channel is the most difficult of all top quark pair decay modes. The channel has the advantage of the largest branching fraction and fully reconstructed events, but the huge background from multijet production makes it difficult to separate signal from background.

DØ has published a measurement of the top quark mass using the all hadronic decay channel in Run I using the template method [118], but not yet published a result with Run II.

2.4.1 Template method

CDF has applied a template method with in situ jet energy calibration to $2.9\ \mathrm{fb}^{-1}$ of data [119]. The event selection requires between six and eight jets and vetoes on large missing transverse momentum and on identified leptons in the event. To further suppress the huge multijet background without top quarks a neural network is employed which uses 13 input variables. These input variables include event shape observables like sphericity, the minimal and maximal two and three jet invariant masses and observables based on the transverse jet energies and the jet angles. Only events above a threshold value of the neural network output are accepted. The threshold has been optimised for the most precise top quark mass result and depends on the number of b tagged jets in the events. Finally, at least one of the six leading jets needs to be identified as b jets using CDFs secondary vertex tagger.

The background remaining after the selection is estimated from data. Jet by jet tag rate functions are measured in events with exactly four jets and parameterised as function of the jet transverse energy, the track multiplicity and number of vertices in the event. These jet tag rates are applied to the selected data with six to eight jets, but before requiring b tagged jets. Corrections for the fact that heavy flavour quarks are produced in pairs and for the signal content in the pre-tag samples are applied. This background model was verified on data with the neural network cut inverted.

The top quark mass is reconstructed in each selected event by performing a constrained kinematic fit to the six leading jets. The quark energies are allowed to vary within the experimental resolution of the measured jet momenta. The two pairs of light quark momenta are constrained to the W boson mass within the W boson width. The invariant mass of the two triples of quarks from the top and antitop quark decays are constrained to agree with the top quark mass within the nominal top quark width, which is assumed as $\Gamma_t = 1.5$ GeV. The top quark mass and the quark energies are free parameters of the fit. The corresponding fit χ^2 can thus be written as follows:

$$\begin{aligned}\chi^2 &= \frac{(M_{q\bar{q}}^{(1)} \quad M_W)^2}{\Gamma_W^2} + \frac{(M_{q\bar{q}}^{(2)} \quad M_W)^2}{\Gamma_W^2} \\ &\quad + \frac{(M_{bq\bar{q}}^{(1)} - m_t^{\text{reco}})^2}{\Gamma_t^2} + \frac{(M_{bq\bar{q}}^{(2)} - m_t^{\text{reco}})^2}{\Gamma_t^2} \\ &\quad + \sum_{i=1}^{6} \frac{(p_T^{i,q} - p_T^{i,\text{jet}})^2}{\sigma_i^2}.\end{aligned} \tag{47}$$

Of the possible jet to parton assignments required for the above fit, only those that assign b tagged jets to b quarks are used. With six jets this still allows thirty permutations for events with one identified b jet and six permutations in presence of two tagged jets. The top quark mass obtained from the jet–parton assignment that yields the lowest χ^2 is used as the reconstructed mass for the event under consideration. To obtain a handle on the jet energy scale, also the W mass is reconstructed in each event. This is done by repeating the above fit with also the W boson mass as a free parameter. Again the jet–parton assignment with the lowest χ^2 is chosen for further analysis.

The distribution of these mass values reconstructed in each of the data events is compared to templates for various nominal top quark masses. These templates are constructed by applying the above fitting procedure to events from top quark pair signal simulation for various nominal top quark masses and to the described background estimate. For the background estimate the reconstructed top quark and

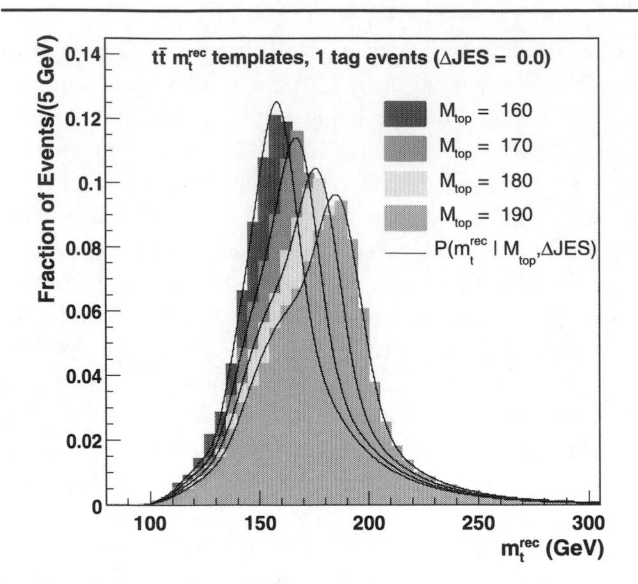

Fig. 18 Signal templates for various nominal top quark masses obtained for the sub-sample of events with exactly one b tagged jet. The lines correspond to the fitted functions used in the likelihood computation [119]

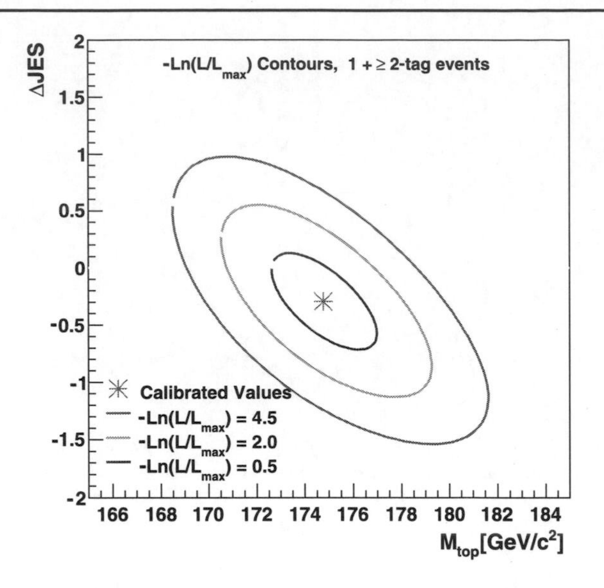

Fig. 19 Top quark mass and jet energy scale shift observed with the template method by CDF in 2.9 fb^{-1} using the all hadronic decay channel. The curves correspond to points of equal likelihood distance from the optimum [119]

W boson masses are entering the distributions with weights computed from the tagging probabilities and the corrections described above. Signal templates are created not only for various top quark mass values but also for a range of jet energy scale shifts, Δ_{JES}. To obtain smooth signal templates the distribution obtained at discrete values of the nominal top quark mass and jet energy scale shift are fitted by functional forms. See Fig. 18 for examples of such templates and their parametrisation.

Now the results obtained in data are compared to these templates with a likelihood that is maximised with respect to the top quark mass, m_t, the jet energy scale shift, Δ_{JES}, and the number of signal and background events. The likelihood consists of a term describing the agreement with the W boson mass templates and of terms for the subsamples containing one or more than one b tagged jets, $\mathcal{L}_1$ and $\mathcal{L}_2$:

$$\mathcal{L}(m_t, \Delta_{\mathrm{JES}}) = \exp\left(-\frac{\Delta_{\mathrm{JES}}^2}{2\sigma_c}\right)\mathcal{L}_1(m_t, \Delta_{\mathrm{JES}})\mathcal{L}_2(m_t, \Delta_{\mathrm{JES}}). \quad (48)$$

The individual terms are constructed very similar to the CDF template method lepton plus jets, see (12).

To estimate the performance of the method, the procedure is applied to ensembles of pseudo-data for various nominal values of the top quark mass and of the jet energy scale shift. The pseudo-data are constructed from the simulated signal and the data based background templates. Then the nominal top quark mass for an ensemble of pseudo-data is compared to the average of the results obtained for each of the pseudo-data in that ensemble. Similarly, a calibration curve for the jet energy scale shift is constructed. These calibration curves show excellent linearity and only very minor overall offsets. In addition the uncertainty estimate is verified with the width of the pull distribution and corrected accordingly.

Systematic uncertainties are estimated as the average effect determined on ensembles of pseudo-data with systematic effects included. The largest single uncertainty is estimated to stem from residual (mass and jet energy dependent) biases from the template parametrisation, not covered by the calibration procedure ($^{+0.8}_{-0.4}$ GeV). Residual effects of the jet energy scale follow with ± 0.5 GeV. This analysis also evaluates the effect of changing the underlying event model from a default [120] to a new model including colour reconnections [72–74] and finds an uncertainty of 0.4 GeV.

In 2.9 fb^{-1} of data CDF determines the top quark mass with in situ jet energy calibration from events in the all hadronic channel to be [119]

$$174.8 \pm 2.4_{\mathrm{stat+JES}}{}^{+1.2}_{-1.0}{}_{\mathrm{syst}}\ \mathrm{GeV}. \quad (49)$$

The measured two-dimensional likelihood is shown in Fig. 19. The jet energy scale shift determined is consistent with CDFs nominal value, but has a smaller total uncertainty.

2.4.2 *Ideogram method*

Also the ideogram method described for the case of lepton plus jet events in Sect. 2.2.2 is well suited to be applied in the all hadronic channel.

CDF applied this method of measuring the top quark mass in up to 1.9 fb^{-1} of data [121, 122]. The first steps of the analysis are very similar to the template based analysis for the all hadronic channel. The selection of events requires exactly six jets. A neural network is used to further suppress background from multijet production with no top quarks. For the signal sample at least two of the jets are required to be identified as b jets using CDFs secondary vertex algorithm.

The background contribution after this selection is computed by applying jet by jet tag rate functions on the data before the b tag requirement. The required tag rate functions are derived on events with four jets, where the signal contamination is negligible.

Then a constrained kinematic fit is performed to simultaneously determine the top quark mass and the W boson mass for the event under consideration. Transfer corrections are applied to correct differences between the jet energy inside the chosen cone radius of 0.4 and originating quark. These transfer corrections optionally implement a jet energy scale shift. The constraints of the fit are that the quark momenta yield identical top quark masses and identical W boson masses in the two decay chains of the event, cf. (47). The result for all possible jet to parton assignments which assign b tagged jets to (anti-)b quarks are kept for the further analysis.

With these numbers in the ideogram method an event-by-event likelihood is computed that describes the probability to observe the reconstructed top quark and W boson masses for the various jet to parton assignments given the true top quark mass and jet energy scale shift. This likelihood consists of probabilities, P_{sig} and P_{bkg}, corresponding to the signal and the background hypothesis, cf. (15). The signal and background probabilities obtained for each of the jet to parton assignment are weighted by their χ^2 probability, w_i, which depends on the jet energy scale applied. In addition CDF adds a term to the signal probabilities to describe events with no correct jet to parton assignment of the selected six jets:

$$\begin{aligned}
&P_{\text{sig}}(x; m_t, \Delta_{\text{JES}}) \\
&\quad = f_{\text{nm}} \sum_{i=1}^{N} w_i(\Delta_{\text{JES}}) S(m_{t,i}, M_{W,i}, \Delta_{\text{JES}}) \\
&\qquad + (1 - f_{\text{nm}}) S_{\text{nm}}(m_{t,i}, M_{W,i}, \Delta_{\text{JES}}) \\
&P_{\text{bkg}}(x; \Delta_{\text{JES}}) = \sum_{i=1}^{N} w_i(\Delta_{\text{JES}}) B(m_{t,i}, M_{W,i}).
\end{aligned} \tag{50}$$

N is the number of possible jet–parton assignments, i.e. 6 for doubly tagged events and 18 for events with three b tagged jets. Δ_{JES} the jet energy scale shift in units of the nominal jet energy scale uncertainty, $m_{t,i}$ and $M_{W,i}$ are the fit results for the jet–parton assignment number i and $1 - f_{\text{nm}}$ is the fraction of events with no correct jet–parton assignment.

The signal probability, S, is written as the convolution of the theoretically expected Breit–Wigner, **BW**, and the experimental smearing represented by a Gaussian, **G**. Here two Breit–Wigner functions are needed, one for the top quark and one for the W boson, and the Gaussian is two-dimensional, constructed including correlations between the top quark and the W boson mass extracted from the kinematic fit:

$$\begin{aligned}
&S(m_{t,i}, M_{W,i}, \Delta_{\text{JES}}) \\
&\quad = \int\int \mathbf{G}(m_{t,i}, M_{W,i}, m', M', \sigma_i) \\
&\qquad \times \mathbf{BW}(m', m_t)\mathbf{BW}(M', M_W)\, \mathrm{d}m'\mathrm{d}M'.
\end{aligned} \tag{51}$$

The background probability, B, and the probability for the signal events with no matching, S_{nm}, are derived from simulation. The final likelihood is then written as the product of the event likelihoods times a prior likelihood for the jet energy scale shift to describe the external calibration

$$\begin{aligned}
&\mathcal{L}(m_t, \Delta_{\text{JES}}) \\
&\quad = \exp\left(-\frac{\Delta_{\text{JES}}^2}{2\sigma_c}\right) \prod_{\text{events}} \mathcal{L}_{\text{evt}}(x; m_t, \Delta_{\text{JES}})
\end{aligned} \tag{52}$$

and maximised with respect to the top quark mass, m_t, and the jet energy scale shift, Δ_{JES}.

The performance of this method is evaluated in ensemble tests. Ensembles of pseudo-data are generated for a grid of nominal top quark masses and jet energy scale shifts using the PYTHIA generator and full CDF detector simulation and reconstruction. The mean response of the analysis on these ensembles is used to correct the top quark mass and jet energy scale shift obtained in data.

Systematic uncertainties are evaluated on ensembles of pseudo-data with the corresponding systematic variation. A variation of initial and final state parameters of the simulation yields the dominating contribution of ± 1.2 GeV, followed by the difference obtained from comparing PYTHIA to HERWIG (± 0.8 GeV). The residual jet energy scale contributes with ± 0.7 GeV.

In 1.9 fb^{-1} of data CDF measures the top quark mass from all hadronic events to be [122]

$$m_t = 165 \pm 4.4_{\text{stat+JES}} \pm 1.9_{\text{syst}}\ \text{GeV}. \tag{53}$$

Figure 20 shows the result with points of equal likelihood distance from the optimum in the top quark mass vs. jet energy scale shift plane.

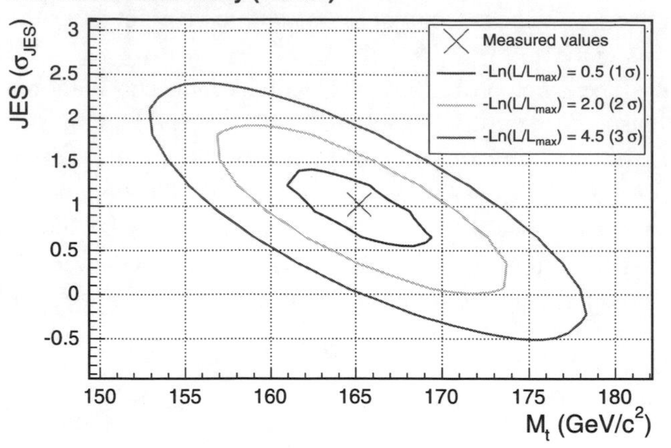

Fig. 20 Top quark mass and jet energy scale shift observed with the ideogram method by CDF in 1.9 fb^{-1} using the all hadronic decay channel. The curves correspond to points of equal likelihood distance from the optimum [122]

2.5 Top quark mass from cross-section

As discussed in Sect. 2.1 the definition of quark masses is inherently ambiguous. The quoted mass values are often considered to be given in the pole mass definition. However, this definition usually yields significant higher order corrections. For Monte Carlo simulations the order of the corrections included in the determination cannot be easily computed due to the presence of parton shower cut offs and modelling of hadronisation.

Deriving the top quark mass from the top quark cross-section measurements avoids using the simulation for calibration and allows the determination of the top quark mass using a well defined mass definition in an understandable approximation.

Both predicted and the measured cross-section depend on the true top quark mass. In the theoretical prediction this dependence stems from the change in phase space with varying top quark mass. For the experimental results this dependence is introduced through the selection efficiencies which vary with the amount of energy available for the decay products.

DØ has compared their cross-section measurement for up to 1 fb^{-1} of data to theoretical predictions as function of the top quark pole mass for various theoretical approximations [124–126].

For the extraction of the top quark mass the dependence of the theoretical predictions and the experimental cross-section results are parametrised with a polynomial. Both the theoretical and the experimental cross-section uncertainties are assumed to be Gaussian to build a likelihood that then allows to find the top quark pole mass that yields the best agreement and its uncertainty. This assumption is justified also for the theoretical uncertainty as it is dominated by PDF uncertainties.

The measured cross-section is compared to several theoretical predictions in Fig. 21: A pure NLO prediction [127], approximate next-to-next-to-leading order [34, 35] and a next-to-leading order with resummation of leading and next-to-leading soft logarithms [32]. All are evaluated with the CTEQ6.6 parton density distribution [128].

In this method statistical and systematic uncertainties are already combined. Using the combined cross-section in $\ell+$jets, dilepton and τ+lepton, DØ derives in 1 fb^{-1} [126]:

$$\begin{aligned}
&\text{NLO [127, 128]} && m_t = 165.5 \pm 6.0\ \text{GeV}\\
&\text{NNLO}_{\text{approx}}\ [34] && m_t = 169.1 \pm 5.6\ \text{GeV}\\
&\text{NNLO}_{\text{approx}}\ [35] && m_t = 168.2 \pm 5.6\ \text{GeV}\\
&\text{NLO} + \text{NLL [32]} && m_t = 167.5 \pm 5.7\ \text{GeV}.
\end{aligned} \tag{54}$$

The results on the top quark pole mass agree well with the world average of direct measurement, but have a much larger uncertainty. As these numbers refer to the same data, the differences between the results of about ± 2 GeV has to be attributed to theoretical differences and may be an indication of the sensitivity of the pole mass definition to higher order corrections.

Recently in a first determination of the $\overline{\text{MS}}$-mass of the top quark has been presented [129]. Based on the data of the same DØ cross-section result [126] a running top quark mass of $\bar{m}_t = 160.0 \pm 3.3$ GeV is extracted using the approximate NNLO prediction [34]. To compare this number to the pole mass result quoted above it has to be converted to the pole mass. This conversions strongly depends on the order used. In leading order no change occurs, while using the NNLO formula yields $m_t = 168.2 \pm 3.6$ GeV. (For simplicity the slightly asymmetric uncertainties were symmetrised by the author.) It should be noted that the $\overline{\text{MS}}$-mass determined from different orders of the $\overline{\text{MS}}$ cross-section calculation change by less than 1 GeV between leading order and

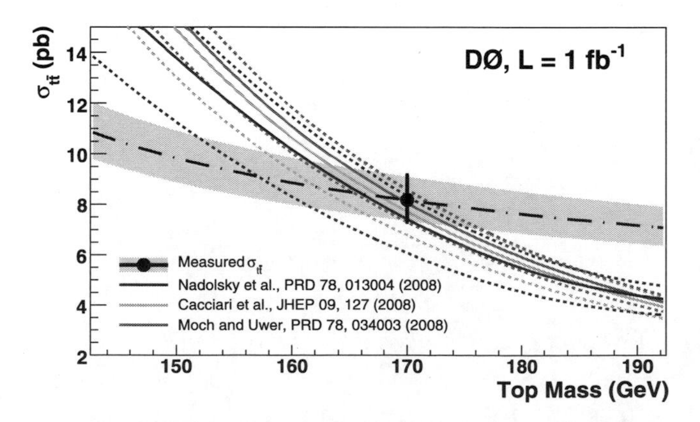

Fig. 21 Comparison of the theoretical top quark pair cross-section to the experimental results ($\ell+$jets, dilepton and $\tau+$lepton) with their top quark mass dependence. *The lines* show parametrised dependence and the error bands [123]

next-to-next-to-leading order, while the corresponding pole mass of the same order changes by nearly 10 GeV.

2.6 Modelling of non-perturbative effects

Modelling non-perturbative effects in simulations is a notoriously difficult task. Beside the effect of hadronisation, which is believed to be well understood since LEP times, in hadron-hadron collisions multiple parton interactions yield a different kind of non-perturbative effects. The uncertainties on the top quark mass due to their modelling uncertainties have only recently been considered.

In the context of hadron collisions at the Tevatron several efforts to tune the default model present in PYTHIA to data [120] resulted in several parameter tunes, known as Tune A, Tune DW, Tune D6, etc. These tunes required large values of parameters describing the colour correlations between the multiple parton interactions and the hard process. This is often interpreted as an implicit hint for the need of colour reconnections, i.e. a modification of the colour flow obtained from the simulation of the hard and semi-hard process and parton showers.

Newer versions of PYTHIA implement an extended model to describe the underlying event including variants of the PYTHIA model with explicit colour reconnection [72, 73]. These are tuned and their influence on the top quark mass measurements is studied with a toy top quark mass analysis [73, 74]. Compared to the variants of the default PYTHIA-model the new model and its variants yields shifts in the extracted top quark mass of up to 1 GeV. The greater portion, namely 0.7 GeV, of this shift can be assigned to differences between the different parton showers used in the old and the new underlying event models. Less than 0.5 GeV is attributed to non-perturbative effects. It should be noted here that additional alternative models of colour reconnection suitable for hadron collisions exists [130, 131], but were not yet studied in the context of the top quark mass.

While the smallness of the non-perturbative portion of the derived shift on the top quark mass confirms the prejudice that non-perturbative effects are expected at the order of (a few times) the confinement scale, Λ_{QCD}, the sizable shift due to parton shower details is very worrying. Only very recent top quark mass measurements include an uncertainty on the colour reconnection effects (see the previous subsections) and confirm the estimated uncertainty due to modelling of non-perturbative effects. The greater shift due to the parton shower differences is still not included.

It has recently been realised that the two types of models differ mainly in their b-quark jet energy scale and that the shape of b-jets does not agree with the models using the p_T-ordered parton shower [132, 133]. The simultaneous determination of the light quark jet energy scale and the b-quark jet energy scale as suggested by [134, 135] may thus help to resolve the issue.

2.7 Combination of top quark mass results

The Tevatron experiments regularly combine their Run I and Run II top quark mass results to take advantage of the increase in statistical power. The most recent combination was performed in July 2010 [90].

For each measurement that enters the combination a detailed break down of errors is performed. Uncertainties that are believed to have correlations of one measurement with any other measurement of the same or the other collaborations are separated so that these correlations can be taken into account in the combination. At this point only measurements from independent dataset or selections are used in the combination. The evaluation of partial correlations, as they would appear when using results form multiple methods on the same dataset and channel, are thereby avoided. The various uncertainty contributions are considered to be either fully correlated or to be uncorrelated.

The average is then computed with the best linear unbiased estimator (BLUE) [89] and yields [90]:

$$m_t = 173.3 \pm 1.1\ \mathrm{GeV}. \qquad (55)$$

Uncertainties on this estimation due to the approximations of the procedure including a cross-check with reduced correlation coefficients was estimated to be much smaller that 0.1 GeV.

2.8 Mass difference between top and antitop quark

According to the CPT-theorem local quantum field theories require the mass of any particle to be equal to that of its antiparticle. A measurement of the mass difference between top and antitop quarks is a unique test of the validity of the CPT-theorem in the quark sector, as other quark masses are more difficult to assess due to hadronisation. Both Tevatron experiments have extended their studies of the top quark mass to determine this mass difference.

2.8.1 Matrix Element method

DØ extends the Matrix Element method using 1 fb^{-1} of data [136]. The analysis selects events with one isolated lepton, transverse missing energy and exactly four jets. At least one of the jets is required to be identified as b-jet.

The Matrix Element method used to measure the top quark mass as described in Sect. 2.2.3 is extended by explicitly keeping a separate dependence on the top quark mass, m_t, and the antitop quark mass, $m_{\bar{t}}$ in (21) to (24). For this the leading order matrix element is rewritten to explicitly depend on m_t and $m_{\bar{t}}$. At the same time the dependence on the overall jet energy scale factor, f_{JES}, is dropped. This yields a likelihood $\mathcal{L}(m_t, m_{\bar{t}}, f_{\text{top}})$ corresponding to (24). The likelihood to observe a mass difference of $\Delta = m_t - m_{\bar{t}}$ is obtained by integrating over all possible average mass values $m_{\text{avg}} = (m_t + m_{\bar{t}})/2$.

$$\mathcal{L}(\Delta) = \int \mathrm{d}m_{\text{avg}}\, \mathcal{L}\big(m_t, m_{\bar{t}}, f_{\text{top}}^{\text{best}}(m_t, m_{\bar{t}})\big). \tag{56}$$

Here $f_{\text{top}}^{\text{best}}(m_t, m_{\bar{t}})$ is the fraction of top quark events fitted for the considered values of m_t and $m_{\bar{t}}$. The measured mass difference Δ is the one that maximises this likelihood.

As for the mass measurements the method is calibrated using pseudo-experiments. The pseudo-data in these tests are constructed from $t\bar{t}$ signal events and $W +$ jets background. For the signal various values of m_t and $m_{\bar{t}}$ were generated with a modified version of PYTHIA that allows to set $m_t \neq m_{\bar{t}}$. It was confirmed that the mass difference measured in the various pseudo datasets is very close to the generated value. The small deviation in the reconstructed value and the derived uncertainty are corrected for in the final result.

Many uncertainties that are important for the mass measurement cancel in the determination of the top antitop quark mass difference. This includes the uncertainties due to the jet energy scale and the uncertainty due to a difference in the detector response between light and b quark jets. Instead the dominating systematic uncertainty in the measurement of the mass difference is that of modelling additional jets in events of top quark pairs. This signal modelling uncertainty is estimated from data and simulation. Using simulation the amount of events with more than four jets is varied to agree with data. In addition, pseudo-experiments are performed in which appropriate data events with more than four jet are added to the standard signal simulation with various amounts. In combination DØ finds an uncertainty of 0.85 GeV due to signal modelling. Because of the necessity to distinguish top from antitop quarks also the uncertainty in the lepton charge determination and the uncertainty in modelling the differences of the calorimeter response for b and $\bar{b}$ jets is evaluated. The latter gives a sizable contribution of

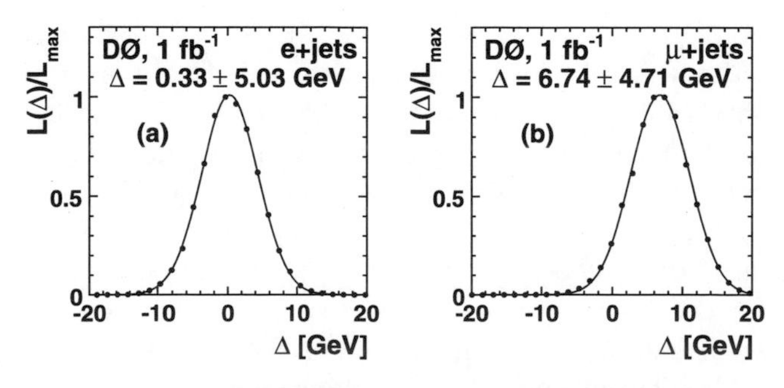

Fig. 22 Normalised likelihood distribution as function of the top to antitop quark mass difference, Δ, as measured by DØ in the electron and muon plus jets channels, separately [136]

0.4 GeV in the DØ result, but its determination is limited by the statistics of the corresponding simulation. Due to the negligence of multijet background in the likelihood computation and its calibration the multijet contamination yields another contribution of 0.4 GeV. The total systematic uncertainty is 1.2 GeV.

DØ applies the described method using 1 fb^{-1} of data in the $e +$ jets and the $\mu +$ jets channel separately. The corresponding likelihood functions are shown in Fig. 22. The two channels are then combined by a weighted average which yields

$$\Delta = 3.8 \pm 3.7\ \text{GeV} \tag{57}$$

in good agreement with the CPT-theorem expectation of zero [136]. As a cross-check the likelihood $\mathcal{L}(m_t, m_{\bar{t}}, f_{\text{top}}^{\text{best}}(m_t, m_{\bar{t}}))$ is integrated over all possible mass differences and used to determine the standard top quark mass. From this the relative difference is obtained to be

$$\Delta/m_t = (2.2 \pm 2.2)\%. \tag{58}$$

2.8.2 Template method

For the measurement of the top quark mass difference [137] CDF has expanded the template method with full reconstruction of the top quark decay products described in Sect. 2.3.1. The event selection requires one isolated energetic lepton, missing transverse energy and at least four energetic jets in the final state. Events are categorised by the charge sign of the lepton and the number of identified b-jets. Only for events with more than one identified b-jets more than four high p_T jets are allowed.

In each event the mass difference is reconstructed with a fit constrained by the top quark decay kinematics. The χ^2 for this fit is described by (11) when replacing m_t^{reco} with 172.5 GeV $\pm\ \Delta/2$. If the top quark decays hadronically and the antitop quark leptonically, the positive sign applies for term including M_{bqq} and the negative sign for the term including $M_{b\ell\nu}$. If the top quark decays leptonically instead, the signs are exchanged so that Δ represents the reconstructed mass difference in all cases: $\Delta = m_t - m_{\bar{t}}$.

The χ^2 is minimised for all possible associations of jets to the four quarks requiring that identified b-jets are only associated to b-quarks. For further analysis the values of $\Delta^{(1)}$ and $\Delta^{(2)}$ corresponding to the association with the smallest and the second smallest χ^2 are used. Events where the lowest χ^2 is larger than 3.0 (9.0) are rejected for events without (with) b-tagged jets.

Simulations are used to compute the expected signal contributions as function of the two reconstructed mass difference values for various nominal mass differences using MADGRAPH + PYTHIA. The contribution of multijet background is modelled using data with loosened lepton identification criteria. The distribution of mass differences reconstructed in W + jets background is modelled with ALPGEN + PYTHIA. Its normalisation is derived from data. Further smaller backgrounds are taken from simulation normalised to the NLO cross-sections.

As for the mass determination (Sect. 2.3.1) the signal simulations are used to derive a probability density, P^{sig}, to reconstruct a pair $\Delta^{(1)}$, $\Delta^{(2)}$ for a signal event with a given nominal mass difference. These two-dimensional functions depend on the lepton charge, which is the reason to separate samples of different lepton charge. Similarly a probability density for background events, P^{bkg}, is computed. Then a likelihood for the six subsamples, $i = 1, \ldots, 6$, is computed as

$$\mathcal{L}_i = \exp\left(-\frac{(b_i - b_i^{\text{e}})^2}{2\sigma_{b_i}^2}\right) \prod_{\text{events}} \frac{s_i P_i^{\text{sig}} + b_i P_i^{\text{bkg}}}{s_i + b_i} \tag{59}$$

with s_i and b_i being the number of signal and background events in the corresponding sample; b_i^{e} and σ_{b_i} are the corresponding background expectations and its error.

The final likelihood, i.e. the product of all $\mathcal{L}_i$, is available only at discrete values of the nominal mass difference. For the final maximisation of the likelihood it is interpolated with polynomial smoothing. CDF found their method to be bias free in pseudo-experiments. A small correction to the derived statistical uncertainty is applied.

Systematic uncertainties are dominated by the signal modelling uncertainty of 0.7 GeV, which is derived by comparing MADGRAPH to pure PYTHIA and by comparing the results of using the HERWIG to the PYTHIA parton shower. The next-to-dominating uncertainties stem from the description of multi hadron interactions and differences in the detector response for b and $\bar{b}$ quarks. The former is derived by verifying the agreement of the simulation with data as function of the number of primary vertices. The latter is determined by comparing the p_T balance in di-b-jet events. These effects yield systematic uncertainties of 0.4 GeV and 0.3 GeV, respectively.

Using 5.6 fb^{-1} of data CDF finds the top quark to antitop quark mass difference to be [137]

$$\Delta = -3.3 \pm 1.4_{\text{stat}} \pm 1.0_{\text{syst}} \tag{60}$$

which agrees with the CPT-theorem at a little less than two standard deviations. The analysis assumes explicitly an average (anti-)top quark mass of 172.5 GeV.

2.9 Conclusions and outlook to LHC

In the Standard Model of particle physics the top quark mass is an a priori unknown parameter. The Tevatron experiments have employed many different techniques to determine its value. Emphasis has been put on the reduction of the experimental uncertainties. The Ideogram and the Matrix Element methods aim at the maximal utilisation of the available experimental information. The simultaneous 'in situ' determination of the jet energy scale along with the top quark mass addresses the dominant systematic uncertainty. Also the next dominant uncertainty, the b-jet energy scale, can in principle be determined in situ, however, more statistics is needed to achieve an improvement over the current uncertainty.

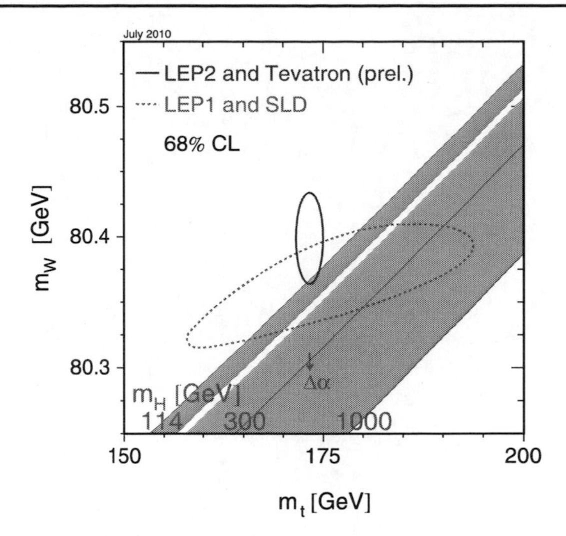

Fig. 23 (Colour online) Mass of the W boson vs. the top quark mass. The smaller (*blue*) contour indicates the one 68% C.L. of the direct measurements, the larger (*red*) contour the indirect results. The expectation in the Standard Model forms diagonal lines that depend on the Higgs boson mass [90]

These results constrain the Standard Model prediction for processes that involve top quarks either as real particles or in virtual loops. Besides the real production of top quarks, electroweak precision data are sensitive to the value of the top quark mass. Even before the discovery of the top quark the electroweak precision data constrained the top quark mass [138]. Now such indirect values are compared with the direct measurements at the Tevatron to verify the consistency of the Standard Model [90, 139]. Figure 23 shows the direct measurement of the W boson and the top quark mass compared to the indirect top quark mass from electroweak precision measurements. It indicates the Standard Model expectation as function of the Higgs mass. Also the Standard Model agreement with the global electroweak results is usually computed. Figure 24 shows the χ^2 fit as function of the Higgs boson mass which also determines the indirect constraints on the allowed Higgs boson mass.

Implicitly, such comparisons assume that the experimentally measured top quark mass values are given in the pole mass scheme. This assumption, however, is not confirmed to the level of the experimental precision of the combined result from direct top quark mass measurements. Therefore conclusions based on such comparisons need to be drawn

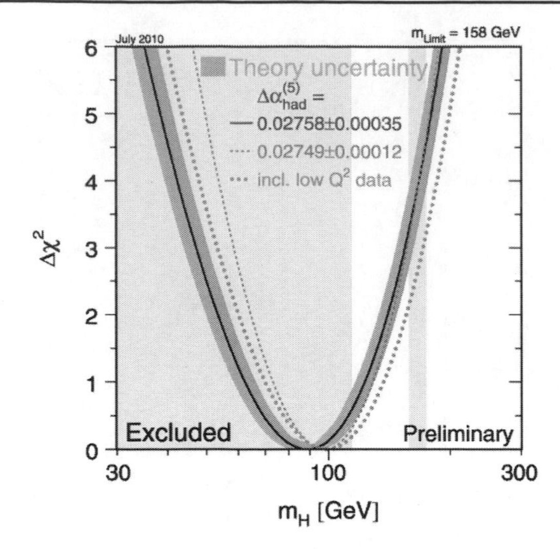

Fig. 24 (Colour online) Quality of the agreement of precision measurements with the Standard Model prediction as function of the assumed Higgs boson mass after a global fit of the parameters to data. *The yellow* region indicates the direct exclusion limit. The central result is shown as a *back line* with *blue band*. Alternative fits are shown in *other colours* [90]

with great care. Hopefully, in the future it will be possible to evaluate the exact top quark mass definition used in the Monte Carlos and used in the calibration of top quark mass measurements. Any bias in the current usage of the value can then be corrected for and the quoted experimental uncertainties can be used in these comparisons unchanged. Until the mass definition of the simulations used for calibration can be determined, it will be an important task to develop top quark mass measurements that are based on well defined mass schemes.

The LHC has started data taking in 2010 at an initial centre-of-mass energy of 7 TeV. First results show a tremendous quality of the data taken by the experiments and top quark pair production was already seen in about 3 pb^{-1} [140, 141]. With the planned LHC luminosity of 1 fb^{-1} the LHC experiments will become competitive to the Tevatron experiments within 2011. Systematic uncertainties on the top quark mass using the methods developed essentially at the Tevatron were expected to be below 2 GeV even before the initial data taking [142–144]. With the performance shown in the first year of running a result comparable to the current Tevatron result can be expected. The statistical uncertainty in these results will still be noticeable.

A truly negligible statistical uncertainty can only be expected after a successful LHC data taking at the design centre-of-mass energy of 14 TeV or close. This energy can be expected only after a long year shutdown. Only then the correspondingly enhanced production cross-section will open the door to methods applicable only at the LHC. The top quark mass determination involving leptonic J/ψ decays from the b-quark jet [145, 146] is expected to be less dependent on the jet energy scale allowing to cross-check the in situ methods. In production with very high transverse top quark momentum the top decay products are clearly separated from the antitop quark decay products. These events are theoretically better understood and allow a mass measurement in a well defined mass definition [63, 147]. At a design LHC this might resolve the puzzle of interpreting the current experimental top quark mass.

3 Interaction properties

The Standard Model fixes the properties of top quark for all three interactions considered in the Standard Model. To establish that the top quark discovered at the Tevatron is in fact the Standard Model top quark it is important to verify the expected properties experimentally and to set limits on possible deviations. This subsection will consider interaction properties, i.e. measurements of top quark properties and possible deviations from the Standard Model that do not assume explicit presence of non-Standard Model particles. First measurements of top quark properties regarding the weak interaction shall be described. Then the verification of the electrical charge is summarised followed by properties of the top quark production through the strong force. Measurements that involve non-Standard Model particles are covered in Sect. 4.

3.1 *W* boson helicity

One of the first properties of the electroweak interaction with top quarks that was measured is that of the W boson helicity states in top quark decays. Within the Standard Model top quarks decay into a W boson and a b quark. To check the expected $V-A$ structure of this weak decay the W boson helicity is investigated. Only left-handed particles are expected to couple to the W boson and thus the W boson can be either left-handed ($-$) or longitudinal (0). For the known b and t quark masses the fractions should be $f_- = 0.3$ and $f_0 = 0.7$, respectively. The right-handed ($+$) contribution is expected to be negligible.

Depending on the W boson helicity ($-$, 0, $+$) the charged lepton in the W boson decay prefers to align with the b quark direction, stay orthogonal or escape in the opposite direction. Several observables are sensitive to the helicity: the transverse momentum of the lepton, the lepton-b-quark invariant mass, M_{lb}, and the angle between the lepton and the b quark directions, $\cos\theta^*$. For best sensitivity at Tevatron energies $\cos\theta^*$ is measured in the W boson rest frame.

All of these observables have been used to measure the W boson helicity fractions at the Tevatron. The most recent and thus most precise results use $\cos\theta^*$ as the observable.

CDF

Analysis based on M_{lb}^2 The lepton-b-quark invariant mass, M_{lb}, was used in an analysis of approximately 700 pb^{-1} [148]. Events with a lepton plus jets signature containing an isolated lepton, missing transverse energy and at least 3 jets with identified b-jets were studied separately for one and two identified jets. In addition events with a dilepton signature containing two identified leptons with opposite electrical charge, missing transverse energy and at least two jets are investigated.

For each selected event the squared invariant lepton-b-quark mass, M_{lb}^2, is computed. In lepton plus jets events with a single identified b-jet the computation is unambiguous, however, the identified b-jet and the lepton are from the same top quark in only half of the cases, see Fig. 25 (left). For events with two identified b-jets a two-dimensional distribution of M_{lb}^2 is constructed. One dimension being M_{lb}^2 computed with the higher energetic b-jet, the other with the lower energetic b-jet. In the dilepton events the two-dimensional histogram is filled twice, i.e. using each of the leptons.

The contribution of background events in the lepton plus jets samples is modelled by ALPGEN $W+b\bar{b}$ events and by multijet events obtained from a control data sample. In the singly tagged sample multijet events contribute 15%, in the doubly tagged sample they are neglected. For the dilepton sample the background is described by about 50% $Z+$jets from ALPGEN, 30% $W+$jets with one jet misidentified as lepton using a single lepton sample and applying the misidentification rate. The final 20% of background are from diboson samples, WW and WZ.

The signal expectation is simulated with ALPGEN and PYTHIA assuming a top mass of $m_t = 175$ GeV for both $V-A$ and $V+A$ coupling to the W boson. A binned log likelihood fit is used to extract the fraction of $V+A$ coupling in the top decay, f_+. The likelihood uses Poisson probabilities to describe the expected number of events in each bin of the M_{lb}^2 distributions. The parameters of the Poisson distributions are taken from the described simulation and are smeared with nuisance parameters for top pair cross-section and the total background contribution in each of three samples. These nuisance parameters are constrained to their nominal values with Gaussian probability distribution.

The procedure was verified using a large number of pseudo-experiments with different nominal contributions from $V+A$ decays. The fit was found to be stable and unbiased. Systematic uncertainties are dominated by the uncertainty on the jet energy scale, followed by uncertainties of the background shape and normalisation as well as the limited Monte Carlo statistics.

The likelihood distributions obtained with this procedure are shown in Fig. 25 (right). The left-handed W boson fraction in the top quark decay is measured to be $f_+ = -0.02 \pm 0.07$ in agreement with the Standard Model expectation of a negligible contribution. The upper limit is computed using Bayesian statistics with a flat prior for f_+ between 0 and 1 and yields $f_+ < 0.09$ at 95% C.L. Results for individual samples differ by at most 1.8 standard deviations.

Analyses based on $\cos\theta^*$ The angle, $\cos\theta^*$, measured in the W boson rest frame between the lepton and the b quark direction yields calculable distributions of this angle for each of the possible W boson helicities $(-, 0, +)$:

$$\begin{aligned} d_0(\theta^*) &= \frac{3}{4}\left(1-\cos^2\theta^*\right), \\ d_\pm(\theta^*) &= \frac{3}{8}\left(1 \pm \cos\theta^*\right). \end{aligned} \tag{61}$$

Measuring $\cos\theta^*$ thus allows to reconstruct the contribution of each of these helicities in top quark decays.

CDF has used this observable in several analyses to measure the W helicity fractions in top decays [149, 150]. In [149] events with an isolated lepton, missing transverse energy and at least four jets including at least one identified b-jet of 1.9 fb^{-1} are investigated. Two different methods of reconstructing the full top pair kinematics are used to then compute the W boson rest frame and $\cos\theta^*$ for each individual event.

One method recovers the unmeasured neutrino momentum from the missing transverse energy and from solving

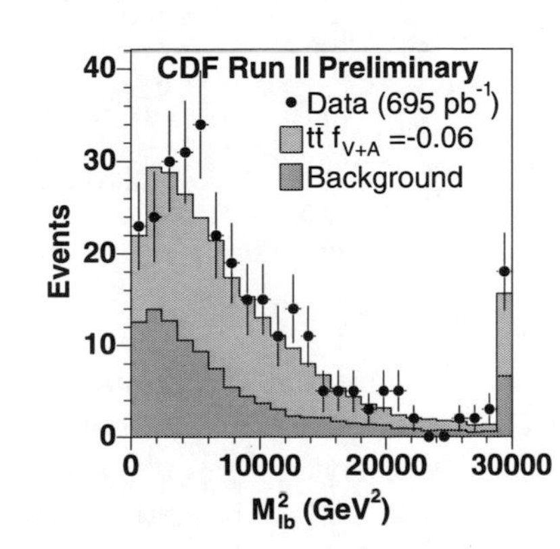

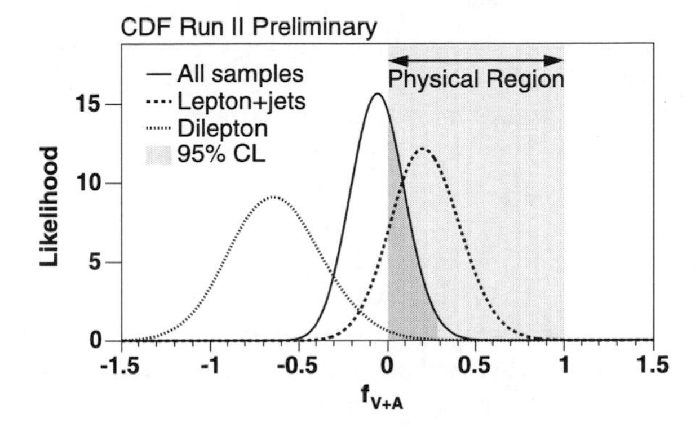

Fig. 25 (Colour online) Distribution of the squared lepton-b-quark invariant mass measured by CDF compared to best fit expectation (*left*). Likelihood distributions as function of the $V+A$ fraction f_{V+A} for the individual and the combined channels. *The shaded yellow* region indicates the allowed 95% confidence range (*right*) [148]

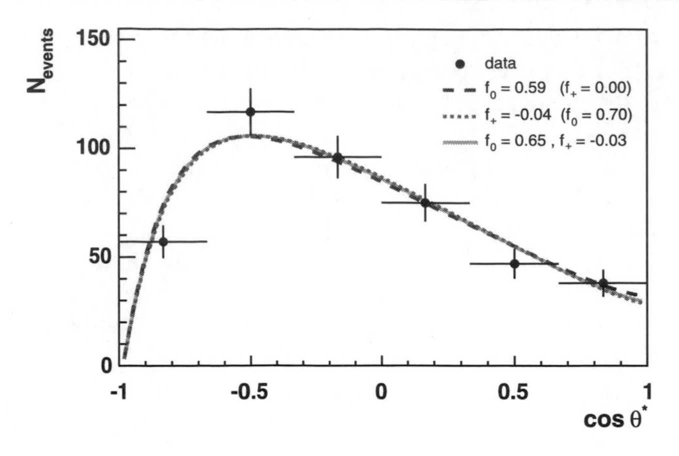

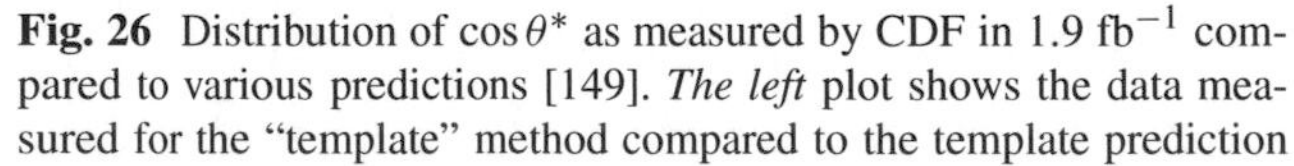

Fig. 26 Distribution of $\cos\theta^*$ as measured by CDF in 1.9 fb^{-1} compared to various predictions [149]. *The left* plot shows the data measured for the "template" method compared to the template prediction with the best fit parameters. *The right* plot shows the de-convoluted data of the "convolution" method compared to theoretical curves for individual helicity types and the best fit combination

the quadratic equation following from the $W \to \ell\nu$ decay kinematics when using the nominal W boson mass. This part of the analysis (for reasons that will become clear below) is called the "convolution" method. The other method (which will be called "template" method) uses a constrained kinematic fit to determine the lepton and parton momenta, where these momenta are allowed to float within the experimental uncertainties of the measured quantities. Constraints are built from the W boson mass and the equality of the top and antitop quark masses constructed from the fitted lepton and parton momenta. Both methods require an association of the measured jets to the partons of the top quark pair topology and use the quality of the constrained fit to select this assignment.

The reconstructed $\cos\theta^*$ distribution is used in two different likelihood fits to determine the helicity fractions f_0 and f_+ either individually, fixing the other one to the Standard Model value, or simultaneously. Both methods require simulation to construct their log likelihood function. PYTHIA and HERWIG are used to simulate Standard Model top quark pair production. Modified versions of HERWIG and MADEVENT + PYTHIA [151] are used to generate samples with varied helicity fractions. The W + jets background is simulated with ALPGEN + HERWIG and normalised to the amount of data before b-tagging and after removing all other background and $t\bar{t}$ signal events. Multijet background is taken from a control sample. Additional minor backgrounds from diboson, Z + jets and single top quark production are taken from simulation normalised according to their theoretical cross-sections.

In the "template" method the background expectations are combined with signal templates for the three different helicity states. The helicity fractions are taken from an unbinned likelihood fit with proper correction for acceptance effects. For this method an additional cut on the scalar sum of all transverse energy was required, $H_T > 250$ GeV.

The "convolution" method uses the signal simulation to derive acceptance functions which are then convoluted with the theoretically predicted number of events in each bin of $\cos\theta^*$. The helicity fractions are then taken from a binned likelihood fit after subtracting the background estimation from data.

Systematic uncertainties for both methods were determined from pseudo-datasets with templates modified according to the systematic effect under consideration. The jet energy scale uncertainty is among the dominating systematic effects in both methods. Only for the f_0 in the two-dimensional fits the initial/final state radiation uncertainties are more important. In general the "convolution" technique yields slightly larger systematic uncertainties. Data are compared to the estimated results for the optimal fit parameters in Fig. 26. For the "convolution" method (right) the data shown are de-convoluted and compared to the pure theory prediction.

The results of the methods are combined using the BLUE method [89] with the statistical correlation between the two methods determined in pseudo-experiments and the systematic uncertainties considered completely correlated. The combined results yields an improvement of about 10% compared to the individual methods. The model independent 2d fit yields

$$\begin{aligned} f_0 &= 0.66 \pm 0.16(\text{stat}) \pm 0.05(\text{syst}) \\ f_+ &= -0.03 \pm 0.06(\text{stat}) \pm 0.03(\text{syst}) \end{aligned} \quad (62)$$

with a correlation of -82% between f_0 and f_+. Upper limits on the positive helicity fraction, f_+ are not set.

In [150] CDF investigates the dilepton channel in 4.8 fb^{-1} to determine the W boson helicity fractions from $\cos\theta^*$. The selection requires two isolated leptons and at least two jets one of which is required to be identified as b-jet.

To reconstruct $\cos\theta^*$ the neutrino momenta are deduced by assuming the top quark pair decay kinematics with an intermediate W boson. Requiring that the momenta of the corresponding decay products are compatible with the W boson mass and a top quark mass of 175 GeV and using the missing transverse energy allows to determine the neutrino momenta up to an 8-fold ambiguity. To account for the experimental resolutions this determination is repeated with appropriately smeared input momenta. Of the eight (smeared) solutions the one that yields the smallest invariant $t\bar{t}$ mass is used. The distribution of measured reconstructed $\cos\theta^*$ values is compared to templates with a binned likelihood function. Templates for the signal are obtained for various values of f_0 and f_+ using a customised HERWIG generator. In addition templates for background processes of diboson, Drell–Yan and fake leptons are added to obtain the complete expectation.

Significant systematic uncertainties come from the jet energy scale and from the uncertainty of the modelling of initial and final state radiation. This special study is also suffering from limited template statistics.

In dilepton events of 4.8 fb^{-1} CDF determines the helicity fractions to be [150]

$$\begin{aligned} f_0 &= 0.78 \pm 0.20(\text{stat}) \pm 0.06(\text{syst}) \\ f_+ &= -0.11 \pm 0.10(\text{stat}) \pm 0.04(\text{syst}). \end{aligned} \quad (63)$$

Additional improvements are expected from relaxing the requirement of identified b-jets.

Analysis using the Matrix Element method In addition to the two analyses with explicit reconstruction of the top quark kinematics described above, CDF has performed an analysis that uses the Matrix Element technique to measure the longitudinal W boson helicity fraction, f_0, in 1.9 fb^{-1} of data [152, 153]. The analysis selects events with one isolated lepton, large missing transverse energy and at least four energetic jets including at least one jet identified as b-jet.

For each selected event a likelihood, $L(f_0, f_+|\{j\})$, to observe the measured quantities, $\{j\}$ (the lepton and jet momenta and the missing transverse energy), is computed as function of the W helicity fractions. This likelihood consists of a signal term which depends on f_0 and f_+, and a background term which is independent of these:

$$\begin{aligned} &L_i(f_0, f_+|\{j\}) \\ &\quad = f_{\text{top}} P_{t\bar{t}}(\{j\}; f_0, f_+) + (1 - f_{\text{top}}) P_{W+\text{jets}}(\{j\}). \end{aligned} \quad (64)$$

The signal and background probabilities, $P_{t\bar{t}}$ and $P_{W+\text{jets}}$, are computed by integrating the differential parton level signal and background cross-sections according to the leading order cross-sections for $q\bar{q} \to t\bar{t}$ and $W + 4$jet production, respectively. These integrations account for the experimental resolutions. The product of likelihoods for all selected events is evaluated for discrete values of f_0 and f_+. At each point the signal fraction, f_{top}, is chosen to minimise the total likelihood. The result for the longitudinal helicity fraction and its statistical uncertainty are taken from the minimum of the log likelihood and its change by 0.5 units.

The method is validated and calibrated using simulation for top pair signal and various background processes. Samples of various nominal f_0, f_+ values were created by reweighting the top pair events according to the expected $\cos\theta^*$ distribution. Applying the above method to various pseudo datasets with various nominal f_0, f_+ values yields a calibration curve. The observed slope of less than one is explained by the incomplete description of signal and background each with only a single leading order matrix element. The final results are corrected with this calibration curve.

The largest systematic uncertainty for this method stems from the uncertainty in simulation. It is estimated by checking the difference between PYTHIA and HERWIG. In addition, uncertainties due to initial and final state radiation, different PDF sets, the jet energy scale uncertainty and various other experimental effects are considered.

Using the described Matrix Element method for 2.7 fb^{-1} and assuming $m_t = 175$ GeV CDF finds

$$\begin{aligned} f_0 &= 0.88 \pm 0.11(\text{stat}) \pm 0.06(\text{syst}) \\ f_+ &= -0.15 \pm 0.07(\text{stat}) \pm 0.06(\text{syst}) \end{aligned} \quad (65)$$

with a correlation of -59%. A variation of the assumed top quark mass shifts the results for f_0 by 0.017 and for f_+ by -0.010 per $+1$ GeV of shift in the top quark mass from the central value [153].

DØ The DØ collaboration has investigated a total of 5.3 fb^{-1} to determine the W boson helicity fractions based on the reconstruction of the decay angle $\cos\theta^*$ in lepton plus jets and in dilepton events [154, 155]. In the lepton plus jets events the hadronic decay is utilised to measure $|\cos\theta^*|$ which helps to measure the longitudinal fraction, f_0.

Events are selected by requiring an isolated lepton, missing transverse energy and at least 4 jets. No second lepton is allowed in the event. Dilepton events are selected with two isolated charged leptons with opposite charge, large missing transverse energy and at least two jets. Some additional cuts are applied to suppress $Z \to \ell\ell$ events in the ee and $\mu\mu$ channels and to ensure a minimal transverse energy in the $e\mu$ channel. All channels use a multivariate likelihood discriminant based on kinematic observables and the neural network b-jet identification to improve the purity of the top quark pair signal.

In lepton plus jets events the decay kinematics of the top quark pairs is reconstructed using a constrained fit that determines the momenta of top quark and the W boson decay products from the measured jets and lepton momenta as well

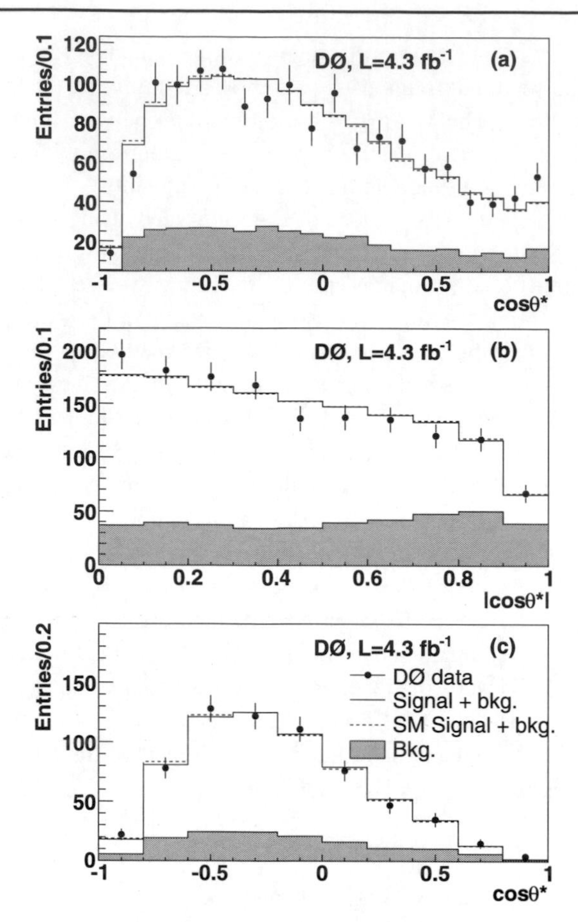

Fig. 27 *W* decay angle distributions as measured in 4.3 fb^{-1} of DØ Run IIb data. *The top* plot shows the distribution obtained from the leptonic decay in the lepton + jet events, *the middle* plot the one obtained from the hadronic decay. *The bottom* plot shows the distribution obtained from dilepton events. In addition to the data shown with error bars all plots contain the best fit prediction shown as full line and the Standard Model as a dashed line histogram. *The shaded* area represents the background contribution [155]

as from the missing transverse energy. Only the four leading jets in p_T are used. The fit requires the momenta of W boson decay products to be consistent with the nominal W boson mass and the momenta of the top quark decay products to yield a top quark mass of 172.5 GeV. Of the 12 possible jet–parton assignments the one with the highest probability of being the correct one is chosen. This probability for each association is computed using the fit χ^2 as well as the output value of the neural network b-tagger for the four jets and its consistency with the light or heavy quark assignment under consideration. From the chosen solution $\cos\theta^*$ is computed from the leptonic side and a second measurement of absolute value from the hadronic side, see Fig. 27 (top and middle).

The kinematics of dilepton events can be solved assuming the top quark mass with a fourfold ambiguity. In addition the two possible assignments of the two leading jets in p_T to the b quarks are considered. For each of these solutions the decay angle $\cos\theta^*$ is computed. To explore the full phase space consistent with the measurements, the measured jet and charged lepton momenta are fluctuated according to the detector resolution and $\cos\theta^*$ is computed for each fluctuation. The average over all solutions and all fluctuations is computed for each jet to find two $\cos\theta^*$ values per event. The resulting distribution for Run IIb data is shown in Fig. 27 (bottom).

The expected distribution for signal top pair events is simulated using ALPGEN + PYTHIA with various $V-A$ to $V+A$ ratios. These samples are then reweighted to form samples corresponding to the three W boson helicity states. Important backgrounds in the selected event sample are W + jets and multijet events in the lepton plus jets channel, WW + jets and Z + jets in the dilepton samples. Backgrounds of a single weak boson with jets are simulated with ALPGEN + PYTHIA, diboson samples are generated with PYTHIA. The multijet contribution is estimated from data for each bin of the $\cos\theta^*$ distribution.

From the data and the estimated signal and background contributions a binned likelihood, $\mathcal{L}(f_0, f_+)$, is computed for the observed data to be consistent with the sum of the backgrounds and the estimates for the three W boson helicity states. The normalisation of the background is kept as a nuisance parameter with a Gaussian constraint to its nominal value. The measured helicity fractions, f_0 and f_+, are those that minimise this likelihood.

Systematic uncertainties are evaluated in ensemble tests. Pseudo-datasets are drawn from models with systematic variations and compared to the standard templates to find the resulting shift in the obtained helicity fractions. Dominating uncertainties stem from the modelling of $t\bar{t}$, which are determined by exchanging the standard ALPGEN + PYTHIA with pure PYTHIA, HERWIG and MC@NLO. Further significant contributions to the systematic uncertainties stem from background modelling, from the limited template statistics, from jet energy scale, jet energy calibration and jet reconstruction. An uncertainty due to the top quark mass uncertainty of 1.4 GeV is also included.

Combining the datasets of Run IIa and Run IIb, which are analysed separately, DØ finds

$$\begin{aligned} f_0 &= 0.669 \pm 0.078(\text{stat}) \pm 0.065(\text{syst}) \\ f_+ &= 0.023 \pm 0.041(\text{stat}) \pm 0.034(\text{syst}). \end{aligned} \tag{66}$$

The correlation between the two numbers is about −0.8. The result shows a good consistency between the Run IIa and Run IIb datasets as well as between the dilepton and lepton plus jet channel. As for the other good agreement with the Standard Model expectation.

3.2 The CKM element V_{tb}

Another aspect of the weak coupling is that of flavour changing charged currents. The Standard Model explains these through the CKM matrix that needs to be determined from experiment. The elements V_{td} and V_{ts} of this matrix have been determined from experiment assuming the Standard Model and allow to infer $0.9990 < |V_{tb}| < 0.09992$ [6, 156] using unitarity of a 3×3 CKM matrix. Physics beyond the Standard Model may invalidate these assumptions and leave $|V_{tb}|$ unconstrained [156]. The value of $|V_{tb}|$ directly influences the single top quark production cross-section and the ratio between top quark decays to b-quarks ($t \to bW$) and to light quarks ($t \to qW$ with $q = d, s$). Both effects have been studied by the Tevatron experiments to constrain the CKM elements related to the top quark.

3.2.1 Single top quark

The single top quark cross-section from Standard Model sources is proportional to $|V_{tb}|^2$. Both experiments have used this relation to convert their cross-section measurements to determinations of the CKM element [157–159]. In addition to the uncertainty of the single top quark cross-section measurement, theoretical uncertainties on the Standard Model cross-section need to be taken into account.

The combination of the CDF and DØ measurements yields a cross-section for single top quark production of $2.76^{+0.58}_{-0.47}$ pb [159] assuming a top quark mass of 170 GeV in modelling the signal efficiencies. To extract the CKM matrix element it is assumed that $|V_{tb}|$ is much larger than $|V_{td}|$ and $|V_{ts}|$, so that the top quark decay is dominated by the decay to Wb and no significant production through d and s quarks in the initial state takes place. No assumption about the unitarity of the CKM matrix is made.

With theses assumptions the analyses performed to determine the cross-section can remain unchanged for the determination of $|V_{tb}|$. Attributing the full deviations of the experimental result from the SM prediction to the value of $|V_{tb}|$ the combined CDF and D0 single top quark production yields [159]:

$$|V_{tb}| = 0.88 \pm 0.07 \quad \text{or} \quad |V_{tb}| > 0.77 \quad \text{at } 95\% \text{ C.L.} \tag{67}$$

3.2.2 Top quark pairs

In top quark pair decays the number of identified b jets is used to measure the branching fraction for $t \to bW$. This fraction can be expressed in terms of the CKM matrix elements

$$R_b = \frac{|V_{tb}|^2}{|V_{td}|^2 + |V_{ts}|^2 + |V_{tb}|^2}, \tag{68}$$

assuming that the top quark decay is restricted to Standard Model quarks.

CDF CDF has investigated 160 pb^{-1} of data using both events with lepton plus jets and dilepton events [160]. For the lepton plus jets sample events are selected by requiring an isolated lepton, missing transverse momentum and four jets. Dilepton events consist of two charged leptons, missing transverse momentum and two jets. Both samples are classified according to the number of jets that are identified as b jets.

The background in the lepton plus jets sample is dominated by W + jets events and multijet events with fake electrons. The multijet background is estimated from data using a range of control samples. The W + jets background is simulated using the ALPGEN + HERWIG generator. In the subsamples with one or more than one identified b-jet, it is normalised using data before b-tagging. The fraction of heavy flavour in these samples is scaled according to the Monte Carlo to data ratio in a control sample of inclusive jet events. For the description of W + jets events without identified b jets, a neural network based on kinematic observables is used, which enriches W + jets background at low values and top quark pair signal at high values. The neural network was trained on ALPGEN + HERWIG samples. The distribution of neural network output values measured in data is compared to the simulation of W + jets and $t\bar{t}$ to fit the signal and background contribution. The shape of the multijet background is included at the rate determined above.

For the dilepton sample the main backgrounds stem from Drell–Yan, diboson and from W + jets events with fake leptons. The Drell–Yan background for ee and $\mu\mu$ is simulated using PYTHIA normalised to the number of Z bosons in a mass window around M_Z. Other electroweak backgrounds are fully taken from simulation. The W + jets backgrounds are taken from the ALPGEN+HERWIG simulation applying lepton fake rates, which are determined in a complementary jet sample. A tag rate probability for generic QCD jets is used to find the contribution of fake lepton events to the various b-tag subsamples.

The distribution of the number of b-tags for top quark pair events depends on the branching fraction, R_b. It is determined from events generated with PYTHIA and passed through full CDF detector simulation (as all simulations above).

Finally, a Poisson likelihood for the observed data to agree with the expectation is constructed as function of R_b. Gaussian functions with nuisance parameters are used to take systematic uncertainties including correlations between the samples and the b-tag bins into account. The dominant uncertainty comes from the background estimate in the 0-tag samples and the b quark identification efficiency.

In the analysed 160 fb^{-1} CDF finds $R_b = 1.12 \pm 0.2(\text{stat})^{+0.14}_{-0.13}(\text{syst})$ [160]. The Feldman–Cousins approach [6, 161] is used to compute a lower limit of

$$R_b > 0.61 \quad \text{or} \quad |V_{tb}| > 0.78 \quad \text{at } 95\% \text{ C.L.} \tag{69}$$

where the conversion to the limit on the CKM element is done assuming three generations and unitary of the CKM matrix, only.

DØ DØ has measured the top quark branching fraction, R_b, in conjunction with the top quark pair cross-section using 0.9 fb^{-1} [162]. Events are selected for the lepton plus jets channel requiring an isolated lepton, missing transverse momentum and at least three jets. In data b jets are identified using a neural network tagger.

Top quark pair signal is simulated with PYTHIA including samples in which one or both top quarks decay to a light quark and a W boson. The dominating W + jets background is simulated using ALPGEN + PYTHIA. Its heavy flavour content of the W + jets background was corrected according to a measurement in a control sample. The fake lepton background from multijet events is fully estimated from data. Additional smaller backgrounds from diboson, single top and Z + jets are simulated using PYTHIA, SINGLETOP and ALPGEN + PYTHIA, respectively, and normalised to their NLO cross-sections. All simulations are passed through the DØ detector simulation and reconstruction. In the simulation tag rate functions, determined on control samples in data, are used to describe the probability for a given jet to be identified as b jet. Figure 28 (left) illustrates the probability to have zero, one or more identified b jets in top quark pair events as function of the top quark branching fraction obtained from simulation.

The event samples are separated by lepton type, number of jets (3 or ≥ 4) and number of identified b jets (0, 1 or ≥ 2). The 0-b-tag sample with four or more jets is further split in bins of a topological likelihood discriminant to obtain additional separation between W + jets background and top quark pair signal events, cf. Fig. 28 (middle).

To simultaneously determine the top quark pair production cross-section, $\sigma_{t\bar{t}}$, and the branching fraction, R_b, a binned likelihood is constructed. Poisson distributions according to the expected event count as function of $\sigma_{t\bar{t}}$ and R_b is used for each sample and discriminant bin. The normalisation for W + jets is fixed globally by subtracting all other backgrounds as well as top quark pair estimates from data. Systematic uncertainties are included using nuisance parameters with Gaussian constraints.

For the determination of R_b the measurement is dominated by the statistical uncertainty. Systematic uncertainties are dominated by the uncertainty of the b tagging efficiency. In contrast to the CDF measurement, due to the global determination of the W + jets normalisation, the uncertainty on the size of this background is no significant source of uncertainty.

In 0.9 fb^{-1} of lepton plus jets data DØ obtains $R_b = 0.97^{+0.09}_{-0.08}(\text{total})$ [162] consistent with the expectation of the Standard Model. In Fig. 28 (right) the observed number of events is compared to expectations for various values of R_b. Limits on R_b obtained using the Feldman–Cousins procedure yield

$$R_b > 0.79 \quad \text{at } 95\% \text{ C.L.} \tag{70}$$

This limit is converted to a limit on the ratio of $|V_{tb}|^2$ to the off-diagonal elements:

$$\frac{|V_{tb}|^2}{|V_{td}|^2 + |V_{ts}|^2} > 3.8 \quad \text{at } 95\% \text{ C.L.} \tag{71}$$

The only assumption entering this limit is that top quarks cannot decay to quarks other than the known Standard

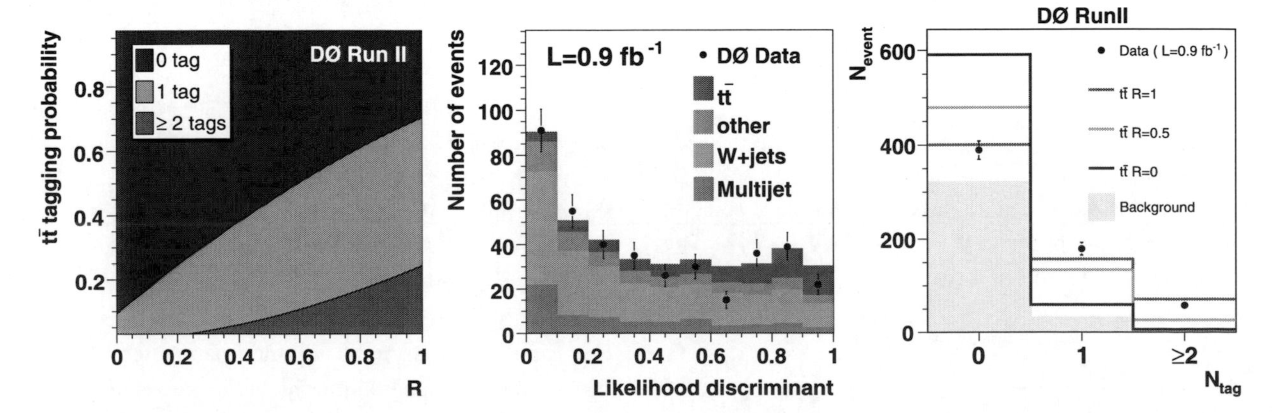

Fig. 28 *Left*: Probability to have zero, one or more identified b jets in top quark pair events as function of the top quark branching fraction, R_b. *Middle*: Data compared to contribution from various backgrounds as function of the topological likelihood discriminant. *Right*: Observed number of events as function of the number of identified b jets compared to expectations for various values of R_b [162]

Model quarks. Thus it is valid even in presence of an additional generation of quarks as long as the b' quark is heavy enough.

3.3 Flavour changing neutral currents

Flavour changing neutral currents (FCNC) do not appear in the SM at tree level and are suppressed in quantum loops [163–166]. However, anomalous couplings could lead to enhancements of FCNC in the top quark sector and their observation would be a clear sign of new physics [167, 168].

The Tevatron experiments have looked for FCNC both in top quark decays [169] and in the production of (single) top quarks [170–172]. Limits on the single top production through anomalous couplings were also set with LEP and HERA data [173–180].

3.3.1 Top quark decay through Z bosons

In an investigation of data with a total luminosity of 1.9 fb^{-1} CDF looks for top quark pairs that show a flavour changing neutral current decay through a Z boson [169]. The analysis aims to identify events in which the Z boson decays leptonically and the second top quark decays through a W boson into hadrons. The event selection thus looks for a pair of leptons and at least four jets. The leptons need to be of the same flavour and have opposite charge. Their invariant mass is required to be within 15 GeV of the Z boson mass. The cuts on the total transverse mass and the transverse energy of the leading and sub-leading jets were optimised in simulation. Events failing these cuts are used as control sample, events passing these cuts are split into events without any identified b jet and events with at least one identified b jet.

The dominant background with this selection stems from Standard Model $Z+$jets production, which is simulated using ALPGEN. Further but much smaller background contributions stem from Standard Model top quark pair production, and diboson production. The signal of top pairs with FCNC decay was simulated using PYTHIA. The events are reweighted to yield helicities of 65% longitudinal and 35% left-handed Z bosons.

To separate signal from background the mass of the W boson is reconstructed from two jets, the top quark mass is reconstructed by adding a third jet and a second top quark mass is reconstructed from the Z boson with the fourth jet. A χ^2 variable is built from the differences of the reconstructed masses to the nominal W boson and top quark masses, respectively. The χ^2 of the jet–parton assignment that yields the lowest χ^2 is used to build a distribution of χ^2 values.

The estimated shapes of the various backgrounds and the signal events are used as templates that are fitted to the distribution measured in data. The main parameters of the fit

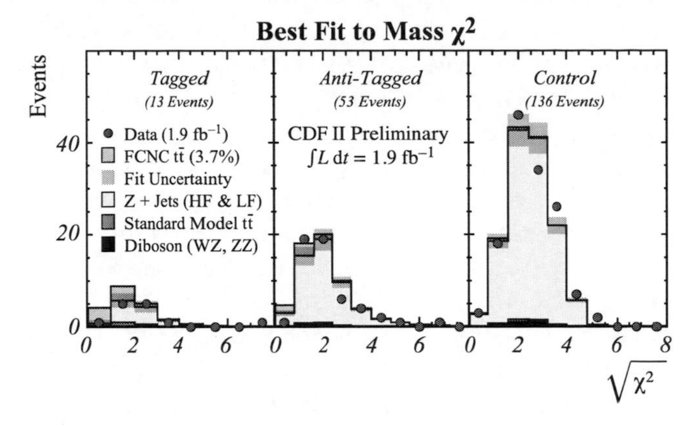

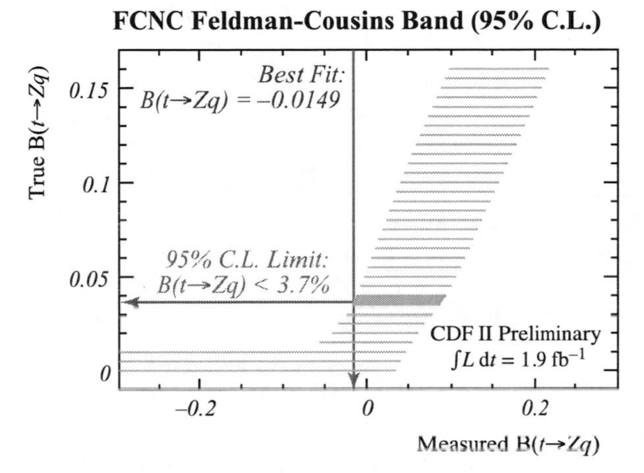

Fig. 29 *Top*: Mass χ^2 distribution observed in the two signal and one control sample compared the best fit expectation plus a signal of 3.7% branching to Z boson. *Bottom*: Feldman–Cousins band of 95% C.L. with the measured branching fraction $\mathcal{B}(t \to Zq)$ [169]

are the branching fraction, $\mathcal{B}(t \to Zq)$, and the normalisation of the dominating $Z+$jets background (in the control sample). Further parameters describe the difference of the background normalisation between the signal and the control samples (with a Gaussian constraint), the b quark identification fraction and the jet energy scale shift. The latter is considered to cover all shape changing affects.

The distribution of observed χ^2 values is shown with the best fit of the signal and background templates in Fig. 29. Data agree well with the Standard Model templates and thus limits on the branching fraction $t \to Zq$ are set. For this the Feldman–Cousins method is applied and yields $\mathcal{B}(t \to Zq) < 3.7\%$ at 95% C.L. [169] The Run I result in addition sets a limit on flavour changing neutral currents in the photon plus jet mode of $\mathcal{B}(t \to \gamma u) + \mathcal{B}(t \to \gamma c) < 3.2\%$ [181].

3.3.2 Anomalous single top quark production

While the above study of top quark decays addresses a flavour changing neutral current through the Z boson, investigations of the production of single top quark events can be used to restrict anomalous gluon couplings.

CDF In an analysis of 2.2 fb^{-1} CDF looks for the production of single top quarks without additional jets, $u(c)+g \rightarrow t$. To select such events with a leptonic top quark decay, one isolated lepton, transverse missing energy and exactly one hadronic jet are required. The jet must be identified as b jet. Additional cuts are used to reduce the backgrounds without a W boson as in the single top quark analyses [170].

To describe the expected background from Standard Model processes diboson and top quark pair events are simulated with PYTHIA and normalised to the NLO cross-sections. Single top quark events are simulated using MADGRAPH + PYTHIA [182, 183]. Finally, processes of weak vector bosons are simulated with ALPGEN + PYTHIA. In these samples the heavy flavour contribution is enhanced according to the findings in a control sample. The total normalisation of the $W+$jets samples is taken from sideband data. The signal of FCNC production of single top quark is simulated using TOPREX + PYTHIA [184].

Due to the large background from $W+$ 1jet data, a neural network is employed to differentiate between FCNC and Standard Model production. Fourteen observables, which each allow a significance of more than 3σ in discriminating signal and background, were chosen as inputs to the neural network. They utilise kinematical properties of the measured quantities and the reconstructed W boson as well as the output of a special flavour separation neural network. The neural network is trained on samples with equal amount of signal and background. It is then applied to the individual signal and background samples to obtain templates for all simulated physics processes considered, see Fig. 30 (left).

To determine the possible contribution of FCNC single top quark production the background templates are added according to their expected contribution and a binned maximum likelihood fit is used to measure the contribution due to FCNC production. Systematic uncertainties are parametrised in the likelihood function with Gaussian constraints. They are dominated by uncertainties on the cross-sections of the background samples normalised to NLO and the selection efficiency for signal events.

CDF finds no significant contribution of FCNC single top quark production in 2.2 fb^{-1} of data. The limit on the allowed production cross-section σ_t^{FCNC} is set using Bayesian statistics with a flat prior for positive cross-sections and yields $\sigma_t^{\text{FCNC}} < 1.8$ pb at 95% C.L. [185]. This cross-section limit is converted to limits on FCNC top quark–gluon coupling constants following [186, 187]. Assuming that only one of the couplings differs from the Standard Model expectation CDF finds $\kappa_{gtu}/\Lambda < 0.018\ \text{TeV}^{-1}$ or $\kappa_{gtc}/\Lambda < 0.069\ \text{TeV}^{-1}$. Expressed in terms of the top quark branching fraction through this processes these limits correspond to $\mathcal{B}(t \rightarrow u+g) < 3.9 \times 10^{-4}$ and $\mathcal{B}(t \rightarrow c+g) < 5.7 \times 10^{-3}$ as shown in Fig. 30 (middle and right). These small limits justify the approximation of pure Standard Model decays made in simulating signal samples above.

DØ DØ has set limits on the FCNC anomalous couplings of the top quark in up to 2.3 fb^{-1} of data [171, 172]. The analyses investigate the singly production of a top quark in association with at least one additional jet. The event selection requires an isolated charged lepton, missing transverse momentum and at least two jets. Exactly one of the jets has to be identified as b jet. With this the selection follows closely the selection used for the measurements of the single top quark production [157, 188–190].

In the most recent analysis the single top quark samples for the SM and the FCNC signal processes are simulated with the SINGLETOP generator [191, 192]. Background contributions from top quark pair, $W+$jets and $Z+$jets production are simulated using ALPGEN + PYTHIA and diboson production is simulated by PYTHIA. All samples are passed through GEANT to simulate the DØ detector and then reconstructed with the standard event reconstruction. The SM samples of single top quarks, top quark pairs, $Z+$jets and dibosons are normalised to the NLO (or bet-

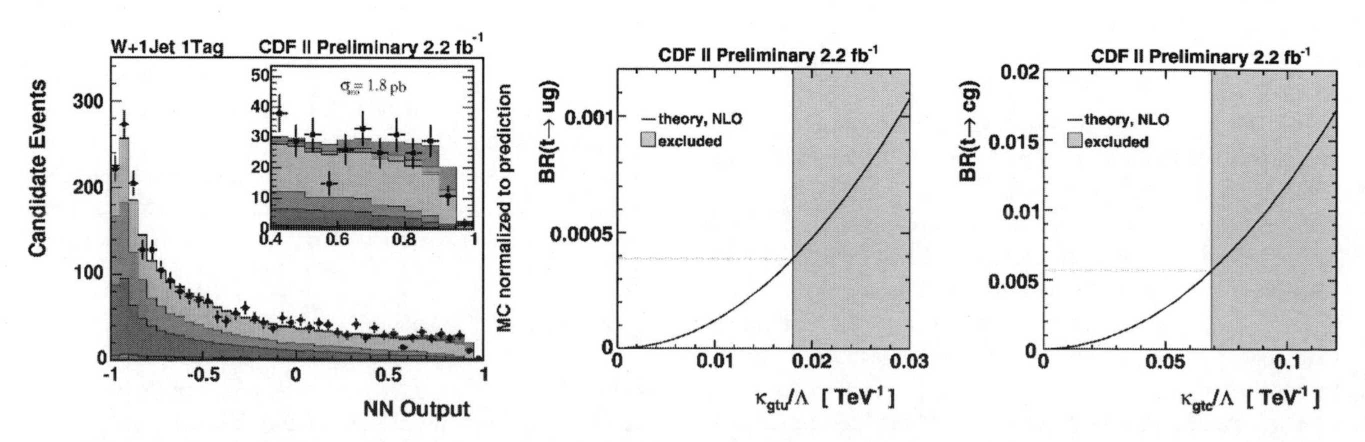

Fig. 30 Neural network output compared to 2.2 fb^{-1} of CDF data (*left*). Upper limits on the anomalous branching fractions derived from the limits on the anomalous couplings to u-quarks (*middle*) and c-quarks (*right*) [170, 185]

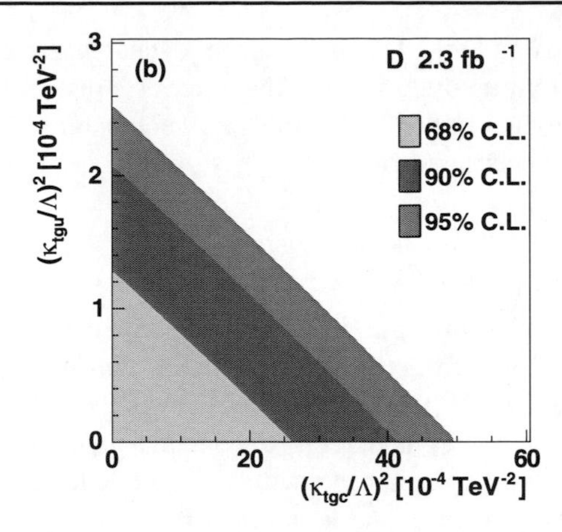

Fig. 31 Exclusion contour of the quadratic FCNC couplings obtained by DØ in a two-dimensional approach using 2.3 fb^{-1} [172]

ter) cross-sections. Events from W + jets production are normalised to data before b-tagging accounting for other simulated backgrounds and multijet background. The multijet background is described using data in which the lepton candidates fail one of the lepton identification cuts.

To separate FCNC from Standard Model prediction the method of a Bayesian neural network (BNN) is employed. A large number of variables is used as input to the neural network. These variables describe object and event kinematics, top quark reconstruction, jet width and angular correlations. For training the BNN the two FCNC processes including tgu and tgc couplings, respectively, are combined into a single training sample. Separate BNNs are trained for each lepton flavour and jet multiplicity. The distribution of BNN outputs observed in data agrees well with the pure SM expectation, thus limits on the allowed anomalous gluon couplings are computed.

These limits are computed using a Bayesian approach. A likelihood for the distribution of neural network outputs observed in data to occur is computed from the events expected in the Standard Model and the FCNC production of single top quark as function of the anomalous gluon couplings κ_{gtu}/Λ and κ_{gtc}/Λ. The likelihood for each bin is based on a Poisson distribution. Systematics are taken into account by smearing the Poisson parameters with a corresponding Gaussian distribution. The dominant uncertainties stem from shape changing effects like those from the jet energy scale and the modelling of b quark identification. In addition normalisation uncertainties for the background simulations and the overall luminosity uncertainty give a significant contribution.

The likelihood is folded with a prior flat in the FCNC cross-sections and exclusion contours are computed as contours of equal probability that contain 95% of the volume. The two-dimensional limits on the squared couplings observed by DØ in 2.3 fb^{-1} [172] are shown in Fig. 31. One-dimensional limits are obtained by integrating over one of the two anomalous couplings and yield $\kappa_{gtu}/\Lambda < 0.013$ TeV^{-1} and $\kappa_{gtc}/\Lambda < 0.057$ TeV^{-1}, corresponding to branching fractions of $\mathcal{B}(t \to u + g) < 2.0 \times 10^{-4}$ and $\mathcal{B}(t \to c + g) < 3.9 \times 10^{-3}$. These DØ results are the currently most stringent branching fraction limits.

3.4 Top quark charge

The top quark's electrical properties should be fixed by its charge. However, in reconstructing top quarks the charges of the selected objects are usually not checked. Thus an exotic charge value of $|q_t| = 4e/3$ is not excluded by standard analyses. Furthermore the amount of photon radiation off the top quark may a priori differ from the expectation based on the top quark charge. The Tevatron experiments have searched for deviations from the SM expectation in both aspects of the top quark charge.

3.4.1 Exotic top quark charge

To distinguish between the Standard Model and the exotic top quark charge it is necessary to reconstruct the charges of the top quark decay products, the W boson and the b quark. The W boson charge can be taken from the charge of the reconstructed lepton, but finding the charge of the b quark is more difficult.

DØ DØ has performed an analysis of ℓ + jets events with at least two b-tagged jets in 370 pb^{-1} using a jet charge technique to determine the charge of the b jets [193]. Semileptonic events are selected following the cross-section analysis by requiring exactly one isolated lepton, transverse missing energy and four or more jets. At least two of the jets must be identified as b jets using a secondary vertex tagging algorithm.

The charge of a jet can be defined as the sum of the charges of all tracks inside the cone of that jet. In this analysis the sum has been weighted with the component of the track momenta transverse to the jet momentum, $p_\perp$:

$$Q_{\text{jet}} := \frac{\sum q_i \cdot p_{\perp,i}^{\kappa}}{\sum p_{\perp,i}^{\kappa}} \quad \text{with } \kappa = 0.6 \tag{72}$$

where the sums run over all tracks, i, within the jet under consideration and q_i is the charge sign of the track i. Because particles may easily escape the jet cone such a jet charge fluctuates strongly from event to event, so only statistical statements can be made. It is crucial to determine the expected distribution of Q_{jet} in the case of b or $\bar{b}$ quark and, because a significant fraction of charm quarks gets flagged by the secondary vertex tagger, also for the c and $\bar{c}$ quarks. These expected distributions, cf. Fig. 32 (left), are derived from dijet data using a tag and probe method.

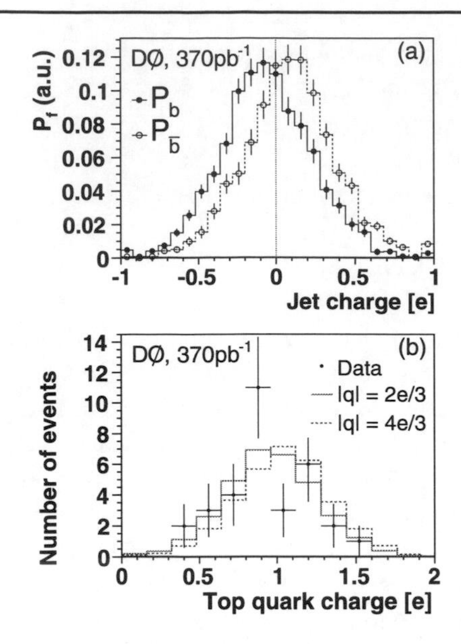

Fig. 32 Expected jet charge distribution for b and anti-b quarks (*left*). Measured absolute top quark charge compared to Standard Model and exotic models (*right*) [193]

To determine the top quark charge an assignment of b-jets to the leptonic or hadronic event side is necessary. This analysis uses the quality of a fit to the $t\bar{t}$ hypothesis, which uses the W boson and top quark masses as constraints, to select the best possible assignment. The jet charge for the b jet on the leptonic (hadronic) side, q_{b_l} (q_{b_h}) is then combined with the charge of the measured lepton q_l to define two top quark charge values per event: $Q_{\text{lep}} = |q_l + q_{b_l}|$ and $Q_{\text{had}} = |-q_l + q_{b_h}|$. The distribution of the measured top quark charges is compared to templates simulated for the Standard Model and the exotic case, where the exotic case has been obtained by inverting the jet charge, see Fig. 32 (right). The top quark pair events are simulated using ALPGEN + PYTHIA. The dominating W + jets background is simulated using ALPGEN + PYTHIA with a normalisation to data. Multijet templates are derived from data alone. All simulated events are passed through full DØ detector simulation.

An unbinned likelihood ratio accounting also for remaining background yields a p-value for the exotic case of 7.8% and a Bayes factor of $B_f = 4.3$ favouring the Standard Model charge scenario [193, 194].

CDF, jet charge The corresponding analysis by CDF investigates 1.5 fb^{-1} with events from the semileptonic and the dileptonic decay channel [195]. The former are selected requiring an isolated charged lepton, missing transverse momentum and at least four jets. Two of the jets are required to be identified b jets using CDFs secondary vertex algorithm. Dilepton events are selected by asking two oppositely charged leptons, missing transverse momentum and at least two jets, one of which needs be identified as b jet.

Compared to the DØ analysis the jet charge is computed slightly differently. Instead of the transverse momentum the scalar product of the jet and the track momentum is used to weigh the measured charges:

$$Q_{\text{jet}} := \frac{\sum q_i \cdot (\boldsymbol{p}_i \cdot \boldsymbol{p}_{\text{jet}})^\kappa}{\sum (\boldsymbol{p}_i \cdot \boldsymbol{p}_{\text{jet}})} \quad \text{with } \kappa = 0.5. \tag{73}$$

Depending on the sign of Q_{jet} the identified b jet is considered to stem from the b or the $\bar{b}$ quark. The purity of this assignment is calibrated on dijet events with two identified b jets. One of the jets is required to contain a muon that serves as tag in the tag and probe method. The resulting purity is corrected for effects due to muons from secondary decays, for B meson mixing and for light or c-quark jets misidentified as b jets.

To compute the charge of the top and antitop quark the jets need to be associated to the leptons. In the semileptonic channel a kinematic fit with constraints on the top quark mass and the W-boson mass is used. The jet–parton association with the lowest χ^2 of this fit is kept. In the dilepton channel the invariant mass of each pair of one lepton and one jet, M_{lb}^2, is computed. The combination which does not produce the largest value of M_{lb}^2 is used. In both channels cuts on χ^2 and M_{lb}^2, respectively, are used to enhance the purity of correct assignments.

Each event can now be classified as Standard Model like or as exotic model like. To obtain a statistical interpretation a likelihood is computed as function of the fraction of Standard Model like signal pairs, f_+. Nuisance parameters that represent the number and purity of signal and background events are optimised for each value of f_+. The systematic uncertainties considered include effects from the choice of parton density function, the uncertainties in the simulation of initial and final state radiation, the jet energy scale and the choice of the generator. All systematic uncertainties are included in the statistical treatment through their effect on the nuisance parameters.

Background predictions are obtained as for other CDF lepton plus jets and dilepton analyses based on a mixture of simulation and data. For this analysis each background is checked for a correlation between the charge of the signal lepton and the jet charge value of the corresponding b jet. Such a correlation could occur for the semileptonic channel from the $b\bar{b}$ background when the lepton from the b decay passes the lepton selection criteria and from single top quark events. In both cases the correlation is found to be small and consistent with zero within uncertainties.

In 1.5 fb^{-1} of data CDF finds the most likely value of the fraction of Standard Model like signal events to be $f_+ = 0.87$. This corresponds to a p-value of 31% [195], see also Fig. 33. Because this is larger than the a priori chosen limiting probability of 1% to falsely reject the Standard Model hypothesis, CDF claims to confirm the Standard Model hypotheses. The confidence limit corresponding

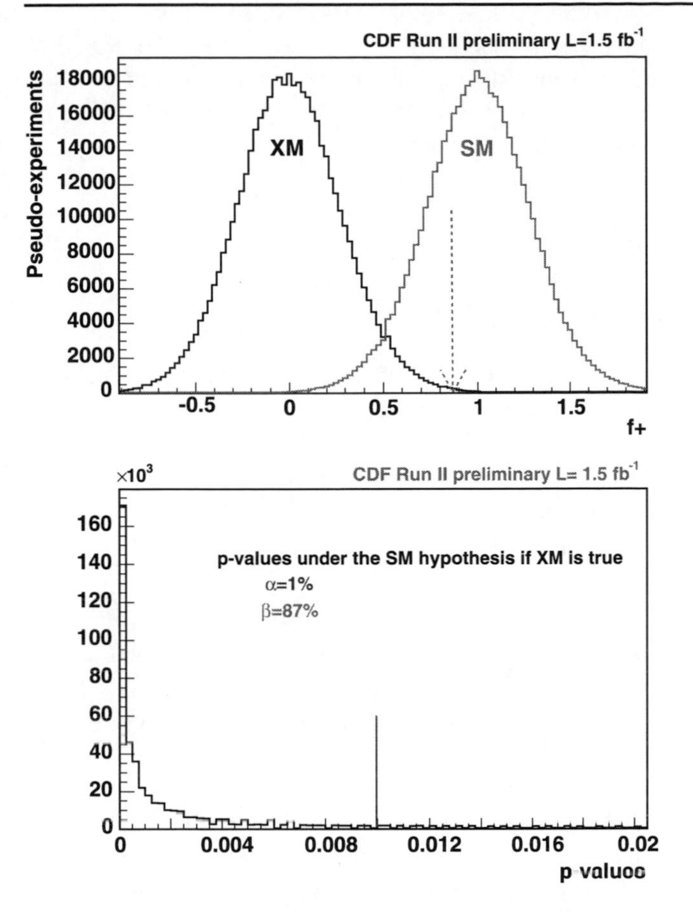

Fig. 33 *Top*: Distribution of the obtained fraction of Standard Model like pairs, f_+, in ensembles with the exotic and the Standard Model top quark charge. Indicated is the value measured in data: $f_+ = 0.87$. *Bottom*: Distribution of p-values for the Standard Model hypothesis to be true obtained in ensembles of the exotic model. Indicated is the a priori chosen limiting probability of incorrectly rejecting the Standard Model if Standard Model is true, 1%, and the corresponding probability of rejecting the Standard Model if the exotic model is true, 87% [195]

to this 1% choice is computed as 87%. CDF computes the Bayes factor to be $2\log(B_f) = 12$, which shows that this analysis with 1.5 fb^{-1} yields a much stronger exclusion of the exotic hypothesis than the DØ analysis of 370 pb^{-1} described above.

CDF, soft lepton tag With more data an alternative method to determine the charge of the b quarks is to concentrate on the leptonic decays of the b quarks. CDF applies this method on 2.7 fb^{-1} [196].

Events are selected requiring an isolated energetic lepton, missing transverse energy and at least four jets. Vetoes are applied against additional isolated energetic leptons, conversion electrons, cosmic muons, and Z bosons. In the four jets at least one soft lepton and one secondary vertex tag must be reconstructed. The soft lepton tag is optimised to select leptonic b quark decays, $b \to \ell \nu X$, but to suppress cascade decays, $b \to c \to \ell \nu X$. With this the charge of the b quark can be deduced from the measured soft lepton charge.

To resolve the ambiguities of assigning the measured jets and energetic leptons to the (anti-)top quark decay products a kinematic fitter is applied. For the computation of the top quark charge the solution with the best fit quality is used. Only solutions which assign the tagged jets to b quarks are considered. Events with a bad fit quality are rejected, if $\chi^2 < 9$ or $\chi^2 < 27$ for events with one or two tagged jet, respectively.

To estimated the expected performance of the analysis, signal events are generated with PYTHIA using EVTGEN [197] for the decays. Background from W +jet events are generated using ALPGEN + PYTHIA correcting the heavy flavour contribution by a factor derived in $W + 1$ jet events. These samples are normalised to the data before requiring any b tags. The contribution of diboson single top, Z +jets and Drell–Yan events is considered using simulation normalised to the theoretical or experimentally measured cross-sections. Contributions from multijet events with fake leptons are estimated from data. The total background estimation is $B = 2.4 \pm 0.8$ events. The purity of the charge determination is determined from these simulations and then calibrated in data using pure $b\bar{b}$ events.

In 2.7 fb^{-1} of data CDF finds a total of 45 events, $N_{\mathrm{SM}} = 29$ of which are reconstructed as SM like and $N_{\mathrm{XM}} = 16$ with the exotic charge value [196]. The statistical significance of this result is obtained by considering a large number of pseudo experiments. Figure 34 shows the expected distribution in terms of the asymmetry

$$A = \frac{1}{D_s} \frac{N_{\mathrm{SM}} - N_{\mathrm{XM}} + B D_b}{N_{\mathrm{SM}} + N_{\mathrm{XM}} - B} \tag{74}$$

where D_s and D_b are the dilution factors obtained in the calibration step for signal and background, respectively. The SM scenario corresponds to positive values, the exotic model to negative values. In 69% of the pseudo-experiments

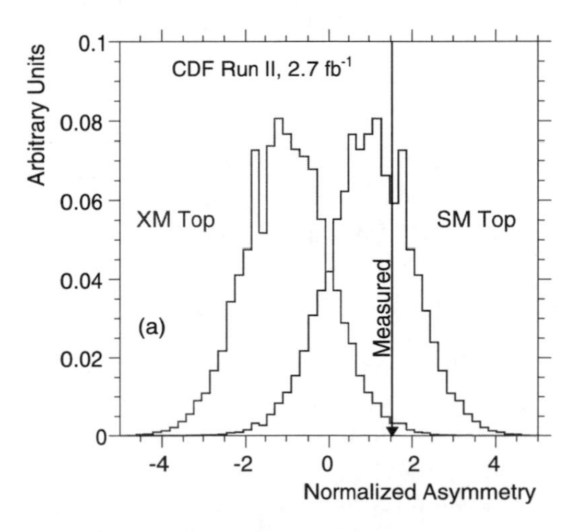

Fig. 34 Distribution of the asymmetry observed in a large number of pseudo experiments for the SM and the exotic charge case. The measured result is indicated by a *vertical line* [196]

assuming the SM the asymmetry turned out to be smaller than the measured value. Only in 0.9% of the pseudo experiments assuming a fully exotic top quark showed an asymmetry larger than the measured value. From these studies the exotic charge model can be excluded at 95% C.L.

3.4.2 Photon radiation of top quarks

The top quark charge is also expected to define the amount of photon radiation off the top quark. The measurement of this process is thus a complementary way of verifying the electrical charge of the top quark. The CDF collaboration has searched for top quark pairs with an additional photon in 1.9 fb^{-1} as part of a more general search [198].

For the search of the $t\bar{t}\gamma$ final state, events are selected that contain an isolated energetic photon, one isolated energetic lepton, large missing transverse energy and at least three jets. One of the jets is required to be identified as b-jet.

With this selection the dominant backgrounds stem from jets misidentified as photons and from light-jets misidentified as b-jets. The former is estimated from data by measuring the photon fake rate in jet events as function of $\not{E}_T$. This fake rate is then applied on events passing the selection without the photon requirement. Also the amount of fake b-jets is determined from data and then applied to the events without the requirement of a b-jet identification. Additional backgrounds including $W\gamma +$jet and diboson events are estimated using MADGRAPH + PYTHIA normalised to NLO cross-sections. MADGRAPH + PYTHIA passed through the full detector simulation and the reconstruction is also used to determine the signal selection efficiencies.

In 1.9 fb^{-1} of data CDF finds a total of 16 events which pass the described selection. The expectation including SM $t\bar{t}\gamma$ events is 11.2 ± 2.2 events. Attributing the full difference between the expectation and the data to the $t\bar{t}\gamma$ process, CDF measures the cross-section for radiation off top quark pair events of

$$\sigma_{t\bar{t}\gamma} = 0.15 \pm 0.08 \text{ pb} \qquad (75)$$

with the quoted uncertainty being dominated by the statistical uncertainty. The result is in agreement with the SM expectation of 0.080 ± 0.011 pb [198].

3.5 Spin correlations

At the level of the hard interaction the spins between the top and the antitop quark are correlated in top quark pair production. Because top quarks decay before they hadronise these correlations are conserved in the weak decay and thus "good" quantum-mechanic observables [199–203]. The degree of correlation depends on the production and decay processes. Its measurement thus probes the details of the production mechanism (when assuming a SM weak decay). The amount of correlation also depends on the reference axes used to define the top and antitop quark spin states. At the Tevatron the spin correlations expected in the SM are largest in the so-called off-diagonal and beam bases.

Both Tevatron experiments have measured the spin correlation of top quark pairs in different channels and using different spin bases.

3.5.1 Dilepton channel

In the dilepton channel both experiments use the normalised double differential cross-section with respect to the angle of flight directions of the two leptons as their observable. The angle of flight, $\cos\theta_\pm$, is measured with respect to the spin quantisation axis chosen in the analysis. This double differential cross-section depends on the spin correlation coefficient κ:

$$\frac{1}{\sigma}\frac{\mathrm{d}^2\sigma}{\mathrm{d}\cos\theta_+ \mathrm{d}\cos\theta_-} = \frac{1}{4}(1 + \kappa\cos\theta_+\cos\theta_-). \qquad (76)$$

In the off-diagonal and beam bases the SM predicts a correlation coefficient κ of about 0.8.

CDF The CDF analysis of the dilepton channel is based on 2.8 fb^{-1} [204]. The event selection requires two leptons of opposite charge, large missing transverse energy and at least two energetic jets. A veto is applied on Drell–Yan events and on special multijet fake configurations.

The analysis selects the off-diagonal basis for the definition of the flight directions. In this basis the spin quantisation axis is defined in the $t\bar{t}$ centre-of-mass frame. It is defined as the direction which deviates from the direction of flight of the (anti-)top quark by the angle ξ in clockwise direction. The value of ξ is computed as function of the top quark velocity, β, and the angle between the top quark and proton flight direction, θ^*:

$$\tan\xi = \sqrt{1-\beta^2}\tan\theta^*. \qquad (77)$$

The directions of flight with respect to the direction defined by ξ are measured in the rest frames of the top and the antitop quark. The analysis uses the angles obtained for the leptons ($\cos\theta_+, \cos\theta_-$) and the (anti-)$b$-quarks ($\cos\theta_b, \cos\theta_{\bar{b}}$).

To reconstruct these angles in an individual event a full kinematic reconstruction of the top quark pair and its decay to $l^+l^-\nu\bar{\nu}b\bar{b}$ is necessary. In the dilepton channel with its two neutrinos this requires the six constraints. The reconstructed W boson and the top quark masses need to be consistent with their nominal values (two constraints each). In addition the sum of the transverse momenta of the reconstructed neutrinos are required to agree with the measured missing transverse energy, $\not{E}_T$. Due to the quadratic nature of the corresponding equations these constraints yield up to four solutions for the unmeasured neutrino momenta, $\boldsymbol{p}_\nu$ and $\boldsymbol{p}_{\bar{\nu}}$.

To further improve the reconstruction a kinematic fit is applied that varies the reconstructed (anti-)b quark energies within the experimental resolution of the b jet measurements and the reconstructed sum of transverse neutrino momenta within the resolution of $\not{E}_T$. Besides these resolutions the kinematic fit includes the probability densities for the distribution of the p_z, the p_T and the invariant mass of the top quark pair in its likelihood. Of the initially up to four solutions for the neutrino momenta, $\boldsymbol{p}_\nu$ and $\boldsymbol{p}_{\bar{\nu}}$, the one which gives the best likelihood is used to compute the flight directions in each event.

Simulations are used to determine the expected outcome of this measurement as function of the spin correlation. For the $t\bar{t}$ signal simulation PYTHIA is used. Because the PYTHIA samples do not contain spin correlations between the generated top quarks, the generated events are weighted proportional to $1 + \kappa \cos\theta_+ \cos\theta_-$ according to the generated values of $\cos\theta_+$ and $\cos\theta_-$. Background of dibosons and Drell–Yan are simulated using PYTHIA and ALPGEN + PYTHIA, respectively. Background due to fake leptons is simulated from data with a single energetic lepton and jets. In these events one jet is artificially interpreted as a fake lepton. Both signal and background templates are smoothened by a polynomial function, where for the signal template the dependence on κ is kept.

The measured two-dimensional distributions of the reconstructed $(\cos\theta_+, \cos\theta_-)$ and $(\cos\theta_b, \cos\theta_{\bar{b}})$ values are now compared to the templates to determine the spin correlation coefficient κ using a likelihood fit. The measured value is corrected slightly according to a calibration performed on ensembles of pseudo-experiments.

In 2.8 fb^{-1} of dilepton top quark pair events CDF measures the spin correlation coefficient in the off-diagonal basis to be

$$\kappa = 0.32^{+0.55}_{-0.78}. \tag{78}$$

The total uncertainty is by far dominated by the statistical error. The leading systematics stems from the uncertainty on relative contribution of signal and background events. At the available statistics this results poses only a very minor constraint on the real spin correlations.

DØ The measurement of the spin correlation by DØ uses dilepton events of up to 4.2 fb^{-1} of data [205]. Events are selected requiring two oppositely charged leptons, large missing transverse energy and at least two jets. Drell–Yan events are vetoed near the Z boson resonance.

In this study the flight directions in (76) are computed using the beam basis. The spin quantisation axes for the beam basis are the directions of the proton and antiproton in the $t\bar{t}$ rest frame. The flight directions with respect to the quantisation axes are measured in the top and antitop quark rest frame, respectively. DØ uses the direction of flight angles for the two measured leptons, $\cos\theta_+$ and $\cos\theta_-$.

For the full reconstruction of the top quark pair kinematics DØ applies the Neutrino Weighting Method described in the context of top quark mass measurement in Sect. 2.3.1. For the spin correlation measurement the weight, W, is computed for a fixed top quark mass of 175 GeV, but scanning possible values of the flight angles to find its dependence on the product $\cos\theta_+ \cos\theta_-$, i.e. for each event $W = W(\cos\theta_+ \cos\theta_-)$. For further analysis the mean, μ, of the weight function in each event is used:

$$\mu = \int x W(x)\, \mathrm{d}x. \tag{79}$$

The distribution of μ values found in the selected events is now compared to templates obtained from simulation. Signal templates for a range of κ values are obtained by reweighting PYTHIA. Templates for the Z/γ events is generated with ALPGEN + PYTHIA, while for diboson processes PYTHIA is used alone. Both background types are normalised to the theoretical cross-sections. To enhance the bad simulation of the Z boson p_T distribution, a reweighting of these events has been applied. The measured spin correlation parameter κ is obtained by a binned likelihood fit of the templates to the measured data.

DØ has studied the performance of this method in ensembles of pseudo experiments using various nominal values of κ. This calibration is used in the Feldman–Cousins procedure [6, 161] to obtain the final results. Using dilepton events DØ determines the spin correlation coefficient in the beam axis

$$\kappa = -0.17^{+0.64}_{-0.53}. \tag{80}$$

The uncertainty is dominated by statistics. The leading systematic uncertainty is found to come from signal modelling. This includes a dependence on the assumed top quark mass as well as differences observed when replacing PYTHIA by ALPGEN + PYTHIA or MC@NLO. The observed spin correlation is in slight tension with the SM expectation of about 0.8.

3.5.2 Lepton plus jets channel

CDF has also determined top quark spin correlations in the semileptonic decay channel [206, 207]. In 5.3 fb^{-1} events with one energetic lepton, large missing transverse momentum and at least four jets are selected; at least one of the jets is required to be identified as b-quark jet.

This analysis uses the beam and the helicity bases. The spin quantisation axes are defined in the $t\bar{t}$ rest frame. For the beam basis the beam direction is used, for the helicity basis the (anti-)top quark directions are taken. Then the angles between the flight directions of the top quark decay products, $b\bar{b}\ell\nu q q'$, and the quantisation axis are reconstructed in the (anti-)top quark rest frame. The analysis considers the lepton direction, $\cos\theta_\ell$, the bottom quark from the hadronic top quark decay, $\cos\theta_b$, and the down-like quark, $\cos\theta_d$.

For the determination of these angles the full event kinematics of the top quark pair decay is reconstructed using a kinematic fitter which constrains the top quark mass to 172.5 GeV. Only events with a reasonable fit quality are considered in the analysis. In the following analysis the distributions of the cos of the reconstructed angles are compared to templates. For the reconstructed value of $\cos\theta_d$ the jet which is closest to the b-jet in the W rest frame is used. This is found to be the down-type quark in about 60% of the cases.

For the analysis of the angles in the helicity basis, samples of the four possible helicity combinations of the top and antitop quark are generated with a modified version of Herwig. From these four combinations, equal helicity and opposite helicity samples are obtained assuming parity and CP conservation. In the analysis of the angles with respect to the beam basis the signal samples are obtained by reweighting PYTHIA events according to the generated decay angles. In addition templates for the expected background are produced, including multijet, $W+$jets and diboson events.

To determine the spin correlation the two-dimensional distribution of $\cos\theta_\ell\cos\theta_d$ vs. $\cos\theta_\ell\cos\theta_b$ is fitted with the described templates using a binned likelihood fit. The contribution of the same spin and the opposite spin templates are allowed to float freely, the background contribution is allowed to float but constrained to the expectation within its errors. The determined fractional contribution of the opposite spin contribution, f_o, is converted to the spin correlation coefficient using $\kappa = 2f_o - 1$.

In 5.3 fb^{-1} CDF determines the spin correlation in the helicity and in the beam bases to be

$$\begin{aligned}\kappa_{\text{helicity}} &= 0.48 \pm 0.48_{\text{stat}} \pm 0.22_{\text{syst}} \\ \kappa_{\text{beam}} &= 0.72 \pm 0.64_{\text{stat}} \pm 0.26_{\text{syst}}.\end{aligned} \tag{81}$$

The measurement is clearly dominated by the limited statistical uncertainty. Of the systematic uncertainties the uncertainty on the signal modelling is by far dominating.

3.6 Charge forward–backward asymmetry

At the Tevatron the initial state of proton antiproton is not an eigenstate under charge conjugation. Thus in principle also the final state may change under this operation. In QCD, however, such a charge asymmetry appears only at next-to-leading order and arises mainly from interference between contributions symmetric and antisymmetric under the exchange of top and antitop quarks [208–212].

Experimentally, CDF and DØ investigated forward–backward asymmetries [213–216]

$$A_{\text{FB}} = \frac{N_{\text{F}} - N_{\text{B}}}{N_{\text{F}} + N_{\text{B}}} \tag{82}$$

where N_{F} and N_{B} are the number of events observed in the forward and backward direction, respectively. The forward and backward directions are either defined in the laboratory frame, i.e. according to the sign of the rapidity of the top quark, y_t, or can be defined in the frame where the top quark pair system rests along the beam axis, i.e. according to the sign of the rapidity difference between top and antitop quark, $\Delta y = y_t - y_{\bar{t}}$. The two different definitions of forward and backward yield two different asymmetries that are labelled $A_{\text{FB}}^{p\bar{p}}$ and $A_{\text{FB}}^{t\bar{t}}$ according to their rest frame of definition. In the Standard Model at NLO asymmetries are expected to be 0.05 and 0.08, respectively [217], but at NNLO significant corrections are predicted for the contributions from $t\bar{t}+X$ [218].

The smallness of the asymmetries expected within the Standard Model make them a sensitive probe for new physics.

CDF The CDF collaboration has investigated up to 5.3 fb^{-1} of data and measures both charge asymmetries defined above from top quark pairs with semileptonic decay [214, 216]. The event selection requires an isolated lepton, missing transverse energy and at least four hadronic jets, one of which must be identified as b jet.

The top and antitop quark kinematics are reconstructed from the jet momenta, the lepton momentum and the missing transverse momentum using mass constraints from the W boson and the top quark. The reconstructed values of these masses are constrained by the nominal values of $M_W = 80.4$ GeV and $m_t = 172.5$ GeV and b tagged jets are assigned to b quarks only [216]. Of the possible jet–parton assignments the one with the best fit probability is taken. The rapidity of the hadronically decayed (anti-)top quark, y_h, is multiplied by minus the charge, Q_ℓ, of the lepton to obtain the top quark rapidity: $y_t = -Q_\ell\, y_h$. For the $t\bar{t}$ frame asymmetry, $A_{\text{FB}}^{t\bar{t}}$ the rapidity difference is computed $\Delta y = y_t - y_{\bar{t}}$ as $\Delta y = Q_\ell\,(y_\ell - y_h)$, with Q_ℓ and y_h as above and y_ℓ being the rapidity of the leptonically decayed (anti-)top quark.

The NLO Standard Model expectation of top quark pair production is done with MCFM [219] and the next-to-leading order generator MC@NLO [220] which contain a small asymmetry. Leading order signal simulation without asymmetry from PYTHIA is used to check for any detector or selection asymmetry. The dominating background events

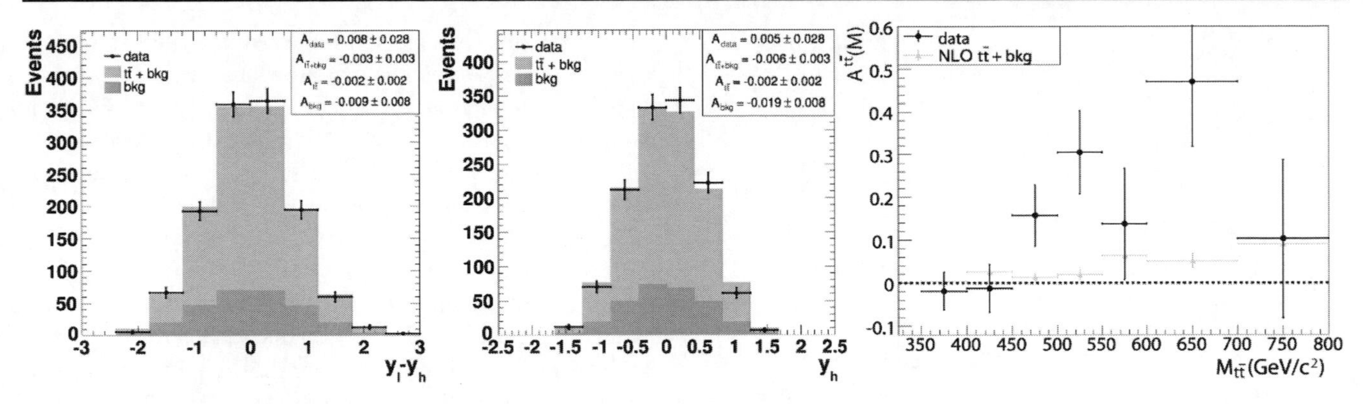

Fig. 35 Distribution of the top quark rapidity (*left*) and rapidity difference (*middle*) as measured by CDF in 5.3 fb^{-1} compared to the Standard Model prediction. *Right*: The functional dependence of the rapidity difference of the top quark pair invariant mass, $M_{t\bar{t}}$ [216]

of $W+\text{jets}$ are simulated with ALPGEN + PYTHIA, diboson backgrounds and single top quark events are simulated with PYTHIA and MADEVENT, respectively. The normalisation of the $W+\text{jets}$ background and contributions from misreconstructed multijet events are estimated from data.

The uncorrected rapidity and rapidity difference distributions measured in data are compared to the expectations in Fig. 35. These distributions differ from the true particle level shape due to acceptance and reconstruction effects. After background subtraction CDF derives the particle level distributions inverting the acceptance efficiencies and migration probability matrices as derived from PYTHIA simulation with zero asymmetry using a reduced number of only four bins. The final asymmetries are computed from these unfolded distributions.

The dominating systematic uncertainties are the background normalisation and shape. For $A_{\text{FB}}^{p\bar{p}}$ the amount of initial and final state radiation contributes significantly, while for $A_{\text{FB}}^{t\bar{t}}$ the jet energy scale is the next leading uncertainty. Further uncertainties from the parton distribution functions, due to colour reconnection and the MC generator are considered.

The final asymmetries measured in the update to 5.3 fb^{-1} of CDF data are [216]

$$\begin{aligned} A_{\text{FB}}^{p\bar{p}} &= 0.150 \pm 0.050_{\text{stat}} \pm 0.024_{\text{syst}} \\ A_{\text{FB}}^{t\bar{t}} &= 0.158 \pm 0.072_{\text{stat}} \pm 0.017_{\text{syst}}. \end{aligned} \tag{83}$$

These values are somewhat larger than the 0.038 and 0.058 expected in the Standard Model at NLO, respectively, but agree within two standard deviations.

CDF also investigated the dependence of the asymmetry on several other topological and kinematic properties. Considering the two ranges of Δy yields:

$$\begin{aligned} A_{\text{FB}}^{t\bar{t}}\big(|\Delta y| < 1.0\big) &= 0.03 \pm 0.12 \quad (\text{SM: } 0.39) \\ A_{\text{FB}}^{t\bar{t}}\big(|\Delta y| \geq 1.0\big) &= 0.61 \pm 0.26 \quad (\text{SM: } 0.123). \end{aligned} \tag{84}$$

Clearly, the deviation from the expectation is driven by the effect at large rapidity differences.

The functional dependence of the asymmetry on the invariant mass of the top quark pairs is shown in Fig. 35 (right). Separating this result in two bins yields

$$\begin{aligned} A_{\text{FB}}^{t\bar{t}}(M_{t\bar{t}} < 450\ \text{GeV}) &= -0.12 \pm 0.15 \quad (\text{SM: } 0.04) \\ A_{\text{FB}}^{t\bar{t}}(M_{t\bar{t}} \geq 450\ \text{GeV}) &= +0.48 \pm 0.11 \quad (\text{SM: } 0.09). \end{aligned} \tag{85}$$

The events with high invariant mass show a deviation 3.4 standard deviations from the NLO prediction obtained with MCFM.

CDF completes their study by verifying that the asymmetries are consistent with CP conservation by separately considering events with positively and negatively charged leptons. The slightly enhanced asymmetry observed in the inclusive measurement thus seems to stem from effects at large rapidity difference and high invariant mass of the top quark pair system.

DØ The DØ collaboration has investigated up to 4.3 fb^{-1} of data to measure $A_{\text{FB}}^{t\bar{t}}$ in semileptonic top quark pair events [213, 215]. The event selection requires exactly one isolated lepton, missing transverse momentum and at least four jets, the hardest of which must have $p_T > 40$ GeV. At least one of the jets is required to be identified as b jet with DØ's neural network tagger.

The top quark pair kinematics is reconstructed by fitting the momenta of the top quark decay products to the measured jet and lepton momenta and the missing transverse energy with constraints on the reconstructed W boson and top quark mass to 80.4 GeV and 172.5 GeV, respectively. Only the b jets and the remaining three leading jets are used. The possible jet–parton assignments are reduced by assigning identified b jets only to b quarks. In the final analysis only the assignment with the best fit probability is used. The rapidity difference with the correct sign is determined from the

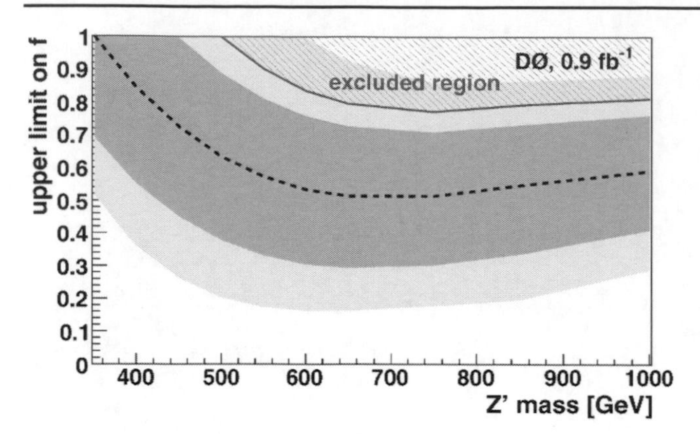

Fig. 36 Limits on a possible fraction, f, of resonant top quark pair production through a Z' boson obtained from the measurement of the forward–backward asymmetry in DØ [213]

rapidities reconstructed of the leptonic and the hadronic side (y_ℓ and y_h) and the lepton charge, Q_ℓ: $\Delta y = Q_\ell\,(y_\ell - y_h)$.

To estimate the dominant background of $W+$jets production a set of observables well described by the simulation is used to construct a likelihood discriminant that does not depend on Δy. The expected shape of top quark pair signal and the $W+$jets background in the distribution of the discriminant and on the asymmetry is determined from MC@NLO and ALPGEN+PYTHIA simulation, respectively, passed through DØ detector simulation and reconstruction. The effect of multijets events on the asymmetry and the discriminant is determined from data that fail the lepton identification. Other backgrounds were checked to have negligible effects.

The final reconstructed asymmetry, $A_{\mathrm{FB}}^{t\bar{t}}$, in signal events is determined by maximising the combined likelihood of the observed discriminant distribution and the distribution of the sign of Δy as function of the signal and background contributions and of the signal asymmetry. The dominant systematic uncertainties on the asymmetry are the jet energy calibration and the asymmetry reconstructed in $W+$jets events. All of them are much smaller than the statistical uncertainty.

In 4.3 fb^{-1} of data DØ finds a final observed asymmetry [215] of

$$A_{\mathrm{FB}}^{t\bar{t}\,\mathrm{obs}} = 0.08 \pm 0.04_{\mathrm{stat}} \pm 0.01_{\mathrm{syst}}. \tag{86}$$

To keep the result model independent and in contrast to the CDF results this number is not corrected for acceptance and resolution effects. Instead it needs to be compared to a theory prediction for the phase space region accepted in this analysis which is corrected for dilution effects. For NLO QCD and the cuts used in this analysis DØ evaluates $A_{\mathrm{FB}}^{t\bar{t}} = 0.01^{+0.02}_{-0.01}$. Thus as for CDF this result corresponds to an asymmetry that is slightly higher than expected in NLO QCD, but not by more than two standard deviations.

In addition to the QCD expectation in the published analysis [213] DØ provides a parameterised procedure to compute the asymmetry expected for an arbitrary model of new physics. As an example the measurement's sensitivity to top quark pair production via a heavy neutral boson, Z', with couplings proportional to that of the Standard Model Z boson is studied. PYTHIA is used to obtain a prediction of this kind of top quark pair production and due to the parity violating decay yields large observable asymmetries of 13 to 35% depending on the assumed Z' boson mass. Limits on the possible fraction of heavy Z' production are determined as function of the Z' boson mass using the Feldman–Cousins approach. These limits are shown in Fig. 36 and can be applied to wide Z' resonance by averaging the appropriate mass range.

3.7 Differential cross-section

Measurements of the differential cross-sections of top quark pair production can be used to verify the production mechanism assumed in the Standard Model. Due to the required unfolding these measurements are especially cumbersome. The CDF collaboration has measured the differential cross-section with respect to the invariant top quark pair mass, $\frac{\mathrm{d}\sigma_{t\bar{t}}}{\mathrm{d}M_{t\bar{t}}}(M_{t\bar{t}})$, using 2.7 fb^{-1} of data [221]. The event selection requests a lepton with high transverse momentum, large missing transverse momentum and at least four jets. At least one of the jets needs to be identified as b-jet.

The invariant mass of the top quark pairs is reconstructed from the four-momenta of the four leading jets in p_T, the four-momentum of the lepton and the missing transverse energy. The z-component of the neutrino is not reconstructed but used as if it was zero [222].

The dominating background in this selection stems from $W+$jets production. Its kinematics simulated with ALPGEN + PYTHIA correcting heavy flavour contribution for differences between data and Monte Carlo. The required normalisation is measured in data before applying the b-jet requirement [223]. Multijet background is extrapolated from data with low missing transverse momentum. The smaller backgrounds of diboson, $Z+$jets and single top quark is fully taken from simulation using PYTHIA, ALPGEN+PYTHIA and MADGRAPH, respectively. All simulated events are passed through the CDF detector simulation and reconstruction.

To obtain the differential cross-section from the background subtracted distribution of observed $M_{t\bar{t}}$ values, acceptance effects and smearing effects from the reconstruction need to be corrected for. The required acceptance correction is computed from signal simulation with PYTHIA. Factors to correct for differences between data and Monte Carlo observed in control samples are applied for the lepton identification and b-jet identification rates. The distortions of the reconstructed distribution are unfolded using the singular value decomposition [224] of the response matrix that is obtained from simulations.

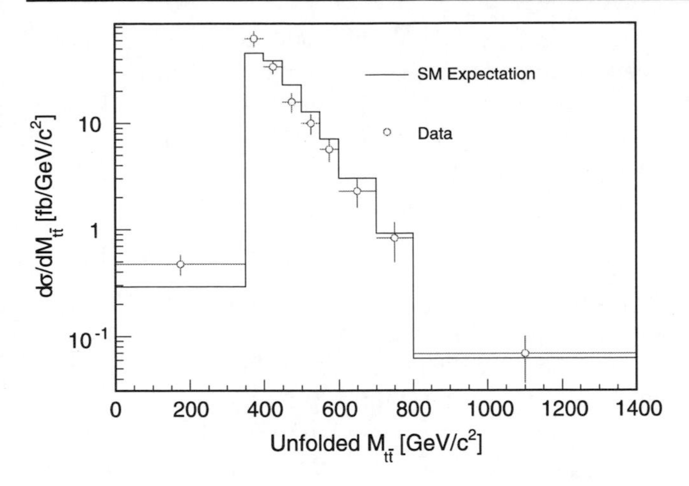

Fig. 37 Differential top quark pair production cross-section measured by CDF in 2.7 fb^{-1} of data using the semileptonic decay mode. Indicated are the total uncertainties for each bin, excluding the overall luminosity uncertainty of 6% [221]

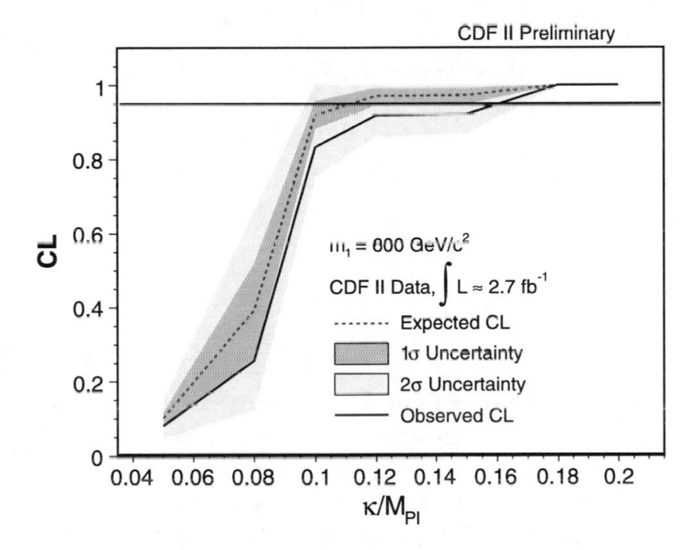

Fig. 38 Expected and observed limit on κ/M_{Pl} in a Randall–Sundrum model obtained from $\text{d}\sigma_{t\bar{t}}/\text{d}M_{t\bar{t}}$ [221, 222]

Relevant systematic uncertainties arise from the background normalisation, the acceptance, parton density distributions, the used Monte Carlo generator and jet energy scale. The relative contributions of the uncertainties strongly depend on $M_{t\bar{t}}$. To reduce the uncertainty on the jet energy scale an in situ calibration of the jet energy scale is performed. This uses the invariant dijet mass reconstructed from the two non-b-tagged jets most consistent with M_W.

The differential cross-section obtained in 2.7 fb^{-1} of data using the semileptonic decay mode is shown in Fig. 37 [221]. The consistency with the Standard Model expectation is computed using Anderson–Darling statistics [225]. The observed p-value is 0.28, showing good agreement with the Standard Model.

Finding no evidence for physics beyond the Standard Model limits on gravitons in a Randall–Sundrum model [226] decaying to top quarks are set using the CL_s method [227, 228]. Signal is modelled with MADGRAPH + PYTHIA assuming a first resonance with a mass of 600 GeV. The Anderson–Darling statistics is used as test statistics in the CL_s method. For the ratio of the warping parameter over the Planck mass CDF finds $\kappa/M_{\text{Pl}} < 0.16$ at 95% C.L, see Fig. 38.

The invariant top quark pair mass was used in further analyses by both CDF and DØ to search for new physics. These results are described in Sect. 4.4.

3.8 Gluon production vs. quark production

Top pair production at the Tevatron with $\sqrt{s} = 1.96$ TeV takes place either through quark antiquark annihilation or through gluon fusion. The former is expected to dominate with the gluon fusion contributing about 15%. Due to the large uncertainties of the large-x gluon density in the proton the exact size of the gluon contribution is rather uncertain [30, 31, 34].

Two properties of the two production processes allow to separate them and to measure their relative contributions. Close to threshold the spin states of the gluon fusion are $J = 0$, $J_z = 0$, while the $q\bar{q}$ annihilation yields $J = 1$, $J_z = \pm 1$ [229]. This yields angular correlations between the charged leptons in the dilepton channel. Alternatively one can exploit the difference in the amount of gluon radiation from quarks and gluons: The gluon fusion processes are expected to contain more particles from initial state radiation. CDF has used both features to measure the gluon fraction of top quark pair production.

3.8.1 Angular correlation methods

Dilepton channel CDF has investigated 2.0 fb^{-1} of data with an event signature of top quark pair dilepton events [230]. The selection requires two oppositely charged leptons, at least one of which must be isolated, and at least two jets. The scalar sum of the lepton and jets transverse energies must exceed 200 GeV. Additional cuts are placed to reject cosmic particles, leptons from photon conversion and Z boson events.

The azimuthal angle between the two leptons is measured in each event. Then a template method is used to measure the fractional contribution of the different production mechanisms. The expected behaviour of signal events is simulated using HERWIG with the top quark mass set to $m_t = 175$ GeV and the CTEQ5L parton distribution function. PYTHIA and MC@NLO are used in systematic studies. Backgrounds are dominated by diboson production and Z boson events with tauonic decay. These are simulated by PYTHIA. In addition the background from events with only one true lepton and

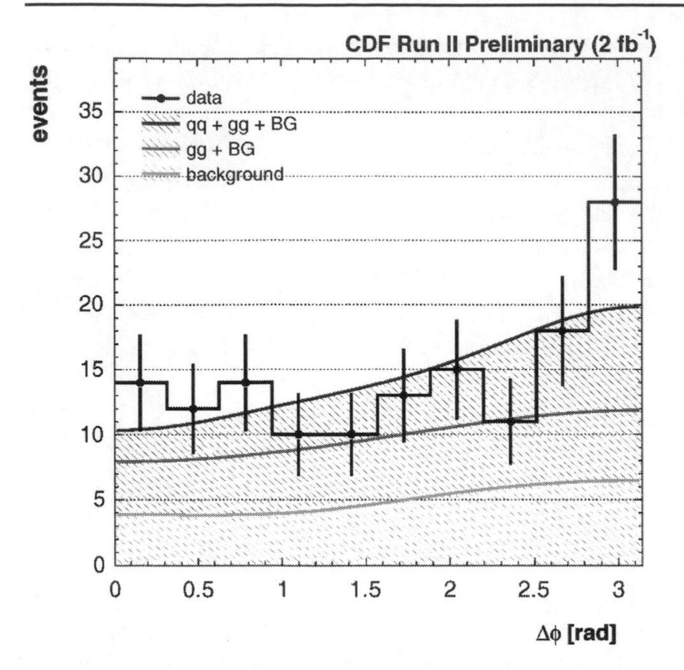

Fig. 39 Distribution of azimuthal angle between the two leptons, $\Delta\phi$, observed in 2.0 fb^{-1} of CDF data, compared to the best fit template curves [230]

a jet misidentified as lepton are described using data. All simulated events are passed through the full CDF detector simulation and reconstruction.

The angular distributions obtained for events produced by $q\bar{q}$ annihilation and gg fusion and the sum of backgrounds are separately fitted with smooth functions, which then serve as signal and background templates.

The measured fraction of top quark pairs produced through gluon fusion is obtained from an unbinned likelihood fit of these templates to the observed data, cf. Fig. 39 (left). Systematic uncertainties include uncertainties on the template shapes, the acceptance differences between $q\bar{q}$ annihilation and gluon fusion, the used matrix element, initial and final state radiation and PDF uncertainties. The uncertainties are determined as function of the nominal gluon production fraction and several of them may contribute up to 10%. All systematic uncertainties are included in the determination of the Feldman–Cousins band, see Fig. 39 (right), which is used to obtain the final result with errors.

In the investigated 2.0 fb^{-1} of dilepton events CDF obtains a gluon fusion fraction of 0.53 ± 0.37 [230]. The total uncertainty is dominated by statistical uncertainties and is not yet able to restrict the theoretical uncertainties on the gluon fusion production.

Lepton plus jets channel Angular correlations are also used by CDF in an analysis of 0.96 fb^{-1} with lepton plus jets events [231]. The events are required to contain one energetic lepton, large missing transverse energy and at least four jets. One of the jets is required to be identified as b jet through the presence of a secondary vertex.

In each event the decay chain of the top quark pair decay is reconstructed from the four leading jets using a kinematic fit with constraints on the W boson and the top quark mass, cf. Sect. 2.3.1. Only jet–parton assignments which associate the tagged b jet(s) with a b quark are considered. The one with the best fit quality is used for further analysis.

From the reconstructed top quark pair decay, eight observables are used to feed a neural network that is trained to distinguish between $q\bar{q}$ and gg production. The observables are the (cosine of the) angle between the top quark momentum and the beam direction of the incoming proton, the top quark velocity and the (cosines of the) six angles of the top quark decay product as defined in the off-diagonal spin basis, cf. Sect. 3.5. The first two observables are contributing about one third of the total sensitivity, each. The six decay angles yield the remaining third.

The neural network is trained separately for events with one and with more than one identified b jet. Simulation of the two signal production processes is done with HERWIG, the dominant background of W + jets is generated with ALPGEN + HERWIG. The generated events are passed through the full detector simulation and reconstruction chain of CDF. To obtain the measured gg production fraction, f_{gg}, templates for the neural network output are constructed from the simulation as function of f_{gg} and a likelihood to observe the measured data as function of f_{gg} is maximised. The Feldman–Cousins approach [6, 161] is applied to restrict the final result to the physically allowed range.

To determine the systematic uncertainties various pseudo-experiments with systematically varied signal and/or background templates are studied. The deviation of f_{gg} obtained in these samples from the standard template result is considered as systematic uncertainty. The dominant uncertainty is found to stem from background shape and composition as well as from the differences between leading and next-to-leading order simulation of the signal.

In 0.96 fb^{-1} of lepton plus jets events CDF is able to set limits on the gg production fraction of $f_{gg} < 0.61$ at 95% C.L. [231]. This method yields independent information, but is not as sensitive as the following method.

3.8.2 *Soft track method*

Another method that CDF applies to measure the fraction of top quark pair production through gluon fusion relies on differences that occur because gluons have a higher probability to radiate than quarks [232]. The analysis is based on top quark pairs with semileptonic decays in 0.96 fb^{-1} of CDF data.

As sensitive observable the average number of soft tracks per event, $\langle N_{\rm trk}\rangle$, with 0.9 GeV $< p_T <$ 2.9 GeV in the central detector region $|\eta| \le 1.1$ is used. In simulation this is shown to have linear relation to the average number of gluons, $\langle N_g\rangle$, in the hard process. This relation is calibrated on two samples with different gluon content: W + 0jets events for low gluon content and dijet for high gluon content. W + 1jet events are used as cross-check.

The dijet event sample for calibration is selected requiring a leading jet with transverse momentum between 80 and 100 GeV and a second recoiling jet with $|\Delta\phi| \geq 2.53$. Vetoes are applied on lepton candidates and missing transverse energy. The W + jets sample is selected requiring an isolated lepton, large missing transverse energy. For the signal top quark pair sample in addition at least four jets are required. At least one of these jets must be identified as b-jet. In the W + jets and top quark pair samples vetoes on additional lepton candidates and on leptons consistent with photon conversion or cosmic rays are applied.

After calibration of the relation between $\langle N_{\text{trk}} \rangle$ and $\langle N_g \rangle$, the fraction of events with a high gluon content, f_g, is determined using a binned likelihood fit. The fit result is corrected according to the expected background contribution. The fraction of events with high gluon content in the background, f_g^{bkg}, is extrapolated from the W + jets sample with up to three jets to the four or more jet sample. The expected amount of background in the selected signal sample is determined following the neural network based method in [223]. The obtained high gluon content in top quark pair events, $f_g^{t\bar{t}}$, in a last step is corrected for the differences in acceptance between the gluon fusion and the $q\bar{q}$ annihilation processes.

The systematic uncertainties of this measurement are dominated by uncertainties of the calibration procedure and were determined by varying the corresponding parameters in the analysis.

In the dataset of 0.96 fb^{-1} CDF determines a gluon fusion fraction of top quark pair production in semileptonic events of $0.07 \pm 0.14_{\text{stat}} \pm 0.07_{\text{syst}}$ [232]. This number corresponds to an upper limit of 0.33 at 95% C.L. well in agreement with the Standard Model expectations. Also this measurement is statistically limited. A combination of this result with the result of the Angular Correlation Method described in the previous section yields an about 10% improvement on the upper limit [231].

3.9 Top quark width and lifetime

The top quark width and its lifetime are related by Heisenberg uncertainty principle. In the Standard Model the top quark width is expected to be 1.34 GeV corresponding to a very short lifetime of about 5×10^{-25} s. Experimentally, these predictions have been challenged for deviations in very different analyses. CDF constrains the lifetime from the distribution of reconstructed top quark mass values and from the distribution of lepton track impact parameters. DØ combines the measured $t \to Wb$ branching fraction and the single top t-channel cross-section to measure the top quark width.

3.9.1 Top quark mass distribution

The limit on the top quark width was obtained by CDF from the distribution of reconstructed top quark mass values from top quark pairs decaying to lepton plus jets in up to 4.3 fb^{-1} of data [233, 234]. After selecting events with one lepton, large missing transverse momentum and at least four jets. One of the jets is required to be identified as b-jet.

In these events the top quark mass is reconstructed using a kinematic fit that determines the four-momenta of the top quark decay productions from the measured jet and lepton momenta and the transverse missing energy. The fit uses constraints that force the W boson decay products to build the W boson mass within the width of the W boson and the reconstructed top and antitop quark masses to be equal within the top quark width. In the ambiguous association of jets to partons identified b-jets are only associated to b-quarks. Of the remaining associations and the two solutions for the neutrino z-momentum, the one with the best χ^2 is used. It was checked that the use of the constraint of the equality of the top quark masses width does not destroy the sensitivity to the true width.

To find the measured value of the top quark width the distribution of top quark masses reconstructed with the best association in each event is compared to parametrised templates with varying nominal width. Templates for top quark pair signal events were generated using PYTHIA with m_t = 172.5 GeV [234]. Background contributions of W + jets are modelled ALPGEN + PYTHIA. Multijet contributions from data with non-isolated leptons. Single top quark and diboson events are simulated with MADGRAPH. The template distributions for discrete values of the nominal top quark width are parametrised to obtain smooth template functions that can now be interpreted as probability densities. The measured top quark width is determined in an unbinned likelihood fit.

Recently, in addition to the reconstructed top quark mass, the invariant mass of the jets assigned to the hadronic W decay is considered as an observable. This allows a simultaneous fit of the top quark width and the jet energy scale [234].

The Feldman–Cousins approach [6, 161] is used to determine the final result excluding the unphysical values of negative widths that may occur in the fit. The jet resolution followed by colour reconnection effects yield the biggest single contribution to the systematic uncertainties. are propagated to the final Feldman–Cousins band by convoluting their effects with the fitted width function.

Including all systematics this CDF analysis of 4.3 fb^{-1} yields an upper limit of the top quark width $\Gamma_t < 7.5$ GeV at 95% C.L. which corresponds to $\tau_t > 8.7 \times 10^{-26}$ s. At 68% C.L. the top quark width is determined as 0.4 GeV $< \Gamma_t <$ 4.4 GeV [234].

3.9.2 *Lepton impact parameter*

The limit from the lepton track impact parameter distribution was obtained by CDF using lepton plus jets events in 318 pb^{-1} of data [235]. Events are selected requiring one isolated lepton, missing transverse energy and at least three jets. At least one of the jets has to be identified as b-jet. The lepton track needs to be reconstructed with at least three R–ϕ positions in the CDF silicon tracker.

The lepton impact parameter, d_0, chosen as observable in this measurement is defined as the smallest distance between the collision point and the lepton track in the transverse projection. The collision point is computed as the position of the beam line in the transverse plane at the reconstructed z position of the primary vertex.

The distribution of lepton impact parameters expected in an ideal detector for various top quark lifetimes is simulated with PYTHIA. The resolution of the CDF detector is measured in Drell–Yan data near the Z boson resonance and used to derive the templates for the real detector expectation. The dominant backgrounds like W + jets consist of prompt leptons, which are described by the zero lifetime template. But the distribution of multijet events, backgrounds with τ leptons and electrons from photon conversions need to be modelled. Multijets and electron conversions are modelled from control samples in data. Backgrounds with τ leptons are modelled using HERWIG.

From these templates a likelihood as function of $c\tau_t$ is built. The maximal likelihood is obtained from the 0 μm template. Systematic uncertainties on the signal and background systematics are computed with correspondingly varied templates and are dominated from the uncertainty on the detector resolution for prompt leptons. The Feldman–Cousins approach is used to determine the observed limit of $c\tau_t < 52.5$ μm at 95% C.L.

3.9.3 *Branching fraction and single top cross-section*

The total width of the top quark can be written as the ratio of the partial width for the decay $t \to Wb$ and the corresponding branching fraction:

$$\Gamma_t = \Gamma(t \to Wb)/\mathcal{B}(t \to Wb). \tag{87}$$

DØ measures the top quark width by relating the partial width and branching fraction to the results of two independent analyses [236]. The branching fraction in the denominator is equal to the branching fraction ratio, R_b, measured from top quark pairs (cf. Sect. 3.2.2). This assumes that the top quark always decays to a W-boson plus a quark. The partial width is derived from the single top production cross-section which is proportional to $\Gamma(t \to Wb)$. When considering only the t-channel cross-section, $\sigma_{t\text{-channel}}$, this proportionality is valid also in the presence of anomalous couplings in the tWb-vertex. Thus the partial width is determined as

$$\Gamma(t \to Wb) = \sigma_{t\text{-channel}} \frac{\Gamma^{\text{SM}}(t \to Wb)}{\sigma^{\text{SM}}_{t\text{-channel}}}, \tag{88}$$

where the superscript SM indicates the Standard Model expectations. As the total width is by definition larger than the partial width, a lower bound of $\Gamma_t > 1.21$ GeV at 95% C.L. can be set from the t-channel production cross-section [190] alone. In the combination of the partial width with the branching fraction systematics are classified and treated as either fully correlated or uncorrelated. They are dominated by uncertainties in the background description and the description of the b-jet identification in the input analyses.

With the combination of the t-channel production cross-section measured in 2.3 fb^{-1} [190] and the branching fraction ratio measured in 1 fb^{-1} [162] DØ obtains a top quark width of $\Gamma_t = 1.99^{+0.69}_{-0.55}$ GeV corresponding to a top quark lifetime of $\tau_t = (3.3^{+1.3}_{-0.9}) \times 10^{-25}$ s [236].

3.10 Outlook to LHC

Within the SM the interaction properties of the top quark are fully defined once the top quark mass is known. The verification of the predicted properties establishes the top quark as the particle expected in the SM. Huge progress has been made in the last years in experimentally verifying the expectations. Measurements of the W boson helicity, the CKM element V_{tb} and searches for FCNC confirm the expected weak interaction properties. The electric charge has been challenged and the strong interaction properties have been verified in differential cross-section measurements, in a verification of the contribution of gluon fusion processes and also in the forward–backward charge asymmetry. First tests of the spin structure and determinations of the top quark width complete the current picture. All current results are compatible with the SM expectations, a single deviation of 3.4σ appears in the charge asymmetry at high $M_{t\bar{t}}$.

Despite the great progress, the precision of the experimental knowledge of interaction properties of the top quark is still limited. Many results have not yet reached a precision of 10%. As a consequence of the experimental progress even the more precise results, like the W boson helicity and the measurement of V_{tb}, are statistically limited.

The near future will see updates of the results with the yet unanalysed Tevatron data to a total 10 fb^{-1}. In addition the LHC schedule envisions to collect 1 fb^{-1} of proton–proton collisions at $\sqrt{s} = 7$ TeV. The LHC experiments will then have about twice as many top quark pairs as the Tevatron and even three times as many single top quarks. For the top quark pairs the production through gluon fusion and for the single top quark the t-channel production will dominate. When systematic uncertainties can be controlled similarly well as at the Tevatron, the LHC results on this dataset will still be limited by statistics.

Only the update of the LHC to the design centre-of-mass energy of 14 TeV will provide sufficiently many top quarks to enter an area of precision measurements in the verification of top quark interactions.

An important exception to these statements is the measurement of the forward–backward asymmetry, which cannot be measured in pp collisions. Instead related asymmetries need to be studied [211, 217, 237–239]. These measurements will probably require large luminosity as only quark annihilation diagrams contribute to the asymmetries.

4 New particles in top quark events

The phenomenology of the top quark may also be altered by particles that are not expected within the Standard Model, but in one of the many models of new physics. Such particles beyond the Standard Model may occur in the top quark production or its decay, depending on the specific model or its parameters. Some models of new physics also contain new particles with signatures that are very similar to the Standard Model top quark. The Tevatron experiments have checked for all these different extensions of the Standard Model in the top quark sector.

This subsection will actually start with a process that is expected in the Standard Model though at very low rate: associated Higgs production. In some models of new physics this process is expected to be enhanced. Then particles beyond the Standard Model in the top quark decay will be discussed. Finally, searches for the production of particles that look like the top quark but are not are described.

4.1 Associated Higgs boson production, ttH

Top quark pair production may be associated by the production of a Higgs boson. For parameters where Higgs bosons dominantly decay to bottom quark pairs, i.e. low Higgs masses, this associated production is a possibility to measure the top quark Yukawa coupling. While the corresponding cross-section in the Standard Model is too low to allow a Higgs discovery in this channel alone, it still contributes to the combination of the Standard Model Higgs searches. In some models including new physics an enhancement of $t\bar{t}H$ production is expected [240–242].

DØ DØ performed an analysis searching for associated Higgs production in events with a lepton (e or μ) missing transverse energy and at least four jets [243]. The analysis uses the scalar sum of transverse momenta, H_T, the number of jets and the number of jets identified as b-jets to discriminate the Standard Model backgrounds and top pair production containing no Higgs from the signal.

Signal events are simulated using PYTHIA. For $t\bar{t}$ production pure PYTHIA simulation was compared to ALPGEN + PYTHIA simulation. Due to the difference between the two simulations a 50% uncertainty was assigned to the contribution of $t\bar{t}b\bar{b}$ through QCD processes. Background from W + jets events is simulated with ALPGEN + PYTHIA and normalised to data. Multijet background was completely estimated from data. Smaller backgrounds are taken from simulation normalised to NLO cross-sections.

In the investigation of 2.1 fb^{-1} the observed data agree with the Standard Model expectations within statistical and systematic uncertainties [243]. To compute limits signal and background contributions are fitted to the data for a background only assumption and for a signal plus background assumption. Limits on $\sigma(t\bar{t}H) \cdot \mathcal{B}(H \to b\bar{b})$ are then derived using the CL$_s$ method [227, 228] for Higgs masses between 105 and 155 GeV.

For $M_H = 115$ GeV the cross-section limit corresponds to about 60 times the Standard Model value. While this allows to exclude unexpectedly large Higgs boson production in association with the top quark, its contribution to the Standard Model Higgs search remains small.

4.2 Charged Higgs boson

Particles beyond the Standard Model in the final state of top quark pair events may alter the branching fractions of the various top quark decay channels and modify the kinematic properties of the final state.

Charged Higgs bosons appear in many extensions of the Standard Model due to the need for an additional Higgs doublet with a separate vacuum expectation value. These models are characterised by the ratio of the vacuum expectation values of the two Higgs doublets, $\tan\beta$. A charged Higgs

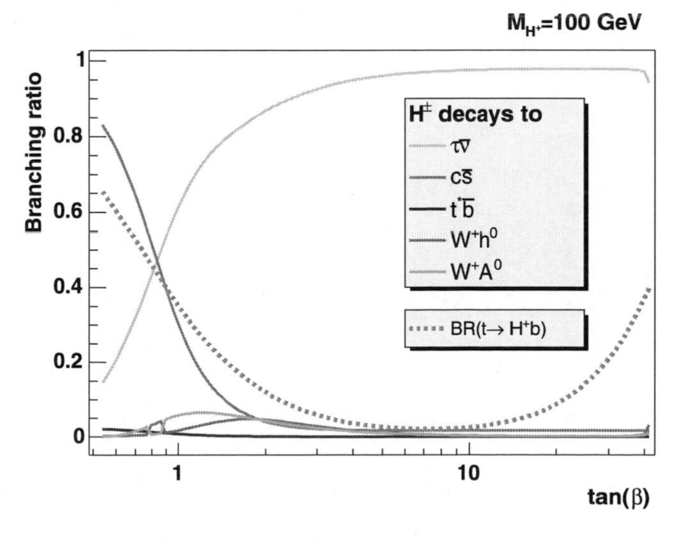

Fig. 40 Charged Higgs boson branching fraction in the MSSM as function of $\tan\beta$ for a low Higgs mass [244]

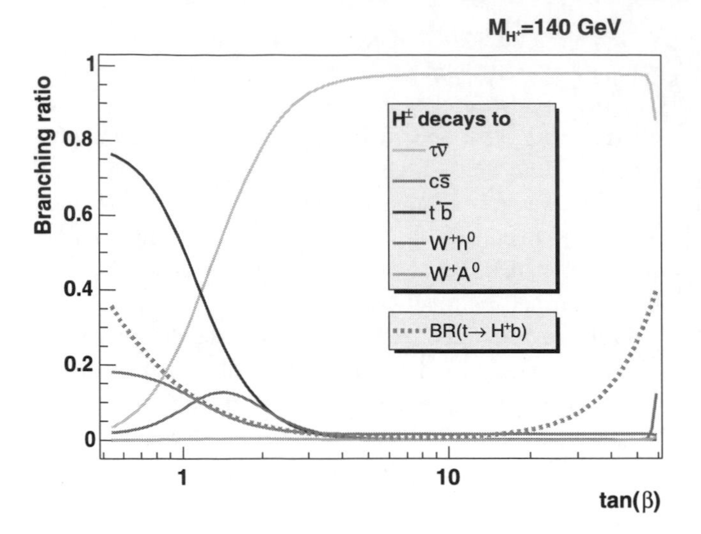

Fig. 41 Charged Higgs boson branching fraction in the MSSM as function of $\tan\beta$ for a high Higgs mass [244]

boson can replace the W boson in top quark decays. Because charged Higgs bosons have different branching fractions than W bosons this alters the branching fractions to the various top quark pair decay channels. If its mass is different from the M_W it also modifies the kinematic properties of the top quark pair final state.

In the Minimal Supersymmetric Standard Model (MSSM) [245] the decay at low $\tan\beta$ is dominated by hadronic decay to $c\bar{s}$ at low Higgs boson masses and to $t^*\bar{b}$ for Higgs boson masses above about 130 GeV. For $\tan\beta$ larger than about 1 a leptonic decay to $\tau\bar{\nu}$ dominates, cf. Figs. 40, 41. The figures also show the expected branching fraction of $t \to H^\pm b$ which is especially large for very low and very high $\tan\beta$ and rather small in the intermediate range.

CDF CDF has performed two analyses with different approaches. An analysis based on 0.2 fb^{-1} uses the CDF $t\bar{t}$ cross-section measurements in various channels and recasts the interpretation to obtain limits on the charged Higgs boson production. A second more recent analysis on 2.2 fb^{-1} investigates the kinematic differences between lepton plus jets events from top quark pair production with Standard Model decay and those including charged Higgs boson decays.

Recast of top quark pair cross-section To obtain limits on a possible charged Higgs boson contribution in top quark decay CDF utilises cross-section measurements performed in the lepton plus jets channel (with exactly one b-tag or two or more b tags), the dilepton channel and the τ plus lepton channel [244]. Care is taken to avoid overlap between the various channels. Beside the Standard Model decays of a top quark through a W boson, four decay modes through the charged Higgs boson are considered: $H^+ \to \bar{\tau}\nu$, $H^+ \to c\bar{s}$, $H^+ \to t^*\bar{b}$ and $H^+ \to W^+h^0 \to W^+b\bar{b}$. The latter has a non-negligible contribution at intermediate values of $\tan\beta$.

Selection efficiencies are taken from simulation of top quark pair events for various masses of the top quark, the charged and neutral Higgs boson, h^0. The simulation takes the dependence of the width of the top quark and the charged Higgs boson into account. The production cross-section is kept at its Standard Model value for $m_t = 175$ GeV: $\sigma_{t\bar{t}} = 6.7 \pm 0.9$ pb.

The event counts observed in data in the four channels are compared to the expectations in three different ways. For specific benchmarks of the MSSM a Bayesian approach is used to set limits on $\tan\beta$. This analysis uses a flat prior on $\log\tan\beta$ within the theoretically allowed range. These limits are computed for various values of the charged Higgs boson mass and five different parameter benchmarks. Figure 42 (left) shows the results for one specific benchmark.

For the high $\tan\beta$ region $H^+ \to \bar{\tau}\nu$ dominates in a large fraction of the MSSM parameter space. Setting the branching fraction of $H^+ \to \bar{\tau}\nu$ to 100%, limits on the charged Higgs contribution to top quark decays are set using Bayesian statistics. A flat prior for $\mathcal{B}(t \to H^+b)$ between 0 and 1 is used. For charged Higgs boson masses between 80 GeV and 160 GeV CDF can exclude $\mathcal{B}(t \to H^+b) > 0.4$ at 95% C.L.

Finally, a more model independent limit is computed by scanning the full range of possible charged Higgs boson decays. For all five $H^\pm$ decay modes considered the branching fraction is scanned in 21 steps, assuring that the sum of branching fractions adds to one. Limits on $\mathcal{B}(t \to H^+b)$ are computed for each combination. The least restrictive limit is quoted. Also this analysis is repeated for various charged Higgs boson masses. The limits obtained in this more general approach, shown in Fig. 42 (right), exclude only very high contributions of charged Higgs bosons to top quark decays of above approximately 0.8 to 0.9, depending on the charged Higgs boson mass.

Investigation of kinematic differences At low $\tan\beta$, where the charged Higgs boson can also decay to $c\bar{s}$, CDF used the invariant dijet mass to search for a possible $H^\pm$ contribution in top quark pair events [246]. Lepton plus jet events are selected requiring at least two of the four leading jets in p_T to be b-tagged. The four leading jets are used in a kinematic fit that requires consistency of the fitted lepton and neutrino momenta with the W boson mass and reconstructed top quark masses to be 175 GeV. The dijet mass of the hadronic W decay remains unconstrained. The jet–parton assignment with the best χ^2 is used and the charged Higgs boson mass is reconstructed from the non-b-tagged of the four leading jets. For events with more than four jets the fifth jet is added to its closest neighbour if their $\Delta R < 1.0$ to improve the dijet mass resolution.

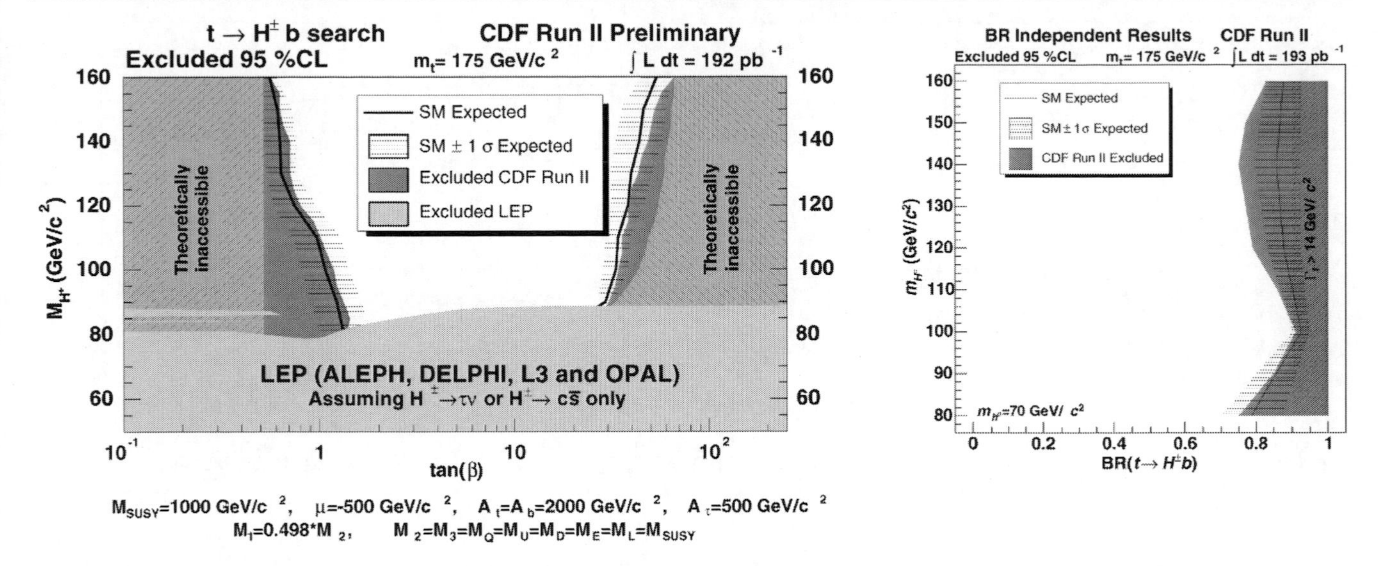

Fig. 42 Results from recasting the CDF top quark cross-section measurements [244]. *Left*: Exclusion region in the MSSM m_H–$\tan\beta$ plane for an example benchmark scenario corresponding to the parameters indicated below the plot. *Right*: Upper limits on $\mathcal{B}(t \to H^+b)$ derived without assumptions on the charged Higgs boson branching fraction

Background events are dominated by top quark pair production with Standard Model decay. Further processes included are $W +$ jets, $Z +$ jets, diboson, single top quark and multijet events. Except for multijet events the backgrounds are estimated from simulation. The normalisation for $W +$jets taken from data, for the others it is taken from theory. The multijet background is fully determined from data.

To determine a possible contribution of charged Higgs boson in the decay of top quark pair production a binned likelihood fit is performed. The likelihood is constructed with templates for the backgrounds and using the branching fraction of top quark to charged Higgs bosons, the number of top quark pair events and the number of background events as parameters. The number of background events is constrained within the uncertainty to the expectation. The observed dijet mass distribution and the fitted background composition is shown in Fig. 43 including a charged Higgs boson contribution of 10%.

Systematic uncertainties are computed by fitting pseudo-data created from systematically varied templates with the standard unshifted templates. The change of the branching ratio due to the systematic variation is taken as systematic uncertainty for each variation considered. These uncertain-

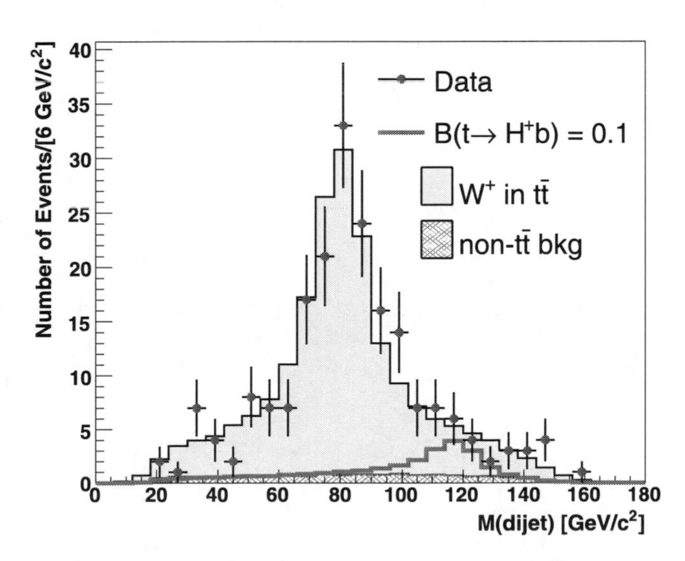

Fig. 43 CDF dijet mass distribution with 120 GeV Higgs boson events assuming $\mathcal{B}(t \to H^+b) = 0.1$. The size of Higgs boson signal corresponds to the expected upper limit branching ratio at 95% C.L. for 120 GeV [246]

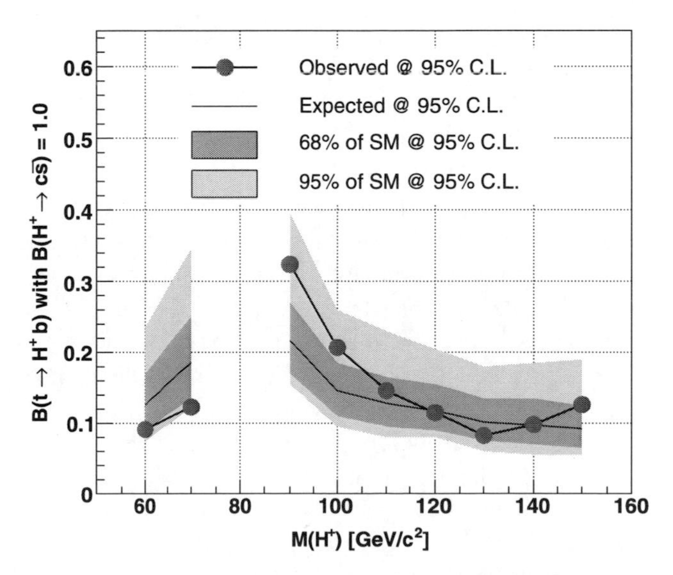

Fig. 44 (Colour online) CDF limits $\mathcal{B}(t \to H^+b)$ observed in 2.2 fb^{-1} data (*red dots*) compared with the expected limit assuming the Standard Model (*black line* with uncertainty bands) [246]

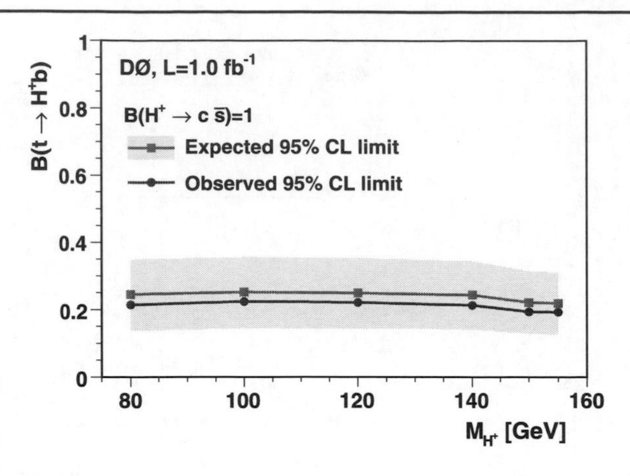

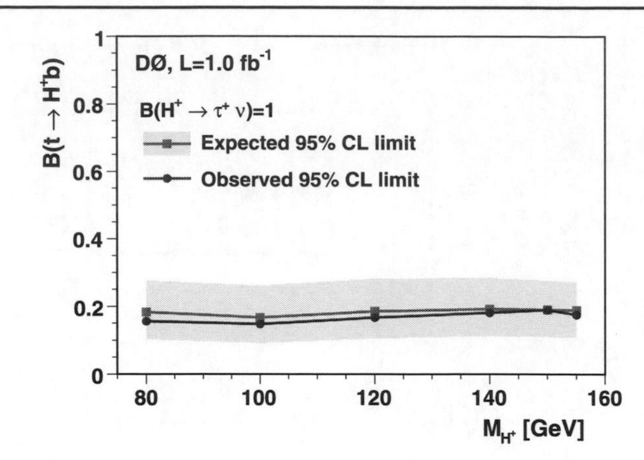

Fig. 45 Limits on the contribution of a charged Higgs boson in top quark decays for a leptophobic model (*left*) and a tauonic model (*right*) obtained in 1 fb^{-1} of DØ data assuming the SM $t\bar{t}$ production cross-section [247]

ties are included to a final likelihood by convoluting a Gaussian distribution to the original likelihood. Systematic uncertainties are dominated by the signal modelling that is derived from replacing the default Pythia sample of $t\bar{t}$ events by a Herwig sample. For Higgs boson masses close to the W boson mass the jet energy scale uncertainty becomes dominant.

For various assumed Higgs boson masses 95% C.L. limits on the branching fraction are determined by integrating the likelihood distribution to 95% of its total. As shown in Fig. 44 limits between about 10% and 30% can be set depending on the mass of the charged Higgs, consistent with the expected limits for pure Standard Model top quark decays. This result is less model dependent than the above CDF limits, but not as strict within the models used above.

DØ The DØ collaboration searched for a light charged Higgs boson contribution in top quark decay by reinterpreting the cross-section measurements in various decay channels and for heavy charged Higgs boson contributing to single top quark production.

Charged Higgs boson in top quark decay In the search for a light charged Higgs boson in top quark decay DØ uses the cross-section analyses for lepton plus jets, dilepton and lepton plus tau decay channels, with lepton referring to e and μ only [126, 247, 248]. The channels are kept disjoint and further separated into subsamples depending on the number of jets, the number of b tags and the lepton type. The number of expected events for each of the subsamples is computed from $t\bar{t}$ simulation with PYTHIA including the Standard Model decays and decays of the top quark to a leptophobic or a tauonic charged Higgs boson and charged Higgs boson masses between 80 and 155 GeV.

A likelihood for the observed data is built each of the two models given the number of expected events as function of the branching fraction $\mathcal{B}(t \to H^{\pm}b)$. The observed $\mathcal{B}(t \to H^{\pm}b)$ is extracted by maximising the likelihood.

In a first iteration the production cross-section is fixed at a value of $\sigma_{t\bar{t}} = 7.48$ pb. Limits are set according to the Feldman–Cousins procedure including systematic uncertainties. The systematic uncertainties in this method are dominated by uncertainties due to the assumed top quark pair cross-section, the luminosity and b-jet identification. The resulting limits obtained with 1 fb^{-1} of data exclude a branching fraction above around 20% for the pure leptophobic model and above 15–20% for the tauonic model, cf. Fig. 45 [247].

In a second iteration the $t\bar{t}$ production cross-section is treated as a free parameter and determined simultaneously with the limits on charged Higgs production, see Fig. 45. This reduces the assumptions made in the determination of the limit. In addition the result is much less sensitive to the luminosity. In this method the description of multijet background becomes the largest systematic uncertainty. Such a two-dimensional fit is only possible for the tauonic decay

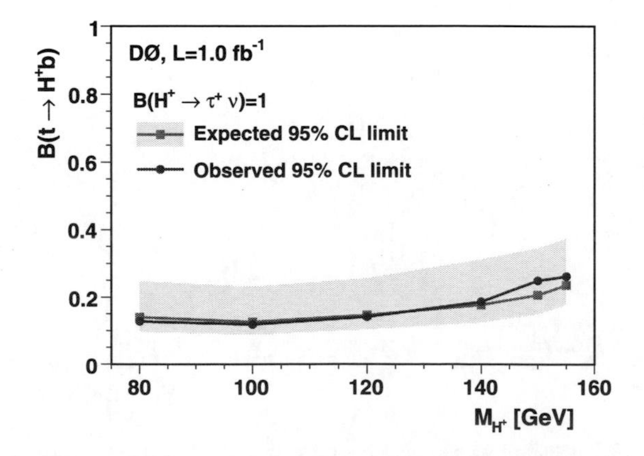

Fig. 46 Limits on the contribution of a charged Higgs boson in top quark decays for the tauonic model obtained in 1 fb^{-1} of DØ data in a simultaneous fit of the SM $t\bar{t}$ production cross-section [247]

model where this analysis includes channels that get enhanced in the presence of a charged Higgs and others that get depleted. In this channel the sensitivity enhances by more than 20% for low Higgs boson masses [247].

DØ further discusses the implication for various supersymmetric models and sets exclusion limits on these models in the plane of model parameters $\tan\beta$ and $M_{H^\pm}$.

Charged Higgs boson in single top quark production Heavy charged Higgs bosons cannot only occur in the decay of top quarks but may contribute to single top quark production. Their signature is identical to Standard Model s-channel single top quark production, but may have a resonant structure in the invariant mass distribution of its decay products, the top and the bottom quarks.

Following their single top quark analysis, DØ selects events with an isolated lepton, missing transverse energy and exactly two jets, one of which is required to be identified as b-jet (Fig. 46) [249]. Background estimation for W + jets and $t\bar{t}$ production is simulated using ALPGEN. Standard Model single top quark production is modelled using SINGLETOP [191, 192]. Charged Higgs boson signal events are simulated with a narrow width for the charged Higgs boson using COMPHEP. Three types of two Higgs doublet models (2HDM) are considered. In the Type I 2HDM one doublet gives mass to all fermions; in the Type II model one doublet gives mass to the u-type quarks (and neutrinos), the other to the d-type quarks and charged leptons. This model is realised in the MSSM. In the Type III 2HDM both doublets contribute to the masses of all fermions. Due to the different couplings the cross section of single top quark production in these 3 models is quite different.

Standard Model and charged Higgs boson production of single top quarks is separated by reconstructing the invariant mass of the two jets and the W boson. This distribution shows good agreement between data and the Standard

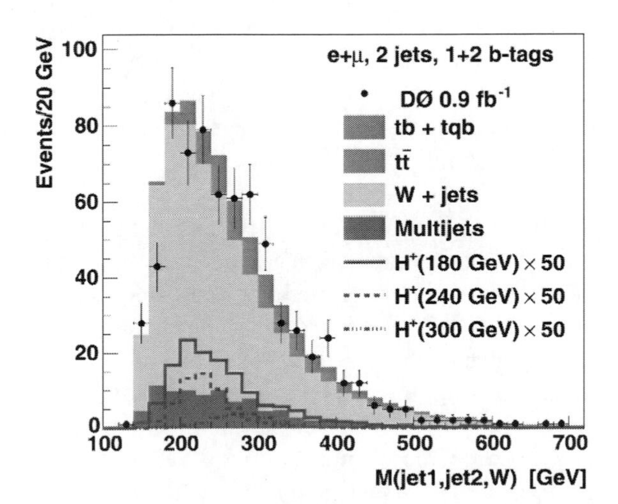

Fig. 47 Observed and expected invariant mass of the W boson and the two leading jets in 0.9 fb^{-1} of DØ data [249]

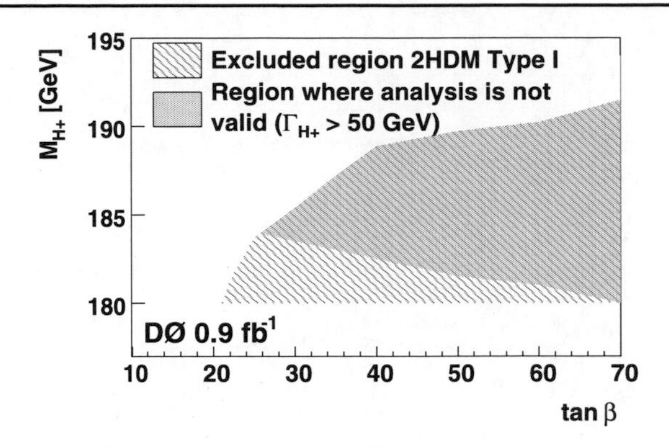

Fig. 48 Exclusion areas derived from 0.9 fb^{-1} of DØ data for Type I two Higgs Doublet Model (2HDM) [249]

Model expectation, see Fig. 47. Bayesian statistics is used to set limits on the allowed cross-section for single top quark production through a charged Higgs boson. For the Type I 2HDM some region in $\tan\beta$ vs. $M_{H^\pm}$ can be excluded, cf. Fig. 48, a significant fraction of phase space is not accessible by the analysis in its current form due to the restriction to small $H^\pm$ decay widths [249].

4.3 Heavy charged vector boson, W'

New charged gauge bosons, W', are expected in extensions of the Standard Model with additional gauge symmetries and in supersymmetric models, see e.g. [6, 245]. Its couplings may be to left-handed fermions, like for the Standard Model W boson, or include right-handed fermions. In general a mixture of these two options is possible. If the W' boson has left-handed couplings, it will have a sizeable interference with the SM W^+ boson [250]. For purely right-handed couplings, a leptonic decay may only occur when the right-handed neutrinos are lighter than the W' boson. In this case the decay to a top and bottom quark is an interesting channel to perform direct searches for such W' bosons.

Both CDF and DØ search for various types of W' bosons decaying to tb pairs in conjunction with their single top quark analyses. The main discriminating observable is the reconstructed invariant mass of the decay products, which was also utilised to search for a heavy charged Higgs boson, cf. Sect. 4.2.

DØ DØ has published a search for a heavy W' boson with decay to top and bottom quarks using 0.9 fb^{-1} [251]. The event selection follows the single top quark analysis and requires one isolated lepton, missing transverse momentum and two or three jets, one of which must be identified as b-jet.

The invariant mass, $\sqrt{\hat{s}}$, of the bottom and the top quark decay products is computed from the measured four-momenta of the leading two jets, the charged lepton and the neutrino. The transverse momentum of the neutrino is

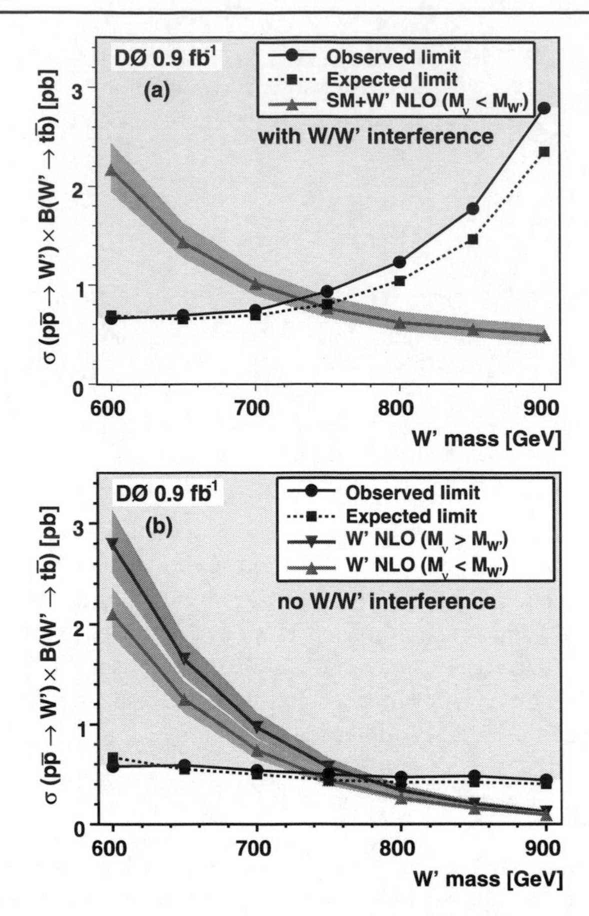

Fig. 49 DØ results on a search for W' boson decaying to top and bottom quark using 0.9 fb^{-1} of data. *Top*: Expected and observed limits on a left-handed W' production cross-section times branching fraction of the decay to top and bottom quark as function of $M_{W'}$ compared to the theory prediction. *Bottom*: Same but for right-handed W' production [251]

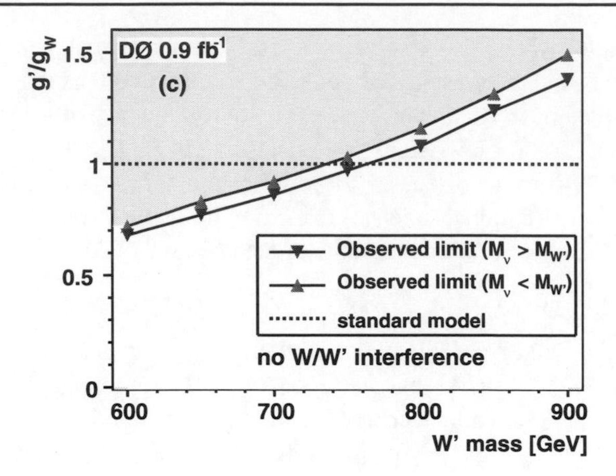

Fig. 50 DØ results on a search for W' boson decaying to top and bottom quark using 0.9 fb^{-1} of data. Limits on the W' boson coupling relative to the Standard Model W-boson coupling [251]

identified with the transverse missing momentum, its z-component inferred by solving $M_W^2 = (p_\ell + p_\nu)^2$ choosing the solution with the smaller $|p_\nu^z|$.

The distributions expected within the Standard Model from a combination of simulation and data. Single top quark and top quark pair production is generated with SINGLE-TOP and ALPGEN + PYTHIA normalised to their theoretical cross-sections. W + jets background is generated with ALPGEN + PYTHIA and normalised to data before b-tagging in such a way that it includes diboson backgrounds. Also the W + heavy flavour fraction is derived from data. Multijet background is fully taken from data. Samples of W' boson events with masses up to 900 GeV are generated in conjunction with the single top quark samples taking interferences with the W boson into account that are present for the left-handed W' bosons. Because of this interference the Standard Model single top quark production in the s-channel is treated as part of the signal in the search for W'_L bosons.

The distribution of reconstructed $\sqrt{\hat{s}}$ measured by DØ agrees with the expectation from the Standard Model. Limits on a possible contribution from W' bosons decaying to top and bottom quarks are derived as a function of $M_{W'}$ assuming couplings like in the Standard Model, though possibly to right-handed fermions. DØ uses the Bayesian approach with a flat non-negative prior on the cross-section times branching fraction. Expected and observed results are shown in Fig. 49. Comparing upper limits on the W' boson cross-section times branching fraction to top and bottom quark to the NLO theory predictions [252] excludes left-handed W' bosons with $M_{W'_L} < 731$ GeV. If only hadronic decays are allowed the right-handed W'_R boson is excluded for $M_{W'_R} < 768$ GeV, when leptonic decays are also possible the limit is 739 GeV.

Without assuming the coupling strength the Bayesian approach is used to determine a limit on the size of this coupling relative to the Standard Model, see Fig. 50. These limits assume no interference between the Standard Model W and the W' bosons.

In computing the above limits systematic uncertainties are included. They include effects due to uncertainties on the integrated luminosity, the theoretical cross-sections, branchings fraction, object identification efficiencies, trigger efficiencies, fragmentation models, jet energy scale and heavy flavour simulation.

CDF In an investigation of 1.9 fb^{-1} of data [253] CDF selects W + jets events requiring one lepton (e, μ) isolated from jets, missing transverse energy and two or three energetic jets. At least one of the jets must be tagged as b-jet.

In these events the neutrino momentum, p_ν, is inferred from the missing transverse momentum and by solving $M_W^2 = (p_\ell + p_\nu)^2$ for the longitudinal component of the neutrino. The W boson mass, M_W, is set to its nominal value, p_ℓ is the measured lepton momentum. In case of complex solutions CDF assigns the real part of the solution to the longitudinal neutrino momentum. The invariant mass of the

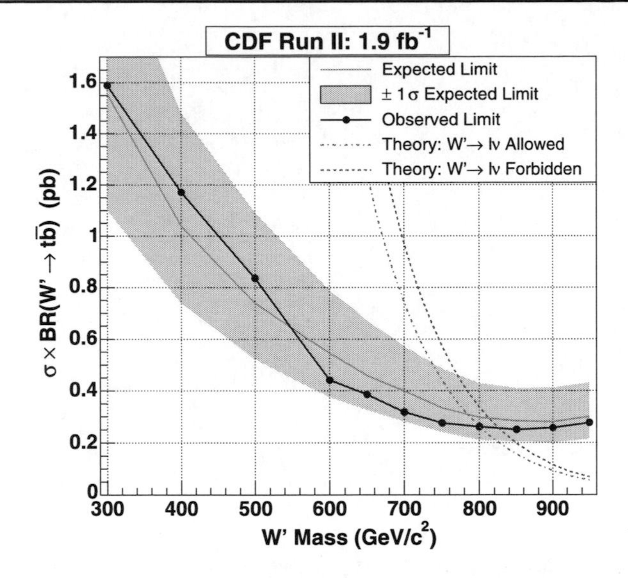

Fig. 51 Limits on W'_R cross-section times branching fraction to tb as function $M_{W'_R}$ compared to theory obtained in the CDF search for W'_R using 1.9 fb^{-1} of data [253]

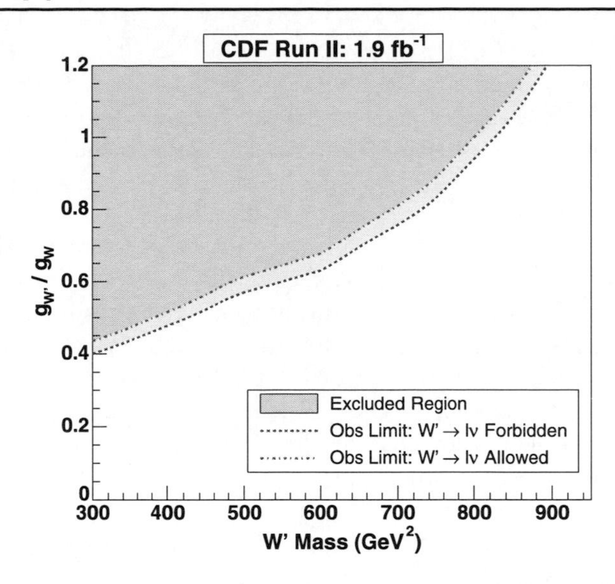

Fig. 52 Limits on the W'_R coupling strength relative the Standard Model coupling as function $M_{W'_R}$ obtained in the CDF search for W'_R using 1.9 fb^{-1} of data [253]

lepton, the neutrino and the two leading jets, M_{WJJ}, is then used as a discriminating observable.

The distribution expected within the Standard Model is computed from a combination of simulation and data. The contribution of events containing a real W boson is taken from simulation. W + jets samples are normalised to data before b-tagging using a scale factor to correct the heavy flavour contribution to fit the observation in $W + 1$ jet data. The other samples are normalised to their theoretical expectation. The identification of heavy flavour jets is estimated from simulation and corrected with a scale factor. Misidentification of light flavour jets is computed from mistag rate functions. The contribution from events without real W bosons are estimated from events with electrons that pass only a subset of the full electron identification and thus are enriched with jets misidentified as electrons.

W' boson signal events are simulated using Pythia for W' boson masses between 300 and 950 GeV with fermion couplings identical to the Standard Model W boson. When the right-handed W' boson is heavier than the right-handed neutrinos, the branching fraction to $\ell\nu$ is corrected according to the additional decay modes.

Limits are constructed according to the CL$_s$ method [227, 228]. Probabilities are computed from pseudo-experiments which are generated including variations due to systematic uncertainties. The dominating systematic uncertainties are the jet energy scale and the scale factor used to account for differences between simulation and data in the b-tagging algorithm of CDF.

Limits on the W'_R boson production cross-section are set as a function of $M_{W'_R}$ assuming the Standard Model coupling strength. These are converted to mass limits by comparison to the corresponding theoretical expectation and yield $M_{W'_R} > 800$ GeV for W'_R bosons which decay leptonically and $M_{W'_R} > 825$ GeV for $M_{\nu_R} > M_{W'_R}$. For the more general case that the W'_R coupling is a priori unknown the W'_R coupling strength, g', relative to the Standard Model coupling, g_W, is constrained. Limits are computed from the above analysis as function of the assumed $M_{W'}$. The observed and expected limits derived by CDF for $M_{W'_R}$ and g'/g_W are shown in Figs. 51 and 52.

4.4 Resonant top quark pair production

Due to the fast decay of the top quark, no resonant production of top quark pairs is expected within the Standard Model. However, unknown heavy resonances decaying to top quark pairs may add a resonant part to the Standard Model production mechanism. Resonant production is possible for massive Z-like bosons in extended gauge theories [254], Kaluza–Klein states of the gluon or Z boson [255, 256], axigluons [257], Topcolour [258, 259], and other theories beyond the Standard Model. Independent of the exact model, such resonant production could be visible in the reconstructed $t\bar{t}$ invariant mass.

CDF CDF has employed several different techniques to search for resonances in the $t\bar{t}$ invariant mass distribution. All analyses use a very similar event selection: an isolated lepton, missing transverse energy and four or more jets. One analysis (the Matrix Element plus Template method) is also applied to the full hadronic channel selecting six or seven jets. Their main difference is the method to reconstruct the $t\bar{t}$ invariant mass distribution.

Constrained fit plus template The method that uses the least assumptions reconstructs the invariant mass using a constrained fit [260] and is performed requiring at least one identified b jet. In the fit the final state lepton and quark momenta are determined from the measured lepton momentum, the missing transverse energy (which is assumed to stem from the unseen neutrino) and the measured jet momenta. Constraints are imposed that require the sum of neutrino and lepton momenta as well as the two light quark momenta to be consistent with the W boson mass. In addition these pairs in combination with one b quark need be consistent with the top quark mass of 175 GeV. The fitted momenta may vary within the experimental resolution of their assigned measurement and the mass constraints are varied within the natural widths of the W boson and the top quark, respectively. The fit thus assumes the lepton plus jets decay topology of top quark pair events.

Of the multiple jet–parton assignments the one with the best χ^2 from the fit is used to compute the top quark pair invariant mass for each event. The expected distribution of this observable is dominated by Standard Model top quark pair events which are simulated using HERWIG. Further backgrounds like $W+$jets, misidentified multijet events, diboson and single top quark events are modelled with combination of simulation and control data.

Templates for a resonant production of top quark pairs are simulated using PYTHIA for resonance masses between 450 and 900 GeV. The resonance couplings are proportional to those of a Standard Model Z boson. The width of this Z' boson was kept at $0.012 M_{Z'}$. The observed invariant top quark pair mass distribution is compared to the distribution expected from the Standard Model in Fig. 53.

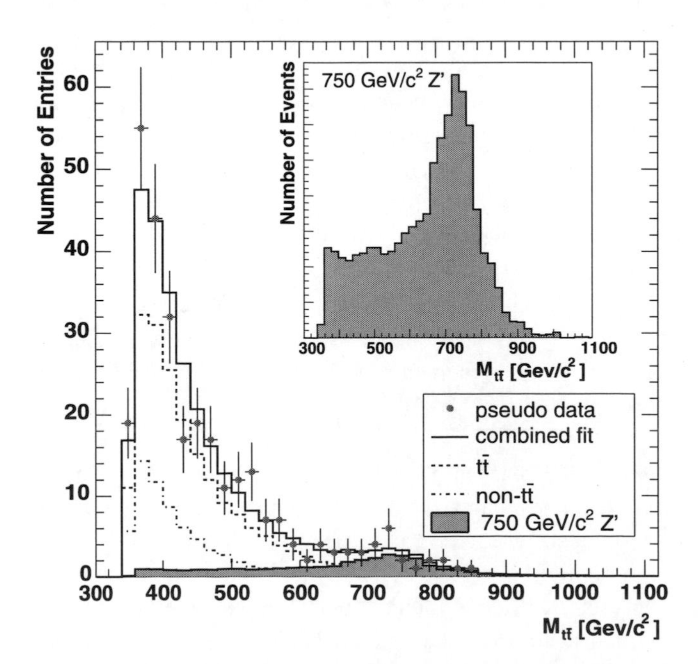

Fig. 53 Distribution of the invariant top quark pair mass reconstructed with a constrained fit in 1 fb^{-1} of CDF data [260]

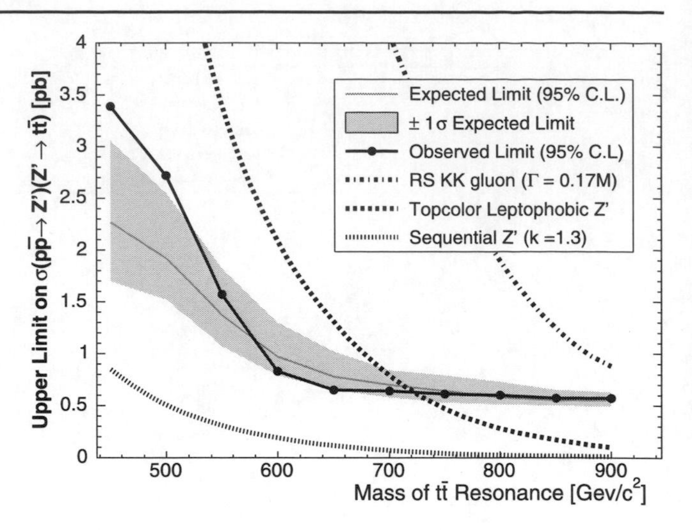

Fig. 54 Limits on the cross-section times branching fraction for resonant top quark pair production observed in 1 fb^{-1} of CDF data [260]. Theoretical curves are shown for various model and used to set mass limit on the corresponding resonance

From the expected distributions CDF constructs a likelihood for the expected bin content of the distribution of the invariant top quark pair mass as function of the resonant production cross-section time branching ratio, $\sigma_X \mathcal{B}$, the number of Standard Model top quark pairs and the number of non-$t\bar{t}$ events. Nuisance parameters with Gaussian constraints are used to implement the effect of systematic uncertainties. These include uncertainties that affect the relative background normalisation and the luminosity, the uncertainty of the jet energy scale and the shape change due to the top quark mass uncertainty. Minor contributions come from varying the PDFs between CTEQ6M [36] and MRST [261] parametrisations and the uncertainty on the strength of initial and final state radiation.

To find the upper limits the maxima of the likelihood as function of $\sigma_X \mathcal{B}$ is integrated to the point where the integral reaches 95% of its area. This is done for each assumed resonance mass between 450 and 900 GeV in 50 GeV steps. Expected and observed limits are shown in Fig. 54. At high resonance masses the observed limits exclude a resonant top quark pair production with $\sigma_X \mathcal{B} > 0.55$ pb at 95% C.L. Production through leptophobic Topcolour assisted Technicolour is excluded for resonance masses up to 720 GeV.

Matrix element plus template The resolution of the reconstructed invariant top quark pair mass can be improved by assuming additional information. In an analysis of 680 pb^{-1} CDF employed the Matrix Element technique to reconstruct the invariant mass distribution that is used to search for resonant production in lepton plus jets events [262]. Following the mass analysis [93] described in Sect. 2.2.3, for each event a probability density, $P(\{\boldsymbol{p}\}|\{\boldsymbol{j}\})$, is computed to find the momenta of the top quark pair decay products (four quarks, charged lepton and neutrino, $\{\boldsymbol{p}\}$) given the

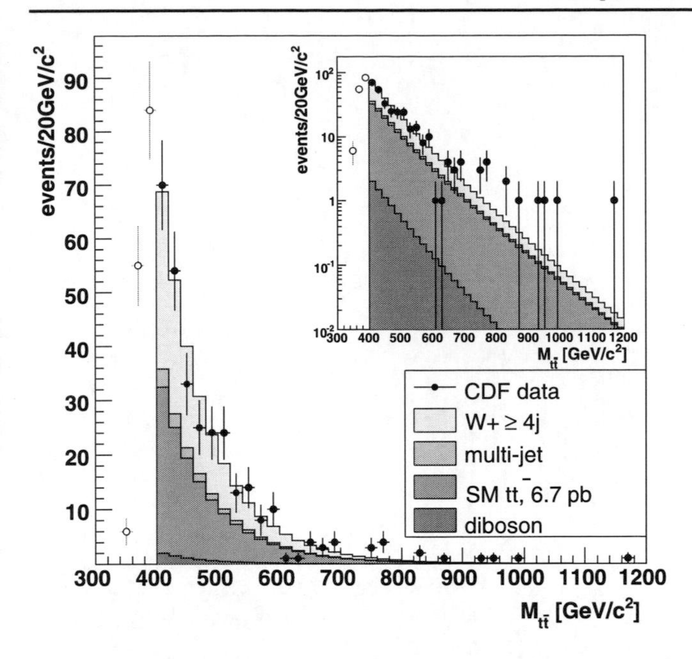

Fig. 55 Distribution of the invariant top quark pair mass reconstructed with the Matrix Element technique in 680 pb^{-1} of CDF data using semileptonic events [262]

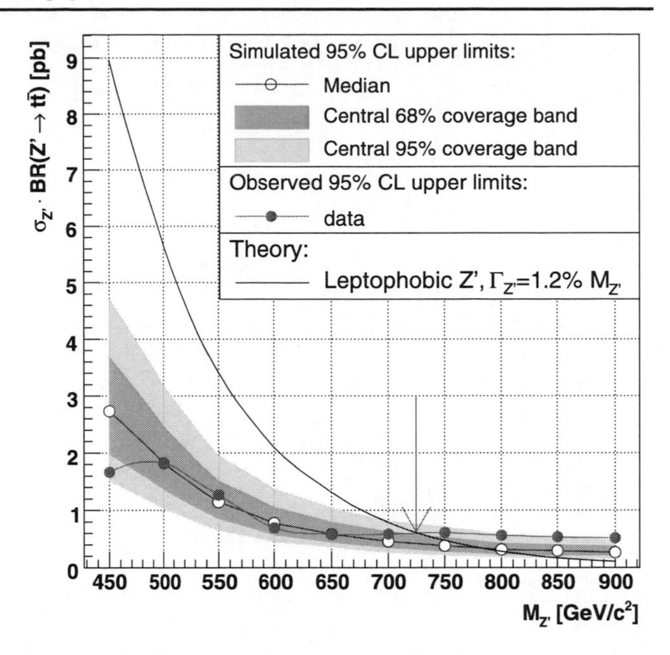

Fig. 56 Expected and observed limits on $\sigma_X \cdot \mathcal{B}(X \to t\bar{t})$ obtained by CDF in 680 pb^{-1} using the Matrix Element technique in lepton plus jets events [262]

observed quantities, $\{j\}$. This probability is computed from the parton density functions, the theoretical Matrix Element for Standard Model top quark pair production and decay and jet transfer functions that fold in the detector resolution. It is converted to a probability density for the top quark pair invariant mass using

$$P_f(M_{t\bar{t}}|\{j\}) = \int \mathrm{d}\{p\}\, P(\{p\}|\{j\})\delta(M_{t\bar{t}} - m(\{p\})). \quad (89)$$

The result for all possible jet–parton assignments is summed, before the mean value is computed and taken as the reconstructed invariant top quark pair mass for the event under consideration. Here the b-tagging information is used to reduce the number of allowed jet–parton assignments. The events are not required to contain b-tagged jets in the event selection.

Template distributions for Standard Model and resonant $t\bar{t}$ processes are derived from PYTHIA, W +jets events from ALPGEN + HERWIG including full detector simulation. As a resonance signal a Z' boson with a width of $1.2\% M_{Z'}$ was generated. The multijet background template was taken from data. The Standard Model $t\bar{t}$ samples and diboson samples are normalised to the theoretical cross-section, while the sum of multijet and W+jet samples are scaled to fit the observed data and depend on the assumed signal contribution. The resulting expected distribution compared to the observed data is shown in Fig. 55.

The possible contribution from a resonant top quark pair production is computed using Bayesian statistics. The posterior probability density is build from a likelihood that implements Poissonian expectations in each bin. The parameters of the Poisson distribution are smeared with Gaussians according to the systematic uncertainties. Finally, the posterior probability density is convoluted with a flat prior in $\sigma_{Z'} \cdot \mathcal{B}(Z' \to t\bar{t})$. The uncertainties of the Standard Model top quark pair production cross section, the jet energy scale and the variation of initial and final state gluon radiation have the largest impact on the resulting limits.

The events observed by CDF in 680 pb^{-1} of data show no evidence for resonant top quark pair production and upper limits are derived for $\sigma_{Z'} \cdot \mathcal{B}(Z' \to t\bar{t})$ as shown in Fig. 56. A comparison to the leptophobic Topcolour assisted Technicolour model yield an exclusion of this model for $M_{Z'} <$ 725 GeV at 95% C.L. Using additional assumptions about the kinematics through the matrix element thus allows to exclude slightly higher Z' boson masses despite using less data.

The Matrix Element plus Template method was also applied in the all hadronic decay channel using 2.8 fb^{-1} of CDF data with six or seven jets [263]. Top quark pair events are enriched with a neural net event selection and b jet identification. The dominant multijet background is described with a data driven method from events without the b identification requirement.

No evidence for resonant top quark pair production is found. CDF computed upper limits on the resonant production cross-section time branching fraction, $\sigma_X \cdot \mathcal{B}(X \to t\bar{t})$, as shown in Fig. 57. In the leptophobic Topcolour assisted Technicolour model resonance masses of $M_{Z'} < 805$ GeV are excluded at 95% C.L.

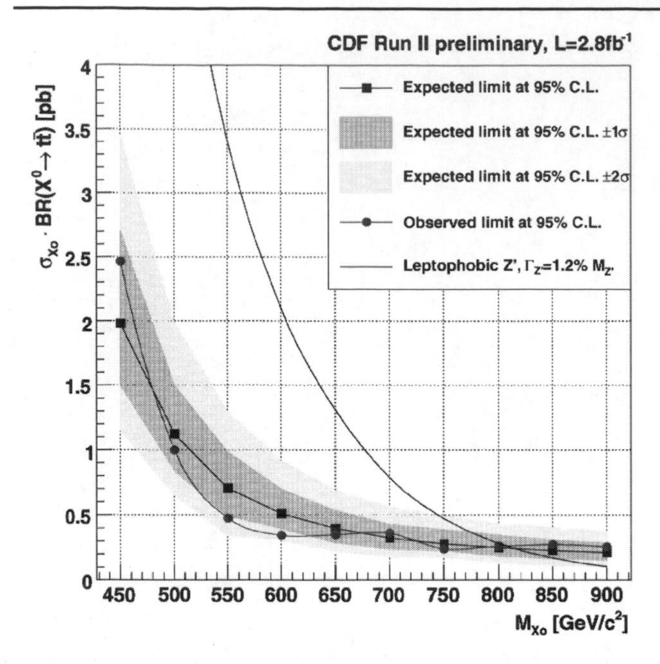

Fig. 57 Expected and observed upper limits on $\sigma_X \cdot \mathcal{B}(X \to t\bar{t})$ obtained by CDF with a Matrix Element technique in 2.8 fb^{-1} of data using all hadronic events [263]

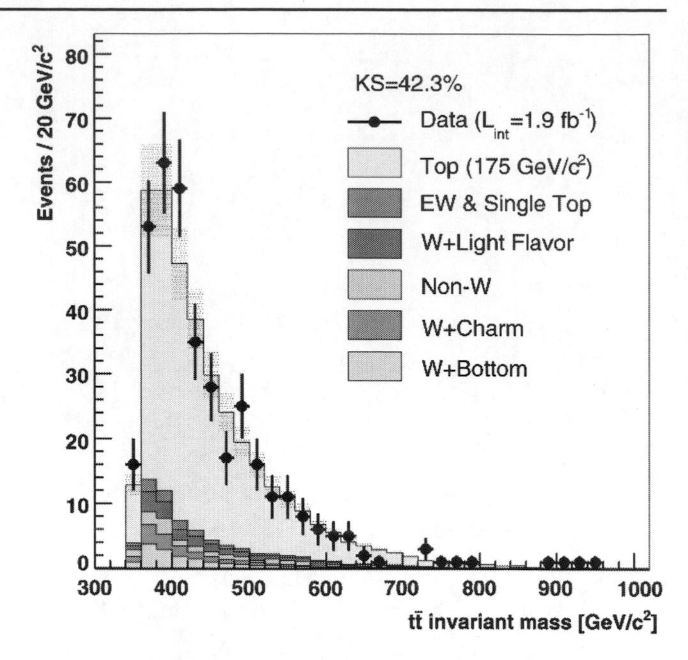

Fig. 58 Invariant top quark pair mass reconstructed with the dynamical likelihood method in 1.9 fb^{-1} of CDF data [264]

Dynamical likelihood method with massive gluon interpretation Another analysis of CDF is based on the dynamical likelihood method (DLM, see also Sect. 2.2.3). It investigates 1.9 fb^{-1} of data with an isolated lepton, missing transverse energy and exactly four reconstructed jets [264]. In contrast to the previously described analysis the invariant top quark pair mass is reconstructed without using the production part of the matrix element in the construction of the probability densities in (89). This avoids a bias towards the Standard Model production mechanism. The resulting distribution and the corresponding Standard Model expectation is shown in Fig. 58.

The distribution is then used to search for a new colour-octet particle, called a massive gluon. An unbinned likelihood fit based on the production matrix elements with and without massive gluon contribution is used to extract the possible coupling strengths of such a massive gluon contributing to the top quark pair production. The likelihood is computed for various masses and widths of the massive gluon. The systematic uncertainties are incorporated in the likelihood calculation. Jet energy scale and top mass uncertainties are the largest contribution to the total uncertainty on the fitted coupling, λ.

The observed data agree with the Standard Model expectation within $\sim 1.7\sigma$. This agreement is cross-checked by reconstructing the top quark p_T-distribution, which is also found to be in agreement with the Standard Model expectation. Limits on the possible coupling strength of a massive gluon, G, contributing to the top quark pair production are

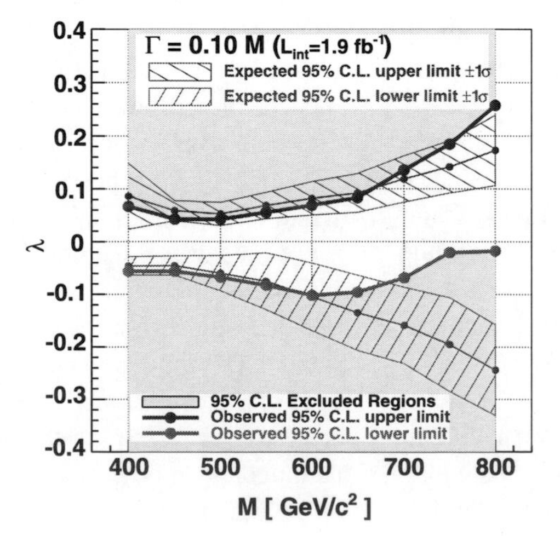

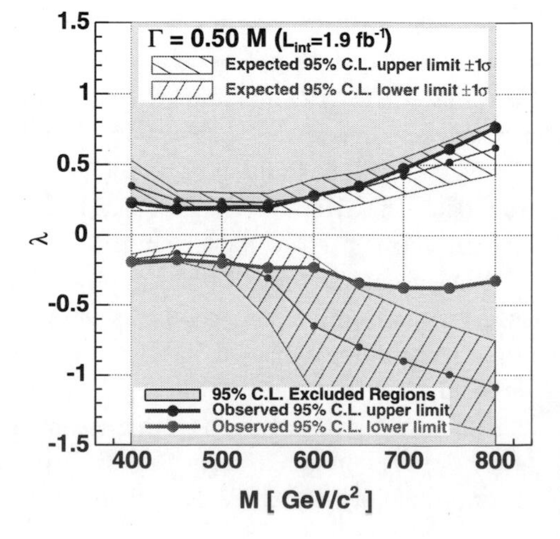

Fig. 59 Limits on the coupling strength of a massive gluon, G, contribution deduced for various masses and two widths [264]

set at 95% C.L. for various values of the width, Γ_G as function of the mass, M_G. Figure 59 shows the expected and observed limits for two choices of the massive gluon width.

DØ DØ investigated the invariant mass distribution of top quark pairs in up to 3.6 fb^{-1} of lepton plus jets events [265–267]. The event selection requires an isolated lepton, transverse missing momentum and at least three jets. At least one of the jets needs to be identified as b jet. Signal simulation is created for various resonance masses between 350 and 1000 GeV. Separated resonance samples were generated with couplings proportional to Standard Model Z boson couplings, with pure vector couplings and with pure axial-vector couplings. The width of the resonances was chosen to be 1.2% of their mass, which is much smaller than the detector resolution.

The top quark pair invariant mass, $M_{t\bar{t}}$, is reconstructed directly from the reconstructed physics objects. A constrained kinematic fit is not applied. Instead the momentum of the neutrino is reconstructed from the transverse missing energy, $\not{E}_T$, which is identified with the transverse momentum of the neutrino and by solving $M_W^2 = (p_\ell + p_\nu)^2$ for the z-component of the neutrino momentum. p_ℓ and p_ν are the four-momenta of the lepton and the neutrino, respectively.

The $t\bar{t}$ invariant mass can then be computed without any assumptions about a jet–parton assignment that is needed in constrained fits. Compared to the constrained fit reconstruction, applied in an earlier analysis, this gives better performance at high resonance masses and in addition allows the inclusion of ℓ+3 jets events. The expected distribution of Standard Model processes and the measured data is shown in Fig. 60. For comparison a resonance with a mass of 650 GeV is shown at the cross-section expected in the Topcolour assisted Technicolour model used for reference.

Cross-sections for resonant production are evaluated with Bayesian statistics using a non-zero flat prior (for positive values) of the resonant top quark pair cross-section time branching fraction, $\sigma_X \cdot B(X \to t\bar{t})$. A Poisson distribution is assumed for the number of events observed in each bin of the likelihood. The prior for the combined signal acceptance and background yields is a multivariate Gaussian with uncertainties and correlations described by a covariance matrix. The measured $\sigma_X \cdot B(X \to t\bar{t})$ correspond to the maximum of the Bayesian posterior probability density, limits are set at the point where the integral of the posterior probability density from zero reaches 95% of its total. Expected limits are obtained by applying the procedure when assuming that the observed result corresponded to the Standard Model expectation. The limits obtained for the Z-like, vector and axial-vector resonances are found to agree with each other, thus these limits are valid for a general (colour neutral) narrow resonance.

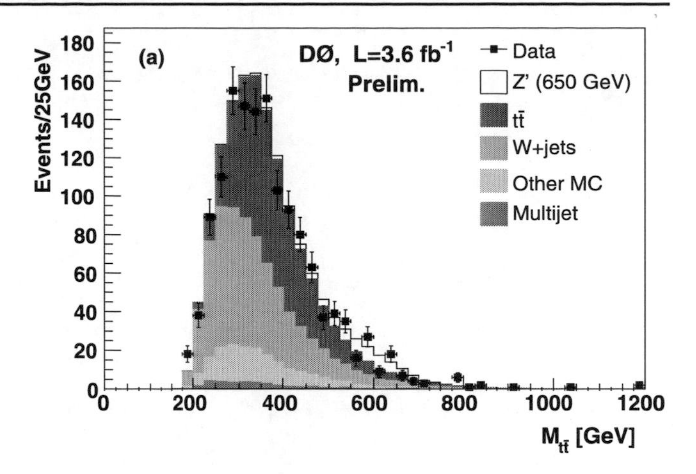

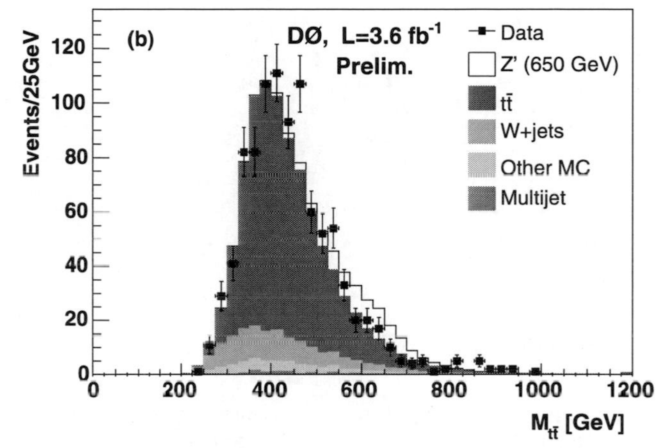

Fig. 60 Expected and observed $t\bar{t}$ invariant mass distribution for the combined (**a**) $\ell + 3$ jets and (**b**) $\ell + 4$ or more jets channels, with at least one identified b jet. Superimposed as white area is the expected signal for a Topcolour assisted Technicolour Z' boson with $M_{Z'} = 650$ GeV [267]

These expected limits were used to optimise major analysis cuts and the b-tag working point. In Fig. 61 (left) the expected limits are used to visualise the effect of the various systematics by including one after another. The lowest curve corresponds to a purely statistical limit. Adding the jet energy uncertainty shows that this uncertainty mainly contributes at medium resonance masses. The various object identification efficiencies and the luminosity are added. They essentially scale like the background shape. Finally the effect of the top quark mass is included and it is most important at low resonance masses [268].

In 3.6 fb^{-1} of DØ data the observed cross-sections are close to zero for all considered resonance masses, as shown in Fig. 61 (right). The largest deviation (around 650 GeV) corresponds to a little more than 1.5 standard deviations. Thus limits are set on the $\sigma_X \cdot B(X \to t\bar{t})$ as function of the assumed resonance mass, M_X. The excluded values range from about 1 pb for low mass resonances to less than 0.2 pb for the highest considered resonance mass of 1 TeV. The

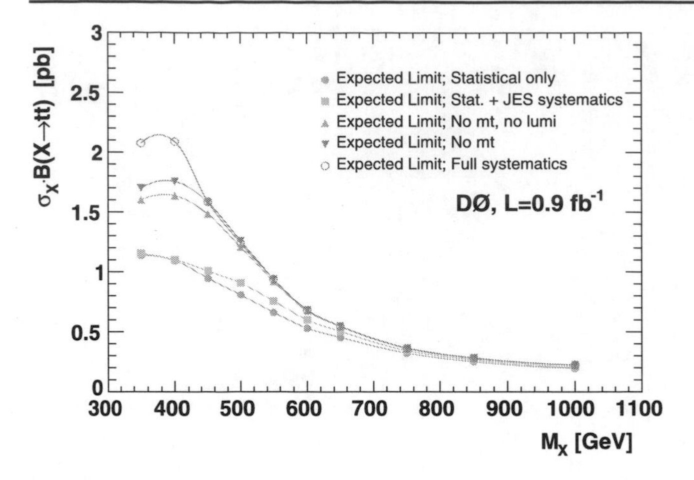

Fig. 61 *Left*: Expected limits $\sigma_X \cdot B(X \to t\bar{t})$ vs. the assumed resonance mass for 0.9 fb^{-1}. *From bottom* to *top* the lines represent the limit expected without systematics, including only JES systematics, excluding selection efficiencies, m_t and luminosity, all except m_t and complete systematics [268]. *Right*: Expected upper limits with complete systematics in 3.6 fb^{-1} compared to the observed cross-section and exclusion limits at 95% C.L. [267]

benchmark Topcolour assisted Technicolour model can be excluded for resonance masses of $M_{Z'} < 820$ GeV at 95% C.L.

4.5 Admixture of stop quarks

A final very fundamental question that may be asked in the context of top quark physics beyond the Standard Model is, whether the events that are considered to be top quarks actually are all top quarks or whether some additional unknown new particle is hiding in the selected data. The top quark's supersymmetric partners, the stop quarks $\tilde{t}_1$ and $\tilde{t}_2$, are possible candidates in such a scenario.

4.5.1 DØ, lepton plus jets

The stop quark decay modes to neutralino and top quark, $\tilde{\chi}_1^0 t$, or through chargino and b quark, $\tilde{\chi}_1^{\pm} b$, both yield a final state with a neutralino, a b quark and a W boson, $\tilde{\chi}_1^0 bW$. The neutralino is the lightest supersymmetric particle in many models and is stable if R-parity is conserved. Then it escapes the detector and the experimental signature of stop quark pair production differs from that of semileptonic top quark pair events only by the additional contribution to the missing transverse energy from the neutralino.

DØ has searched for a contribution of such stop quark pair production in the semileptonic channel in data with 0.9 fb^{-1} [269, 270]. Semileptonic events were selected following the corresponding $t\bar{t}$ cross-section analysis by looking for isolated leptons (e and μ), missing transverse energy and four or more jets. At least one of the jets was required to be identified as b-jet using DØ's neural network algorithm.

To describe the Standard Model expectation a mixture of data and simulation is employed. The description of top quark pair production (and of further minor backgrounds) is taken fully from simulation normalised to the corresponding theoretical cross-sections. For W + jets the kinematics is taken from simulation, but the normalisation is taken from data. Multijet background is fully estimated from data. As signal the lighter of the two stop quarks, $\tilde{t}_1$, is considered. Production of $\tilde{t}_1\bar{\tilde{t}}_1$ is simulated for various combinations of stop quark and chargino masses, $m_{\tilde{t}_1}, m_{\tilde{\chi}_1^{\pm}}$. For the sake of this analysis the stop quark mass was assumed to be less or equal to the top quark mass. The neutralino mass $m_{\tilde{\chi}_1^0} = 50$ GeV was chosen to be close to the experimental limit.

To detect a possible contribution of stop quark pairs the differences between stop quark pair events and Standard Model top quark pair production kinematic variables are combined into a likelihood, $\mathcal{L} = P_{\text{stop}}/(P_{\text{stop}} + P_{\text{SM}})$. The kinematic variables considered include the transverse momentum of the (leading) b jet, distances between leading b jet and lepton or leading other jet. Additional variables were reconstructed by applying a constrained fit. In this fit reconstructed physics objects (lepton, missing transverse energy and jets) are assigned to the decay products of an assumed semileptonic top quark pair event and the measured quantities are allowed to vary within their experimental resolution to fulfil additional constraints. It was required that the W boson mass is consistent with the invariant mass of the jets assigned to the two light quarks as well as with the invariant mass of the lepton with the neutrino. The masses of the reconstructed top quark pairs were constrained to be equal. Of the possible jet–parton assignments only the one with the best χ^2 was used. From the constrained fits observables the angle between the b-quarks and the beam axis in the

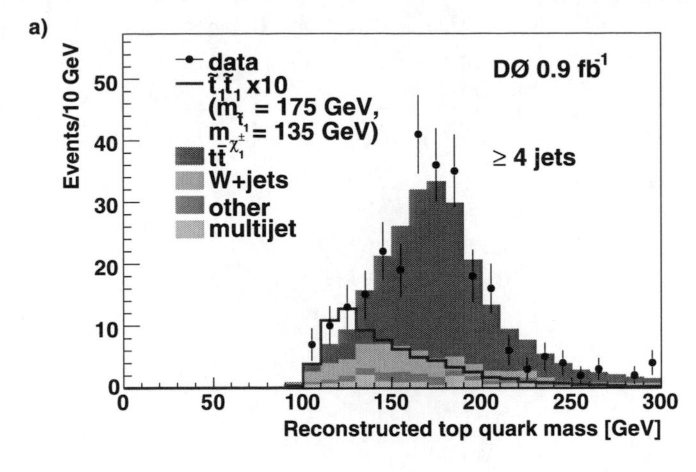

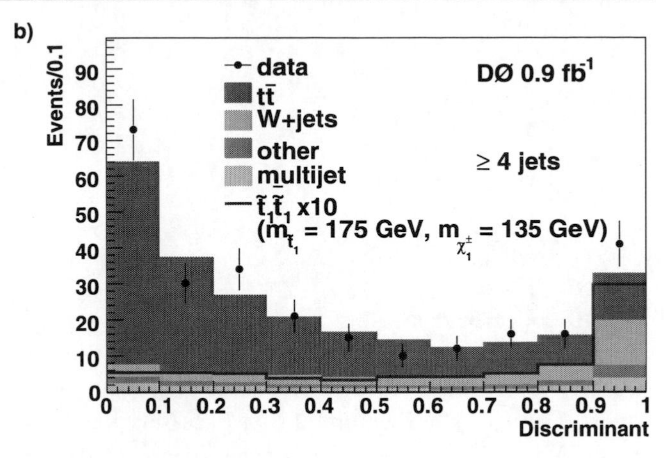

Fig. 62 *Left*: Expected and observed distribution for the reconstructed top quark mass. *Right*: Expected likelihood distribution for Standard Model and signal compared to data. In both plots only events with four or more jets are shown. The stop quark contribution corresponds to ten times the expectation in the MSSM with $m_{\tilde{t}_1} = 175$ GeV, $m_{\tilde{\chi}_1^\pm} = 135$ GeV [270]

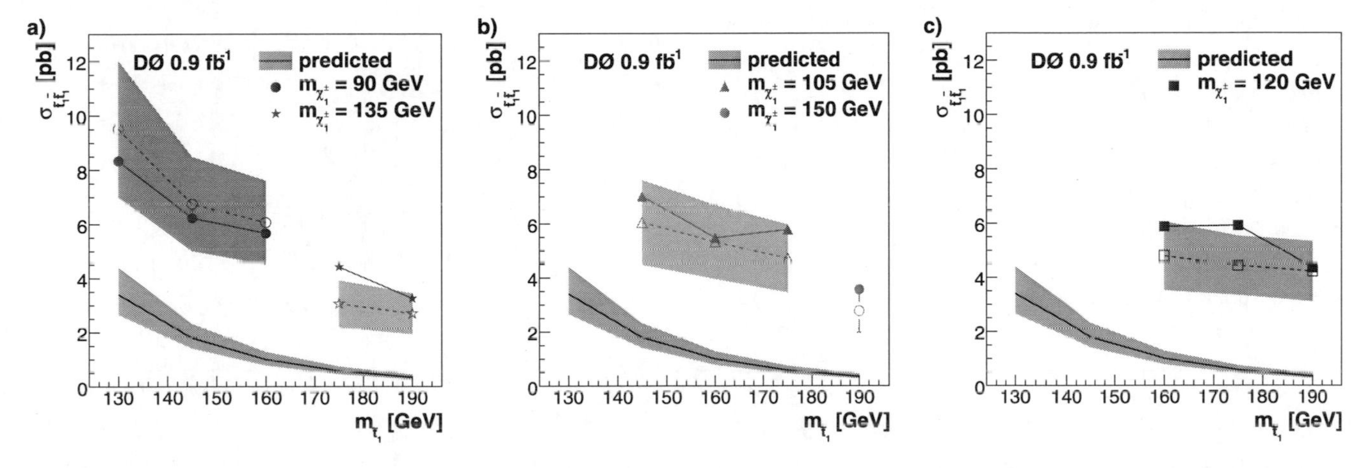

Fig. 63 Expected and observed limits on the stop quark pair production cross-section compared to the expectation in the various MSSM parameter sets [270]

$b\bar{b}$ rest frame, the $b\bar{b}$ invariant mass, the distances between the b's and the same-side or opposite-side W bosons and the reconstructed top quark mass are considered. The likelihood was derived for each $m_{\tilde{t}_1}$, $m_{\tilde{\chi}_1^\pm}$ combination separately and the selection of variables used has been optimised each time. Figure 62 shows the separation power of the reconstructed top quark mass and the full likelihood for the case of $m_{\tilde{t}_1} = 175$ GeV, $m_{\tilde{\chi}_1^\pm} = 135$ GeV and the comparison to the observed data.

To determine limits on the possible contribution of stop quark pair production in the selected channel Bayesian statistics is employed using a non-zero flat prior for positive values of the stop quark pair cross-section. A Poisson distribution is assumed for the number of events observed in each bin of the likelihood. The prior for the combined signal acceptance and background yields is a multivariate Gaussian with uncertainties and correlations described by a covariance matrix. The systematic uncertainty is dominated by the uncertainties on the theoretical cross-section of top quark pair production, on the selection efficiencies and the luminosity determination. Figure 63 shows the expected and observed limits on the stop quark pair production cross-section compared to the MSSM prediction for various values of $m_{\tilde{t}_1}$ and $m_{\tilde{\chi}_1^\pm}$. The theoretically expected stop quark signal cross-section in the MSSM is smaller than the experimental limits for all parameter points considered.

4.5.2 CDF, dilepton

CDF has searched for a contribution of stop quarks in the dilepton channel using up to 2.7 fb^{-1} of data [271]. The dilepton event signature was chosen to cover chargino decay modes to $\ell + \nu$ that do not involve an intermediate W boson and thus may not have a corresponding hadronic decay to build semileptonic events.

$$\begin{aligned}
\tilde{\chi}_1^{\pm} &\to \tilde{\chi}_1^0 + H^{\pm} \to \tilde{\chi}_1^0 + \ell + \nu \\
\tilde{\chi}_1^{\pm} &\to \ell + \tilde{\nu}_\ell \qquad \to \tilde{\chi}_1^0 + \ell + \nu \\
\tilde{\chi}_1^{\pm} &\to \tilde{\ell} + \nu \qquad \to \tilde{\chi}_1^0 + \ell + \nu \\
\tilde{\chi}_1^{\pm} &\to \tilde{\chi}_1^0 + G^{\pm} \to \tilde{\chi}_1^0 + \ell + \nu.
\end{aligned} \tag{90}$$

CDF collected data using an inclusive high-p_T trigger and selects events with two leptons, one of which needs to be isolated from calorimeter energies not associated to that lepton. The events are required to have missing transverse energy and at least two jets. A Z boson veto is applied for ee and $\mu\mu$ events. To suppress the leading background, Standard Model top quark pair events, from the selected event a cut in the plane of H_T vs. Δ plane is applied, where Δ is the product of the azimuthal angles between the leading jets and the two leptons: $\Delta = \Delta\phi(\text{jet}_1, \text{jet}_2)\Delta\phi(\ell_1, \ell_2)$. This cut reduces top quark pair production by a factor of 2, but reduces stop quark by approximately 12% only.

The Standard Model background expectation is modelled using simulation and control data. Simulation of top quark pair production and other minor backgrounds are normalised to their NLO cross-section. Z + jets samples are normalised to control data of low missing transverse energy near the Z-pole separately for events with and without b-tags. To model events with faked leptons that may stem from other top quark pair decay channels, from W + jets or from multi-jet events, parametrised lepton fake rates are derived from a large sample of generic jets. These fake rates are applied to events with lepton plus electron or muon like events to find the contribution of fakes in the signal region.

To describe signal events stop quark pair production is simulated with various combinations of neutralino, chargino and stop quark masses lighter than the top quark mass. Generated samples are interpolated through a template morphing technique to obtain any combination of masses within the generated range.

The reconstructed stop quark mass is used to distinguish a stop quark signal from Standard Model backgrounds including top quark pair production. The stop quark mass, $m_{\tilde{t}_1}$, is determined following an extension of the dilepton neutrino weighting technique.

b-jets are assigned to their proper lepton based on jet-lepton invariant mass quantities. A correct assignment is reached in 85% to 95% of the cases with both b-jets being the leading 2 jets. Neutralino and neutrino are considered as a single, though massive, pseudo particle. For given ϕ directions of the pseudo-particles the particle momenta are determined with a fit to the measured quantities using constraints on the assumed pseudo-particle mass, the assumed chargino mass and the equality of the two stop quark masses.

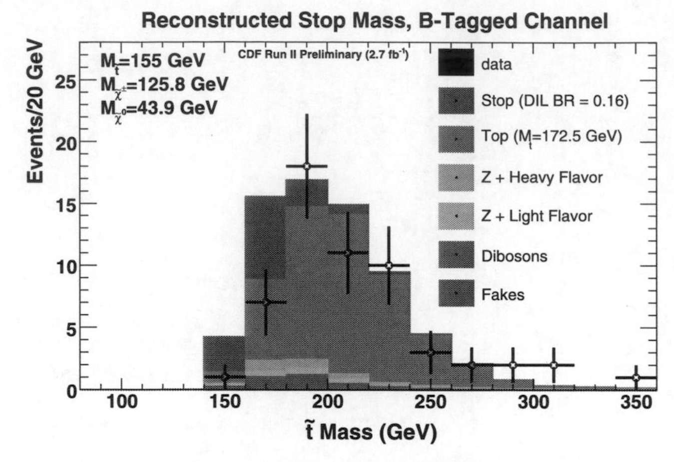

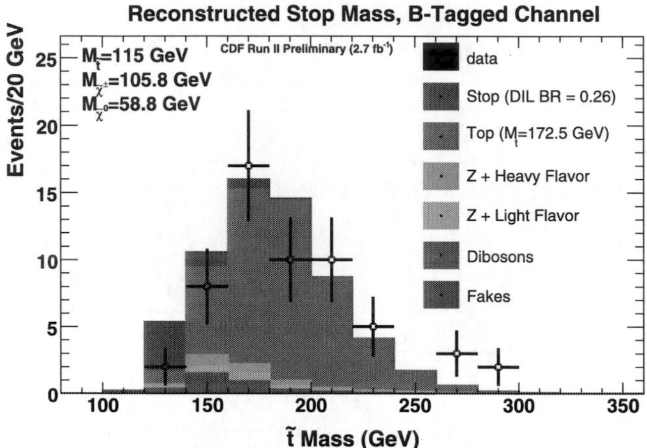

Fig. 64 Distribution of reconstructed stop quark masses in data and simulation for two choices of the stop quark, chargino and neutralino masses [271]

The reconstructed stop quark mass is computed as weighted average of the fitted stop quark masses, where the average is computed over all values of ϕ, with weights of $e^{-\chi^2}$. The expected and observed distributions for two choices of parameters are shown in Fig. 64.

The combination of reconstructed stop quark mass templates from the signal and the various background components is fitted to data. Systematic uncertainties enter the fit through nuisance parameters, signal and background contributions are allowed to vary within their rate uncertainties and the shape may vary according to CDFs morphing technique. The ratio of the likelihoods is used to do the limit-setting according to the CL$_s$ technique [227, 228].

Depending on the dilepton branching ratio limits are set in the stop quark vs. neutralino mass plane. Figure 65 shows the results for two choices of parameters. These limits are derived using only very few assumptions, these are (a) $\tilde{\chi}_1^0$ is the LSP and $\tilde{q}$, $\tilde{\ell}$ and $\tilde{\nu}$ are heavy, (b) $m_{\tilde{t}_1} \lesssim m_t$ and (c) $m_{\tilde{\chi}_1^{\pm}} < m_{\tilde{t}_1} - m_b$. Thus the limits are valid over a large range of SUSY parameter space.

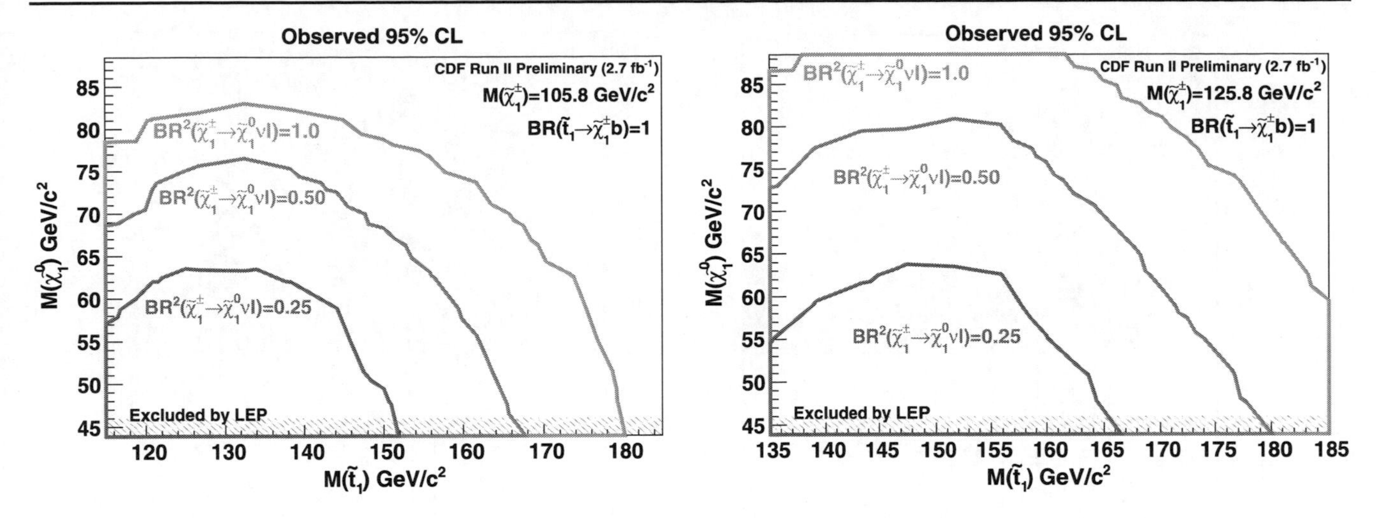

Fig. 65 CDF observed 95% C.L. limits in the stop mass vs. chargino mass plane. *The left* plot shows limits for a chargino mass of 105.8 GeV, *the right* for 125.8 GeV [271]

4.6 Heavy top-like quark, t'

Another class of hypothetical particles that may hide in samples usually considered as top quarks are new heavy quarks, in this context usually called t' quark. These new particles a considered to decay to Wq and thus show a signature very similar to that of top quarks.

Heavy top-like quarks appear in a large number of new physics models: A fourth generation of fermions [272], Little Higgs models [273] and more named in [274]. Strong bounds are placed on such models by electroweak precision data, but for special parameters the effects of the fourth generation particles on electroweak observables compensate. Among other settings a small mass splitting between the fourth u-type quark, t', and its isospin partner, b', is preferred, i.e. $m_{b'} + M_W > m_{t'}$ [272]. Especially, when the new top-like quark is very heavy, it should be distinguishable from Standard Model top production in kinematic observables. A search for such heavy top-like quarks has been performed by both Tevatron experiments.

CDF The CDF collaboration has repeatedly analysed their samples of lepton plus jets events to search for a new top-like quark. The published result uses 760 pb^{-1} [274] and the preliminary updated results 4.6 fb^{-1} of data [275]. The analyses consider pair production of a new quark heavier than the top quark and with a subsequent decay to Wq. The event selection requires exactly one isolated lepton, large missing transverse energy and at least four energetic jets.

The dominant Standard Model processes that pass this selection are top quark pair production which is simulated using PYTHIA, $W+$jets events which are simulated using ALPGEN + HERWIG or PYTHIA and normalised to data and multijet events which are modelled from data with reversed lepton identification. Minor backgrounds like $Z+$jets and diboson events are considered to be described by the $W+$ jets simulation. The simulation of $t'\bar{t}'$ signal is performed using PYTHIA.

As the t' quark decay chain is the same as the top quark decay chain a constrained fit to the kinematic properties is performed. The momenta of the quarks and leptons from the $t^{(\prime)}$ quark and the following W boson decays are fitted to the observed transverse momenta. The constraints require that the W boson decay products form the nominal W mass and that the decay products of the $t^{(\prime)}$ quark yield the same mass on the hadronic and the leptonic side. Of the 12 different jet–parton assignments CDF chooses the one with the best χ^2 of the fit. The corresponding $t^{(\prime)}$ mass from the fit, M_{Reco}, is used as one observable to separate signal from background. The second observable is the total transverse energy, H_T, i.e. the sum of transverse energies of the observed jets, the lepton and the missing transverse energy. The choices explicitly avoid imposing b-quark tagging requirements. The expected and observed individual distributions are shown in Fig. 66.

The signal and background shapes in the two-dimensional M_{Reco}–H_T plane are used to construct a likelihood for the observed data as function of the assumed t' quark cross-section, $\sigma_{t'}$. Then Bayesian statistics is employed to compute expected and observed limits on $\sigma_{t'}$ that are shown in Fig. 67.

Systematic uncertainties are implemented through nuisance parameters that are constrained in the fit with a Gaussian function to their nominal value within their expected uncertainty. The jet energy scale is named as one of the largest uncertainties. In addition uncertainties on the Q^2 scale used in simulating $W+$jets, on initial state and final state radiation, on the multijet background determination, the integrated luminosity, the lepton identification, the par-

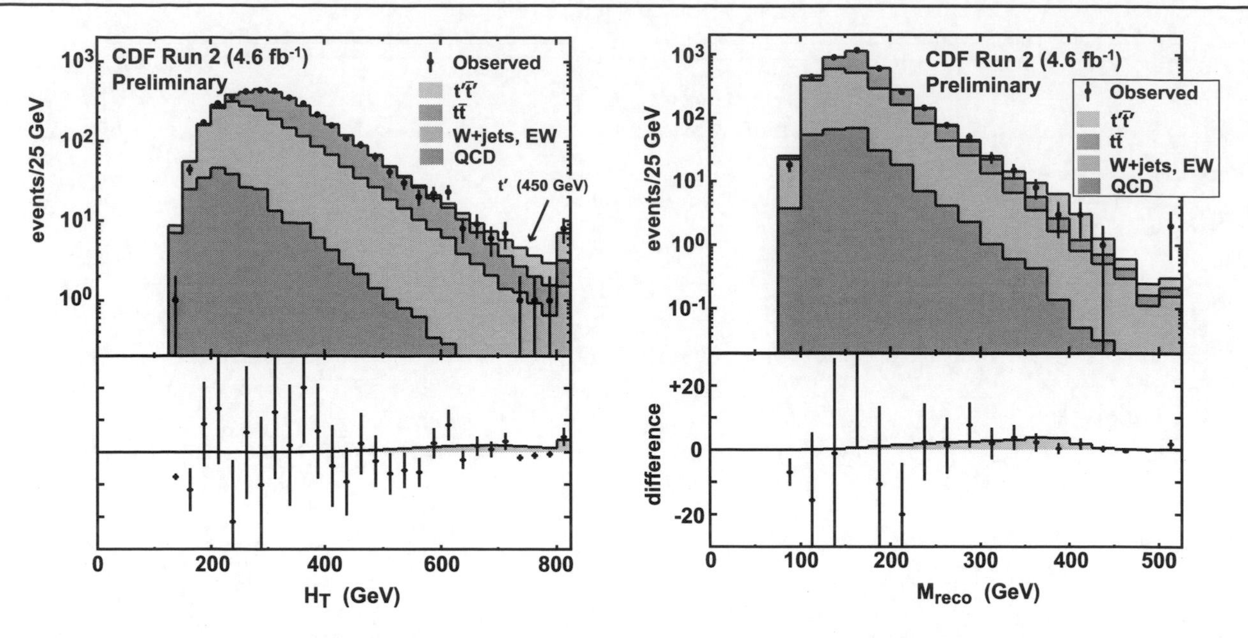

Fig. 66 Expected and observed distribution of the total transverse energy, H_T, and the reconstructed $t^{(\prime)}$ mass in 4.6 fb^{-1} of CDF data including a hypothetical signal at $m_{t'} = 450$ GeV [275]

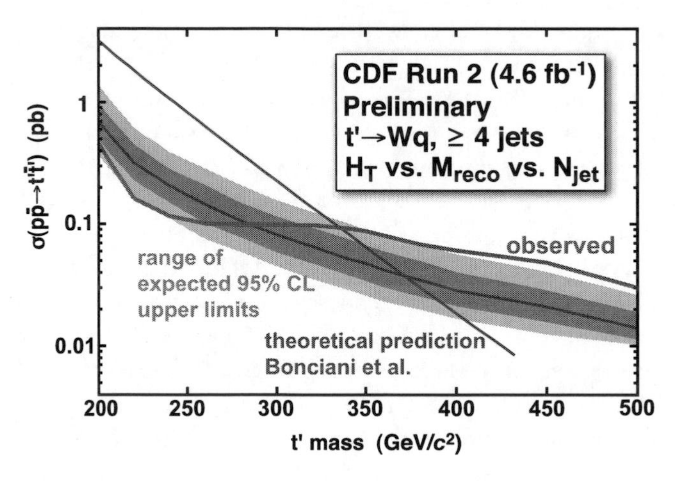

Fig. 67 Expected and observed limits on the cross-section of a new top-like quark determined from 4.6 fb^{-1} of CDF data [275]. The shaded bands show the expected one and two sigma variation on the expected upper limit

ton density functions and the expected t' quark cross-section as function of the t' quark mass are considered.

The limits on t' quark pair production cross-section, $\sigma_{t'}$, determined in this search for a new top-like heavy quark are compared to the theoretical prediction [31]. Assuming $t' \to Wq$ CDF concludes that a t' pair production can be excluded for $m_{t'} < 335$ GeV at 95% C.L. However, for masses of $m_{t'} \simeq 350$ GeV the observed limit is worse than the expectation by about two standard deviations, which indicates a surplus of data for that range.

DØ The DØ collaboration has searched for a heavy top-like quark with a decay to a W-boson and a quark using 4.3 fb^{-1} of data [276]. Events are selected which show an isolated lepton, large missing transverse energy and at least four jets.

The dominating backgrounds from Standard Model processes are top quark pair and $W+$jets production, which are simulated using ALPGEN + PYTHIA. The $t\bar{t}$ simulation and additional smaller backgrounds from diboson and Z-boson production are normalised to the expected Standard Model cross-sections. The normalisation of the $W+$jets contribution is determined below. The contribution of multijet background faking a lepton is derived from a sample with loosened lepton selection. Signal simulation is obtained from PYTHIA for 13 different t' mass values using a fixed decay width of 10 GeV.

Also DØ uses a reconstructed t'-quark invariant mass and the scalar sum of transverse momenta to distinguish the t'-quark signal from the Standard Model backgrounds. The t'-quark invariant mass is derived in a full event reconstruction assuming the expected decay $t'\bar{t}' \to \ell\nu bq\bar{q}\bar{b}$. In this reconstruction the momenta of the assumed decay products are fitted to the reconstructed measured objects applying constraints that enforce the mass of the intermediate W-bosons and equality of the initial t' and $\bar{t}'$ masses. All possible associations of the four leading jets in p_T to the quarks are considered. To select the best association in addition to the fit χ^2 a $\Delta\chi^2$ term is computed that prefers low transverse momenta for the reconstructed t'-quark. The t'-mass reconstructed, m_{fit}, with the association that minimises $\chi^2 + \Delta\chi^2$ is taken for further analysis.

Two-dimensional distributions of H_T vs. m_{fit} are used to fit the data compositions with free $W+$jets and signal

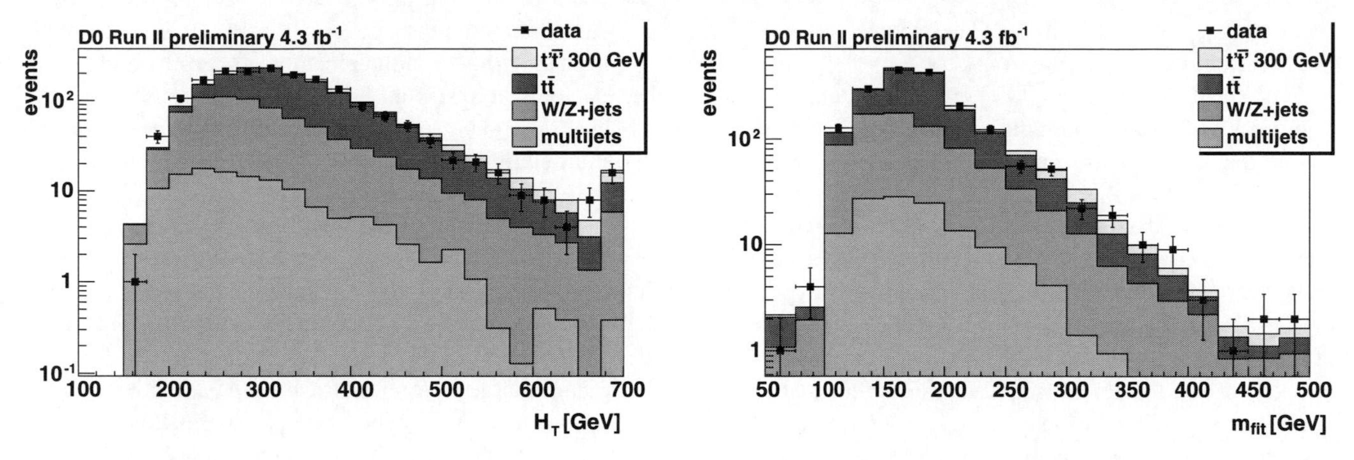

Fig. 68 Expected and observed distribution of the total transverse energy, H_T, and the reconstructed $t^{(\prime)}$ mass in 4.3 fb^{-1} of DØ data including a hypothetical signal at $m_{t'} = 300$ GeV [276]

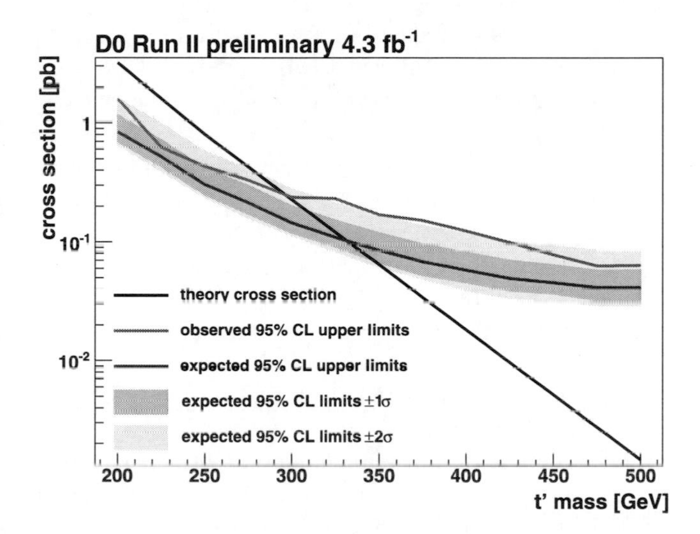

Fig. 69 Expected and observed limits for the production of a t'-quark obtained in 4.3 fb^{-1} of DØ data. The shaded bands indicate the one and two sigma deviations from the expected limit [276]

normalisations, cf. Fig. 68. This is done for all 13 t'-quark mass hypotheses and for a background only hypothesis. DØ uses the CL$_s$ method [227, 228] with a Poisson likelihood as the test statistics to determine the cross-section limits. DØ includes several systematic uncertainties affecting the normalisation of the fixed background contributions as well as shape changing effects in the limit calculation. Of the normalisation uncertainties the luminosity uncertainty gives the largest single contribution.

The limits on a t'-quark production cross-section observed by DØ in 4.3 fb^{-1} [276] and the corresponding expectations are shown in Fig. 69 as function of the hypothetical t' mass. By comparing to the theoretical prediction [31] DØ excludes heavy top-like quarks with masses less than 296 GeV at 95% C.L. Over almost the full range the observed limit stays behind the expected limit indicating a surplus of data in the t'-signal range. Correspondingly the expected mass exclusion of 330 GeV is not reached.

4.7 Outlook to LHC

Due to the high mass of the top quark it has been speculated that the top quark may play a special role in particle physics. Many models of new physics involve new particles that may occur in top quark production or decay. The Tevatron experiments have investigated a wide range of options for such new particles in top quark events and also looked for particles with signatures very similar to that of top quarks. So far no significant deviation has been found.

In the near future the LHC is scheduled to deliver at least 1 fb^{-1} at $\sqrt{s} = 7$ TeV. With such a dataset the LHC experiments will have roughly twice as many top quark pairs and three times as many single top quarks as the Tevatron with the expected 10 fb^{-1}. The benefit of this data, however, depends on the process in which the new particle appears.

Searches for new particles that occur in the top quark decay, like e.g. a charged Higgs boson, will benefit directly from the increased dataset and yield results competitive to those of the Tevatron. Searches for new particles that occur in the production like, Z' or W' bosons or a t' quark, will benefit from the larger centre-of-mass energy. The cross-sections of such processes, however, scale with the corresponding parton luminosity, which increases much more for the gluon than for the quark luminosity. Particles like the Z' boson that can by produced only by quark-antiquark annihilation thus will have less signal events in 1 fb^{-1} than the Tevatron will have in the expected 10 fb^{-1}. Such analyses will still extend the Tevatron results to the phase space areas opened with the higher centre-of-mass energy. Finally, the search for a boson produced in the single top quark s-channel, like the W', will suffer from the comparably moderate increase of the corresponding cross-section and thus require more statistics at the LHC to become competitive to the Tevatron results.

With the design LHC at $\sqrt{s} = 14$ TeV the larger production cross-section and the enlarged phase space will again extend the results significantly. An ATLAS study assuming collisions at $\sqrt{s} = 14$ TeV and a luminosity of 1 fb^{-1} [144] has considered the potential for discovering resonant top quark pair production through a narrow Z' boson. A significant degradation of the selection efficiency is expected at high top pair invariant masses because the top quark decay products join into the same jet more and more often. Given the performance visible in the first results shown at conferences from LHC experiments searches for new physics in top quark events will greatly advance at each new energy reached by the LHC after collecting a moderate luminosity of e.g. 1 fb^{-1}.

5 Conclusions

More than fifteen years after the discovery of the top quark the experimental verification of the properties expected in the Standard Model has reached a first level of maturity. With the luminosity delivered by the Tevatron accelerator in its Run II to the CDF and DØ experiments many top quark properties have been challenged and contributions for new particles have been searched for.

The top quark mass measurements, with the elaborate statistical methods of the Matrix Element technique, have now reached a statistical precision of less than 1 GeV and experimental systematics of 1 GeV making the top quark mass the most precisely known mass of all quarks. This precision significantly exceeds the precision goals of about 3 GeV of the Tevatron Run II programme set for 2 fb^{-1} [47, 277]. Despite the low branching fraction of the dileptonic channel, even results in this channel alone now have achieved an experimental precision exceeding this goal. In order to reach these small uncertainties, it was important to constrain the jet energy scale, leading to the dominating uncertainty, in situ to data. Unfortunately, the experimental precision on the top quark mass is currently not matched by a corresponding theoretical understanding. The uncertainty of the mass definition used in the simulations that the experiments apply to calibrate their measurements is not known better than to the order of 1 GeV. This complicates the comparison of the top quark mass results, e.g. with other electroweak precision data, and is a fundamental problem also for the LHC programme. Discussions on these issues have started between the experimental, the generator and the theoretical community in order to collect the information needed to overcome this issue. The development of methods that precisely determine the mass in well defined definitions is an important problem in current top quark physics.

Many interaction properties have been challenged by the Tevatron experiment in the recent years. Production properties such as the quark and gluon induced production rates, the electrical charge and the various decay properties did not show significant deviation from the expectation and thus confirm that we observe the top quark expected in the SM. The $V - A$ structure of the weak top quark decay has been tested in the W boson helicity measurements to the 5% level and weak decays through flavour changing neutral currents are constrained to be below the 4% level. The largest deviation (3.4σ) from the SM was found in the forward–backward charge asymmetry at high $m_{t\bar{t}}$. While the sheer number of tests documents a great progress the precision of these measurements is generally limited by statistics. Searches for new particles have been performed in events with top quark like signatures. Neither specific searches for supersymmetric top quark partners, for charged Higgs bosons nor searches for W' bosons, t' quarks or generic resonances found any significant deviations from the SM expectations. Also these studies are limited by the statistics available so far.

The Tevatron is planned to be shutdown in autumn 2011. Until then it is expected to increase the total integrated luminosity for the CDF and DØ experiments to more than 10 fb^{-1}. The analyses of the Tevatron take advantage of several years of optimising the reconstruction of the recorded events. Thus further updates of existing analyses, but also investigation of yet untested properties can be expected. In addition the LHC has started to operate at $\sqrt{s} = 7$ TeV well above that of the Tevatron centre-of-mass energy. It is expected to deliver an integrated luminosity of at least 1 fb^{-1} throughout 2011. Due to the increased collision energy at the LHC the top production cross-sections are significantly higher than at the Tevatron. With 1 fb^{-1} the LHC experiments will have collected about twice as many top quark pairs and three times as many single top quarks than the Tevatron experiments at 10 fb^{-1}.

Given the impressive performance of the LHC experiments shown at recent conferences it is natural to assume that the systematic uncertainties can be controlled comparably well as at the Tevatron experiments, if not better. We will thus see a plethora of competitive results from the LHC experiments. Clearly some searches profit from the increased phase space and will be dominated by the LHC. There are also results where the Tevatron has an advantage, like the charge asymmetry and the single top quark W' search, however, most results will be comparable. Also at the LHC many will be statistically limited which makes it useful to foresee a combination of the results.

After an update to the design LHC with proton–proton collisions at $\sqrt{s} = 14$ TeV the field will be taken over by the LHC experiments. With a top quark pair cross-section which is 100 times higher than at the Tevatron and dominant backgrounds like $W+$jets and $Z+$jets increasing only by factors of about 10 the LHC will become a real top quark factory. Again many searches for new particles will profit from the enlarged phase space. Measurements of interaction properties and searches for new particles in top quark decays will become limited by systematics after only a short running at design luminosity. At the same time with the huge expected statistics new types of measurement methods on rare subsamples will become feasible. For the top quark mass methods using leptonic J/ψ decays or using events with very high top quark momentum will have very different experimental and theoretical uncertainties. Such alternative methods with different systematic uncertainties will contribute to the combination of results. And may even help to solve the puzzle of the meaning of the top quark mass.

Acknowledgements The author would like to thank his colleagues in the DØ and CDF collaborations for their support in the preparation of this review. Special thanks belong to Frank Fiedler, Lina Galtieri, Alexander Grohsjean, Klaus Hamacher, Michele Weber and Wolfgang Wagner for valuable discussions about experimental details and to Werner Bernreuther, André Hoang, Sven Moch and Peter Uwer for detailed support and suggestions on theoretical questions. Finally, I would like to thank Stefan Tapprogge for valuable suggestions in the preparation of this work.

References

1. S.L. Glashow, Partial symmetries of weak interactions. Nucl. Phys. **22**, 579 (1961)
2. J. Goldstone, A. Salam, S. Weinberg, Broken symmetries. Phys. Rev. **127**, 965 (1962)
3. S. Weinberg, A model of leptons. Phys. Rev. Lett. **19**, 1264 (1967)
4. H. Fritzsch, M. Gell-Mann, H. Leutwyler, Advantages of the color octet gluon picture. Phys. Lett. B **47**, 365 (1973)
5. P.W. Higgs, Broken symmetries, massless particles and gauge fields. Phys. Lett. **12**, 132 (1964)
6. K. Nakamura et al. (Particle Data Group), Review of particle physics. J. Phys. G **37**, 075021 (2010)
7. P. de Bernardis et al. (Boomerang), A flat universe from high-resolution maps of the cosmic microwave background radiation. Nature **404**, 955–959 (2000). arXiv:astro-ph/0004404
8. D.N. Spergel et al. (WMAP), First year Wilkinson Microwave Anisotropy Probe (WMAP) observations: Determination of cosmological parameters. Astrophys. J. Suppl. Ser. **148**, 175 (2003). arXiv:astro-ph/0302209
9. P. Astier et al. (SNLS), The supernova legacy survey: Measurement of Ω_M, Ω_Λ and w from the first year data set. Astron. Astrophys. **447**, 31–48 (2006). arXiv:astro-ph/0510447
10. D.N. Spergel et al. (WMAP), Wilkinson Microwave Anisotropy Probe (WMAP) three year results: Implications for cosmology. Astrophys. J. Suppl. Ser. **170**, 377 (2007). arXiv:astro-ph/0603449
11. F. Abe et al. (CDF), Observation of top quark production in $\bar{p}p$ collisions. Phys. Rev. Lett. **74**, 2626–2631 (1995). arXiv:hep-ex/9503002
12. S. Abachi et al. (D0), Observation of the top quark. Phys. Rev. Lett. **74**, 2632–2637 (1995). hep-ex/9503003
13. R.S. Chivukula, B.A. Dobrescu, H. Georgi, C.T. Hill, Top quark seesaw theory of electroweak symmetry breaking. Phys. Rev. D **59**, 075003 (1999). arXiv:hep-ph/9809470
14. B.A. Dobrescu, C.T. Hill, Electroweak symmetry breaking via top condensation seesaw. Phys. Rev. Lett. **81**, 2634–2637 (1998). arXiv:hep-ph/9712319
15. A. Quadt, Top quark physics at hadron colliders. Eur. Phys. J. C **48**, 835–1000 (2006)
16. M. Beneke et al., Top quark physics. arXiv:hep-ph/0003033
17. W. Wagner, Top quark physics in hadron collisions. Rep. Prog. Phys. **68**, 2409–2494 (2005). arXiv:hep-ph/0507207
18. R. Kehoe, M. Narain, A. Kumar, Review of top quark physics results. Int. J. Mod. Phys. A **23**, 353–470 (2008). arXiv:0712.2733
19. R. Demina, E.J. Thomson, Top quark properties and interactions. Annu. Rev. Nucl. Part. Sci. **58**, 125–146 (2008)
20. M.-A. Pleier, Review of top quark properties measurements at the tevatron. Int. J. Mod. Phys. A **24**, 2899–3037 (2009). arXiv:0810.5226
21. W. Bernreuther, Top quark physics at the LHC. J. Phys. G **35**, 083001 (2008). arXiv:0805.1333
22. J.R. Incandela, A. Quadt, W. Wagner, D. Wicke, Status and prospects of top-quark physics. Prog. Part. Nucl. Phys. **63**, 239–292 (2009). arXiv:0904.2499
23. F. Deliot, D. Glenzinski, Top quark physics at the tevatron. arXiv:1010.1202. Submitted to Rev. Mod. Phys.
24. CDF, A. Heinson (D0), Observation of single top quark production at the Tevatron collider. Mod. Phys. Lett. A **25**, 309–339 (2010). arXiv:1002.4167
25. Y. Fukuda et al. (Super-Kamiokande), Evidence for oscillation of atmospheric neutrinos. Phys. Rev. Lett. **81**, 1562–1567 (1998). arXiv:hep-ex/9807003
26. Q.R. Ahmad et al. (SNO), Measurement of the rate of $\nu_e + d \to p + p + e^-$ interactions produced by ^{8}B solar neutrinos at the Sudbury Neutrino Observatory. Phys. Rev. Lett. **87**, 071301 (2001). arXiv:nucl-ex/0106015
27. Q.R. Ahmad et al. (SNO), Direct evidence for neutrino flavor transformation from neutral-current interactions in the Sudbury Neutrino Observatory. Phys. Rev. Lett. **89**, 011301 (2002). arXiv:nucl-ex/0204008
28. N. Cabibbo, Unitary symmetry and leptonic decays. Phys. Rev. Lett. **10**, 531–533 (1963)
29. M. Kobayashi, T. Maskawa, CP violation in the renormalizable theory of weak interaction. Prog. Theor. Phys. **49**, 652–657 (1973)
30. N. Kidonakis, R. Vogt, Next-to-next-to-leading order soft-gluon corrections in top quark hadroproduction. Phys. Rev. D **68**, 114014 (2003). hep-ph/0308222
31. M. Cacciari, S. Frixione, M.L. Mangano, P. Nason, G. Ridolfi, The $t\bar{t}$ cross-section at 1.8 TeV and 1.96 TeV: A study of the systematics due to parton densities and scale dependence. J. High Energy Phys. **04**, 068 (2004). hep-ph/0303085
32. M. Cacciari, S. Frixione, M.L. Mangano, P. Nason, G. Ridolfi, Updated predictions for the total production cross sections of top and of heavier quark pairs at the Tevatron and at the LHC. J. High Energy Phys. **09**, 127 (2008). arXiv:0804.2800
33. S. Moch, P. Uwer, Theoretical status and prospects for top-quark pair production at hadron colliders. Phys. Rev. D **78**, 034003 (2008). arXiv:0804.1476
34. S. Moch, P. Uwer, Heavy-quark pair production at two loops in QCD. Nucl. Phys. B, Proc. Suppl. **183**, 75–80 (2008). arXiv:0807.2794
35. N. Kidonakis, R. Vogt, The theoretical top quark cross section at the Tevatron and the LHC. Phys. Rev. D **78**, 074005 (2008). arXiv:0805.3844
36. J. Pumplin et al., New generation of parton distributions with uncertainties from global QCD analysis. J. High Energy Phys. **07**, 012 (2002). arXiv:hep-ph/0201195
37. B.W. Harris, E. Laenen, L. Phaf, Z. Sullivan, S. Weinzierl, The fully differential single top quark cross-section in next

to leading order QCD. Phys. Rev. D **66**, 054024 (2002). arXiv:hep-ph/0207055
38. Z. Sullivan, Understanding single-top-quark production and jets at hadron colliders. Phys. Rev. D **70**, 114012 (2004). arXiv:hep-ph/0408049
39. N. Kidonakis, Single top production at the Tevatron: Threshold resummation and finite-order soft gluon corrections. Phys. Rev. D **74**, 114012 (2006). arXiv:hep-ph/0609287
40. A.D. Martin, R.G. Roberts, W.J. Stirling, R.S. Thorne, Physical gluons and high E_T jets. Phys. Lett. B **604**, 61–68 (2004). arXiv:hep-ph/0410230
41. J.H. Kühn, Acta Phys. Pol. **12**, 347 (1981)
42. http://www-d0.fnal.gov/Run2Physics/top/top_public_web_pages/top_feynman_diagrams.html
43. http://www-bdnew.fnal.gov/operations/rookie_books/rbooks.html
44. http://www-d0.fnal.gov/runcoor/RUN/run2_lumi.html
45. V. Rusu, Tevatron operation and physics. FNAL Physics Advisory Commitee, http://www.fnal.gov/directorate/program_planning/Mar2009PACPublic/PACMarch09AgendaPublic.htm, March, 2009
46. F. Abe et al. (CDF), The CDF detector: an overview. Nucl. Instrum. Methods A **271**, 387–403 (1988)
47. R. Blair et al. (CDF-II), The CDF-II detector: Technical design report. FERMILAB-PUB-96-390-E, 1996
48. D.E. Acosta et al. (CDF), Measurement of the J/ψ meson and b-hadron production cross sections in $p\bar{p}$ collisions at $\sqrt{s} = 1960$ GeV. Phys. Rev. D **71**, 032001 (2005). arXiv:hep-ex/0412071
49. A. Sill (CDF), CDF Run II silicon tracking projects. Nucl. Instrum. Methods A **447**, 1–8 (2000)
50. A. Bardi et al., The CDF online silicon vertex tracker. Nucl. Instrum. Methods A **485**, 178–182 (2002)
51. A.A. Affolder et al. (CDF), Intermediate silicon layers detector for the CDF experiment. Nucl. Instrum. Methods A **453**, 84–88 (2000)
52. D. Acosta et al. (CDF-II), A time-of-flight detector in CDF-II. Nucl. Instrum. Methods A **518**, 605–608 (2004)
53. S. Abachi et al. (D0), The D0 detector. Nucl. Instrum. Methods A **338**, 185–253 (1994)
54. V.M. Abazov et al. (D0), The upgraded D0 detector. Nucl. Instrum. Methods A **565**, 463–537 (2006). arXiv:physics/0507191
55. A. Heinson, http://www-d0.fnal.gov/Run2Physics/top/top_public_web_pages/top_dzero_detector.html
56. V.M. Abazov et al. (D0), Measurement of the $t\bar{t}$ production cross section in $p\bar{p}$ collisions at $\sqrt{s} = 1.96$ TeV using kinematic characteristics of lepton + jets events. Phys. Rev. D **76**, 092007 (2007). arXiv:0705.2788
57. A. Abulencia et al. (CDF), Measurement of the $t\bar{t}$ production cross section in $p\bar{p}$ collisions at $\sqrt{s} = 1.96$ TeV using lepton + jets events with jet probability b^- tagging. Phys. Rev. D **74**, 072006 (2006). arXiv:hep-ex/0607035
58. G. Blazey et al., in *QCD and weak boson physics in Run II*, ed. by U. Baur, R.K. Ellis, D. Zeppenfeld (2000), FERMILAB-PUB-00-297
59. D.E. Acosta et al. (CDF), Measurement of the $t\bar{t}$ production cross section in $p\bar{p}$ collisions at $\sqrt{s} = 1.96$ TeV using lepton plus jets events with semileptonic B decays to muons. Phys. Rev. D **72**, 032002 (2005). arXiv:hep-ex/0506001
60. T. Scanlon, b-tagging and the search for neutral supersymmetric Higgs bosons at D0. FERMILAB-THESIS-2006-43
61. E. Laenen, Top quark in theory. arXiv:0809.3158, 2008
62. S. Fleming, A.H. Hoang, S. Mantry, I.W. Stewart, Factorization approach for top mass reconstruction at high energies. arXiv:0710.4205, 2007
63. A.H. Hoang, I.W. Stewart, Top mass measurements from jets and the Tevatron top-quark mass. Nucl. Phys. B, Proc. Suppl. **185**, 220–226 (2008). arXiv:0808.0222
64. DELPHI, Mass effects in the PYTHIA generator. DELPHI 2003-061. PHYS 932, June, 2003
65. T. Aaltonen et al. (CDF), First simultaneous measurement of the top quark mass in the lepton + jets and dilepton channels at CDF. Phys. Rev. D **79**, 092005 (2009). arXiv:0809.4808
66. CDF, Combined template-based top quark mass measurement in the lepton + jets and dileptons channels using 2.7 fb^{-1} of data. CDF Note 9578, Oct., 2008
67. CDF, Simultaneous template-based top quark mass measurement in the lepton + jets and dileptons channels including m_{T2}. CDF Note 9679. http://www-cdf.fnal.gov/physics/new/top/2009/mass/TMT_p19_public/, Mar., 2009
68. CDF, Combined template-based top quark mass measurement in the lepton + jets and dileptons channels using 5.6 fb^{-1} of data. CDF Note 10273, Aug., 2010
69. A.A. Affolder et al. (CDF), Measurement of the $t\bar{t}$ production cross section in $p\bar{p}$ collisions at $\sqrt{s} = 1.8$ TeV. Phys. Rev. D **64**, 032002 (2001). arXiv:hep-ex/0101036. Phys. Rev. D **65**, 039902 (2002), Erratum
70. M.L. Mangano, M. Moretti, F. Piccinini, R. Pittau, A.D. Polosa, ALPGEN, a generator for hard multiparton processes in hadronic collisions. J. High Energy Phys. **07**, 001 (2003). arXiv:hep-ph/0206293
71. T. Sjostrand, S. Mrenna, P. Skands, PYTHIA 6.4 physics and manual. J. High Energy Phys. **05**, 026 (2006). arXiv:hep-ph/0603175
72. M. Sandhoff, P. Skands, Colour annealing: A toy model of colour reconnections. FERMILAB-CONF-05-518-T, in Les Houches 'Physics at TeV Colliders' (2005) SM and Higgs Working Group: Summary report, hep-ph/0604120, 2005
73. P. Skands, D. Wicke, Non-perturbative QCD effects and the top mass at the Tevatron. Eur. Phys. J. C **52**, 133–140 (2007). hep-ph/0703081
74. D. Wicke, P.Z. Skands, Non-perturbative QCD effects and the top mass at the Tevatron. Nuovo Cimento **123B**, 1–8 (2008). arXiv:0807.3248
75. D0, Measurement of the top quark mass in the lepton + jets channel using DØ Run II data. D0 note 4574-CONF, August, 2004
76. D0, Measurement of the top quark mass in the lepton + jets channel using DØ Run II data: The low bias template method. D0 note 4728-CONF, 2005
77. P. Abreu et al. (DELPHI), Measurement of the W pair cross-section and of the W mass in e^+e^- interactions at 172 GeV. Eur. Phys. J. C **2**, 581–595 (1998)
78. P. Abreu et al. (DELPHI), Measurement of the mass of the W boson using direct reconstruction at $\sqrt{s} = 183$ GeV. Phys. Lett. B **462**, 410–424 (1999)
79. V.M. Abazov et al. (D0), Measurement of the top quark mass in the lepton + jets channel using the ideogram method. Phys. Rev. D **75**, 092001 (2007). arXiv:hep-ex/0702018
80. V.M. Abazov et al. (D0), Measurement of the $t\bar{t}$ production cross section in $p\bar{p}$ collisions at $\sqrt{s} = 1.96$ TeV using kinematic characteristics of lepton + jets events. Phys. Lett. B **626**, 45–54 (2005). arXiv:hep-ex/0504043
81. Y. Peters, K. Hamacher, D. Wicke (D0), Precise tuning of the b fragmentation for the D0 Monte Carlo. FERMILAB-TM-2425-E, D0 Note 5229
82. V.M. Abazov et al. (D0), A precision measurement of the mass of the top quark. Nature **429**, 638–642 (2004). arXiv:hep-ex/0406031
83. K. Kondo, T. Chikamatsu, S.H. Kim, Dynamical likelihood method for reconstruction of events with missing momentum. 3: Analysis of a CDF high p_T $e\mu$ event as $t\bar{t}$ production. J. Phys. Soc. Jpn. **62**, 1177–1182 (1993)
84. R.H. Dalitz, G.R. Goldstein, Test of analysis method for top-antitop production and decay events. Proc. R. Soc. Lond. Ser. A, Math. Phys. Sci. **455**, 2803–2834 (1999). arXiv:hep-ph/9802249

85. D0, Measurement of the top quark mass in the lepton + jets final state with the matrix element method. Phys. Rev. D **74**, 092005 (2006). hep-ex/0609053
86. V.M. Abazov et al. (D0), Precise measurement of the top quark mass from lepton + jets events at D0. Phys. Rev. Lett. **101**, 182001 (2008). arXiv:0807.2141
87. D0, Measurement of the top quark mass in the lepton + jets channel using the matrix element method in 2.2 fb^{-1} of DØRun II data. D0 note 5750-CONF, July, 2008
88. D0, Measurement of the top quark mass in the lepton + jets channel using the matrix element method in 3.6 fb^{-1} of DØRun II data. D0 note 5877-CONF, March, 2009
89. L. Lyons, D. Gibaut, P. Clifford, How to combine correlated estimates of a single physical quantity. Nucl. Instrum. Methods A **270**, 110 (1988)
90. The Tevatron Electroweak Working Group for the CDF and D0 Collaborations, Combination of CDF and D0 results on the mass of the top quark. arXiv:1007.3178, 2010
91. A. Abulencia et al. (CDF), Precise measurement of the top quark mass in the lepton + jets topology at CDF II. Phys. Rev. Lett. **99**, 182002 (2007). arXiv:hep-ex/0703045
92. CDF, Top mass measurement using matrix element method and lepton + jets channel. CDF Conf. Note 9725, Apr., 2009
93. A. Abulencia et al. (CDF), Measurement of the top quark mass with the dynamical likelihood method using lepton plus jets events with b-tags in $p\bar{p}$ collisions at $\sqrt{s} = 1.96$ TeV. Phys. Rev. D **73**, 092002 (2006). hep-ex/0512009
94. CDF, Top quark mass measurement using the dynamical likelihood template medhod in the lepton plus jets channel at CDF Run II. CDF Note 9135, Dec., 2007
95. CDF, Top mass measurement in the lepton + jets channel using a matrix element method with Quasi-Monte Carlo integration and in situ jet calibration with 2.7 fb^{-1}. CDF Conf. Note 9427, July, 2008
96. CDF, Top mass measurement in the lepton + jets channel using a matrix element method with Quasi-Monte Carlo integration and in situ jet calibration with 3.2 fb^{-1}. CDF Conf. Note 9692, Feb., 2009
97. T. Aaltonen et al. (CDF), Top quark mass measurement in the lepton plus jets channel using a modified matrix element method. Phys. Rev. D **79**, 072001 (2009). arXiv:0812.4469
98. CDF, Top mass measurement in the lepton + jets channel using a matrix element method with Quasi-Monte Carlo integration and in situ jet calibration with 5.6 fb^{-1}. CDF Conf. Note 10191, June, 2010
99. T. Aaltonen et al. (CDF), Top quark mass measurement in the lepton + jets channel using a matrix element method and in situ jet energy calibration. Phys. Rev. Lett. **105**, 252001 (2010). arXiv:1010.4582
100. R. Kleiss, W.J. Stirling, Top quark production at hadron colliders: Some useful formulae. Z. Phys. C **40**, 419–423 (1988)
101. A. Abulencia et al. (CDF, Run II), Measurement of the top quark mass in $p\bar{p}$ collisions at $\sqrt{s} = 1.96$ TeV using the decay length technique. Phys. Rev. D **75**, 071102 (2007). arXiv:hep-ex/0612061
102. T. Aaltonen et al. (CDF), Measurements of the top-quark mass using charged particle tracking. Phys. Rev. D **81**, 032002 (2010). arXiv:0910.0969
103. CDF, Measurement of the top quark mass with 2.7 fb^{-1} of CDF RunII data in the lepton + jets channel using only leptons. CDF Note 9683, July, 2009
104. CDF, Measurement of the top quark mass from the lepton p_T in the $t\bar{t} \rightarrow$ dilepton channel using b-tagging at 2.8 fb^{-1}. CDF Note 9831, July, 2009
105. CDF, Lepton + jets and dilepton combined measurement of the top quark mass from the leptons' p_T using b-tagging at 2.8 fb^{-1}. CDF Note 9881, Aug., 2009
106. T. Aaltonen et al. (CDF), Measurement of the top quark mass at CDF using the 'neutrino ϕ weighting' template method on a lepton plus isolated track sample. Phys. Rev. D **79**, 072005 (2009). arXiv:0901.3773
107. G. Corcella et al., HERWIG 6.5: An event generator for hadron emission reactions with interfering gluons (including supersymmetric processes). J. High Energy Phys. **01**, 010 (2001). arXiv:hep-ph/0011363
108. D0, Measurement of the top quark mass in dilepton final states via neutrino weighting. D0 note 5746-CONF, July, 2008
109. V.M. Abazov et al. (D0), Measurement of the top quark mass in final states with two leptons. Phys. Rev. D **80**, 092006 (2009). arXiv:0904.3195
110. V.M. Abazov et al. (D0), Measurement of the top quark mass in the dilepton channel. Phys. Lett. B **655**, 7 (2007). arXiv:hep-ex/0609056
111. R.H. Dalitz, G.R. Goldstein, The decay and polarization properties of the top quark. Phys. Rev. D **45**, 1531–1543 (1992)
112. D0, Measurement of the top quark mass in the dilepton channel using the matrix weigthing method at DØ. D0 note 5463-CONF, Aug., 2007
113. D0, Measurement of the mass of the top quark in $e\mu$+jets final states at DØ with 5.3 fb^{-1}. D0 note 6104-CONF, Aug., 2010
114. A. Abulencia et al. (CDF), Top quark mass measurement from dilepton events at cdf ii. Phys. Rev. Lett. **96**, 152002 (2006). hep-ex/0512070
115. A. Abulencia et al. (CDF), Top quark mass measurement from dilepton events at CDF II with the matrix-element method. Phys. Rev. D **74**, 032009 (2006). hep-ex/0605118
116. CDF, Measurement of the top quark mass in the dilepton channel using a matrix element method with 1.8 fb^{-1}. CDF Note 8951, Aug., 2007
117. D0, Measurement of the top quark mass in the electron-muon channel using the matrix element method with 3.6 fb^{-1}. D0 note 5897-CONF, Mar., 2009
118. V.M. Abazov et al. (D0), Measurement of the top quark mass in all-jet events. Phys. Lett. B **606**, 25–33 (2005). hep-ex/0410086
119. T. Aaltonen et al. (The CDF), Measurement of the top quark mass and $p\bar{p}$–$t\bar{t}$ cross section in the all-hadronic mode with the CDFII detector. Phys. Rev. D **81**, 052011 (2010). arXiv:1002.0365
120. R.D. Field (CDF), The underlying event in hard scattering processes. arXiv:hep-ph/0201192, CDF Note 6403; further recent talks available from webpage http://www.phys.ufl.edu/~rfield/cdf/, 2002
121. T. Aaltonen et al. (CDF), Measurement of the top-quark mass in all-hadronic decays in p anti-p collisions at CDF II. Phys. Rev. Lett. **98**, 142001 (2007). arXiv:hep-ex/0612026
122. CDF, Measurement of the top quark mass with in situ jet energy calibration in the all-hadronic channel using the ideogram method with 1.9 fb^{-1}. CDF Note 9265, Mar., 2008
123. V.M. Abazov et al. (D0), Top quark mass extraction from $t\bar{t}$ cross section measurements. DØ Note 5742 conf, 2008
124. V.M. Abazov et al. (D0), Measurement of the $t\bar{t}$ production cross section in $p\bar{p}$ collisions at $\sqrt{s} = 1.96$ TeV. Phys. Rev. Lett. **100**, 192004 (2008). arXiv:0803.2779
125. V.M. Abazov et al. (D0), Measurement of the ttbar production cross section and top quark mass extraction using dilepton events in ppbar collisions. Phys. Lett. B **679**, 177–185 (2009). arXiv:0901.2137
126. V.M. Abazov et al. (D0), Combination of $t\bar{t}$ cross section measurements and constraints on the mass of the top quark and its decays into charged Higgs bosons. Phys. Rev. D **80**, 071102 (2009). arXiv:0903.5525
127. W. Beenakker, H. Kuijf, W.L. van Neerven, J. Smith, QCD corrections to heavy quark production in p anti-p collisions. Phys. Rev. D **40**, 54–82 (1989)

128. P.M. Nadolsky et al., Implications of CTEQ global analysis for collider observables. Phys. Rev. D **78**, 013004 (2008). arXiv:0802.0007
129. U. Langenfeld, S. Moch, P. Uwer, Measuring the running top-quark mass. Phys. Rev. D **80**, 054009 (2009). arXiv:0906.5273
130. J. Rathsman, A generalised area law for hadronic string reinteractions. Phys. Lett. B **452**, 364–371 (1999). hep-ph/9812423
131. B.R. Webber, Colour reconnection and Bose-Einstein effects. J. Phys. G **24**, 287–296 (1998). arXiv:hep-ph/9708463
132. A.B. Galtieri, MC4LHC Readiness Workshop, CERN http://indico.cern.ch/conferenceOtherViews.py?view=standard&confId=74601, March, 2010
133. D.E. Acosta et al. (CDF), Study of jet shapes in inclusive jet production in $p\bar{p}$ collisions at $\sqrt{s} = 1.96$ TeV. Phys. Rev. D **71**, 112002 (2005). arXiv:hep-ex/0505013
134. F. Fiedler, Independent measurement of the top quark mass and the light- and bottom-jet energy scales at hadron colliders. Eur. Phys. J. C **53**, 41–48 (2008). arXiv:0706.1640
135. F. Fiedler, A. Grohsjean, P. Haefner, P. Schieferdecker, The matrix element method and its application to measurements of the top quark mass. Nucl. Instrum. Methods A **624**, 203–218 (2010). arXiv:1003.1316
136. V.M. Abazov (D0) et al., Direct measurement of the mass difference between top and antitop quarks. Phys. Rev. Lett. **103**, 132001 (2009). arXiv:0906.1172
137. CDF, Measurement of top quark and anti-top quark mass difference in the lepton + jets channel. CDF Note 10173, June, 2010
138. LEP Electorweak Working Group, A combination of preliminary electroweak measurements and constraints on the standard model. LEPEWWG/94-02, ALEPH 94-121 PHYSIC 94-105, DELPHI 94-110 PHYS 427, L3 Note 1631, Opal TN245, July, 1994
139. The ALEPH, DELPHI, L3, OPAL, SLD Collaboration, The LEP Electroweak Working Group, The SLD Electroweak Heavy Flavour Groups, Precision electroweak measurements on the Z resonance. Phys. Rep. **427**, 257 (2006). arXiv:hep-ex/0509008
140. G. Aad et al. (Atlas), Measurement of the top quark-pair production cross section with ATLAS in pp collisions at $\sqrt{s} = 7$ TeV. arXiv:1012.1792
141. V. Khachatryan et al. (CMS), First measurement of the cross section for top-quark pair production in proton–proton collisions at $sqrt(s) = 7$ TeV. Phys. Lett. B **695**, 424–443 (2011). arXiv:1010.5994
142. ATLAS, ATLAS: detector and physics performance technical design report. Vol. 2. CERN-LHCC-99-15
143. G.L. Bayatian et al. (CMS), CMS technical design report, volume II: Physics performance. J. Phys. G **34**, 995–1579 (2007). CERN-LHCC-2006-021, CMS-TDR-008-2
144. G. Aad et al. (ATLAS), Expected performance of the ATLAS experiment—detector, trigger and physics. arXiv:0901.0512
145. A. Kharchilava, Top mass determination in leptonic final states with J/ψ. Phys. Lett. B **476**, 73–78 (2000). arXiv:hep-ph/9912320
146. R. Chierici, A. Dierlamm, Determination of the top mass with exclusive events $t \to Wb \to l\nu J/\psi X$. CERN-CMS-NOTE-2006-058, 2006
147. A.H. Hoang, I.W. Stewart, Top-mass measurements from jets and the Tevatron top mass. Nuovo Cimento B **123**, 1092–1100 (2008)
148. A. Abulencia et al. (CDF), Search for $V + A$ current in top quark decay in $p\bar{p}$ collisions at $\sqrt{s} = 1.96$ TeV. Phys. Rev. Lett. **98**, 072001 (2007). arXiv:hep-ex/0608062
149. T. Aaltonen et al. (CDF), Measurement of W-boson helicity fractions in top-quark decays using $\cos\theta^*$. Phys. Lett. B **674**, 160–167 (2009). arXiv:0811.0344
150. CDF, W boson helicity measurement in $t\bar{t}$ dilepton channel at cdf. CDF Conf. Note 10333, Nov., 2010
151. J. Alwall et al., MadGraph/MadEvent v4: the new web generation. J. High Energy Phys. **09**, 028 (2007). arXiv:0706.2334
152. CDF, Measurements of W boson fractions in top quark decay to lepton + jets events using a matrix element analysis technique with 1.9 fb^{-1} of data. CDF Conf. Note 9144, Dec., 2007
153. CDF, Measurements of W boson fractions in top quark decay to lepton + jets events using a matrix element analysis technique with 2.7 fb^{-1} of data. CDF Conf. Note 10004, Nov., 2009
154. V.M. Abazov et al. (D0), Measurement of the W boson helicity in top quark decay at D0. Phys. Rev. D **75**, 031102 (2007). arXiv:hep-ex/0609045
155. V.M. Abazov et al. (D0), Measurement of the W boson helicity in top quark decays using 5.4fb^{-1} of $p\bar{p}$ collision data. Phys. Rev. D **83**, 032009 (2011). arXiv:1011.6549
156. J. Alwall et al., Is $V_{tb} \simeq 1$? Eur. Phys. J. C **49**, 791–801 (2007). arXiv:hep-ph/0607115
157. V.M. Abazov et al. (D0), Observation of single top-quark production. Phys. Rev. Lett. **103**, 092001 (2009). arXiv:0903.0850
158. T. Aaltonen et al. (CDF), First observation of electroweak single top quark production. Phys. Rev. Lett. **103**, 092002 (2009). arXiv:0903.0885
159. CDF, D0, T. E. W. Group, Combination of CDF and D0 measurements of the single top production cross section. arXiv:0908.2171
160. D. Acosta et al. (CDF), Measurement of $B(t \to Wb)/B(t \to Wq)$ at the Collider Detector at Fermilab. Phys. Rev. Lett. **95**, 102002 (2005). hep-ex/0505091
161. G.J. Feldman, R.D. Cousins, A unified approach to the classical statistical analysis of small signals. Phys. Rev. D **57**, 3873–3889 (1998). arXiv:physics/9711021
162. V.M. Abazov et al. (D0), Simultaneous measurement of the ratio $\mathcal{B}(t \to Wb)/\mathcal{B}(t \to Wq)$ and the top quark pair production cross section with the D0 detector at $\sqrt{s} = 1.96$ TeV. Phys. Rev. Lett. **100**, 192003 (2008). arXiv:0801.1326
163. S.L. Glashow, J. Iliopoulos, L. Maiani, Weak interactions with lepton-hadron symmetry. Phys. Rev. D **2**, 1285–1292 (1970)
164. J.L. Diaz-Cruz, R. Martinez, M.A. Perez, A. Rosado, Flavor changing radiative decay of THF t quark. Phys. Rev. D **41**, 891–894 (1990)
165. G. Eilam, J.L. Hewett, A. Soni, Rare decays of the top quark in the standard and two Higgs doublet models. Phys. Rev. D **44**, 1473–1484 (1991). Phys. Rev. D **59** 039901 (1999), Erratum
166. B. Mele, S. Petrarca, A. Soddu, A new evaluation of the $t \to c$ H decay width in the standard model. Phys. Lett. B **435**, 401–406 (1998). arXiv:hep-ph/9805498
167. H. Fritzsch, t quarks may decay into Z bosons and charm. Phys. Lett. B **224**, 423 (1989)
168. J.A. Aguilar-Saavedra, Top flavour-changing neutral interactions: Theoretical expectations and experimental detection. Acta Phys. Pol. A **35**, 2695–2710 (2004). arXiv:hep-ph/0409342
169. T. Aaltonen et al. (CDF), Search for the flavor changing neutral current decay $t \to Zq$ in $p\bar{p}$ collisions at $\sqrt{s} = 1.96$ TeV. Phys. Rev. Lett. **101**, 192002 (2008). arXiv:0805.2109
170. T. Aaltonen et al. (CDF), Search for top-quark production via flavor-changing neutral currents in $W + 1$ jet events at CDF. Phys. Rev. Lett. **102**, 151801 (2009). arXiv:0812.3400
171. V.M. Abazov et al. (D0), Search for production of single top quarks via flavor-changing neutral currents at the Tevatron. Phys. Rev. Lett. **99**, 191802 (2007). arXiv:hep-ex/0702005
172. V.M. Abazov et al. (D0), Search for flavor changing neutral currents via quark-gluon couplings in single top quark production using 2.3 fb^{-1} of $p\bar{p}$ collisions. Phys. Lett. B **693**, 81–87 (2010). arXiv:1006.3575
173. A. Heister et al. (ALEPH), Search for single top production in e^+e^- collisions at $\sqrt{s}$ up to 209 GeV. Phys. Lett. B **543**, 173–182 (2002). arXiv:hep-ex/0206070

174. J. Abdallah, et al. (DELPHI), Search for single top production via FCNC at LEP at $\sqrt{s} = 189$ GeV–208 GeV. Phys. Lett. B **590**, 21–34 (2004). arXiv:hep-ex/0404014
175. P. Achard et al. (L3), Search for single top production at LEP. Phys. Lett. B **549**, 290–300 (2002). arXiv:hep-ex/0210041
176. G. Abbiendi et al. (OPAL), Search for single top quark production at LEP2. Phys. Lett. B **521**, 181–194 (2001). arXiv:hep-ex/0110009
177. S. Chekanov et al. (ZEUS), Search for single-top production in ep collisions at HERA. Phys. Lett. B **559**, 153–170 (2003). arXiv:hep-ex/0302010
178. A. Aktas et al. (H1), Search for single top quark production in ep collisions at HERA. Eur. Phys. J. C **33**, 9–22 (2004). arXiv:hep-ex/0310032
179. H1, Search for single top quark production in ep collisions at HERA. Contributed paper to EPS2007, abstract 776, H1prelim-07-163, 2007
180. F.D. Aaron et al. (H1), Search for single top quark production at HERA. arXiv:0904.3876
181. F. Abe et al. (CDF), Search for flavor-changing neutral current decays of the top quark in $p\bar{p}$ collisions at $\sqrt{s} = 1.8$ TeV. Phys. Rev. Lett. **80**, 2525–2530 (1998)
182. T. Stelzer, W.F. Long, Automatic generation of tree level helicity amplitudes. Comput. Phys. Commun. **81**, 357–371 (1994). arXiv:hep-ph/9401258
183. F. Maltoni, T. Stelzer, MadEvent: Automatic event generation with MadGraph. J. High Energy Phys. **02**, 027 (2003). arXiv:hep-ph/0208156
184. S.R. Slabospitsky, L. Sonnenschein, TopReX generator (version 3.25): Short manual. Comput. Phys. Commun. **148**, 87–102 (2002). arXiv:hep-ph/0201292
185. CDF, Combination of CDF single top searches with 2.2 fb^{-1} of data. CDF Note 9251, Mar., 2008
186. J.J. Liu, C.S. Li, L.L. Yang, L.G. Jin, Next-to-leading order QCD corrections to the direct top quark production via model-independent FCNC couplings at hadron colliders. Phys. Rev. D **72**, 074018 (2005). arXiv:hep-ph/0508016
187. L.L. Yang, C.S. Li, Y. Gao, J.J. Liu, Threshold resummation effects in direct top quark production at hadron colliders. Phys. Rev. D **73**, 074017 (2006). arXiv:hep-ph/0601180
188. V.M. Abazov et al. (D0), Multivariate searches for single top quark production with the D0 detector. Phys. Rev. D **75**, 092007 (2007). arXiv:hep-ex/0604020
189. V.M. Abazov et al. (D0), Evidence for production of single top quarks. Phys. Rev. D **78**, 012005 (2008). arXiv:0803.0739
190. V.M. Abazov et al. (D0), Measurement of the t-channel single top quark production cross section. Phys. Lett. B **682**, 363–369 (2010). arXiv:0907.4259
191. E.E. Boos, V.E. Bunichev, L.V. Dudko, V.I. Savrin, A.V. Sherstnev, Method for simulating electroweak top-quark production events in the NLO approximation: SingleTop event generator. Phys. At. Nucl. **69**, 1317–1329 (2006)
192. E. Boos et al. (CompHEP), CompHEP 4.4: Automatic computations from Lagrangians to events. Nucl. Instrum. Methods A **534**, 250–259 (2004). arXiv:hep-ph/0403113
193. V.M. Abazov et al. (D0), Experimental discrimination between charge $2e/3$ top quark and charge $4e/3$ exotic quark production scenarios. Phys. Rev. Lett. **98**, 041801 (2007). arXiv:hep-ex/0608044
194. D0, Clarification on the D0 measurement of the top quark charge. http://www-d0.fnal.gov/Run2Physics/WWW/results/final/TOP/T06D/extra/topQ.htm, April, 2007
195. CDF, First CDF measurement of the top quark charge using the top decay products. CDF Note 8967, Aug., 2007
196. CDF, Exclusion of exotic top-like quark with $-4/3$ electric charge using soft lepton tags. CDF Note 9939, Jan., 2010
197. D.J. Lange, The EvtGen particle decay simulation package. Nucl. Instrum. Methods A **462**, 152–155 (2001)
198. T. Aaltonen et al. (CDF), Searching the inclusive $\ell\gamma \not{E}_T +$ b-quark signature for radiative top quark decay and non-standard-model processes. Phys. Rev. D **80**, 011102 (2009). arXiv:0906.0518
199. I.I.Y. Bigi, Y.L. Dokshitzer, V.A. Khoze, J.H. Kühn, P.M. Zerwas, Production and decay properties of ultraheavy quarks. Phys. Lett. B **181**, 157 (1986)
200. V.D. Barger, J. Ohnemus, R.J.N. Phillips, Spin correlation effects in the hadroproduction and decay of very heavy top quark pairs. Int. J. Mod. Phys. A **4**, 617 (1989)
201. T. Stelzer, S. Willenbrock, Spin correlation in top quark production at hadron colliders. Phys. Lett. B **374**, 169–172 (1996). arXiv:hep-ph/9512292
202. W. Bernreuther, A. Brandenburg, Z.G. Si, P. Uwer, Top quark spin correlations at hadron colliders: Predictions at next-to-leading order QCD. Phys. Rev. Lett. **87**, 242002 (2001). arXiv:hep-ph/0107086
203. W. Bernreuther, Z.-G. Si, Distributions and correlations for top quark pair production and decay at the Tevatron and LHC. Nucl. Phys. B **837**, 90–121 (2010). arXiv:1003.3926
204. CDF, A measurement of the $t\bar{t}$ spin correlation coefficient in 2.8 fb^{-1} dilepton candidates. CDF Note 9824, June, 2009
205. V.M. Abazov et al. (D0), Spin correlations in $t\bar{t}$ production in dilepton events. DØNote 5950 conf, July, 2009
206. CDF, Measurement of tt halicity fraction and spin correlation using reconstructed lepton + jets events. CDF Note 10211, Jan., 2010
207. CDF, Measurement of tt halicity fraction and spin correlation using reconstructed lepton + jets events. CDF Note 10211, July, 2010
208. F. Halzen, P. Hoyer, C.S. Kim, Forward–backward asymmetry of hadroproduced heavy quarks in QCD. Phys. Lett. B **195**, 74 (1987)
209. P. Nason, S. Dawson, R.K. Ellis, The one particle inclusive differential cross-section for heavy quark production in hadronic collisions. Nucl. Phys. B **327**, 49–92 (1989). Nucl. Phys. B **335**, 260 (1990), Erratum
210. W. Beenakker, W.L. van Neerven, R. Meng, G.A. Schuler, J. Smith, QCD corrections to heavy quark production in hadron hadron collisions. Nucl. Phys. B **351**, 507–560 (1991)
211. J.H. Kühn, G. Rodrigo, Charge asymmetry of heavy quarks at hadron colliders. Phys. Rev. D **59**, 054017 (1999). arXiv:hep-ph/9807420
212. M.T. Bowen, S.D. Ellis, D. Rainwater, Standard model top quark asymmetry at the Fermilab Tevatron. Phys. Rev. D **73**, 014008 (2006). arXiv:hep-ph/0509267
213. V.M. Abazov et al. (D0), First measurement of the forward–backward charge asymmetry in top quark pair production. Phys. Rev. Lett. **100**, 142002 (2008). arXiv:0712.0851
214. T. Aaltonen et al. (CDF), Forward–backward asymmetry in top quark production in $p\bar{p}$ collisions at $\sqrt{s} = 1.96$ TeV. Phys. Rev. Lett. **101**, 202001 (2008). arXiv:0806.2472
215. V.M. Abazov et al. (D0), Measurement of the forward–backward production asymmetry of t and $\bar{t}$ quarks in $p\bar{p} \to t\bar{t}$ events. DØ Note 6062 conf, July, 2010
216. T. Aaltonen et al. (CDF), Evidence for a mass dependent forward–backward asymmetry in top quark pair production. arXiv:1101.0034
217. O. Antunano, J.H. Kühn, G. Rodrigo, Top quarks, axigluons and charge asymmetries at hadron colliders. Phys. Rev. D **77**, 014003 (2008). arXiv:0709.1652
218. S. Dittmaier, P. Uwer, S. Weinzierl, NLO QCD corrections to $t\bar{t}$ + jet production at hadron colliders. Phys. Rev. Lett. **98**, 262002 (2007). arXiv:hep-ph/0703120

219. J.M. Campbell, R.K. Ellis, An update on vector boson pair production at hadron colliders. Phys. Rev. D **60**, 113006 (1999). arXiv:hep-ph/9905386
220. S. Frixione, B.R. Webber, Matching NLO QCD computations and parton shower simulations. J. High Energy Phys. **06**, 029 (2002). arXiv:hep-ph/0204244
221. T. Aaltonen et al. (CDF), First measurement of the $t\bar{t}$ differential cross section $d\sigma/dm_{t\bar{t}}$ in $p\bar{p}$ collisions at $\sqrt{s}=1.96$ TeV. Phys. Rev. Lett. **102**, 222003 (2009). arXiv:0903.2850
222. CDF, Measurement of the $t\bar{t}$ differential cross section, $d\sigma/dM_{t\bar{t}}$ in 2.7 fb^{-1} of data. Public analysis webpage http://www-cdf.fnal.gov/physics/new/top/2008/tprop/dXs_27fb/webpage/dXs27fb_Public.htm, Nov., 2008
223. D.E. Acosta et al. (CDF), Measurement of the cross section for $t\bar{t}$ production in $p\bar{p}$ collisions using the kinematics of lepton + jets events. Phys. Rev. D **72**, 052003 (2005). arXiv:hep-ex/0504053
224. A. Hocker, V. Kartvelishvili, SVD approach to data unfolding. Nucl. Instrum. Methods A **372**, 469–481 (1996). arXiv:hep-ph/9509307
225. T.W. Anderson, D.A. Darling, Asymptotic theory of certain "goodness of fit" criteria based on stochastic processes. Ann. Math. Stat. **23**(2), 193–212 (1952)
226. L. Randall, R. Sundrum, A large mass hierarchy from a small extra dimension. Phys. Rev. Lett. **83**, 3370–3373 (1999). arXiv:hep-ph/9905221
227. T. Junk, Confidence level computation for combining searches with small statistics. Nucl. Instrum. Methods A **434**, 435–443 (1999). arXiv:hep-ex/9902006
228. A. Read, Workshop on confidence limits, CERN, Geneva, Switzerland, 17–18 Jan. 2000: Proceedings. CERN-2000-005
229. T. Arens, L.M. Sehgal, Azimuthal correlation of charged leptons produced in $p\bar{p}\to t\bar{t}+$. Phys. Lett. B **302**, 501–506 (1993)
230. CDF, Measurements of the gluon fusion fraction in $t\bar{t}$ production using azimuthal correlation of charged leptons. CDF Conf. Note 9432, July, 2008
231. T. Aaltonen et al. (CDF), Measurement of the fraction of $t\bar{t}$ production via gluon-gluon fusion in $p\bar{p}$ collisions at $\sqrt{s}=1.96$-TeV. Phys. Rev. D **79**, 031101 (2009). arXiv:0807.4262
232. T. Aaltonen et al. (CDF), First measurement of the fraction of top quark pair production through gluon-gluon fusion. Phys. Rev. D **78**, 111101 (2008). arXiv:0712.3273
233. T. Aaltonen et al. (CDF), First direct bound on the total width of the top quark in $p\bar{p}$ collisions at $\sqrt{s}=1.96$ TeV. Phys. Rev. Lett. **102**, 042001 (2009). arXiv:0808.2167
234. CDF, A measurement of the top quark width using the template method in the lepton plus jets channel with 4.3 fb^{-1}. CDF Note 10035, Jan., 2010
235. CDF, First direct limit on the top quark lifetime. CDF Note 8104, Feb., 2006
236. V.M. Abazov et al. (D0), Determination of the width of the top quark. arXiv:1009.5686
237. J.H. Kühn, G. Rodrigo, Charge asymmetry in hadroproduction of heavy quarks. Phys. Rev. Lett. **81**, 49–52 (1998). arXiv:hep-ph/9802268
238. Y.-k. Wang, B. Xiao, S.-h. Zhu, One-side forward–backward asymmetry in top quark pair production at CERN large hadron collider. Phys. Rev. D **82**, 094011 (2010). arXiv:1008.2685
239. B. Xiao, Y.-K. Wang, Z.-Q. Zhou, S.-h. Zhu, Edge charge asymmetry in top pair production at the LHC. arXiv:1101.2507
240. A. Stange, S. Willenbrock, Yukawa correction to top quark production at the Tevatron. Phys. Rev. D **48**, 2054–2061 (1993). arXiv:hep-ph/9302291
241. T.-F. Feng, X.-Q. Li, J. Maalampi, The anomalous Higgs—top couplings in the MSSM. Phys. Rev. D **69**, 115007 (2004). arXiv:hep-ph/0310247
242. J.A. Aguilar-Saavedra, Light Higgs boson discovery in the Standard Model and beyond. J. High Energy Phys. **12**, 033 (2006). arXiv:hep-ph/0603200
243. D0, Search for the Standard Model Higgs boson in the $t\bar{t}h\to t\bar{t}b\bar{b}$ channel. D0 note 5739-conf, July, 2008
244. A. Abulencia et al. (CDF), Search for charged Higgs bosons from top quark decays in $p\bar{p}$ collisions at $\sqrt{s}=1.96$ TeV. Phys. Rev. Lett. **96**, 042003 (2006). hep-ex/0510065. http://www-cdf.fnal.gov/physics/new/top/2005/ljets/charged_Higgs/Higgs/V2/HiggsAnalysis_publicV2.html
245. S.P. Martin, A supersymmetry primer. arXiv:hep-ph/9709356
246. T. Aaltonen et al. (CDF), Search for charged Higgs bosons in decays of top quarks in $p-\bar{p}$ collisions at $\sqrt{s}=1.96$ TeV. Phys. Rev. Lett. **103**, 101803 (2009). arXiv:0907.1269
247. V.M. Abazov et al. (D0), Search for charged Higgs bosons in top quark decays. Phys. Lett. B **682**, 278–286 (2009). arXiv:0908.1811
248. V.M. Abazov et al. (D0), Search for charged Higgs bosons in decays of top quarks. Phys. Rev. D **80**, 051107 (2009). arXiv:0906.5326
249. V.M. Abazov et al. (D0), Search for charged Higgs bosons decaying to top and bottom quarks in $p\bar{p}$ collisions. Phys. Rev. Lett. **102**, 191802 (2009). arXiv:0807.0859
250. E. Boos, V. Bunichev, L. Dudko, M. Perfilov, Interference between W' and W in single-top quark production processes. Phys. Lett. B **655**, 245–250 (2007). arXiv:hep-ph/0610080
251. V.M. Abazov et al. (D0), Search for W' Boson resonances decaying to a top quark and a bottom quark. Phys. Rev. Lett. **100**, 211803 (2008). arXiv:0803.3256
252. Z. Sullivan, Fully differential W' production and decay at next-to- leading order in QCD. Phys. Rev. D **66**, 075011 (2002). arXiv:hep-ph/0207290
253. T. Aaltonen et al. (CDF), Search for the production of narrow tb resonances in 1.9 fb^{-1} of ppbar collisions at $\sqrt{s}=1.96$ TeV. Phys. Rev. Lett. **103**, 041801 (2009). arXiv:0902.3276
254. A. Leike, The phenomenology of extra neutral gauge bosons. Phys. Rep. **317**, 143 (1999). arXiv:hep-ph/9805494
255. B. Lillie, L. Randall, L.-T. Wang, The bulk RS KK-gluon at the LHC. J. High Energy Phys. **09**, 074 (2007). arXiv:hep-ph/0701166
256. T.G. Rizzo, Testing the nature of Kaluza-Klein excitations at future lepton colliders. Phys. Rev. D **61**, 055005 (2000). arXiv:hep-ph/9909232
257. L.M. Sehgal, M. Wanninger, Forward–backward asymmetry in two jet events: Signature of axigluons in $p\bar{p}$ collisions. Phys. Lett. B **200**, 211 (1988)
258. C.T. Hill, S.J. Parke, Top production: Sensitivity to new physics. Phys. Rev. D **49**, 4454–4462 (1994). arXiv:hep-ph/9312324
259. R.M. Harris, C.T. Hill, S.J. Parke, Cross section for topcolor $Z'(t)$ decaying to $t\bar{t}$. arXiv:hep-ph/9911288, 1999
260. T. Aaltonen et al. (CDF), Limits on the production of narrow $t\bar{t}$ resonances in $p\bar{p}$ collisions at $\sqrt{s}=1.96$ TeV. Phys. Rev. D **77**, 051102 (2008). arXiv:0710.5335
261. A.D. Martin, R.G. Roberts, W.J. Stirling, R.S. Thorne, Parton distributions and the LHC: W and Z production. Eur. Phys. J. C **14**, 133–145 (2000). arXiv:hep-ph/9907231
262. T. Aaltonen et al. (CDF), Search for resonant $t\bar{t}$ production in $p\bar{p}$ collisions at $\sqrt{s}=1.96$ TeV. Phys. Rev. Lett. **100**, 231801 (2008). arXiv:0709.0705
263. CDF, Search for resonant $t\bar{t}$ production in $p\bar{p}$ collisions at $\sqrt{s}=$ 1.96 TeV. CDF Note 9844, July, 2009
264. T. Aaltonen et al. (CDF), Search for new color-octet vector particle decaying to $t\bar{t}$ in $p\bar{p}$ collisions at $\sqrt{s}=1.96$ TeV. Phys. Lett. B **691**, 183–190 (2010). arXiv:0911.3112
265. V.M. Abazov et al. (D0), Search for $t\bar{t}$ resonances in the lepton plus jets final state in $p\bar{p}$ collisions at $\sqrt{s}=1.96$ TeV. Phys. Lett. B **668**, 98–104 (2008). arXiv:0804.3664

266. V.M. Abazov et al. (D0), Search for $t\bar{t}$ resonances in the lepton + jets final state in $p\bar{p}$ collisions at $\sqrt{s} = 1.96$ TeV. DØ Note 5600 conf, 2008
267. V.M. Abazov et al. (D0), Search for $t\bar{t}$ resonances in the lepton + jets final state in $p\bar{p}$ collisions at $\sqrt{s} = 1.96$ TeV. DØ Note 5882 conf, 2009
268. D. Wicke, Top BSM at D0. Nuovo Cimento B **123**, 1269–1277 (2008). arXiv:0807.0188
269. V.M. Abazov et al. (D0), Search for scalar top admixture in the $t\bar{t}$ lepton + jets final state at $\sqrt{s} = 1.96$ TeV in 1 fb^{-1} of DØ data. DØ Note 5438 Conf, 2007
270. V.M. Abazov et al. (D0), Search for admixture of scalar top quarks in the $t\bar{t}$ lepton + jets final state at $\sqrt{s} = 1.96$ TeV. Phys. Lett. B **674**, 4–10 (2009). arXiv:0901.1063
271. CDF, Search for pair production of stop quarks mimicking top event signatures. CDF Conf. Note 9439, July, 2008
272. G.D. Kribs, T. Plehn, M. Spannowsky, T.M.P. Tait, Four generations and Higgs physics. Phys. Rev. D **76**, 075016 (2007). arXiv:0706.3718
273. T. Han, H.E. Logan, B. McElrath, L.-T. Wang, Loop induced decays of the little Higgs: $H \rightarrow gg, \gamma\gamma$. Phys. Lett. B **563**, 191–202 (2003). arXiv:hep-ph/0302188
274. T. Aaltonen et al. (CDF), Search for heavy top-like quarks $t' \rightarrow Wq$ using lepton plus jets events in 1.96 TeV proton–antiproton collisions. Phys. Rev. Lett. **100**, 161803 (2008). arXiv:0801.3877
275. CDF, Search for heavy top $t' \rightarrow Wq$ in lepton plus jets events in 4.6 fb^{-1}. CDF Conf. Note 10110, Mar. (2010)
276. D0, Search for a fourth generation t' quark that decays to W boson + jet. D0 note 5892-CONF, July (2010)
277. D. Amidei et al. (TeV-2000 Study Group), Future electroweak physics at the Fermilab Tevatron: Report of the TeV-2000 Study Group. SLAC-REPRINT-1996-085